**BASIC
ELECTRICAL
ENGINEERING**

Balabanian and LePage: ELECTRICAL SCIENCE, Book II: Dynamic Networks

Belove and Drossman: SYSTEMS AND CIRCUITS FOR ELECTRICAL ENGINEERING TECHNOLOGY

Belove, Schachter, and Schilling: DIGITAL AND ANALOG SYSTEMS, CIRCUITS AND DEVICES

Bracewell: THE FOURIER TRANSFORM AND ITS APPLICATIONS

Cannon: DYNAMICS OF PHYSICAL SYSTEMS

Chirlian: BASIC NETWORK THEORY

Desoer and Kuh: BASIC CIRCUIT THEORY

Fitzgerald, Higginbothan, and Grabel: BASIC ELECTRICAL ENGINEERING

Ghausi: PRINCIPLES AND DESIGN OF LINEAR ACTIVE CIRCUITS

Hammond and Gehmlich: ELECTRICAL ENGINEERING

Hayt and Hughes: INTRODUCTION TO ELECTRICAL ENGINEERING

Hayt and Kemmerly: ENGINEERING CIRCUIT ANALYSIS

Hilburn and Johnson: MANUAL OF ACTIVE FILTER DESIGN

Huelsman: DIGITAL COMPUTATIONS IN BASIC CIRCUIT THEORY

Huelsman: THEORY AND DESIGN OF ACTIVE RC CIRCUITS

Huelsman and Allen: INTRODUCTION TO THE THEORY AND DESIGN OF ACTIVE FILTERS

Jong: METHODS OF DISCRETE SIGNAL AND SYSTEM ANALYSIS

Kalman, Falb, and Arbib: TOPICS IN MATHEMATICAL SYSTEM THEORY

Lewis, Reynolds, Bergseth, and Alexandro: LINEAR SYSTEMS ANALYSIS

Liu and Liu: LINEAR SYSTEMS ANALYSIS

Papoulis: THE FOURIER INTEGRAL AND ITS APPLICATION

Peatman: DESIGN OF DIGITAL SYSTEMS

Ruston and Bordogna: ELECTRIC NETWORKS: Functions, Filters, Analysis

Sage: METHODOLOGY FOR LARGE SCALE SYSTEMS

Schwartz and Friedland: LINEAR SYSTEMS

Temes and LaPatra: INTRODUCTION TO CIRCUIT SYNTHESIS

Timothy and Bona: STATE SPACE ANALYSIS

Truxal: INTRODUCTORY SYSTEM ENGINEERING

Tuttle: CIRCUITS

A. E. FITZGERALD, SC.D.

Late Vice President for Academic Affairs
and Dean of the Faculty
Northeastern University

DAVID E. HIGGINBOTHAM, S.M.

Late Professor and Chief
Electrical Engineering Section
United States Coast Guard Academy

ARVIN GRABEL, SC.D.

Professor of Electrical Engineering
Northeastern University

BASIC ELECTRICAL ENGINEERING

**CIRCUITS
ELECTRONICS
MACHINES
CONTROLS**

Fifth Edition

McGraw-Hill, Inc.

New York St. Louis San Francisco Auckland Bogotá
Caracas Lisbon London Madrid Mexico Milan
Montreal New Delhi Paris San Juan Singapore
Sydney Tokyo Toronto

BASIC
ELECTRICAL
ENGINEERING

9 10 11 12 13 DOR DOR 9 9 8 7 6 5 4 3 2

This book was set in Times Roman by Progressive Typographers.
The editors were Frank J. Cerra and J. W. Maisel;
the designer was Nicholas Krenitsky;
the production supervisor was Dominick Petrellese.
New drawings were done by J & R Services, Inc.
Printed and bound by Impresora Donneco Internacional S. A.
de C. V. a division of R. R. Donnelley and Sons, Inc.

Library of Congress Cataloging in Publication Data

Fitzgerald, Arthur Eugene, date
    Basic electrical engineering.

    (McGraw-Hill series in electrical engineering : Net-
works and systems)
    Includes index.
    1. Electric engineering. 2. Electronics.
I. Higginbotham, David E., joint author. II. Grabel,
Arvin, joint author. III. Title. IV. Series.
TK145.F53 1981   621.3   80-19420
ISBN 0-07-021154-X
ISBN 0-07-021155-8 (solutions manual)

Manufactured in Mexico.

CONTENTS

7 THE PHYSICAL FOUNDATIONS
OF SEMICONDUCTOR DEVICES   327

8 ELEMENTARY AMPLIFIER STAGES   395

9 MULTISTAGE AMPLIFIERS   477

The objective of this book—as it was for the four prior editions—is to introduce undergraduate engineering and science students to the essentials of electrical engineering. By stressing the fundamental physical concepts and the diversity of applications, we have attempted to convey both the substance and flavor of the discipline. The scope and treatment in this volume may also serve as a valuable adjunct to the continuing education of professionals in fields akin to engineering.

This book has been extensively rewritten and a substantial amount of new material has been added. As in previous editions, the text is divided into four major sections: circuit theory, electronics, electromechanical energy conversion, and analog and digital control systems. Sufficient description is given to the physical principles in each of these areas to make applications meaningful. The treatment is carried beyond a bare introduction so that the reader can make the transition to more advanced, specialized texts without undue intellectual embarrassment. Increased emphasis is given to the exciting applications made possible by operational amplifiers, microprocessors, and other integrated circuits and semiconductor devices which are part of current engineering practice and which will continue to influence future development.

The book is organized to provide a maximum of flexibility without loss of continuity, so that a variety of one- or two-semester courses can be presented. Thus, the individual faculty member can readily adapt the material to meet the needs of his or her students. An instructor's guide is available, one section of which describes several possible arrangements of the material for courses of different length, content, and sequence.

A chapter on the Laplace transform has been added; coverage of this material is optional. In the study of circuit theory, some professors may wish to introduce the concepts of the natural response prior to their treatment of steady-state ac circuits. They may do so by first covering Sections 3-1 to 3-4 and Chapter 4 before returning to complete those sections in Chapter 3 that are important to them. Similarly, there is subject matter in Chapters 2 and 6 that can be skipped if the sections in electronics, machines, and controls which employ these techniques are also omitted.

The parts of the book relating to electronics and machines can be taken up interchangeably. The introductory chapters of these parts (Chapter 7 for electronics and Chapters 13 and 14 for machines) present the underlying physical concepts of the respective fields. These chapters may be supplemented by as many of the analytical chapters that follow as are appropriate in view of technical objectives. Furthermore, in the electronics section, it is possible to treat digital systems prior to analog systems.

The final chapter serves to introduce a variety of methods by which the performance of physical systems is controlled. By focusing on control aspects, the integration of many of the concepts used in the preceding chapters is effected.

The importance of electrical and electronic instrumentation is considerable. Rather than isolating this topic in a separate section, instrumentation is integrated throughout the book, particularly in the sections on circuit theory and electronics. In addition, a separate section on transducers is included in Chapter 17.

The book contains over 700 problems and 150 illustrative examples. The vast majority of these either are new or have been reworked and most have been classroom tested. Solutions to problems at the end of each of the chapters appear in the instructor's guide and answers to selected problems are given at the end of the text.

We have had the benefit of valuable advice and suggestions from the many professors and practicing engineers who used earlier editions of this book as either instructor or student. All these individuals have influenced this edition, so to them we express our thanks and appreciation.

The late A. E. Fitzgerald was the guiding force behind this book for over three decades. A truly professional gentleman, he had a distinguished career both as teacher and writer and as a practicing engineer. The intellect and spirit of my teacher, colleague, and friend appears on every page of this book.

Also apparent throughout this edition is the influence of the late David E. Higginbotham. The insight, sensitivity, and experience of this outstanding teacher have contributed immeasurably.

Arvin Grabel

BASIC
ELECTRICAL
ENGINEERING

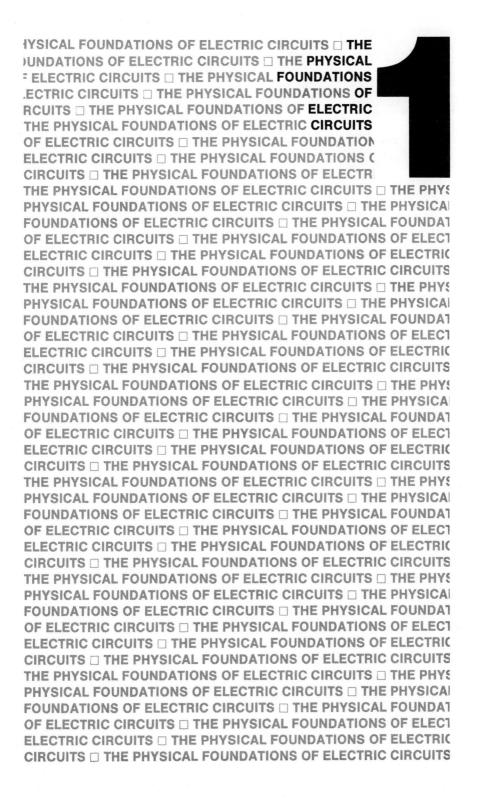

lectrical engineers have contributed significantly to technological achievements during the twentieth century. These contributions have resulted in the widespread use of electric energy and the advantages it affords. Electric energy has a number of desirable attributes not possessed by alternative energy forms. Among these attributes are the ease with which electric energy is converted to and from other forms of energy, the ability to distribute it over large areas, and the speed with which it is transported.

Economies result when large quantities of energy are generated at one location and distributed to sites where the energy is needed. Ease of transportation to geographically separate regions is a significant factor in realizing these economies. Information, in the form of electric signals, can be propagated at great speed. As a result the effectiveness of long-range communication systems, data-processing systems, and control systems is enhanced. In many cases it is the speed at which the energy is transferred that makes the system feasible.

As an end in itself, electric energy is not generally useful. In most applications, other forms of energy are converted to electric energy at the input and reconverted at the output. Between input and output, the electric energy is processed and transmitted to the desired location in a manner appropriate for use. Motors are the converters which provide the mechanical energy to operate the machinery in a factory. In turn, the electric energy to drive the motor is derived by converting mechanical energy in a steam-driven turbogenerator at the power plant. Similarly, the information in the sound and picture in a television system is converted to electric signals at the studio by means of microphones and vidicons. Reconversion to optical and acoustical information in the receiver is accomplished by cathode-ray tubes and loudspeakers. The availability of devices that readily perform the interconversion of energy is crucial to the use of electric energy. These conversion devices, coupled with the generally small size of electric components, form the basis for the use of electric energy in a variety of control, communication, and instrumentation systems.

The study of electrical engineering becomes one of investigating the characteristics and uses of devices and systems for energy conversion, processing, and transfer. Such devices and systems impact on all professional branches of engineering and science. Most instrumentation and control systems are, in part, electrical or electronic in nature. The incorporation of the digital computer in such systems increases the utilization of electrical instrumentation. A variety of engineers deal with motors and power distribution in building systems, process control, and the design of manufacturing facilities. Medical scientists use electric devices in diagnostics, in prosthetics, and for environmental control. The aforementioned applications have, in turn, a vital impact on the economic, structural, and behavioral aspects of organizations and are, therefore, of vital concern to managers and executives.

For convenience, a basic study of electrical engineering is traditionally divided into treatments of circuit theory, of electronics and electronic
devices, of energy conversion and electromechanical conversion devices,
and of control devices and systems. This chapter will lay the groundwork
for such a study by developing the elementary concepts and defining the
basic terms. The typical reader already will have encountered much of this
material. Thus, for many, it will simply be a coordinated review reflecting
the general viewpoint of the remainder of the book.

1-1 ENERGY AND INFORMATION

The earliest uses of electric energy were the telegraph, telephone, and electric lighting. Both the telephone and telegraph have as their objective the
transfer of information from one location to another. In the latter, the sender
codes the message into electric impulses (the familiar Morse code representation of the alphabet), whereupon it is transmitted, received, and decoded.
Similarly, telephone transmission requires the conversion of the speaker's
voice to a corresponding electric signal which can be faithfully reproduced at
the listener's receiver. In both cases it is the form of the electric energy,
which contains the information, that is most important. The amount of energy expended need be only what is necessary to recover the information at
the receiving end.

To be effective in lighting buildings and streets, the incandescent lamp
requires that electric energy be available at the point of use. The initial suc-
cess of electric lighting resulted from incorporation of the lamp in a total
system for the generation and distribution of the energy needed. Here the
important factor is the quantity of energy transferred; its form is of secondary importance since no information transfer occurs.

As illustrated by these early applications, the uses of electric energy can
be classified as systems that process information and those that process energy. Modern systems for communication, computation, and control are
generally considered to process information; electric power-generating, conversion, and distribution systems are evidently energy processors. However, both classes of systems are concerned with the transfer of electric energy, i.e., with doing work by delivering energy to the right place, in the
right form, and at sufficient power for the intended purposes. The general
import of the preceding sentence can be gathered by thinking of the words
work, energy, and *power* as they are used in ordinary speech. However,
these terms, so basic to engineering, must be considered in their more precise and quantitative meanings.

Work is done when something is moved against a resisting force. For example, work is done when a weight is lifted against the pull of gravity. Quantitatively, the work done is found by multiplying the force applied by the distance through which the force moves. The unit of work in the meter-
kilogram-second (SI) system of units is the *joule* (abbreviated J). It is the
work done when a force of one *newton* (0.225 lbf.) acts through a distance of
one meter. The English unit of work is the *foot-pound*. ($1 \text{ J} = 1 \text{ N} \cdot \text{m} =$

0.738 ft-lbf.)[1] The SI system of units is used predominantly in this book. The English system is used only where dictated by common industrial usage.

Energy is the capacity for doing work; another way to think of work is as a transfer of energy. Mechanical energy is measured in the same units as work. When a mass is elevated, energy is expended by the human body or by some hoisting device. This mass, on the other hand, acquires *potential energy*. That is, by virtue of its elevation above the ground, the mass can do work. It can raise another weight by means of a pulley; or it can be allowed to fall, as in a pile driver, transferring its energy when it hits the pile at the bottom. In this case, the potential energy at the beginning of the fall would change into *kinetic energy* as the bottom was neared. That is, the body would have the ability to do work because of its motion. At impact the energy of a pile driver is transferred to the pile.

A general principle applicable to all physical systems is the *principle of conservation of energy,* which states that energy is neither created nor destroyed; it is merely changed in form. Energy can be changed into heat, light, or sound; it may be mechanical energy of position or of motion; it may be stored in a battery or in a spring; but it is not created or destroyed. The twentieth century interconversion of matter and energy is a refinement which does not affect the usefulness of the law for nonrelativistic processes.

For practical purposes we are very much concerned with the *rate* of doing work or otherwise delivering energy. This rate is called *power.* In SI units, power is measured in *watts* (abbreviated W), one watt equaling one joule per second. In English units, power is measured in foot-pounds per second. *Horsepower,* equivalent to 550 ft-lb/s (1 hp = 746 W), is still sometimes used to indicate power, particularly motor performance. Thus, from the definition of power, if W is the work accomplished or energy expended or delivered in time t_o seconds, the average power for that period is

$$P = \frac{W}{t_o} \tag{1-1}$$

just as we get our average speed on a trip by dividing the distance traveled by the time required. We are also interested in the *instantaneous power,* however, just as we are interested in the momentary reading of the speedometer of the car. The expression for instantaneous power may be written

$$p = \frac{dW}{dt} \tag{1-2}$$

where the lowercase letter p indicates an instantaneous quantity and the term dW/dt is, of course, the usual mathematical notation for the time rate at which work is being done. Equation (1-2) can be rewritten as Eq. (1-3) to show the relationship between energy and power.

$$W = \int p \, dt \tag{1-3}$$

---

[1] In general, units are abbreviated in this book when they follow numerical value or a symbolic equation; otherwise, they are spelled out. A listing of abbreviations is given in Appendix A.

The average power, expressed in Eq. (1-1), can now be obtained from the evaluation of Eq. (1-3) during a desired time interval. This relation is given in Eq. (1-4) and can be recognized as the mean-value theorem:

$$P = \frac{1}{t_2 - t_1} \int_{t_1}^{t_2} p \, dt \tag{1-4}$$

where $t_2 - t_1$ represents the time interval over which the average is taken.

Because of the intimate relationship between power and energy, we often find energy expressed in such units as *wattseconds* or *kilowatthours* (which equals $1,000 \times 3,600$, or 3.6×10^6 wattseconds). A wattsecond is, of course, the same measurement as a joule; the picture conveyed is the amount of energy delivered in one second by a constant power of one watt.

1-2 FUNDAMENTAL ELECTRICAL QUANTITIES

Electric energy transfer occurs because of the actions of *electric charges*. An indication of the quantity of electricity, electric charge is usually denoted by Q or q and is measured in *coulombs* (abbreviated C). As a convention generally followed in this book, capital letters are used to represent quantities that do not vary with time; lowercase letters refer to time-varying quantities. While we can perceive of a wire or a similar object carrying a charge, it is difficult to visualize the charge itself being divorced from the object. The nature of electric charges is best understood in terms of the effects they produce. Justification for the existence of charged particles is based on a wealth of experimental evidence and the theoretical models which predict observed behavior.

One of the first phenomena observed in the study of electric charges was that there are two kinds: positive and negative. *Protons* are considered positive charges, *electrons* negative. The charge on an electron, sometimes called the *electronic charge,* is -1.602×10^{-19}C so that nearly 6.3×10^{18} electrons are required to form one coulomb.

A most significant effect of an electric charge is that it can produce a force. Specifically, a charge will repel other charges of the same sign; it will attract other charges of the opposite sign. Since the charge on an electron is negative, any charge which attracts (or is attracted by) an electron is accordingly a positive charge; the charge on a proton is an example. Note that the force of attraction or repulsion is felt equally by each of the charges or charged particles. The magnitude of the force between two charged bodies is proportional to the product of the charges and inversely proportional to the square of the distance between them. That is, the force F between two charged bodies having charges Q_1 and Q_2 is given by *Coulomb's law* as

$$F = k \frac{Q_1 Q_2}{d^2} \tag{1-5}$$

where d is the distance between the charges and k is a constant depending on the medium surrounding the charges. The direction of the force is along the line connecting the two charges. Note that this is the same sort of relation that governs the forces of gravity between two masses. Such relations are referred to as *inverse-square laws.*

The situation defined in Eq. (1-5) can be described by saying that there is a region of influence in the neighborhood of an electric charge wherein a force will be exerted when another charge is introduced. The force will grow progressively weaker as the new charge is placed in more remote positions. Such a region of influence is called a *field.* The field set up by the presence of electric charges is an *electric field.* Since we are for the moment dealing with charges at rest, it can be called an *electrostatic field.* Note again the general similarity to gravity and gravitational forces. When we say, for example, that a space vehicle has left the earth's gravitational field, we mean that the force of gravity acting on the vehicle has become negligibly small. Obviously, a similar statement can be made about a charged particle with respect to an electric field.

The electric field is defined at a point as the force per unit-positive charge. That is, the electric field at any point is the force, in magnitude and direction, which would act on a unit-positive charge at that point. Contributions to the total field at any point are made by all the charges that are close enough to have any influence.

We are now in a position to think about work and energy transfer in connection with electric forces. Suppose we move a positive charge in an electric field in a direction opposite to that of the field, that is, against the force acting on it due to other electric charges. For example, if the field were caused by the existence of a nearby negative charge, we would move the positive charge farther from it. Now work would be done in moving the charge against the forces acting on it, just as work is done in lifting a weight in the earth's gravitational field. Moreover, the law of conservation of energy applies; that is, the particle would now be in a position of higher potential, just as the lifted weight would possess greater potential energy. Consequently, we may think of storing energy by means of the field and subsequently transferring this energy to do work.

An important electrical quantity, *potential difference* or *voltage,* is defined as the work per unit-positive charge in moving a charge between two points in the field. Mathematically, this is expressed as

$$V = \frac{W}{Q} \qquad \text{or} \qquad W = VQ \tag{1-6}$$

where V is the potential difference in volts and W is the work done in transporting a charge Q between two points a and b. The letter E is also used to represent voltage; in this book, both are used. Where a preference for E or V exists because of common usage or tradition, that preference is followed. Since work is the force F times the distance l between a and b and the strength of the electric field $\mathscr{E}$ is the force per unit charge, Eq. (1-6) is often given as

$$V = \frac{Fl}{Q} = \mathscr{E}l \tag{1-7}$$

It is assumed, in Eq. (1-7), that the force (field) acts in the direction of the line defined by a and b. If the energy increases in moving from a to b, a voltage rise exists in the direction a to b. Conversely, a voltage drop exists when the same positive charge loses energy in moving from b to a. It is evident that the voltage drop from b to a has the same value as the rise from a to b. However, we note that the voltage drop between b and a is the negative of the voltage rise from b to a.

We are familiar with the devices for obtaining useful work from weights moving to positions of lower potential in the earth's field. Perhaps the most useful one to think of is a waterwheel to obtain work from a stream of water going over a falls. In more-or-less analogous ways we can obtain work from a stream of charges moving under the influence of electric forces to a position of lower potential.

The idea of the previous paragraph leads us to the conclusion that for engineering purposes we are primarily interested in charges in motion and the resultant energy transfer. We are particularly (although not exclusively) interested in those situations where the motion is confined to a definite path formed by materials such as copper and aluminum which, as experience has shown, are good *conductors* of electricity. By way of contrast, other materials, such as Bakelite, mica, glass, and polyethylene, are known to be extremely poor conductors. They are called *insulators* and are used to confine the electricity to the specific conducting paths by constituting barriers to departure from these paths, called *circuits*.

The rate of motion of charge in a circuit is called *current*. Imagine that we are standing at a point in a circuit watching the charge go by. Suppose we see it pass at the uniform rate of Q coulombs every t seconds. The rate then has the steady value

$$I = \frac{Q}{t} \tag{1-8}$$

Often the rate at which the charges flow changes with time, so that the current also changes in value. Thus, the instantaneous current i in a circuit may be written

$$i = \frac{dq}{dt} \quad \text{or} \quad q = \int i \, dt \tag{1-9}$$

The unit of current is the *ampere* (abbreviated A). One ampere exists when the charge flows at the rate of one coulomb per second. We must specify the direction as well as the size of the current. Historically, as a result of Benjamin Franklin's experiments with lightning, a positive current was conceived of as a stream of positive charges. We now know that currents in ordinary conductors consist of the displacement of electrons, but the convention has not changed. Positive current is by definition the direction of flow of positive charges, which is opposite to the direction of electron flow.

In a *direct current* (dc), the flow of charges is all in one direction for the period of time under consideration. Figure 1-1, for example, shows a graph

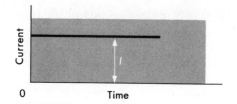

FIGURE 1-1 Graph of a steady direct current.

of a direct current as a function of time; more specifically, it shows a *steady* direct current, for the magnitude is constant at the value *I*.

In an *alternating current* (*ac*), the charges flow first in one direction and then in the other, repeating this cycle with a definite frequency. The variation of current with time is often as is shown in Fig. 1-2, where the solid wave shows one complete cycle of values and the dotted extensions denote the current continuing to follow this cyclic pattern. The ordinary house current in the United States, for example, usually has this waveform and alternates at the frequency of 60 cycles per second. The unit of frequency is the hertz (abbreviated Hz and equal to one cycle per second).

A second type of field which is extremely important in energy transfer is the *magnetic field* which exists in the vicinity of current-carrying elements. Magnetic fields cause forces to be exerted on other current-carrying elements and on magnetic materials. Because it can exert a force and consequently produce work, energy can be stored in the magnetic field. Both the electric field and the magnetic field exist simultaneously; the former is caused by the presence of charge, the latter by the motion of charge. A simi-

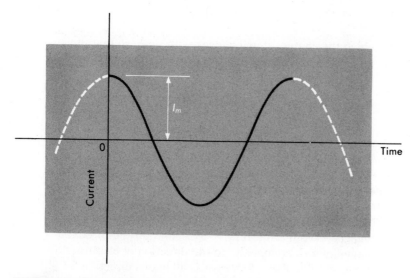

FIGURE 1-2 Graph of alternating current.

lar situation is found in mechanical systems. A mass is capable of storing energy because of its position (potential energy) or by virtue of its motion (kinetic energy).

1-3 ELECTRIC POWER AND ENERGY

The work done or energy transferred in an electric circuit or part of a circuit is expressed in Eq. (1-7) as the product of voltage and charge. If work is being done at a constant rate and the total charge Q is moved through a voltage of V volts in t seconds, then the power, or work per unit time, is

$$P = \frac{VQ}{t} \quad \text{W or J/s} \tag{1-10}$$

Now, from a practical viewpoint, we are more interested in the current than we are in the charge. A more useful form of Eq. (1-10) is obtained by use of Eq. (1-8); that is,

$$P = VI \quad \text{W} \tag{1-11}$$

In a particular case, the power is that given out or absorbed by the portion of the circuit across which V is measured; this circuit element must also carry I amperes. If both the voltage v and current i vary with time, the power p will also vary instant by instant. The instantaneous power is then

$$p = vi \quad \text{W} \tag{1-12}$$

Both Eqs. (1-11) and (1-12) follow naturally from the definitions of voltage and current, for the product ei has the dimensions

$$\frac{\text{Energy}}{\text{Charge}} \frac{\text{charge}}{\text{time}} = \frac{\text{energy}}{\text{time}}$$

which is the meaning of power. In terms of basic mathematical definitions, Eq. (1-12) becomes

$$\frac{dw}{dq} \frac{dq}{dt} = \frac{dw}{dt}$$

Where V and I both remain constant over the time t seconds, the total energy given out or absorbed is

$$W = VIt \quad \text{Ws or J} \tag{1-13}$$

Example 1-1

A 12-V battery is to be charged by a constant current of 20 A supplied to it for 10 h. Of the energy supplied to the battery, seventy percent is converted to chemical energy and stored. The remaining 30 percent of the energy is converted to heat and lost to the surrounding environment. Determine

a The cost of charging the battery if electricity costs 5 cents/kWh

b The amount of chemical energy stored

SOLUTION **a** From Eq. (1-13), the total energy is

$$W = 12 \times 20 \times 10 = 2,400 \text{ Wh} = 2.40 \text{ kWh}$$

The cost is

$$2.40 \times 5.0 = 12 \text{ cents}$$

b The charging process has an efficiency of 70 percent as only that amount of the supplied energy is converted to useful chemical energy. The amount of chemical energy is

$$W_{\text{chem}} = 0.7 \text{ W} = (0.7)(2,400) = 1,680 \text{ Wh} = 1,680 \times 3,600$$
$$= 6,048,000 \text{ J}$$

At this point we have introduced the principal electrical quantities with which we shall be dealing. They are summarized in Table 1-1, together with their more usual unit of measurement and its abbreviation. For some purposes these units may be inconveniently small or large. To denote larger or smaller units, a series of prefixes to the name of the basic unit is used, a multitude of zeros thereby being avoided before or after the decimal point. These prefixes, with their abbreviations in parentheses, are given in Table 1-2; a listing of them may also be found in Appendix A. Thus, in heavy-power industrial or public-utility circuits we often speak of kilovolts (kV), kilowatts (kW), and megawatts (MW). In lower-power-level electronic and communication circuits, however, we often deal in millivolts (mV) and microamperes (μA). As we shall see, these prefixes are also used with the units for the circuit constants.

We are now in the position to consider how the electrical phenomena

TABLE 1-1 SUMMARY OF PRINCIPAL ELECTRICAL QUANTITIES

ELECTRICAL QUANTITY	SYMBOL	UNIT (SI SYSTEM)	RELATED EQUATION	MECHANICAL ANALOGY	HYDRAULIC ANALOGY
Charge	q, Q	Coulomb (C)	. . .	Position	Volume
Current	i, I	Ampere (A)	$i = dq/dt$	Velocity	Flow
Potential difference, or voltage	e, E or v, V	Volt (V)	$v = dw/dq$	Force	Head, or pressure
Power	p, P	Watt (W)	$p = vi$	Power	Power
Energy, or work	w, W	Joule, or wattsecond (J, or Ws)	$w = \int v\, dq$ or $w = \int vi\, dt$	Energy, or work	Energy, or work

TABLE 1-2 PREFIXES USED WITH ELECTRICAL UNITS

FOR LARGER QUANTITIES		FOR SMALLER QUANTITIES	
Kilo (k)	10^3 units	Milli (m)	10^{-3} units
Mega or meg (M)	10^6 units	Micro (μ)	10^{-6} units
Giga (G)	10^9 units	Nano (n)	10^{-9} units
Tera (T)	10^{12} units	Pico (p)	10^{-12} units

that occur in physical systems can be modeled. Electric circuits, whose components are called *circuit elements,* are the usual method by which systems are represented. Different kinds of circuit elements are used to indicate energy dissipation, energy storage, and as a consequence of conservation of energy, the supplied energy. In the succeeding five sections of this chapter (Secs. 1-4 through 1-8), the various kinds of elements are introduced. A keystone in the logical, coordinated analysis of the circuits comprised of these elements lies in their mathematical and physical description.

1-4 RESISTANCE

The circuit element used to represent energy dissipation is most commonly described by requiring the voltage across the element be directly proportional to the current through it. Mathematically, the voltage is

$$v = Ri \quad \text{V} \tag{1-14}$$

where i is the current in amperes. The constant of proportionality R is the *resistance* of the element and is measured in ohms (abbreviated Ω). The voltage-current relation expressed in Eq. (1-14) is known as *Ohm's law.* A physical device whose principal electrical characteristic is resistance is called a *resistor.*

Electric resistance is comparable to pipe friction in the hydraulic analog and also to friction in a mechanical system. Resistance or friction directly opposes the current, water flow, or motion, and the energy dissipated in overcoming this opposition appears as heat. Since an electric charge gives up energy when passing through a resistance, the voltage v in Eq. (1-14) is a voltage drop in the direction of the current. Alternatively, v is a voltage rise in the direction opposite to the current. The conventional diagrammatic representation of a resistance, together with designations of the current direction and voltage polarity, is shown in Fig. 1-3. The plus and minus signs denote decrease of potential, and hence a voltage drop, from left to right (or plus to minus). Note that in Fig. 1-3, the resistance exists between two terminals a and b. A *terminal* is a point at which an electric connection may be made to other elements or devices.

The power dissipated by the resistance may be determined from Eq. (1-12) combined with Eq. (1-14):

$$p = vi = (Ri)i = i^2R = v\frac{v}{R} = \frac{v^2}{R} \quad \text{W} \tag{1-15}$$

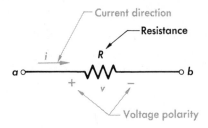

FIGURE 1-3 Schematic representation of resistance.

Equation (1-14) gives the voltage across a resistance in terms of its current. A reciprocal relationship giving the current in terms of the voltage is often of equal or greater value in a particular case. As a result, Ohm's law is often expressed as

$$i = Gv \quad A \tag{1-16}$$

where

$$G = 1/R \tag{1-17}$$

Reciprocal resistance G is called *conductance* and is measured in *mhos*. (The SI unit for conductance is the siemens, but mho is more frequently used. The symbol $\mho$ is used as an abbreviation.) Conductance power becomes

$$p = vi = v(Gv) = v^2G = i\frac{i}{G} = \frac{i^2}{G} \quad W \tag{1-18}$$

The schematic representation depicted in Fig. 1-3 is that of an ideal resistance. Practical resistors are subject to the limitations imposed by the "real" world, such as power-handling capacity, manufacturing tolerance, and temperature variation. Thus the 50-Ω 10 percent 2-W resistor we purchase at the store refers to an element whose nominal value of resistance is 50 Ω at a specified temperature (usually 25°C). Manufacturing variations may be ±10 percent of nominal value so that the resistance range lies between 45 and 55 Ω. The 2-W power rating is the maximum dissipation permissible without doing permanent damage to the resistor. The import of the power rating is that the allowable current is also limited. For the 50-Ω 2-W resistor, this current, from Eq. (1-15), is 0.2 A.

Resistance values also vary with environmental conditions, with temperature being most significant. Dependent on the material used, values of resistance may either increase or decrease as temperature increases. For example, carbon-composition resistors exhibit a decrease in resistance with an increase in temperature, while the resistance of metallic-wire and silicon-film resistors increases with increasing temperature. Although deviations from nominal values are factors with which the designer must cope, the variation of resistance with temperature is often used to advantage in instrumentation and control.

1-5 CAPACITANCE

The second electric circuit element is used to represent charge storage and, consequently, energy stored in the electric field. As expressed quantitatively in Eq. (1-19), the current through the element is proportional to the derivative of voltage across it.

$$i = C \frac{dv}{dt} \tag{1-19}$$

Solving Eq. (1-19) for the voltage yields

$$v = \frac{1}{C} \int i\,dt \tag{1-20}$$

and, with the use of Eq. (1-19),

$$v = \frac{q}{C} \quad \text{and} \quad q = Cv \tag{1-21}$$

The proportionality constant C expresses the charge-storing property of the element and is called the *capacitance* of the element. With q in coulombs and v in volts, the capacitance is in *farads* (abbreviated F). Because a farad is physically a large unit, C is frequently expressed in microfarads (equal to 10^{-6} farad and abbreviated μF) or in picofarads (equal to 10^{-12} farad and abbreviated pF). A *capacitor* is a physical element which exhibits the property of capacitance.

The schematic representation of capacitance in which current and voltage reference directions are indicated is shown in Fig. 1-4. In this figure and in Eqs. (1-19) and (1-20), a voltage drop exists in the direction of the current. Charge flow from a higher potential to a lower potential, i.e., from plus to minus, signifies that energy can be removed from the circuit and stored. The capacitive effect may be thought of as opposing a change in current.

The power associated with a capacitance is

$$p = vi = Cv \frac{dv}{dt} \quad \text{W} \tag{1-22}$$

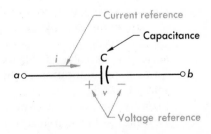

FIGURE 1-4 Schematic representation of capacitance.

and the energy is

$$w = \int p\,dt = \int Cv\frac{dv}{dt}\,dt = \int Cv\,dv = \frac{1}{2}Cv^2 \quad \text{J} \tag{1-23}$$

The value of the energy stored in the capacitance, given in Eq. (1-23), is dependent only on the voltage magnitude and not on the manner of reaching that magnitude.

The stored energy is returned to the circuit as the voltage is reduced to zero. For example, if a capacitance is discharged by placing a resistance across it, a current persists in the resistance until the stored energy is dissipated as heat in the resistance. When a *short circuit*, i.e., an element of zero resistance, is placed across the terminals of the capacitance, the stored energy is radiated to the surrounding environment. Practically speaking, however, no zero resistance elements exist, so that the energy is converted to heat or any spark produced (as might occur when you walk across a woolen carpet and reach for a metal doorknob).

In one sense, the behavior of a capacitance can be likened to that of a mechanical spring. Whether it is elongated or compressed, the potential energy stored in a spring is a function of position only. This energy can reappear when the spring returns to its equilibrium condition. The analogy that may be drawn is then between charge and position, energy in the electric field and potential energy.

Example 1-2

The current in a 0.1-F capacitance is shown in Fig. 1-5a. The curve, called a *waveform*, is that of a pulse having a value of 4 A for times between 0 and 0.5 s and zero thereafter. Sketch the waveforms of the capacitance voltage v, charge q, power p, and stored energy w as functions of time.

SOLUTION The desired results are displayed in Fig. 1-5. The voltage (Fig. 1-5b) is found from Eq. (1-20). The charge (Fig. 1-5c) is, from Eq. (1-21), Cv. The power and energy (Fig. 1-5d and e) are determined from Eqs. (1-22) and (1-23), respectively. It is of note that after the current pulse has ceased to exist, the capacitance remains charged, a voltage exists across its terminals, and energy is stored in its electric field.

To obtain the voltage and energy stored from the current and power, we can use graphical integration. Thus, v is the area under the current pulse, scaled by the value of capacitance; and w is the area under the power waveform. For the duration of the current pulse, its area increases linearly with time, resulting in the voltage waveform of Fig. 1-5b. Similarly, the area defined by the power waveform increases parabolically between 0 and 0.5 s. Beyond $t = 0.5$ s, p is zero and no further contribution to the area occurs. This results in w in Fig. 1-5e remaining constant at a value equal to the area of the triangle in Fig. 1-5d. The graphical integration technique may already

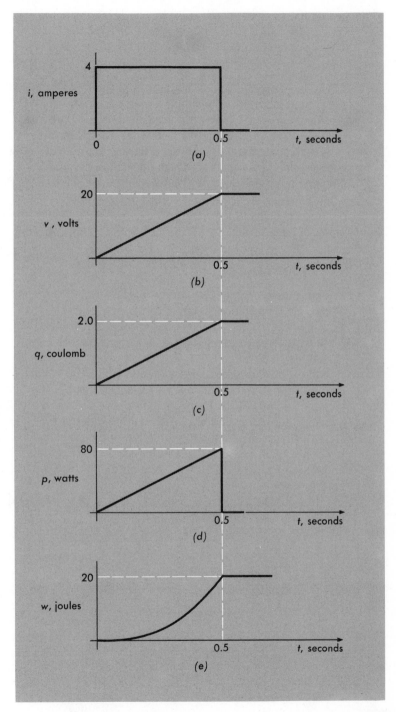

FIGURE 1-5 Waveforms of (a) current, (b) voltage, (c) charge, (d) power, and (e) stored energy in Example 1-2.

be familiar to students of mechanics. It is often used to relate acceleration to velocity or velocity to displacement in dynamics and to relate shear and bending moment (moment-shear diagrams) in elementary structural analysis.

In its simplest form a capacitor is constructed by forming a pair of metal plates separated by an insulating material (dielectric). Practical capacitors employ a variety of geometries and dielectrics to provide a wide range of capacitance values, usually from a few picofarads to one farad. However, as a given capacitor cannot store an arbitrary amount of energy, the maximum voltage which may exist across the capacitor terminals is limited. Exceeding the maximum voltage rating causes permanent damage to the capacitor as a result of dielectric breakdown. For physical dimensions no larger than one's thumb, it is usual that the larger the capacitance values, the smaller the maximum voltage ratings.

1-6   INDUCTANCE

The circuit element used to represent the energy stored in a magnetic field is defined by the relation

$$v = L \frac{di}{dt} \tag{1-24}$$

Equation (1-24) describes a situation in which the voltage across the element is proportional to the time rate of change of current through it. The constant of proportionality L is the *self-inductance,* or simply the *inductance* of the element, and is measured in *henrys* (abbreviated H). The voltage v in Eq. (1-24) is a voltage drop in the direction of the current and can be considered to oppose an increase in current. Figure 1-6 depicts the schematic representation of an inductance and its associated reference direction for current and voltage polarity.

If the voltage across the inductance is known, Eq. (1-24) may be rewritten so that the current is determined as

$$i = \frac{1}{L} \int v \, dt \tag{1-25}$$

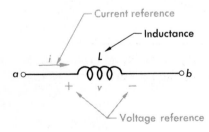

FIGURE 1-6   Schematic representation of inductance.

Equation (1-25) shows that inductance current depends not on the instantaneous value of the voltage but on its past history, that is, on the integral or sum of the volt-second products for all time prior to the time of interest. For many applications, where knowledge of an inductance current following a switching process (usually occurring at an arbitrary time called $t = 0$) is desired, Eq. (1-25) may be written in the form

$$i = \frac{1}{L} \int_0^t v \, dt + i(0) \qquad (1\text{-}26)$$

where $i(0)$ is the current existing at the time of switching and is a measure of the past history of the inductance prior to the switching process.

Because its effect is to oppose changes in the velocity of flow of charge, inductance is analogous to mass or inertia in a mechanical system and to the mass of liquid in hydraulics. Inductance prevents the current from changing instantly [notice that Eq. (1-24) requires an infinite voltage to cause an instantaneous change in current], just as the mass of an automobile prevents it from stopping or starting instantaneously.

The power associated with the inductive effect in a circuit is

$$p = vi = Li \frac{di}{dt} \qquad \text{W} \qquad (1\text{-}27)$$

and the energy is

$$w = \int p \, dt = \int Li \frac{di}{dt} \, dt = \int Li \, di = \frac{1}{2} Li^2 \qquad \text{J} \qquad (1\text{-}28)$$

Unlike the resistive energy, which goes into heat, the inductive energy is stored in the same sense that kinetic energy is stored in a moving mass. It can be seen from Eq. (1-28) that its value is dependent only on current magnitude and not on the manner of reaching that magnitude. The stored inductive energy reappears in the circuit as the current is reduced to zero. For example, if a switch is opened in a current-carrying inductive circuit, the current decreases rapidly, but not instantaneously. In accordance with Eq. (1-24), a relatively high voltage appears across the separating contacts of the switch, and an arc may form. The arc makes it possible for the stored energy to be dissipated as heat in the arc and circuit resistances.

Example 1-3

The current in a 10-H inductance is shown in Fig. 1-7a. Sketch the waveforms for the voltage v, the instantaneous power p, and the energy stored w as functions of time.

SOLUTION   The waveforms depicted in Fig. 1-7b, c, and d are the desired results. The voltage waveform (Fig. 1-7b) is obtained from Eq. (1-24). Recalling that the derivative represents the slope of the function, the voltage

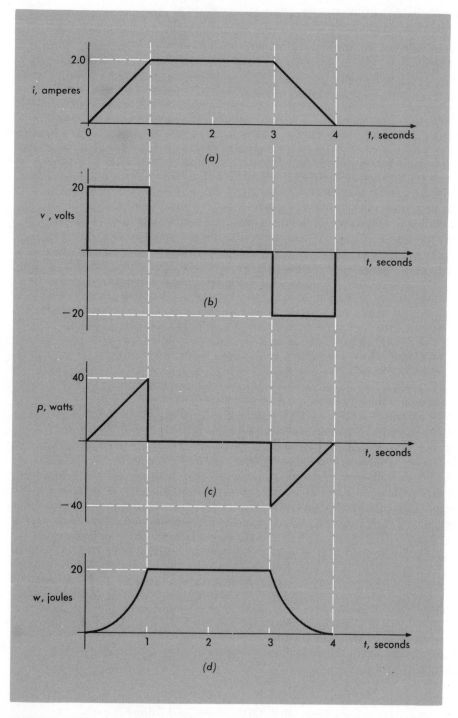

FIGURE 1-7   Waveforms of (a) current, (b) voltage, (c) power, and (d) stored energy in Example 1-3.

is zero whenever the current is constant (the 1 to 3-s time interval). The power waveform in Fig. 1-7c is found by the use of Eq. (1-27); note that the waveform is zero when the voltage is zero, a consequence of a constant current. Graphical integration of the power waveform or the use of Eq. (1-28) yields the stored energy illustrated in Fig. 1-7d. Although both the voltage and power are zero between 1 and 3 s, energy is stored in the inductance during that time.

The self-inductance of a circuit is intimately associated with the magnetic field linking the circuit. The self-inductance voltage may be thought of as the voltage induced in the circuit by a magnetic field produced by the circuit current. Since a magnetic field exists in the region around the current which produces it, there is also a possibility that a voltage may be induced in other circuits linked by the field. Two circuits linked by the same magnetic field are said to be *coupled* to each other. The circuit element used to represent magnetic coupling is shown in Fig. 1-8 and is called *mutual inductance*. It is represented by the symbol M and, like self-inductance, is measured in henrys. The voltampere relationship is one which gives the voltage induced in one circuit by a current in another and is

$$v_2 = M \frac{di_1}{dt} \tag{1-29}$$

A similar equation can, of course, be written giving a voltage v_1 induced by a current i_2. The two dots, called *polarity markings,* in Fig. 1-8 are used to indicate the direction of the magnetic coupling between the two coils; note from the figure and from Eq. (1-29) that by matching the dots, the directions of current and voltage drops are made to correspond with those of Fig. 1-9 and Eq. (1-24) for the self-inductance.

If currents are present in both coupled circuits, voltages of self-inductance and mutual inductance are induced in each circuit. The self-induced voltages have the directions shown in Fig. 1-6; the mutually induced voltages follow the pattern of Fig. 1-8. Figure 1-9 shows two such elements. Each coil is characterized by its own self-inductance; the combination has a

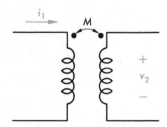

FIGURE 1-8 Schematic representation of mutual inductance.

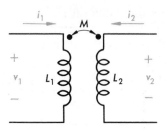

FIGURE 1-9 Mutual inductance with current in both coils.

mutual inductance indicating coupling between the coils. The voltampere relationships for this case combine Eqs. (1-24) and (1-29) and are

$$v_1 = L_1 \frac{di_1}{dt} + M \frac{di_2}{dt} \qquad (1\text{-}30)$$

and
$$v_2 = L_2 \frac{di_2}{dt} + M \frac{di_1}{dt} \qquad (1\text{-}31)$$

Coupling between two closed circuits permits the transfer of energy between the circuits through the medium of the mutual magnetic field. This phenomenon is the basis on which all transformers operate; at times it is an annoyance, creating extraneous coupling between supposedly isolated circuits (causing interference from power lines to nearby telephone lines, for example).

1-7 INDEPENDENT SOURCES

Each of the elements described in the previous three sections is capable of removing energy from the system by storage in a field or by dissipation. In order that energy be conserved, we must introduce elements capable of supplying energy. Such elements are called *sources* and deliver energy at either a specified voltage or a specified current.

The schematic representation of an ideal voltage source is given in Fig. 1-10*a*. The voltage drop from *a* to *b* in Fig. 1-10*a* is independent of the load that may be connected to the source and is determined exclusively by the function *v*(*t*). The circuit symbol shown in Fig. 1-10*b* is frequently used to indicate an ideal voltage source that presents a constant voltage *V* at its terminals. In particular, battery voltages are so depicted. When a load is connected to the source terminals, as in Fig. 1-10*c*, a current exists in the source in the direction of the voltage rise. A current direction from minus to plus (− to +) signifies the source is delivering energy to the load. The amount of current is determined by the power required by the load. Changes in load power are accompanied by corresponding changes in current as the energy is delivered at the voltage value which defines the source. It is also possible for the current in the source to exist in the direction of the voltage drop. In this situ-

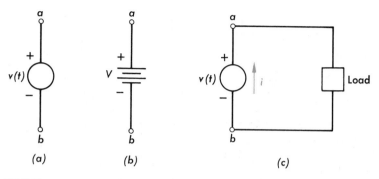

FIGURE 1-10 Circuit symbols for (a) an ideal voltage source, (b) a battery or ideal voltage source with constant terminal voltage, and (c) a voltage source with load.

ation the voltage source is extracting energy from the system as is the case when a battery is charged.

An ideal current source representation is shown in Fig. 1-11a; the arrow in the figure points in the direction of positive current. An ideal current source is characterized by a current which is independent of load connections at its terminals. The value of the source at any instant of time is given only by the function $i(t)$. As illustrated in Fig. 1-11b, when supplying power to a load, the source current is in the direction of the voltage rise. The value of the voltage rise varies with the power requirements of the load; the power is provided at the specified current of the source $i(t)$.

An analogous situation exists in the behavior of certain mechanical systems. The use of a rotating shaft supplying energy to a mechanical load can be accomplished by driving the load at either a specified torque or a specified angular velocity. In the former case, angular velocity adjusts to satisfy load requirements; in the latter, it is the torque that adjusts to accommodate the load power. The voltage and current sources may be considered as analogous to the torque and angular velocity, respectively.

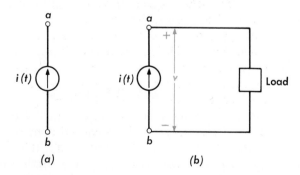

FIGURE 1-11 (a) Circuit symbol for an ideal current source. (b) A current source with load.

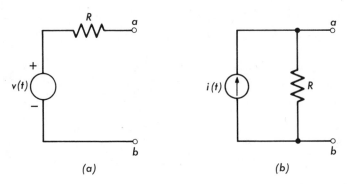

FIGURE 1-12   Representations of practical (a) voltage source and (b) current source.

The values of $v(t)$ and $i(t)$ in Figs. 1-10a and 1-11a are not affected by the loads connected to them or by the electrical conditions that exist elsewhere in the systems in which they are employed. For this reason, the ideal voltage and current sources described to this juncture are referred to as independent sources. Sometimes, in engineering practice, the terms *constant voltage source* and *constant current source* are used even when the source quantities are functions of time. However, in the hope of avoiding needless confusion in this text, constant voltage and constant current sources denote only those cases for which the values of source voltage and source current do not vary with time. It is also important to note that the source values of $v(t)$ and $i(t)$ in independent sources are not arbitrary. They depend on the characteristics of the energy conversion systems used in the realization of the sources. Thus, the terminal voltage of a battery is a function of the internal electrochemical action.

An implied characteristic of an ideal voltage source is that no limit exists on the current and, consequently, the power the source can deliver. A similar statement applies to the voltage across and the power delivered by a current source. Yet, we know that the battery used to power a transistor radio cannot start an automobile, because the radio battery is incapable of delivering the high current required for ignition. When we account for the current, voltage, and power limitations, practical sources are most often represented by the ideal source-resistance combinations depicted in Fig. 1-12. A detailed, analytic description of practical sources is given in Sec. 2-3.

1-8 DEPENDENT SOURCES

A second important class of electrical sources is characterized by the fact that the source voltage or current depends on the current or voltage elsewhere in the circuit. Sources which exhibit this dependency are called *controlled* or *dependent sources*. Both controlled current and controlled voltage sources exist; since a voltage or current can serve as the control quantity, four types of dependent sources are possible. These four are represented schematically in Fig. 1-13. The sources, designated by diamonds in Fig.

1-13, act between terminals c and d, with control provided at terminals a and b. The plus and minus indications in Fig. 1-13a and b are the reference polarities for the voltage sources. The direction of positive current is given by the arrows in Fig. 1-13c and d. The proportionality constants A, r, and g indicate the relation between the control quantities and the strength of the sources; in practice they are important parameters of the physical devices used to realize controlled sources.

The behavior of dependent sources may be demonstrated by their comparison with independent ones. For example, at terminals c-d, the voltage-controlled current source of Fig. 1-13c acts in the same manner as does the independent current source in Fig. 1-11a. That is, in both cases, power supplied to a load connected to the source terminals is independent of that load. One principal difference illustrated in Fig. 1-13c is that the source current gv_1 can be altered by a change in the circuit conditions which affect v_1. Thus, decreasing v_1 serves to decrease the strength of the source, while increasing v_1 increases the source current. In Fig. 1-11c, the value of $i(t)$ is independent of all circuit conditions.

Another major difference is that four terminals are required to define a controlled source; whereas only two are needed for an independent source. Of the four dependent source terminals, one pair provides the control, and the second pair exhibits the properties of the source. Often terminals b and d in Fig. 1-13 are common, so that only three terminals may be necessary to

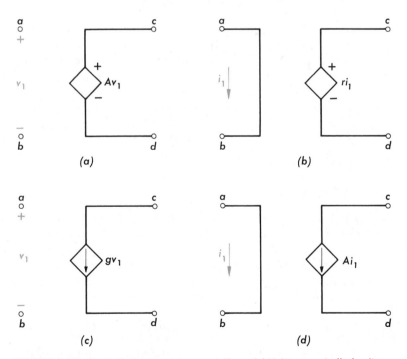

FIGURE 1-13 Controlled-source representations. (a) Voltage-controlled voltage source, (b) current-controlled voltage source, (c) voltage-controlled current source, and (d) current-controlled current source.

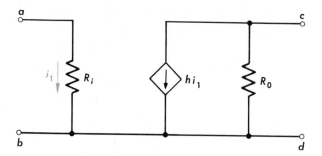

FIGURE 1-14 Nonideal current-controlled current source which can serve as transistor model.

represent a controlled source as a circuit element. The advantage gained from the use of a three- or four-terminal circuit element is that the control terminals may be connected to components that are remote from those connected to the source terminals. The controlled source may be thought of as providing "action at a distance" so that various circuit elements may be physically isolated from one another.

The importance and widespread application of dependent sources is underscored by the fact that transistors and operational amplifiers are the most common physical devices used to obtain controlled-source characteristics. In the transistor, the output current can be considered proportional to the input current, so that it may be modeled as a current-controlled current source. (As described in Chap. 8, transistors are also often represented as voltage-controlled current source.) Operational amplifiers display voltage-controlled voltage source characteristics since the output voltage is directly related to input voltage. It is sometimes convenient to consider electric generators, in which the voltage induced in one winding depends on the current in a second winding, as current-controlled voltage sources. Because physical devices only approximate ideal sources, their representations include additional circuit elements which reflect practical limitations. As an example, the circuit in Fig. 1-14 can be used as a model for the transistor.

1-9 KIRCHHOFF'S LAWS

The basic laws that electric circuits obey follow rationally from the nature of the electrical quantities defined in earlier sections of this chapter. They lead directly to methods for the systematic analysis of electric circuits. These laws are known as *Kirchhoff's laws,* and they describe the relationships among circuit voltages and circuit currents that must be satisfied. The first of these laws is Kirchhoff's current law (KCL):

1 *The algebraic sum of all the currents directed away from a node is zero.*

As used in this book, a *node* is defined as a point where three or more connections to elements or sources are made. While Kirchhoff's current law is equally valid for a two-connection point, its application merely states the

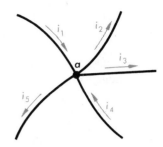

FIGURE 1-15 A node in a circuit.

fact that the two element currents are equal and may therefore be described by a single current.

When this law is applied, currents directed away from the node are considered positive and those directed toward it negative (or vice versa, as long as one is consistent). In Fig. 1-15, the equation at node a is

$$-i_1 + i_2 + i_3 - i_4 + i_5 = 0$$

which is the equivalent of saying that all the charges flowing toward the junction must also flow away from it. Violation of the current law would result in a "stream of amperes" from node a into free space, an evident impossibility.

Example 1-4

In Fig. 1-16, a portion of a circuit is shown for which the following voltages and current are known: $v_1 = 10\epsilon^{-2t}$ volts, $v_2 = 2\epsilon^{-2t}$ volts, and $i_3 = 6\epsilon^{-2t}$. Find v_4.

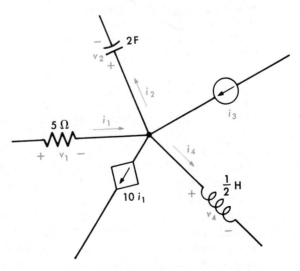

FIGURE 1-16 Circuit for Example 1-4.

SOLUTION The KCL equation is

$$-i_1 + i_2 - i_3 + i_4 + 10i_1 = 0$$

Current i_3 is known and currents i_1 and i_2 are found from the voltampere relations for the resistance and capacitance, respectively. From Eq. (1-14),

$$i_1 = \frac{10\epsilon^{-2t}}{5} = 2\epsilon^{-2t} \quad \text{A}$$

and from Eq. (1-19)

$$i_2 = 2\frac{d}{dt}(2\epsilon^{-2t}) = -8\epsilon^{-2t} \quad \text{A}$$

The current i_4 becomes

$$\begin{aligned} i_4 &= i_1 - i_2 + i_3 - 10i_1 \\ &= 2\epsilon^{-2t} + 8\epsilon^{-2t} + 6\epsilon^{-2t} - 20\epsilon^{-2t} = -4\epsilon^{-2t} \quad \text{A} \end{aligned}$$

The voltage across the inductance v_4 is determined by use of Eq. (1-24) and is

$$v_4 = \frac{1}{2}\frac{d}{dt}(-4\epsilon^{-2t}) = 4\epsilon^{-2t} \quad \text{V}$$

The second of Kirchhoff's laws is the voltage law (KVL) and is

2 *The algebraic sum of the voltage drops taken in a specified direction around a closed path is zero.*

A closed path in an electric circuit is called a *loop* or *mesh*. The second law is a consequence of the energy-conservation principle and is equivalent to striking a balance by equating energy input to energy output. In writing KVL equations, one may go around the path in either direction and sum the voltage drops provided that the voltage across each element in the path is accounted for once and only once. (The algebraic sum of the voltage rises can also be used in applying KVL as long as one is consistent within a particular equation.)

In general, the solution of circuit problems involves the determination of currents and voltages in certain circuit elements when the currents or voltages in other elements are given. Assignment of current and voltage variables is arbitrary, although an orderly engineering process is usually followed to reduce the chance of error. However, for the three elements described in Secs. 1-4, 1-5, and 1-6, the direction of current and voltage polarity cannot both be chosen arbitrarily. As given in Fig. 1-3 and Eqs. (1-14) and (1-16) for the resistance, Fig. 1-4 and Eqs. (1-19) and (1-20) for the capacitance, and Fig. 1-6 and Eqs. (1-24) and (1-25) for the inductance, selection of a current direction determines the voltage polarity. Conversely, specification of the voltage polarity requires a current direction consistent with the defining volt-ampere relations for each of the three elements.

Example 1-5

Figure 1-17 depicts a portion of a circuit in which $v_1 = 12 \sin 2t$, $i_2 = 4 \sin 2t$, and $i_3 = 2 \sin 2t - 4 \cos 2t$. Determine i_4.

SOLUTION   For the mesh shown in the figure, starting at point a, the KVL equation is

$$-v_1 + v_2 - v_3 - 5v_2 + v_4 = 0$$

The voltage v_1 is known, and v_2 and v_3 are determined from Eqs. (1-24) and (1-14), respectively, as

$$v_2 = 2 \frac{d}{dt} (4 \sin 2t) = 16 \cos 2t \quad \text{V}$$

$$v_3 = 5(2 \sin 2t - 4 \cos 2t) = 10 \sin 2t - 20 \cos 2t \quad \text{V}$$

The voltage v_4 is then

$$
\begin{aligned}
v_4 &= v_1 - v_2 + v_3 + 5v_2 \\
&= 12 \sin 2t - 16 \cos 2t + 10 \sin 2t - 20 \cos 2t + 80 \cos 2t \\
&= 22 \sin 2t + 44 \cos 2t \quad \text{V}
\end{aligned}
$$

By use of Eq. (1-19), i_4 is determined as

$$i_4 = \frac{1}{4} \frac{d}{dt} (22 \sin 2t + 44 \cos 2t) = 11 \cos 2t - 22 \sin 2t \quad \text{A}$$

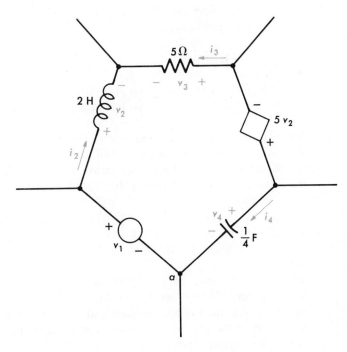

FIGURE 1-17 Circuit for Example 1-5.

It should be noted that the voltage polarities for v_2 and v_3 must be as indicated in Fig. 1-17, since the current directions for i_2 and i_3 are specified.

Electric circuit problems are thus seen to be the determination of a cause-effect relationship in a network made up of one or more energy sources and one or more circuit elements. The cause is usually the voltage or current sources exciting the circuit; the desired effect is often the voltage or current existing in a certain portion of the network. Solution is achieved through application of Kirchhoff's current and voltage laws and the volt-ampere relationships of the three circuit elements. The formal body of knowledge organizing the use of these rules, and hence expediting the solution of specific problems, is called *circuit theory*. Succeeding chapters will deal with circuit theory in some detail.

PROBLEMS

1-1 An incandescent lamp uses electric energy at the rate of 100 W when supplied by a constant 120-V source. Determine
a The current in the lamp.
b The electric charge flowing through the circuit per hour.
c The cost of operation for one year assuming the lamp is on 10 h/day and the cost of electric energy is 6.0 cents/kWh.

1-2 A 12-V lead storage battery is used to supply a 120-W load.
a Determine the load current.
b If the battery is rated at 60 A-h, how long can the load be supplied?
c What is the total charge stored in the battery?

1-3 The voltage and current waveforms for the electric device shown in Fig. 1-18*a* are depicted in Fig. 1-18*b*.
a Sketch a curve of the device power as a function of time.
b What is the average power dissipated by the device during the interval $0 \le t \le 4$ s.

1-4 The trapezoidal waveforms for current and voltage shown in Fig. 1-19 approximate those which exist in electronic switches used in digital computers.
a Sketch the power as a function of time.
b Determine the average power during the first 100 ns.
c Determine the average voltage and current during the first 100 ns.
d Compare the product of the average voltage and current with the result obtained in (*b*) and explain any difference.

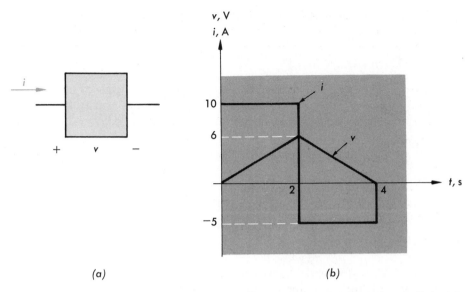

FIGURE 1-18 (a) Device and (b) waveform for Prob. 1-3.

1-5 The current and voltage in the device shown in Fig. 1-18a are

$$v(t) = 20 \cos t \qquad i(t) = 5 \cos t$$

a Sketch the instantaneous power.
b Determine the average power dissipated during one cycle ($0 \leq t \leq 2\pi$).

1-6 Repeat Prob. 1-5 for
a $v(t) = 10 \cos t \qquad i(t) = 2 \cos (t + 60°)$
b $v(t) = 15 \cos t + 5 \cos 2t \qquad i(t) = 2 \cos t$

1-7 Repeat Prob. 1-5 for
a $v(t) = 5 - 5 \cos 2t \qquad i(t) = 2 + 2 \cos 2t$
b $v(t) = 50 \cos 2t \qquad i(t) = 5 \sin 2t$

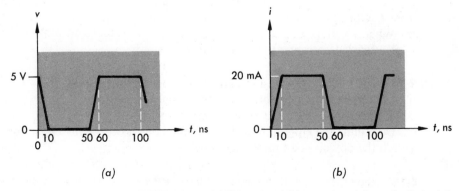

FIGURE 1-19 (a) Voltage and (b) current waveforms for Prob. 1-4.

1-8 The voltage and current in an electrical network are given by

$$v(t) = 100\epsilon^{-3t} \qquad i(t) = 2(1 - \epsilon^{-3t})$$

a Sketch $v(t)$ and $i(t)$.
b Sketch the instantaneous power.
c Determine the total energy dissipated during the interval $0 \leq t \leq \frac{1}{3}$.
d What is the total energy dissipated during the first second?

1-9 Repeat Prob. 1-8 for
a $v(t) = 10\epsilon^{-1.5t} \qquad i(t) = 1 - \epsilon^{-1.5t}$
b $v(t) = 20\epsilon^{-t} \qquad i(t) = t\epsilon^{-t}$

1-10 Find the resistance in ohms of
a The incandescent lamp of Prob. 1-1.
b The load in Prob. 1-2.

1-11 A resistor whose power rating is 0.5 W has a resistance of 10 kΩ.
a What is the maximum dc current in the resistor for which the power rating is not exceeded?
b A current, $i(t) = (2 + \sqrt{2} \sin 10^6 t) \times 10^{-3} A$, is supplied to the resistor. Sketch the voltage and instantaneous power for one cycle.
c Determine the average power dissipated in one cycle, when the resistor is supplied by the current given in b.

1-12 The resistance of a certain platinum resistance-thermometer is given by

$$R_T = R_0(1 + 5.42 \times 10^{-2} T - 4.17 \times 10^{-6} T^2)$$

where R_T = resistance at temperature T
       R_o = resistance at $T = 0°C$
and
       T = temperature, °C

a For $R_o = 25 \, \Omega$, sketch R_T as a function of T for the range $0° \leq T \leq 600°C$.
b When a voltage of 6 V is applied to the thermometer, a current of 15 mA is measured. Determine the temperature.

1-13 The voltage across a 100-μF capacitor is $120\sqrt{2} \sin 377t$.
a Determine the capacitor current as a function of time.
b Sketch the current and voltage waveforms.
c Determine the equation for the instantaneous power in the capacitor and sketch the power as a function of time.

1-14 Repeat Prob. 1-13 for $v(t) = 12(1 - \epsilon^{-10^3 t})$.

1-15 The voltage across a 50-pF capacitor is given by

$$v(t) = \begin{cases} 10^8 t & 0 \le t \le 50 \text{ ns} \\ 7.5 - 5 \times 10^7 t & 50 \le t \le 150 \text{ ns} \\ 0 & t > 150 \text{ ns} \end{cases}$$

a Determine the equation for the capacitor current as a function of time.
b Sketch the energy stored in the capacitor as a function of time.

1-16 The current in a 10-mH inductance is

$i(t) = 0.1414 \sin 377t$

a Determine the voltage across the inductance.
b Sketch the voltage and current waveforms.
c Determine the equation for, and sketch, the instantaneous power.

1-17 Repeat Prob. 1-16 for $i(t) = 10^{-3}(1 - \epsilon^{-10^6 t})$

1-18 The current through a 500-μH inductance is

$$i(t) = \begin{cases} 10^4 t & 0 \le t \le 2 \ \mu s \\ 2.5 \times 10^{-2} - 2.5 \times 10^3 t & 2 \le t \le 10 \ \mu s \\ 0 & t \ge 10 \ \mu s \end{cases}$$

a Determine the equation for the voltage across the inductance as a function of time.
b Sketch the energy stored in the inductance as a function of time.

1-19 The currents $i_1(t)$ and $i_2(t)$ in the mutually coupled coils shown in Fig. 1-9 are

$i_1(t) = 10 \cos 377t \qquad i_2(t) = 5 \cos 377t$

The element values are: $L_1 = 10$ H, $L_2 = 5$ H, $M = 5$ H.
Determine the voltages $v_1(t)$ and $v_2(t)$.

1-20 Repeat Prob. 1-19 for $i_1(t) = 5(1 - \epsilon^{-t})$ and $i_2(t) = 10(1 - \epsilon^{-t})$.

1-21 In the circuit of Fig. 1-20, the source voltages are $V_1 = 45$ V and $V_2 = 15$ V. Determine:
a The current from a to b
b The voltage drop V_{ac}
c The power supplied by V_1
d The power absorbed by the segment bc

1-22   In the circuit of Fig. 1-20, $V_{ac} = 12$ V and $V_{bd} = 21$ V. Determine the values of V_1 and V_2.

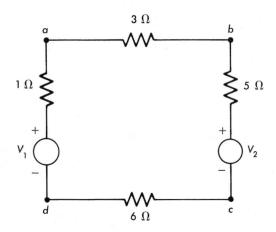

FIGURE 1-20   Circuit for Prob. 1-21 and 1-22.

1-23   In the circuit of Fig. 1-21, the source currents are $I_A = 3$ A and $I_B = 5$ A. Determine:
a   The voltage V
b   The currents I_1, I_2, and I_3
c   The power supplied by I_B

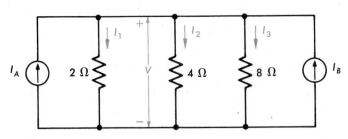

FIGURE 1-21   Circuit for Prob. 1-23.

1-24   For the circuit segment shown in Fig. 1-22,

$$i_1 = 2 \text{ A} \qquad i_3(t) = 10\epsilon^{-t} \qquad \text{and} \qquad i_4(t) = 5 \cos 2t$$

Find v_1, v_2, v_3, v_4, i_2 and i_5.

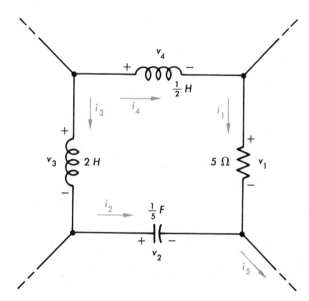

FIGURE 1-22   Circuit segment for Prob. 1-24.

1-25   For the circuit segment shown in Fig. 1-23, find i_A and v_{ab}.

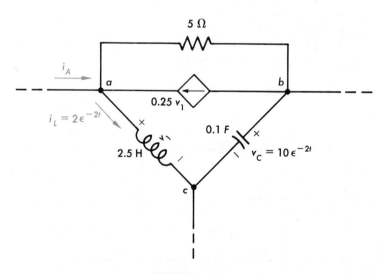

FIGURE 1-23   Circuit segment for Prob. 1-25.

1-26 For the circuit segment shown in Fig. 1-24, find $v_o(t)$.

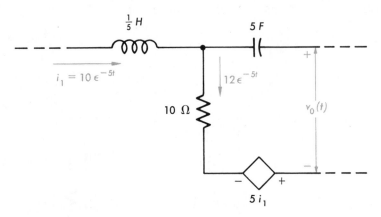

FIGURE 1-24 Circuit segment for Prob. 1-26.

1-27 For the circuit shown in Fig. 1-25, determine the power supplied by the 6-V source.

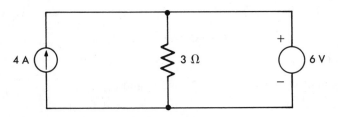

FIGURE 1-25 Circuit for Prob. 1-27.

1-28 For the circuit segment shown in Fig. 1-26, determine $v(t)$.

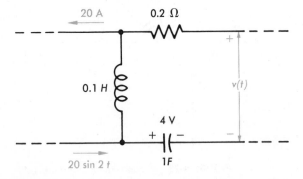

FIGURE 1-26 Circuit segment for Prob. 1-28.

1-29   In the circuit of Fig. 1-27, $v(t) = 5\epsilon^{-t}$ volts. Use Kirchhoff's laws and the volt-ampere relations for the elements to determine the source current $i(t)$.

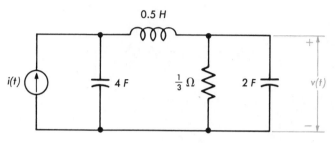

FIGURE 1-27   Circuit for Prob. 1-29.

1-30   Repeat Prob. 1-29 for $v(t) = 10$ V.

1-31   Find the source voltage $v(t)$ in the circuit of Fig. 1-28 when the current $i(t) = -20\epsilon^{-2t}$.

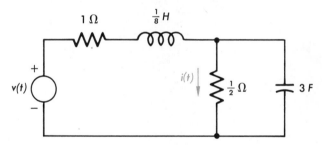

FIGURE 1-28   Circuit for Prob. 1-31.

1-32   Repeat Prob. 1-31 for $i(t) = 24 \sin (2t/3)$.

1-33   In the circuit of Fig. 1-29, $V_o = 5$ V. Determine I_S.

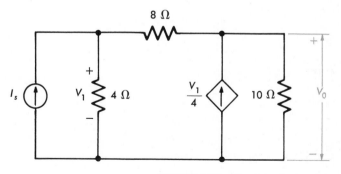

FIGURE 1-29   Circuit for Prob. 1-33.

1-34   The circuit in Fig. 1-30 is often used to represent an operational amplifier stage. In the circuit of Fig. 1-30, the power supplied by the 2.5 V source is 1.25 mW. Determine the voltage V_o.

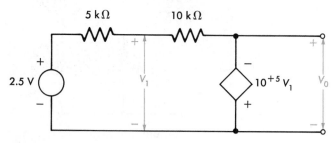

FIGURE 1-30   Circuit for Prob. 1-34.

2

CIRCUIT ANALYSIS-RESISTIVE NETWORKS

irchhoff's laws, used in conjunction with the element definitions described in the previous chapter, are the basis upon which circuit problems are solved. The purposes of this chapter are twofold: first, to develop the orderly processes by which the equations needed for the solution of network problems may be readily written and solved; second, to investigate techniques which permit networks to be represented by relatively simple equivalent circuits. In order to focus attention on the network rather than on the complexities of the mathematics, we will use resistive networks excited by constant current and voltage sources to illustrate the techniques.

2-1 KIRCHHOFF'S LAWS AND THE MEASUREMENT OF CURRENT, VOLTAGE, AND RESISTANCE

Electric circuits are combinations of elements in which at least one mesh exists. In general, circuits contain several loops and nodes and are excited by one or more sources. The response (or responses) is the current or voltage in one of the circuit elements. As a rule, known quantities are the element values. The unknown quantities include the voltages across current sources; the currents in voltage sources; and the currents and voltages associated with the resistances, capacitances, and inductances in the circuit.

The tools available to determine unknown quantities are Kirchhoff's current-law equations (KCL), Kirchhoff's voltage-law equations (KVL), and element volt-ampere relations (in this chapter, Ohm's law). When applied to circuits composed of the elements described in Chap. 1, these relations give rise to a system of equations which is linear. In a linear system of equations, the variables (unknowns) appear only to the first power, so that the resultant voltage and current responses are linear functions of each of the excitations. A specific response is, consequently, the sum of the responses to the individual excitations. This property, known as the *principle of superposition,* is the fundamental property of linear systems. (For further discussion, see Sec. 2-7.) All three kinds of equations are used simultaneously to solve circuit problems. In this section, these equations are developed for some simple circuits which are the models of the basic instruments used to measure current, voltage, and resistance.

The *D'Arsonval galvanometer* is the primary indicating instrument used to measure direct currents and voltages. The schematic representation of the meter movement of a D'Arsonval instrument is shown in Fig. 2-1. The wire coil is situated such that it is free to rotate in the field of the permanent magnet. When a current exists in the coil, a torque is produced which causes a deflection of the pointer whose resultant position is directly proportional to the average (dc) current in the coil. Typically, current values needed to effect full-scale deflection in the fine wire coils used lie in the range from

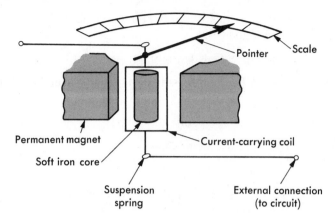

FIGURE 2-1 Representation of a D'Arsonval movement.

10 μA to 1 mA. To be useful in measuring the wide variation of current and voltage values encountered in practice, the galvanometer must be augmented by additional circuit elements. These circuit modifications result in *ammeters* to measure current and *voltmeters* to indicate voltage.

An ammeter capable of indicating a full-scale current in excess of the coil current is depicted in Fig. 2-2. As shown in the figure, a resistance R_S, called a *shunt*, is placed across the meter movement R_M. The currents I_0 and I_{MO} are the full-scale ammeter current and full-scale coil current, respectively. The purpose of the shunt is to permit most of the current to pass through it so that a specified ammeter current I_0 produces a coil current I_{MO}. The selection of the value of R_S must be based on the current I_0 and on the

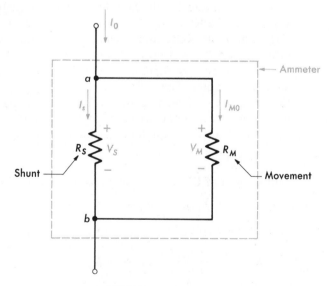

FIGURE 2-2 Circuit diagram of an ammeter.

characteristics of the meter movement. Application of the current law at node a gives

$$-I_O + I_S + I_{MO} = 0 \tag{2-1}$$

The KVL equation, starting at a and traversing the loop in a clockwise direction, is

$$V_M - V_S \quad \text{or} \quad V_M = V_S \tag{2-2}$$

When we use Ohm's law,

$$V_M = I_{MO} R_M \quad \text{and} \quad V_S = I_S R_S \tag{2-3}$$

Solving Eq. (2-1) for I_S and by combination of Eqs. (2-2) and (2-3), the value of R_S becomes

$$R_S = \frac{V_S}{I_S} = \frac{R_M I_{MO}}{I_O - I_{MO}} \tag{2-4}$$

Equation (2-4) indicates the value of R_S necessary for the ammeter to read full-scale for a current I_O and a meter movement having resistance R_M and coil current I_{MO}. In most laboratory ammeters, the value of R_S can be changed; each particular value of R_S results in a different value of I_O. Therefore, a given meter movement can provide an ammeter with a wide range of current scales.

It is important to note that the voltage drops across R_M and R_S are equal as is indicated in Eq. (2-2). In Fig. 2-2, both R_S and R_M are connected between nodes a and b. Circuit connections for which KVL requires the voltage across each element be the same, independent of the element values, are referred to as *parallel connections* or *parallel circuits*.

Also of interest is the fact that the KCL equation at node b is not used. If we write the KCL equation at b, we discover that it is the negative of Eq. (2-1). Thus, the KCL equations at a and b are not independent of one another and only one is needed to describe circuit behavior.

The meter movement can be augmented, as shown in Fig. 2-3, to construct a voltmeter. The resistance R_V is in *series* with R_M since KCL requires that the current in each be identical. The inclusion of R_V permits a full-scale deflection to occur for a voltage value of V_O rather than the meter voltage drop V_M. The KVL equation starting at a is

$$V_R + V_M - V_O = 0 \tag{2-5}$$

Substitution of Ohm's law relations for R_S and R_V in Eq. (2-5) yields

$$I_{MO} R_M + I_{MO} R_V - V_O = 0 \tag{2-6}$$

which, when solved for R_V, gives

$$R_V = \frac{V_O - I_{MO} R_M}{I_{MO}} = \frac{V_O}{I_{MO}} - R_M \tag{2-7}$$

The value of R_V in Eq. (2-7) is that necessary to convert the galvanometer to an instrument whose full-scale deflection occurs for a voltage V_O impressed

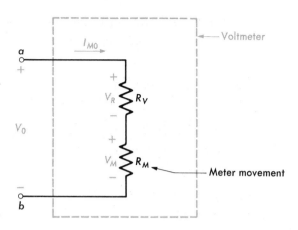

FIGURE 2-3 Voltmeter circuit schematic.

across terminals a and b. Practical voltmeters permit the value of R_V to be adjusted to accommodate a range of voltage scales.

The circuit in Fig. 2-4 illustrates how the galvanometer can be used to measure resistance. When used as shown in Fig. 2-4, the circuit is an *ohm-meter;* as the unknown resistance R_X is in series with the meter movement, the circuit is often called a series ohmmeter. The purpose of R_O is to limit the current in the meter to I_{MO} when R_X is zero, i.e., when terminals a and b are short-circuited. With R_X in the circuit, the KVL equation is

$$-V_B + I_M R_O + I_M R_M + I_M R_X = 0 \tag{2-8}$$

Solving Eq. (2-8) for I_M yields

$$I_M = \frac{V_B}{R_O + R_M + R_X} \tag{2-9}$$

With R_X equal to 0, Eq. (2-9) reduces to Eq. (2-6) with V_O set equal to V_B so that R_O is determined by Eq. (2-7). The scale on an ohmmeter is calibrated in ohms, although the deflection of the pointer is produced by the current I_M in the meter. The scale is nonlinear, however, because the current produced is not proportional to R_X. In practical ohmmeters, shunts may be placed in parallel with the meter to accommodate different resistance scales. Often, when

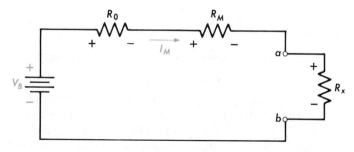

FIGURE 2-4 Schematic for a series ohmmeter.

accuracies in the order of 3 to 5 percent are permissible, the ammeter, volt-meter, and ohmmeter are combined in a single instrument called a *multi-meter*.

Example 2-1

A D'Arsonval galvanometer is to be used for an ammeter, a voltmeter, and a series ohmmeter. A current of 1 mA produces full-scale deflection and meter resistance is 50 Ω. A 3-V battery is used in the ohmmeter. Determine the values of

a R_S needed for an ammeter with 1.0-A full-scale reading

b R_V for a voltmeter having 10-V full-scale deflection

c R_0 for the ohmmeter

d What value of R_X produces half-scale deflection on the ohmmeter?

SOLUTION **a**   For the ammeter, the value of R_S is given by Eq. (2-4) as

$$R_S = \frac{50 \times 10^{-3}}{1 - 10^{-3}} = 0.05005 \; \Omega$$

b   From Eq. (2-7), the value of R_V is

$$R_V = \frac{10}{10^{-3}} - 50 = 9{,}950 \; \Omega$$

c   For the ohmmeter, by use of Eq. (2-7),

$$R_0 = \frac{3}{10^{-3}} - 50 = 2{,}950 \; \Omega$$

d   Half-scale deflection corresponds to a current of 0.5 mA. Using the value of R_0 found in part *c*, Eq. (2-9) gives

$$0.5 \times 10^{-3} = \frac{3}{2{,}950 + 50 + R_X}$$

from which

$$R_X = 3{,}000 \; \Omega$$

2-2   DIRECT APPLICATION OF KIRCHHOFF'S LAWS

The solution of multimode, multimesh circuit problems requires that the independent variables (unknowns) be identified and that a number of inde-pendent equations equal to the number of unknowns be written. The inde-pendent equations are obtained by the use of Ohm's law, KCL, and KVL for the circuit resistances. Stated without proof, the number of independent equations of each type is available:

1 The number of independent element volt-ampere equations is equal to the number of resistances.

2 The number of independent KCL equations is equal to one less than the number of nodes.

3 The number of independent KVL equations is equal to the number of independent loops. An independent loop is one whose KVL equation includes at least one voltage not included in other equations.

The direct-application method involves the assignment of appropriate current and voltage variables, the writing of equations of the three types listed above, and the solution of these equations for the desired quantities. The assignment of unknown variables is arbitrary; however, certain practices are usually followed in the interest of organization and simplicity. Resistance voltages and currents are chosen as shown in Fig. 1-3 in order that the Ohm's law equations can be written with positive signs. The series circuit and parallel circuit also simplify the assignment of variables. As the current in a series circuit is the same in each element, only one current variable need be assigned. Similarly, only one voltage variable is identified for the parallel circuit as the voltage across each element is identical. The following example serves to illustrate the method employed.

Example 2-2

Find the unknown voltages and currents in the circuit of Fig. 2-5.

SOLUTION Assign the reference polarities for the unknown voltages and directions for the unknown currents as shown in Fig. 2-5. Note that the 5-A source, the 3-Ω and 6-Ω resistances are in parallel and require identification of only one unknown voltage, V_1. Similarly, the 12-Ω resistance and 24-V source are in series and have the same current I_3. The currents I_1 and I_2 and

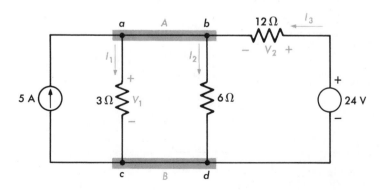

FIGURE 2-5 Circuit for Example 2-2.

the voltage V_2 are selected in accordance with reference directions in Fig. 1-3.

The Ohm's law relations for the three resistances are

$$V_1 = 3I_1 \qquad V_1 = 6I_2 \qquad \text{and} \qquad V_2 = 12I_3$$

The KCL equations are written next. The circuit has two nodes A and B and only one independent equation is necessary. To highlight the fact that only two nodes exist, the portion of the circuit between ab and cd can be redrawn as shown in Fig. 2-6. Circuits are traditionally drawn as depicted in Fig. 2-5; points a and b are each at the same electrical potential and are, therefore, part of the same node. A similar statement can be made for points c and d being part of node B. Then, at node A

$$-5 + I_1 + I_2 - I_3 = 0$$

The KVL equation for the single independent loop, clockwise from b to d and back to b, is

$$-V_2 + 24 - V_1 = 0$$

The five simultaneous equations for the five variables can be solved by any appropriate method. Substitution of the Ohm's law relation into either the KVL or the KCL equation will eliminate the voltage or current variables. If the Ohm's law equations are substituted into the KCL equation, the result is

$$-5 + \frac{V_1}{3} + \frac{V_1}{6} - \frac{V_2}{12} = 0$$

Simultaneous solution of the above equation with the KVL equation yields

$$V_1 = 12 \text{ V} \qquad \text{and} \qquad V_2 = 12 \text{ V}$$

The currents, from Ohm's law, are found to be

$$I_1 = 4 \text{ A} \qquad I_2 = 2 \text{ A} \qquad I_3 = 1 \text{ A}$$

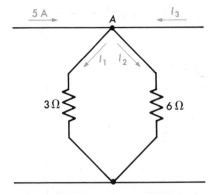

FIGURE 2-6 Portion of Figure 2-5 emphasizing the two nodes.

Example 2-2 illustrates the use of the direct-application method in a straightforward and explicit manner. The process and resulting set of simultaneous equations can be made more compact if the Ohm's law relations for the resistances were written explicitly into either the KVL or KCL equations. In Example 2-2, this would permit the KCL equation expressed in terms of the voltage variables to be written directly. Thus, the circuit performance would be described by two variables rather than by the five variables in the example. The Ohm's law equations are still required; however, they are implicitly included in writing the equations.

Example 2-3

Determine the unknown currents and voltages in the circuit of Fig. 2-7.

SOLUTION The reference polarities for the variables are those indicated in the figure. The KCL equation at A is

$$-I_1 + I_2 + I_3 = 0$$

The Ohm's law relations for the resistances will be included in the KVL equations for the two loops. As a result, the voltage variables are eliminated. For mesh 1, the KVL equation is

$$-12 + 2I_1 + 4I_2 = 0$$

and for mesh 2, we have

$$-4I_2 + 6I_3 - 8I_1 = 0$$

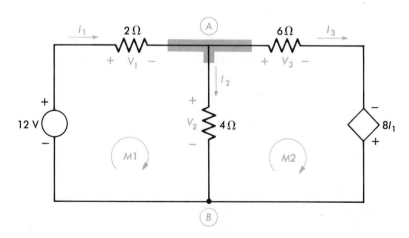

FIGURE 2-7 Circuit of Example 2-3.

Simultaneous solution of the three equations is facilitated when they are written in standard form as

$$-I_1 + I_2 + I_3 = 0$$
$$2I_1 + 4I_2 + 0I_3 = 12$$
$$-8I_1 - 4I_2 + 6I_3 = 0$$

One systematic method for solving multivariable, linear simultaneous equations is through the use of determinants and Cramer's rule. For a set of three equations

$$a_{11}x_1 + a_{12}x_2 + a_{13}x_3 = y_1$$
$$a_{21}x_1 + a_{22}x_2 + a_{23}x_3 = y_2$$
$$a_{31}x_1 + a_{32}x_2 + a_{33}x_3 = y_3$$

in three variables, x_1, x_2, and x_3, any variable may be found as a quotient of two determinants. For example,

$$x_1 = \frac{\begin{vmatrix} y_1 & a_{12} & a_{13} \\ y_2 & a_{22} & a_{23} \\ y_3 & a_{32} & a_{33} \end{vmatrix}}{\begin{vmatrix} a_{11} & a_{12} & a_{13} \\ a_{21} & a_{22} & a_{23} \\ a_{31} & a_{32} & a_{33} \end{vmatrix}} \quad \text{and} \quad x_2 = \frac{\begin{vmatrix} a_{11} & y_1 & a_{13} \\ a_{21} & y_2 & a_{23} \\ a_{31} & y_3 & a_{33} \end{vmatrix}}{\begin{vmatrix} a_{11} & a_{12} & a_{13} \\ a_{21} & a_{22} & a_{23} \\ a_{31} & a_{32} & a_{33} \end{vmatrix}}$$

The determinants may be evaluated by any appropriate method, and the process may be extended to include equation sets with any number of variables. (Standard subroutines are available for the solution of linear simultaneous equations by computers, including programmable hand calculators. Students are encouraged to use these techniques and to develop their own methods.)

For the set of equations in this problem, I_3 is

$$I_3 = \frac{\begin{vmatrix} -1 & 1 & 0 \\ 2 & 4 & 12 \\ -8 & -4 & 0 \end{vmatrix}}{\begin{vmatrix} -1 & 1 & 1 \\ 2 & 4 & 0 \\ -8 & -4 & 6 \end{vmatrix}} = \frac{-144}{-12} = 12 \text{ A}$$

The other currents may be similarly determined as $I_1 = 10$ A and $I_2 = -2$ A. The voltages, from Ohm's law, are $V_1 = 20$ V, $V_2 = -8$ V, and $V_3 = 72$ V. The significance of the negative sign associated with I_2 is that the assumed direction specified at the outset of the problem is opposite to the actual direction of the current. Thus, I_2 was chosen as being directed from A to B; in reality, the positive direction of I_2 is from B to A. Note that the voltage drop V_2 is wholly consistent with both the assumed and actual directions of I_2. A negative voltage drop from A to B is simply a positive drop from B to A.

The method of direct application of fundamental laws is useful in that it focuses attention on those laws and their importance in circuit analysis. It also provides data regarding the number and types of variables and equations which are required for the solution of network problems. With an organized approach, it gives insight into the manner of selecting variables and writing equations. However, a limitation of the direct-application method is that the number of variables increases rapidly as the circuits become more complex. Circuits containing only a few more elements than the circuits in Figs. 2-5 and 2-7 require several additional variables, thereby increasing the number of equations to be solved simultaneously. Such a solution often becomes cumbersome. The nodal and mesh analyses developed in Secs. 2-4 to 2-6 are techniques which reduce the number of circuit variables and thus decrease the number of equations to be solved simultaneously.

2-3 SOURCE EQUIVALENCE AND CONVERSION

Practical sources can only approximate the ideal sources introduced in Sec. 1-7. First presented in Fig. 1-12, the circuits redrawn as part of Fig. 2-8 are convenient representations of practical physical sources. Each is composed of an ideal source and a resistance. The series resistance R_0 in the voltage source of Fig. 2-8a serves to limit the current that can be delivered to the load. It also indicates that some of the energy in the physical source is dissipated internally. For the current source shown in Fig. 2-8b, the role of the

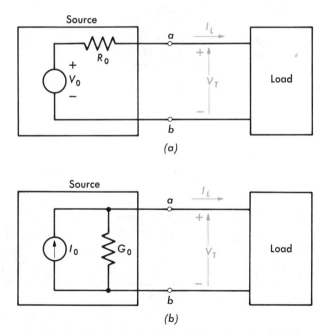

FIGURE 2-8 Practical sources with loads: (a) voltage source, (b) current source.

parallel conductance G_O is analogous to that of R_O in the voltage source. The load current I_L and terminal voltage V_T are the electrical quantities that can be measured directly. The relation between them is useful in describing the circuit behavior of a practical source.

The KVL equation for the loop in the voltage source is

$$-V_O + I_L R_O + V_T = 0 \qquad \text{(2-10)}$$

and solving for I_L gives

$$I_L = -\frac{1}{R_O} V_T + \frac{V_O}{R_O} \qquad \text{(2-11)}$$

The expression in Eq. (2-11) is the volt-ampere or terminal characteristic of the source. As shown in Fig. 2-9a, it is a straight line of slope $-1/R_O$. The maximum current that can be delivered to the load is V_O/R_O and occurs for V_T equal to zero, a situation which corresponds to a zero-resistance load. Because zero resistance indicates a short-circuit connection between terminals a and b, V_O/R_O is referred to as the short-circuit current I_{sc}. For open-circuit conditions (infinite load resistance), I_L is zero and V_T equals V_O, the open-circuit voltage V_{oc}.

In the current source represented in Fig. 2-8b, the KCL equation at a is

$$-I_O + G_O V_T + I_L = 0 \qquad \text{(2-12)}$$

Rearrangement of the terms in Eq. (2-12) gives

$$I_L = -G_O V_T + I_O \qquad \text{(2-13)}$$

The volt-ampere characteristic given by Eq. (2-13) is the straight line of slope $-G_O$ and intercept I_O depicted in Fig. 2-9b. Note that I_O is the short-circuit current and the open-circuit voltage is I_O/G_O. Comparison of Eq. (2-13) and Fig. 2-9b with Eq. (2-11) and Fig. 2-9a indicates that they are identical if

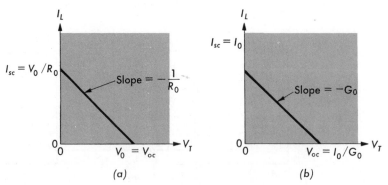

(a)   (b)

FIGURE 2-9   Volt-ampere characteristic of (a) voltage-source and (b) current source.

$$\frac{V_O}{R_O} = I_O \quad \text{or} \quad \frac{V_{oc}}{R_O} = I_{sc} \quad \text{or} \quad V_{oc} = I_{sc}R_O \tag{2-14}$$

and $\quad R_O = \dfrac{1}{G_O} \tag{2-15}$

The import of Eqs. (2-14) and (2-15) is that the circuits in Fig. 2-8 are equivalent and either may be used to represent the same practical source. The student is cautioned to recognize that equivalent does not mean identical. Equivalence refers to the fact that both the voltage-source and current-source representations exhibit the same terminal characteristic and each produces the same effect in the load. Electrical measurements at terminals ab cannot be used to distinguish between the two representations. They are not, however, identical as is seen from open-circuit conditions. No current exists in R_O in the voltage source; while from Eq. (2-12), I_O must exist in G_O when ab remains unconnected. Internally, both representations are different; externally, the same effects are produced so equivalence is achieved.

Often, to facilitate the solution of circuit problems, it is convenient to represent voltage sources as current sources or current sources as voltage sources. To *convert* a voltage source in series with a resistance to a current source in parallel with a resistance (conductance), or vice versa, Eqs. (2-14) and (2-15) are used.

Example 2-4

Convert the current source in Fig. 2-10a to a voltage source and the voltage source in Fig. 2-10b to a current source.

SOLUTION In the circuit of Fig. 2-10a, $I_O = I_{sc} = 50$ mA and $R_O = 1$ kΩ. From Eq. (2-14) $V_{oc} = 50 \times 10^{-3} \times 10^3 = 50$ V. For the circuit of Fig. 2-10b, $V_{oc} = 120$ V and $R_O = 4$ Ω. By the use of Eq. (2-14) $I_{sc} = \frac{120}{4} = 30$ A and from Eq. (2-15) $G_O = \frac{1}{4}$ ℧. The converted circuits are shown in Fig.

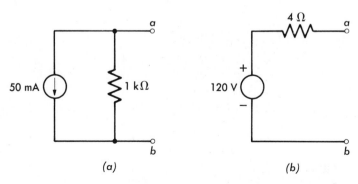

(a)   (b)

FIGURE 2-10 Source representations for Example 2-4.

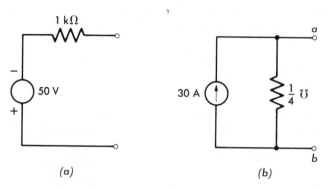

FIGURE 2-11 Source conversions of the circuits in (a) Fig. 2-10a, and (b) Fig. 2-10b.

2-11. Note that the voltage polarity in Fig. 2-11a corresponds to the voltage rise in the direction of the source current in Fig. 2-10a. Similarly, the current direction in Fig. 2-11b is in the direction from minus to plus in the voltage source of Fig. 2-10b.

2-4 NODAL ANALYSIS

Nodal analysis is a formulation of circuit equations in which a set of voltage variables is selected that implicitly satisfies the KVL relations. The circuit can then be described completely by the necessary number of KCL equations whose solution permits the determination of the current and voltage in every circuit element. The basis upon which the set of variables is selected is to make use of the one node in the circuit at which a KCL equation need not be written. By selection of that node as a reference (datum), *node-voltage* variables are defined as the voltage drops from each of the remaining nodes to the reference node. That this set of variables inherently satisfies Kirch-

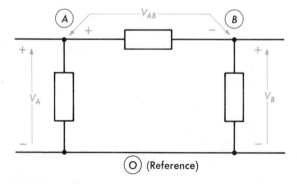

FIGURE 2-12 Circuit segment to demonstrate KVL in terms of node voltages.

hoff's voltage law is demonstrated for the circuit portion shown in Fig. 2-12. For the loop in the circuit, Kirchhoff's voltage law is

$$-V_A + V_{AB} + V_B = 0 \qquad (2\text{-}16)$$

where the node-voltages V_A and V_B are the voltage drops from A to O and B to O, respectively. But, in terms of the node-voltages,

$$V_{AB} = V_{AO} + V_{OB} = V_A - V_B \qquad (2\text{-}17)$$

Substitution of Eq. (2-17) for Eq. (2-16) results in the required identity.

The use of nodal analysis is developed through a study of the circuit in Fig. 2-13. There are three nodes in the circuit; with node C as reference, V_A and V_B are the node-voltage variables. The two independent KCL equations are written at nodes A and B and are

$$A: \qquad -I_1 + V_A G_1 + (V_A - V_B)G_2 = 0 \qquad (2\text{-}18)$$

and

$$B: \qquad I_3 + V_B G_3 + (V_B - V_A)G_2 = 0 \qquad (2\text{-}19)$$

At each node, it is assumed that positive currents are leaving. It is left to the student as an exercise to show that a current $(V_A - V_B)G_2$ leaving A is consistent with a current $(V_B - V_A)G_2$ leaving B. Rearranging the terms in Eqs. (2-18) and (2-19) gives

$$A: \qquad (G_1 + G_2)V_A - G_2 V_B = I_1 \qquad (2\text{-}20)$$

and

$$B: \qquad -G_2 V_A + (G_2 + G_3)V_B = -I_3 \qquad (2\text{-}21)$$

Examination of Eqs. (2-20) and (2-21) shows a pattern which will permit equations of this type to be written readily after inspection. In Eq. (2-20), written at node A, the coefficient of V_A is the positive sum of the conductances connected to node A; the coefficient of V_B is the negative sum of the

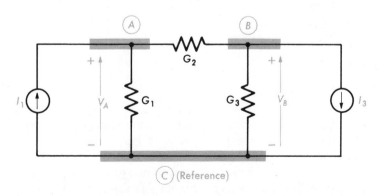

FIGURE 2-13 Circuit used to illustrate nodal analysis.

conductances connected between nodes A and B; the right-hand side of the equation is the sum of the current sources feeding into node A. Now consider Eq. (2-21), written for node B. An analogous situation exists: The coefficient of V_B is the positive sum of the conductances connected to node B; the coefficient of V_A is the negative sum of the conductances connected between nodes B and A; the right-hand side of the equation is the sum of the current sources feeding into node B. (Note that I_3 is directed away from node B.) The fact that these two equations are similar in structure is not a coincidence! The similarity follows from the KCL equations and the manner in which the voltage variables are selected.

The formal procedure for writing nodal or node-voltage equations is outlined in the following steps:

1 Convert all voltage sources in series with resistances to current sources in parallel with conductances, as outlined in Sec. 2-3. The circuit is then redrawn.

2 Select a reference node O and identify the node-voltage variables V_A, V_B, . . . , V_N as the voltage drops from nodes A, B, . . . , N to node O. The choice of a reference is arbitrary; its selection is often based on convenience.

3 Write the KCL equations at nodes A, B, . . . , N in terms of the node-voltage variables. For circuits which contain no dependent sources, the resultant set of equations is of the form

$$A: \quad G_{AA}V_A - G_{AB}V_B - \cdots - G_{AN}V_N = I_A$$
$$B: \quad -G_{AB}V_A + G_{BB}V_B - \cdots - G_{BN}V_N = I_B$$
$$\cdots \cdots \cdots \cdots \cdots \cdots \cdots \cdots$$
$$N: \quad -G_{AN}V_A - G_{BN}V_B - \cdots + G_{NN}V_N = I_N$$

where G_{JJ} = sum of all conductances connected to node J
G_{JK} = sum of all conductances connected between nodes J and K
I_J = sum of all current sources entering node J

4 Solve the equations for the desired node voltages. Other voltages and currents in the circuit are determined by application of Kirchhoff's voltage law and Ohm's law.

Example 2-5

Use nodal analysis to determine the voltage across the 5 Ω resistance and the current in the 12-V source in Fig. 2-14a.

SOLUTION The procedure outlined for writing nodal equations is followed. First, the 12-V source and its 4-Ω series resistance is converted to a 3-A source in parallel with a 0.25-℧ conductance. The redrawn circuit is

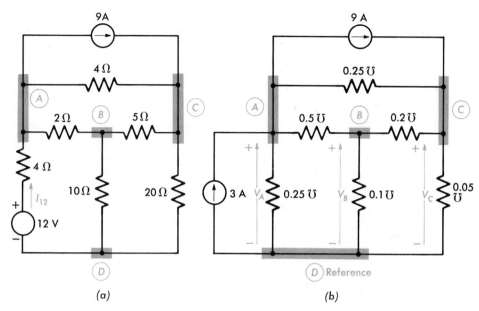

FIGURE 2-14 (a) Circuit for Example 2-5; and (b) circuit of Fig. 2-14a with source conversion.

shown in Fig. 2-14b. Next, node D is selected as reference and variables V_A, V_B, and V_C are selected. The nodal equations are

A: $(0.25 + 0.5 + 0.25)V_A - 0.5V_B - 0.25V_C = 3 - 9$

B: $-0.5V_A + (0.5 + 0.1 + 0.2)V_B - 0.2V_C = 0$

C: $-0.25V_A - 0.2V_B + (0.2 + 0.25 + 0.05)V_C = 9$

and by rearrangement

A: $1.0V_A - 0.5V_B - 0.25V_C = -6$

B: $-0.5V_A + 0.8V_B - 0.2V_C = 0$

C: $-0.25V_A - 0.2V_B + 0.5V_C = 9$

Simultaneous solution of the three equations yields $V_A = 4$ V, $V_B = 8.33$ V, and $V_C = 23.3$ V.

   The voltage across the 5-Ω resistance is $V_B - V_C = 8.33 - 23.3 = -15$ V; the negative sign indicates that node C is at a higher potential than is node B with respect to the reference node D. To obtain the current in the 12-V source, the circuit of Fig. 2-14a is used as the current I_{12} does not appear in Fig. 2-14b. However, between nodes A and D both circuits are equivalent, so that V_A in both circuits is the same. Between A and D in Fig. 2-14a, the KVL expression is $V_A = 4 = -4I_{12} + 12$ or $I_{12} = 2A$.

The node-voltage method is a well-ordered method for solving network problems using the KCL equations as a base. In this method, the number of simultaneous equations which must be solved is equal to one less than the number of network nodes.

2-5 MESH ANALYSIS

An alternative formulation of circuit equations based on a set of current variables is *mesh analysis*. Current-law relations are inherently satisfied by the variables selected so that circuits are described only by the necessary number of independent voltage-law equations. The solution of the mesh or loop equations obtained by this method allows the current and voltage in each circuit element to be determined. Rather than assign a current variable to each element or series combination of elements, the technique employed in mesh analysis is to identify a current variable that is associated with each loop in the circuit. These variables, called *mesh currents,* exist in each and every element in the mesh. Thus, at a node contained in a loop, the mesh current both enters and leaves the node, guaranteeing that the KCL equation is satisfied. As more than one mesh current may exist in a particular element, the actual current becomes the algebraic sum of the mesh currents. The method of approach for mesh analysis is developed through the use of the circuit in Fig. 2-15.

In Fig. 2-15, the mesh current variables I_1 and I_2 are selected for the left-hand and right-hand loops, respectively. Mesh current I_1 exists in each element in the loop *abda;* that is, I_1 exists in R_1, R_2, V_B, and V_A in the direction indicated. Similarly, for the loop *bcdb,* I_2 exists in R_3, V_C, V_B, and R_2. Since the actual current in each element is the sum of the mesh currents in it, the current in the series combination V_B, R_2 is $I_1 - I_2$ directed from *b* to *d.* Note that I_1 and I_2 both enter and leave node *b;* Kirchhoff's current law at *b* is thereby satisfied.

The results of writing the KVL equations for the two loops in Fig. 2-15

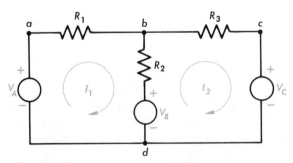

FIGURE 2-15 Circuit illustrating mesh analysis.

are

$1:$ $\quad I_1R_1 + (I_1 - I_2)R_2 + V_B - V_A = 0$ $\hspace{3cm}$ (2-22)

$2:$ $\quad I_2R_3 + V_C - V_B + (I_2 - I_1)R_2 = 0$ $\hspace{3cm}$ (2-23)

The terms in Eqs. (2-22) and (2-23) can be rearranged as

$1:$ $\quad (R_1 + R_2)I_1 - R_2I_2 = V_A - V_B$ $\hspace{3cm}$ (2-24)

$2:$ $\quad -R_2I_1 + (R_2 + R_3)I_2 = V_B - V_C$ $\hspace{3cm}$ (2-25)

The pattern displayed in Eqs. (2-24) and (2-25) is similar to those obtained in Eqs. (2-20) and (2-21) in nodal analysis. In Eq. (2-24), the coefficient of I_1 is the sum of the resistances in the mesh; the coefficient of I_2 is the negative sum of the resistances common to both meshes. The term $V_A - V_B$ represents the sum of the voltage-source rises taken in the direction of I_1. Analogous statements can be made about the terms in Eq. (2-25).

It is not coincidence that the pattern of Eqs. (2-24) and (2-25) occurs; rather it is a result of the selection of the mesh currents I_1 and I_2. Although other choices are permissible, selection of the mesh currents in the same direction, with each contained within an elementary loop, causes the equations to have the form displayed. An elementary loop is readily identified as one of the "windowpanes" of the entire circuit.

The formal procedure for writing mesh equations is given in the following step-by-step outline:

1 Convert each current-source parallel-conductance combination to a voltage source in series with resistance by the method given in Sec. 2-3. For convenience, the circuit is redrawn.

2 Select a mesh current variable for each elementary loop. It is usual to choose all currents in the same direction; a clockwise direction is traditional.

3 Write a KVL equation for each loop in the direction of the mesh current for the loop. The resultant form, for circuits which contain no dependent sources, is

$1:$ $\quad R_{11}I_1 - R_{12}I_2 - \cdots - R_{1N}I_N = V_1$
$2:$ $\quad -R_{12}I_1 + R_{22}I_2 - \cdots - R_{2N}I_N = V_2$

$\cdots \cdots \cdots \cdots \cdots \cdots \cdots \cdots \cdots \cdots \cdots \cdots$

$N:$ $\quad -R_{1N}I_1 - R_{2N}I_2 - \cdots - R_{NN}I_N = V_N$

where R_{JJ} = sum of all resistances contained in mesh J
$\hspace{1.5cm}$ R_{JK} = sum of all resistances common to both meshes J and K
$\hspace{1.5cm}$ V_J = sum of the source-voltage rises in mesh J, taken in the direction of I_J

4 Solve the set of equations for the desired mesh currents. Other currents and voltages in the circuit can be obtained by use of Ohm's law and Kirchhoff's current law.

Example 2-6

Use mesh analysis to determine the currents in the 6-Ω, 8-Ω, and 10-Ω resistances in the circuit of Fig. 2-16a.

SOLUTION   The 2-A source in parallel with the 12-Ω resistance is first converted to a 24-V source in series with a 12-Ω resistance. In the redrawn circuit of Fig. 2-16b, the mesh currents I_1, I_2, and I_3 are selected as indicated. The KVL equations for the meshes are

1: $(12 + 8 + 20)I_1 - 8I_2 - 20I_3 = 24$
2: $-8I_1 + (8 + 4 + 6)I_2 - 6I_3 = 0$
3: $-20I_1 - 6I_2 + (20 + 6 + 10)I_3 = 0$

or

1: $40I_1 - 8I_2 - 20I_3 = 24$
2: $-8I_1 + 18I_2 - 6I_3 = 0$
3: $-20I_1 - 6I_2 + 36I_3 = 0$

Simultaneous solution of the equations yields $I_1 = 1.125$ A, $I_2 = 0.75A$, and $I_3 = 0.75$ A. The current in the 6-Ω resistance, from c to d, is $I_2 - I_3 = 0.75 - 0.75 = 0$; in the 8-$\Omega$ resistance, the current from a to c is $I_1 - I_2 = 1.125 - 0.75 = 0.375$ A. The mesh current $I_3 = 0.75$ A is the current in the 10-Ω resistance.

   The circuit depicted in Fig. 2-16 is a *Wheatstone bridge*, which is a useful instrument in the accurate measurement of resistance. When the bridge is balanced as it is in this example, no current exists in the branch between c and d. In general, R_M represents the resistance of a galvanometer, R_1 and R_2 are known precision resistances, R_3 is a known variable resistance, and R_4 is the unknown resistance whose value we wish to determine. The process used is to adjust R_3 until no current is detected by the galvanometer, i.e., until balance has been achieved. The balance condition, satisfied in this example, requires that $R_1/R_3 = R_2/R_4$. As R_1, R_2, and R_3 are known,

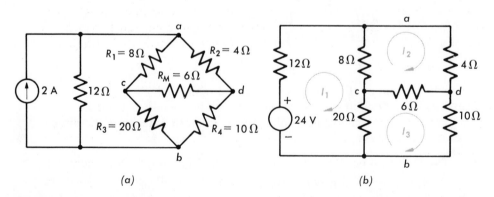

(a)   (b)

FIGURE 2-16   Circuit of Example 2-6.

R_4 is readily evaluated. Because resistance values are influenced by a variety of nonelectrical physical factors, the Wheatstone bridge is the keystone of many measuring instruments. Strain gauges which measure stress in structural members, temperature measuring systems employing thermocouples and thermistors, and hot-wire anemometers are typical of the instruments which utilize bridge circuits.

From the discussions in this and the preceding section, it is seen that the node-voltage and loop-current methods complement each other. The node-voltage method uses KCL equations written with unknown voltages as variables; the number of necessary equations is one less than the number of network nodes. In the loop-current method, KVL equations are written in terms of unknown currents; the number of necessary equations here is equal to the number of independent loops. A decision to use one method instead of the other is usually based on the number of equations needed for each method.

2-6 LOOP AND NODE EQUATIONS WITH CONTROLLED SOURCES

Two basic properties of controlled sources allow the direct extension of the node-voltage method of Sec. 2-4 and the loop-current method of Sec. 2-5 to circuits containing controlled sources. The first property is that at its terminals a controlled source acts in the same manner as does an independent source. Therefore, source conversion and the application of the Kirchhoff voltage and current laws are treated identically for both types of sources.

The second property is that the strength of a controlled source depends on the value of a voltage or current elsewhere in the network. To include the effect of this dependence in the KCL or KVL equations requires that a constraint equation be written. The constraint equation is used to relate the control voltage or current for the controlled source to the voltage or current variables selected for the node-voltage or loop-current equations written. The combination of the constraint equations with the appropriate KCL or KVL equations results in the node-voltage or loop-current equations which describe the entire circuit. The following examples are used to illustrate the two methods for circuits which contain controlled sources.

Example 2-7

Determine the voltage across the 0.8-℧ conductance in the circuit of Fig. 2-17 by the use of nodal analysis.

SOLUTION Select node B as reference and identify the voltage variables V_A and V_C as is indicated in Fig. 2-17. As node B is the reference node, the volt-

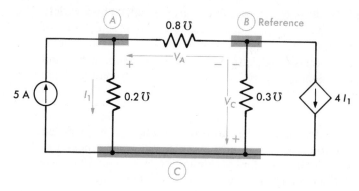

FIGURE 2-17 Circuit of Example 2-7.

age across the 0.8-℧ conductance is simply V_A. The KCL equations at nodes A and C, written by treating the dependent source as any other source, are

A:　　　$(0.8 + 0.2)V_A - 0.2V_C = 5$

C:　　$-0.2V_A + (0.2 + 0.3)V_C = -5 + 4I_1$

Next, the constraint equation for the controlled source is written. This requires that I_1 be expressed in terms of the variables V_A and V_C. From Fig. 2-17 it is seen that

$$I_1 = 0.2(V_A - V_C)$$

Substitution of the constraint equation into the original KCL equations and combining similar terms gives

A:　　　$1.0V_A - 0.2V_C = 5$

C:　　$-1.0V_A + 1.3V_C = -5$

Simultaneous solution of these equations gives $V_A = 5V$.

Example 2-8

Find the currents I_1 and I_2 in the circuit of Fig. 2-18.

SOLUTION The loop-current method will be used. First, the voltage-controlled current source and its parallel resistance must be converted into a voltage-controlled voltage source and series resistance. The converted source has a value $(0.5V_1)(10) = 5V_1$ volts; the series resistance is 10 Ω. The converted circuit is shown in Fig. 2-19.

The KVL equations for the circuit of Fig. 2-19 are first written by treating all sources identically. They are

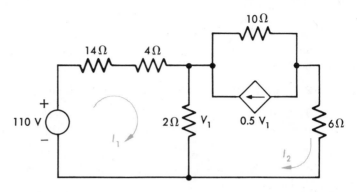

FIGURE 2-18 Circuit of Example 2-8.

$$(14 + 4 + 2)I_1 - 2I_2 = 110$$
$$-2I_1 + (2 + 10 + 6)I_2 = -5V_1$$

for the left-hand and right-hand loops, respectively. The relationship between V_1 and the variables I_1 and I_2 is the constraint equation

$$V_1 = (I_1 - I_2) \times 2$$

Combination of the constraint and the original loop equations, after the rearrangement of terms, results in

$$20I_1 - 2I_2 = 110$$
$$8I_1 + 8I_2 = 0$$

from which

$$I_1 = 5 \text{ A} \qquad \text{and} \qquad I_2 = -5 \text{ A}$$

Examples 2-7 and 2-8 illustrate the applicability of the loop-current and node-voltage methods to circuits containing controlled sources. The principal features of the method described in these examples can be summarized by the four steps useful in writing the desired circuit equations.

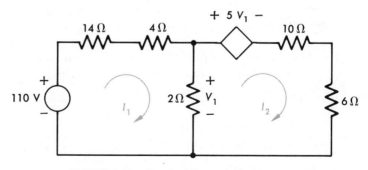

FIGURE 2-19 Circuit of Figure 2-18 after source conversion.

1   Write the loop or node equations, treating all sources as independent sources. This step includes converting all voltage sources and series resistances to current sources and parallel resistances in nodal analysis. The opposite conversion applies in mesh analysis.

2   Identify the controlled sources and write a constraint equation for each.

3   Combine the constraint equations with the equations obtained in step 1, rearranging terms where necessary, so that the resultant set of equations is in a form convenient for solution.

4   Solve the resultant set of equations for the unknown current or voltage variables.

2-7 SUPERPOSITION AND LINEARITY

In any electric circuit, the element voltages and currents are the responses (effects) produced by the applied sources, which are regarded as the excitations (causes). For a linear network containing several independent sources, each response can be considered to be the sum of several components, each component produced by one of the individual sources. The principle of superposition states that the current or voltage in any branch produced by several sources acting simultaneously is the algebraic sum of the currents or voltages produced in that branch by each source acting separately. For resistive circuits of the type described in this chapter, the superposition principle is a consequence of the linear current-voltage relation in a resistance.

Superposition is applicable to a wide range of systems which occur, for example, in the study of mechanical vibrations, fluid mechanics, and operations research. In general terms, the principle may be stated as follows: For a system in which excitations x_1 and x_2 produce responses y_1 and y_2, respectively, application of x_1 and x_2 results in a response of y_1 and y_2. Symbolically, this is written as

$$
\begin{aligned}
x_1 &\longrightarrow y_1 \\
x_2 &\longrightarrow y_2 \\
x_1 + x_2 &\longrightarrow y_1 + y_2
\end{aligned}
\tag{2-26}
$$

where the arrow indicates the cause-effect relation. For practical purposes the relation described in Eq. (2-26) also serves as the definition of a linear system. Therefore, all linear systems obey superposition, and physical systems for which superposition is satisfied are linear.[1]

An alternative to Eq. (2-26) is to express the response in terms of the applied excitation as shown in Eq. (2-27).

$$
\begin{aligned}
y_1 &= g\,x_1 \\
y_2 &= g\,x_2 \\
y_1 + y_2 &= g(x_1 + x_2) = g\,x_1 + g\,x_2
\end{aligned}
\tag{2-27}
$$

---

[1] The conditions in Eq. (2-26) are necessary but not sufficient conditions for linearity. A more rigorous definition is $Ax_1 + Bx_2 \rightarrow Ay_1 + By_2$, where A and B are constants.

The symbol g in Eq. (2-27) represents an *operator;* that is, g designates the mathematical operation performed on the excitation that produces the response. In resistive circuits, g is simply a real number and is indicative of the fact that voltages and currents in resistances are related by Ohm's law. On the other hand, for a capacitor, $i = (C)(dv/dt)$ and the operation performed on the voltage excitation is that of differentiation, scaled by the value C. Differentiation is a linear operation because $(d/(du)(f_1 + f_2) = df_1/du + df_2/du$. Thus, circuits that comprise R, L, and C as defined in Chap. 1 are linear systems and satisfy superposition.

The principle of superposition can be used in the solution of circuit problems. Each *independent source* acting on the network is considered separately; the other independent sources are suppressed. The total response is computed by algebraically summing the component responses produced by the individual sources. To suppress a source means to reduce the strength of the source to zero. For a voltage source, suppression requires that the source-voltage be made zero, a short-circuit condition. Current sources are open-circuited to reduce their effect to zero.

Example 2-9

Determine I_2 in the circuit of Fig. 2-20 by use of the superposition principle.

SOLUTION The 3-A and 9-V sources are suppressed in turn. The terms V'_A, V'_B, I'_1, and I'_2 are the components of V_A, V_B, I_1, and I_2 produced by the 9-V source. Double-primed quantities identify the corresponding components caused by the 3-A source. The circuit of Fig. 2-20 with the 3-A current source suppressed is depicted in Fig. 2-21a. For reasons which will be apparent later in this example, a literal solution precedes numeric evaluation.

At node B, the KCL equation is

$$-\frac{V'_A}{3} + \frac{V'_B}{R_3} + \frac{V'_B}{R_4} = 0$$

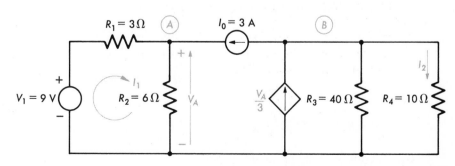

FIGURE 2-20 Circuit of Example 2-9.

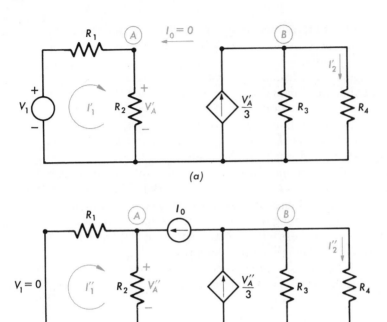

FIGURE 2-21   Circuit of Figure 2-20 with (a) current source suppressed, and (b) voltage source suppressed.

from which

$$V'_B = \frac{V'_A}{3} \frac{R_3 R_4}{R_3 + R_4}$$

and

$$I'_2 = \frac{V'_B}{R_4} = \frac{V'_A}{3} \frac{R_3}{R_3 + R_4}$$

For the left-hand side of the mesh, KVL gives

$$- V_1 + I'_1 R_1 + I'_1 R_2 = 0$$

so that

$$I'_1 = \frac{V_1}{R_2 + R_2}$$

and

$$V'_A = R_2 I'_1 = V_1 \frac{R_2}{R_1 + R_2}$$

Substitution of numeric values gives

$$V'_A = 9\frac{6}{3 + 6} = 6V \qquad \text{and} \qquad I'_2 = \frac{6}{3}\frac{40}{40 + 10} = 1.6 \text{ A}$$

Figure 2-21b displays the circuit in Fig. 2-20 with the 9-V source suppressed. Nodal equations at A and B are

A: $\qquad \dfrac{V''_A}{R_1} + \dfrac{V''_A}{R_2} = I_0$

B: $\qquad \dfrac{V''_B}{R_3} + \dfrac{V''_B}{R_4} = \dfrac{V''_A}{3} - I_0$

Solution of these equations yields

$$V''_A = \frac{R_1 R_2}{R_1 + R_2} I_0 \qquad V''_B = \frac{I_0 R_3 R_4}{R_3 + R_4}\left[\frac{R_1 R_2}{3(R_1 + R_2)} - 1\right]$$

Numeric evaluation results in

$$V''_A = \frac{3 \times 6}{3 + 6} I_0 \qquad V''_B = \frac{3 \times 40 \times 10}{40 + 10}\left[\frac{3 \times 6}{3(3 + 6)} - 1\right] = -8 \text{ V}$$

$$I''_2 = \frac{V''_B}{10} = -0.8 \text{ A}$$

The total current I_2 is $I'_2 + I''_2 = 1.6 - 0.8 = 0.8$ A.

In the course of solving this problem, three observations can be made based on the analysis of the circuit in Fig. 2-21a. The first of these involves the circuit segment containing node B in which the source current $V'_A/3$ enters the node and leaves through R_3 and R_4. The resistances R_3 and R_4 form a *current divider*, the relation for which is given by the expression for I'_2. The second observation arises in the mesh defined by I'_1. Here, the source voltage V_1 is distributed across R_1 and R_2 which form the *voltage divider* described by the expression for V'_A. Both current and voltage dividers are used in practical circuits. Often only a single voltage source or current source is used to provide power for an entire system. Different sections of the system may require excitation at voltage and current levels different from that of the source. Voltage dividers and current dividers are simple and effective circuits used to provide the different voltage and current values required.

The third feature relates to the nature of the dependent source. In Fig. 2-21a, no resistance couples the circuit portions containing nodes A and B. The independent source is applied to the segment containing A, yet a voltage appears at B. The response is produced at node B, because energy is transferred to that circuit segment by means of the dependent source. The amount of power delivered to the node B segment is, however, controlled by the electrical activity in the mesh containing node A. This is the "action-at-a-distance" analogy described in Sec. 1-8.

It is also important to note that in the solution only the independent sources are suppressed. A dependent source has zero value only when its control voltage or current is zero. In Fig. 2-21, the source $V_A/3$ is zero if V_A is zero; as neither V_A' nor V_A'' becomes zero, the effect of the dependent source must be included in evaluating I_2' and I_2''.

2-8 THÉVENIN'S THEOREM

We think of a system as a combination of components organized to accomplish a desired objective. The necessity for knowing how the various components interact and how they affect system performance is apparent. One such method was the use of the superposition principle to obtain each source's contribution to the total response. In order to focus attention on the behavior of a specific element or group of elements, it is convenient to describe the other portions of the system by simple, equivalent representations.

Figure 2-22 depicts a circuit in which the elements are portioned into two segments, N_1 and N_2. To determine the effect of N_1 on N_2 and often to adjust N_2 for optimum performance, we wish to represent N_1 by an equivalent circuit. The heuristic approach which follows establishes the nature of the equivalent that is used. In Fig. 2-22, consider N_2 to appear as a zero-value resistance (short circuit) at terminals ab. Then, V is zero and I is the current in the short circuit I_{sc}. When N_2 is open-circuited, corresponding to $R \rightarrow \infty$, I is zero and the voltage across ab is the open-circuit voltage V_{oc}. As the circuit N_1 is linear, V and I must be linearly related. Thus, if a graph of I versus V is plotted, it is a straight line and passes through the coordinates $(0, I_{sc})$ and $(V_{oc}, 0)$. This type of characteristic is displayed in Fig. 2-9 as the volt-ampere characteristic of a practical source. Consequently, the network segment N_1 can be represented as shown in Fig. 2-23a as a voltage source in series with a resistance. The circuit depicted in Fig. 2-23a is a *Thévenin equivalent* circuit; source conversion results in the *Norton equivalent* of Fig. 2-23b. The Thévenin equivalent is the essence of *Thévenin's theorem,* stated for resistive networks as follows:

> At a pair of terminals ab, any linear network can be represented by an equivalent circuit composed of a voltage source in series with a resistance. The source voltage is the voltage V_{oc} measured across the open-circuited terminals ab. The resistance value R_0 is the ratio of the open-circuit voltage to the current I_{sc} that exists between the terminals ab when these are short-circuited.

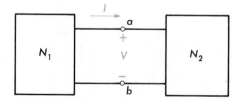

FIGURE 2-22   Network partitioned into two segments.

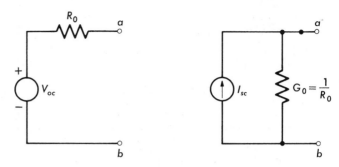

FIGURE 2-23 (a) Thévenin equivalent circuit, and (b) Norton equivalent circuit.

Thévenin's theorem can be more formally justified by demonstrating that an arbitrary linear network has the same volt-ampere characteristic at a pair of terminals as a practical source. In Fig. 2-24a, N represents a linear network which contains independent sources; these sources establish the open-circuit voltage, at terminals ab. By virtue of the open circuit, the current I_L is zero. Figure 2-24b shows the same network with a voltage source equal to the open-circuit voltage V_{oc} applied to the terminals. Since the source voltage exactly balances the network open-circuit voltage, I_L is still zero. By superposition, suppressing V_{oc} results in a current I_{sc} as depicted in Fig. 2-24c. Suppression of all independent sources in N requires that the current produced by V_{oc} be identical to I_{sc} in the direction indicated in Fig. 2-24d. From Ohm's law, the ratio of V_{oc} to I_{sc} is the equivalent resistance R_0 of the network N with independent sources suppressed, as viewed from terminals ab. In Fig. 2-24e, an additional source V_1 is connected in series with V_{oc}. As a result of V_1 a current I_L now exists, the total contribution of all other sources being zero. With all sources suppressed other than V_1 (Fig. 2-24f), the current I_L is given by Ohm's law as

$$I_L = \frac{V_1}{R_0}$$

as N is represented by its equivalent resistance. From Fig. 2-24e,

$$V_T = V_{oc} - V_1$$

and then

$$I_L = -\frac{V_T}{R_0} + \frac{V_{oc}}{R_0}$$

which is the volt-ampere characteristic given in Eq. (2-11).

To obtain the Thévenin equivalent of a circuit, two of the three parameters V_{oc}, I_{sc}, and R_0 must be determined. The interrelation among these quantities is given in Eqs. (2-15) and (2-16). The resistance R_0 can be obtained either from V_{oc}/I_{sc} or by direct determination of the equivalent resistance of the network when all independent sources are suppressed.

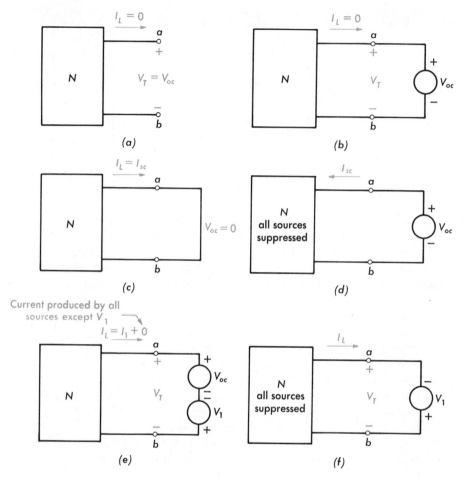

FIGURE 2-24   Development of Thévenin's theorem.

Example 2-10

For the circuit in Fig. 2-25, determine

a   The value of R which will absorb the greatest power from the circuit

b   The amount of the power

SOLUTION   To determine power, it is necessary to know both the current I and the resistance R. Therefore, a relation giving I as a function of R must be found.

Since the resistance R is to be studied, it will be removed and a Thévenin equivalent circuit formed of the remainder of the circuit. The circuit with R removed is shown in Fig. 2-26a with the 48-V source 24-Ω re-

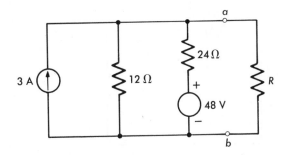

FIGURE 2-25   Circuit of Example 2-10.

sistance combination converted to a 2-A source in parallel with 24-Ω. The node equation at a is used to obtain the open-circuit voltage V_{oc} and is

$$V_{oc} \left(\tfrac{1}{12} + \tfrac{1}{24} \right) = 3 + 2 \qquad \text{or} \qquad V_{oc} = 40 \text{ V}$$

Next, the short-circuit current is determined from the circuit in Fig. 2-26b. As $V = 0$, no current exists in either the 12- of 24-Ω resistances, so that KCL at a gives

$$I_{sc} = 3 + 2 = 5 \text{ A}$$

The Thévenin resistance R_0 equals $V_{oc}/I_{sc} = 40/5 = 8 \; \Omega$.

The circuit in Fig. 2-27 displays the Thévenin equivalent to which the resistance R is reconnected. The KVL equation around the loop is

$$-40 + 8I + RI = 0 \qquad \text{or} \qquad I = \frac{40}{8 + R}$$

The power dissipated in the resistance R, from Eq. (1-16), is

$$P = I^2 R = \frac{1,600R}{(8 + R)^2}$$

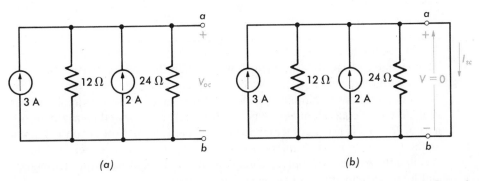

FIGURE 2-26   Circuits to calculate (a) open-circuit voltage, and (b) short-circuit current in Example 2-10.

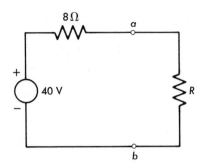

FIGURE 2-27 Thévenin equivalent of circuit in Figure 2-25.

To find the value of R for maximum power, form dP/dR and equate it to zero. This yields

$$\frac{dP}{dR} = \frac{(8 + R)^2 \times 1,600 - 1,600R \times 2 \times (8 + R)}{(8 + R)^4} = 0$$

Solving for R gives

$$R = 8 \ \Omega$$

To demonstrate that $R = 8$ results in maximum power, consider the following physical argument: For $R = 0$, $P = I^2R = 0$; for $R \to \infty$, P is also zero as no power is delivered to an open circuit. As P is positive in the resistance (a resistance dissipates power), the power for $R = 8$ must be a maximum. To verify this result, the student can evaluate d^2P/dR^2 for $R = 8 \ \Omega$ and show that it is negative.

For $R = 8 \ \Omega$, the current is

$$I = \frac{40}{8 + 8} = 2.5 \ \text{A}$$

and the maximum power is

$$P_{\text{max}} = (2.5)^2 \times 8 = 50 \ \text{W}$$

This example illustrates the facility with which the characteristics of a network with respect to a single circuit element may be determined by Thévenin's theorem. Notice that the load resistance R for maximum power is equal to the equivalent resistance of the source as viewed from the terminals of R. This is not a coincidence but is generally true; the result is known as the *maximum power transfer theorem*. The equivalent source resistance is called the *output resistance*, and the process of equating output resistance and load resistance is known as *resistance matching*.

Example 2-11

Obtain a Thévenin equivalent at terminals ab for the circuit shown in Fig. 2-28a.

SOLUTION   The open-circuit voltage is determined first. At node a, the KCL equation is

$$-I + I_1 - 99I = 0 \quad \text{or} \quad I_1 = 100I$$

The KVL equation for the left-hand mesh is

$$-11 \times 10^{-3} + 2,200I + 220I_1 = 0$$

Solving for I_1 gives

$$I_1 = 45.5 \ \mu A$$

and

$$V_{oc} = 220I_1 = 10.0 \text{ mV}$$

When ab are short-circuited as shown in Fig. 2-28b, $V_{ab} = 0$, so that $I_1 = 0$. At a, the KCL equation is

$$-I - 99I + I_{sc} = 0 \quad \text{or} \quad I_{sc} = 100I$$

Application of KVL to the outside loop gives

$$-11 \times 10^{-3} + I \times 2,200 = 0$$

from which $I = 5 \ \mu A$ and $I_{sc} = 500 \ \mu A$.
The equivalent resistance is

$$R_0 = \frac{V_{oc}}{I_{sc}} = \frac{10 \times 10^{-3}}{500 \times 10^{-6}} = 20 \ \Omega$$

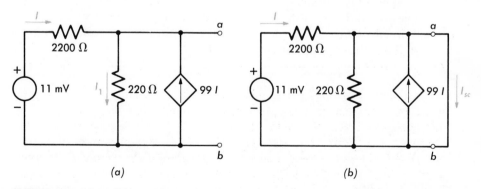

(a)   (b)

FIGURE 2-28   Circuits for computing (a) open-circuit voltage, and (b) short-circuit current for Example 2-11.

The principal value of Thévenin's theorem is that it permits a simple representation of a complex network at an output terminal pair and, as a result, allows ready calculation of the effect of an output load on the network or, conversely, the effect the load will have on the terminal behavior of the network. In our everyday experience, the concept of the Thévenin equivalent allows us to represent the source of electric energy in our houses by the alternating voltage measured at the receptacle (plug) without having to consider the entire generation and distribution system. Similarly, in our stereo system, the Thévenin equivalent permits us to match the speakers to the amplifier output.

2-9 NETWORK SIMPLIFICATION

The node-voltage and loop-current methods of Secs. 2-4 to 2-6 are logical refinements of the method of direct application of fundamental laws. In all cases, they provide reasonably direct solution of the voltage and current in each branch of a circuit. Many circuit problems, however, require solution of a single unknown voltage or current; other problems are concerned with the effect produced on a single source by a complex network. Both situations have in common the fact that attention is focused on a single element (or group of elements) in the circuit. Accordingly, any techniques which make it possible to reduce the complexity of the network surrounding the element of interest are of value in the solution of these types of problems. The principle of superposition, developed in Sec. 2-7, is one such method which focuses on the effects produced by individual circuit sources. A second technique is Thévenin's theorem, described in the previous section, through which the concepts of an equivalent source and equivalent resistance were introduced.

Network reduction is one technique for simplifying network complexity. The technique usually involves recognition of certain network configurations which permit ready consolidation of the configuration components; it may involve sources or elements, although each must be handled in a somewhat different fashion. The circuits of Fig. 2-29 are series circuits. The circuit of Fig. 2-29a shows a series combination of voltage sources. From the KVL equation, it is evident that

$$V_{eq} = V_1 - V_2 + \cdots + V_n$$

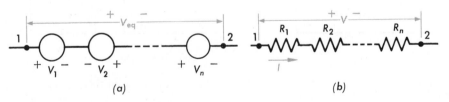

FIGURE 2-29   Series combination of (a) sources and (b) resistances.

FIGURE 2-30   Equivalent series combinations of Fig. 2-29a and b.

and, as a consequence, the *equivalent-source representation* of Fig. 2-30a may be used in place of the several sources of Fig. 2-29a. Apparently, then, voltage sources in series add algebraically since reference polarities must be included. Figure 2-29b shows a series combination of resistances. Recall from Sec. 2-1 that each component in a series circuit carries the same current. The KVL equation for this figure is then

$$V = IR_1 + IR_2 + \cdots + IR_n = I(R_1 + R_2 + \cdots + R_n)$$

or

$$V = IR_{eq}$$

where

$$R_{eq} = R_1 + R_2 + \cdots + R_n \tag{2-28}$$

Resistances in series can thus be replaced by an *equivalent resistance* (see Fig. 2-30b), which is the sum of the component resistances.

Parallel combinations of sources and resistances are shown in Fig. 2-31a and b. The resistive elements are described by their conductances. In parallel circuits, voltage is the common quantity and currents are added. Thus, the equivalent source I_{eq} of Fig. 2-32a is, from Kirchhoff's current law,

$$I_{eq} = I_1 - I_2 + \cdots + I_n$$

As with series voltage sources, parallel current sources are added (taking directions into account). The resistances of Fig. 2-31b are combined by writing the KCL equation at point A.

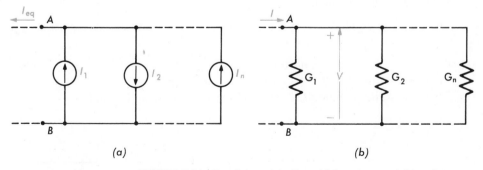

FIGURE 2-31   Parallel combination of (a) sources and (b) resistances.

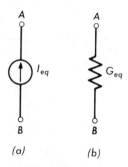

FIGURE 2-32 Equivalent parallel combinations of Fig. 2-31a and b.

$$I = VG_1 + VG_2 + \cdots + VG_n$$
$$= V(G_1 + G_2 + \cdots + G_n)$$

or

$$I = VG_{eq}$$

where

$$G_{eq} = G_1 + G_2 + \cdots + G_n \tag{2-29}$$

Conductances in parallel may thus be replaced by an equivalent conductance (see Fig. 2-32b), which is the sum of the component conductances. Equation (2-29) can be expressed in terms of the resistance, recognizing that $G = 1/R$, since

$$R_{eq} = \frac{1}{(1/R_1) + (1/R_2) + \cdots + (1/R_n)} \tag{2-30}$$

One configuration commonly encountered is the parallel combination of two resistances. Equation (2-29) applied to this configuration gives

$$G_{eq} = \frac{1}{R_1} + \frac{1}{R_2} = \frac{R_1 + R_2}{R_1 R_2}$$

The equivalent resistance is, from Eq. (2-30),

$$R_{eq} = \frac{R_1 R_2}{R_1 + R_2} \tag{2-31}$$

Networks containing combinations of series- and parallel-resistance configurations can often be reduced by continuing applications of Eqs. (2-28) to (2-31). The technique is illustrated in the following example.

Example 2-12

The circuit shown in Fig. 2-33 is that of a two-terminal network. The terminals x and y are connected to a source v. The four loads are connected to the

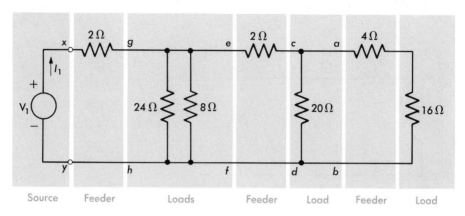

FIGURE 2-33 Circuit for Example 2-12.

source by means of feeders, the load and feeder resistances being given on the diagram. To facilitate a study of the source requirements, replace the load-feeder circuit by a single equivalent resistance. Determine the necessary resistance value.

SOLUTION   In applying network simplification, the resistances should be combined by starting at the point of the network that is most removed from the source and working to the source.

The 4- and 16-Ω resistances to the right of the terminals ab are in series. The equivalent resistance looking to the right of ab is then

$$R_{ab} = 4 + 16 = 20 \ \Omega$$

The equivalent resistance R_{ab} is in parallel with the 20-Ω load just to the left of terminals ab. The equivalent resistance to the right of terminals cd is, from Eq. (2-31),

$$R_{cd} = \frac{(20)(20)}{20 + 20} = 10 \ \Omega$$

The equivalent resistance R_{cd} is now in series with the 2-Ω feeder resistance; the equivalent resistance looking to the right of terminals ef is then

$$R_{ef} = 2 + 10 = 12 \ \Omega$$

The equivalent resistance R_{ef} is in parallel with the 8- and 24-Ω resistances. The equivalent resistance looking to the right of terminals gh is, from Eq. (2-29),

$$G_{gh} = \frac{1}{12} + \frac{1}{8} + \frac{1}{24} = \frac{1}{4} \ \mho$$

or

$$R_{gh} = 4 \ \Omega$$

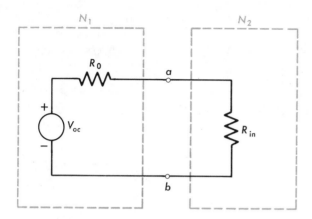

FIGURE 2-34   Equivalent representation of Figure 2-22.

The equivalent resistance R_{gh} is in series with the 2-Ω resistance. The equivalent resistance R_{xy}, which will completely replace the entire network as far as the terminals x and y are concerned, is then

$$R_{xy} = 2 + 4 = 6 \ \Omega$$

The resistance R_{xy} in Example 2-12 is the equivalent resistance of the network viewed from the source terminals. In Fig. 2-33, by use of Ohm's law, R_{xy} is the ratio of the source voltage V_1 to the current in the source I_1 and is often called the *input resistance*. The ratio of voltage to current at a pair of terminals for a network containing no independent sources, or in which the independent sources are suppressed, is also called the *driving-point resistance* of the network at the terminals. The *driving-point conductance* is similarly defined as the ratio of current to voltage at the terminal pair. As described in the analysis corresponding to Fig. 2-24d, the Thévenin resistance is also a driving-point resistance. The driving-point resistance allows a complex network to be represented at a terminal pair by a single equivalent resistance. Thus, for the circuit described in Fig. 2-22, the network segment N_1 is replaced by its Thévenin equivalent; and N_2, if it contains no independent sources, is replaced by its input resistance R_{in}. The equivalent representation of the circuit in Fig. 2-22 is depicted in Fig. 2-34; the interaction between N_1 and N_2 is clearly displayed and readily computed.

The driving-point resistance in Example 2-12 was computed by a succession of series- and parallel-combination reductions. Circuits which contain dependent sources, however, require the ratio of voltage to current be calculated to obtain the driving-point resistance as illustrated in Example 2-13.

Example 2-13

Find the equivalent resistance for the 11-mV source in Example 2-11 and Fig. 2-28a.

SOLUTION   The equivalent resistance is the ratio of the source voltage to the current I in it. In Example 2-11, I_1 was found to be 45.5 μA and $I_1 = 100I$. Then,

$$I = \frac{I_1}{100} = \frac{45.5 \times 10^{-6}}{100} = 0.455 \ \mu A$$

The input resistance R_{in} is

$$R_{in} = \frac{V}{I} = \frac{11 \times 10^{-3}}{0.455 \times 10^{-6}} = 24.2 \ k\Omega$$

There are certain network configurations which cannot be resolved by series-parallel combinations alone. These configurations may often be handled by the use of a Y-Δ *transformation*. This transformation permits three resistors which make a Y configuration to be replaced by three in a Δ configuration, or vice versa. The circuits of Fig. 2-35a and b are Δ and Y networks, respectively. If these networks are to be equivalent, the resistance between any pair of terminals must be the same in the Y as in the Δ. Three simultaneous equations may be written expressing this equivalence of terminal resistances. Thus, considering terminals x and y, the equivalent Δ resistance is the resistance R_c in parallel with the series combination of R_a and R_b, and the equivalent Y resistance is the series combination of R_1 and R_2. Expressed algebraically,

$$R_{xy} = R_1 + R_2 = \frac{R_c(R_a + R_b)}{(R_a + R_b) + R_c} \qquad (2\text{-}32)$$

Two similar equations may be written for the other two terminal pairs.

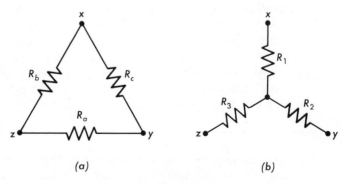

(a)   (b)

FIGURE 2-35 Δ and Y equivalent circuits

The resulting three equations may be solved simultaneously for the Δ values R_a, R_b, and R_c or the Y values R_1, R_2, and R_3. The results are

$$R_1 = \frac{R_b R_c}{R_a + R_b + R_c} \tag{2-33}$$

$$R_2 = \frac{R_a R_c}{R_a + R_b + R_c} \tag{2-34}$$

$$R_3 = \frac{R_a R_b}{R_a + R_b + R_c} \tag{2-35}$$

or

$$R_a = \frac{R_1 R_2 + R_2 R_3 + R_3 R_1}{R_1} \tag{2-36}$$

$$R_b = \frac{R_1 R_2 + R_2 R_3 + R_3 R_1}{R_2} \tag{2-37}$$

$$R_c = \frac{R_1 R_2 + R_2 R_3 + R_3 R_1}{R_3} \tag{2-38}$$

Example 2-14

Use network reduction to determine the voltage V in the circuit of Fig. 2-36.

SOLUTION   The three-resistance segment formed by the two 4-Ω resistances and the 8-Ω resistance between nodes a, b, and c is a Δ network. Its Y equivalent is found by the use of Eqs. (2-33) to (2-35):

$$R_1 = \frac{4 \times 8}{4 + 4 + 8} = 2 \; \Omega$$

$$R_2 = \frac{4 \times 8}{4 + 4 + 8} = 2 \; \Omega$$

$$R_3 = \frac{4 \times 4}{4 + 4 + 8} = 1 \; \Omega$$

The replacement of the Δ by its equivalent Y and the conversion of the 48-A 3-Ω source is shown in Fig. 2-37. The combination of the two pairs of series resistances and the conversion the 144-V 4-Ω source to its current-source representation is given in Fig. 2-38. Successive reductions and combinations are shown in Fig. 2-39a, b, and c. The final result is $V = 54$ V. Notice that network reduction has resulted in the Thévenin equivalent; the entire network in Fig. 2-36 can be replaced at terminals cd by the circuit shown in Fig. 2-39c.

The concept of equivalent resistance which often results in circuit simplification is a useful tool in circuit analysis. Certain circuits lend themselves to configurational simplification such as series and parallel combinations and

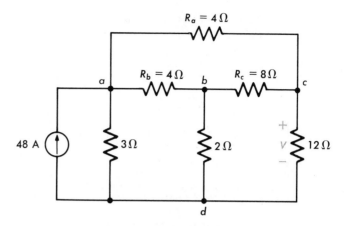

FIGURE 2-36 Circuit of Example 2-14.

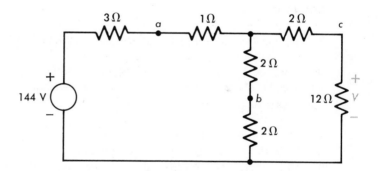

FIGURE 2-37 Circuit of Fig. 2-36 after Δ-Y and source conversion.

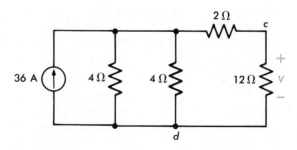

FIGURE 2-38 Circuit of Fig. 2-36 after simplification.

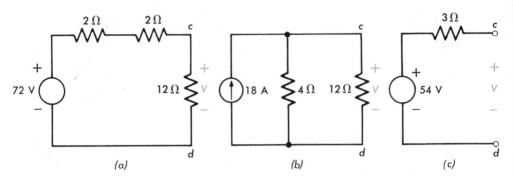

FIGURE 2-39 Successive simplifications of circuit of Fig. 2-38: (a) After parallel combination and source conversion; (b) after a second parallel combination and source conversion.

Δ-Y conversion. As a general rule, the equivalent resistance can always be obtained from the ratio of voltage to current at a terminal pair or, if independent sources exist in the network, by the use of Thévenin's theorem (the ratio of open-circuit voltage to short-circuit current).

PROBLEMS

2-1 A 1.0-mA current produces full-scale deflection in a D'Arsonval movement having a resistance of 25 Ω.
a Construct an ammeter whose full-scale deflection is 5.0 A.
b Use the meter movement to construct a voltmeter with a full-scale deflection of 50 V.

2-2 Use the meter movement in Prob. 2-1 as
a A voltmeter having a full-scale deflection of 300 V
b An ammeter whose full-scale deflection is 10 A.

2-3 a A voltmeter having a 100-V scale is constructed using a 50-Ω, 100-μA D'Arsonval movement. Determine the value of the series resistance.
b If the maximum error in the meter reading is 1 percent of full-scale deflection, what is the maximum error in a voltmeter reading of 40 V?

2-4 The circuits shown in Fig. 2-40 are often used to determine the value of an unknown resistance R_x. Called the voltmeter-ammeter method, the technique employs a test source V_T and an ammeter A and voltmeter V. The unknown resistance is computed from the ratio of the voltmeter reading to the ammeter reading. We wish to investigate the accuracy of each of the circuits when a voltmeter whose resistance is 10 kΩ and an ammeter whose resistance is 1.0 Ω are used with a test source of 10 V.
a For $R_x = 10$ Ω, determine the ammeter and voltmeter readings for each circuit.
b For each circuit, determine the measured resistance and compute the error in this resistance value. Explain any differences.

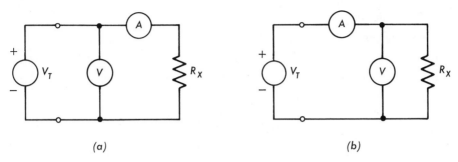

FIGURE 2-40 Circuits for determining resistance by voltmeter-ammeter method.

2-5 Repeat Prob. 2-4 for $R_x = 2,000\ \Omega$.

2-6 Repeat Prob. 2-4 for $R_x = 100\ \Omega$.

2-7 In the series ohmmeter in Fig. 2-4, the resistance R_o is adjusted to produce full-scale deflection in the meter when R_x is zero. For a meter having 1-mA full-scale deflection and a resistance of 75 Ω, determine
a The value of R_o for $V_B = 3$ V
b The value of R_x that produces half-scale deflection.

2-8 A series ohmmeter is constructed as shown in Fig. 2-4. The meter has a resistance of 50 Ω and a full-scale deflection of 100 μA, and the test voltage is 1.5 V.
a Adjust the value of R_o for full-scale deflection when $R_x = 0$.
b Plot a calibration curve for the ohmmeter; that is, a curve of meter reading as a function of R_x.
c If the meter accuracy is ± 3 percent of full-scale deflection, what values if any of R_x can be measured to an accuracy of ± 10 percent?

2-9 The meter movement in Prob. 2-8 is employed in the circuit shown in Fig. 2-41 in which the resistance R_p is used to change the range of resistance that can be measured. For $R_p = 5\ \Omega$, repeat Prob. 2-8.

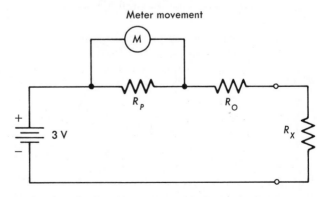

FIGURE 2-41 Ohmmeter circuit for Prob. 2-9.

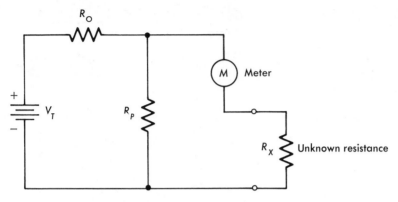

FIGURE 2-42 Shunt-ohmmeter for Prob. 2-10.

2-10 A 50-Ω milliameter which deflects full scale for a current of 0.5 mA is used in the *shunt-ohmmeter* in Fig. 2-42. This type of circuit is useful in measuring smaller resistance values than is usually convenient for series ohmmeters. The resistance R_o is adjusted for full-scale deflection when R_x is zero; R_p is used to alter the range of resistances that can be accurately measured. For $R_p = 45$ and $V_T = 3$ V, determine
a The value of R_o
b The value of R_x which causes half-scale deflection

2-11 An ohmmeter of the type shown in Fig. 2-42 is to be constructed using a 75-Ω milliameter whose full-scale deflection is 1.0 mA and a 1.5-V test source.
a Determine R_o and R_p so that $R_x = 1,000$ Ω produces half-scale deflection.
b If the milliameter accuracy is ± 3 percent of full-scale deflection, what is the largest value of R_x that can be measured with an accuracy of ± 10 percent?

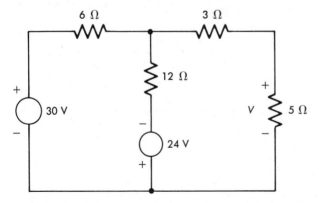

FIGURE 2-43 Circuit for Prob. 2-12.

2-12 Use the direct application of Kirchhoff's laws to find the voltage V across the 5-Ω resistor in the circuit of Fig. 2-43.

2-13 Find the current I in the circuit of Fig. 2-44. Use the direct application of the fundamental laws.

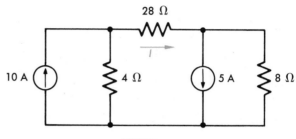

FIGURE 2-44 Circuit for Prob. 2-13.

2-14 The circuit shown in Fig. 2-45 is that of a transistor amplifier stage. By direct application of Kirchhoff's laws, determine
a The currents I_1 and I_2
b The power supplied by the 0.3-V source
c The power dissipated by the 4-kΩ resistance

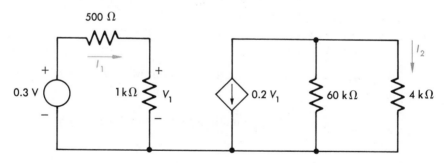

FIGURE 2-45 Circuit for Prob. 2-14.

2-15 The circuit in Fig. 2-46 is a simplified model of a system used to detect the presence of gaseous impurities caused by leaks in a piping system. The sensing element is R_2 whose resistance varies with impurity concentration. The shaded portion is the representation of an operational amplifier stage whose output V_o can be measured, displayed or used to drive an alarm. Assume

$$R_2 = 2 \times 10^3(1 + 50.4 \times 10^{-3}\eta - 4.24 \times 10^{-5}\eta^2)$$

where η is the impurity concentration in parts per million (ppm).
a Plot a curve of V_o as a function of η for the range $0 \le \eta \le 10^3$.
b A measurement indicates $V_o = 9$ V; determine the concentration from **a**.

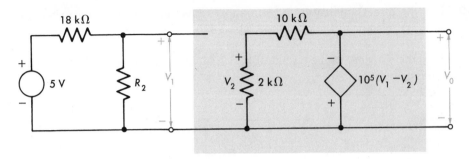

FIGURE 2-46 Circuit for Prob. 2-15.

2-16 A circuit model of an amplifier stage, called an emitter-follower, is depicted in Fig. 2-47. Determine the voltage V_o.

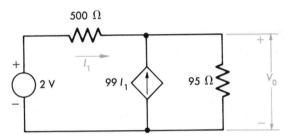

FIGURE 2-47 Circuit for Prob. 2-16.

2-17 Use the node-voltage method to determine the voltage V in the circuit of Fig. 2-43.

2-18 Use the node-voltage method to determine the current I in the circuit of Fig. 2-44.

2-19 Use the node-voltage method to determine the voltage across the 12-Ω resistance in Fig. 2-36.

2-20 a Use the node-voltage method to determine the voltage V_B in the circuit of Fig. 2-48.
b From the result in a, find I_3.

2-21 Find the voltage across the 8-A current source in the circuit of Fig. 2-49 by the node-voltage method. (*Hint:* The 4-V source has no series resistance and cannot be converted into a current-source equivalent. Select node D as the reference and write the node-voltage equations in the usual manner. Then, recognition that V_C is known leads to the desired solution.)

2-22 Use nodal analysis to find the voltage V_{cd} in the circuit of Fig. 2-16a.

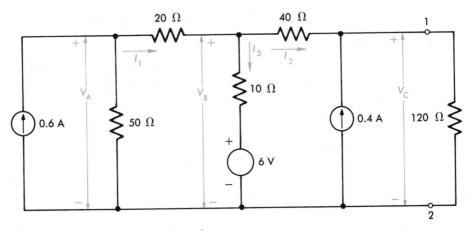

FIGURE 2-48   Circuit for Prob. 2-20.

2-23   Use mesh analysis to find the current *I* in the circuit of Fig. 2-44.

2-24   Determine the current in the 12-Ω resistance in Fig. 2-43 by the use of mesh analysis.

2-25   Use mesh analysis to determine the currents I_1, I_2, and I_3 in the circuit of Fig. 2-48.

2-26   Find the voltage across the 12-Ω resistance in the circuit of Fig. 2-36. Use mesh analysis.

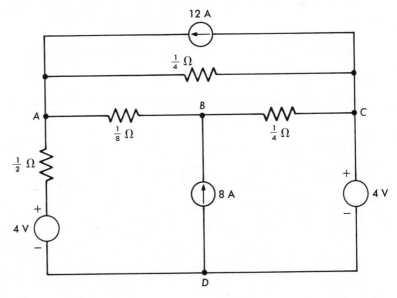

FIGURE 2-49   Circuit for Prob. 2-21.

2-27   Find the power supplied by the 9-A source in the circuit of Fig. 2-14a. Use mesh analysis.

2-28   Use mesh analysis to determine the current in the $\frac{1}{2}$-Ω resistance in the circuit of Fig. 2-49. (*Hint:* The 8-A source cannot be converted into a voltage-source equivalent. Select mesh currents in the usual manner and recognize that the current in the branch B-O is known.)

2-29   A strain gauge is essentially a Wheatstone bridge, as shown in Fig. 2-50. The resistance $R + \Delta R$ is the sensing element attached to the structural member under test. To eliminate temperature effects, a dummy gauge having an unstressed resistance R is used as one arm of the bridge. The two remaining resistors R_1 and R_2 are made equal so that when the gauge is unstrained the bridge is balanced. Applying a force to the structural member causes the sensing element to both elongate and constrict increasing the gauge resistance to $R + \Delta R$. The change in resistance is related to the elongation of the gauge by

$$\frac{\Delta R}{R} = K \frac{\Delta L}{L}$$

where ΔL = change in gauge length
L = unstrained length
K = gauge constant which depends on the gauge material
a   For $R = R_1 = R_2 = 120 \ \Omega$, and $V_T = 12$ V, derive an equation which relates V_{ab} to R and ΔR.
b   What value of ΔR is needed to make $V_{ab} = 1.0$ mV?
c   In a bonded-wire strain gauge, $K = 2$. What is the minimum value of $\Delta L/L$ that can be measured if the minimum value of V_{ab} that can be detected is 100 μV?
d   A typical value of K for a semiconductor gauge is 150. Repeat c for this type of strain gauge.

2-30   Use nodal analysis to find the voltage across the 5-Ω resistance in the circuit of Fig. 2-51.

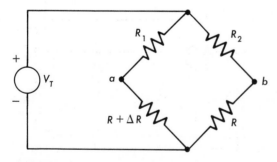

FIGURE 2-50   Strain gauge for Prob. 2-29.

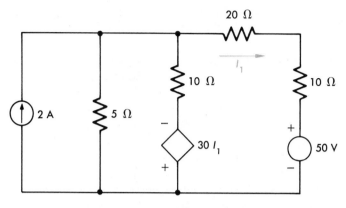

FIGURE 2-51 Circuit for Prob. 2-30.

2-31 Use nodal analysis to find the voltage across the 12-Ω resistance in the circuit of Fig. 2-52.

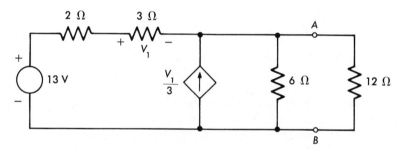

FIGURE 2-52 Circuit for Prob. 2-31.

2-32 Find the current in the 2-Ω resistance in the circuit of Fig. 2-53 by the use of nodal analysis.

2-33 Find the voltage across the 5-Ω resistance in the circuit of Fig. 2-51 by the use of mesh analysis.

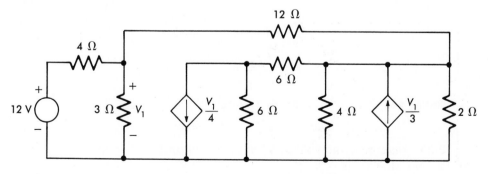

FIGURE 2-53 Circuit for Prob. 2-32.

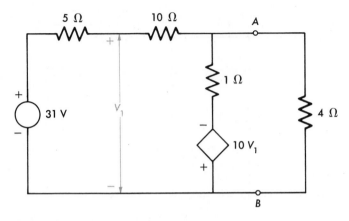

FIGURE 2-54 Circuit for Prob. 2-34.

2-34 a Use mesh analysis to determine the power dissipated by the 4-Ω resistance in the circuit of Fig. 2-54.
b What is the power supplied by the 31-V source?

2-35 The circuit shown in Fig. 2-55 is the representation of a transistor amplifier stage. Determine, by use of mesh analysis,
a The current I_o
b The power supplied by the 1.0-mV source
c The power dissipated in the 2-kΩ resistance
d The power in the dependent source (Is it supplied or absorbed?).

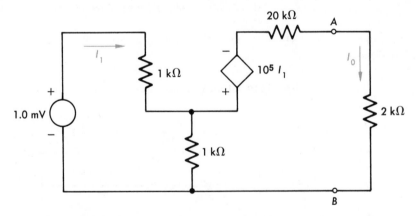

FIGURE 2-55 Transistor amplifier stage for Prob. 2-35.

2-36 Determine the voltage V_o in the circuit of Fig. 2-56 by
a Nodal analysis
b Mesh analysis.

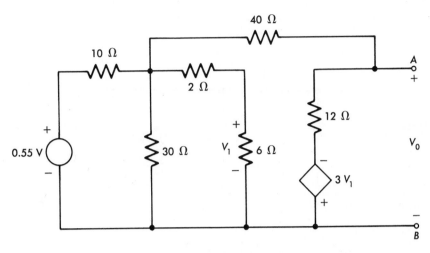

FIGURE 2-56   Circuit for Prob. 2-36.

2-37   In the Wheatstone bridge circuit of Fig. 2-57, the resistance R_M represents a D'Arsonval galvanometer and V_2 is the output of a thermocouple. The thermocouple is a device which produces a voltage proportional to the temperature difference between the two junctions formed when a pair of dissimilar metals are joined.

a   For $R_1 = R_2 = 100$ Ω, $R_3 = R_4 = 400$ Ω, $V_1 = 6$ V, $R_M = 20$ Ω, and $V_2 = 1.5$ V, determine the current I_{ab}. Use the principle of superposition.

b   What value of R_3 is needed to make $I_{ab} = 0$, assuming all other component values are those given in a?

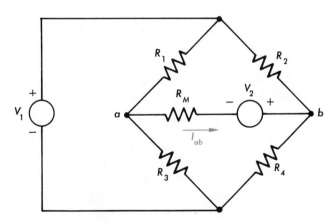

FIGURE 2-57   Wheatstone bridge circuit for Prob. 2-37.

2-38 Use the principle of superposition to find the value of I needed to cause $V_2 = 0$ in the circuit of Fig. 2-58.

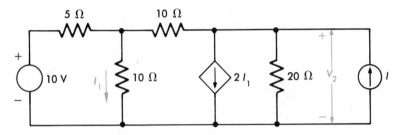

FIGURE 2-58 Circuit for Prob. 2-38.

2-39 a Determine the Thévenin equivalent circuit as viewed by the resistor R in the circuit of Fig. 2-59.
b What value of R is required if the power dissipated by R is to be maximum?
c What is the value of the power in b?

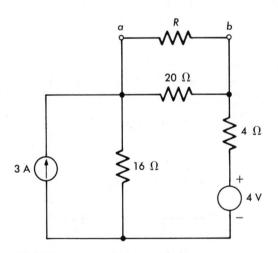

FIGURE 2-59 Circuit for Prob. 2-39.

2-40 Use the Thévenin equivalent circuit to find the voltage across the 12-Ω resistance in the circuit of Fig. 2-36.

2-41 a Obtain the Thévenin equivalent of the portion of the circuit to the left of terminals 1-2 in Fig. 2-48.
b Use the result in a to find the current in the 120-Ω resistance.

2-42 a Determine the Thévenin equivalent at terminals *A-B* for the circuit in Fig. 2-52.
b Use the result in *a* to find the voltage across the 12-Ω resistance.

2-43 Find the current in the 4-Ω resistance in the circuit of Fig. 2-54 by the use of the Thévenin equivalent circuit.

2-44 Find the voltage across the 2-kΩ resistance in the circuit of Fig. 2-55 by use of Thévenin's theorem.

2-45 In the circuit of Fig. 2-57, the parameter values are: $R_1 = 4\ \Omega$, $R_2 = 6\ \Omega$, $R_3 = 12\ \Omega$, $R_4 = 3\ \Omega$, $R_M = 2\ \Omega$, $V_1 = 12$ V, and $V_2 = 1.5$ V. Determine the current I_{ab} by first removing the branch *ab* and replacing the remainder of the bridge circuit by its Thévenin equivalent.

2-46 a Obtain a Thévenin equivalent circuit at terminals *A-B* for the circuit of Fig. 2-56.
b What is the maximum power that can be provided to a resistance *R* connected to terminals *A-B*?

2-47 Find the resistance R_{ab} in the circuit of Fig. 2-60.

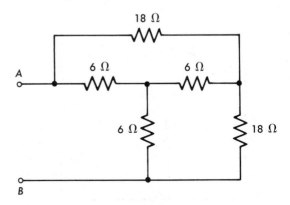

FIGURE 2-60 Circuit for Prob. 2-47.

2-48 Determine I_2 in the circuit of Fig. 2-48 by use of network reduction.

2-49 Find the resistance seen by the 31-V source in the circuit of Fig. 2-54.

2-50 In the circuit of Fig. 2-55, determine the resistance seen by the 1 mV source.

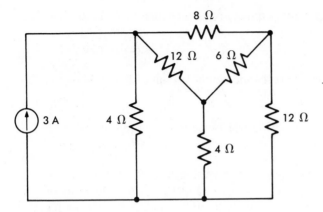

FIGURE 2-61   Circuit for Prob. 2-51.

2-51   Find the voltage across the 8-Ω resistance in the circuit of Fig. 2-61. Use network reduction and source conversion.

TEADY-STATE CIRCUIT ANALYSIS □ **SINUSOIDAL**
TATE CIRCUIT ANALYSIS □ SINUSOIDAL **STEADY-**
IRCUIT ANALYSIS □ SINUSOIDAL STEADY-**STATE**
NALYSIS □ SINUSOIDAL STEADY-STATE **CIRCUIT**
□ SINUSOIDAL STEADY-STATE CIRCUIT **ANALYSIS**
SINUSOIDAL STEADY-STATE CIRCUIT ANALYSIS □
STEADY-STATE CIRCUIT ANALYSIS □ SINUSOIDAl
-STATE CIRCUIT ANALYSIS □ SINUSOIDAL STEAD
CIRCUIT ANALYSIS □ SINUSOIDAL STEADY-STATI
ANALYSIS □ SINUSOIDAL STEADY-STATE CIRCUIT ANALYSIS □ SINI
SINUSOIDAL STEADY-STATE CIRCUIT ANALYSIS □ SINUSOIDAL STE
STEADY-STATE CIRCUIT ANALYSIS □ SINUSOIDAL STEADY-STATE C
-STATE CIRCUIT ANALYSIS □ SINUSOIDAL STEADY-STATE CIRCUIT
CIRCUIT ANALYSIS □ SINUSOIDAL STEADY-STATE CIRCUIT ANALYS
ANALYSIS □ SINUSOIDAL STEADY-STATE CIRCUIT ANALYSIS □ SINI
SINUSOIDAL STEADY-STATE CIRCUIT ANALYSIS □ SINUSOIDAL STE
STEADY-STATE CIRCUIT ANALYSIS □ SINUSOIDAL STEADY-STATE C
-STATE CIRCUIT ANALYSIS □ SINUSOIDAL STEADY-STATE CIRCUIT
CIRCUIT ANALYSIS □ SINUSOIDAL STEADY-STATE CIRCUIT ANALYS
ANALYSIS □ SINUSOIDAL STEADY-STATE CIRCUIT ANALYSIS □ SINI
SINUSOIDAL STEADY-STATE CIRCUIT ANALYSIS □ SINUSOIDAL STE
STEADY-STATE CIRCUIT ANALYSIS □ SINUSOIDAL STEADY-STATE C
-STATE CIRCUIT ANALYSIS □ SINUSOIDAL STEADY-STATE CIRCUIT
CIRCUIT ANALYSIS □ SINUSOIDAL STEADY-STATE CIRCUIT ANALYS
ANALYSIS □ SINUSOIDAL STEADY-STATE CIRCUIT ANALYSIS □ SINI
SINUSOIDAL STEADY-STATE CIRCUIT ANALYSIS □ SINUSOIDAL STE
STEADY-STATE CIRCUIT ANALYSIS □ SINUSOIDAL STEADY-STATE C
-STATE CIRCUIT ANALYSIS □ SINUSOIDAL STEADY-STATE CIRCUIT
CIRCUIT ANALYSIS □ SINUSOIDAL STEADY-STATE CIRCUIT ANALYS
ANALYSIS □ SINUSOIDAL STEADY-STATE CIRCUIT ANALYSIS □ SINI
SINUSOIDAL STEADY-STATE CIRCUIT ANALYSIS □ SINUSOIDAL STE
STEADY-STATE CIRCUIT ANALYSIS □ SINUSOIDAL STEADY-STATE C
-STATE CIRCUIT ANALYSIS □ SINUSOIDAL STEADY-STATE CIRCUIT
CIRCUIT ANALYSIS □ SINUSOIDAL STEADY-STATE CIRCUIT ANALYS
ANALYSIS □ SINUSOIDAL STEADY-STATE CIRCUIT ANALYSIS □ SINI

3

n the previous chapter resistive networks excited by constant voltages and currents were treated. This chapter introduces the analysis of circuits that are excited by time-varying sources and that contain the energy-storage elements, capacitance and inductance. In particular, attention is focused on the steady-state response of networks to sinusoidal (ac) excitations. Sinusoidal excitations are important for several reasons. The time functions generated by virtually all large-capacity generating systems are sinusoids. Sinusoidal signals of adjustable frequency are one of the most commonly used signals for testing electronic equipment. Most periodic functions can be represented as a sum of sinusoids; techniques developed through such a representation can be extended to include aperiodic functions.

Because of the many conveniences afforded, the algebra of complex numbers is used to solve steady-state ac circuits problems. This approach has been universal since the method was introduced by C. P. Steinmetz at the turn of the twentieth century. Students who feel unfamiliar with complex-number manipulation may wish to review the material in Appendix B.

The initial sections of this chapter deal with the nature of network behavior and the response of circuits to exponential excitations. The use of exponential functions provides a logical foundation for both the ac steady-state analysis treated in the remaining sections of this chapter and the excitation-response techniques presented in Chap. 4.

Those readers who wish to proceed directly to the more general analysis of circuits to time-varying excitations in Chap. 4 may do so by first studying Secs. 3-1 to 3-3. The remainder of the material in this chapter may be treated after Chaps. 4 and 5 without loss of continuity.

3-1 THE NATURE OF NETWORK RESPONSES

The response of a physical system to an applied excitation is determined by three factors: the excitation, the kind of elements the system comprises and how these elements are interconnected, and the past history of the system. As a qualitative illustration of the effect each factor has on the response, the system depicted in Fig. 3-1 is considered. In Fig. 3-1, the surface of mass M is in contact with a stationary surface, where D represents the friction between the surfaces. The excitation is the force $F(t)$; the response is the velocity $u(t)$. From our experience we know the velocity depends on the characteristics of the force. These include the force's intensity, length of time of application, and variation with time. Similarly, the size of the mass and amount of friction affect the velocity no matter what the excitation is. The effect of past history on the response is evident in the answers to the following: Is the system moving when the force is applied? If so, in what direction and how fast?

The qualitative discussion in the previous paragraph is the basis for the

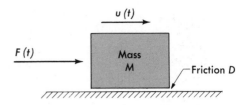

FIGURE 3-1 Mass-friction system.

quantitative description of the system response. Consider the situation of the mass-friction system in Fig. 3-1 being accelerated from an initial velocity u_0 to velocity u_1 by means of a constant force. The resultant velocity response is shown in Fig. 3-2. The time at which the force is applied is arbitrarily assigned; for convenience, it is often selected as $t = 0$. Prior to the application of the force, the system exists in its initial state, corresponding to u_0. For times greater than zero, the total response is considered to have two components which describe the new state of the system and the manner by which the change from the initial state occurs. The response component indicative of the new state is the *forced response.* The excitation and the effects it produces in the system elements determine the forced response. The forced response is called the *steady-state response,* particularly when the excitation is constant or when it is repetitive in time.

The *natural response* describes the transition between the initial state and the new forced response. The form of the natural response is determined almost exclusively by the system elements and their interaction. Because it describes the time interval between the initial state and forced response, the natural response is frequently referred to as the *transient response.*

The past history is incorporated into the total response by means of boundary conditions, which, for physical reasons, are usually *initial conditions* evaluated at the time the excitation is applied. The initial conditions determine the amplitude of the natural response and reflect the degree of mismatch between the original state and forced response.

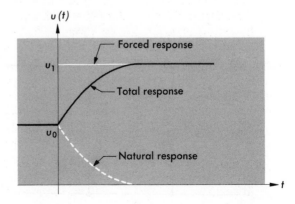

FIGURE 3-2 Velocity response of mass-friction system.

For many applications, the solution for the forced response is all that is required. This is frequently the case when excitations are periodic or when they are applied for lengthy durations. In these instances, the transient period is short enough, when compared with other times involved, to render unnecessary, for practical purposes, the solution for the natural response.

3-2 SINGLE-ELEMENT RESPONSES TO EXPONENTIAL EXCITATIONS

Sources of the form $A\epsilon^{st}$ are used to investigate network responses for the principal reason that a wide variety of excitations encountered in practice can be represented by exponentials. The factor s multiplying the time variable t in ϵ^{st} is a general variable which can assume real, imaginary, or complex values. Since the exponent is dimensionless and t is measured in seconds, the dimension of s is accordingly $1/s$ and is the reason s is often called the *complex-frequency variable*. For a specific excitation, s will, of course, have a specific numerical value.

Many excitations are capable of being expressed by exponentials. The time-invariant source (dc) is obtained simply from $A\epsilon^{st}$ by setting $s = 0$. In Chap. 4, pulses are shown to be composed of a number of time-invariant components. Similarly, the periodic functions discussed in Sec. 3-12 are composed of sinusoidal components, with each sinusoid expressed as an exponential function as described in Sec. 3-4. Therefore, exponential representation is applicable to both simple and complex excitations and is useful in the evaluation of network responses.

To demonstrate the use of $A\epsilon^{st}$ as an excitation, the relationship between voltage and current for the three circuit elements is developed. In the circuit of Fig. 3-3, the current source $i(t)$ is used to establish a voltage $V\epsilon^{st}$ in each of the circuit elements. The current through each element is determined from the appropriate volt-ampere relations (see Chap. 1). Thus

$$i_G(t) = Gv = GV\epsilon^{st} = I_G\epsilon^{st} \qquad (3\text{-}1)$$

$$i_C(t) = C\frac{dv(t)}{dt} = CsV\epsilon^{st} = I_C\epsilon^{st} \qquad (3\text{-}2)$$

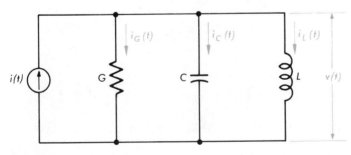

FIGURE 3-3 GLC parallel circuit.

$$i_L(t) = \frac{1}{L} \int v(t) \, dt = \frac{1}{Ls} V\epsilon^{st} = I_L\epsilon^{st} \tag{3-3}$$

Each of the currents given in Eqs. (3-1) to (3-3) is an exponential whose exponent is the same as that for the voltage across the elements. The amplitude of each current can then be related to the voltage amplitude as

$$I_G = GV \qquad \text{or} \qquad \frac{I_G}{V} = G \tag{3-4}$$

$$I_C = CsV \qquad \text{or} \qquad \frac{I_C}{V} = Cs \tag{3-5}$$

$$I_L = \frac{1}{Ls} V \qquad \text{or} \qquad \frac{I_L}{V} = \frac{1}{Ls} \tag{3-6}$$

The relations in Eqs. (3-5) and (3-6) have the form of the Ohm's law relation of Eq. (3-4). As each term is the ratio of a current to a voltage, the quantities G, Cs, and $1/Ls$ have the dimension of mhos. The term G represents a conductance; a more general term used to define the ratio of current to voltage is *admittance*, the symbol of which is $Y(s)$. It is important to note that the admittance is a function of the complex-frequency variable s.

For many applications, the ratio of voltage to current at a pair of terminals, called the *impedance*, is useful. The impedance is designated by the symbol $Z(s)$ and has the dimension of ohms; at a pair of terminals $Z(s)$ is the reciprocal of $Y(s)$ and is similarly a function of s.

Table 3-1 shows the symbolic representation and defining relations for the three circuit elements. The time-domain volt-ampere relations are shown in Table 3-1a. In Table 3-1b the elements are represented by their impedances and admittances for the case of exponential voltages and currents. As the impedances and admittances are, in general, functions of s and are used to relate the amplitudes of the exponential voltages and currents, these volt-ampere relations are called the *frequency-domain representations* of the elements.

The source current $i(t)$ in Fig. 3-3 can be determined by the application of Kirchhoff's current law. Thus,

$$i(t) = i_G(t) + i_C(t) + i_L(t) = Gv(t) + C\frac{dv(t)}{dt} + \frac{1}{L} \int v(t) \, dt \tag{3-7}$$

Substitution of Eqs. (3-1) to (3-3) in Eq. (3-7) and combining terms give

$$i(t) = \left(G + Cs + \frac{1}{Ls} \right) V\epsilon^{st} = I\epsilon^{st} \tag{3-8}$$

Again, in Eq. (3-8), it should be noted that the source current is also an exponential function having the same exponent as the voltage, so that one may conclude that exponential excitations produce exponential responses. The ratio of I/V is the circuit admittance $Y(s)$; the circuit impedance $Z(s)$ is the ratio of V/I. These are

TABLE 3-1 Element volt-ampere relationship with (a) time-varying excitation, (b) exponential excitation, and (c) sinusoidal excitation

	RESISTANCE (CONDUC-TANCE)	INDUCTANCE	CAPACITANCE	PASSIVE NETWORKS

a Time-varying excitation in which $v = v(t)$ and $i = i(t)$

Symbol and equations

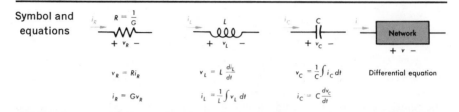

$$v_R = Ri_R$$

$$i_R = Gv_R$$

$$v_L = L\frac{di_L}{dt}$$

$$i_L = \frac{1}{L}\int v_L\, dt$$

$$v_C = \frac{1}{C}\int i_C\, dt$$

$$i_C = C\frac{dv_C}{dt}$$

Differential equation

b Exponential excitation in which $v = Ve^{st}$ and $i = Ie^{st}$

Impedance (ohm) symbol and equation

$$V_R = RI_R$$

$$V_L = sLI_L$$

$$V_C = \frac{1}{sC}I_C$$

$$V = ZI$$

Admittance (mho) symbol and equation

$$I_R = GV_R$$

$$I_L = \frac{1}{sL}V_L$$

$$I_C = sCV_C$$

$$I = YV$$

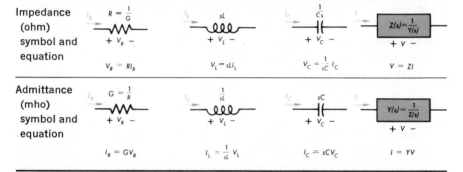

c Sinusoidal excitation in which $v = V_m \cos(\omega t + \theta)$ and $i = I_m \cos(\omega t + \alpha)$

Impedance (ohm) symbol and equations

$$V_R = RI_R$$

$$V_L = j\omega L\, I_L = jX_L\, I_L$$

$$V_C = \frac{1}{j\omega C}I_C = jX_C$$

$$V = ZI = (R + jX)I$$

Admittance (mho) symbol and equations

$$I_R = GV_R$$

$$I_L = \frac{1}{j\omega L}V_L = jB_L V_L$$

$$I_C = j\omega C V_C = jB_C V_C$$

$$I = YV = (G + jB)V$$

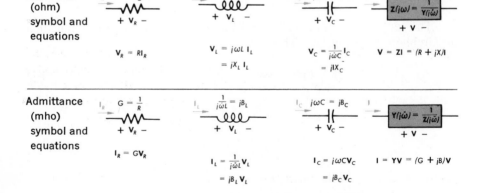

$$Y(s) = \frac{I}{V} = G + Cs + \frac{1}{Ls} \qquad Z(s) = \frac{V}{I} = \frac{1}{G + Cs + 1/Ls} \qquad (3\text{-}9)$$

It should be noted in Eq. (3-9) that $Y(s)$, which represents the three elements in parallel, can be obtained by addition of the admittance of each element in the same manner as parallel conductances are combined in the resistive networks of Chap. 2. In Table 3-1a, the general network is represented by a differential equation of the type in Eq. (3-7) in the time domain. For exponential excitation, $Z(s)$ and $Y(s)$ indicated in Table 3-1b are used to relate the values of V and I by means of an algebraic equation.

Table 3-1c presents a series of entries for the case where $s = j\omega$; the significance of these entries will be discussed in Sec. 3-4 where the forced response to sinusoidal excitation is introduced. For the moment we will simply note that the case of $s = j\omega$ will be of considerable significance.

Networks drawn using the symbols of Tables 3-1b and 3-1c are called *transformed networks*. These networks play a significant role in the further study of circuit theory. The next section of this chapter is devoted to the examination of the forced response of circuits to exponential excitations and leads directly to the important special case of sinusoidal excitations. In Chaps. 4 and 5 the use of exponential functions and transformed networks are employed again in the total solution of network responses.

3-3 FORCED RESPONSE WITH EXPONENTIAL EXCITATION

In this section, the forced component of a network response resulting from an exponential excitation $A\epsilon^{st}$ is determined. The forced component is that portion of the response produced by the particular excitation applied.

The circuits of Fig. 3-4 are used to indicate how the forced response is determined. The circuit in Fig. 3-4a is described by the use of Kirchhoff's voltage law as

$$v(t) = Ri(t) + L\frac{di(t)}{dt} + \frac{1}{C}\int i(t)\, dt \qquad (3\text{-}10)$$

As indicated in Sec. 3-2, an exponential excitation causes an exponential response, both having the same value of s. Thus, the forced component of response is $i(t) = I\epsilon^{st}$. To determine I, Table 3-1b is used; each element is represented by its impedance, resulting in the transformed network of Fig. 3-4b. The KVL expression for the circuit in Fig. 3-4b becomes

$$V = RI + LsI + \frac{1}{Cs}I \qquad (3\text{-}11)$$

Solving for I in Eq. (3-11) yields

$$I = \frac{V}{R + Ls + 1/Cs} = \frac{V}{Z(s)} \qquad (3\text{-}12)$$

It should be noted that the denominator in Eq. (3-12) is the expression for the

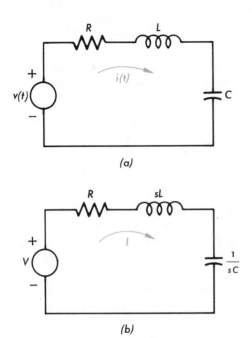

FIGURE 3-4 RLC series circuit (a) in the time
domain and (b) transformed into the s domain.

impedance of the circuit and can be obtained by the addition of each series
impedance. The time function corresponding to the frequency-domain
response in Eq. (3-12) is

$$i(t) = I\epsilon^{st} = \frac{V}{R + Ls + 1/Cs} \epsilon^{st} \tag{3-13}$$

The steps necessary to determine the response are summarized as
follows:

1 Express the excitation in the form $A\epsilon^{st}$.

2 Transform the network by the use of Table 3-1b.

3 Write the appropriate Kirchhoff's law expressions and solve for the
frequency-domain response.

4 Convert the frequency-domain response of (3) to the time function by
multiplication by ϵ^{st}. The value of s used is that of the excitation given in (1).

Example 3-1

Determine the forced component of the current through the capacitor in the
circuit of Fig. 3-5a for $i(t)$ equal to

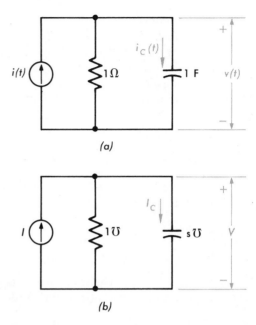

FIGURE 3-5 Circuit of Example 3-1 (a) in the time domain and (b) transformed to the s domain.

a $10\epsilon^{-2t}$

b   10 A

SOLUTION   **a**   The excitation is in exponential form with $s = -2$ so that the response is also an exponential. The transformed network based on the admittances in Table 3-1b is shown in Fig. 3-5b, and V and I may be related by use of Kirchhoff's current law as

$$I = \frac{V}{1} + sV$$

from which

$$V = \frac{I}{1 + s}$$

The current I_C through the capacitor is

$$I_C = sV = \frac{s}{1 + s} I$$

Substitution of $s = -2$ and $I = 10$ gives

$$I_C = 20$$

from which

$$i_C(t) = 20\epsilon^{-2t}$$

b The constant excitation can be expressed as $10\epsilon^{0t}$ so that substitution of $s = 0$ in I_C in part a gives $I_C = 0$ and consequently $i_C(t) = 0$. The significance of this result is that it indicates the capacitor behaves as an open circuit in the dc steady state. This can be further verified by considering the impedance of the capacitor for $s = 0$ which tends toward infinity. (Infinite impedance is used to signify an open circuit.)

Example 3-2

For the circuit of Fig. 3-6a, determine the voltage across the inductor for $e(t)$ equal to

a $10\epsilon^{-2t}$

b 10 V

SOLUTION **a** The transformed network is shown in Fig. 3-6b. The KVL expression for the loop in Fig. 3-6b is

$$V = 1I + sI = (1 + s)I \qquad \text{or} \qquad I = \frac{V}{1 + s}$$

The voltage V_L is sI so that

$$V_L = \frac{s}{1 + s} V$$

Substitution of $s = -2$ and $V = 10$ gives

$$V_L = \frac{-2}{1 - 2} \times 10 = +20$$

from which

$$v_L(t) = 20\epsilon^{-2t}$$

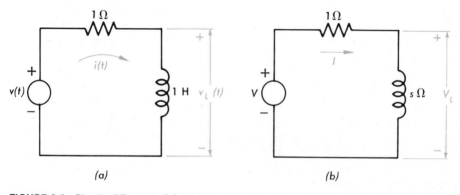

(a) (b)

FIGURE 3-6 Circuit of Example 3-2 (a) in the time domain and (b) transformed to the s domain.

b   The constant excitation is represented by $10\epsilon^{0t}$; substitution of $s = 0$ in the expression for V_L in part a yields $V_L = 0$ and, consequently, $v_L(t) = 0$. The short-circuit behavior of the inductance in the dc steady state is indicated by the zero value of inductance voltage. From Table 3-1b, it is seen that for $s = 0$, the impedance of the inductor is zero, zero impedance signifying a short circuit.

The numeric results for $i_C(t)$ and $v_L(t)$ in Examples 3-1 and 3-2 are identical and highlight the dualism of series and parallel circuits. Two electric circuits are said to be the *duals* of each other if the KVL equations of one circuit are numerically the same as the KCL equations of the second circuit. This is the case in these examples; the network functions relating the response and excitation (I_C/I in Example 3-1 and V_L/V in Example 3-2) are identical functions of s. Therefore, for excitations which are numerically equivalent ($10\epsilon^{-2t}$ and $10\epsilon^{0t}$), the respective responses should also be numerically equivalent. The concept of duality therefore requires both element and configurational correspondence. If network B is the dual of network A, then all values of L, R, and C in network A are replaced by numerically equal values of C, G, and L, respectively, in network B. Similarly, all voltage and current sources in A are numerically equal to the respective current and voltage sources in B; elements in series (parallel) in A correspond to elements in parallel (series) in B.

The concept of duality can be extended to other nonelectric physical systems by means of analogs. For example, in a dynamical system a mass M is governed by Newton's law which states

$$F(t) = M \frac{du}{dt} \quad \text{or} \quad u(t) = \frac{1}{M} \int F(t) \, dt \tag{3-14}$$

where $F(t)$ is the force on the mass and $u(t)$ is the velocity. The form of the relations in Eq. (3-14) is identical to that for inductance and capacitance given in Table 3-1a. Two analogies are possible; the first is called the *force-voltage analog*, the second the *force-current analog*. If the force-voltage analog is used, the velocity and mass analogs are the current and inductance, respectively. Table 3-2 is used to identify the analogous electric quantities for a mechanical system.

The significance of the analogs in Table 3-2 is that by making the quantities in either of the two right-hand columns numerically identical to those of the mechanical systems, the response of the analogous electric system will be numerically equivalent to that of the corresponding mechanical system. Thus, the mechanical-system characteristics can be investigated by means of an equivalent electrical network. This principle is the basis for use of electronic analog computers. Other systems such as those encountered in thermodynamics and fluid flow can also be represented by analogous electric networks.

TABLE 3-2 Mechanical-electrical analogs

QUANTITY	SYMBOL	DEFINING RELATION	FORCE-CURRENT ANALOG	FORCE-VOLTAGE ANALOG
Force	$\longrightarrow F(t)$	....	$i(t)$	$v(t)$
Velocity	$\longrightarrow u(t)$	....	$v(t)$	$i(t)$
Mass		Newton's law $$F = M \frac{du}{dt}$$	C	L
Compliance = $\dfrac{1}{\text{stiffness}}$	C_m	Hooke's law $$F = \frac{1}{C_m} \int u\, dt$$	L	C
Viscous friction	D	$F = Du$	G	R

3-4 SINUSOIDAL EXCITATION

Most of the world's electric energy is generated and distributed by means of voltages and currents which vary sinusoidally with time. Furthermore, the electronic signals present in many instrumentation, control, and communication systems are often conveniently represented as sinusoids. Thus, sine and cosine functions are an important class of excitation in electrical engineering.

The waveform of a cosine function representing a current is shown in Fig. 3-7; the equation for the current is given in Eq. (3-15).

$$i(t) = I_m \cos \omega t \tag{3-15}$$

In Eq. (3-15), $i(t)$ is the *instantaneous value* of the current at any time t, and I_m is the maximum value or amplitude. The cosine function is *periodic;* that is, it repeats itself every T seconds. In general, a function which satisfies

$$f(t) = f(t + T) \tag{3-16}$$

for all values of t is said to be periodic with *period T*. Two other periodic functions commonly encountered are the square-wave and sawtooth wave depicted in Fig. 3-8. The portion of each waveform existing for one period in Fig. 3-7 and 3-8 is called a *cycle*. The number of cycles of the waveform per second is the *frequency* measured in *hertz* (Hz). Evidently,

$$f = \frac{1}{T} \quad \text{Hz} \tag{3-17}$$

For the cosine wave in Fig. 3-7, the segment $0ab$ represents one cycle.

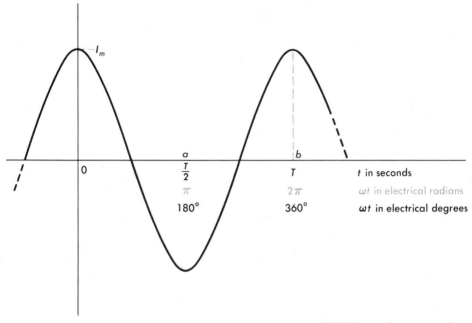

FIGURE 3-7 Current waveform.

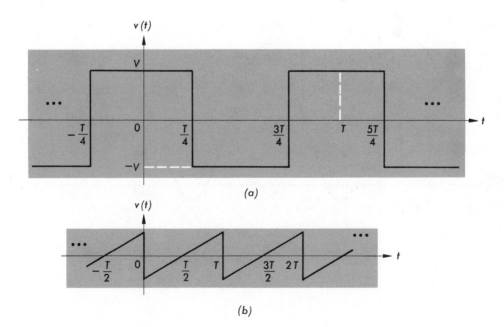

FIGURE 3-8 Periodic waveforms: (a) square-wave and (b) saw-tooth wave.

When Eq. (3-15) is plotted as a function of ωt, one cycle corresponds to an electrical angle of 2π radians or 360 degrees. Thus, the *angular frequency* ω can be expressed as

$$\omega = \frac{2\pi}{T} = 2\pi f \tag{3-18}$$

from which Eq. (3-15) can be expressed in the alternative forms

$$i(t) = I_m \cos \omega t = I_m \cos 2\pi ft = I_m \cos \frac{2\pi}{T} t \tag{3-19}$$

In Fig. 3-7, the time origin or time reference axis is chosen at the point where the current is a maximum positive value. In general, the origin can be chosen arbitrarily at any point on the wave and is selected for greatest convenience in the specific problem at hand. In many circuits, the current and voltage waves do not go through zero or their maxima at the same time but are displaced from one another by a *phase angle*. Figure 3-9 shows current and voltage waves which neither are in phase with one another nor have their maxima at zero time; the equations of these waves are

$$v = V_m \cos (\omega t + \alpha) \tag{3-20}$$

$$i = I_m \cos (\omega t + \beta) \tag{3-21}$$

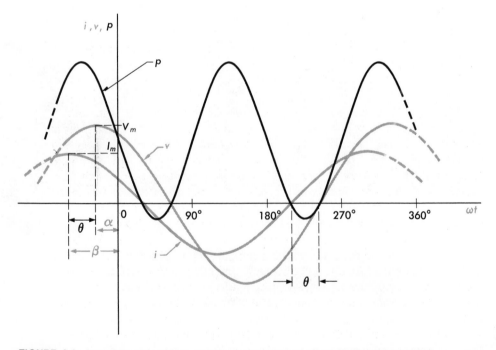

FIGURE 3-9 Instantaneous voltage, current, and power in a sinusoidally excited circuit.

The phase angle between the two waves is θ. The voltage wave, which reaches a peak after the current wave, is said to *lag* the current by the angle θ; or conversely, the current, which reaches its peak at an earlier time than does the voltage, *leads* the voltage by the angle θ. It should be noted that the quantity ωt in the argument of the cosine is dimensionally in radians; to be strictly correct, therefore, the values of α and β in Eqs. (3-20) and (3-21) should also be in radians. For example, the voltage v might be

$$v = 155 \cos \left(377t + \frac{\pi}{6} \right)$$

Common usage, however, specifies the phase angles in degrees, and the voltage v then becomes

$$v = 155 \cos (377t + 30°)$$

In calculating instantaneous values of the wave, this dimensional inconsistency should be noted. The normal application of the voltage equation is such that the specification of the phase angle in degrees is more helpful in calculations.

The term *sinusoidal excitation* as used in this book means excitation whose waveform is sinusoidal, regardless of when in time the maximum value of the excitation occurs. Thus the voltage and current functions described by Eqs. (3-20) and (3-21) are called *sinusoidal functions*. Circuits excited with sinusoidal currents and voltages are often called *alternating-current (ac) circuits;* circuits excited with constant currents and voltages are similarly called *direct-current (dc) circuits.*

If the current and voltage functions of Eqs. (3-20) and (3-21) represent the current in and voltage across a portion of an ac circuit, the instantaneous power in that portion of the circuit is, from Eq. (1-13), the product vi. The product curve $p = vi$ is also shown in Fig. 3-9, from which it should be noted that although v and i are negative during half of each cycle, the power p is predominantly positive. (The case of $\theta = 90°$ is an exception.) Thus, alternating currents and voltages are effective means for the transfer of power and energy. During a part of the power cycle equal to the phase angle θ between current and voltage, the power becomes negative and hence the direction of energy flow is reversed. Compared with the dc case, this reversal is, of course, a disadvantage, as is the fact that the instantaneous power transfer is not constant in magnitude. Practically speaking, however, these disadvantages are outweighed by the fact that alternating currents and voltages permit the use of transformers. Transformers make possible electric generation at the most economical generator voltage, power transfer at the most economical transmission voltage, and power utilization at the most economical and efficacious voltage for the particular utilization device. The resulting advantages are so great that electricity is almost universally generated and transmitted in ac form. If direct current is desired for a particular application, it is obtained by conversion from alternating current at or very near the point of utilization.

The representation of sinusoids by exponential functions is based on Euler's identity, which states

$$\epsilon^{j\theta} = \cos \theta + j \sin \theta \tag{3-22}$$

where j is used to express the imaginary number $\sqrt{-1}$. Substitution of $-\theta$ for θ in Eq. (3-22) yields

$$\epsilon^{-j\theta} = \cos \theta - j \sin \theta \tag{3-23}$$

Algebraic manipulation of Eqs. (3-22) and (3-23) gives

$$\cos \theta = \frac{\epsilon^{j\theta} + \epsilon^{-j\theta}}{2} \tag{3-24}$$

$$\sin \theta = \frac{\epsilon^{j\theta} + \epsilon^{-j\theta}}{2j} \tag{3-25}$$

Equations (3-24) and (3-25) demonstrate that sinusoids can be expressed by two exponential functions, each of which has an imaginary exponent. Thus an excitation of the form

$$v(t) = V_m \cos (\omega t + \phi) \tag{3-26}$$

where V_m is the peak amplitude and ϕ is the phase angle as shown in Fig. 3-10, can be written as

$$v(t) = \frac{V_m}{2} [\epsilon^{j(\omega t + \phi)} + \epsilon^{-j(\omega t + \phi)}] \tag{3-27}$$

The value of s is the portion of the exponent which multiplies t; it is then convenient to express $v(t)$ as

$$v(t) = \frac{V_m}{2} \epsilon^{j\phi} \epsilon^{j\omega t} + \frac{V_m}{2} \epsilon^{-j\phi} \epsilon^{-j\omega t} = \mathbf{V}_1 \epsilon^{j\omega t} + \mathbf{V}_2 \epsilon^{-j\omega t} \tag{3-28}$$

where $\mathbf{V}_1 = (V_m/2)\epsilon^{j\phi}$ and $\mathbf{V}_2 = (V_m/2)\epsilon^{-j\phi}$.

The quantities $\mathbf{V}_1$ and $\mathbf{V}_2$ are voltages represented by complex numbers and as such are written in boldface type. As we shall see it is frequently convenient to express voltages, currents, and voltage-current or current-voltage ratios as complex numbers. The use of boldface characters to indicate such representation will be followed in the remainder of the book. (The reader who is unfamiliar with complex numbers may refer to Appendix B.)

The voltage function of Eq. (3-26) is a cosine function. It should be noted, however, that the results are perfectly general, since any sine, cosine, or combination of sine and cosine waves of the same frequency may be written as a cosine wave with a phase angle. The trigonometric identities associated with this representation are

$$\sin (\omega t + \phi) = \cos \left(\omega t + \phi - \frac{\pi}{2} \right) \tag{3-29}$$

and

$$A \cos \omega t + B \sin \omega t = \sqrt{A^2 + B^2} \cos \left(\omega t - \tan^{-1}\frac{B}{A}\right) \qquad (3\text{-}30)$$

where, for dimensional consistency, $\tan^{-1} B/A$ is expressed in radians.

From Eq. (3-22), $\cos \theta$ is the real part of $\epsilon^{j\theta}$ so that the voltage given in Eq. (3-26) can be written in an alternative form as

$$v(t) = V_m \cos (\omega t + \phi) = \text{Re}\{V_m \epsilon^{j(\omega t+\phi)}\} = \text{Re}\{\mathbf{V}_m \epsilon^{j\omega t}\} \qquad (3\text{-}31)$$

where the term Re is used to denote the real part. Similarly, a sine function can be expressed as the imaginary part (Im) of an exponential as given in Eq. (3-32).

$$v(t) = V_m \sin (\omega t + \alpha) = \text{Im}\{V_m \epsilon^{j(\omega t+\alpha)}\} = \text{Im}\{\mathbf{V}_m \epsilon^{j\omega t}\} \qquad (3\text{-}32)$$

3-5 FORCED RESPONSE WITH SINUSOIDAL EXCITATION

The method outlined in Sec. 3-3 can also be used to determine the sinusoidal steady-state response. Consider the circuit of Fig. 3-10a which is excited by a voltage source $v(t) = V_m \cos \omega t$. The voltage source is expressed as the sum of two exponentials as given in Eq. (3-27) and leads to the circuit shown

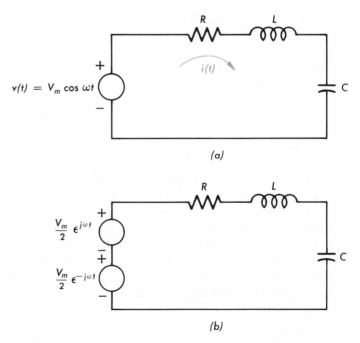

FIGURE 3-10 RLC circuit with excitation represented as (a) a sinusoid and (b) the sum of two exponentials.

in Fig. 3-10b. Each of the sources in Fig. 3-10b is an exponential for which the value of s for $v_1(t)$ is $j\omega$, that for $v_2(t)$ is $(-j\omega)$. The forced response $i(t)$ can be determined by the use of superposition. As both $v_1(t)$ and $v_2(t)$ are of the form $A\epsilon^{st}$, the circuit is described by the transformed network of Fig. 3-4b. The components of the response are then obtained from Eq. (3-12), for which

$$\mathbf{I}_1 = \frac{V_m/2}{R + j\omega L + 1/j\omega C} \qquad \text{and} \qquad \mathbf{I}_2 = \frac{V_m/2}{R - j\omega L + 1/(-j\omega C)} \qquad (3\text{-}33)$$

The component $\mathbf{I}_1$ is the frequency-domain response caused by $v_1(t)$ so that $s = j\omega$ is used in Eq. (3-12). Similarly, $s = -j\omega$ in the determination of $\mathbf{I}_2$ as it is the response produced by $v_2(t)$. Each of the terms in Eq. (3-33) can be expressed in exponential form (see Appendix B) as

$$\mathbf{I}_1 = I_1 \epsilon^{j\theta_1} \qquad \text{and} \qquad \mathbf{I}_2 = I_2 \epsilon^{j\theta_2} \qquad (3\text{-}34)$$

where

$$I_1 = I_2 = \frac{V_m/2}{\sqrt{R^2 + (\omega L - 1/\omega C)^2}} \qquad (3\text{-}35)$$

and

$$\theta_1 = -\theta_2 = -\tan^{-1}\frac{\omega L - 1/\omega C}{R} \qquad (3\text{-}36)$$

The corresponding time functions are

$$i_1(t) = I_1 \epsilon^{j\theta_1} \epsilon^{j\omega t} \qquad \text{and} \qquad i_2(t) = I_1 \epsilon^{-j\theta_1} \epsilon^{-j\omega t} \qquad (3\text{-}37)$$

The total response is the superposition of $i_1(t)$ and $i_2(t)$ and becomes

$$i(t) = i_1(t) + i_2(t) = I_1[\epsilon^{j(\omega t + \theta_1)} + \epsilon^{-j(\omega t + \theta_1)}] \qquad (3\text{-}38)$$

Substitution of Eqs. (3-35) and (3-36) and use of Eq. (3-24) give

$$i(t) = \frac{V_m}{\sqrt{R^2 + (\omega L - 1/\omega C)^2}} \cos(\omega t + \theta_1) = I_m \cos(\omega t + \theta_1) \qquad (3\text{-}39)$$

The total response in Eq. (3-39) is characterized by its amplitude I_m and phase angle θ_1. By virtue of the expressions in Eqs. (3-35) and (3-36), both $\mathbf{I}_1$ and $\mathbf{I}_2$ contain the same information about the amplitude and phase of the response. This redundancy affords the convenience of reducing the computation required since the total response can be determined from only one component of the response. Thus, for the purpose of determining the sinusoidal steady-state response, only one component of the form of $v_1(t)$ in Fig. 3-10b need be considered as the circuit excitation. Consequently, an excitation function

$$(V_m \epsilon^{j\phi})\epsilon^{j\omega t} = \mathbf{V}_m \epsilon^{j\omega t}$$

producing a response

$$(I_m \epsilon^{j\theta})\epsilon^{j\omega t} = \mathbf{I}_m \epsilon^{j\omega t}$$

can be used to imply a sinusoidal excitation

$$v(t) = V_m \cos(\omega t + \phi)$$

and its response

$$i(t) = I_m \cos(\omega t + \theta)$$

The justification is the pragmatic one that use of one of the components of Eq. (3-33) provides all the information required to write Eq. (3-39). The terms $\mathbf{V}_m$ and $\mathbf{I}_m$ are used to specify both the amplitude and phase of a sinusoid. As such they are generally called *phasors*.

The basis for the use of a single exponential term can also be demonstrated by considering the response components $\mathbf{I}_1$ and $\mathbf{I}_2$ in Eq. (3-33). These values are conjugates, as was expected since the values of s for the two excitations are conjugates. Consequently, i_1 and i_2 are conjugates and, by recalling that the sum of a number and its conjugate is twice the real part of the number,

$$i_1 + i_2 = 2\mathrm{Re}\, \mathbf{I}_1 \epsilon^{j\omega t} = I_m \cos(\omega t + \theta_1) \tag{3-40}$$

Equation (3-40) indicates that only $\mathbf{I}_1$ need be evaluated in order to specify the amplitude and phase of the sinusoidal response. Thus, a single exponential of the form $\mathbf{V}_m \epsilon^{j\omega t}$ can be used to indicate an excitation $V_m \cos(\omega t + \theta)$. Alternatively, the cosine function may be expressed as in Eq. (3-31). Verification that the use of the real part of $\epsilon^{j\omega t}$ for a cosine wave excitation can be used to evaluate the response is left to the interested student as an exercise.

The use of a single exponential function for which $s = j\omega$ to imply sinusoidal excitation and response leads to the volt-ampere relations given in Table 3-1c. Because sinusoidally excited networks are common, it is convenient to identify the frequency-domain representations of the elements separately. The differences between Tables 3-1b and 3-1c are the use of boldface symbols for current and voltage [see Eq. 3-28)] and the substitution of $j\omega$ for s. Both differences indicate sinusoidal excitation and sinusoidal response.

Note that the general impedance and admittance functions $Z(s)$ and $Y(s)$ of Table 3-1b are replaced by the complex functions $\mathbf{Z}(j\omega)$ and $\mathbf{Y}(j\omega)$ in Table 3-1c. Since $\mathbf{Z}$ and $\mathbf{Y}$ are complex numbers, they may be expressed in terms of their real and imaginary components. Thus the impedance is

$$\mathbf{Z} = R + jX \tag{3-41}$$

and the admittance is

$$\mathbf{Y} = G + jB \tag{3-42}$$

where R and G are the real components of $\mathbf{Z}$ and $\mathbf{Y}$ and are already defined as resistance and conductance. The imaginary components of impedance and admittance are called *reactance* and *susceptance*, respectively. From the definition of Eq. (3-41) and Table 3-1c, it is seen that inductive and capacitive reactances are

$$X_L = \omega L \quad \text{and} \quad X_C = -\frac{1}{\omega C} \tag{3-43}$$

respectively. Similarly, inductive and capacitive susceptances are

$$B_L = -\frac{1}{\omega L} \quad \text{and} \quad B_C = \omega C \tag{3-44}$$

The minus signs in Eqs. (3-43) and (3-44) come from the definitions of the parameters and the fact that $1/j = -j$. In more complex networks, the sign of the reactance or susceptance is used to indicate whether a network behaves inductively or capacitively at the excitation frequency.

The values of X and B in Eqs. (3-41) and (3-42) are consistent in describing inductive or capacitive behavior. A positive value of X and a negative value of B indicates inductive behavior. Capacitive effects are indicated when X is negative and B is positive.

The representation of the elements as given in Table 3-1c and the use of $\mathbf{A}\epsilon^{j\omega t}$ to imply sinusoidal excitation allows the response to be determined as follows:

1 Express the sinusoidal excitation in the form $\mathbf{A}\epsilon^{j\omega t}$.

2 Transform the network by the use of Table 3-1c.

3 Write the appropriate Kirchhoff-law expressions and solve for the frequency-domain response.

4 Convert the frequency-domain response of (3) to the time function by multiplication by $\epsilon^{j\omega t}$ as given in (1).

5 Express the time function in (4) by its implied sinusoidal function. If the excitation in (1) is a cosine (sine) wave, the response is similarly a cosine (sine) wave.

In many problems, steps (4) and (5) are omitted as sinusoidal excitation implies a sinusoidal response at the same frequency.

Example 3-3

Determine the forced component of the current in the inductor in the circuit of Fig. 3-11a for $i(t) = 15 \cos 3t$.

SOLUTION   The function $15\epsilon^{j3t}$ is used to imply the sinusoidal excitation. The transformed network for $s = j3$ is shown in Fig. 3-11b and is based on the admittance given in Table 3-1c. The use of KCL at node A relates $\mathbf{V}$ and $\mathbf{I}$ by

$$\mathbf{I} = 3\mathbf{V} + j6\mathbf{V} - j2\mathbf{V}$$

from which

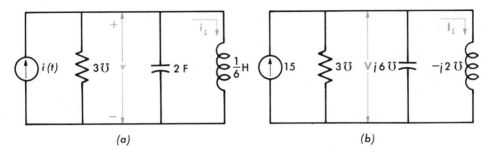

FIGURE 3-11 Circuits for Example 3-3: (a) In the time domain and (b) the transformed circuit.

$$V = \frac{I}{3 + j6 - j2} = \frac{I}{3 + j4}$$

The current in the inductor is

$$I_L = -j2V = \frac{-j2I}{3 + j4}$$

Substitution of $I = 15$ gives

$$I_L = 6\epsilon^{-j143.1°}$$

The corresponding time function is $I_L\epsilon^{j3t} = 6\epsilon^{j(3t-143.1°)}$ which implies

$$i_L(t) = 6 \cos (3t - 143.1°)$$

Example 3-4

For the circuit of Fig. 3-12a, determine the steady-state voltage across the inductor for $v(t) = 12 \sin (2t + 30°)$.

SOLUTION The function $12\epsilon^{j30°}\epsilon^{j2t}$ is used to represent the sinusoidal excitation. Figure 3-12b is the transformed network of Fig. 3-12a for $s = j2$ and

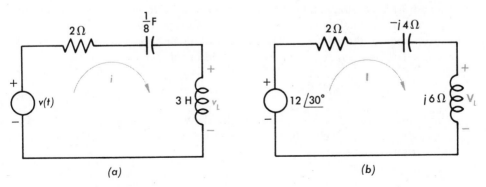

FIGURE 3-12 Circuits for Example 3-4: (a) Time domain and (b) frequency-domain representations.

is based on the impedances given in Table 3-1c. The KVL expression for the loop is

$$\mathbf{V} = 2\mathbf{I} - j4\mathbf{I} + j6\mathbf{I}$$

from which

$$\mathbf{I} = \frac{\mathbf{V}}{2 + j2}$$

The inductor voltage is

$$\mathbf{V}_L = j6\mathbf{I} = \frac{j6\mathbf{V}}{2 + j2}$$

and substitution of $\mathbf{V} = 12\epsilon^{j30°}$ gives

$$\mathbf{V}_L = 18\sqrt{2}\ \epsilon^{j75°} = 18\sqrt{2}\ \underline{/75°}$$

The corresponding time function is $\mathbf{V}_L\epsilon^{j2t} = 18\sqrt{2}\epsilon^{j75°}\epsilon^{j2t}$ which implies

$$v_L(t) = 18\sqrt{2}\ \sin{(2t + 75°)}$$

Note that $v_L(t)$ is a sine function, a direct result of the fact that the excitation function is a sine function.

3-6   RMS VALUES, POWER, AND REACTIVE POWER

Sinusoidal voltages and currents are completely specified by Eqs. (3-20) and (3-21). These equations, however, do not readily permit the effectiveness of such currents and voltages in performing power transfer to be readily assessed. In addition, as the average value over a cycle of a sinusoid is zero and maximum amplitudes do not accurately reflect power, an alternative specification of current and voltage strength is necessary.

Historically, the basis for such specification evolved from the comparison of the heating effect in a resistance of both direct and alternating currents. With a direct current of I amperes, the power expended as heat in a resistance is constant, independent of time, and equal to I^2R. For a current i which varies periodically with time, the power is i^2R; this power is a function of time. If the heating effects of a periodic current and a direct current are to be compared, the average heating power of the periodic current must be considered. This average power over a cycle of the current wave is

$$\frac{1}{T}\int_0^T i^2R\ dt$$

and this average must be compared with the value of the dc power, I^2R. The periodic current which will deliver the same average power as the direct current I must evidently be

$$\sqrt{\frac{1}{T}\int_0^T i^2\ dt}$$

because both quantities, when squared and multiplied by the resistance R, give the average power.

The *effective value* of the periodically varying current i is given by the expression

$$I_{eff} = \sqrt{\frac{1}{T} \int_0^T i^2 \, dt} \quad A \tag{3-45}$$

A similar expression with currents replaced by voltages is the effective value of a periodic voltage. Effective values are also known as *root-mean-square* (abbreviated *rms*) values because of the manner in which they are defined. The designation root-mean-square, said backward, outlines the process of evaluation: First the ordinates of the wave are *squared,* then the *mean* ordinate of the squared wave is found, and lastly the square *root* of this mean ordinate is taken.

For a sinusoidal current, the effective value is

$$I_{eff} = \sqrt{\frac{1}{T} \int_0^T I_m^2 \cos^2 \frac{2\pi t}{T} \, dt}$$

$$= \frac{I_m}{\sqrt{2}} = 0.707 I_m \tag{3-46}$$

Although the discussion leading up to Eq. (3-45) is based on heating effect alone, the usefulness of rms values for expressing the effectiveness of currents and voltages is not so restricted. Fundamentally, the purpose of any electric circuit is the transfer of power and energy, both of which are proportional to the product of voltage and current or, for fixed circuit parameters, to the square of the current or voltage alone. Root-mean-square values are therefore so suitable that they are used almost invariably for specifying magnitudes of alternating currents and voltages. Thus a 110-V house-lighting circuit refers to an rms voltage of 110 V, and a 6-V transformer refers to an rms voltage of 6 V. Another advantage is that the most common ac instrument movements inherently indicate effective values.

Because of the common uses to which effective values are put, the subscript *eff* is usually omitted; alternating voltages or currents referred to by the symbols E, V and I are effective values. The symbols E_m, V_m and I_m are retained for maximum values. The time functions of Eqs. (3-20) and (3-21) may then be written

$$v = \sqrt{2} \, V \cos(\omega t + \alpha) \tag{3-47}$$

and

$$i = \sqrt{2} \, I \cos(\omega t + \beta) \tag{3-48}$$

The definition of rms value given in Eq. (3-45) is applicable to all periodic waveforms. When applied to nonsinusoidal periodic waveforms, the rms value is often used as a means of comparison of such a wave with a sinusoidal signal.

Example 3-5

As alternating current is almost universally used to provide electric power, conversion of ac to dc becomes necessary in applications where the devices used operate with direct current. One method of ac-to-dc conversion employs a rectifier, a device in which the current in the rectifier is nonzero during the positive half-cycle of the sinusoid. In certain control applications, such as speed-control of a motor, the duration during which the current is nonzero is adjusted to effect the control. The typical waveform that results is shown in Fig. 3-13, where the wave is sinusoidal between $\frac{\pi}{3}$ and π radians and is zero for the remainder of the cycle. For this waveform, determine

a   the average value

b   the rms value

SOLUTION   **a**   The average value is given by

$$I_{av} = \frac{1}{T} \int_0^T i(t) \, dt = \frac{1}{2\pi} \int_0^{2\pi} i(t) \, dt$$

As $i(t)$ is zero between 0 and $\frac{\pi}{3}$ and π and 2π, the integral reduces to

$$I_{av} = \frac{1}{2\pi} \int_{\pi/3}^{\pi} 10 \sin \omega t \, d(\omega t) = 2.39 \text{ A}$$

The average value is the dc equivalent of the rectified current.

b   The rms value is given by Eq. (3-45) which becomes

$$I_{eff} = \sqrt{\frac{1}{2\pi} \int_{\pi/3}^{\pi} 10^2 \sin^2 \omega t \, d(\omega t)} = 4.48 \text{ A}$$

Note that the base is the entire period in both part a and part b, even though the current is zero over a substantial part of the period.

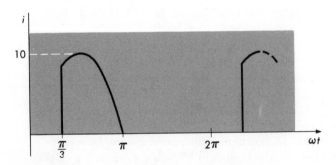

FIGURE 3-13   Current waveform in Example 3-5.

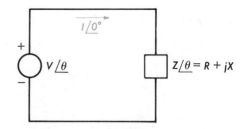

FIGURE 3-14 Ac network.

The use of effective values of current and voltage permits the average power to be determined from phasor quantities. Consider the circuit of Fig. 3-14 for which a voltage

$$v(t) = \sqrt{2}\, V \cos{(\omega t + \theta)}$$

is applied to the network and produces a current

$$i(t) = \sqrt{2}\, I \cos{\omega t}$$

The corresponding phasors are

$$V/\underline{\theta} \text{ and } I/\underline{0°}$$

so that the equivalent impedance of the network is

$$Z/\underline{\theta} = R + jX$$

The instantaneous power p supplied to the network by the source is, from Eq. (1-13),

$$p = vi = \sqrt{2}\, V \cos{(\omega t + \theta)}\sqrt{2}\, I \cos{\omega t} \tag{3-49}$$

The voltage, current, and power waves are shown in Fig. 3-15.

As pointed out in Sec. 3-4, the practical usefulness of alternating currents and voltages as transporters of energy depends on the fact that the time-average value of this pulsating power is not zero. The time-average power P is used to specify the energy-delivering ability of the alternating current and voltage.

By use of trigonometric transformations, Eq. (3-49) may be rewritten as

$$p = VI \cos{\theta}(1 + \cos{2\omega t}) - VI \sin{\theta} \sin{2\omega t} \tag{3-50}$$

In this equation, the time-average values of the terms $\cos{2\omega t}$ and $\sin{2\omega t}$ are zero. Only the term $VI \cos{\theta}$ contributes to the average. The average power P supplied to the circuit is, therefore,

$$P = VI \cos{\theta} \tag{3-51}$$

where θ is the phase angle between the voltage and current.

Of the three circuit elements present in the general RLC circuit, only the resistance absorbs a net amount of energy or a definite average power. The

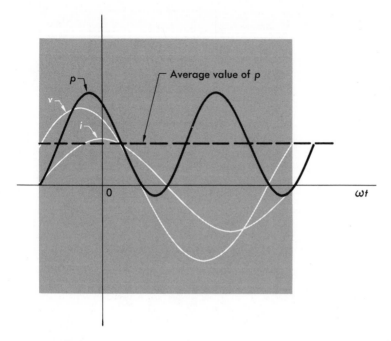

FIGURE 3-15 Instantaneous voltage, current, and power in an ac circuit.

instantaneous power taken by the resistance is

$$p_R = iv_R = i(iR)$$
$$= \sqrt{2}\,I \cos \omega t \, \sqrt{2} \, IR \cos \omega t$$
$$= I^2R(1 + \cos 2\omega t) \quad \text{W} \tag{3-52}$$

which is given by the dotted curve of Fig. 3-16. Since the second term in parentheses in Eq. (3-52) drops out when an average is taken, the average power taken by the entire circuit is

$$P = I^2R \quad \text{W} \tag{3-53}$$

a result which might have been written immediately from the definition of rms current.

The other two circuit elements, inductance and capacitance, affect the instantaneous power but do not contribute to the average power. When the current through an inductance is increasing, energy is transferred from the circuit to the magnetic field, but this energy is returned when the current is decreasing. Similarly, when the voltage across a capacitance is increasing, energy is transferred from the circuit to the electric field, but this energy is returned when the voltage is decreasing. These effects may be seen analytically by writing the expression for the instantaneous power p_X taken by the reactance. For an inductance,

$$p_X = iv_L = iL\frac{di}{dt}$$
$$= -\sqrt{2}\,I\cos\omega t\,\sqrt{2}\,\omega\,LI\sin\omega t$$
$$= -I^2 X_L \sin 2\omega t \tag{3-54}$$

For a capacitance,

$$p_X = iv_C = i\frac{1}{C}\int i\,dt$$

$$= \sqrt{2}\,I\cos\omega t\,\sqrt{2}\left(\frac{1}{\omega C}\right) I\sin\omega t$$

$$= -I^2 X_C \sin 2\omega t \tag{3-55}$$

(Recall that the value of capacitive reactance is negative.)

The instantaneous power p_X for an inductive circuit is given by the dashed curve of Fig. 3-16. Addition of the dashed curve and the dotted curve gives the total instantaneous power p, shown by the solid curve.

The shuttling back and forth of energy indicated by the curve of p_X represents an undesirable condition from the viewpoint of accomplishing net energy transfer. It contributes nothing to this transfer, yet it constitutes just as much a loading of the equipment as if it did. The amplitude of the power oscillation,

$$Q = I^2 X = VI\sin\theta \tag{3-56}$$

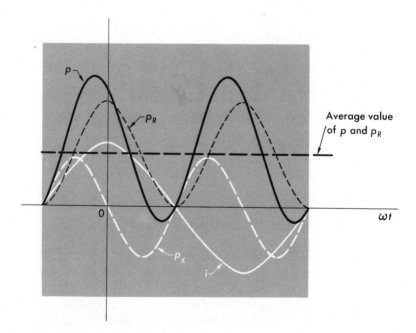

FIGURE 3-16   Total instantaneous power p, and the portions p_R taken by the resistance and p_x taken by the reactance.

is a convenient measure of this undesirable feature. It is called *reactive power* or *wattless power* and is measured in reactive voltamperes (abbreviated *var* from the initials of *voltamperes reactive;* 1,000 var = 1 *reactive kilovoltampere* or *kilovar,* abbreviated *kvar*). It is a much-used term in the power-generation, -distribution, and -utilization fields. In order to emphasize the distinction between reactive power Q and average power P, the latter is often referred to as the *active* or *real power.*

In the circuit of Fig. 3-14, whose current and voltage waves are shown in Fig. 3-15, the current lags the voltage. The equivalent reactance of this circuit is evidently inductive. Inductive reactive power is, by convention, defined as positive reactive power. Since the instantaneous power oscillations in a capacitance are of opposite sign to those of an inductance, capacitive reactive power is negative. Since capacitive reactance is a negative number, the quantity I^2X in Eq. (3-56) inherently gives the correct sign for reactive power absorbed. Thus, an inductor draws positive reactive power from the system, and a capacitor draws negative reactive power. A capacitor may then be considered as a source of positive reactive power, a concept which is particularly useful in power engineering. This viewpoint will be considered in further detail in Sec. 3-11.

Of the two equivalent expressions given by Eqs. (3-51) and (3-53) for active power P in an ac circuit, the latter is the same as in a dc circuit, but the former differs by the reduction factor cos θ. This reduction factor appears because, in the general case, only a portion of the circuit voltage appears across the equivalent resistance. The term cos θ is called the *power factor.* Similarly, the term sin θ appearing in the reactive-power expression, Eq. (3-56), is called the *reactive factor.* A circuit in which the current lags the voltage (i.e., an inductive circuit) is said to have a *lagging power factor.* A circuit in which the current leads the voltage (i.e., a capacitive circuit) is said to have a *leading power factor.*

The term VI which appears in Eqs. (3-51) and (3-56) is called the *voltamperes* (VA) of the circuit:

$$VA = VI \qquad \text{voltamperes (VA)}$$
$$= \frac{VI}{1,000} \qquad \text{kilovoltamperes (kVA)} \tag{3-57}$$

This quantity is evidently a combination of P and Q, for

$$\sqrt{P^2 + Q^2} = \sqrt{(VI)^2(\sin^2 \theta + \cos^2 \theta)}$$
$$= VA \tag{3-58}$$

Power factor is thus the ratio of power to voltamperes, and reactive factor is the ratio of reactive power to voltamperes.

Voltamperes is a significant quantity because ac apparatus, such as generators, transformers, and cables, is generally rated in voltamperes rather than in watts. The allowable output is limited by heating and hence by the losses in the device; these losses in turn are determined by the voltage and current and are almost independent of the power factor. Consequently, the amount of electrical equipment installed to supply a given load is essentially determined by the voltamperes of that load rather than by the power alone. On the other hand, the boiler and turbine sizes and fuel requirements in a generating station (turbine size and water requirements in a hydroelectric station) are determined essentially by the power output and not by the voltamperes. It is completely logical, therefore, to find electrical rate structures dependent on both power and either voltamperes or reactive power. All three quantities are of economic importance.

Active power, reactive power and voltamperes can also be obtained directly from the phasor values of voltage and current. The "complex power" **S**, whose magnitude is the voltamperes, is a convenient way to accomplish the relationship and is given as

$$\mathbf{S} = \mathbf{V}\mathbf{I}^* = S\underline{/\theta} = P + jQ \tag{3-59}$$

In Eq. (3-59) the term $\mathbf{I}^*$ is used because the power factor depends on the phase angle between V and I. The relationships given in Eqs. (3-51) and (3-56) are obtained directly from S as

$$P = \text{Re } \mathbf{S} = S \cos \theta = VI \cos \theta \tag{3-60}$$

and

$$Q = \text{Im } \mathbf{S} = S \sin \theta = VI \sin \theta \tag{3-61}$$

3-7 THE PHASOR METHOD

The *phasor method* is the technique for solving circuits problems when the excitations are sinusoids of the same frequency and the forced or steady-state responses are desired. (For cases in which the excitation frequencies are different, superposition may be used.) The method is basically one in which the time functions are transformed to the phasor representations of the sinusoids as described in Secs. 3-2 to 3-5. For convenience, these representations are repeated here. Current and voltage of the forms

$$i = \sqrt{2} I \cos (\omega t + \alpha) \qquad \text{and} \qquad v = \sqrt{2} V \cos (\omega t + \beta)$$

are transformed to

$$\mathbf{I} = I\underline{/\alpha} \qquad \text{and} \qquad \mathbf{V} = V\underline{/\beta} \tag{3-62}$$

The phasor quantities $\mathbf{I}$ and $\mathbf{V}$ have magnitudes equal to the rms values of the original functions and angles given by the argument of the cosine function at $t = 0$.

The relationships between current and voltage for the three circuit elements—resistance, capacitance, and inductance—follow the defining equations for the transformed network. Relationships involving impedance and admittance parameters are as shown in Table 3-1c. The phasor volt-ampere equations are

$$\mathbf{V} = R\mathbf{I} \qquad \text{or} \qquad \mathbf{I} = G\mathbf{V} \tag{3-63}$$

for the resistance;

$$\mathbf{V} = j\omega L \mathbf{I} \quad \text{or} \quad \mathbf{I} = \frac{1}{j\omega L} \mathbf{V} = -j \frac{1}{\omega L} \mathbf{V} \tag{3-64}$$

for the inductance;

$$\mathbf{V} = \frac{1}{j\omega C} \mathbf{I} = -j \frac{1}{\omega C} \mathbf{I} \quad \text{or} \quad \mathbf{I} = j\omega C \mathbf{V} \tag{3-65}$$

for the capacitance; and

$$\mathbf{V} = \mathbf{IZ} \quad \text{or} \quad \mathbf{I} = \mathbf{YV} \tag{3-66}$$

for the general case where $\mathbf{Z}$ and $\mathbf{Y}$ respectively are impedance and admittance phasor operators giving the appropriate current-voltage phasor relations. The term *phasor operators* is used for $\mathbf{Z}$ and $\mathbf{Y}$ because they are complex quantities which act to change the magnitude and angle of the associated current and voltage phasors. The KVL and KCL equations are also written in phasor form.

The several voltage and current phasors in a network can be shown on a *phasor diagram* in which each phasor is represented by its corresponding directed arrow drawn to scale. Phasor diagrams are often useful as a visual check of the algebraic solution of a problem, especially since the KVL and KCL equations can be shown as graphical addition.

Example 3-6

Find the current i in the circuit of Fig. 3-17a. Only the forced response is needed. Use the phasor method and draw a phasor diagram showing the circuit voltages and current.

SOLUTION The excitation and circuit elements are shown transformed in Fig. 3-17b; impedance parameters are used. The circuit equations are

$$\mathbf{V}_R = 3\mathbf{I}$$

$$\mathbf{V}_C = -j4\mathbf{I}$$

and

$$\mathbf{V}_R + \mathbf{V}_C = 10\underline{/0°}$$

from which

$$\mathbf{I} = 2\underline{/53.1°} \quad \text{and} \quad i = \sqrt{2}\, 2 \cos (2t + 53.1°) \quad \text{A}$$

The phasor diagram is shown in Fig. 3-17b.

The phasor diagram of Fig. 3-17c shows three voltage phasors and one current phasor. Notice that the sum of $\mathbf{V}_R$ and $\mathbf{V}_C$ adds to form the source

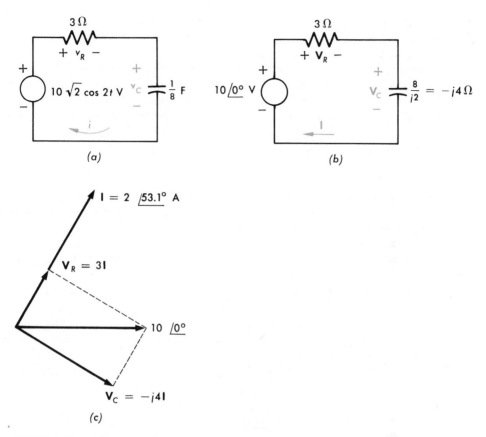

FIGURE 3-17 (a) Circuit of Example 3-6, (b) phasor representation, and (c) phasor diagram for Example 3-6.

voltage phasor, providing a visual check on the KVL equation. Notice also that $\mathbf{V}_R$ lies along $\mathbf{I}$ (they are said to be *in phase*) and the $\mathbf{V}_C$ is 90° behind $\mathbf{I}$ (they are said to be *in quadrature* with $\mathbf{I}$ *leading* $\mathbf{V}_C$ or, conversely, $\mathbf{V}_C$ *lagging* $\mathbf{I}$). In general, the following observations can be made regarding the element phasor voltage-current relationships and the Kirchhoff-law equations:

1 The current in a resistance is in phase with the resistance voltage.

2 The current in an inductance lags the inductance voltage by 90°.

3 The current in a capacitance leads the capacitance voltage by 90°.

4 Circuit voltages and currents can be added graphically in accordance with Kirchhoff's laws through the use of phasor diagrams.

The circuit of Fig. 3-17a is in the time domain; element values are in ohms, henrys, and farads, and voltages and currents are functions of time. In the transformed circuit of Fig. 3-17b, element values are impedances in ohms and voltages and currents are expressed as phasors. For many problems the information of Fig. 3-17b is in the problem statement: admittance or impedance values are given directly; the excitation is given as an rms-valued phasor and the desired result is the rms value of the response and its phase angle with respect to either the excitation or some arbitrary time. In these cases, only the phasor representation and transformed network are used.

Example 3-7

Use a phasor diagram to find the value of the source current **I** required to establish a voltage $\mathbf{V} = 10\underline{/0^\circ}$ V in the circuit of Fig. 3-18.

SOLUTION   The currents $\mathbf{I}_R$, $\mathbf{I}_C$, and $\mathbf{I}_L$ are found from Eqs. (3-63) to (3-65) and are

$$\mathbf{I}_R = 0.3\mathbf{V} = 3\underline{/0^\circ}$$

$$\mathbf{I}_C = j0.6\mathbf{V} = j6 = 6\underline{/90^\circ}$$

$$\mathbf{I}_L = -j0.2\mathbf{V} = -j2 = 2\underline{/-90^\circ}$$

The phasor diagram showing the voltage **V**, the three element currents, and their sum is shown in Fig. 3-19. From Kirchhoff's current law,

$$\mathbf{I} = \mathbf{I}_R + \mathbf{I}_C + \mathbf{I}_L = 5\underline{/53.1^\circ}\text{ A}$$

The two previous examples demonstrate several aspects of the utility of the phasor method. Firstly, the determination of the desired response is carried out by use of purely algebraic manipulation. Secondly, because the KVL and KCL equations are algebraic, nearly all the techniques developed in Chap. 2 for resistive circuits are applicable to steady-state ac circuits. For example, in Example 3-6, the capacitor and resistor are in series; the impedance parameters shown in Fig. 3-17b are added directly, in the same manner as resistors in series are combined. Similarly, the parallel admittances in Example 3-8 can be combined exactly as parallel conductances are combined.

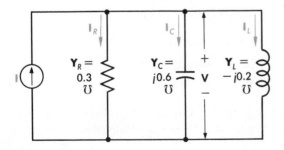

FIGURE 3-18   Circuit of Example 3-7.

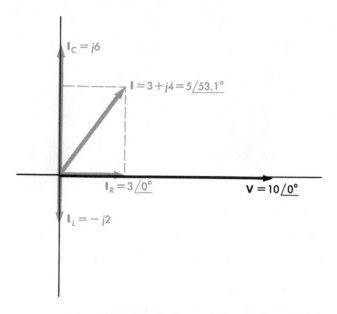

FIGURE 3-19 Phasor diagram for Example 3-7.

Thus, in Fig. 3-20,

$$\mathbf{Z} = \frac{\mathbf{V}}{\mathbf{I}} = R + jX$$

is the impedance at a pair of terminals with the plus sign indicating an equivalent representation of a resistance R in series with a reactance X as shown in Fig. 3-21a. Similarly,

$$\mathbf{Y} = \frac{\mathbf{I}}{\mathbf{V}} = \frac{1}{\mathbf{Z}} = G + jB$$

suggests an equivalent representation of Fig. 3-21b showing a conductance G in parallel with a susceptance B. In general, we can observe that impedances in series and admittances in parallel add, always keeping in mind that the addition performed involves complex numbers.

The next three sections show further that the resistive techniques developed in Chap. 2 apply to ac circuits using the phasor method.

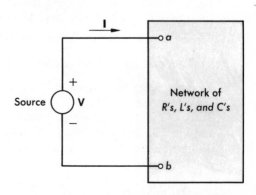

FIGURE 3-20 Network.

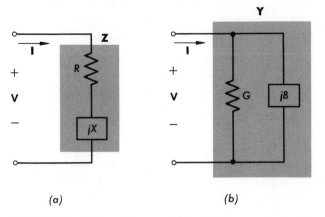

FIGURE 3-21 (a) Impedance and (b) admittance equivalent-circuit representation.

3-8 MESH AND NODAL ANALYSES

In the majority of systems for the transmission and distribution of large amounts of power, more than one source of energy supplies the network and important load centers are fed over at least two (and frequently more) transmission and distribution routes. In most electronic systems, both the systems themselves and the devices which are the system components contain many circuit branches. As a result, system equivalent circuits in both areas generally contain several loops and nodes. Mesh and nodal analyses, presented in Secs. 2-4 to 2-6, are frequently used to analyze these networks. They have the obvious advantage of being well-organized and systematic methods of analysis. Application of nodal and mesh analyses for ac circuits follows the procedures outlined in Chap. 2. However, the constant voltages and currents in dc circuits are replaced by phasor voltages and currents in ac circuits. Similarly, resistances and conductances are replaced by the complex quantities for impedance and admittance.

Example 3-8

Figure 3-22a is the equivalent circuit of a load supplied from two sinusoidal sources of the same frequency. The impedances $\mathbf{Z}_1$ and $\mathbf{Z}_2$ are the feeder line impedances and $\mathbf{Z}_L$ is the load impedance. For the values shown, determine

a The load voltage using nodal analysis

b The feeder line currents $\mathbf{I}_1$ and $\mathbf{I}_2$ and the load current $\mathbf{I}_L$

SOLUTION **a** The voltage source-series impedance combinations $\mathbf{V}_1 - \mathbf{Z}_1$ and $\mathbf{V}_2 - \mathbf{Z}_2$ are converted to current sources and parallel admittances as indicated in Fig. 3-22b. The load is also represented by its admittance. The node equation is

$$\mathbf{V}_L[(0.6 - j0.8) + (0.04 - j0.03) + (0.5 - j0.866)]$$
$$= 240\underline{/-53.13°} + 240\underline{/-63.43°}$$

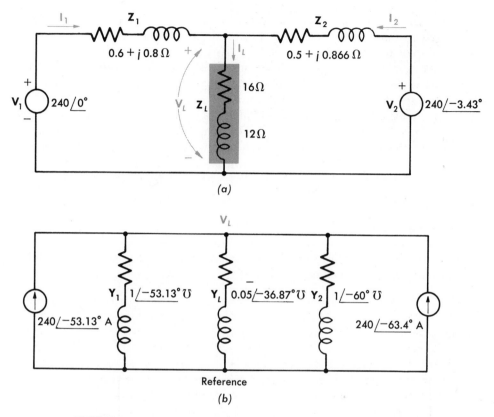

FIGURE 3-22 (a) Circuit for Example 3-8 and (b) circuit after source transformations.

Solution of this equation for $\mathbf{V}_L$ gives

$$\mathbf{V}_L = 233.9\underline{/-2.19°} \text{ V}$$

b The load current $\mathbf{I}_L$ is

$$\mathbf{I}_L = \frac{\mathbf{V}_L}{\mathbf{Z}_L} = \frac{233.9\underline{/-2.19°}}{16 + j12} = 11.70\underline{/-39.06°} \text{ A}$$

In Fig. 3-22a, the feeder line currents are

$$\mathbf{I}_1 = \frac{\mathbf{V}_1 - \mathbf{V}_L}{\mathbf{Z}_1} \quad \text{and} \quad \mathbf{I}_2 = \frac{\mathbf{V}_2 - \mathbf{V}_L}{\mathbf{Z}_2}$$

Evaluation yields

$$\mathbf{I}_1 = \frac{240 - (233.9\underline{/-2.19°})}{0.6 + j0.8} = 10.90\underline{/+2.12°} \text{ A}$$

$$\mathbf{I}_2 = \frac{240\underline{/3.43°} - (233.9\underline{/-2.19°})}{0.5 + j0.0866} = 7.859\underline{/-102.3°} \text{ A}$$

If the circuit in Fig. 3-22a were constructed and measurements made with ac meters, these instruments would give the proper indication for the magnitudes of the currents and voltages. However, use of magnitudes only in making subsequent calculations leads to serious errors. Note, in particular, that while $I_1 + I_2 = I_L$, the sum of the magnitudes of I_1 and I_2 does not give the correct result for I_L. It is essential to use complex quantities to perform the necessary evaluations. In the laboratory, or in the field, additional measurements are made so that appropriate phase angle information is obtained. Often, the active power is measured from which the phase angle is evaluated from the power factor. Sometimes, it is convenient to measure the phase angle directly by means of an oscilloscope or phase-meter.

Example 3-9

The circuit in Fig. 3-23a is the equivalent circuit of an electronic amplifier. Determine the forced component of the output voltage v_2 for

$$i(t) = 10^{-4}\sqrt{2} \cos 2 \times 10^7 t \quad \text{A}$$

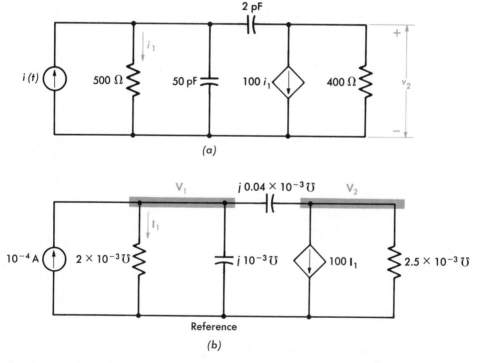

(a)

(b)

FIGURE 3-23 (a) Time domain and (b) frequency-domain representations of amplifier circuit in Example 3-9.

SOLUTION   The transformed circuit of Fig. 3-23b is obtained using the admittances in Eqs. (3-63) to (3-65) and Table 3-1. For the circuit in Fig. 3-23b the nodal equations are

$$\mathbf{V}_1(2 \times 10^{-3} + j10^{-3} + j0.04 \times 10^{-3}) - j0.04 \times 10^{-3}\mathbf{V}_2 = 10^{-4}$$

$$-j0.04 \times 10^{-3}\mathbf{V}_1 + \mathbf{V}_2(+j0.04 \times 10^{-3} + 2.5 \times 10^{-3}) = -100\mathbf{I}_1$$

The constraint equation for the controlled source is, from Ohm's law,

$$\mathbf{I}_1 = 2 \times 10^{-3}\mathbf{V}_1$$

Substitution of the constraint equation and rearrangement of terms yields

$$\mathbf{V}_1 \times 10^{-3}(2 + j1.04) - \mathbf{V}_2 \times 10^{-3}(j0.04) = 10^{-4}$$

$$\mathbf{V}_1 \times 10^{-3}(200 - j0.04) + \mathbf{V}_2 \times 10^{-3}(2.5 + j0.04) = 0$$

Solution for $\mathbf{V}_2$ gives

$$\mathbf{V}_2 = 1.70\underline{/-245°} \text{ V}$$

The corresponding time function is

$$v_2(t) = 1.70\sqrt{2} \cos (2 \times 10^7 t - 245°)$$

Example 3-10

Determine the voltage $\mathbf{V}_{AB}$ in the circuit of Fig. 3-24 by use of mesh analysis.

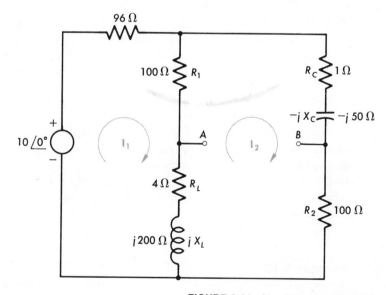

FIGURE 3-24 Circuit for Example 3-10.

SOLUTION The mesh equations for the two loops are

$$\mathbf{I}_1(96 + 100 + 4 + j200) - \mathbf{I}_2(100 + 4 + j200) = 10\underline{/0°}$$

$$-\mathbf{I}_1(100 + 4 + j200) + \mathbf{I}_2(4 + j200 + 100 + 1 - j50 + 100) = 0$$

Simultaneous solution for $\mathbf{I}_1$ and $\mathbf{I}_2$ yields

$$\mathbf{I}_1 = 5.053 \times 10^{-2}\underline{/-0.736°} \text{ A} \qquad \mathbf{I}_2 = 4.484 \times 10^{-2}\underline{/25.60°} \text{ A}$$

The voltage $\mathbf{V}_{AB}$ is

$$(\mathbf{I}_1 - \mathbf{I}_2)(4 + j200) - \mathbf{I}_2 \times 100 = \mathbf{V}_{AB}$$

from which

$$\mathbf{V}_{AB} = 0$$

The circuit analyzed in Example 3-10 is one form of *ac* bridge, and its behavior is similar to a Wheatstone bridge. When $\mathbf{V}_{AB} = 0$, the bridge is balanced and becomes a useful method for experimentally determining many parameters quite accurately. A common arrangement is to fix the values of R_1 and R_2, their ratio being a multiplication factor which permits bridge usage over a wide range of parameter values. The branch containing R_L and X_L is the unknown to be determined and R_C and X_C are adjusted to achieve balance. Bridge structures similar to that in Fig. 3-24 are widely used in practice; one application is in the use of inductive-type strain gauges.

3-9 THÉVENIN'S THEOREM

We have now seen that the basic methods of writing network current and voltage equations are generally the same for resistive and ac circuits with one exception: Phasor representation of currents, voltages, and impedances must be used for ac circuits instead of the simpler algebraic values of currents, voltages, and resistances used for resistive circuits. Such replacements immediately make available for ac circuit analysis the Y-Δ transformations of Sec. 2-9, the superposition principle of Sec. 2-7, and the Thévenin equivalent of Sec. 2-8.

With specific reference to Thévenin's theorem, the steps outlined in Sec. 2-8 are followed also for the ac case. The open-circuit voltage $\mathbf{V}_0$ and equivalent impedance $\mathbf{Z}_0$ must now be found in complex rectangular or polar form. The process will be demonstrated by means of two specific examples, which at the same time will further illustrate the application of dc strategy to ac problems.

Example 3-11

Find the Thévenin equivalent of the circuit of Fig. 3-25a.

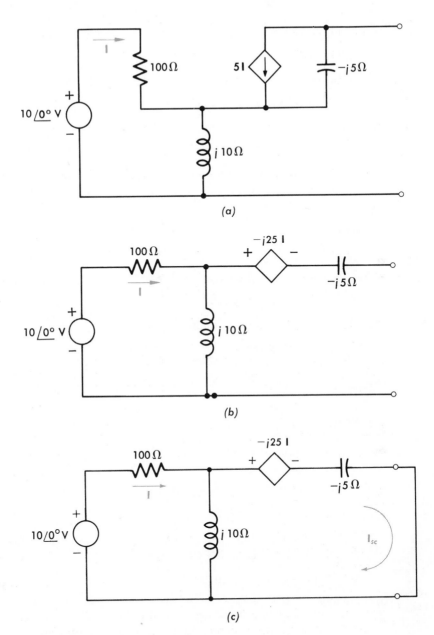

FIGURE 3-25 (a) Circuit of Example 3-11, (b) circuit to calculate open-circuit voltage, and (c) circuit to calculate short-circuit current.

SOLUTION The Thévenin impedance will be computed as the ratio of open-circuit voltage to short-circuit current. The controlled source is converted to its voltage-source equivalent as shown in Fig. 3-25b. The open-circuit voltage V_0 can now be calculated. By noting that the current in the right-hand loop is zero because of the open-circuit, the KVL equation for the left-hand loop is

$$(100 + j10)\mathbf{I} = 10\underline{/0°}$$

The value of $\mathbf{V}_0$ is

$$\mathbf{V}_0 = -(-j25\mathbf{I}) + j10\mathbf{I} = j35\mathbf{I}$$

from which

$$\mathbf{V}_0 = 3.48\underline{/84.3°} \text{ V}$$

The short-circuit current is found from the circuit of Fig. 3-25c, which is the circuit of Fig. 3-25b with the output terminals short-circuited. The mesh equations are

$$(100 + j10)\mathbf{I} - j10\mathbf{I}_{sc} = 10\underline{/0°}$$

$$-j10\mathbf{I} + \mathbf{I}_{sc}(j10 - j5) = -(-j25\mathbf{I})$$

After rearrangement, the equations are solved for $\mathbf{I}_{sc}$ and give

$$\mathbf{I}_{sc} = 0.600\underline{/+31.0°} \text{ A}$$

The Thévenin impedance is then

$$\mathbf{Z}_0 = \frac{\mathbf{V}_0}{\mathbf{I}_{sc}} = \frac{3.48\underline{/84.3}}{0.600\underline{/31.0}} = 5.80\underline{/53.3°} \text{ } \Omega$$

Example 3-12

a   Find the Thévenin equivalent of the circuit of Fig. 3.26a to the left of terminals A-B, and

b   Use the Thévenin equivalent to determine the active power absorbed by the 3 + j4 Ω load impedance.

SOLUTION   **a**   First the open-circuit voltage is found using the circuit of Fig. 3-26b. The KVL equation for the loop is

$$\mathbf{I}(5 + j10 + 3 - j4) = 20\underline{/90°}$$

from which

$$\mathbf{I} = 2.00\underline{/53.1} \text{ A}$$

The value of $\mathbf{V}_0$ is

$$\mathbf{V}_0 = (2.00\underline{/53.1})(3 - j4) = 10\underline{/0°}$$

The Thévenin impedance can be found by suppressing the voltage source and combining impedances. Figure 3-26c is used to perform the calculation. The 3-Ω resistance and ($-j4$ Ω) reactance are in series; this combination is in parallel with the series combination of the 5-Ω resistance and $j10$ Ω reac-

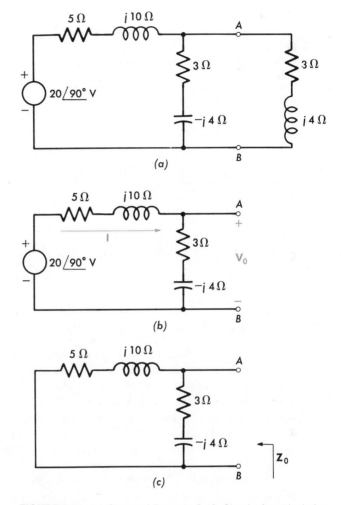

FIGURE 3-26 (a) Circuit of Example 3-12. Circuits for calculating (b) open-circuit voltage and (c) Thévenin impedance.

tance. Therefore,

$$\mathbf{Z}_0 = \frac{(3 - j4)(5 + j10)}{(3 - j4) + (5 + j10)} = 5.59\underline{/-26.6°} \; \Omega$$

b   The circuit of Fig. 3-27 shows the Thévenin equivalent to which the $(3 + j4)$ impedance is connected. The current in the loop is

$$\mathbf{I}_L = \frac{10\underline{/0°}}{(5.59\underline{/-26.6°}) + 3 + j4} = 1.22\underline{/-11.2°} \; A$$

The power is $I_L^2 R$ so that

$$P = (1.22)^2 \times 3 = 4.47 \; W$$

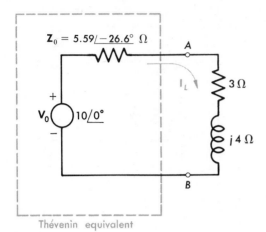

Thévenin equivalent

FIGURE 3-27 Thévenin equivalent of circuit in Figure 3-26a.

3-10 MAXIMUM-POWER CONDITIONS

A circuit for the utilization of ac power and energy may consist of a source with an external network connected to its terminals. The external network may consist simply of the load connected directly to the source terminals, or it may consist of the load together with a coupling network interposed between the source and the load. In either case, the complete external network can be simplified by network reduction to a single equivalent impedance between the two source terminals. This single equivalent impedance is the input, or driving-point, impedance of the external network.

Input impedance is a valuable quantity for judging the effect of the external network upon the source, for when the analysis is concerned with the behavior of the source itself, it is desirable to represent the external network in the simplest possible manner. The source itself may be represented, in accordance with Thévenin's theorem (and subject to the assumptions of linearity underlying that theorem), by a voltage $\mathbf{V}_o$ in series with an impedance $\mathbf{Z}_o$. Accordingly, the source and its external circuit may be reduced to the simple representation of Fig. 3-28.

In the circuit of Fig. 3-28, let

$$\mathbf{Z}_o = R_o + jX_o = Z_o\underline{/\theta_o}$$

and

$$\mathbf{Z}_i = R_i + jX_i = Z_i\underline{/\theta_i}$$

The magnitude of the current I is

$$I = \frac{V_o}{\sqrt{(R_o + R_i)^2 + (X_o + X_i)^2}}$$

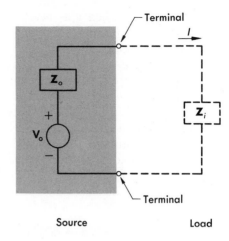

FIGURE 3-28 Equivalent circuit for source and its load.

The power input to the external circuit is then

$$P_i = I^2 R_i = \frac{V_o^2 R_i}{(R_o + R_i)^2 + (X_o + X_i)^2} \tag{3-67}$$

$$= \frac{V_o^2 Z_i \cos \theta_i}{(Z_o \cos \theta_o + Z_i \cos \theta_i)^2 + (Z_o \sin \theta_o + Z_i \sin \theta_i)^2} \tag{3-68}$$

Now consider that the source voltage and impedance are fixed but that some adjustments are possible in the impedance of the external circuit. It is desired to transfer the maximum power P_i to this external circuit. The conditions for maximum power transfer depend on the nature of the adjustments which can be made in the external impedance. If only the reactance X_i can be varied, inspection of Eq. (3-67) shows that maximum P_i is obtained when

$$X_i = -X_o \tag{3-69}$$

If only the resistance R_i can be varied, the process of equating dP_i/dR_i to zero from Eq. (3-67) shows that P_i is a maximum when

$$R_i = \sqrt{R_o^2 + (X_o + X_i)^2} \tag{3-70}$$

If both R_i and X_i can be varied independently, the optimum values are given by Eq. (3-69) and the condition

$$R_i = R_o \tag{3-71}$$

the latter being obtained by substituting Eq. (3-69) in Eq. (3-70).

A not uncommon practical condition, however, is that the magnitude Z_i can be varied but the impedance angle θ_i must remain fixed. The process of equating dP_i/dZ_i to zero from Eq. (3-68) then shows maximum P_i to be obtained when the magnitude

$$Z_i = Z_o \tag{3-72}$$

Thus, when it is necessary to obtain the largest possible amount of power from a source, an advantage is gained if the circuit constants can be adjusted to bring about equality of impedances. This adjustment is called *impedance matching*. Moreover, when the reactive component of input impedance can be made equal and opposite to the reactance of the source, optimum energy-transfer conditions are obtained. Impedance matching is an important consideration in electronic and communications devices, where a low level of power is generally dealt with. For power devices, such as generators, motors, and transmission lines, however, impedance matching is never a guiding principle because of the associated low efficiency. If Z_o and Z_i are purely resistive, for example, the power lost as heat in the source equals that delivered to the external network when impedances are matched.

3-11 THE VOLT-AMPERE METHOD

An approach to solving *ac* circuits problems, particularly for circuits involved in power distribution, is called the *volt-ampere method*. When it is applied, the power and reactive power taken by the individual circuit elements are computed and their respective sums are equated to the power and reactive power of the source. This procedure is the equivalent of writing Kirchhoff's voltage law in phasor form. Translations from powers and reactive powers to currents, voltages, and phase angles, as well as the converse translations, are accomplished by means of Eqs. (3-51) and (3-56). This method may be particularly convenient when power and power factor or voltamperes are known at one point and corresponding values are desired at other points. When only currents, voltages, and impedances or admittances are involved, however, it may be a needlessly artificial approach. The volt-ampere method may be used in the solution of multiple-source problems as well as for singly excited systems.

The relations among power, reactive power, and voltamperes can be remembered conveniently by means of the volt-ampere triangle shown in Fig. 3-29. This triangle shows graphically the relationships presented in Eqs.

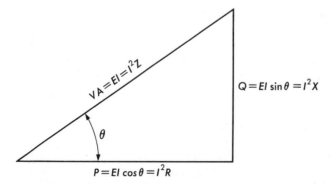

FIGURE 3-29 Volt-ampere triangle.

(3-51), (3-53), (3-56), and (3-58). The additional relationship

$$Q = P \tan \theta \qquad (3\text{-}73)$$

can also be derived from the triangle. Also note that the voltamperes of a group of circuit elements must be combined by adding the powers and reactive powers separately, taking care to use the appropriate signs for the reactive powers.

Example 3-13

Figure 3-30 is the equivalent circuit of a load supplied from two generators over lines with the impedances shown. The load requires 10 kW at 0.80 power-factor lagging. Source G_1 operates at a terminal voltage of 460 V and supplies 5 kW at 0.80 power-factor lagging.

Find values of V_2 and V_L and power and reactive power for G_2.

SOLUTION This example is one which may conveniently be handled by the volt-ampere method.

$$I_1 = \frac{P_1}{V_1 \cos \theta_1} = \frac{5{,}000}{(460)(0.80)} = 13.6 \text{ A}$$

$$P_{r1} = P_1 - I_1^2 R_1 = 5{,}000 - (13.6)^2(1.4) = 4{,}740 \text{ W}$$

$$Q_{r1} = Q_1 - I_1^2 X_1 = 5{,}000 \tan (\cos^{-1} 0.80) - (13.6)^2(1.6)$$
$$= 3{,}450 \text{ var}$$

$$V_L = \frac{\text{VA}}{\text{current}} = \frac{\sqrt{(4{,}740)^2 + (3{,}450)^2}}{13.6} = 432 \text{ V}$$

Since the load requires 10,000 W and 10,000 tan $(\cos^{-1} 0.80)$ or 7,500 var,

$$P_{r2} = 10{,}000 - 4{,}740 = 5{,}260 \text{ W}$$

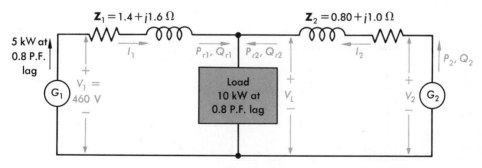

FIGURE 3-30 Simple power system for Example 3-13.

and

$$Q_{r2} = 7,500 - 3,450 = 4,050 \text{ var}$$

$$I_2 = \frac{\text{VA}}{\text{voltage}} = \frac{\sqrt{(5,260)^2 + (4,050)^2}}{432} = 15.4 \text{ A}$$

$$P_2 = P_{r2} + I_2^2 R_2 = 5,260 + (15.4)^2(0.80) = 5,450 \text{ W}$$

$$Q_2 = Q_{r2} + I_2^2 X_2 = 4,050 + (15.4)^2(1.0) = 4,290 \text{ var}$$

$$V_2 = \frac{\text{VA}}{\text{current}} = \frac{\sqrt{(5,450)^2 + (4,290)^2}}{15.4} = 451 \text{ V}$$

As pointed out in Sec. 3-6, the electrical equipment that must be installed to supply a given load is determined by the volt-ampere requirements of the load and hence is directly affected by load power factor. The usual industrial load on a power system operates at a lagging power factor, and in many cases this power factor is low enough so that improvement at the load is economically justifiable. Such power-factor improvement or correction is accomplished by connecting a bank of capacitors in parallel with the load, the size of the bank being so chosen that the power factor of the parallel combination reaches the desired value. Capacitors for this purpose are rated in kilovoltamperes or reactive kilovoltamperes. The method is illustrated in Example 3-14.

Example 3-14

As shown in Fig. 3-31, a 230-V 60-Hz generator is used to supply two loads, L_1 and L_2. The power requirement for L_1 is 10 kW at 0.8 power-factor lagging; that for L_2 is 10 kVA at 0.6 power-factor lagging. A capacitor is added as indicated to provide power-factor correction.

a Determine the value of capacitance required for the generator to operate at unity power factor.

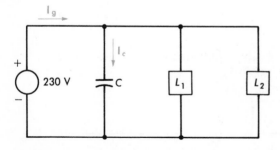

FIGURE 3-31 Circuit for Example 3-14.

b Find the magnitude of the generator current with and without the capacitor.

SOLUTION **a** The capacitor provides negative reactive voltamperes. To make the generator operate at unity power factor, the value of Q_C must exactly balance the total Q of the two loads. For L_1

$$P_{L1} = 10 \text{ kW}$$

$$Q_{L1} = P_{L1} \tan \theta = 10 \times \tan (\cos^{-1} 0.8) = 7.5 \text{ kvar}$$

and for L_2

$$P_{L2} = 10 \times 0.6 = 6 \text{ kW}$$

$$Q_{L2} = 10 \times \sin (\cos^{-1} 0.6) = 8 \text{ kvar}$$

The total load absorbs

$$P_T = 10 + 6 = 16 \text{ kW}$$

$$Q_T = 7.5 + 8 = 15.5 \text{ kvar}$$

Thus, $Q_C = -Q_{\text{tot}} = -15.5$ kvar

$$I_C = \frac{|Q_C|}{V} = \frac{15,500}{230} = 67.4 \text{ A}$$

$$X_C = \frac{Q_C}{I_C^2} = \frac{-15,500}{67.4^2} = -3.41 \ \Omega$$

from which

$$C = \frac{-1}{2\pi f X_C} = \frac{-1}{2\pi \times 60 \times -3.41} = 7.78 \times 10^{-3} F = 7,780 \ \mu F$$

b Without the capacitor, the generator provides the total voltamperes required by the two loads so that

$$kVA_{\text{gen}} = \sqrt{P_T^2 + Q_T^2}$$

$$= \sqrt{16^2 + 15.5^2} = 22.3 \text{ kVA}$$

Then,

$$I_g = \frac{22,300}{230} = 96.9 \text{ A}$$

When the capacitor is added, the reactive voltamperes at the generator is zero. Then,

$$kVA_{\text{gen}} = P_T$$

and

$$I_g = \frac{16,000}{230} = 69.6 \text{ A}$$

Note that the power-factor correction significantly reduces the current the generator must supply.

3-12   FOURIER SERIES

The phasor method of circuit analysis can be extended to include nonsinusoidal, periodic functions. The properties and characteristics of periodic waveforms are introduced and defined in Sec. 3-4. Almost all such functions can be expressed as a sum of sine and cosine terms called the *Fourier series*. The mathematical theorem associated with Fourier series states that a periodic function $f(t)$ can be written in the form

$$f(t) = a_0 + a_1 \cos \omega t + a_2 \cos 2\omega t + \cdots + b_1 \sin \omega t$$
$$+ b_2 \sin 2\omega t + \cdots \quad (3\text{-}74)$$

or

$$f(t) = \sum_{n=0}^{\infty} a_n \cos n\omega t + \sum_{n=1}^{\infty} b_n \sin n\omega t \quad (3\text{-}75)$$

where $\omega = 2\pi/T$ and is the *fundamental angular frequency*, or 2π times the frequency of the original periodic wave. The term a_0 is the *average ordinate* or *dc* component of the wave. The term

$$a_1 \cos \omega t + b_1 \sin \omega t$$

is the *fundamental component* and has the same frequency and period as the original wave. The remaining terms, taken as pairs of the form

$$a_n \cos n\omega t + b_n \sin n\omega t$$

give the *nth-harmonic* component of the function.

The values of the a and b coefficients can be determined as follows. The coefficient a_0 is the average value of $f(t)$ and is

$$a_0 = \frac{1}{T} \int_0^T f(t)\, dt \quad (3\text{-}76)$$

To find a_m, first multiply each term in Eq. (3-74) by $\cos m\omega t$ and integrate the result from 0 to T. By this operation, Eq. (3-74) becomes

$$\int_0^T f(t) \cos m\omega t\, dt = \sum_{n=1}^{n=m-1} \int_0^T a_n \cos n\omega t \cos m\omega t\, dt$$

$$+ \int_0^T a_m \cos^2 m\omega t\, dt$$

$$+ \sum_{n=m+1}^{\infty} \int_0^T a_n \cos n\omega t \cos m\omega t\, dt$$

$$+ \sum_{n=1}^{\infty} \int_0^T b_n \sin n\omega t \cos m\omega t\, dt \quad (3\text{-}77)$$

All the integrals on the right-hand side of Eq. (3-77) are zero, except for the term

$$\int_0^T a_m \cos^2 m\omega t \; dt = \frac{a_m T}{2}$$

Thus

$$a_m = \frac{2}{T} \int_0^T f(t) \cos m\omega t \; dt \qquad m \neq 0 \tag{3-78}$$

In a similar fashion, it can be shown that

$$b_m = \frac{2}{T} \int_0^T f(t) \sin m\omega t \; dt \tag{3-79}$$

Example 3-15

The periodic waveform shown in Fig. 3-32 is called a *square wave*. It is useful in representing an excitation which is switched on for one-half the cycle, and switched off one-half the cycle. In Fig. 3-32, the constant excitation V_m is applied for $0 < t < T/4$ and $3T/4 < t < T$ during the first period. During the interval $T/4 < t < 3T/4$, the excitation is switched off. Square waves are encountered in adjustable-frequency motor drives (Sec. 17-4), are used to set timing sequences in digital computers and control systems (Sec. 17-6), and are often generated by means of multivibrators (Sec. 11-4).

Determine the Fourier series for the square wave in Fig. 3-32.

SOLUTION The value of a_o can be found from Eq. (3-76) and is

$$a_o = \frac{1}{T} \left(\int_0^{T/4} V_m \; dt + \int_{3T/4}^T V_m \; dt \right) = \frac{V_m}{2}$$

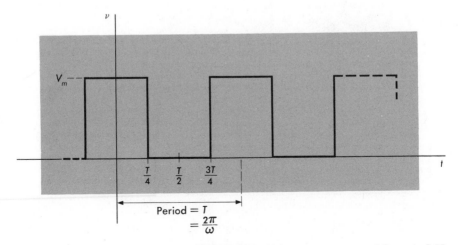

FIGURE 3-32 Voltage square wave of Example 3-15.

The values of a_1, a_2, etc., are found from Eq. (3-78); thus

$$a_m = \frac{2}{T} \left(\int_0^{T/4} V_m \cos m\omega t \, dt + \int_{3T/4}^T V_m \cos m\omega t \, dt \right)$$

$$= 0 \qquad m \text{ even}$$

$$= \frac{2V_m}{m\pi} \qquad m \text{ odd}$$

(The algebraic sign is + for $m = 1$ and changes for each successive term.)
The b coefficients are, from Eq. (3-79),

$$b_m = \frac{2}{T} \left(\int_0^{T/4} V_m \sin m\omega t \, dt + \int_{3T/4}^T V_m \sin m\omega t \, dt \right) = 0$$

The Fourier series is then written

$$v = \frac{V_m}{2} + \frac{2V_m}{\pi} \cos \omega t - \frac{2V_m}{3\pi} \cos 3\omega t + \frac{2V_m}{5\pi} \cos 5\omega t - \cdots \qquad (3\text{-}80)$$

Exact representation of the nonsinusoidal, periodic wave requires an infinite number of terms in the Fourier series. Notice, however, that the amplitude of the harmonics in Eq. (3-80) decreases progressively as the order of the harmonic increases, indicating that a good approximation can be obtained with comparatively few terms. A particularly severe test of approximate representation is offered by the square wave because of its flat top and infinitely sloped edges. Figure 3-33 presents a comparison between the square wave and its representation by the first five terms of its Fourier series given in Eq. (3-80).

The Fourier series can also be expressed as a sum of exponentials. The

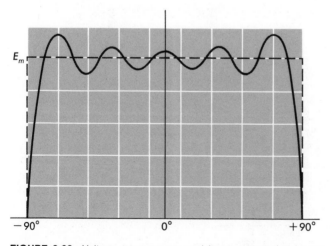

FIGURE 3-33 Voltage square wave and its representation by the first five terms of its Fourier series.

kth harmonic is

$$a_k \cos k\omega t + b_k \sin k\omega t$$

From Eq. (3-30), this term can be rewritten as

$$\sqrt{a_k^2 + b_k^2} \cos (k\omega t - \tan^{-1}b_k/a_k) = c_k \cos (k\omega t + \theta_k)$$

Expressing the cosine term as a sum of exponentials [see Eqs. (3-24) to (3-28)], the kth harmonic becomes

$$\mathbf{C}_k \epsilon^{jk\omega t} + \mathbf{C}_k^* \epsilon^{-jk\omega t}$$

where

$$\mathbf{C}_k = \frac{c_k}{2} \epsilon^{j\theta_k}$$

When each harmonic in the series of Eq. (3-75) is written as a sum exponential, the result is

$$f(t) = \sum_{-\infty}^{\infty} \mathbf{C}_n \epsilon^{jn\omega t} \tag{3-81}$$

Equation (3-81) is called the *exponential form* of the Fourier series. The summation from $-\infty$ to $+\infty$ is a consequence of the two exponentials in the representation of cos $k\omega t$, namely $\epsilon^{+jk\omega t}$ and $\epsilon^{-jk\omega t}$. The relation used to obtain $\mathbf{C}_n$ is

$$\mathbf{C}_n = \frac{1}{T} \int_0^T f(t) \epsilon^{-jn\omega t} \, dt \tag{3-82}$$

The derivation of Eq. (3-82) is similar to that used to arrive at Eqs. (3-78) and (3-79).

3-13   NETWORK RESPONSE TO PERIODIC FUNCTIONS

The fact that any periodic waveform can be written as a Fourier series leads to the conclusion that knowledge of how to solve problems involving sinusoidal excitations permits the solution of problems with any periodic excitation. The Fourier series for the excitation is written, and each term of the series is considered as a separate source. The principle of superposition then states that the total response is the sum of the responses produced by each individual component. As indicated in the preceding section, many periodic functions can be approximately represented by a relatively few terms of the Fourier series, greatly reducing the computational effort. The decision to include or to ignore a specific term is one requiring engineering judgment and depends mainly on how accurately the results must be known and also on the accuracy with which the data are presented.

The Fourier-series representation of a periodic excitation and the superposition principle allows the phasor technique to be used to determine the system response. Each frequency component of the response is produced by the corresponding harmonic of the excitation; the sum of these responses is the Fourier series of the system response. Implicit in the use of the phasor technique is that the response being determined is the forced or steady-state response.

Example 3-16

A voltage source having the waveform of Fig. 3-32 and Example 3-15 is applied to the RC network of Fig. 3-34. Let the maximum value of the voltage V_m be 10 V and the period T be 2π s.

a   Find the first four nonzero terms of the Fourier series of $v_C(t)$.

b   Draw to scale the four components found in part a. Add these to get the approximate waveform of $v_C(t)$.

SOLUTION   **a**   The general Fourier series for the source voltage is given in Eq. (3-80). If V_m is 10 V and the fundamental angular frequency $\omega = 2\pi/T = 1$ rad/s, the series becomes

$$v(t) = 5 + 6.36 \cos t - 2.12 \cos 3t + 1.27 \cos 5t \qquad (3\text{-}83)$$

when only the first four nonzero terms are used.

The network of Fig. 3-34 is transformed into a phasor network as shown in Fig. 3-35. Note that the capacitive impedance is expressed in terms of ω since the frequency of each term in the Fourier series is different.

A general phasor expression for $\mathbf{V}_C$ will now be developed. The current in the circuit is

$$\mathbf{I} = \frac{\mathbf{V}}{2 + 1/j\omega}$$

and the voltage $\mathbf{V}_C$ is

$$\mathbf{V}_C = \frac{1}{j\omega} \mathbf{I}$$

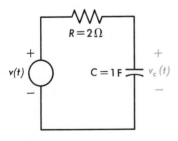

FIGURE 3-34   Circuit of Example 3-16.

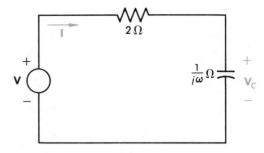

FIGURE 3-35 Circuit of Figure 3-34 transformed.

or

$$\mathbf{V}_C = \frac{\mathbf{V}}{1 + j2\omega} \tag{3-84}$$

Each Fourier-series term will now be treated separately.

1 The average or dc component of Eq. (3-83) is 5 V. For the dc case, $\omega = 0$, and from Eq. (3-84),

$$V_{C\text{dc}} = V_{\text{dc}} = 5 \text{ V}$$

2 The fundamental component of Eq. (3-83) is $v_1 = 6.36 \cos t$, which transforms into $\mathbf{V}_1 = (6.36/\sqrt{2})\underline{/0°}$. From Eq. (3-84),

$$\mathbf{V}_{C1} = \frac{(6.36/\sqrt{2})\underline{/0°}}{1 + j2} = (2.84/\sqrt{2})\underline{/-63.4°} \text{ V}$$

and the corresponding time function is

$$v_{C1} = 2.84 \cos(t - 63.4°) \quad \text{V}$$

3 The third-harmonic ($\omega = 3$) component is $v_3 = -2.12 \cos 3t$, which transforms into $\mathbf{V}_3 = -(2.12/\sqrt{2})\underline{/0°}$, from which

$$\mathbf{V}_{C3} = \frac{(2.12/\sqrt{2})\underline{/0°}}{1 + j(2)(3)} = -(0.35/\sqrt{2})\underline{/-80.5°} \text{ V}$$

and

$$v_{C3} = -0.35 \cos(3t - 80.5°) \quad \text{V}$$

4 The fifth-harmonic ($\omega = 5$) component is $v_5 = 1.27 \cos 5t$, which transforms into $\mathbf{V}_5 = (1.27/\sqrt{2})\underline{/0°}$, from which

$$\mathbf{V}_{C5} = \frac{(1.27/\sqrt{2})\underline{/0°}}{1 + j(2)(5)} = (0.13/\sqrt{2})\underline{/-84.5°} \text{ V}$$

and

$$v_{C5} = 0.13 \cos(5t - 84.5°) \quad \text{V}$$

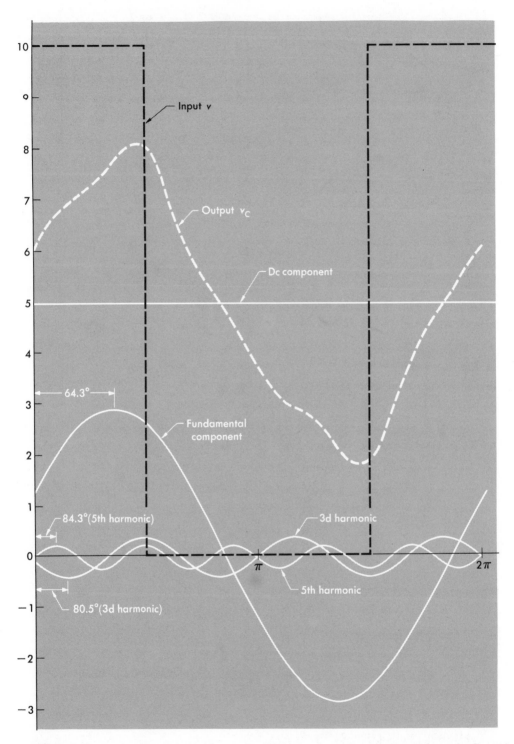

FIGURE 3-36 Dc, fundamental, third-harmonic, fifth-harmonic, and resultant waveforms for Example 3-16.

The Fourier series for v_C is then

$$v_C = 5 + 2.84 \cos (t - 63.4°) - 0.35 \cos (3t - 80.5°)$$
$$+ 0.13 \cos (5t - 84.5°) \qquad \text{V}$$

Note particularly that, while the phasor method is employed to determine each frequency component, the individual time functions must be used in forming the series.

b The several components of part a are plotted in Fig. 3-36. The general shape of $v_C(t)$ is evident (it is a series of exponentials), although more terms would more nearly produce the exact result.

The method used in the solution of Example 3-16 is applicable to all linear systems for which the excitation is represented by a Fourier series. The technique is also applied to determine the response of systems which are simultaneously excited by a number of sinusoidal signals of differing frequencies as in radio and television receivers which receive all stations within range.

PROBLEMS

3-1 The current in a 2.0-H inductance is $5\epsilon^{-t}$. Find
a The impedance of the inductance
b The admittance of the inductance
c The voltage across the inductance

3-2 The voltage across a 0.1-F capacitor is $2.0\epsilon^{-5t}$. Find
a The impedance and admittance of the capacitor
b The current in the capacitor.

3-3 The voltage across the parallel combination of a 5-Ω resistance and 0.2-F capacitance is $10\epsilon^{-2t}$. Determine
a The current in each element
b The total current that is supplied to the parallel combination
c The admittance of the parallel combination

3-4 The current in the series combination of a 10-Ω resistance and 0.1-H inductance is $3.0\epsilon^{-20t}$. Determine
a The voltage across each element
b The voltage across the series combination
c The impedance of the series combination

3-5 The current $i(t)$ in the circuit of Fig. 3-37 is given by $i(t) = I\epsilon^{st}$.
a Draw the transformed network
b Determine the source voltage $v(t)$

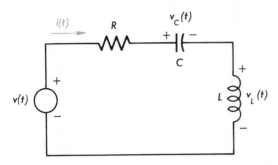

FIGURE 3-37 Circuit for Prob. 3-5.

3-6 In the circuit of Fig. 3-38, the element values are: $R = 2\,\Omega$, $C = 0.5$ F, and $L = 3$ H.
a Draw the transformed network.
b For $i_L(t) = 12\epsilon^{-t/2}$, find the voltage $v(t)$ and the source current $i(t)$.

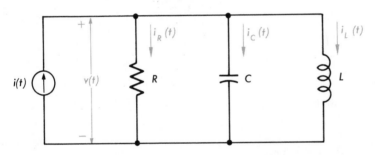

FIGURE 3-38 Circuit for Prob. 3-6.

3-7 In the circuit of Fig. 3-38, the element values are: $R = 4\,\Omega$, $C = 1$ F, and $L = 2$ H. The current $i_R(t) = 3\epsilon^{-t}$. Find the source current $i(t)$.

3-8 The element values in the circuit of Fig. 3-37 are: $R = 10\,\Omega$, $C = 0.1$ F, and $L = 1.0$ H. The capacitor voltage $v_c(t)$ is $5\epsilon^{-10t}$. Determine the source voltage $v(t)$.

3-9 In circuit of Fig. 3-37: $v(t) = 12\epsilon^{-3t}$, $R = 6\,\Omega$, $C = \frac{1}{6}$ F, and $L = \frac{1}{3}$ H. Determine $v_L(t)$.

3-10 In circuit of Fig. 3-38: $i(t) = 12\epsilon^{-3t}$, $R = \frac{1}{6}\,\Omega$, $C = \frac{1}{3}$ F, and $L = \frac{1}{6}$ H. Find the capacitor current $i_c(t)$.

3-11 The differential equation relating the force $F(t)$ and the velocity $u(t)$ for the spring-mass-friction system shown in Fig. 3-39 is

$$F(t) = M\frac{du}{dt} + Du + \frac{1}{C_m}\int u\,dt$$

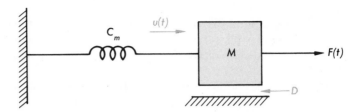

FIGURE 3-39   Mechanical system for Prob. 3-11.

The parameter values are: $M = 10$ kg, $D = 2$ kg/s, and $C_m = 4$ N/m.
a Draw the equivalent electric network using the force-current analog.
b For $F(t) = 20\epsilon^{-t/4} N$, determine $u(t)$.

3-12 For the system described in Prob. 3-11,
a Draw the equivalent electric network using the force-voltage analog.
b For $F(t) = 20\epsilon^{-t/4} N$, determine $u(t)$.
c Repeat *b* for $F(t) = 10\epsilon^{-t/8}$.

3-13 An alternating current is given by the equation

$$i(t) = 10^{-3} \cos (3.14 \times 10^5 t - 30°)$$

Determine:
a The frequency of the current
b The time t_1 when the wave reaches its first maximum value
c The time t_2 when the wave reaches its first minimum value
d The time between two consecutive maxima

3-14 An alternating voltage has a frequency of 60 Hz, a peak amplitude of 169 V, and a value at $t = 0$ of 102 V.
a Determine the equation of the voltage.
b What is the first time the voltage is zero?
c What is the amplitude at $t = \frac{1}{30}$ s?

3-15 Repeat Prob. 3-6 for $i_L(t) = 2 \cos t/3$.

3-16 Repeat Prob. 3-8 for $v_c(t) = 10 \cos t$.

3-17 Repeat Prob. 3-10 for $i(t) = 12 \cos (3t + 45°)$.

3-18 Repeat Prob. 3-9 for
a $v(t) = 10 \cos (2t - 30°)$
b $v(t) = 10 \cos \sqrt{6}\, t$.

3-19 The current in the series combination of a 50-Ω resistance and 10-mH inductance is $10^{-2} \cos 5 \times 10^3 t$.
a What is the impedance of the series combination?

b Determine the voltage across the series combination.
c Draw a phasor diagram showing the current, voltage across each element, and the voltage across the series combination.

3-20 The voltage across the parallel combination of a 100-kΩ resistance and a 0.02-μF capacitor is 50 cos (5 $\times$ $10^4 t$ + 26.6°).
a Determine the admittance of the parallel combination.
b Draw a phasor diagram showing the current in each element, the voltage across the parallel combination, and the current that supplies the parallel circuit.

3-21 Determine the rms and average values of the voltage waveform in Fig. 3-40a.

3-22 Determine the rms and average values of the full-wave rectified sinusoidal waveform in Fig. 3-40b.

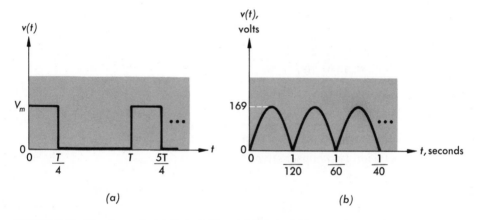

(a)   (b)

FIGURE 3-40 Waveforms for (a) Prob. 3-21 and (b) Prob. 3-22.

3-23 The waveform in Fig. 3-41a is often referred to as a "sawtooth"; for this waveform find the rms and average values.

3-24 Find the rms and average values for the triangular waveform displayed in Fig. 3-41b.

3-25 The voltage $v(t)$ and current $i(t)$ measured at terminals AB in Fig. 3-42 are given by the equations

$$v(t) = 100 \sqrt{2} \cos 10^3 t \qquad i(t) = 10 \sqrt{2} \cos (10^3 t + 60°)$$

a Determine the average power, reactive power, and voltamperes taken by the network.
b Determine the equivalent resistance and reactance of the network as seen at terminals AB.

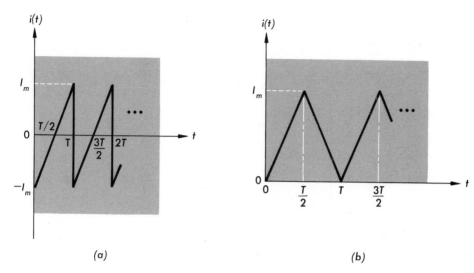

FIGURE 3-41   Triangular waveforms for (a) Prob. 3-23 and (b) Prob. 3-24.

c   At the excitation frequency, what value of capacitance or inductance is required to realize the reactance obtained in *b*?

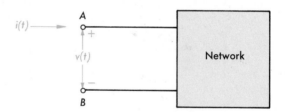

FIGURE 3-42   Network for Prob. 3-25.

3-26   Repeat Prob. 3-25 for

$v(t) = 10^{-3} \cos (2\pi \times 10^5 t + 30°)$
$i(t) = 2 \times 10^{-3} \cos (2\pi \times 10^5 t - 23.1°)$

3-27   Repeat Prob. 3-25 for

$v(t) = 220 \sqrt{2} \cos (377t - 15°)$ $i(t) = 10 \cos (377t - 60°)$

3-28   Repeat Prob. 3-25 for

$v(t) = 4 \cos 3t + 3 \sin 3t$ $i(t) = 3 \cos 3t - 4 \sin 3t$

3-29   Determine the voltage $v_2(t)$ in the circuit of Fig. 3-43 using mesh analysis.

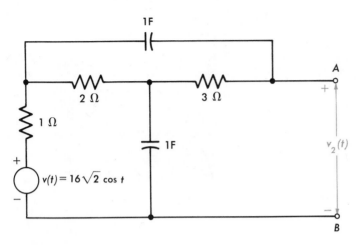

FIGURE 3-43 Circuit for Prob. 3-29.

3-30 Use mesh analysis to find the current I_2 in the circuit of Fig. 3-44.

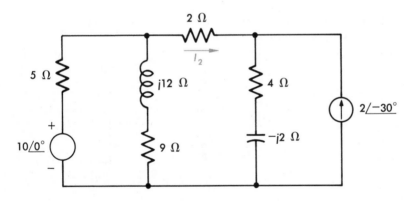

FIGURE 3-44 Circuit for Prob. 3-30.

3-31 Repeat Prob. 3-30 using nodal analysis.

3-32 Repeat Prob. 3-29 using nodal analysis.

3-33 Use nodal analysis to determine the voltage $v_2(t)$ in the circuit of Fig. 3-45.

3-34 Repeat Prob. 3-33 using mesh analysis.

3-35 The ac equivalent of a Wheatstone bridge, depicted in Fig. 3-46, is widely used in a variety of instrumentation schemes. In Fig. 3-46, V is a fixed-frequency sinusoidal signal; Z_1, Z_2 and Z_3 are known impedances; and

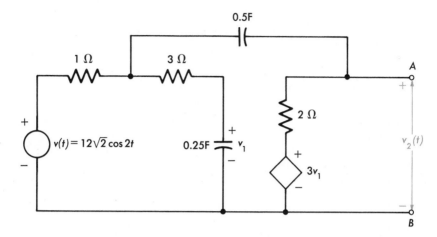

FIGURE 3-45   Circuit for Prob. 3-33.

Z_4 is the unknown impedance. The detector, often called a *null-detector*, is represented by the impedance Z_m. Derive the balance condition, that is, the condition for which the current in Z_m is zero.

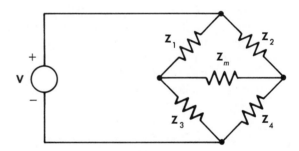

FIGURE 3-46   Circuit for Prob. 3-35.

3-36   One form of the bridge structure in Fig. 3-46, useful in measuring an inductance and its series resistance is shown in Fig. 3-47. The resistance ratio R_1/R_2 can be varied and serves as a multiplier; the resistance R limits the current. The resistance-capacitance combination $R_p - C_p$ is adjusted to achieve a balanced condition. From the values of C_p and R_p and the multiplier, the unknown inductance L_s and series resistance R_s are determined.

a   For $R_1 = R_2 = 1$ kΩ, $R = 500$ Ω, $R_m = 50$ Ω, and $v(t) = 10\sqrt{2}$ cos $2\pi \times 10^3$ t, determine the values of R_p and C_p required to achieve a balance for $L_s = 10$ mH and $R_s = 10$ Ω.

b   Repeat *a* for $R_1 = 10$ kΩ.

c   Repeat *a* for $R_1 = 100$ Ω.

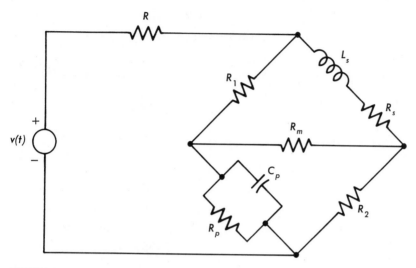

FIGURE 3-47 Ac-bridge circuit for Prob. 3-36.

3-37 Often, the values of L_s and R_s in Fig. 3-47 represent other physical quantities such as pressure, displacement, and strain. In these instances the nominal values of L_s and R_s change when subjected to the force which produces a change in pressure, or causes a displacement or elongation. The bridge is initially balanced with no external force applied. The inductance change produced by the external force causes an imbalance in the bridge and the resultant detector current is used to measure the effect of the change.

a For $R = R_1 = R_2 = 1 \text{ k}\Omega$, $R_m = 100\Omega$, determine C_p and R_p so that the bridge is balanced for $L_s = 20 \text{ mH}$ and $R_s = 30 \ \Omega$. The frequency of the test signal is 10 kHz.

b Find the amplitude of the test signal if the rms current in the $L_s - R_s$ arm of the bridge is to be 1.0 mA at balance.

c Derive a relationship between the change in inductance ΔL and the current in the detector. Assume R_s remains constant.

d If $\Delta L = K \ \Delta P$, where ΔP is the change in pressure in N/m² and K is a constant of proportionality in H-m²/N, determine the value of K if an rms detector current of 1.0 μA is to correspond to a change in pressure of 10^4 N/m².

3-38 The capacitance value of a parallel-plate capacitor can be varied by (1) changing the area of the plates, (2) changing the distance between the plates, and (3) changing the dielectric constant ϵ of the material between the plates. The first method is commonly employed as the tuning mechanism in a radio. The second method is used in capacitor microphones and is readily adapted to measure force, displacement, velocity, acceleration, and pressure. The third method is often used to detect small quantities of trace elements in a dielectric material, particularly in a gas. As capacitance is directly

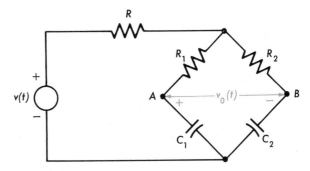

FIGURE 3-48 Differential capacitance-bridge for Prob. 3-38.

proportional to the dielectric constant, the presence of trace elements alters the capacitance value.

The circuit shown in Fig. 3-48 is employed to detect changes in capacitance. The bridge is initially balanced, generally with $C_1 = C_2$ and $R_1 = R_2$. The capacitances C_1 and C_2 most often have identical geometries and are filled with the same gaseous dielectric. Subsequently, C_2 is opened and the dielectric material which contains the trace element fills the region between the plates. The voltage $v_0(t)$ that results from the imbalance is amplified and the amplified signal is recorded or displayed on an oscilloscope or digital voltmeter.

Consider the system in Fig. 3-48 for which $R = R_1 = R_2 = 1$ kΩ, and the test signal $v(t) = 10\sqrt{2} \cos 2\pi \times 10^4 \, t$. The minimum voltage $v_0(t)$ that can be detected is 10-μV rms.

a What is the minimum change in capacitance that can be detected if $C_1 = C_2 = 1{,}000$ pF?
b The relation between C_2 and trace-element concentration η (in parts per million) is $C_2 = 10^{-9} (1 + 4\eta)$. What is the concentration that corresponds to the change in capacitance in a?

3-39 Find the voltage V_2 in the circuit of Fig. 3-49 using nodal analysis.

3-40 Repeat Prob. 3-39 using mesh analysis.

3-41 In the circuit of Fig. 3-50, determine, using nodal analysis,
a $v_0(t)$
b The impedance seen by the voltage source

3-42 Repeat Prob. 3-41 using mesh analysis. (*Hint:* The 600-Ω resistance and the reactance of the 100-pF capacitance can be combined to form a single impedance.)

3-43 A *wattmeter,* an instrument which measures power, contains two coils as shown in Fig. 3-51. The current coil serves as an ammeter, the voltage

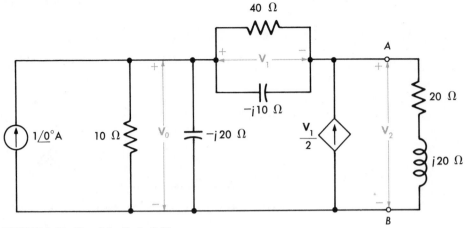

FIGURE 3-49 Circuit for Prob. 3-39.

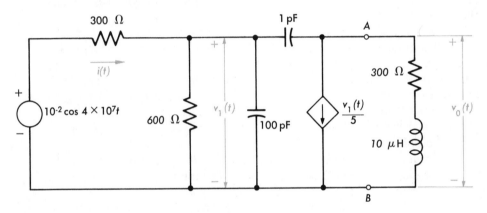

FIGURE 3-50 Circuit for Prob. 3-41.

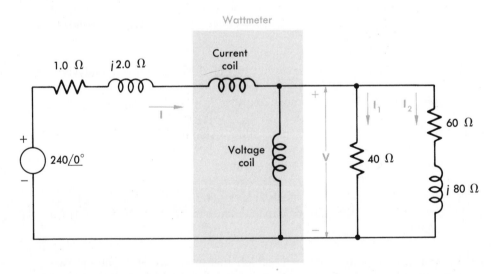

FIGURE 3-51 Wattmeter circuit for Prob. 3-43.

coil acts as a voltmeter. The deflection of the indicator (pointer) is proportional to the product of the current, voltage, and power factor. Thus, the wattmeter can be calibrated to read $VI \cos \theta$. For the circuit of Fig. 3-51, determine

a The wattmeter reading
b The sum of the power dissipated by the 40- and 60-Ω resistances. Assume the current and voltage coil impedances have negligible effect on circuit performance.

3-44 A load impedance is driven by a 100-V source and takes 7.5 A. A wattmeter indicates the power dissipated is 500 W.
a What is the power factor of the load?
b If the excitation frequency is 50 Hz, determine the two possible combinations of elements which comprise the load.

3-45 a In the circuit of Fig. 3-52, determine the reading of a wattmeter if the current coil is connected between AB and the voltage coil is connected between BC.
b Repeat a for the current and voltage coil connections between DE and EF, respectively.
c How much power does each generator supply?

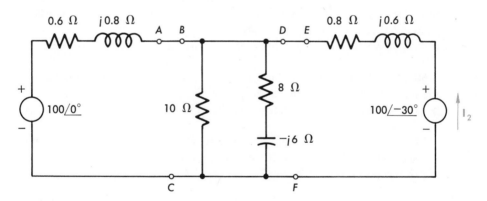

FIGURE 3-52 Circuit for Prob. 3-45.

3-46 Determine the current $\mathbf{I}_A$ in the circuit of Fig. 3-53.

3-47 a Obtain a Thévenin equivalent at terminals AB in the circuit of Fig. 3-43.
b What impedance $\mathbf{Z}_L$, when connected to AB, produces maximum power in $\mathbf{Z}_L$?
c What is the value of this maximum power?

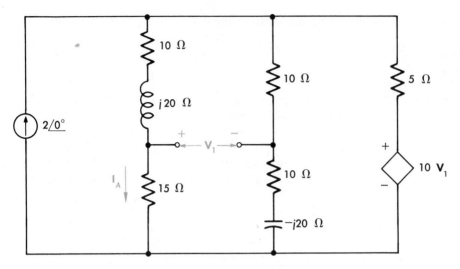

FIGURE 3-53   Circuit for Prob. 3-46.

3-48   a   Represent the network at terminals AB in Fig. 3-45 by a Thévenin equivalent.
b   What impedance $\mathbf{Z}_L$ connected to AB produces maximum power in $\mathbf{Z}_L$?
c   Determine the value of the maximum power.

3-49   a   Obtain a Thévenin equivalent of the portion of the network seen by the 2-Ω resistance in Fig. 3-44.
b   Use the result in a to determine the current in the 2-Ω resistance.

3-50   a   Obtain a Thévenin equivalent of the portion of the network to the left of terminals AB in the circuit of Fig. 3-49.
b   Use the result in a to determine the voltage $\mathbf{V}_2$.

3-51   a   Obtain a Thévenin equivalent of the portion of the network to the right of terminals AC in the circuit of Fig. 3-52.
b   Determine the power supplied by the $100\underline{/0°}$ V source.

3-52   a   For the portion of the network to the left of terminals DF in Fig. 3-52, obtain a Thévenin equivalent.
b   Find the current I_2 using the result in a.

3-53   Determine the voltage $v_o(t)$ in the circuit of Fig. 3-50 by first obtaining the Thévenin equivalent at terminals AB.

3-54   Determine the voltamperes, power, and reactive power provided by the $240\underline{/0°}$ V- source in the circuit of Fig. 3-51.

3-55   In the circuit of Fig. 3-54, find
a   The current $\mathbf{I}_2$

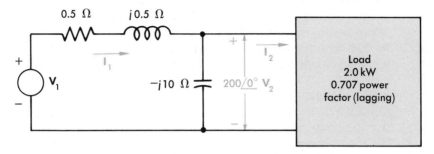

FIGURE 3-54 Circuit for Prob. 3-55.

b The current I_1 and the source voltage V_1
c The power, reactive power, and voltamperes provided by the source

3-56 Two loads are connected in parallel across a 13,200-V line. Load A draws 200 kW at 0.6 power factor (lagging); load B takes 200 kVA at 0.8 power factor (lagging).
a Find the power, reactive power, voltamperes, and line current for the combined load.
b If the frequency of excitation is 60 Hz, what value of capacitance placed across the load is required for the line current to be in phase with the line voltage?

3-57 Three parallel loads are supplied by a 440-V line. Load A draws 10 kVA at a power factor of 0.80 (leading); load B takes 15 kW at a 0.50 lagging power factor, and load C operates at unity power factor and dissipates 5 kW.
a Determine the power, reactive power, voltamperes and line current of the combined load.
b What value of capacitance is required for the line to operate at unity power factor if the excitation frequency is 50Hz?
c Repeat b if the line is to operate at 0.90 power factor (leading).

3-58 The schematic diagram of a generator supplying two loads over a distribution system is depicted in Fig. 3-55. Load A draws 8 kW at a leading

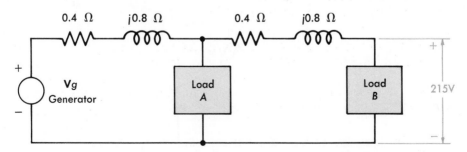

FIGURE 3-55 Loads and distribution system for Prob. 3-58.

power factor of 0.80; load B draws 10 kW at a 0.6 power factor (lagging). The terminal voltage at load B is 215 V. Determine
a   The terminal voltage at load A
b   The voltamperes, power factor, and terminal voltage of the generator

3-59   Two generators supply a 10-kW load at 0.80 power factor (lagging) through a distribution system as illustrated in Fig. 3-56. Find
a   The generator voltages V_1 and V_2
b   The voltamperes and power factor of generator V_1.
Generator V_2 supplies 5 kW at a 0.6 power factor (lagging)

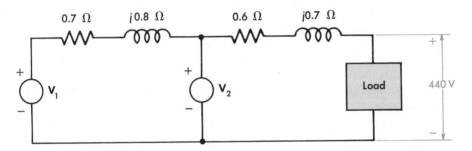

FIGURE 3-56   Circuit for Prob. 3-59.

3-60   Determine the Fourier series for the waveform shown in Fig. 3-40a.

3-61   a   Determine the Fourier series for the waveform shown in Fig. 3-40b.
b   How many harmonics are needed if the amplitude of the last term is to be at least 5 percent of the fundamental?

3-62   Determine the Fourier series for the waveform displayed in Fig. 3-41a.

3-63   Repeat Prob. 3-61 for the waveform depicted in Fig. 3-41b.

3-64   The output of a half-wave rectifier, approximated by the first three terms of its Fourier series as

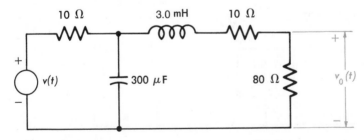

FIGURE 3-57   Circuit for Prob. 3-64.

$$v(t) = \frac{V_m}{\pi} \left(1 + \frac{\pi}{2} \cos 377t + \frac{2}{3} \cos 754t + \cdots \right)$$

is applied to the circuit shown in Fig. 3-57. Determine the output voltage $v_o(t)$.

3-65 Repeat Prob. 3-64 if the input $v(t)$ to the circuit in Fig. 3-57 is the full-wave rectified waveform shown in Fig. 3-40b.

3-66 The waveform displayed in Fig. 3-32 whose Fourier series is given in Eq. (3-80) is the voltage $v(t)$ applied to the circuit shown in Fig. 3-58. The amplitude and period of the square wave are 10 V and 1 ms, respectively. Sketch the output waveform $v_o(t)$. Only consider frequency components through the fifth harmonic.

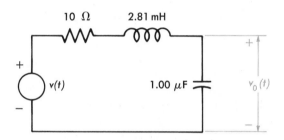

FIGURE 3-58 Circuit for Prob. 3-66.

3-67 The triangular current waveform in Fig. 3-41b, having an amplitude $I_m = 10$ mA and a period 1 ms, is the current $i(t)$ applied to the circuit of Fig. 3-59. Sketch the output waveform $v_o(t)$. Consider only the first four harmonics in the Fourier series.

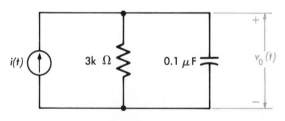

FIGURE 3-59 Circuit for Prob. 3-67.

THE RESPONSE OF SINGLE NETWORKS □ **THE**
ONSE OF SINGLE NETWORKS □ THE **RESPONSE**
SE OF SINGLE NETWORKS □ THE RESPONSE **OF**
INGLE NETWORKS □ THE RESPONSE OF **SINGLE**
ORKS □ THE RESPONSE OF SINGLE **NETWORKS**
THE RESPONSE OF SINGLE NETWORKS □ THE R
RESPONSE OF SINGLE NETWORKS □ THE RESP(
OF SINGLE NETWORKS □ THE RESPONSE OF SII
SINGLE NETWORKS □ THE RESPONSE OF SINGL
NETWORKS □ THE RESPONSE OF SINGLE NETWORKS □ THE RESP(
THE RESPONSE OF SINGLE NETWORKS □ THE RESPONSE OF SINGL
RESPONSE OF SINGLE NETWORKS □ THE RESPONSE OF SINGLE NI
OF SINGLE NETWORKS □ THE RESPONSE OF SINGLE NETWORKS □
SINGLE NETWORKS □ THE RESPONSE OF SINGLE NETWORKS □ TH
NETWORKS □ THE RESPONSE OF SINGLE NETWORKS □ THE RESP(
THE RESPONSE OF SINGLE NETWORKS □ THE RESPONSE OF SINGL
RESPONSE OF SINGLE NETWORKS □ THE RESPONSE OF SINGLE NI
OF SINGLE NETWORKS □ THE RESPONSE OF SINGLE NETWORKS □
SINGLE NETWORKS □ THE RESPONSE OF SINGLE NETWORKS □ TH
NETWORKS □ THE RESPONSE OF SINGLE NETWORKS □ THE RESP(
THE RESPONSE OF SINGLE NETWORKS □ THE RESPONSE OF SINGL
RESPONSE OF SINGLE NETWORKS □ THE RESPONSE OF SINGLE NI
OF SINGLE NETWORKS □ THE RESPONSE OF SINGLE NETWORKS □
SINGLE NETWORKS □ THE RESPONSE OF SINGLE NETWORKS □ TH
NETWORKS □ THE RESPONSE OF SINGLE NETWORKS □ THE RESP(
THE RESPONSE OF SINGLE NETWORKS □ THE RESPONSE OF SINGL
RESPONSE OF SINGLE NETWORKS □ THE RESPONSE OF SINGLE NI
OF SINGLE NETWORKS □ THE RESPONSE OF SINGLE NETWORKS □
SINGLE NETWORKS □ THE RESPONSE OF SINGLE NETWORKS □ TH
NETWORKS □ THE RESPONSE OF SINGLE NETWORKS □ THE RESP(
THE RESPONSE OF SINGLE NETWORKS □ THE RESPONSE OF SINGL
RESPONSE OF SINGLE NETWORKS □ THE RESPONSE OF SINGLE NI
OF SINGLE NETWORKS □ THE RESPONSE OF SINGLE NETWORKS □
SINGLE NETWORKS □ THE RESPONSE OF SINGLE NETWORKS □ TH
NETWORKS □ THE RESPONSE OF SINGLE NETWORKS □ THE RESP(
THE RESPONSE OF SINGLE NETWORKS □ THE RESPONSE OF SINGL
RESPONSE OF SINGLE NETWORKS □ THE RESPONSE OF SINGLE NI
OF SINGLE NETWORKS □ THE RESPONSE OF SINGLE NETWORKS □
SINGLE NETWORKS □ THE RESPONSE OF SINGLE NETWORKS □ TH
NETWORKS □ THE RESPONSE OF SINGLE NETWORKS □ THE RESP(

4

he response of networks to time-varying sources is considered in this chapter. In particular, the response when the excitation is suddenly applied or suddenly changed is examined. The two components of the response described in Sec. 3-1, the forced and natural responses needed to make the gradual adjustment in the response of systems which contain energy-storage elements, are determined by the traditional techniques for solving the linear differential equations which arise.

To represent the abrupt changes in excitation, encountered when a switch is opened or closed and in individual sequences of pulses, a class of excitation, often called *singularity functions*, is introduced. This type of representation leads to the step and impulse functions. The methods for evaluating the natural and forced component of the response are also applied to these excitations.

4-1 THE FORCED RESPONSE

The *forced response* is that component of the system response that indicates the effects produced by the applied excitation. Our intuition and experience lead to the conclusion that the form of the forced response is virtually the same as the form of the excitation. In Sec. 3-2 this conclusion was reinforced for the case of exponential excitation. Thus, constant (dc) excitations, for which $s = 0$, produced constant responses; and sinusoidal responses result when sinusoidal (ac) excitations are applied, the case for $s = j\omega$. With the aid of Table 3-1, it was shown that the strength of the excitation and the circuit parameters determined the strength of the forced response. However, the form of this response was an exponential having the same exponent as the excitation. This knowledge is the underlying factor in developing the methods used to determine the forced response to more general excitations.

Consider the *RL* circuit of Fig. 4-1, excited by a voltage source $v(t) = Kt$ applied at $t = 0$ when the switch is closed. The KVL equation for the circuit is

$$Ri_f(t) + L\frac{di_f(t)}{dt} = v(t) = Kt \tag{4-1}$$

where $i_f(t)$ is the forced response of the circuit. Equation (4-1) is valid only for $t > 0$; that is, for all times after the switch is closed and the excitation is applied. As such, the solution for $i_f(t)$ can describe only the forced response for $t > 0$.

The search for the function $i_f(t)$ which satisfies Eq. (4-1) for all $t > 0$ is conducted in a methodical manner based on our previous experience. In addition to finding the solution for $i_f(t)$, the method indicates the thought process to be followed in other cases. Our experience suggests that our first choice for $i_f(t)$ be of the same form as the excitation; that is, assume $i_f(t) =$

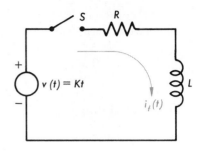

FIGURE 4-1 *RL* circuit.

At. The desired solution is obtained by determination of the value of A. Substitution of At for $i_f(t)$ in Eq. (4-1) gives

$$RAt + LA = Kt$$

Since no value of A exists which satisfies this equation for all values of t, the assumed solution is wrong. However, as Eq. (4-1) contains $i_f(t)$ and its derivative, let us assume that the form of $i_f(t)$ is that of the excitation and its derivative. Therefore, let

$$i_f(t) = At + B \qquad (4\text{-}2)$$

so that determination of A and B leads to the desired solution. Substitution of Eq. (4-2) in Eq. (4-1) yields

$$RAt + RB + L\frac{d}{dt}(At + B) = Kt$$

which, after rearrangement, results in

$$RAt + RB + LA = Kt \qquad (4\text{-}3)$$

Comparison of terms on the left- and right-hand sides of Eq. (4-3) gives

$$RA = K \quad \text{and} \quad RB + LA = 0$$

from which

$$A = \frac{K}{R} \quad \text{and} \quad B = -\frac{LK}{R^2}$$

The forced response then becomes

$$i_f(t) = \frac{K}{R}t - \frac{LK}{R^2} \qquad (4\text{-}4)$$

The method of solution which results in Eq. (4-4) is called the *method of undetermined coefficients* and is a standard technique used to obtain the particular solution (forced response) of a linear differential equation. This class of differential equations arises most frequently in circuit theory. In this book it will be dealt with almost exclusively. In addition, as the circuit parameters are constants, the resultant class of equations is called *linear differential equations with constant coefficients.*

In the general case, following the reasoning leading to Eq. (4-4), it can be shown that *the forced response is a function which contains the excitation and all its unique derivatives.* The use of this form for the response ensures the proper balance of the two sides of the defining differential equation obtained from the KVL and KCL relations for the circuit.

Example 4-1

For the circuit in Fig. 4-2, determine the forced response for $v_f(t)$.

SOLUTION The differential equation, obtained by use of KCL, is

$$2v_f + \frac{1}{2}\frac{dv_f}{dt} = 12te^{-t}$$

The form of the response is assumed to be of the form of the excitation and its unique derivatives. As

$$\frac{d}{dt}(te^{-t}) = -te^{-t} + \epsilon^{-t}$$

$v_f(t)$ is taken as $Ate^{-t} + Be^{-t}$. Note that derivatives beyond the first are redundant, as additional derivatives result in terms te^{-t} and ϵ^{-t}. The coefficients A and B to be determined include the additional derivatives. Substitution for $v_f(t)$ in the original equation gives

$$2Ate^{-t} + 2Be^{-t} + \frac{1}{2}\frac{d}{dt}(Ate^{-t} + Be^{-t}) = 12te^{-t}$$

from which

$$\frac{3}{2}Ate^{-t} + \left(\frac{A}{2} + \frac{3B}{2}\right)\epsilon^{-t} = 12te^{-t}$$

Comparison of terms gives

$$\frac{3}{2}A = 12 \quad\text{and}\quad \frac{A}{2} + \frac{3B}{2} = 0$$

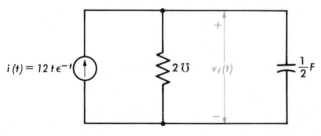

$i(t) = 12\,te^{-t}$ $2\,\mho$ $v_f(t)$ $\frac{1}{2}F$

FIGURE 4-2 Circuit of Example 4-1.

Simultaneous solution for A and B yields

$\qquad A = 8 \qquad$ and $\qquad B = -\frac{8}{3}$

resulting in

$\qquad v_f(t) = 8t\epsilon^{-t} - \frac{8}{3}\epsilon^{-t}$

Example 4-2

Find the forced component of the response $i_f(t)$ for the circuit shown in Fig. 4-3.

SOLUTION   The KVL equation is

$$2i_f(t) + \frac{1}{1/2} \int i_f(t)\, dt = 10 \sin 2t$$

or, by differentiating both sides of the equation,

$$2\frac{di_f(t)}{dt} + 2i_f(t) = 20 \cos 2t$$

To determine the forced response, assume $i_f(t) = A \cos 2t + B \sin 2t$, which, when substituted into the differential equation, gives

$\qquad -4A \sin 2t + 4B \cos 2t + 2A \cos 2t + 2B \sin 2t = 20 \cos 2t$

Equating coefficients yields

$\qquad 2A + 4B = 20 \qquad$ and $\qquad -4A + 2B = 0$

The values of A and B, by simultaneous solution, are

$\qquad A = 2 \qquad$ and $\qquad B = 4$

These values make

$\qquad i_f(t) = 2 \cos 2t + 4 \sin 2t \qquad$ A

[By use of the trigonometric identities in Eqs. (3-29) and (3-30), $i_f(t)$ can be

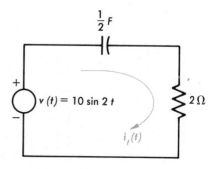

FIGURE 4-3   *RC* circuit of Example 4-2.

expressed as

$$i_f(t) = 2\sqrt{5} \sin (2t + 26.6°) = 2\sqrt{5} \cos (2t - 63.4°) \quad \text{A}$$

which is the same result that is obtained if the method of Sec. 3-2 is used with Table 3-1c.]

4-2 THE NATURAL RESPONSE

In Sec. 3-1, the response of a mass-friction system was shown to gradually change from its initial state to its final state when the excitation was changed. Similarly, if the force is removed suddenly, the velocity does not decrease to zero immediately. The velocity decreases gradually, and the mass comes to rest only after all the kinetic energy stored is dissipated as heat by the friction. This gradual transition from the response prior to the change in excitation to the response produced by the excitation is character-istic of all practical systems which contain energy-storage components. The portion of the system response indicative of this transition is the *natural response*. In this section, one method used to determine the natural response is developed.

The circuit in Fig. 4-4 consists of a conductance G in parallel with a capacitor C that is charged to a voltage V_o at $t = 0$. As a result of the charge, the capacitor has energy stored at $t = 0$. Since there is no external source present, there can be no forced response and only the natural response exists. The voltage $v(t)$ must, therefore, indicate the dissipation of the stored capacitive energy as heat in the conductance.

The KCL equation for the circuit of Fig. 4-4 is

$$C \frac{dv_n(t)}{dt} + Gv_n(t) = 0 \tag{4-5}$$

The voltage is written as $v_n(t)$ to emphasize that it is a function of time. If Eq. (4-5) is to hold for all values of time, then $v_n(t)$ must be a function whose derivative has the same form as the function. In Secs. 3-2 to 3-4, we have already seen that the exponential is one such function. Accordingly, assume that

$$v_n(t) = K\epsilon^{st} \quad \text{V} \tag{4-6}$$

is a solution of Eq. (4-5) and where K and s are constants to be determined.

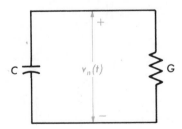

FIGURE 4-4 *GC* parallel circuit.

Substitution of Eq. (4-6) into Eq. (4-5) gives

$$C \frac{d}{dt} (K\epsilon^{st}) + GK\epsilon^{st} = 0$$

or

$$K\epsilon^{st}(sC + G) = 0 \tag{4-7}$$

Equation (4-7) is satisfied if either $K = 0$ or if

$$sC + G = 0 \tag{4-8}$$

The first case ($K = 0$) is a solution and corresponds to an initially uncharged capacitance. The case defined by Eq. (4-8) is the more general solution from which the value of s is determined as

$$s = -\frac{G}{C} \tag{4-9}$$

Use of Eq. (4-9) with Eq. (4-6) gives

$$v_n(t) = K\epsilon^{-Gt/C} \tag{4-10}$$

Of note is that in Eq. (4-8), $sC + G$ represents the admittance $Y(s)$ of the parallel combination of the capacitance and conductance. Thus, Eq. (4-9) is obtained by setting $Y(s) = 0$. The value of s given by Eq. (4-9) is often called the *natural frequency* of the network. In many circuits, the natural frequencies are obtained by computing the values of s which satisfy $Y(s) = 0$.

The value of the constant K is found by determining the value of the response at a particular time (usually, $t = 0$). This value, usually called an *initial condition*, is determined from the application of Kirchhoff's laws and other physical laws to the circuit. Section 4-4 deals with the methods for determining these initial conditions. Equation (4-10) shows that $K = V_o$, where V_o is the value of $v_n(t)$ at $t = 0$. In this case, only $v_n(t)$ is evaluated as there is no forced response.

Example 4-3

Given the *RL* circuit of Fig. 4-5:

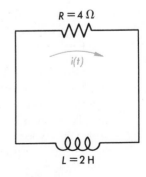

$R = 4\,\Omega$

$i(t)$

$L = 2\,H$

FIGURE 4-5 *RL* network.

a Find the natural response $i_n(t)$ in literal form.

b Find the numerical values of $i_n(t)$ if $i(0) = 5$ A.

SOLUTION **a** The KVL equation is

$$L \frac{di_n(t)}{dt} + Ri_n(t) = 0$$

This equation has the same form as Eq. (4-5), and the solution has the form of Eq. (4-10), with C replaced by L and G replaced by R. Thus,

$$i_n(t) = K\epsilon^{-Rt/L}$$

The natural frequency $s = -R/L$ also represents the value of s obtained from $Z(s) = 0$, where $Z(s)$ is the impedance of the circuit.

An alternative method which helps to justify the assumed exponential solution is as follows: If $i_n(t)$ is a continuous function, it is permissible to rewrite the differential equation as

$$L \frac{di_n(t)}{i_n(t)} + R \, dt = 0$$

or

$$\frac{di_n(t)}{i_n(t)} = - \frac{R}{L} \, dt$$

Integrating both sides of the equation results in

$$ln \; i_n(t) - ln \; K = - \frac{R}{L} t$$

where $(- ln \; K)$ is the constant of integration. Recalling that the difference of logarithms corresponds to division, taking the antilog of both sides yields

$$\frac{i_n(t)}{K} = \epsilon^{-Rt/L} \quad \text{or} \quad i_n(t) = K\epsilon^{-Rt/L}$$

As is evident, this is the same result as previously obtained.

b With the values of R and L given in the problem and the fact that $i(0) = 5$,

$$i(0) = 5 = K\epsilon^{-(4/2)(0)} = K \quad \text{and} \quad i(t) = 5\epsilon^{-2t} \quad \text{A}$$

Equation (4-10) and the results of Example 4-3 both show that an exponential is the form of the solution. These are by no means a pair of isolated cases. In general, the natural component of the response of a linear system can be expressed as either an exponential or a sum of exponentials. To consider the exponential in some detail, a normalized function will be chosen in which ratios rather than specific numbers will be considered. If a current function

$$i(t) = I_0\epsilon^{-at}$$

exists, it can be rewritten in normalized form as

$$\frac{i(t)}{I_0} = \epsilon^{-t/\tau} \tag{4-11}$$

where $\tau = 1/a$.

The ratio of $i(t)/I_0$ as a function of t/τ is plotted in Fig. 4-6 and tabulated in Table 4-1. The curve shows a typical decaying exponential. Note that, while theoretically an infinite time is required for $i(t)/I_0$ to go to zero, at $t/\tau = 5$ the value of the ratio is 0.007—less than 1 percent of the original value of I_0—and that, for most engineering purposes, the natural response

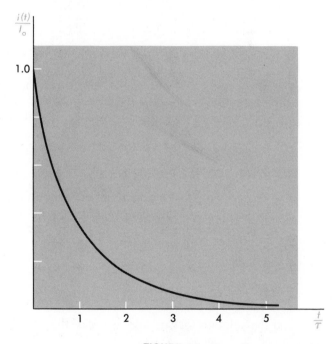

FIGURE 4-6 Normalized exponential.

TABLE 4-1 NORMALIZED VALUES OF
EXPONENTIAL FUNCTION

t/τ	$\dfrac{i(t)}{I_0} = \epsilon^{-t/\tau}$
0	1.000
1	0.368
2	0.135
3	0.050
4	0.018
5	0.007

has reached a negligible value. The value τ is thus seen to be a measure of the length of time required for the natural response to die out. The value τ is called the *time constant* of the network. Networks with long time constants can be expected to take longer for the natural response to decay than those with short time constants; in other words, networks with long time constants take longer to readjust from the effects of a change in stimulus than those with short time constants.

For the circuits of Fig. 4-4, the time constant, from Eq. (4-11), is $-C/G$. As $R = 1/G$, the time constant of an RC circuit is most often written as

$$\tau = RC \tag{4-12}$$

Similarly, from Eq. (4-3), the time constant of an RL circuit is

$$\tau = \frac{L}{R} \tag{4-13}$$

While Eqs. (4-12) and (4-13) are obtained for simple two-element circuits, those results have more general applicability. The value of R need not be that of a single resistance, but is, in general, the equivalent resistance measured from the terminals of the capacitance or inductance. (The techniques for evaluating the equivalent resistance described in Chap. 2 are therefore applicable.)

4-3 THE TOTAL RESPONSE

The *total response* of a system to an excitation that is suddenly applied or changed consists of the sum of the forced and natural responses. The individual response components are determined by the methods described in the previous two sections. Thus, for the circuit shown in Fig. 4-7, the KCL equation is

$$C\frac{dv}{dt} + Gv = i_s(t) \tag{4-14}$$

and the solution for $v(t)$, consisting of the natural component $v_n(t)$ and the forced component $v_f(t)$, is given as

$$v(t) = v_f(t) + v_n(t) \tag{4-15}$$

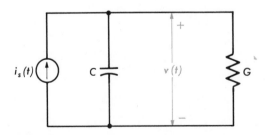

FIGURE 4-7 *GC* circuit with current-source excitation.

Substitution of Eq. (4-15) into Eq. (4-14) yields

$$C \frac{d}{dt} [v_f(t) + v_n(t)] + G[v_f(t) + v_n(t)] = i_s(t)$$

Rearrangement of terms results in

$$\left[C \frac{dv_f(t)}{dt} + Gv_f(t) \right] + \left[C \frac{dv_n(t)}{dt} + Gv_n(t) \right] = i_s(t) \tag{4-16}$$

From Sec. 4-1, the equation for the forced response is identified as

$$C \frac{dv_f(t)}{dt} + Gv_f(t) = i_s(t) \tag{4-17}$$

Consequently, the natural response terms described in Sec. 4-2 are given by

$$C \frac{dv_n}{dt} + Gv_n(t) = 0 \tag{4-18}$$

Note that the natural response terms are obtained from the equation which contains no source as is also the case for the circuits in Figs. 4-4 and 4-5. Networks used to evaluate the natural response contain no independent voltage or current sources and are called *source-free networks*. To obtain the natural, or force-free, response of a network, independent sources must be suppressed; that is, voltage sources are short-circuited and current sources are open-circuited.

The procedure for obtaining the total response is then one which simply requires writing the differential equation for the circuit and then identifying the component responses. The forced response is obtained by use of the method in Sec. 4-1, or for applicable cases, the techniques in Secs. 3-2 to 3-5. The natural response terms are evaluated by suppressing the excitation and forming the force-free equation. The following example illustrates the procedure.

Example 4-4

For the circuit shown in Fig. 4-7, $i_s(t) = 12te^{-t}$, $C = \frac{1}{2}F$ and $G = 2 \, \mho$. The capacitor voltage at $t = 0$ is -3 V. Determine the complete response for $v(t)$.

SOLUTION The differential equation is given in Eq. (4-14), which for the values given becomes

$$\frac{1}{2} \frac{dv}{dt} + 2v = 12te^{-t}$$

The forced component $v_f(t)$ is given in Eq. (4-17) whose solution is given in Example 4-1 as

$$v_f(t) = 8te^{-t} - \tfrac{8}{3}\epsilon^{-t}$$

The natural response is obtained by suppressing the current source and is described by Eq. (4-18). As Eq. (4-18) is the same as Eq. (4-5), the solution for $v_n(t)$ is given by Eq. (4-10) and is

$$v_n(t) = K\epsilon^{-Gt/C} = K\epsilon^{-4t} \quad \text{V}$$

The total solution becomes

$$v(t) = v_f(t) + v_n(t) = 8t\epsilon^{-t} - \tfrac{8}{3}\epsilon^{-t} + K\epsilon^{-4t} \quad \text{V}$$

The evaluation of K follows from the given initial condition

$$v(0) = -3 = 0 - \tfrac{8}{3} + K$$

from which

$$K = -\tfrac{1}{3}$$

and

$$v(t) = 8t\epsilon^{-t} - \tfrac{8}{3}\epsilon^{-t} - \tfrac{1}{3}\epsilon^{-4t} \quad \text{V}$$

Note that the initial condition must be evaluated using the complete response and not simply the natural component of the response.

Mathematically, the procedure described in Example 4-4 is one classical technique for solving linear differential equations. The natural response is obtained from the *homogeneous equation* and describes force-free behavior. As indicated in Examples 4-3 and 4-4, the natural response depends only on circuit parameters and is, therefore, a characteristic of the circuit. The forced component of the response is called the *particular solution* or *particular integral* and represents the system behavior to the specific excitation. From a mathematical viewpoint, evaluation of the constant K appearing in the natural response requires only that the total response be known at any one instant of time. In the following section, constraints on the physical behavior of energy-storage elements are developed which serve as the basis for the initial conditions used to compute the value of K.

4-4 INITIAL CONDITIONS

Examples 4-3 and 4-4 show the need for evaluating circuit currents and voltages and their derivatives at specific instants of time. This section deals with the formal evaluation of these boundary or initial conditions. The tools used in this evaluation are Kirchhoff's laws and the principles of continuity of capacitive voltage and inductive current. The results are used to determine the coefficients of the natural-response terms.

Since most switching processes are assumed for convenience to occur at $t = 0$, it is appropriate to introduce terminology which permits identification of phenomena occurring immediately prior to or immediately after the switching takes place. The symbol (0^-) will be used to indicate time immediately before switching and the symbol (0⁺) is used for the instant of time immediately after switching.

The capacitance is an energy-storage element whose defining equations are given in Sec. 1-5. The volt-ampere relationship is

$$i_C = C\frac{dv_C}{dt} \qquad (4\text{-}19)$$

and the stored energy is

$$w_C = \tfrac{1}{2}Cv_C^2 \qquad (4\text{-}20)$$

Equations (4-19) and (4-20) show that a sudden change in capacitance voltage (that is, $dv_C/dt = \infty$) must be accompanied by an infinite current. Furthermore, such a change will result in a change of stored energy in zero time, requiring an infinite power source. Accordingly, the following principle can be stated: *In the absence of an infinite current, the voltage across a capacitance cannot change instantaneously* [or $v_C(0^-) = v_C(0^+)$].

Example 4-5

In the circuit of Fig. 4-8, the switch S_1 has been closed for a long time prior to $t = 0$. At $t = 0$, switch S_2 is closed. Find $v_C(0^+)$ and $i_C(0^+)$, the capacitance voltage and current, respectively, at the instant after the switching has been completed.

SOLUTION   With the switches in the position shown in the figure and assuming that they have been in this position long enough for the natural response resulting from the switching process to die out, the capacitance has charged up to 10 V and there is no current. Thus, $v_C(0^-) = 10$ V, and $i_C(0^-) = 0$.

At $t = 0$, the switching process connects the second 10 Ω resistance into the circuit. From the principle of the continuity of capacitance voltage,

$$v_C(0^+) = v_C(0^-) = 10 \text{ V}$$

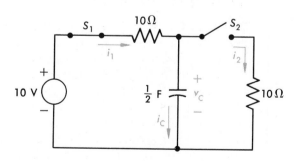

FIGURE 4-8   Circuit of Example 4-5.

To find $i_C(0^+)$, KVL and KCL equations are written for the network at $t = 0^+$. These are

$$10 - 10i_1(0^+) - v_C(0^+) = 0$$

from which $i_1(0^+) = 0$,

$$v_C(0^+) - 10i_2(0^+) = 0$$

which yields $i_2(0^+) = 1$ A. Consequently, from KCL,

$$i_1(0^+) = i_2(0^+) + i_C(0^+)$$

Thus

$$i_C(0^+) = -1 \text{ A}$$

The inductance is an energy-storage element whose defining equations are given in Sec. 1-6. These are the volt-ampere relationship

$$v_L = L \frac{di_L}{dt} \tag{4-21}$$

and the stored energy

$$w_L = \tfrac{1}{2} L i_L^2 \tag{4-22}$$

Note from Eqs. (4-21) and (4-22) that an instantaneous change in inductance current is accompanied by an infinite voltage and power. From this fact, the principle of continuity of inductance current can be formulated: *In the absence of an infinite voltage, the current in an inductance cannot change instantaneously* [or $i_L(0^-) = i_L(0^+)$].

Example 4-6

The switch S in the circuit of Fig. 4-9 has been in position 1 for a long time. At $t = 0$, it is switched to position 2.

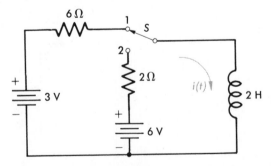

FIGURE 4-9 Circuit of Example 4-6.

a Determine the inductance current $i(t)$ for $t \geq 0$.

b Find the values of $v_L(0^-)$, $v_L(0^+)$, and $\dfrac{di}{dt}(0^+)$.

SOLUTION **a** For $t \geq 0$, the switch is in position 2 and the circuit becomes that shown in Fig. 4-10. The differential equation, based on KVL, is

$$2\frac{di}{dt} + 2i = 6$$

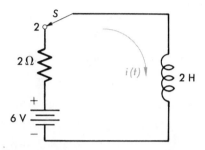

FIGURE 4-10 Circuit of Fig. 4-9 when switch S is in position 2.

The forced-response i_f is assumed to be constant I, so that

$$2\frac{d}{dt}(I) + 2I = 6$$

or

$$0 + 2I = 6 \quad \text{and} \quad I = 3 \text{ A}$$

Note that this value is equal to the applied voltage divided by the resistance. It follows from the result in Example 3-2 where the inductance was shown to behave as a short-circuit to constant excitations.

The natural response is obtained from the force-free equation

$$2\frac{di_n}{dt} + 2i_n = 0$$

whose solution is $K\epsilon^{st}$. Substitution gives

$$2Ks\epsilon^{st} + 2K\epsilon^{st} = 0$$

from which

$$2s + 2 = 0$$

and

$$s = -1$$

The total solution is then

$$i(t) = i_f(t) + i_n(t) = 3 + K\epsilon^{-t}$$

To evaluate K, the value of $i(0^+)$ must be determined. For $t = 0^-$, with the switch in position 1 for a long time, it can be assumed that the natural response which resulted when the 3-V source was applied has had sufficient time to decay. The inductor current is then constant; its value is

$$i(0^-) = \tfrac{3}{6} = 0.5 \text{ A}$$

Again, the short-circuit behavior of the inductance to constant excitation is used. By continuity of inductance current $i(0^+) = 0.5$ A. Then, using the total solution,

$$i(0^+) = 0.5 = 3 + K$$

and $K = -2.5$.
The total solution, valid only for $t \geq 0$, is

$$i(t) = 3 - 2.5\epsilon^{-t}$$

b As the inductance is a short-circuit at $t = 0^-$, its voltage is zero. Thus, $v_L(0^-) = 0$. At $t = 0^+$, from the circuit of Fig. 4-10, the KVL equation is

$$v_L(0^+) + i(0^+) \times 2 = 6$$

and

$$v_L(0^+) = 6 - (0.5) \times 2 = 5 \text{ V}$$

As $v_L = L\dfrac{di}{dt}$,

$$\frac{di}{dt}(0^+) = \frac{v_L(0^+)}{L} = \frac{5}{2} = 2.5 \frac{A}{s}$$

4-5 THE RLC CIRCUIT

In this section, circuits containing both kinds of energy-storage elements are examined. The methods for determining the response of such circuits follow directly from the single storage-element circuits of the previous sections. Consider the RLC parallel circuit shown in Fig. 4-11 in which an excitation (not shown) has been removed at $t = 0$. Assume that energy is stored in both the capacitance and inductance at the instant the source is removed. The response for $t \geq 0$ is just the natural response as no excitation is present. By application of KCL, the differential equation is

$$\frac{1}{R}v(t) + C\frac{dv(t)}{dt} + \frac{1}{L}\int v(t)\,dt = 0$$

Differentiating and rearranging gives

$$C\frac{d^2v(t)}{dt^2} + \frac{1}{R}\frac{dv(t)}{dt} + \frac{1}{L}v(t) = 0 \tag{4-23}$$

As in the RL or RC case, assume the natural response is of the form $K\epsilon^{st}$.

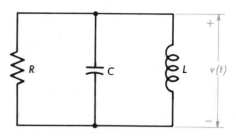

FIGURE 4-11 *RLC* parallel circuit.

Substitution into Eq. (4-23) yields

$$CKs^2\epsilon^{st} + \frac{1}{R} Ks\epsilon^{st} + \frac{K}{L} \epsilon^{st} = 0$$

or

$$K\epsilon^{st} \left(Cs^2 + \frac{s}{R} + \frac{1}{L} \right) = 0 \tag{4-24}$$

As $K = 0$ represents the condition of zero energy storage, the solution of Eq. (4-24) is derived from

$$Cs^2 + \frac{s}{R} + \frac{1}{L} = 0$$

from which

$$s = \frac{-\dfrac{1}{R} \pm \sqrt{\dfrac{1}{R^2} - \dfrac{4C}{L}}}{2C} = -\frac{1}{2RC} \left(1 \pm \sqrt{1 - \frac{4R^2C}{L}} \right) \tag{4-25}$$

There are two values of s which arise from Eq. (4-25), each of which satisfies Eq. (4-24). As a consequence, two terms $K_1\epsilon^{s_1t}$ and $K_2\epsilon^{s_2t}$, where s_1 and s_2 are given in Eq. (4-25) and K_1 and K_2 are constants, arise in the natural response. Physically and mathematically this is not unexpected. First, constraints on both the capacitance voltage and inductance current exist and are independent of one another. As such, two constants are needed if both constraints are to be simultaneously satisfied. Mathematically, Eq. (4-23) is a second-order differential equation as the second derivative is present. To obtain $v(t)$ from d^2v/dt^2, two integrations are needed, each of which requires a constant of integration. Thus, as each exponential term satisfies Eq. (4-23), their sum also must satisfy Eq. (4-23) and the natural response becomes

$$v(t) = K_1\epsilon^{s_1t} + K_2\epsilon^{s_2t} \tag{4-26}$$

Two boundary conditions are required; usually one is the value of $v(t)$ at $t = 0$, the second being the value of $\dfrac{dv(t)}{dt}$ evaluated at $t = 0$. Application of the principles of continuity of inductance current and capacitive voltage to obtain these terms is shown in Example 4-7.

The values of s obtained from Eq. (4-25) depend on the term $1 - \dfrac{4R^2C}{L}$ (the discriminant of the quadratic equation $Cs^2 + \dfrac{s}{R} + \dfrac{1}{L} = 0$). Three cases exist and are

1 $1 - \dfrac{4R^2C}{L} > 0$; s_1 and s_2 are real, negative, and unequal.

2 $1 - \dfrac{4R^2C}{L} < 0$; s_1 and s_2 are complex conjugates.

3 $1 - \dfrac{4R^2C}{L} = 0$; s_1 and s_2 are real, negative, and equal.

The effect of each of these cases on the response of RLC circuits is described in the following examples.

Example 4-7

The switch S in the circuit of Fig. 4-12 is closed at $t = 0$. No energy is stored prior to the application of the excitation. Find $v(t)$ for $t \geq 0$ if C is

a $\frac{1}{9}F$

b $\frac{1}{4}F$

c $\frac{1}{8}F$

SOLUTION The KCL equation is, for $t \geq 0$,

$$\frac{v}{2} + \frac{1}{2} \int v \, dt + C \frac{dv}{dt} = 10$$

Differentiating and rearranging gives

$$C \frac{d^2v}{dt^2} + \frac{1}{2} \frac{dv}{dt} + \frac{1}{2} v = 0$$

which is Eq. (4-23) for the R, L, and C values given.

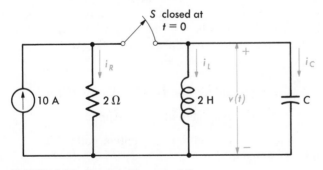

FIGURE 4-12 Circuit for Example 4-7.

a   For $C = \frac{1}{9}F$, the values of s from Eq. (4-25) are

$$s = -\frac{1}{2 \times 2(\frac{1}{9})}\left(1 \pm \sqrt{1 - \frac{4 \times 2^2 \times \frac{1}{9}}{2}}\right)$$

or

$$s_1 = -1.5 \quad \text{and} \quad s_2 = -3$$

The natural response is, from Eq. (4-26),

$$v_n(t) = K_1\epsilon^{-1.5t} + K_2\epsilon^{-3t} \quad \text{V}$$

As the effective excitation in the second-order differential equation is zero [after Eq. (4-23)], the forced response is zero. Physically, this follows from the fact that the inductor is a short-circuit to a dc excitation and, consequently, the voltage across it is zero. The parallel nature of the circuit also makes $v(t) = 0$. The total response is then just the natural response.

To determine K_1 and K_2, initial conditions are used. As no energy is stored before the switch is closed, $i_L(0^-)$ and $v(0^-)$ are both zero. Continuity of inductance current and capacitance voltage require that

$$i_L(0^+) = 0 \quad \text{and} \quad v(0^+) = 0$$

At $t = 0^+$, the KCL equation requires that

$$i_R(0^+) + i_L(0^+) + i_C(0^+) = 10$$

The resistance current is zero because the voltage across it is zero. Therefore,

$$0 + 0 + i_C(0^+) = 10$$

and $i_C(0^+) = 10$ A.
From $i_C = C\dfrac{dv}{dt}$,

$$\frac{dv}{dt}(0^+) = \frac{i_C(0^+)}{C} = \frac{10}{\frac{1}{9}} = 90 \text{ V/s}$$

Two equations involving K_1 and K_2 result from the use of the values of $v(0^+)$ and $\dfrac{dv}{dt}(0^+)$.

$$v(0^+) = 0 = K_1 + K_2$$

and, from $\dfrac{dv}{dt} = s_1 K_1 \epsilon^{s_1 t} + s_2 K_2 \epsilon^{s_2 t}$,

$$\frac{dv}{dt}(0^+) = 90 = -1.5K_1 - 3K_2$$

Simultaneous solution gives

$$K_1 = 60 \quad K_2 = -60$$

which makes

$$v(t) = v_n(t) = 60\epsilon^{-1.5t} - 60\epsilon^{-3t}$$

b   For $C = \frac{1}{4}F$, the values of s are

$$s = -\frac{1}{2 \times 2(\frac{1}{4})} \left(1 \pm \sqrt{1 - \frac{4 \times 2^2 \times \frac{1}{4}}{2}} \right)$$

or $s_1 = -1 + j1$ and $s_2 = -1 - j1$.
The response becomes

$$v(t) = K_1\epsilon^{-t}\epsilon^{+jt} + K_2\epsilon^{-t}\epsilon^{-jt} \quad \text{V}$$

which can be rewritten as

$$v(t) = \epsilon^{-t}[K_1\epsilon^{jt} + K_2\epsilon^{-jt}] \quad \text{V}$$

Expressing ϵ^{jt} and ϵ^{-jt} in terms of Euler's theorem (see Appendix B or Sec. 3-4), and by use of Eq. (3-30), $v(t)$ has the form

$$v(t) = K\epsilon^{-t} \cos(t - \theta)$$

where K and θ are the constants to be evaluated.
    From part a, $v_C(0^+) = 0$ and $i_C(0^+) = 10$ A. As a result

$$\frac{dv}{dt}(0^+) = \frac{10}{\frac{1}{4}} = 40 \text{ V/s}$$

Substitution into $v(t)$ and $\dfrac{dv}{dt}$ yields the following simultaneous equations:

$$v(0^+) = 0 = K \cos(-\theta) = K \cos \theta \quad \text{①}$$

$$\frac{dv}{dt} = -K\epsilon^{-t} \cos(t - \theta) - K\epsilon^{-t} \sin(t - \theta)$$

$$\frac{dv}{dt}(0^+) = 40 = -K \cos(-\theta) - K \sin(-\theta)$$

or

$$40 = -K \cos \theta + K \sin \theta \quad \text{②}$$

From ① , $\cos \theta = 0$ so that

$$\theta = \frac{\pi}{2}$$

and, by use of ② with $\theta = \dfrac{\pi}{2}$,

$$40 = K$$

and $K = 40$.
The solution is then

$$v(t) = 40\epsilon^{-t} \cos\left(t - \frac{\pi}{2}\right) = 40\epsilon^{-t} \sin t \quad \text{V}$$

c When $C = \frac{1}{8}F$, the values of s_1 and s_2 are

$$s = -\frac{1}{2 \times 2(\frac{1}{8})} \left(1 \pm \sqrt{1 - \frac{4 \times 2^2 \times \frac{1}{8}}{2}}\right)$$

and $s_1 = s_2 = -2$.

A special case exists when the values of s_1 and s_2 are equal. The form of the response for this case, stated without proof, is

$$v(t) = K_1 t \epsilon^{-2t} + K_2 \epsilon^{-2t}$$

From a, $v(0^+) = 0$ and $i_C(0^+) = 10$ A, so that

$$\frac{dv}{dt}(0^+) = \frac{10}{\frac{1}{8}} = 80 \text{ V/s}$$

$$v(0^+) = 0 = K_2$$

Since $K_2 = 0$,

$$\frac{dv}{dt} = -2K_1 t \epsilon^{-2t} + K_1 \epsilon^{-2t} \quad \text{V/s}$$

and, at $t = 0^+$

$$80 = 0 + K_1$$

or

$$K_1 = 80$$

The total response is then

$$v(t) = 80t\epsilon^{-2t} \quad \text{V}$$

The forms of the responses given in Example 4-7b and c are the general forms for the cases involving a pair of complex-conjugate or equal roots. Thus, for complex roots $s_1 = -\alpha + j\beta$ and $s_2 = -\alpha - j\beta$, it is convenient to express the form of the natural response as

$$v_n(t) = K\epsilon^{-\alpha t} \cos(\beta t - \theta) \quad \text{V} \tag{4-27}$$

Similarly, for $s_1 = s_2 = -\alpha$, the natural response becomes

$$v_n(t) = K_1 t\epsilon^{-\alpha t} + K_2 \epsilon^{-\alpha t} \quad \text{V} \tag{4-28}$$

Example 4-8

The switch S in the series RLC circuit of Fig. 4-13 is closed at $t = 0$. Find the complete response for $v_c(t)$ for capacitance values of

a $\frac{25}{9} F$ **b** $0.5F$ **c** $1.0F$

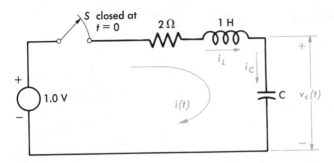

FIGURE 4-13 Circuit for Example 4-8.

SOLUTION When the switch is closed, the KVL equation is

$$2i + 1 \times \frac{di}{dt} + \frac{1}{C} \int i \, dt = 1.0$$

This equation can be rewritten in terms of the capacitor voltage v_c by recognizing that

$$v_c = \frac{1}{C} \int i \, dt$$

and, as this is a series circuit,

$$i(t) = i_c(t) = C \frac{dv_c}{dt}$$

The differential equation, in terms of v_c, becomes

$$2C \frac{dv_c}{dt} + C \frac{d^2 v_c}{dt^2} + v_c = 1.0$$

The forced response v_{cf} is of the form of the excitation and is a constant V_o. Substitution of V_o for v_{cf} gives

$$0 + 0 + V_o = 1.0$$

and

$$V_o = 1.0 = v_{cf}$$

The natural response is obtained from the force-free equation which is

$$2C \frac{dv_{cn}}{dt} + C \frac{d^2 v_{cn}}{dt^2} + v_{cn} = 0$$

Assuming v_{cn} is of the form $K\epsilon^{st}$, the values of s are determined from

$$K\epsilon^{st}[2Cs + Cs^2 + 1] = 0$$

from which

$$Cs^2 + 2Cs + 1 = 0$$

and

$$s = \frac{-2C \pm \sqrt{(2C)^2 - 4C}}{2C} = -1 \left(1 \pm \sqrt{1 - \frac{1}{C}} \right)$$

a   For $C = \frac{25}{9} F$, the values of s_1 and s_2 are

$$s_1 = -0.2 \qquad \text{and} \qquad s_2 = -1.8$$

The natural response is

$$v_{cn}(t) = K_1 \epsilon^{-0.2t} + K_2 \epsilon^{-1.8t} \qquad \text{V}$$

and the total response is

$$v_c(t) = 1.0 + K_1 \epsilon^{-0.2t} + K_2 \epsilon^{-1.8t} \qquad \text{V}$$

The initial conditions needed to evaluate K_1 and K_2 are based on

$$v_c(0^-) = 0 \qquad i_L(0^-) = i(0^-) = 0$$

By the continuity principle,

$$v_c(0^+) = 0 \qquad \text{and} \qquad i_L(0^+) = 0$$

As $i_L(0^+) = i_c(0^+)$,

$$\frac{dv_c}{dt}(0^+) = \frac{i_c(0^+)}{C} = 0$$

Forming $\dfrac{dv_c}{dt}$ as

$$\frac{dv_c}{dt} = -0.2K_1 \epsilon^{-0.2t} - 1.8K_2 \epsilon^{-1.8t} \qquad \text{V/s}$$

and evaluating $\dfrac{dv_c}{dt}(0^+)$ and $v_c(0^+)$ results in

$$v_c(0^+) = 0 = 1 + K_1 + K_2$$

$$\frac{dv_c}{dt}(0^+) = 0 = -0.2K_1 - 1.8K_2$$

Simultaneous solution gives

$$K_1 = -1.125 \qquad \text{and} \qquad K_2 = 0.125$$

The total solution is

$$v_c(t) = 1.0 - 1.125\epsilon^{-0.2t} + 0.125\epsilon^{-1.8t} \qquad \text{V}$$

b   For $C = 0.5F$, the values of s obtained from the quadratic equation are

$$s_1 = -1 + j1 \qquad \text{and} \qquad s_2 = -1 - j1$$

The form of the natural response is given in Eq. (4-27) and is

$$v_{cn}(t) = K\epsilon^{-t} \cos (t - \theta) \qquad \text{V}$$

The total solution becomes

$$v_c(t) = 1.0 + K\epsilon^{-t} \cos(t - \theta) \quad \text{V}$$

The initial conditions, from a, are $v_c(0^+) = 0$ and $\dfrac{dv_c}{dt}(0^+) = 0$. Forming these functions and evaluating at $t = 0^+$ yields

$$v_c(0^+) = 0 = 1 + K \cos \theta$$

$$\frac{dv_c}{dt}(0^+) = 0 = -K \cos \theta + K \sin \theta$$

Simultaneous solution results in

$$K = -\sqrt{2} \quad \text{and} \quad \theta = \frac{\pi}{4}$$

Substitution of these values into the total response gives

$$v_c(t) = 1.0 - \sqrt{2}\epsilon^{-t} \cos\left(t - \frac{\pi}{4}\right) \quad \text{V}$$

c   For $C = 1.0F$, the values of s_1 and s_2 are both equal to -1. The natural response, whose form is given in Eq. (4-28), is

$$v_{cn}(t) = K_1 t\epsilon^{-t} + K_2 \epsilon^{-t} \quad \text{V}$$

The complete response is

$$v_c(t) = 1.0 + K_1 t\epsilon^{-t} + K_2 \epsilon^{-t} \quad \text{V}$$

Evaluating both v_c and $\dfrac{dv_c}{dt}$ at $t = 0^+$, and using their values obtained in a results in

$$v_c(0^+) = 0 = 1 + K_2$$

$$\frac{dv_c}{dt}(0^+) = 0 = K_1 - K_2$$

These equations yield

$$K_1 = K_2 = -1$$

from which

$$v_c(t) = 1 - t\epsilon^{-t} - \epsilon^{-t} = 1 - \epsilon^{-t}(t + 1) \quad \text{V}$$

The responses obtained for the three cases in Example 4-8 are graphed in Fig. 4-14. These cases are often referred to as overdamped (a), underdamped (b), and critically damped (c). A system that is overdamped responds slowly to any change in excitation. As seen in Fig. 4-14, the response given by curve a has not reached its steady-state value (1.0 V) as have the responses indicated by curves b and c. The underdamped system in

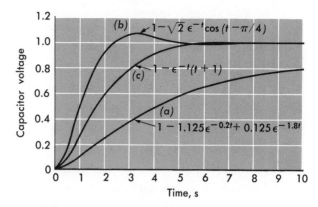

FIGURE 4-14 Responses obtained for circuit of Fig. 4-13 in Example 4-8: (a) overdamped, (b) underdamped, and (c) critically damped cases.

b responds most quickly of the three. However, the speed of response is accompanied by overshoot; that is, the response exceeds and then oscillates about 1.0 V as it gradually approaches the steady-state value. The critically damped system's response, curve c, represents the speediest response possible without any overshoot.

Many practical systems are designed to give slightly underdamped responses. In such systems, the improvement in the speed of response is sufficiently important so as to outweigh the disadvantages of the overshoot that results. However, while further underdamping results in increased speed, the amount of overshoot also increases markedly. Consequently, in system design, the trade-off between advantages and disadvantages usually restricts the overshoot to be less than 10 percent, as is the case in Fig. 4-14.

4-6 THE RESPONSE OF MORE GENERAL NETWORKS

In this section, networks containing several energy-storage elements are examined. The response to suddenly applied or changed excitations of such networks consists of both natural and forced components. Each component response serves the same function as it does in the simpler systems discussed previously. The differential equation which describes the network is obtained by application of Kirchhoff's laws and the element volt-ampere relations. The general form of this equation is

$$a_n \frac{d^n x}{dt^n} + a_{n-1} \frac{d^{n-1} x}{dt^{n-1}} + \cdots + a_1 \frac{dx}{dt} + a_o x = f(t) \qquad (4\text{-}29)$$

where $f(t)$ = excitation
$x(t)$ = response

a coefficients = functions of the circuit parameters (R, L, C and the controlled sources)

The forced response $x_f(t)$ has the form of $f(t)$ and all unique derivatives of $f(t)$. To obtain $x_f(t)$, one substitutes the assumed response into Eq. (4-29) and adjusts the response amplitudes to effect a balance between the left- and right-hand sides of Eq. (4-29).

By setting $f(t) = 0$ in Eq. (4-29), as given in Eq. (4-30), the force-free equation is obtained, from which the natural response $x_n(t)$ is determined.

$$a_n \frac{d^n x_n}{dt^n} + a_{n-1} \frac{d^{n-1} x_n}{dt^{n-1}} + \cdots + a_1 \frac{dx_n}{dt} + a_o x_n = 0 \qquad (4\text{-}30)$$

In previous sections the assumed exponential solution for $x_n(t)$ led to an algebraic polynomial in s from which the values of s are computed. Equations (4-8) and (4-24) are examples of such equations, called the *characteristic equation* of the network. The characteristic equation is obtained by substitution of the general exponential $x_n = K\epsilon^{st}$ into Eq. (4-30) and gives

$$K\epsilon^{st}(a_n s^n + a_{n-1} s^{n-1} + \cdots + a_1 s + a_o) = 0$$

so that

$$a_n s^n + a_{n-1} s^{n-1} + \cdots + a_1 s + a_o = 0 \qquad (4\text{-}31)$$

Equation (4-31) is the characteristic equation for a network whose force-free equation is Eq. (4-30). It can be found directly by replacing the nth derivative of the dependent variable in the force-free equation by s^n. Thus Eq. (4-31) can be written from Eq. (4-30) without the necessity of the intervening substitution of $K\epsilon^{st}$ in Eq. (4-30).

The roots of the characteristic equation are the powers to which the various exponentials in the natural response are raised. Since the roots of an algebraic equation with real coefficients are either real or occur in conjugate pairs (that is, $a \pm jb$), the exponential powers have these forms. Also note that the general characteristic equation of the form of Eq. (4-31) has n roots. This means that the natural response has n exponentials. If the roots of Eq. (4-31) are expressed as $p_1, p_2, \ldots, p_n$, the natural response is

$$x_n(t) = K_1 \epsilon^{p_1 t} + K_2 \epsilon^{p_2 t} + \cdots + K_n \epsilon^{p_n t} \qquad (4\text{-}32)$$

When the characteristic equation has equal roots or complex roots, these pairs are most conveniently expressed by the forms given in Eqs. (4-27) and (4-28).

The total response is

$$x(t) = x_f(t) + x_n(t) = x_f(t) + K_1 \epsilon^{p_1 t} + K_2 \epsilon^{p_2 t} + \cdots + K_n \epsilon^{p_n t} \qquad (4\text{-}33)$$

The constants $K_1, K_2, \ldots, K_n$ are all evaluated from initial or boundary conditions within the network. In the usual case, these are the values of $x(t)$ and $(n - 1)$ of its derivatives at $t = 0$. As only the volt-ampere relations for energy-storage elements involve derivatives and integrals, an nth-order system contains no fewer than n inductances and capacitances. The principles of continuity of inductance current and capacitance voltage are used to develop the necessary boundary conditions.

For high-order systems, the classical method of solution described in this section can become numerically cumbersome. To obtain the n roots of the characteristic equation, computer programs are readily available. The evaluation of the constants $K_1, K_2, \ldots, K_n$ involves solving n simultaneous equations in addition to developing the values of $n - 1$ derivatives. The use of LaPlace transform techniques described in Chap. 5 often provides a more efficient means for obtaining the required solutions. In addition, computer methods, particularly state-variable techniques, are available to solve high-order differential equations.

4-7 STEP FUNCTIONS

In the previous sections of this chapter, the response of systems to suddenly applied or changed excitations is considered. The opening or closing of a switch is used to represent these changes in excitation. For example, the switch S in Fig. 4-13 is closed to indicate the application of the 1.0-V source to the *RLC* series circuit. In similar fashion the square-wave voltage in Fig. 3-32 can be considered as a constant excitation which is periodically switched on and off. The circuit of Fig. 4-15 can be used to represent this process. It is assumed that the switch S is opened and closed externally, i.e., by means independent of the circuit elements. Implicit in this form of representation is that the direct voltage source V and the switching operation exist for all time $(-\infty < t < +\infty)$. The output waveform $v(t)$, therefore, also exists for all time. The concept of switching a source on and off is useful, particularly if the restrictions of source existence for all time and the periodicity of switching are eliminated. In most systems, the excitation is applied at a specific time, usually considered at $t = 0$ for convenience, and knowledge of the response is desired from the time of application $(t \geq 0)$. Practically, a specific excitation lasts for a finite amount of time which is from the time it is applied until the time it is shut down (switched off) or replaced by a new excitation.

What is desired is a technique incorporating the idea of switching on an excitation and its use in representing sources having finite duration, and which allows the response to be determined by standard methods. The

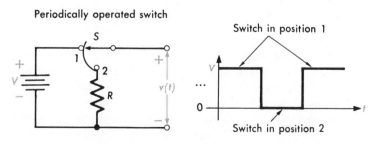

FIGURE 4-15 Square-wave generation by periodic switching.

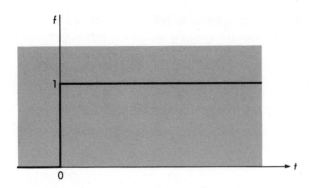

FIGURE 4-16 Unit-step function.

unit-step function or *unit step,* shown in Fig. 4-16, is the basic excitation used to this end. It is a function which is zero for $t < 0$ and unity for $t > 0$, switching from zero to unity at $t = 0$. The mathematical symbol for the unit-step function is $u(t)$; thus

$$u(t) = \begin{cases} 0 & t < 0 \\ 1 & t > 0 \end{cases} \tag{4-34}$$

The physical significance of the unit step is the turning on of something at $t = 0$ which had been previously zero.

The notation of Eq. (4-34) can be extended to indicate other excitations which are turned on at $t = 0$. Three such functions are

$$v_1(t) = 10u(t) \quad V \tag{4-35}$$

$$v_2(t) = (115\sqrt{2} \cos 377t)u(t) \quad V \tag{4-36}$$

$$v_3(t) = 20\epsilon^{-t}u(t) \quad V \tag{4-37}$$

The waveforms of the sources expressed in Eqs. (4-35) to (4-37) are shown in Fig. 4-17. As indicated in Fig. 4-17a, $v_1(t)$ is a step function whose amplitude is 10 V. The expression in Eq. (6-36) is used to represent a constant voltage applied at $t = 0$. Similarly, the waveforms in Fig. 4-17b and c are a sinusoid and an exponential, respectively, which are turned on at $t = 0$. In general, the notation

$$v(t) = f(t)u(t) \tag{4-38}$$

is used to represent a source that is applied at $t = 0$.

The reference time $t = 0$ is arbitrary and generally selected for convenience. For many applications, excitations are applied at times other than $t = 0$. For example, the reference time in a radar system is the time a pulse is transmitted. The received pulse occurs at some later time T, the time difference being proportional to the distance the observed object is from the transmitter. The receiver which processes this pulse for the purpose of display is therefore subject to an excitation at $t = T$ rather than $t = 0$. A second

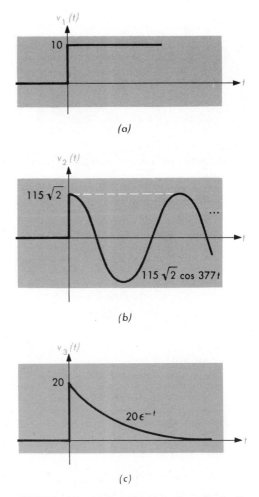

FIGURE 4-17 (a) Step function of amplitude 10; (b) $(115\sqrt{2} \cos 377t) u(t)$; (c) $20\epsilon^{-t}u(t)$.

example is that of an automated factory in which a sequence of operations is performed. The reference time is taken as the start of the sequence. The electric signals, used to control and actuate the automatic equipment, are applied at different times for each operation in the sequence. To indicate the application of an excitation at time $t = T$, it is useful to introduce the *delayed unit step* $u(t - T)$, depicted in Fig. 4-18 and expressed in Eq. (4-39):

$$u(t - T) = \begin{cases} 1 & t > T \\ 0 & t < T \end{cases} \qquad (4\text{-}39)$$

When the unit step or delayed step is applied to a circuit, the response is determined by use of the methods developed earlier in this chapter. This response depends on the particular excitation, the circuit elements and their

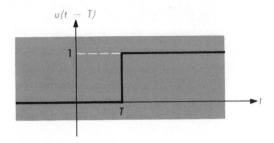

FIGURE 4-18 Delayed unit-step function.

interconnection, and the initial state of the circuit. All three factors are important as the step function causes the system response to change from its initial state. The nature of the step function, when it is the only excitation, is to ensure that no energy is stored in the circuit prior to its application. For $t < 0$, the step function has zero value so that a voltage source appears as a short circuit and a current source as an open circuit. With no effective excitation applied to the circuit for $t < 0$, there can be no initial voltage across a capacitor and no initial current through an inductor at $t = 0$.

Example 4-9

Determine the voltage $v_o(t)$ in the circuit of Fig. 4-19 when

   a $v(t) = 0.25u(t)$

   b $v(t) = 0.25u(t - 1)$

SOLUTION The first step in the solution is to obtain the differential equation which relates the response variable to the excitation and circuit elements. The KCL expression at node A is

$$-6v_1 + \frac{1}{10}v_o(t) + \frac{1}{20}\frac{dv_o(t)}{dt} = 0$$

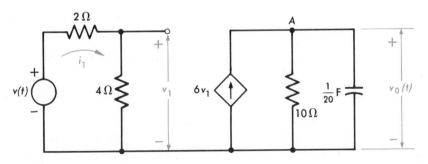

FIGURE 4-19 Circuit for Example 4-9.

The constraint equation for the controlled source is obtained from the KVL expression for the loop indicated by the current i_1 and is

$$v(t) = 6i_1 \qquad \text{and} \qquad v_1 = 4i_1$$

so that

$$v_1 = \tfrac{2}{3}v(t)$$

Substitution of the constraint equation into the KCL expression results in the differential equation for the circuit, which is

$$\frac{1}{10}\,v_o(t) + \frac{1}{20}\frac{dv_o(t)}{dt} = 4v(t)$$

a For $v(t) = 0.25u(t)$, there is no excitation for $t < 0$ and consequently no initial voltage on the capacitor. For $t > 0$, $v(t) = 0.25$, the constant excitation applied to the circuit, and the differential equation becomes

$$\frac{1}{10}\,v_o(t) + \frac{1}{20}\frac{dv_o(t)}{dt} = 1.0$$

The forced component of $v_o(t)$ is a constant V_o which is

$$\tfrac{1}{10}V_o + 0 = 1 \qquad \text{or} \qquad V_o = v_{of}(t) = 10$$

The characteristic equation is

$$\frac{1}{10} + \frac{s}{20} = 0 \qquad \text{or} \qquad s = -2$$

from which the natural response $v_{on}(t)$ is

$$v_{on}(t) = K\epsilon^{-2t} \qquad \text{V}$$

The total response is the sum of $v_{on}(t)$ and $v_{of}(t)$ and is

$$v_o(t) = 10 + K\epsilon^{-2t}$$

With no initial energy stored and the principle of capacitance-voltage continuity, $v_o(0^+) = 0$. Setting $t = 0$ and substituting give

$$0 = 10 + K \qquad \text{or} \qquad K = -10$$

and

$$v_0(t) = 10(1 - \epsilon^{-2t})u(t) \qquad \text{V}$$

The use of $u(t)$ in the expression for $v_o(t)$ follows the notation of Eqs. (4-35) to (4-37), indicating that this is the response for $t > 0$ and that the response for $t < 0$ is zero.

b When $v(t) = 0.25u(t - 1)$, no excitation is applied prior to $t = +1$. As all circuit conditions are the same as in part a, except that the excitation is applied at $t = 1$ rather than $t = 0$, the response must be the same except that it begins at $t = 1$. Thus,

$$v_0(t) = 10[1 - \epsilon^{-2(t-1)}]u(t - 1) \qquad \text{V}$$

Justification for the result shown is simply that the selection of the reference time, $t = 0$, is arbitrary. In both parts a and b the same excitation is applied to the same circuit having the same initial state. Therefore, the response from the time the excitation is applied must be identical in both cases. The waveforms in Fig. 4-20 show identical responses delayed in time.

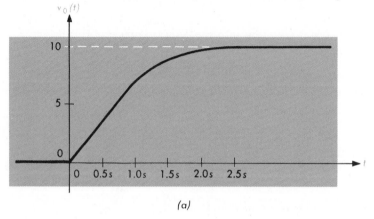

(a)

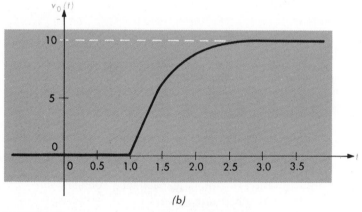

(b)

FIGURE 4-20 Response waveforms for (a) step input and (b) delayed-step input.

4-8 PULSES AS EXCITATIONS

A widely used excitation in control, communication, and computer systems is the rectangular pulse whose waveform is illustrated in Fig. 4-21. It is used to describe a constant excitation that is turned on at $t = 0$ and switched off T seconds later. The pulse duration is the time the excitation is on and is indicated by the interval $0 < t < T$ in Fig. 4-21. In most electric systems, pulses represent excitations that are applied for short periods of time. Typical pulse durations are as long as a few tenths of one second, encountered in some control systems, and are as short as a few tenths of a nanosecond (10^{-9} s) encountered in high-speed digital computers.

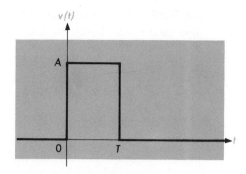

FIGURE 4-21 Rectangular pulse.

The pulse can be expressed mathematically as

$$v(t) = \begin{cases} 0 & -\infty < t < 0 \\ A & 0 < t < T \\ 0 & T < t < +\infty \end{cases} \qquad (4\text{-}40)$$

A convenient way of describing a pulse, and one which is useful in circuit analysis, is by the sum of two step functions. Figure 4-22 shows the development of the pulse of Fig. 4-21 as the sum of step functions. The first

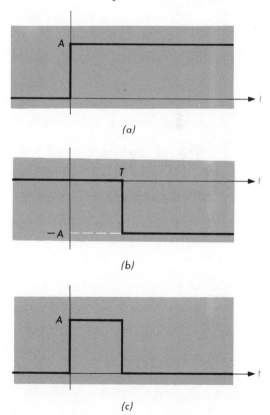

FIGURE 4-22 Representation of rectangular pulse by step and delayed-step responses.

step function, $Au(t)$ shown in Fig. 4-22a, is the value of the pulse in the time interval $0 < t < T$. By the addition of the second step, $-Au(t - T)$, the effect of the first step is canceled for $t > T$. The negative step is shown in Fig. 4-22b; Fig. 4-22c represents the sum of the waveforms in Fig. 4-22a and b which is the pulse to be represented. By use of this development, the pulse defined in Eq. (4-40) can be expressed as

$$v(t) = Au(t) - Au(t - T) \tag{4-41}$$

To determine the response of a circuit to a pulse excitation, each component in Eq. (4-41) is considered as a separate source. By use of the principle of superposition, the response to each component is determined and the circuit response obtained by summing the component responses. The method is described in the following example.

Example 4-10

The circuit of Fig. 4-19 and Example 4-9 is excited by the pulse shown in Fig. 4-23. Determine the voltage $v_o(t)$ and plot the response.

SOLUTION The steps required for solution are

1 Determine the differential equation for the circuit.

2 Express the pulse as the sum of two step functions.

3 Obtain the response to each step function.

4 Add the results in c to form the total response.

The differential equation is determined in Example 4-9 as

$$\frac{1}{10} v_o(t) + \frac{1}{20} \frac{dv_o(t)}{dt} = 4v(t)$$

By use of Eq. (4-41), the pulse in Fig. 4-23 is

$$v(t) = 0.25u(t) - 0.25u(t - 1) \qquad \text{V}$$

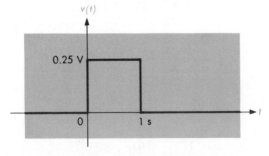

FIGURE 4-23 Pulse excitation for Example 4-10.

The response component to the first step $v_{o1}(t)$, given in Example 4-9, part a, is

$$v_{o1}(t) = 10(1 - \epsilon^{-2t})u(t) \quad V$$

The delayed step is the negative of the delayed step in Example 4-9. In a linear system for which an excitation $x(t)$ produces a response $y(t)$, application of $Kx(t)$, where K is a constant, results in a response $Ky(t)$. The response $v_{o2}(t)$ produced by the delayed step is therefore the negative of that given in Example 4-9, part b, and is

$$v_{o2}(t) = -10[1 - \epsilon^{-2(t-1)}]u(t - 1) \quad V$$

Note that no initial energy storage need be considered in determining $v_{o2}(t)$. When superposition is used, each excitation function produces one component of the response; while it is being considered, all other excitations are considered zero. Thus, when the delayed-step input is applied, the first step function is considered to be zero (nonexistent) and does not contribute any initial energy storage. For $t < 1$, the delayed step has zero value, also causing the initial energy storage to be zero.

The total response, the sum of $v_{o1}(t)$ and $v_{o2}(t)$, is

$$v_0(t) = 10(1 - \epsilon^{-2t})u(t) - 10[1 - \epsilon^{-2(t-1)}]u(t - 1) \quad V$$

and is depicted in Fig. 4-24b. The response during the interval $0 < t < 1$ is attributed only to $v_{o1}(t)$ (Fig. 4-24a), as the delayed step has yet to be applied. For $t > 1$, the response is caused by both step functions. However, as the value of t increases, the component $v_{o1}(t)$ approaches a constant value of 10 V, so that the time dependence for large values of t is most nearly caused by $v_{o2}(t)$ shown in Fig. 4-24a. The physical nature of the circuit also gives support to the output waveform. At $t = 0$, the pulse is applied and the capacitance begins to charge exponentially. The excitation goes to zero at $t = 1$, and the partially charged capacitor discharges through the resistance, returning the circuit to its relaxed state.

When an excitation is composed of the sum of step and delayed step functions, superposition is used to obtain the response. It is the inherent nature of the superposition principle and the step function which permits the initial energy storage to be considered zero in determining each component of the response. Physically, however, when a delayed step is applied, there is energy stored in the circuit caused by the first function applied. The results obtained in Example 4-10 can be used to justify that the condition of capacitance-voltage continuity is satisfied.

The response $v_o(t)$ shown in Fig. 4-24b is the voltage across the capacitor in the circuit of Fig. 4-19. For $t < 0$, no excitation is applied and $v_o(t)$ is zero. For $0 < t < 1$, the response is just $v_{o1}(t)$. The value of $v_{o1}(0^+)$ is zero, so that capacitance-voltage continuity is achieved. Just prior to the application of the delayed step, $t = 1^-$, the voltage across the capacitor is $v_{o1}(1^-) = 8.65$

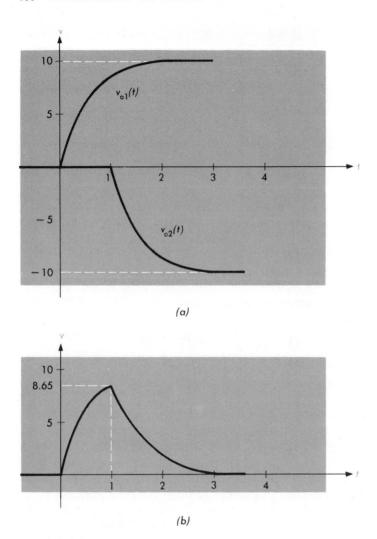

FIGURE 4-24 Response waveforms for Example 4-10: (a) step and delayed-step responses; (b) total response.

V. At $t = 1^+$, after the delayed step is applied, $v_{o1}(1^+) = 8.65$ V as $v_{o1}(t)$ is continuous. The values of $v_{o2}(1^-)$ and $v_{o2}(1^+)$ are both zero for the same reason that $v_{o1}(0^-)$ and $v_{o1}(0^+)$ are zero (see Fig. 4-24b). Thus, each component of the response displays capacitance-voltage continuity so that the total response must also satisfy the continuity property. By a similar analysis, it is readily shown that this method also satisfies inductance-current continuity.

The method of representing a pulse by two step functions can be extended to a group of pulses, or *pulse train*. Figure 4-25 is a pulse train composed of three pulses. Each pulse can be expressed by a step and delayed step as given in Eq. (4-41); the resultant expression for the pulse train is sim-

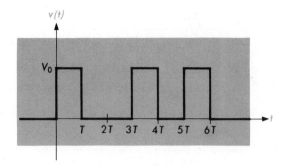

FIGURE 4-25 Pulse train.

ply the sum of the individual pulses. For the waveform in Fig. 4-25 this becomes

$$v(t) = V_o u(t) - V_o u(t - T) + V_o u(t - 3T) - V_o u(t - 4T)$$
$$+ V_o u(t - 5T) - V_o u(t - 6T) \quad (4\text{-}42)$$

The response of a circuit to the pulse train is determined by the technique used in Example 4-10. However, six individual response components must be determined, two for each pulse in the train.

The decomposition of pulse-type waveforms into a number of step functions is commonly used. The major advantage lies in the fact that a circuit has the same response to both a step function and a delayed step, the only difference being the specific time the response begins.

4-9 THE IMPULSE FUNCTION

Short pulses are used as the trigger signals which start a sequence of operations in a digital computer, form part of the sound and picture synchronization signal in motion pictures and television, and are used to initiate timing sequences in digital control systems. A short pulse is one whose duration is small compared with the time constants of the network to which it is applied. In Sec. 4-2, the time constants are shown to be the reciprocals of the values of s which satisfy the characteristic equation of the network and determine the exponents in the expression for the natural response.

To illustrate the effect a short pulse has on circuit response, it is convenient to use the circuit whose pulse response is obtained in Example 4-10. The circuit is redrawn in Fig. 4-26, and the total response given by

$$v_o(t) = 10(1 - \epsilon^{-2t})u(t) + 10[1 - \epsilon^{-2(t-1)}]u(t - 1) \quad (4\text{-}43)$$

The circuit excitation is a pulse expressed as $0.25u(t) - 0.25u(t - 1)$, and the time constant of the circuit $\tau = \frac{1}{2}$ s. If the duration of the pulse is shortened, only the delayed step portion of the excitation is changed to indicate a smaller delay. Similarly, the response $v_o(t)$ is altered only in the delayed term, again indicating a pulse of shorter duration. Increasing the amplitude

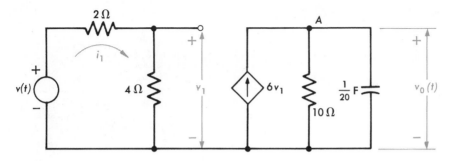

FIGURE 4-26   Circuit to which short pulse is applied.

of the excitation pulse by a factor of K causes an increase in the amplitude of the response by a factor of K.

With the background provided in the preceding paragraph, consider the following pulse excitations applied to the circuit of Fig. 4-26:

(a) $v(t) = 0.25u(t) - 0.25u(t - 0.025)$

(b) $v(t) = 0.50u(t) - 0.50u(t - 0.0125)$

(c) $v(t) = 1.00u(t) - 1.00u(t - 0.00625)$ (4-44)

Each of the pulses shown in Fig. 4-27 has a duration much shorter than the circuit time constant. The respective responses, based on Eq. (4-41), become:

(a) $v_o(t) = 10(1 - \epsilon^{-2t})u(t) + 10[1 - \epsilon^{-2(t-0.025)}]u(t - 0.025)$

(b) $v_o(t) = 20(1 - \epsilon^{-2t})u(t) + 20[1 - \epsilon^{-2(t-0.0125)}]u(t - 0.0125)$

(c) $v_o(t) = 40(1 - \epsilon^{-2t})u(t) + 40[1 - \epsilon^{-2(t-0.00625)}]u(t - 0.00625)$ (4-45)

These responses are shown in Fig. 4-28a, b, and c, respectively, where the time scale is selected to display five time constants, essentially showing the complete discharge of the capacitor. In each case, when the pulse is applied the capacitor begins to charge. However, since the pulse duration is considerably shorter than the time constant, the excitation becomes zero before the capacitor can fully charge. The capacitor voltage then exponentially discharges to zero. On the time scale used, each response appears to be that obtained for the simple discharge of a capacitor with an initial voltage across it (see Sec. 4-2). When the responses are plotted on the same curve sheet, as shown in Fig. 4-29, they are virtually indistinguishable. This comes about because as the pulse duration is decreased, the pulse amplitude is increased by the same percentage so that the area under the pulse remains constant.

In the circuit of Fig. 4-26, the controlled current source provides the excitation to the capacitor. As the amplitude of this source is directly proportional to $v(t)$ its waveform is the same as that for $v(t)$. The area under the current pulse of the controlled source has the dimensions of amperes times seconds, or coulombs, which is the unit of charge. In effect, this current pulse

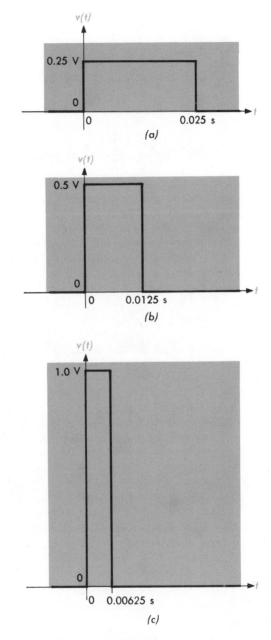

FIGURE 4-27 Pulse waveforms.

has the physical significance of transferring a charge to the capacitor in a very short time. This charge is related to the voltage across the capacitor at the instant the pulse is turned off by the equation $v = q/C$.

In Example 4-9, the voltage-controlled current source is seen to have a strength of $4v(t)$ amperes. Table 4-2 is useful in comparing the two values of

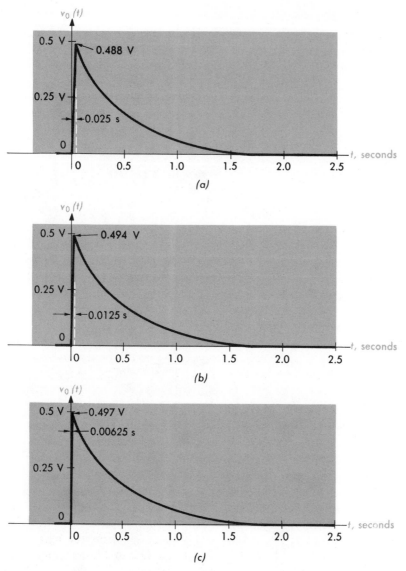

FIGURE 4-28 Response of circuit of Fig. 4-26 to each of the pulse excitations of Fig. 4-27.

capacitor voltage calculated at the instant the pulse is turned off, first by considering charge transfer, and second by means of Eq. (4-45). Comparison of the two calculations in Table 4-2 indicates that both methods result in essentially the same voltage. In addition, the capacitor discharges for all three pulses, as shown in Fig. 4-29, are virtually identical. Both of these factors demonstrate that the effect of the short pulse is to place an initial voltage across the capacitor and then to allow the capacitor to discharge through the

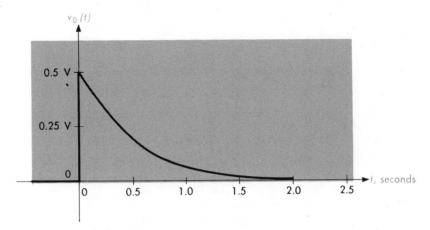

FIGURE 4-29 The responses displayed in Fig. 4-28 shown in relation to each other.

resistor. The system response then appears to be simply the discharge of the charged capacitor in a circuit having no external excitation. (This is true for all times after the pulse is turned off.) With no external excitation, the response is simply the natural response; the initial voltage is used to evaluate the necessary constants.

To represent the effects of pulses of short duration and the responses they produce, it is convenient to introduce the concept of the *unit-impulse function* $\delta(t)$. Mathematically, the unit impulse is defined as

$$\int_{-\infty}^{\infty} \delta(t) \, dt = 1$$

$$\delta(t) = 0 \qquad t \neq 0$$

(4-46)

The interpretation of Eq. (4-46) is that of a function which is zero everywhere except at $t = 0$ and which has an area of unity. To satisfy this condition, $\delta(t)$ becomes infinite at $t = 0$. The symbolic representation of $\delta(t)$ is shown in Fig. 4-30 in which the arrowhead indicates $\delta(t) \rightarrow \infty$ at $t = 0$. The meaning of the impulse in physical systems is that it represents the area under a short voltage or current pulse. Its use is in the fact that it generally reduces the computation required in obtaining the response.

TABLE 4-2 CAPACITOR VOLTAGE AT INSTANT OF SWITCHING

CURRENT OF CONTROLLED SOURCE 4 $v(t)$	PULSE DURATION, s	CAPACITOR VOLTAGE FROM $V = q/C$	CAPACITOR VOLTAGE FROM EQ. (4-45)
1.00	0.025	0.50	0.488
2.00	0.0125	0.50	0.494
4.00	0.00625	0.50	0.497

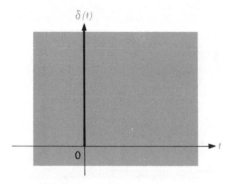

FIGURE 4-30 The unit impulse.

The impulse function can also be used to represent short pulses whose areas differ from unity and which occur at times other than $t = 0$. The interpretation of

$$v(t) = A\, \delta(t - T) \qquad (4\text{-}47)$$

is that $\delta(t - T)$ is zero everywhere except at $t = T$, so that the voltage impulse occurs at time T, and the area under $v(t)$ is A.

To determine the response of a circuit to an impulse, two steps are required. The first involves obtaining the form of the natural response by the methods developed in this chapter. The second step is to evaluate the initial capacitor voltages and inductor currents. This is accomplished by recognizing the fact that a current impulse applied to a capacitance provides an initial voltage of

$$v_C(0^+) = \frac{q}{C} \qquad (4\text{-}48)$$

where the value of q is the area under the impulse function. Similarly, a voltage impulse applied to an inductance provides an initial current whose value is

$$i_L(0^+) = \frac{\lambda}{L} \qquad (4\text{-}49)$$

where λ is the area of the voltage impulse.

The impulse function is used as both a theoretical and experimental tool in characterizing the properties of networks. As a theoretical tool it permits the characterization of networks in terms of the natural response of the system. This response is dependent only on the network elements and their interconnection. By the use of short pulses, impulse functions are approximated experimentally and permit laboratory determination of the natural response in a straightforward manner. In addition, techniques are available (see Chap. 5; others are beyond the scope of this book) which result in the ability to determine system response to nearly all excitation functions once the impulse response is known.

4-10 NETWORK FUNCTIONS AND THE TRANSFORMED NETWORK

The circuits discussed in the previous sections of this chapter require only one KVL or KCL equation for description. Application of Kirchhoff's laws to multinode and multimesh networks results in a system of simultaneous differential equations, as is illustrated by considering the circuit shown in Fig. 4-31. Identification of the voltage variables $v_1(t)$ and $v_2(t)$ in Fig. 4-31 gives rise to the following nodal equations:

$$\frac{v_1(t)}{R_1} + C\frac{d}{dt}[v_1(t) - v_2(t)] = i(t) \tag{4-50}$$

$$C\frac{d}{dt}[v_2(t) - v_1(t)] + \frac{1}{L}\int v_2(t)\,dt + \frac{v_2(t)}{R_2} = 0 \tag{4-51}$$

The simultaneous solution of the coupled differential equations for $v_2(t)$ involves (1) differentiating Eq. (4-50), (2) solving Eq. (4-51) for $[dv_1(t)]/dt$ and differentiating the result to obtain $[d^2v_1(t)]/dt^2$, and (3) substituting $[dv_1(t)]/dt$ and $[d^2v_1(t)]/dt^2$ into the differential form of Eq. (4-50). Performing these operations and combining terms gives

$$\left(\frac{1}{R_1} + \frac{1}{R_2}\right)\frac{d^2v_2(t)}{dt^2} + \left(\frac{1}{R_1R_2C} + \frac{1}{L}\right)\frac{dv_2(t)}{dt} + \frac{v_2(t)}{R_1LC} = \frac{d^2i(t)}{dt^2} \tag{4-52}$$

Clearing fractions yields

$$LC(R_1 + R_2)\frac{d^2v_2(t)}{dt^2} + (R_1R_2C + L)\frac{dv_2(t)}{dt} + R_2v_2(t)$$

$$= LCR_1R_2\frac{d^2i(t)}{dt^2} \tag{4-53}$$

The solution of the differential equation in Eq. (4-53) results in the response $v_2(t)$. In a similar fashion, the variable $v_2(t)$ can be eliminated from Eqs. (4-50) and (4-51) to give a differential equation involving only the variable

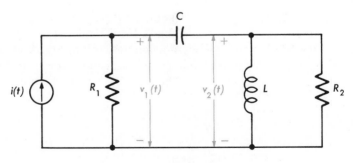

FIGURE 4-31   Multinode *RLC* network.

$v_1(t)$. The result is

$$LC(R_1 + R_2)\frac{d^2v_1(t)}{dt^2} + (R_1R_2C + L)\frac{dv_1(t)}{dt} + R_2\frac{dv_1(t)}{dt}$$

$$= LCR_1R_2\frac{d^2i(t)}{dt^2} + LR_1\frac{di(t)}{dt} + R_1R_2i(t) \quad (4\text{-}54)$$

As is evident in the analysis just concluded, a great deal of manipulation is involved in simply obtaining the differential equation to be solved. The procedure is facilitated by utilizing the transformed networks introduced in Secs. 3-2 and 3-3 and is related to the operational methods developed by Sir Oliver Heaviside at the turn of this century.

Justification of the use of transformed networks and the impedance concept is based on three factors first encountered in Secs. 4-2 and 4-6. These are:

1 The natural response is a sum of terms, each of which is $K\epsilon^{st}$. In Sec. 3-2 we observed that a voltage (current) of the form ϵ^{st} produced a current (voltage) of the form ϵ^{st} in each of the circuit elements.

2 The characteristic equation is obtained from the homogeneous differential equation by replacing d^n/dt^n with the algebraic variable s^n. Thus, differentiation corresponds to multiplication by s; considering integration by the inverse of differentiation, integration corresponds to division by s. Consequently, the volt-ampere relations for L and C can be expressed as indicated in Table 3-1b.

3 Letting $Z(s) = 0$ for a voltage-source excitation and letting $Y(s) = 0$ for a current-source excitation results in the characteristic equation (see Sec. 4-2). This observation is not an isolated case but arises from the fact that the natural response depends on the elements in the circuit and their interconnection. As the impedance and admittance of a network at a pair of terminals is one measure of the effect of network elements, it is reasonable to expect that the impedance and admittance are intimately related to the natural response.

The transformed network of the circuit in Fig. 4-31, obtained by use of the relations given in Table 3-1b, is depicted in Fig. 4-32. The nodal equations for the frequency-domain network in Fig. 4-32 are written by treating admittances in the same manner as conductances are treated in the resistive networks described in Chap. 2. The nodal equations are

$$V_1\left(\frac{1}{R_1} + sC\right) - sC\,V_2 = I \quad (4\text{-}55)$$

$$-sC\,V_1 + V_2\left(\frac{1}{R_2} + Cs + \frac{1}{Ls}\right) = 0 \quad (4\text{-}56)$$

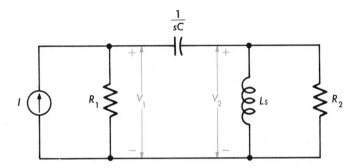

FIGURE 4-32 Transformed network of circuit in Fig. 4-31.

Solution of Eqs. (4-55) and (4-56) for V_1 and V_2 by Cramer's rule yields, after rearranging terms,

$$V_1 = \frac{LCR_1R_2s^2 + LR_1s + R_1R_2}{LC(R_1 + R_2)\,s^2 + (R_1R_2C + L)\,s + R_2}\,I \tag{4-57}$$

$$V_2 = \frac{LCR_1R_2s^2}{LC(R_1 + R_2)\,s^2 + (R_1R_2C + L)\,s + R_2}\,I \tag{4-58}$$

Equations (4-57) and (4-58) can be rewritten as

$$[LC(R_1 + R_2)\,s^2 + (R_1R_2C + L)\,s + R_2]\,V_1$$
$$= [LCR_1R_2s^2 + LR_1s + R_1R_2]\,I \tag{4-59}$$

$$[LC(R_1 + R_2)\,s^2 + (R_1R_2C + L)\,s + R_2]\,V_2 = LCR_1R_2\,s^2\,I \tag{4-60}$$

By recalling that s corresponds to d/dt, comparison of Eqs. (4-59) and (4-60) with Eqs. (4-53) and (4-54) indicate they are identical. Note that the function of s multiplying both V_1 and V_2 in Eqs. (4-59) and (4-60) is the same, it being the characteristic equation for the circuit. Thus, the form of the natural response is the same in all elements of the circuit. In Eqs. (4-59) and (4-60), the homogeneous equation is obtained by setting I to zero, the condition in the frequency domain corresponding to the source-free network. In addition, in Fig. 4-32, the admittance $Y(s)$ seen by the source I is

$$Y(s) = \frac{I}{V_1} = \frac{LC(R_1 + R_2)\,s^2 + (R_1R_2C + L)\,s + R_2}{LCR_1R_2\,s^2 + LR_1\,s + R_1R_2} \tag{4-61}$$

By making $Y(s) = 0$, the characteristic equation of the network is obtained.

The techniques used to develop Eqs. (4-55) to (4-61) are indicative of the utility of the transformed network in the solution of network problems. A principal benefit that accrues from frequency-domain analysis is that the resultant network equations are algebraic rather than the differential equations obtained in the time domain. Thus, the impedance concept in Table 3-1b leading to the transformed network, allows extension of the techniques developed in Chap. 2 to circuits which contain inductance and capacitance as well as resistance.

A second important benefit of transformed networks in the determination of circuit responses is the introduction of the concept of the *network,* or *system, function.* The network function $H(s)$ is defined as the ratio of the response variable to the excitation variable in the transformed network and is given by

$$H(s) = \frac{A(s)}{B(s)} = \frac{\text{transformed network response variable}}{\text{transformed network excitation variable}} \qquad (4\text{-}62)$$

The network functions for the response variables V_1 and V_2 given in Eqs. (4-57) and (4-58) are

$$H_1(s) = \frac{V_1}{I} = \frac{LCR_1R_2\ s^2 + LR_1\ s + R_1R_2}{LC(R_1 + R_2)\ s^2 + (R_1R_2C + L)\ s + R_2} \qquad (4\text{-}63)$$

$$H_2(s) = \frac{V_2}{I} = \frac{LCR_1R_2\ s^2}{LC(R_1 + R_2)\ s^2 + (R_1R_2C + L)\ s + R_2} \qquad (4\text{-}64)$$

The network function $H_1(s)$ in Eq. (4-63) is the ratio of V_1 to I, both of which are defined at the same terminal pair. Because the dimension of $H_1(s)$ is impedance (ohms), it is often referred to as the *driving-point impedance* of the network. Similarly, the unit of $H_2(s)$ is ohms; however, as V_2 and I are defined at different pairs of terminals, such functions are called *transfer impedances.* The network function is often referred to as the *transfer function* because this function is a measure of how the system processes a signal as it is transported or transferred from input to output.

The characteristic equation from which the natural frequencies are determined is given by the denominators of $H_1(s)$ and $H_2(s)$ in Eqs. (4-63) and (4-64). Both $H_1(s)$ and $H_2(s)$ are of the form given in Eq. (4-62); consequently, the natural frequencies are given by the values of s for which $B(s)$ is zero. As $H(s)$ becomes infinite when $B(s)$ is zero, values of s which cause $H(s)$ to tend toward infinity are called *poles.* Thus, the natural frequencies of a system are the poles of the system function.

Values of s which cause $A(s)$ to be zero are called *zeros* as these values cause $H(s)$ to be zero. When a network is excited at the frequency of a zero, the forced response is zero. That is, for an excitation of the form $\epsilon^{s_0 t}$, the forced response is zero if the system function $H(s)$ has a zero at $s = s_0$.

Both the form of the natural response and the forced response of a system can be obtained from the transfer function of the network. The method is illustrated in Example 4-11 in which the applicability of the standard network analysis techniques of Chap. 2 are also demonstrated.

Example 4-11

For the circuit of Fig. 4-33a, determine
a The transfer function which can be used to relate the output $v_2(t)$ to the input $v_s(t)$

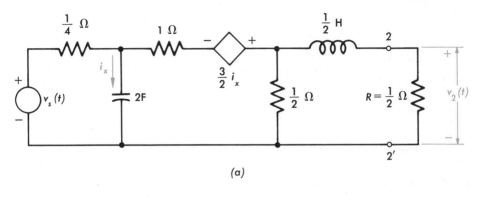

(a)

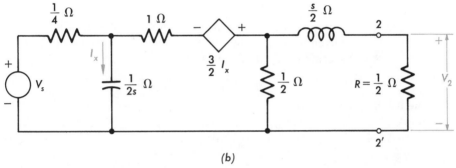

(b)

FIGURE 4-33 (a) Circuit and (b) transformed network for Example 4-11.

b The differential equation which relates $v_2(t)$ and $v_s(t)$
c The form of the natural response
d The total response for $v_s(t) = 6u(t)$
e The total response when $v_s(t)$ is a pulse having an amplitude of 6 V and a duration of 5 s
f The total response for $v_s(t) = 6\epsilon^{-t/3}u(t)$

SOLUTION **a** The transfer function is defined in terms of the transformed network variables only. Therefore, the circuit in Fig. 4-33a is replaced by its frequency-domain equivalent using Table 3-1b and is shown in Fig. 4-33b. To obtain the transfer function $H(s) = V_2/V_s$, the method employed is to first obtain a Thévenin equivalent of the portion of the circuit to the left of terminals 2-2' in Fig. 4-33b. Subsequently, the load resistor R is connected to the Thévenin equivalent circuit, from which the transfer function is determined. In computing the Thévenin equivalent, mesh analysis is employed to determine the open-circuit voltage V_o and nodal analysis is employed to determine the short-circuit current. The student should note that this approach is not the most efficient method for solving this problem. However, the approach illustrates the three major circuit-analysis techniques as applied to transformed networks, which is one of the purposes of this example.

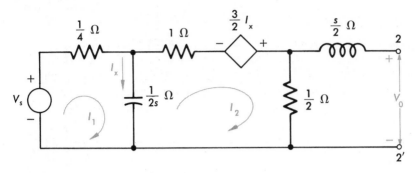

FIGURE 4-34 Transformed network to obtain open-circuit voltage in Example 4-11.

The circuit used to determine the open-circuit voltage is displayed in Fig. 4-34. No current exists in the inductive branch by virtue of the open-circuit between terminals 2 and 2′ making

$$V_o = \tfrac{1}{2} I_2 \tag{1}$$

The mesh equations are

$$-V_s + I_1 \left(\frac{1}{4}\right) + (I_1 - I_2) \frac{1}{2s} = 0 \tag{2}$$

$$(I_2 - I_1) \frac{1}{2s} + I_2(1) - \frac{3}{2} I_x + I_2 \left(\frac{1}{2}\right) = 0 \tag{3}$$

The constraint equation for the dependent source is

$$I_x = I_1 - I_2 \tag{4}$$

Substituting (4) into (3) and combining and rearranging terms gives

$$I_1 \left(\frac{1}{4} + \frac{1}{2s}\right) - \frac{1}{2s} I_2 = V_s$$

$$-I_1 \left(\frac{1}{2s} + \frac{3}{2}\right) + I_2 \left(1 + \frac{1}{2} + \frac{3}{2} + \frac{1}{2s}\right) = 0$$

from which

$$I_2 = \frac{\begin{vmatrix} \dfrac{1}{4} + \dfrac{1}{2s} & V_s \\[2ex] -\left(\dfrac{1}{2s} + \dfrac{3}{2}\right) & 0 \end{vmatrix}}{\begin{vmatrix} \dfrac{1}{4} + \dfrac{1}{2s} & -\dfrac{1}{2s} \\[2ex] -\left(\dfrac{1}{2s} + \dfrac{3}{2}\right) & 3 + \dfrac{1}{2s} \end{vmatrix}} = \frac{4(1 + 3s) V_s}{6s + 7}$$

By use of (1),

$$V_o = \frac{2(1 + 3s) V_s}{6s + 7} \tag{5}$$

The short-circuit current I_{sc} is determined by use of nodal analysis for the circuit in Fig. 4-35a. Figure 4-35b is the circuit of Fig. 4-35a in which both voltage-source–series-impedance combinations are converted to their current-source parallel-admittance equivalents. As is indicated in Fig. 4-35a,

$$I_{sc} = \frac{2}{s} V_B \tag{6}$$

The nodal equations are, for the indicated variables V_A and V_B,

$$-4V_s + 4V_A + 2s V_A + 1(V_A - V_B) + \tfrac{3}{2} I_x = 0 \tag{7}$$

$$-\frac{3}{2} I_x + 1(V_B - V_A) + 2V_B + \frac{2}{s} V_B = 0 \tag{8}$$

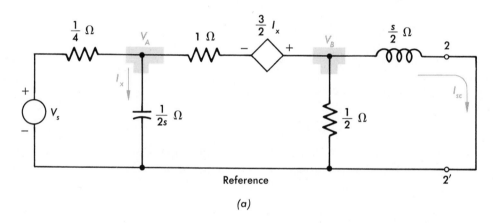

(a)

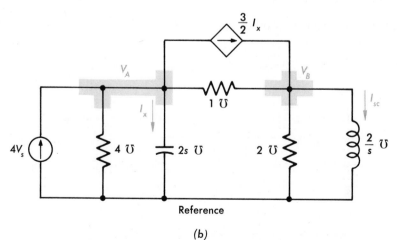

(b)

FIGURE 4-35   (a) Transformed network to obtain short-circuit current in Example 4-11; (b) Circuit of Fig. 4-35a after source conversions.

The constraint equation is expressible as

$$I_x = 2s \, V_A$$

which, when substituted into (7) and (8) gives, after rearrangement of terms,

$$V_A(4 + 2s + 1 + 3s) - 1 \, V_B = 4 \, V_s \tag{9}$$

$$- V_A(1 + 3s) + V_B \left(1 + 2 + \frac{2}{s}\right) = 0 \tag{10}$$

Solution for V_B from (9) and (10) yields

$$V_B = \frac{\begin{vmatrix} 5 + 5s & 4V_s \\ -(1 + 3s) & 0 \end{vmatrix}}{\begin{vmatrix} 5 + 5s & -1 \\ -(1 + 3s) & 3 + \dfrac{2}{s} \end{vmatrix}} = \frac{4s(1 + 3s) \, V_s}{12s^2 + 24s + 10} \tag{11}$$

By use of (11), the value of I_{sc} in (6) is

$$I_{sc} = \frac{4(1 + 3s) \, V_s}{6s^2 + 12s + 5} \tag{12}$$

The Thévenin impedance Z_o is then

$$Z_o = \frac{V_o}{I_{sc}} = \frac{6s^2 + 12s + 5}{2(6s + 7)}$$

The circuit in Fig. 4-36 depicts the Thévenin equivalent circuit in which the $\frac{1}{2}$-Ω resistance R is reconnected to terminals 2 and 2'.

For the circuit in Fig. 4-36, the output V_2 is

$$V_2 = \frac{R}{R + Z_o} V_o$$

which, upon substitution of values and forming the ratio V_2/V_s, yields

$$H(s) = \frac{V_2}{V_s} = \frac{\frac{1}{3}(1 + 3s)}{s^2 + 3s + 2} \tag{13}$$

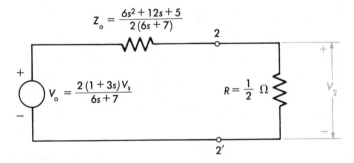

FIGURE 4-36 Thévenin equivalent circuit of Fig. 4-33b.

b The frequency-domain relationship between V_2 and V_s in (13) is also expressible as

$$(s^2 + 3s + 2) V_2 = \tfrac{1}{3} (1 + 3s) V_s \tag{14}$$

Recalling that s corresponds to differentiation with respect to t, (14) becomes

$$\frac{d^2 v_2(t)}{dt^2} + 3 \frac{dv_2(t)}{dt} + 2v_2(t) = \frac{1}{3} v_s(t) + \frac{dv_s(t)}{dt}$$

c The natural frequencies are the poles of $H(s)$; that is, the values of s which cause the denominator of $H(s)$ in (13) to be zero. These roots are $s = -1$ and $s = -2$, thus making the natural response

$$v_{2n}(t) = K_1 \epsilon^{-t} + K_2 \epsilon^{-2t} \quad \text{V}$$

d The natural response is that given in part **c** of this problem. The excitation is of the form $6\epsilon^{ot}$ for $t \geq 0$ so that the forced response is an exponential with $s = 0$ of the form $V_2 \, \epsilon^{ot}$. Then by the use of the transfer function,

$$V_2 = H(s) V_s$$

evaluated at $s = 0$, or

$$V_2 = \frac{\frac{1}{3}}{2} \times 6 = 1$$

Thus, for $t \geq 0$,

$$v_{2f}(t) = 1\epsilon^{ot} = 1$$

and the total solution becomes

$$v_2(t) = v_{2f}(t) + v_{2n}(t) = 1 + K_1 \epsilon^{-t} + K_2 \epsilon^{-2t} \quad \text{V} \tag{15}$$

As (15) is valid only for $t \geq 0$, it is convenient to express $v_2(t)$ as

$$v_2(t) = [1 + K_1 \epsilon^{-t} + K_2 \epsilon^{-2t}] u(t) \quad \text{V}$$

where the multiplier $u(t)$ denotes that $v_2(t)$ exists for $t \geq 0$ and is zero for $t < 0$.

Because the excitation is a step function, there can be no energy stored in the system at $t = 0^-$. Consequently, the voltage across the capacitance and the current in the inductance are zero at $t = 0^-$. By continuity principles, $v_c(0^+)$ and $i_L(0^+)$, indicated in the circuit of Fig. 4-37 which is the circuit of Fig. 4-33a valid at $t = 0^+$, are also zero. The values of $v_c(0^+)$ and $i_L(0^+)$, employed in conjunction with the circuit of Fig. 4-37, are used to determine the initial conditions $v_2(0^+)$ and $[dv_2(0^+)]/dt$ needed to evaluate K_1 and K_2.

In Fig. 4-37, v_2 is simply

$$v_2 = \tfrac{1}{2} i_L$$

so that

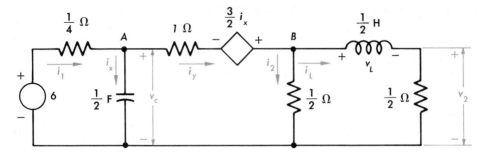

FIGURE 4-37 Circuit of Fig. 4-33a valid at t = 0⁺.

$$\frac{dv_2}{dt} = \frac{1}{2}\frac{di_L}{dt}$$

As $i_L(0^+) = 0$, $v_2(0^+) = 0$.

From the volt-ampere relation for the inductance,

$$\frac{di_L}{dt} = \frac{v_L}{L} \quad \text{and} \quad \frac{dv_2}{dt} = \frac{1}{2}\frac{v_L}{\frac{1}{2}} = v_L$$

For the right-hand loop, the KVL expression is

$$v_L = v_2 - \tfrac{1}{2} i_2 = 0 \tag{16}$$

Evaluation of (16) at $t = 0^+$ gives

$$v_L(0^+) = \frac{1}{2} i_2(0^+) \quad \text{or} \quad \frac{dv_2}{dt}(0^+) = \frac{1}{2} i_2(0^+)$$

The KCL equations at nodes A and B, evaluated at $t = 0^+$, are

$$-i_y(0^+) + i_2(0^+) + i_L(0^+) = 0 \tag{17}$$

$$-i_1(0^+) + i_x(0^+) + i_y(0^+) = 0 \tag{18}$$

As $i_L(0^+) = 0$, (17) becomes

$$i_y(0^+) = i_2(0^+)$$

The KVL expression for the left-hand mesh as $v_c(0^+) = 0$ gives

$$-6 + \tfrac{1}{4} i_1(0^+) = 0 \quad \text{or} \quad i_1(0^+) = 24 \text{ A}$$

Substitution of $i_1(0^+)$ in (18) gives

$$i_x(0^+) + i_y(0^+) = 24 \quad \text{or} \quad i_x(0^+) + i_2(0^+) = 24 \text{ A} \tag{19}$$

From the KVL equation for the center loop, recalling that $v_c(0^+) = 0$ and $i_y(0^+) = i_2(0^+)$,

$$1i_2(0^+) - \tfrac{3}{2} i_x(0^+) + \tfrac{1}{2} i_2(0^+) = 0$$

or

$$i_2(0^+) = i_x(0^+) \tag{20}$$

Use of (20) in conjunction with (19) yields

$$2i_2(0^+) = 24 \qquad \text{or} \qquad i_2(0^+) = 12 \text{ A}$$

which results in

$$\frac{dv_2}{dt}(0^+) = \frac{1}{2} 12 = 6 \text{ V/s}$$

The two initial conditions are

$$\frac{dv_2}{dt}(0^+) = 6 \text{ V/s} \qquad \text{and} \qquad v_2(0^+) = 0$$

Substitution of these values in (15) and in the time derivative of (15) gives, respectively,

$$K_1 + K_2 = -1$$

$$-K_1 - 2K_2 = 6$$

Simultaneous solution yields

$$K_1 = 4 \qquad \text{and} \qquad K_2 = -5$$

from which

$$v_2(t) = (1 + 4\epsilon^{-t} - 5\epsilon^{-2t})u(t) \qquad \text{V}$$

e As described in Sec. 4-8, a pulse can be decomposed into a step function and delayed step function. For the 6-V pulse whose duration is 5 s,

$$v_s(t) = 6u(t) - 6u(t - 5) \qquad \text{V}$$

The use of superposition for each component of the excitation results in the component responses. Thus, for the $6u(t)$ term, the corresponding output is given in part d. For the $-6u(t - 5)$ component, the output is the negative of that given in part d, delayed by 5 s. Therefore,

$$v_2(t) = (1 + 4\epsilon^{-t} - 5\epsilon^{-2t})u(t) - (1 + 4\epsilon^{-(t-5)} - 5\epsilon^{-2(t-5)})u(t - 5) \qquad \text{V}$$

f The forced response for the excitation $6\epsilon^{-t/3}$ is

$$V_2 = 6H(-\tfrac{1}{3})$$

That is, the forced response is of the form $V_2\epsilon^{-t/3}$ and V_2/V_s is the transfer function $H(s)$ evaluated at $s = -\tfrac{1}{3}$, the s value of the excitation. For the $H(s)$ given by (13),

$$H(-\tfrac{1}{3}) = 0$$

The s value of the excitation is the same as a zero of $H(s)$, so no forced component exists. The total response is then

$$v_2(t) = (K_1\epsilon^{-t} + K_2\epsilon^{-2t})u(t) \qquad \text{V}$$

because the form of the natural response is independent of the excitation.

The circuit in Fig. 4-37 is applicable at $t = 0^+$ for the excitation $6\epsilon^{-t/3} u(t)$. As a consequence, the values of $v_2(0^+)$ and $[dv_2(0^+)]/dt$ are those derived in part d. The coefficients K_1 and K_2 are then obtained from

$$K_1 + K_2 = 0 \qquad -K_1 - 2K_2 = 6$$

from which

$$K_1 = +6 \qquad \text{and} \qquad K_2 = -6$$

The resultant response becomes

$$v_2(t) = (6\epsilon^{-t} - 6\epsilon^{-2t})u(t) \qquad \text{V}$$

The methods illustrated in the example just concluded indicate that the use of transformed network is of considerable benefit in determining the network function. From the system function, obtained by standard algebraic circuit analysis, the form of both the forced and natural response components is readily obtained. However, inclusion of initial conditions remains a moderately involved process. The Laplace transformation, described in the succeeding chapter, is a widely used method in which the initial conditions are included in the transformed network. As a result, the total solution is derived by an integrated process.

PROBLEMS

4-1 For the circuit of Fig. 4-38, determine the forced component of the responses $v(t)$ and $i_L(t)$ when
a $i(t) = 4$ A
b $i(t) = 2t$ A
c $i(t) = 4 + 2t$ A

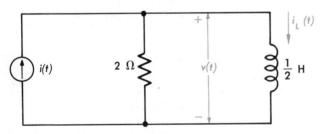

FIGURE 4-38 Circuit for Prob. 4-1.

4-2 In the circuit of Fig. 4-3, determine the forced component of the current when
a $v(t) = 5\epsilon^{-2t}$ V
b $v(t) = 10$ V
c $v(t) = 10 + 5\epsilon^{-2t}$ V

4-3   Determine the forced component of the responses $v_c(t)$ and $i_c(t)$ in the circuit of Fig. 4-39 when $v(t) = 12te^{-t}$ V.

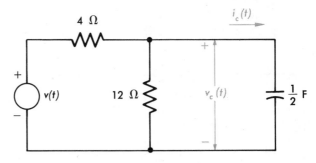

FIGURE 4-39   Circuit for Prob. 4-3.

4-4   In the circuit of Fig. 4-40, determine the forced component of the responses $v_o(t)$ and $i(t)$ when
a $v(t) = 0.1$ V
b $v(t) = 0.2\epsilon^{-10^5 t} \cos 10^5 t$ V

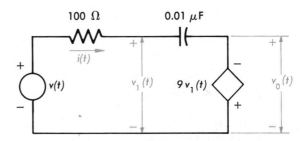

FIGURE 4-40   Circuit for Prob. 4-4.

4-5   In the circuit of Fig. 4-41, find the forced component of $v(t)$ when
a $i(t) = 10^{-3}$ A
b $i(t) = 10^{-3}t$ A
c $i(t) = 10^{-3} - 10^{-3}t$ A

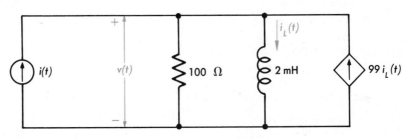

FIGURE 4-41   Circuit for Prob. 4-5.

4-6 Determine the forced component of the response $v_o(t)$ in the circuit of Fig. 4-42 for both $v(t) = 5$ V and $v(t) = 5 \cos t$ V when
a $R_1 = R_2 = 2\ \Omega$
b $R_1 = 1\ \Omega$, $R_2 = 2\ \Omega$
c $R_1 = 2\ \Omega$, $R_2 = 1\ \Omega$

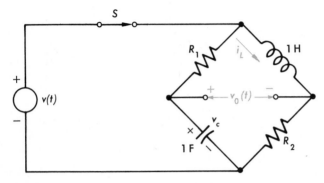

FIGURE 4-42 Circuit for Prob. 4-6.

4-7 In the circuit of Fig. 4-38, determine the natural component of the response $v(t)$.

4-8 The voltage source $v(t)$ in the circuit of Fig. 4-39 is suppressed [$v(t) = 0$].
a Determine the natural responses $i_c(t)$ and $v_c(t)$.
b What is $i_c(0)$ if $v_c(0) = 6$ V?

4-9 The voltage source $v(t)$ in the circuit of Fig. 4-40 is suppressed.
a Determine the natural response $v_o(t)$. Assume $v_o(0) = -9$ V.
b Sketch the response in *a* for five time constants.

4-10 The current source in Fig. 4-41 is suppressed [$i(t) = 0$].
a Determine the natural response $v(t)$ assuming $i_L(0) = 10$ mA.
b Sketch the response in *a* for a duration of five time constants.

4-11 The switch S in Fig. 4-42 is opened at $t = 0$. Determine the natural response $v_o(t)$ for $t \geq 0$ when
a $R_1 = R_2 = 1\ \Omega$
b $R_1 = 2\ \Omega$ and $R_2 = 1\ \Omega$
c $R_1 = 0.5\ \Omega$ and $R_2 = 1\ \Omega$
In each case, assume $v_c(0) = 5$ V and $i_L(0) = 5$ A.

4-12 In the circuit of Fig. 4-38

$$i(t) = 5 \text{ A} \qquad t < 0$$
$$= 12\epsilon^{-t} \qquad t \geq 0$$

Determine $v(0^+)$, $i_L(0^+)$, $(dv/dt)\ (0^+)$, and $(di_L/dt)\ (0^+)$.

4-13 In the circuit of Fig. 4-39,

$$v(t) = 12 \text{ V} \qquad t < 0$$
$$= 6 \cos t \qquad t \geq 0$$

Determine $v_c(0^+)$, $i_c(0^+)$, $(dv_c/dt)(0^+)$, and $(di_c/dt)(0^+)$.

4-14 In the circuit of Fig. 4-40,

$$v(t) = 0 \qquad t < 0$$
$$= 0.2t \qquad t \geq 0$$

Find $v_o(0^+)$ and $(dv_o/dt)(0^+)$.

4-15 The switch S in Fig. 4-42 has been closed for a long time with $v(t) = 3$ V. At $t = 0$, the switch is opened. Determine $v_o(0^+)$, $(dv_o/dt)(0^+)$, $(dv_c/dt)(0^+)$, and $(di_L/dt)(0^+)$ when
a $R_1 = R_2 = 3 \ \Omega$
b $R_1 = 2 \ \Omega$, $R_2 = 4 \ \Omega$
c $R_1 = 4 \ \Omega$, $R_2 = 2 \ \Omega$

4-16 The current $i(t)$ in the circuit of Fig. 4-38 is given in Prob. 4-12. Determine the total response $v(t)$.

4-17 The voltage $v(t)$ in the circuit of Fig. 4-39 is given in Prob. 4-13. Determine the total response for both $v_c(t)$ and $i_c(t)$.

4-18 The value of current source in Fig. 4-41 is zero for $t < 0$. Determine the total response $v(t)$ for each of the three excitations given in Prob. 4-5, assumed to be applied at $t = 0$.

4-19 **a** The excitation in the circuit of Fig. 4-40 is given in Prob. 4-14. Determine $v_o(t)$ for $t \geq 0$.
b Repeat a for the excitation given in Prob. 4-4b. Assume $v(t) = 0$ for $t < 0$.

4-20 The switch S in Fig. 4-42 has been closed for a long time with $v(t) = 3$ V. At $t = 0$, the switch is opened; and at $t = 0.5$ s, the switch is closed again.
a Determine $v_o(t)$ for $t \geq 0$.
1   $R_1 = R_2 = 3 \ \Omega$.
2   $R_1 = 2 \ \Omega$ and $R_2 = 4 \ \Omega$.
3   $R_1 = 4 \ \Omega$ and $R_2 = 2 \ \Omega$.
b For each of the cases given in a, what is the average power dissipated by R_2 in the interval $0 \leq t \leq 0.5$ s?

4-21 The switch S in the circuit of Fig. 4-43 has been open for a long time and is instantaneously closed at $t = 0$. Find $v(t)$ for $t \geq 0$ and sketch the waveform.

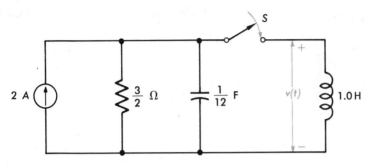

FIGURE 4-43 Circuit for Prob. 4-21.

4-22 In the circuit of Fig. 4-44, the switch S has been in position 1 for a long time and at $t = 0$ it is instantaneously moved to position 2. Determine $i(t)$ for $t \geq 0$ and sketch its waveform.

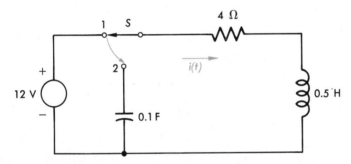

FIGURE 4-44 Circuit for Prob. 4-22.

4-23 In the circuit of Fig. 4-45, the switch S has been open for a long time and is closed at $t = 0$. Find $v_o(t)$ and $i(t)$ for $t \geq 0$.

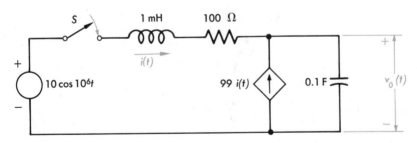

FIGURE 4-45 Circuit for Prob. 4-23.

4-24 **a** The switch S in the circuit of Fig. 4-46 is opened at $t = 0$ having been closed for a long time. Determine $v_o(t)$ for $t \geq 0$.
b How long must the switch remain open for the voltage $v_o(t)$ to be less than 1 percent of its value at $t = 0$?

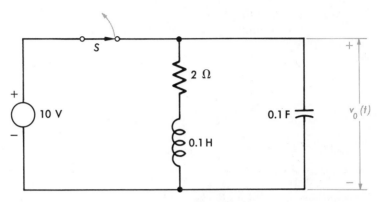

FIGURE 4-46 Circuit for Prob. 4-24.

4-25 In the circuit of Fig. 4-38, the source current $i(t)$ is the unit-step function $u(t)$. Find $v(t)$.

4-26 a Determine $v_o(t)$ in the circuit of Fig. 4-40 for $v(t) = u(t)$.
b Repeat a for $v(t) = u(t - 10^{-3})$.

4-27 a Determine $v(t)$ in the circuit of Fig. 4-41 when $i(t) = 10^{-3} \cos 10^{3}tu(t)$.
b Repeat a when $i(t) = tu(t - 1)$.

4-28 In the circuit of Fig. 4-42, $R_1 = 4 \ \Omega$, $R_2 = 2 \ \Omega$ and $v(t)$ is given by

$$v(t) = 8u(t) - 8u(t - 1)$$

Determine $v_o(t)$ and sketch the waveform.

4-29 The current $i(t)$ in the circuit of Fig. 4-47 is a pulse whose amplitude is 10 A and duration is 0.4 s. Assuming that the pulse begins at $t = 0$, determine $v_o(t)$ for $R = 0.2 \ \Omega$, 0.25 Ω, and 0.5 Ω. Sketch the output waveform for each case.

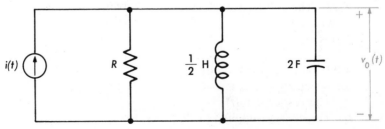

FIGURE 4-47 Circuit for Prob. 4-29.

4-30 Express the "staircase" waveform shown in Fig. 4-48, often encountered in computer control and instrumentation systems, as a sum of step functions.

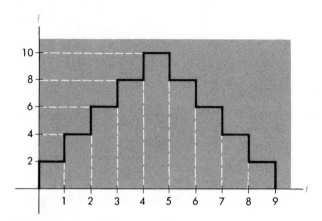

FIGURE 4-48 "Staircase" waveform for Prob. 4-30.

4-31 The voltage waveform in Fig. 4-49 is applied to the circuit in Fig. 4-46. Determine the output voltage $v_o(t)$.

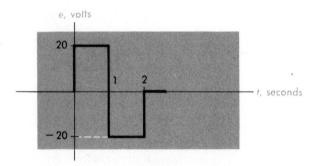

FIGURE 4-49 Voltage waveform for Prob. 4-31.

4-32 The voltage waveform in Fig. 4-49 is applied to an RLC series circuit in which $R = 2\ \Omega$, $L = 2$ H, and $C = 1$ F. Determine the current in the series circuit.

4-33 The voltage waveform in Fig. 4-23 is applied to the circuit in Fig. 4-42 for which $R_1 = 2\ \Omega$ and $R_2 = 0.5\ \Omega$. Determine:
a The response $v_o(t)$
b The energy supplied for $0 \le t \le 1$ s.

4-34 **a** The source voltage $v(t)$ in Fig. 4-39 is

$$v(t) = 4\delta(t)$$

Find $v_c(t)$.
b Repeat a for $v(t) = 4\delta(t - 3)$.

4-35 **a** A unit impulse of voltage is applied to the series combination of a 20-Ω resistance and a 10-mH inductance. Determine the current $i(t)$ in the series combination.
b The voltage applied to the RL circuit in a is

$$v(t) = \delta(t) + \delta(t - T)$$

Find the minimum value of T so that, for practical purposes, the response $i(T)$ is zero.

4-36 The voltage waveform shown in Fig. 4-50 is applied to the series RL circuit of Prob. 4-35.
a Determine and sketch the current in the series circuit as a function of time.
b Select values of T_1 and T_2 so that the response obtained in a approximates the impulse response of the circuit.

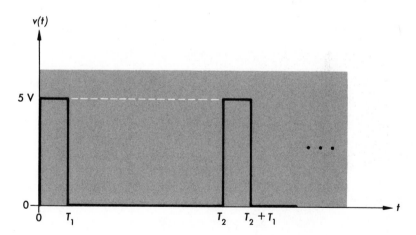

FIGURE 4-50 Voltage waveform for Prob. 4-36.

4-37 Determine the network functions (V_1/V) (s) and (V_o/V) (s) in the circuit of Fig. 4-40 by first replacing the elements by their impedances.

4-38 Determine the driving point admittance $Y(s)$ for the circuit of Fig. 4-41.

4-39 a For the circuit of Fig. 4-42, determine the transfer function (V_o/V) (s).
b Determine the poles and zeros of the transfer function obtained in a.
c What is the form of the natural response of $v_o(t)$?

4-40 Transform the circuit shown in Fig. 4-51 using impedance parameters.
a Determine the voltage-transfer function (V_o/V) (s).
b Determine the driving-point impedance $Z(s)$ seen by the voltage source.
c What is the form of the natural response of $v_o(t)$?
d For an excitation $v(t) = 12\epsilon^{-0.1t}$, find the forced component of $v_o(t)$.

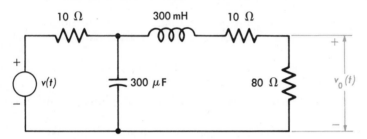

FIGURE 4-51 Circuit for Prob. 4-40.

4-41 For the circuit of Fig. 4-40, use the transformed network to obtain the response $v_o(t)$ to an excitation $v(t) = 12u(t)$.

4-42 Repeat Prob. 4-41 for $v(t) = 12\epsilon^{-t}u(t)$.

4-43 For the transformed network of Fig. 4-52 determine, using nodal analysis,

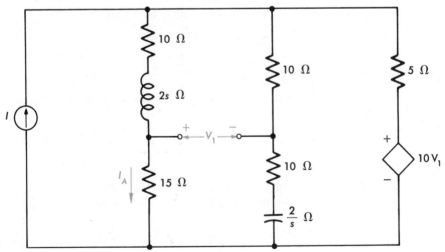

FIGURE 4-52 Circuit for Prob. 4-43.

a The network function I_A/I
b The network function V_1/I
c The form of the natural response for $i_A(t)$
d The forced response component of $v_1(t)$ for $i(t) = 10\epsilon^{-t}$.

4-44 a For the circuit shown in Fig. 4-53 transform the network using admittance parameters.
b Employ the result in a to obtain the transfer function (V_2/I) (s) using nodal analysis.
c Determine the response $v_2(t)$ to an excitation $i(t) = 10^{-3}u(t)$.
d Repeat c for $i(t) = 10^{-3}[u(t) - u(t - 10^{-5})]$ and sketch the result.

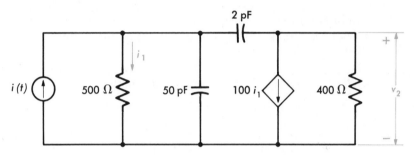

FIGURE 4-53 Circuit for Prob. 4-44.

4-45 Transform the network in Fig. 4-52 and obtain a Thévenin equivalent of the circuit portion to the left of the 40-Ω resistance.

4-46 Repeat Prob. 4-43 using mesh analysis.

4-47 a For the circuit of Fig. 4-53, obtain the Thévenin equivalent seen by the 15-Ω resistance.
b Use the result in a to determine the system function I_A/I.

4-48 For the transformed network in Fig. 4-54, determine the voltage V_c in terms of I, V, and the circuit parameters by use of nodal analysis.

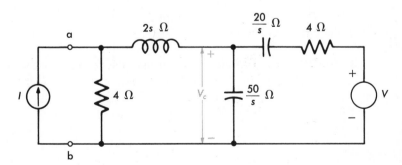

FIGURE 4-54 Circuit for Prob. 4-47.

4-49 a Repeat Prob. 4-48 by use of mesh analysis.
b For what value or values of s is the forced component of the response $v_c(t) = 0$?

4-50 Obtain a Thévenin equivalent for the portion of the network to the right of terminals a and b in Fig. 4-54.

4-51 The driving-point impedance of a network is

$$Z(s) = \frac{s + 4}{s}$$

a Construct a network which has the given $Z(s)$. *Hint:* Express $Z(s)$ as the sum of two terms and use the element volt-ampere relations.
b For the circuit obtained in a, determine the current response to an applied unit-step voltage.

4-52 The driving-point admittance of a network is

$$Y(s) = \frac{2s + 3}{s}$$

a Synthesize a network having the given $Z(s)$. (See *Hint* in Prob. 4-51.)
b Determine the voltage response to an applied unit-step current.

4-53 The driving-point impedance of a network is

$$Z(s) = \frac{s + 2}{s + 1}$$

a Synthesize a circuit having the given $Z(s)$.
b Determine the current response to a voltage applied to the circuit $v(t) = 10\epsilon^{-t}u(t)$.

4-54 Repeat Prob. 4-52 for

$$Y(s) = \frac{s^2 + 4s + 2}{s + 1}$$

4-55 Determine the driving-point admittance seen by the current source I in the circuit of Fig. 4-32.

4-56 Determine the driving-point impedance seen by the source V_s in the circuit of Fig. 4-33b. Use mesh analysis.

4-57 Repeat Prob. 4-56 using nodal analysis.

5

THE LAPLACE TRANSFORM

xponential functions were introduced in Chap. 3 to represent several commonly encountered excitations. One of the advantages gained from the use of exponentials was that the differential equations describing the networks under examination were transformed into algebraic equations. This process considerably simplified the determination of the forced response of the network. Again, in Chap. 4, the natural response was shown to be the sum of exponential terms. By virtue of the exponential form for the natural response, each term in the summation was obtained by algebraic solution of the characteristic equation. At the turn of this century, Oliver Heaviside was the first to develop an algebraic technique for obtaining the total response of a network. The "operational calculus" he developed has been since supplanted by the similar Laplace transform method treated in this chapter. A form of exponential transform, in which many of the ideas introduced in the previous chapters are incorporated, the Laplace transformation provides a systematic algebraic approach for determination of both the forced and natural components of a network response, simultaneously. In addition to an examination of the properties of the Laplace transform, the concept of a transfer function is introduced and its use in solving circuits problems is developed.

5-1 DEFINITION OF THE LAPLACE TRANSFORM

We have already encountered several transformations used in circuit analysis in the previous chapters. Thévenin's theorem, Δ-Y transformations, and source conversions are all examples, each of which involves the representation of a network by an equivalent circuit. The benefit of the equivalent circuit is that both the formulation and solution of the problem are facilitated. As we have seen, this type of circuit transformation often simplifies the calculations required to obtain the solution.

The exponential representation of excitations is essentially a mathematical transformation to which we were introduced in Sec. 3-2. The use of Table 3-1b for exponential excitation and Table 3-1c for sinusoidal excitation transforms the time-domain description of a circuit by differential equations to the frequency-domain characterization by algebraic equations. The process used for obtaining the solution consists of three steps:

1 Transform the equations from the time domain to the frequency domain.

2 Then manipulate the frequency-domain quantities to obtain the frequency-domain equivalent of the desired result.

3 Complete the process by taking the inverse transform; i.e., transform the equivalent result back into the time domain.

The solution of circuits problems by Laplace transformation methods incorporates the time-domain to frequency-domain conversion described in the previous paragraph. Of principal importance is that the differential equations which arise are transformed into algebraic equations which are more easily manipulated. In addition, the Laplace transformation affords us two other advantages over the classical method given in Chap. 4. First, the total response is determined as a single entity; i.e., the natural and forced components are obtained simultaneously and these remain readily identifiable. Second, initial conditions are incorporated within the transform and thus are inherently included in the response. This often results in eliminating the need to evaluate high-order derivatives to arrive at the solution.

The Laplace transform is defined by the relation given in Eq. (5-1).

$$\mathcal{L}[f(t)] = F(s) = \int_0^\infty f(t)\epsilon^{-st}\, dt \tag{5-1}$$

The statement $\mathcal{L}[f(t)] = F(s)$ is read as $F(s)$ is the Laplace transform of $f(t)$. In Eq. (5-1), s is the complex-frequency variable $\sigma + j\omega$. The frequency-domain function $F(s)$ corresponds to the time-domain function $f(t)$. The Laplace transform of $f(t)$ exists provided the integral in Eq. (5-1) converges. That is, $f(t)$ is Laplace transformable when

$$\int_0^\infty f(t)\epsilon^{-st}\, dt < \infty$$

Physical limitations virtually ensure that all functions encountered in practical systems are transformable. Mathematically, ϵ^{-st}, for $\sigma > 0$, decreases more rapidly than nearly all the functions $f(t)$ with which engineers and scientists deal and so guarantees the convergence of the integral.

The limits of integration used in Eq. (5-1) indicate that we are only interested in $f(t)$ for $t \geq 0$. It is usually assumed that $f(t)$ is zero for $t < 0$, a situation which corresponds with our concern for the behavior of systems beginning with the time an excitation is applied.

Example 5-1

Determine the Laplace transform for the functions

a $f_1(t) = 1 \qquad t \geq 0$
$\qquad\quad\; = 0 \qquad t < 0$

b $f_2(t) = \epsilon^{-at}$

SOLUTION **a** The function described defines the *unit-step function* $u(t)$ (see Sec. 4-7). The Laplace transform is obtained from the evaluation of the integral in Eq. (5-1) for the given function. Thus,

$$F_1(s) = \int_0^\infty \epsilon^{-st}\, dt = -\frac{1}{s}\,\epsilon^{-st}\, \Big|_0^\infty$$

and

$$F_1(s) = \mathcal{L}[u(t)] = \frac{1}{s}$$

b $\quad F_2(s) = \int_0^\infty \epsilon^{-at}\epsilon^{-st}\,dt = \frac{-1}{s+a}\,\epsilon^{-(s+a)t}\,\Big|_0^\infty$

and

$$F_2(s) = \mathcal{L}[\epsilon^{-at}] = \frac{1}{s+a}$$

Example 5-2

Evaluate $\mathcal{L}\left[\dfrac{df(t)}{dt}\right]$, assuming $f(t)$ is transformable.

SOLUTION   In this example we are to obtain the transform of the time-derivative of a function $f(t)$. From the definition in Eq. (5-1),

$$\mathcal{L}\left[\frac{df(t)}{dt}\right] = \int_0^\infty \frac{df(t)}{dt}\,\epsilon^{-st}\,dt$$

By the use of $\int u\,dv = uv - \int v\,du$, where

$$u = \epsilon^{-st} \qquad \text{and} \qquad dv = df(t)$$

and, from which

$$du = -s\epsilon^{-st}\,dt \qquad \text{and} \qquad v = f(t)$$

one obtains

$$\int_0^\infty u\,dv = f(t)\epsilon^{-st}\,\Big|_0^\infty - \int_0^\infty f(t)[-s\epsilon^{-st}\,dt]$$

$$= -f(0^+) + s\int_0^\infty f(t)\epsilon^{-st}\,dt$$

Recognition of $\int_0^\infty f(t)\epsilon^{-st}\,dt$ is $\mathcal{L}[f(t)] = F(s)$; the desired result is

$$\mathcal{L}\left[\frac{df(t)}{dt}\right] = sF(s) - f(0^+)$$

The use of $f(0^+)$ to designate the evaluation of $f(t)\epsilon^{-st}$ at the lower limit is consistent with the notation used in Chap. 4 to signify that $f(t)$ is defined for $t \geq 0$.

By recalling $i_C = C\dfrac{dv_C}{dt}$, $v_L = L\dfrac{di_L}{dt}$, and the principle of continuity of capacitance voltage and inductance current, the significance of the term $f(0^+)$ is that the initial condition is automatically included in the Laplace transform of the derivative. For the cases where $f(0^+)$ is zero, indicating no initial energy storage in a physical system,

$$\mathcal{L}\left[\frac{df}{dt}\right] = sF(s)$$

Thus, multiplication by s in the frequency domain corresponds to differentiation in the time domain.

5-2 FUNDAMENTAL PROPERTIES

Examples 5-1 and 5-2 illustrate the technique for determining the Laplace transform from the defining integral. Application of the method to the solution of circuits problems involves the three steps outlined in the previous section. It is useful to consider the circuit of Fig. 5-1 both to demonstrate the method of solution and to investigate the two fundamental properties of the Laplace transform needed to obtain the solution. In the circuit of Fig. 5-1, we wish to find $i(t)$ for $t \geq 0$ given that $i(0^+)$ is 5 A. The KVL equation for the loop is

$$4i(t) + 2\frac{di(t)}{dt} = 0 \tag{5-2}$$

The first step is to transform Eq. (5-2) from the time to the frequency domain by means of the Laplace transform. By use of Eq. (5-1),

$$\mathcal{L}\left[4i(t) + 2\frac{di(t)}{dt}\right] = \int_0^\infty \left[4i(t) + 2\frac{di(t)}{dt}\right] \epsilon^{-st}\, dt$$

$$= 4\int_0^\infty i(t)\epsilon^{-st}\, dt$$

$$+ 2\int_0^\infty \frac{di(t)}{dt}\epsilon^{-st}\, dt \tag{5-3}$$

From the result of Example 5-2 and Eq. (5-1), Eq. (5-3) is rewritten as

$$\mathcal{L}\left[4i(t) + 2\frac{di(t)}{dt}\right] = 4I(s) + 2[sI(s) - i(0^+)] \tag{5-4}$$

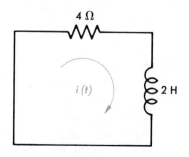

FIGURE 5-1 *RL* circuit.

Embodied in the derivation of Eq. (5-4) is the first fundamental property of the Laplace transform, namely that the Laplace transform is a linear operation. Mathematically, the linearity principle states that if $f_1(t)$ and $f_2(t)$ have Laplace transforms $F_1(s)$ and $F_2(s)$, respectively, then

$$\mathcal{L}[Af_1(t) + Bf_2(t)] = AF_1(s) + BF_2(s) \tag{5-5}$$

where A and B are independent of s and t.

Returning to the solution for $i(t)$, the left side of Eq. (5-4) is the Laplace transform of the left side of Eq. (5-2). Substituting $i(0^+) = 5$ A, the frequency-domain KVL equation is

$$4I(s) + 2sI(s) - 10 = 0 \tag{5-6}$$

The second step in the process is the manipulation of Eq. (5-6) to obtain $I(s)$ as

$$I(s) = \frac{10}{2s + 4} = \frac{5}{s + 2} \tag{5-7}$$

The third step is the transformation of the frequency-domain value of $I(s)$ in Eq. (5-7) to the time-domain function $i(t)$. The process, called the *inverse Laplace transform*, is denoted by

$$\mathcal{L}^{-1}[I(s)] = i(t) \tag{5-8}$$

or, for the general functions $F(s)$ and $f(t)$,

$$\mathcal{L}^{-1}[F(s)] = f(t) \tag{5-9}$$

To obtain the inverse Laplace transform, the second fundamental property is needed. This property is the uniqueness of the Laplace transform; that is, only one function $f(t)$ exists for which the Laplace transform is $F(s)$ and, conversely, only one $F(s)$ exists for which $f(t)$ is the inverse Laplace transform. Uniqueness defines the concept of the *transform-pair* and is stated in Eq. (5-10) as

$$\mathcal{L}[f(t)] = F(s) \longrightarrow \mathcal{L}^{-1}[F(s)] = f(t) \tag{5-10}$$

In Eq. (5-10), the arrow signifies that the statement to the left of the arrow implies that the statement to the right of the arrow is true.

Because of Eq. (5-10), the inverse Laplace transform of Eq. (5-7) is obtained by the use of the result in Example 5-1*b* and Eq. (5-8) as

$$i(t) = \mathcal{L}^{-1}\left[\frac{5}{s + 2}\right] = 5\epsilon^{-2t} \qquad \text{for } t \geq 0 \tag{5-11}$$

The circuit in Fig. 5-1 is also shown in Fig. 4-5 for Example 4-3. As expected, the solution in Eq. (5-11) is identical to the classically obtained solution in Example 4-3. Note, however, that no arbitrary constant is evaluated when the Laplace transform method is used. The initial condition is introduced by the transformation of Eq. (5-2) at the beginning of the solution process and, as such, is inherently part of the final solution.

In Examples 5-1 and 5-2, the transforms were obtained by evaluation of the defining integral. However, in determining $i(t)$ in the circuit of Fig. 5-1, only the results of these examples are used. The uniqueness and linearity of the Laplace transform enabled us to use the concept of the transform pair and the relations we had already obtained. In effect, Examples 5-1 and 5-2 served as entries in a table of transforms and their corresponding time functions. Efficient solution of circuits problems by use of the Laplace transform method relies on the creation of just such a table of transform pairs. Fortunately, the widespread use of the Laplace transform has led to the extensive tabulation of transform pairs. Table 5-1 contains a short table of functions most commonly encountered and is sufficient for the material treated in this book. The use of Table 5-1 is straightforward; given $f(t)$, $F(s)$ is the corresponding Laplace transform and vice versa. Thus, the first and last steps in the problem-solving process involve the use of the table. First, the table is used to transform the circuit equations from the time domain to the frequency domain. After manipulation, step (2), the table is used to obtain the inverse transform of the result.

TABLE 5-1 TABLE OF TRANSFORM PAIRS

$f(t)$		$F(s)$
1 $df(t)/dt$		$sF(s) - f(0^+)$
2 $\dfrac{d^2f(t)}{dt^2}$		$s^2F(s) - sf(0^+) - \dfrac{df}{dt}(0^+)$
3 $\dfrac{d^nf(t)}{dt^n}$		$s^nF(s) - s^{n-1}\dfrac{df}{dt}(0^+) - s^{n-2}\dfrac{d^2f}{dt^2} - \cdots - \dfrac{df^{n-1}}{dt^{n-1}}(0^+)$
4 $g(t) = \displaystyle\int_0^\infty f(t)\,dt$		$\dfrac{F(s)}{s} + \dfrac{g(0^+)}{s}$
5 $u(t)$	the unit step-function	$1/s$
6 $\delta(t)$	the unit impulse function	1
7 t		$1/s^2$
8 $t^{n-1}/(n-1)!$	n is an integer	$1/s^n$
9 ϵ^{-at}		$1/(s+a)$
10 $t\epsilon^{-at}$		$1/(s+a)^2$
11 $t^{n-1}\epsilon^{-at}$		$(n-1)!/(s+a)^n$
12 $\sin\omega t$		$\omega/(s^2+\omega^2)$
13 $\cos\omega t$		$s/(s^2+\omega^2)$
14 $\sin(\omega t + \theta)$		$[s\sin\theta + \omega\cos\theta]/(s^2+\omega^2)$
15 $\cos(\omega t + \theta)$		$[s\cos\theta - \omega\sin\theta]/(s^2+\omega^2)$
16 $\epsilon^{-at}\sin\omega t$		$\omega/[(s+a)^2+\omega^2]$
17 $\epsilon^{-at}\cos\omega t$		$(s+a)/[(s+a)^2+\omega^2]$
18 $t\epsilon^{-at}\sin\omega t$		$2\omega(s+a)/[(s+a)^2+\omega^2]^2$
19 $t\epsilon^{-at}\cos\omega t$		$[(s+a)^2-\omega^2]/[(s+a)^2+\omega^2]^2$

It is assumed that all $f(t)$ exist for $t \geq 0$ and $f(t) = 0$ for $t < 0$. Each of the functions from **7** to **19** can be considered as being multiplied by $u(t)$.

Example 5-3

Determine the inverse Laplace transform of

$$F(s) = \frac{3s + 15}{(s + 1)^2 + (3)^2}$$

SOLUTION   In order to use Table 5-1, the function $F(s)$ must be manipulated so that it corresponds to the appropriate forms listed in the table. As
$F(s)$ appears similar to the form given by transform pair **17**, it can be expressed as

$$F(s) = \frac{3(s + 1) + 12}{(s + 1)^2 + (3)^2} = \frac{3(s + 1)}{(s + 1)^2 + (3)^2}$$

$$+ \frac{12}{(s + 1)^2 + (3)^2}$$

$$= F_1(s) + F_2(s)$$

The form of $F_1(s)$ corresponds to **17** in Table 5-1; $F_2(s)$ can be rewritten in the
form of **16** as

$$F_2 = \frac{4(3)}{(s + 1)^2 + (3)^2}$$

By use of Eq. (5-5),

$$\mathcal{L}^{-1}[F(s)] = \mathcal{L}^{-1}[F_1(s) + F_2(s)] = 3\epsilon^{-t} \cos 3t + 4\epsilon^{-t} \sin 3t$$

or

$$f(t) = \epsilon^{-t}[3 \cos 3t + 4 \sin 3t]$$

By recalling

$$A \cos x + B \sin x = \sqrt{A^2 + B^2} \cos \left(x - \tan^{-1} \frac{B}{A} \right)$$

$$= \sqrt{A^2 + B^2} \sin \left(x + \tan^{-1} \frac{A}{B} \right)$$

the result can be expressed as

$$f(t) = 5\epsilon^{-t} \cos (3t - 53.1°) = 5\epsilon^{-t} \sin (3t + 36.9°)$$

5-3   THE PARTIAL-FRACTION EXPANSION

In most circuit problems, the functions whose inverse transforms give the
time-domain response are often more complicated than the entries in Table
5-1. To make use of the tabulated transform pairs, algebraic manipulation is
necessary so that a particular $F(s)$ will correspond to one of the tabulated entries. Sometimes the process is simple, as is the case in Example 5-3. More
often, $F(s)$ must be decomposed into a sum of component functions whose

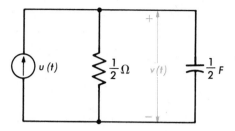

FIGURE 5-2 Parallel *RC* circuit.

inverse Laplace transforms are tabulated. The resultant response is the sum of the inverse transforms of the component functions. The technique most frequently used to decompose $F(s)$ into simple functions is the *partial-fraction expansion* (Heaviside expansion theorem).

To introduce the method consider the parallel *RC* circuit in Fig. 5-2 which is excited by a current source whose strength is the unit step-function $u(t)$. The KCL equation is

$$-u(t) + 2v(t) + \frac{1}{2}\frac{dv(t)}{dt} = 0 \tag{5-12}$$

The Laplace transform of Eq. (5-12) is, by use of Table 5-1,

$$-\frac{1}{s} + 2V(s) + \frac{1}{2}[s\ V(s) - v(0^+)] = 0 \tag{5-13}$$

As the excitation is zero for all $t < 0$, there is no initial energy stored in the capacitor and, consequently, $v(0^+) = 0$. Equation (5-13) becomes

$$\left(2 + \frac{s}{2}\right)V(s) = \frac{1}{s} \tag{5-14}$$

from which

$$V(s) = \frac{2}{s(s + 4)} \tag{5-15}$$

To decompose $V(s)$ by partial-fraction expansion is to express $V(s)$ as a sum of terms as given in Eq. (5-16)

$$V(s) = \frac{K_1}{s} + \frac{K_2}{s + 4} \tag{5-16}$$

The partial-fraction expansion in Eq. (5-16) indicates that $V(s)$ consists of a sum of terms, each of which is a factor of the denominator. The values of K_1 and K_2 are determined by combining the individual fractions by means of the lowest common denominator and comparing the resultant numerator coefficients with those in Eq. (5-15). Thus,

$$V(s) = \frac{K_1(s + 4) + K_2 s}{s(s + 4)} = \frac{s(K_1 + K_2) + 4K_1}{s(s + 4)} \tag{5-17}$$

Comparison of Eq. (5-17) with Eq. (5-15) yields

$$K_1 + K_2 = 0 \quad \text{and} \quad 4K_1 = 2$$

from which

$$K_1 = 0.5 \quad \text{and} \quad K_2 = -0.5$$

Substitution of these values in Eq. (5-16) results in

$$V(s) = \frac{0.5}{s} + \frac{-0.5}{s + 4}$$

whose inverse Laplace transform is, from Table 5-1,

$$v(t) = 0.5 - 0.5\epsilon^{-4t} = 0.5(1 - \epsilon^{-4t})\,u(t) \tag{5-18}$$

The multiplier $u(t)$ in Eq. (5-18) indicates that $v(t) = 0$ for $t < 0$.

The development which leads from Eq. (5-15) to Eq. (5-18) is the basis for the partial-fraction expansion of an arbitrary function. In general, the $F(s)$ whose inverse transform is required can be expressed as the ratio of two polynomials, as indicated in Eq. (5-19).

$$F(s) = \frac{A(s)}{B(s)} = \frac{a_m s^m + a_{m-1} s^{m-1} + \cdots + a_1 s + a_0}{s^n + b_{n-1} s^{n-1} + \cdots + b_1 s + b_0} \tag{5-19}$$

Furthermore, the denominator of $F(s)$ can be factored as

$$B(s) = (s + \alpha_1)(s + \alpha_2) \cdots (s + \alpha_n) = \prod_{i=1}^{n} (s + \alpha_i) \tag{5-20}$$

where each value $s = -\alpha_i$ is a root of $B(s)$. The symbol $\prod$ in Eq. (5-20) indicates the product of terms in a manner similar to the use of Σ to designate the sum of terms.

To develop the general method first consider $F(s)$ for which $A(s)$ is of lower degree than $B(s)$; that is, $m < n$. In addition, it is assumed that $B(s)$ has distinct roots (each of the α_i is different). For this situation, $F(s)$ can be expanded as

$$F(s) = \frac{A(s)}{B(s)} = \frac{K_1}{s + \alpha_1} + \frac{K_2}{s + \alpha_2} + \cdots + \frac{K_n}{s + \alpha_n}$$

$$= \sum_{i=1}^{n} \frac{K_i}{s + \alpha_i} \tag{5-21}$$

Evaluation of the K_i in the manner of Eq. (5-17) requires the simultaneous solution of n equations, an extremely cumbersome task at best. An alternative method to evaluate K_1 is to multiply both sides of Eq. (5-21) by $(s + \alpha_1)$ which gives

$$\frac{(s + \alpha_1)\,A(s)}{B(s)} = K_1 + (s + \alpha_1)\left[\frac{K_2}{s + \alpha_2} + \cdots + \frac{K_n}{s + \alpha_n}\right] \tag{5-22}$$

By setting $s = -\alpha_1$, the right-hand side of Eq. (5-22) is zero except for K_1 so that

$$K_1 = \frac{(s + \alpha_1)A(s)}{B(s)} \bigg|_{s=-\alpha_1} \tag{5-23}$$

Repetition of the process for each root yields the value of K_i for each term in Eq. (5-21). The general result can then be expressed as

$$K_i = \frac{(s + \alpha_i)A(s)}{B(s)} \bigg|_{s=-\alpha_i} \tag{5-24}$$

Example 5-4

Obtain the inverse Laplace transform of the functions

a $F_1(s) = \dfrac{8(s + 2)}{(s + 1)(s + 3)(s + 5)}$

b $F_2(s) = \dfrac{4(s + 1)}{s(s^2 + 2s + 2)}$

SOLUTION **a** Express $F_1(s)$ as

$$F_1(s) = \frac{K_1}{s + 1} + \frac{K_2}{s + 3} + \frac{K_3}{s + 5}$$

From Eq. (5-24),

$$K_1 = (s + 1)F_1(s) \bigg|_{s=-1} = \frac{8(s + 2)}{(s + 3)(s + 5)} \bigg|_{s=-1}$$

$$= \frac{8(-1 + 2)}{(-1 + 3)(-1 + 5)} = 1$$

$$K_2 = (s + 3)F_1(s) \bigg|_{s=-3} = \frac{8(s + 2)}{(s + 1)(s + 5)} \bigg|_{s=-3}$$

$$= \frac{8(-3 + 2)}{(-3 + 1)(-3 + 5)} = 2$$

$$K_3 = (s + 5)F_1(s) \bigg|_{s=-5} = \frac{8(s + 2)}{(s + 1)(s + 3)} \bigg|_{s=-5}$$

$$= \frac{8(-5 + 2)}{(-5 + 1)(-5 + 3)} = -3$$

These values give

$$F_1(s) = \frac{1}{s + 1} + \frac{2}{s + 3} - \frac{3}{s + 5}$$

for which the inverse transform is

$$f_1(t) = \epsilon^{-t} + 2\epsilon^{-3t} - 3\epsilon^{-5t}$$

b   The roots of the denominator of $F_2(s)$ are $s = 0$, $-1 \pm j1$. The partial-fraction expansion of $F_2(s)$ is

$$F_2(s) = \frac{K_1}{s} + \frac{K_2}{s + 1 - j1} + \frac{K_3}{s + 1 + j1}$$

The values of K_1, K_2, and K_3 are found by use of Eq. (5-24) as

$$K_1 = sF_2(s) \Big|_{s=0} = \frac{4(s + 1)}{s^2 + 2s + 2} \Big|_{s=0} = 2$$

$$K_2 = (s + 1 - j1)F_2(s) \Big|_{s=-1+j1} = \frac{4(s + 1)}{s(s + 1 + j1)} \Big|_{s=-1+j1}$$

$$= \frac{2}{-1 + j1} = \sqrt{2}\epsilon^{-j135°}$$

$$K_3 = (s + 1 + j1)F_2(s) \Big|_{s=-1-j1} = \frac{4(s + 1)}{s(s + 1 - j1)} \Big|_{s=-1-j1}$$

$$= \frac{2}{-1 - j1} = \sqrt{2}\epsilon^{+j135°}$$

The resulting $F_2(s)$ becomes

$$F_2(s) = \frac{2}{s} + \frac{\sqrt{2}\,\epsilon^{-j135°}}{s + 1 - j1} + \frac{\sqrt{2}\,\epsilon^{+j135°}}{s + 1 + j1}$$

From Table 5-1,

$$f_2(t) = 2 + \sqrt{2}\,\epsilon^{-j135°} \cdot \epsilon^{(-1+j1)t} + \sqrt{2}\,\epsilon^{+j135°} \cdot \epsilon^{(-1-j1)t}$$

Rearrangement of the terms gives

$$f_2(t) = 2 + \sqrt{2}\,\epsilon^{-t}[\epsilon^{+j(t-135°)} + \epsilon^{-j(t-135°)}]$$

The use of Euler's identity, $\cos x = \frac{1}{2}(\epsilon^{jx} + \epsilon^{-jx})$, permits $f_2(t)$ to be expressed as

$$f_2(t) = 2 + 2\sqrt{2}\,\epsilon^{-t} \cos (t - 135°)$$

In Example 5-4b, the values of K_2 and K_3 are conjugates. The two terms in the partial-fraction expansion associated with the conjugate pair of roots are, therefore, also conjugates. This is not an isolated case but always occurs when $F(s)$ has real coefficients. As a consequence, for a pair of terms of the form

$$\frac{K_1}{s + \alpha - j\beta} + \frac{K_2}{s + \alpha + j\beta}$$

K_1 and K_2, the conjugate of K_1, are expressible as

$$K_1 = A\epsilon^{j\theta} \qquad K_2 = A\epsilon^{-j\theta}$$

The inverse transform follows from Example 5-4b and is

$$\mathscr{L}^{-1}\left[\frac{K_1}{s + \alpha - j\beta} + \frac{K_2}{s + \alpha + j\beta}\right] = 2A\epsilon^{-\alpha t} \cos (\beta t + \theta) \tag{5-25}$$

The result given in Eq. (5-25) can also be obtained by use of entries (**16**) and (**17**) in Table 5-1. Verification of this equivalence is left to the student as an exercise.

For the case where $A(s)$ is of the same degree as or higher degree than $B(s)$, that is, $m \geq n$, Eq. (5-19) can be rewritten as

$$\frac{A(s)}{B(s)} = [A_m s^m + A_{m-1} s^{m-1} + \cdots + A_1 s + A_0] + \frac{A_1(s)}{B(s)} \tag{5-26}$$

Equation (5-26) is obtained from Eq. (5-19) by long division; the bracketed term is the quotient, and $A_1(s)/B(s)$ is the remainder. The remainder can be expanded into partial fractions as in Eq. (5-21). In the next section of this chapter, several properties of the Laplace transform are developed which allow the inverse transform of the quotient to be obtained. However, while Eq. (5-26) is a general result, in most physical systems the degree of $A(s)$ rarely exceeds the degree of $B(s)$. As illustrated in Example 5-5, the transform pairs in Table 5-1 are sufficient to obtain the inverse transform in these cases.

Example 5-5

Obtain

$$f(t) = \mathcal{L}^{-1} \left[\frac{s^2 + 7s + 14}{s^2 + 3s + 2} \right] = \mathcal{L}^{-1}[F(s)]$$

SOLUTION   Performing the required long division and factoring the denominator yields

$$F(s) = 1 + \frac{4s + 12}{(s + 1)(s + 2)} = 1 + \frac{A_1(s)}{B(s)}$$

The partial-fraction expansion of $A_1(s)/B(s)$ is

$$\frac{A_1(s)}{B(s)} = \frac{4s + 12}{(s + 1)(s + 2)} = \frac{K_1}{s + 1} + \frac{K_2}{s + 2}$$

Evaluation of K_1 and K_2 gives

$$K_1 = \left. \frac{4s + 12}{s + 2} \right|_{s=-1} = 8$$

and

$$K_2 = \left. \frac{4s + 12}{s + 1} \right|_{s=-2} = -4$$

which result in

$$F(s) = 1 + \frac{8}{s + 1} - \frac{4}{s + 2}$$

The inverse transform is obtained with use of Table 5-1 as

$$f(t) = \mathscr{L}^{-1}[F(s)] = \delta(t) + 8\epsilon^{-t} - 4\epsilon^{-2t}$$

The term $\delta(t)$ is the unit-impulse function described in Section 4-9.

The last case to be investigated is when $B(s)$ contains a factor $(s + \alpha_i)^k$. In this situation $B(s)$ is said to have a repeated root $-\alpha_i$ of multiplicity k. The partial-fraction expansion with repeated roots is illustrated by consideration of the function

$$F(s) = \frac{A(s)}{(s + a)^3} = \frac{K_1}{(s + a)^3} + \frac{K_2}{(s + a)^2} + \frac{K_3}{(s + a)} \tag{5-27}$$

in which the degree of $A(s)$ is no greater than 2. Multiplication of both sides of Eq. (5-27) by $(s + a)^3$ gives

$$(s + a)^3 F(s) = K_1 + K_2(s + a) + K_3(s + a)^2 \tag{5-28}$$

Setting $s = -a$ in Eq. (5-28) yields

$$K_1 = (s + a)^3 F(s) \Big|_{s=-a} \tag{5-29}$$

K_2 is obtained by differentiating both sides of Eq. (5-28) with respect to s and setting $s = -a$. The result is

$$\frac{d}{ds}[(s + a)^3 F(s)] = K_2 + 2K_3(s + a) \tag{5-30}$$

and for $s = -a$

$$K_2 = \frac{d}{ds}[(s + a)^3 F(s)] \Big|_{s=-a} \tag{5-31}$$

Differentiating both sides of Eq. (5-30) and setting $s = -a$ determines the value of K_3 as

$$K_3 = \frac{1}{2}\frac{d^2}{ds^2}[(s + a)^3 F(s)] \Big|_{s=-a} \tag{5-32}$$

For the general case of repeated roots $F(s)$ is expanded as

$$F(s) = \frac{A(s)}{(s + \alpha)^k} = \frac{K_1}{(s + \alpha)^k} + \frac{K_2}{(s + \alpha)^{k-1}} + \cdots + \frac{K_k}{s + \alpha} \tag{5-33}$$

The constants K_1 to K_k are evaluated in a fashion similar to the development in Eqs. (5-27) to (5-32). The general term has the form

$$K_\nu = \frac{1}{(\nu - 1)!}\frac{d^{\nu-1}}{ds^{\nu-1}}[(s + \alpha)^k F(s)] \Big|_{s=-\alpha} \tag{5-34}$$

Example 5-6

Determine the inverse Laplace transform of

$$F(s) = \frac{s + 2}{s(s + 1)^3}$$

SOLUTION The expansion for $F(s)$ is

$$F(s) = \frac{K_0}{s} + \frac{K_1}{(s + 1)^3} + \frac{K_2}{(s + 1)^2} + \frac{K_3}{s + 1}$$

By use of Eq. (5-24)

$$K_0 = sF(s)\Big|_{s=0} = \frac{s + 2}{(s + 1)^3}\Big|_{s=0} = 2$$

The coefficients K_1, K_2, and K_3 are determined from Eqs. (5-29), (5-30), and (5-31), respectively, as

$$K_1 = (s + 1)^3 F(s)\Big|_{s=-1} = \frac{s + 2}{s}\Big|_{s=-1} = -1$$

$$K_2 = \frac{d}{ds}\left[(s + 1)^3 F(s)\right]\Big|_{s=-1} = \frac{d}{ds}\left[\frac{s + 2}{s}\right]\Big|_{s=-1}$$

$$= \frac{-2}{s^2}\Big|_{s=-1} = -2$$

$$K_3 = \frac{1}{2}\frac{d^2}{ds^2}\left[(s + 1)^3 F(s)\right]\Big|_{s=-1} = \frac{1}{2}\frac{d^2}{ds^2}\left[\frac{s + 2}{s}\right]\Big|_{s=-1}$$

$$= \frac{1}{2}\left[\frac{4}{s^3}\right]\Big|_{s=-1} = -2$$

These values of the K terms yield

$$F(s) = \frac{2}{s} - \frac{1}{(s + 1)^3} - \frac{2}{(s + 1)^2} - \frac{2}{(s + 1)}$$

for which the inverse transform, obtained by use of entries **5, 11, 10,** and **9,** respectively, is

$$f(t) = 2 - \frac{t^2}{2}\epsilon^{-t} - 2t\epsilon^{-t} - 2\epsilon^{-t} = 2 - \epsilon^{-t}\left(\frac{t^2}{2} + 2t + 2\right)$$

5-4 ADDITIONAL PROPERTIES OF THE LAPLACE TRANSFORM

The utility of Table 5-1 in the solution of networks problems is based on both the linearity and the uniqueness of the Laplace transformation. The partial-fraction expansion described in the previous section is one technique which makes the use of Table 5-1 convenient. Several other properties of the

TABLE 5-2   PROPERTIES OF THE LAPLACE TRANSFORM

PROPERTY	TIME DOMAIN	FREQUENCY DOMAIN
1 Time delay	$f(t - T)u(t - T)$	$\epsilon^{-sT}F(s)$
2 Periodic function $f(t) = f(t + nT)$	$f(t)$ $0 \le t \le T$	$F(s)/[1 - \epsilon^{Ts}]$ where $$F(s) = \int_0^T f(t)\epsilon^{-st}dt$$
3 Time scaling	$f(at)$	$\dfrac{1}{a}F(s/a)$
4 Frequency differentiation (multiplication by t)	$tf(t)$	$-\dfrac{dF(s)}{ds}$
5 Frequency integration (division by t)	$\dfrac{f(t)}{t}$	$\displaystyle\int_s^\infty F(s)ds$
6 Frequency shifting	$f(t)\epsilon^{-at}$	$F(s + a)$
7 Initial-value theorem	$\text{Lim } f(t) = f(0^+)$ $t \to 0$	$\text{Lim } sF(s)$ $s \to \infty$
8 Final-value theorem	$\text{Lim } f(t)$ $t \to \infty$ where limit exists	$\text{Lim } sF(s)$ $s \to 0$

Laplace transform are also useful in adapting the table of transforms to the functions which arise in physical systems. In this section these properties, listed in Table 5-2, are briefly described.

The first two entries in Table 5-2 deal with functions whose description involves a time delay. The function $f(t - T)u(t - T)$ is the notation employed to indicate that $f(t)u(t)$ is delayed by T seconds. Such a situation is illustrated in Fig. 5-3. The term $u(t - T)$ is the *delayed unit-step function*, first introduced in Sec. 4-7, whose definition is

$$u(t - T) = 1 \qquad t \ge T$$
$$= 0 \qquad t < T \tag{5-35}$$

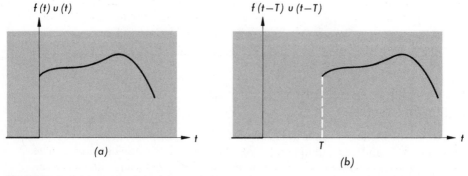

FIGURE 5-3   (a) An arbitrary function and (b) the function in a delayed by T seconds.

To describe the first entry in Table 5-2, consider the Laplace transform of the delayed unit-step function. From the definition in Eq. (5-1)

$$\mathcal{L}[u(t - T)] = \int_0^\infty u(t - T)\epsilon^{-st} \, dt = \int_T^\infty 1 \cdot \epsilon^{-st} \, dt$$

$$= -\frac{1}{s} \, \epsilon^{-st} \, \Big|_T^\infty$$

Evaluation at the indicated limits gives

$$\mathcal{L}[u(t - T)] = \left(\frac{1}{s}\right) \epsilon^{-sT} = \mathcal{L}[u(t)]\epsilon^{-sT} \tag{5-36}$$

In Eq. (5-36), the effect of the delay in the application of the step function by T seconds is reflected in the ϵ^{-sT} term which multiplies the Laplace transform of the unit-step function. In the general case, delay in the time domain is indicated by the ϵ^{-sT} factor in the frequency domain.

The Laplace transform of a general periodic function is given in the second entry in Table 5-2. Consecutive cycles can be represented by the function $f(t)$ and the function $f(t)$ that is time-delayed by one period. For example, the first three cycles of a periodic waveform of period T are expressible in $f(t)u(t)$, $f(t - T)u(t - T)$, and $f(t - 2T)u(t - 2T)$, respectively. The corresponding Laplace transforms are $F(s)$, $F(s)\epsilon^{-sT}$, and $F(s)\epsilon^{-2sT}$. In the transform of the entire periodic function, the factor $1/(1 - \epsilon^{Ts})$ is the sum of the infinite geometric series that results when $f(t + nT)$ is expressed by delay functions.

The time-scaling property **(3)** in Table 5-2 is often convenient in dealing with the numeric values one encounters in practice. The term $f(at)$ indicates time normalization; that is, if the range of the variable t is measured in microseconds, the normalized variable at can be measured in seconds. In the Laplace transform, $F(s/a)$ exhibits the corresponding frequency normalization which translates the range of the variable s from MHz to Hz for the variable s/a.

The time-frequency dualism is the basis for the next three entries, **(4)** to **(6)**, in the table. We have seen that differentiation in the time domain becomes multiplication by s in the frequency domain. In **(4)**, multiplication by t in the time domain results in differentiation in the frequency domain. Similar reasoning results in the integral relation in entry **(5)**.

Dualism is also indicated in the frequency-shifting property in entry **(6)**. A time shift of T seconds in $f(t)$ resulted in multiplication of $F(s)$ by ϵ^{-sT}. A frequency shift, indicated by $F(s + a)$ leads to multiplication by ϵ^{-at} in the time domain.

Example 5-7

Find $\mathcal{L}[t\epsilon^{-t}]$.

SOLUTION

$$\mathcal{L}[t\epsilon^{-t}] = \mathcal{L}[tf(t)]$$

where $f(t) = \epsilon^{-t}$. By use of entry (**4**) in Table 5-2

$$\mathcal{L}[tf(t)] = -\frac{dF(s)}{ds}$$

and

$$F(s) = \mathcal{L}[\epsilon^{-t}] = \frac{1}{s+1}$$

Performing the requisite differentiation,

$$-\frac{dF(s)}{ds} = \frac{1}{(s+1)^2}$$

so that

$$\mathcal{L}[t\epsilon^{-t}] = \frac{1}{(s+1)^2}$$

which is entry (**10**) in Table 5-1.

Alternatively, the frequency-shifting property can be used to evaluate the Laplace transform.

$$\mathcal{L}[t\epsilon^{-t}] = \mathcal{L}[\epsilon^{-t}f(t)] = F(s+1)$$

and where $f(t) = t$.

By use of entry (**7**) in Table 5-1,

$$\mathcal{L}[t] = \frac{1}{s^2} = F(s)$$

which causes

$$\mathcal{L}[t\epsilon^{-t}] = F(s+1) = \frac{1}{(s+1)^2}$$

The two alternative methods of solution indicate the manner in which the properties can be used to evaluate both the Laplace transform and the inverse transform.

The last two entries in Table 5-2, referred to as the *initial-value theorem* and *final-value theorem,* also display the dualism of the time and frequency domains. To obtain the initial value $f(0^+)$, s is allowed to approach ∞. Similarly, as t approaches ∞, the final value of $f(t)$ is evaluated by setting $s = 0$. The use of the theorems is illustrated in Example 5-8.

Example 5-8

The function

$$I(s) = \frac{5}{s + 2}$$

is the frequency-domain response of the RL circuit of Fig. 5-1. Determine $f(0^+)$ and $\lim_{t \to \infty} f(t)$.

SOLUTION   From the initial-value theorem

$$i(0^+) = \lim_{s \to \infty} sI(s) = \lim_{s \to \infty} \frac{5s}{s + 2} = 5 \text{ A}$$

and from the final-value theorem

$$\lim_{t \to \infty} i(t) = \lim_{s \to 0} sI(s) = \lim_{s \to 0} \frac{5s}{s + 2} = 0$$

As expected, these values are identical with those found in Sec. 5-2.

5-5   TRANSFORMED NETWORKS

In Examples 5-1 and 5-2, the Laplace transform was introduced in the solution of circuits problems only after the differential equations which characterize these circuits were written. This section deals with the use of *transformed networks*, first introduced in Secs. 3-2 and 3-3, as a means to convert networks directly from the time domain to the frequency domain. As a result, the need to explicitly write the circuit differential equations is eliminated. Circuit theory problems can then be formulated in the frequency domain and effective use can be made of the standard circuit analysis techniques described in Chaps. 2 and 3.

The transformed network equivalents for the three elements R, L, and C are based on the Laplace transforms of their respective volt-ampere characteristics. These relations are

$$\mathcal{L}[v(t) = Ri(t)] \longrightarrow V(s) = RI(s) \tag{5-37}$$

$$\mathcal{L}\left[v(t) = L\frac{di(t)}{dt}\right] \longrightarrow V(s) = sLI(s) - Li(0^+) \tag{5-38}$$

$$\mathcal{L}\left[v(t) = \frac{1}{C}\int_{-\infty}^{t} i(t)\, dt\right] \longrightarrow V(s) = \frac{1}{Cs}I(s) + \frac{v(0^+)}{s} \tag{5-39}$$

By the use of Table 5-1,

$$\int_{-\infty}^{0} i(t)\, dt = q(0^-)$$

and $q(0^-)/C$ is $v(0^-)$. By continuity of capacitance voltage, $v(0^-) = v(0^+)$, which is the term appearing in Eq. (5-39).

Equivalent circuit representations for the relations in Eqs. (5-37) to (5-39) are shown in Fig. 5-4. Corresponding pairs of terminals for the time-domain and frequency-domain representations in Fig. 5-4 are indicated by 1 and 2. Equivalence between time and frequency domains exists only between corresponding terminal pairs. Thus, for the inductance in Fig. 5-4d, the initial current is included in the transformed network by means of the voltage source $Li(0^+)$. The circuit in Fig. 5-4e is obtained from that in Fig.

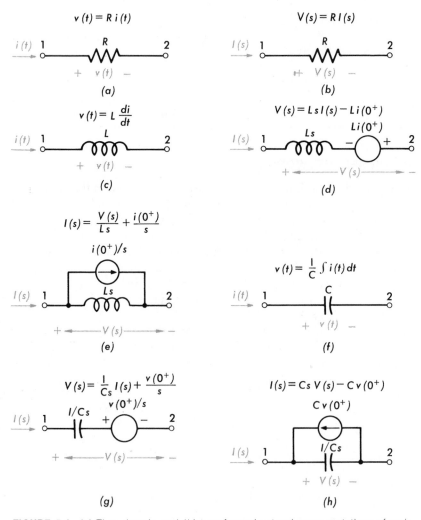

FIGURE 5-4 (a) Time-domain and (b) transformed-network representations of resistance; (c) time-domain and (d) and (e) transformed-network representations of inductance; (f) time-domain and (g) and (h) transformed-network representations of capacitance.

5-4d by converting the voltage source $Li(0^+)$ in series with the impedance Ls to its current source-parallel impedance equivalent. The initial capacitance voltage is indicated by the voltage source $v(0^+)/s$ in Fig. 5-4g. Source conversion of the circuit in Fig. 5-4g results in the circuit representation in Fig. 5-4h.

Comparison of Fig. 5-4 with the entries in Table 3-1 indicates that the inclusion of the initial condition sources in the Laplace transform representation is the major difference between them. The impedance $Z(s)$ and admittance $Y(s)$ for each of the elements, however, is that given in Table 3-1. Thus, sL is the impedance of the inductance and $1/Cs$ is the impedance of the capacitance. Their respective admittances are $1/Ls$ and Cs.

The transformed network representation of current and voltage sources are simply the Laplace transforms of the time functions which define the source current and voltage as illustrated in Fig. 5-5. Dependent-source representations are shown in Fig. 5-6, and follow directly from the transformed networks for independent sources.

The use of transformed networks in the solution of networks problems utilizes the following procedure.

1 Transform the circuit into the frequency domain by use of the representations given in Figs. 5-4 to 5-6.

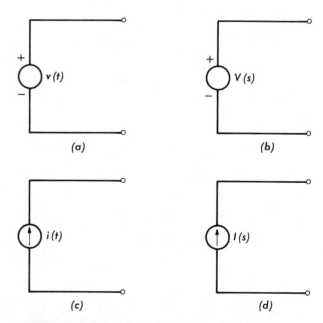

FIGURE 5-5 (a) Time-domain and (b) frequency-domain representations of a voltage source; (c) time-domain and (d) frequency-domain representations of a current source.

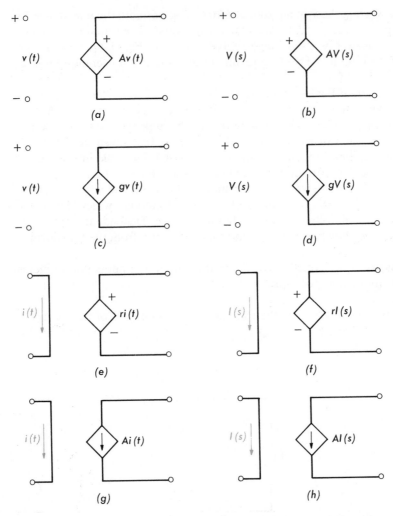

FIGURE 5-6 (a) Time- and (b) frequency-domain representation of a voltage-controlled voltage source; (c) time- and (d) frequency-domain representation of a voltage-controlled current source; (e) time- and (f) frequency-domain representation of a current-controlled voltage source; (g) time- and (h) frequency-domain representation of a current-controlled current source.

2 Formulate the appropriate KVL and KCL equations for the resultant transformed network.

3 Algebraically solve for the desired network response in the frequency domain.

4 Obtain the time-domain response by evaluation of the inverse Laplace transform of the frequency-domain response.

The following examples illustrate the use of the procedure and indicate the applicability of the traditional methods of nodal and mesh analysis.

Example 5-9

Determine $v_2(t)$ for $t \geq 0$ for the circuit in Fig. 5-7.

SOLUTION As there is no initial energy storage in the capacitors, the first step in the solution process gives the transformed network as shown in Fig. 5-8. The nodal equations (step 2) are written as

$$\frac{V_1(s)}{500} + 50 \times 10^{-12}sV_1(s) + 2 \times 10^{-12}s[V_1(s) - V_2(s)]$$

$$- \frac{\sqrt{2} \times 10^{-4}s}{s^2 + 4 \times 10^{14}} = 0$$

$$2 \times 10^{-12}s[V_2(s) - V_1(s)] + \frac{V_2(s)}{400} + 100I_1(s) = 0$$

The constraint equation for the dependent source is

$$I_1(s) = V_1(s)/500$$

Combination of the three equations yields

$$V_1(s)[2 \times 10^{-3} + 52 \times 10^{-12}s] - V_2(s)[2 \times 10^{-12}s]$$
$$= \sqrt{2} \times 10^{-4}s/(s^2 + 4 \times 10^{14})$$

$$- V_1(s)[2 \times 10^{-12}s - 0.2] + V_2(s)[2.5 \times 10^{-3} + 2 \times 10^{-12}s] = 0$$

The third step is to obtain the simultaneous solution for $V_2(s)$, which, after rearrangement of terms, is

$$V_2(s) = \frac{2\sqrt{2} \times 10^6(s - 10^{11})s}{(s^2 + 4 \times 10^{14})(s^2 + 53.4 \times 10^8s + 5 \times 10^{16})}$$

or

$$V_2(s) = \frac{2\sqrt{2} \times 10^6 s(s - 10^{11})}{(s - j2 \times 10^7)(s + j2 \times 10^7)(s + 0.94 \times 10^7)(s + 53.3 \times 10^8)}$$

To obtain $v_2(t)$, the last step in the solution process, $V_2(s)$ is expanded in partial fractions as

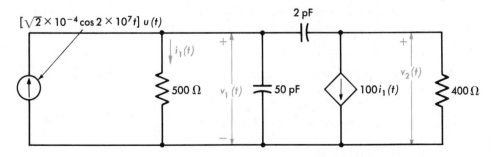

FIGURE 5-7 Circuit for Example 5-9.

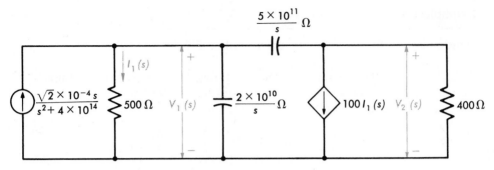

FIGURE 5-8 Transformed network of circuit in Fig. 5-7.

$$V_2(s) = \frac{K_1}{s - j2 \times 10^7} + \frac{K_1^*}{s + j2 \times 10^7} + \frac{K_2}{s + 0.94 \times 10^7}$$

$$+ \frac{K_3}{s + 53.3 \times 10^8}$$

Application of Eq. (5-24) for the factors which contain K_1, K_2, and K_3 gives

$$K_1 = 1.20\underline{/-245°} \qquad K_2 = 1.022 \qquad K_3 = -0.0105$$

By use of Table 5-1 and Eq. (5-25) for the conjugate-pair terms,

$$v_2(t) = [2.40 \cos (2 \times 10^7 t - 245°) + 1.022\epsilon^{-0.94 \times 10^7 t}$$
$$- 0.0105\epsilon^{-53.3 \times 10^8 t}]u(t)$$

The circuit of Fig. 5-7 is identical to that shown in Fig. 3-23a and analyzed in Example 3-9. The cosine term in the response $v_2(t)$ obtained in this example is, as expected, the same as the sinusoidal steady-state response evaluated in Example 3-9.

Example 5-10

In the circuit shown in Fig. 5-9, the switch S has been open for a long time. At $t = 0$, the switch is closed. Determine the currents $i_1(t)$ and $i_2(t)$.

SOLUTION   With the switch open, the circuit is in the steady state at $t = 0^-$ and no voltage drop exists across the 1-H inductance. The current $i_1(0^-)$ is

$$i_1(0^-) = \frac{110}{3 + 8} = 10 \text{ A}$$

The transformed network for the circuit of Fig. 5-9 is given in Fig. 5-10. The mesh equations are

$$- \frac{110}{s} + (3 + s)I_1(s) - 10 + 8[I_1(s) - I_2(s)] = 0$$

$$8[I_1(s) - I_2(s)] + (6 + 2s)I_2(s) = 0$$

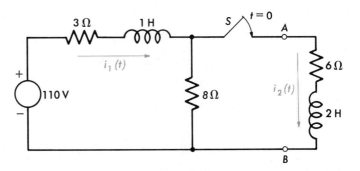

FIGURE 5-9 Circuit for Example 5-10.

Rearrangement of terms gives

$$I_1(s)[11 + s] - 8I_2(s) = \frac{110}{s} + 10$$

$$-8I_1(s) + I_2(s)[14 + 2s] = 0$$

After simultaneous solution and factoring,

$$I_1(s) = \frac{10(s + 11)(s + 7)}{s(s + 3)(s + 15)}$$

$$I_2(s) = \frac{40(s + 11)}{s(s + 3)(s + 15)}$$

Expansion of $I_1(s)$ and $I_2(s)$ into partial fractions results in

$$I_1(s) = \frac{K_1}{s} + \frac{K_2}{s + 3} + \frac{K_3}{s + 15}$$

$$I_2(s) = \frac{K_4}{s} + \frac{K_5}{s + 3} + \frac{K_6}{s + 15}$$

Evaluation of the K terms, using Eq. (5-24), gives

$$K_1 = \tfrac{154}{9} \qquad K_2 = -\tfrac{80}{9} \qquad K_3 = \tfrac{16}{9}$$

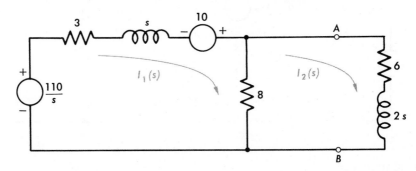

FIGURE 5-10 Transformed network of circuit in Fig. 5-9.

and

$$K_4 = \tfrac{88}{9} \qquad K_5 = -\tfrac{80}{9} \qquad K_6 = -\tfrac{8}{9}$$

The corresponding time functions then become

$$i_1(t) = \tfrac{154}{9} - \tfrac{80}{9}\,\epsilon^{-3t} + \tfrac{16}{9}\,\epsilon^{-15t}$$

$$i_2(t) = \tfrac{88}{9} - \tfrac{80}{9}\,\epsilon^{-3t} - \tfrac{8}{9}\epsilon^{-15t}$$

The expressions for $i_1(t)$ and $i_2(t)$ are valid for $t \geq 0$.

Thévenin's theorem is also applicable in the frequency domain. Both the open-circuit voltage $V_o(s)$ and the equivalent impedance $Z_o(s)$, shown in Fig. 5-11, are, as indicated, functions of the frequency-variable s. The method of obtaining the Thévenin equivalent requires the evaluation of two of the three quantities $V_o(s)$, $I_{sc}(s)$, and $Z_o(s)$ by any of the usual circuit techniques.

Example 5-11

Determine $i_2(t)$ for $t \geq 0$ in the circuit of Fig. 5-9.

SOLUTION Thévenin's theorem will be used. The transformed network is given in Fig. 5-10 as developed in Example 5-10. The open-circuit voltage is evaluated from the circuit of Fig. 5-11a for which

$$V_o(s) = I(s) \times 8$$

The KVL equation for the loop is

$$-\frac{110}{s} + (3 + s)I(s) - 10 + 8I(s) = 0$$

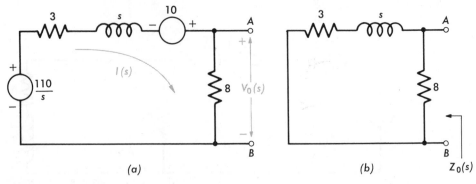

(a) (b)

FIGURE 5-11 Circuits used to compute (a) the open-circuit voltage and (b) the equivalent impedance in Example 5-11.

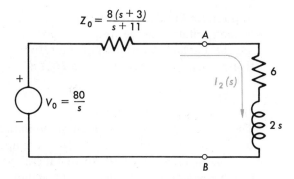

FIGURE 5-12 Thévenin equivalent of circuit in Fig. 5-10.

from which

$$I(s) = \frac{(110/s) + 10}{11 + s}$$

and, after rearrangement,

$$V_0(s) = \frac{80(s + 11)}{s(s + 11)} = \frac{80}{s}$$

The Thévenin impedance is computed from the circuit of Fig. 5-11*b* as the parallel combination of the 8-Ω resistance and the $(3 + s)$-ohm impedance.

$$Z_0(s) = \frac{8(3 + s)}{8 + 3 + s} = \frac{8(3 + s)}{11 + s}$$

Figure 5-12 is the circuit of Fig. 5-10 in which the portion to the left of terminals A-B is replaced by its Thévenin equivalent. The current $I_2(s)$ is then

$$I_2(s) = \frac{80/s}{[8(3 + s)/(11 + s)] + (6 + 2s)} = \frac{40(s + 11)}{s(s^2 + 18s + 45)}$$

Upon factoring,

$$I_2(s) = \frac{40(s + 11)}{s(s + 3)(s + 15)}$$

which is the same result as obtained in Example 5-9. Thus,

$$i_2(t) = \tfrac{88}{9} - \tfrac{80}{9}\,\epsilon^{-3t} - \tfrac{8}{9}\epsilon^{-15t}$$

5-6 NETWORK FUNCTIONS

System performance often requires that component circuits respond appropriately to a variety of excitations. The use of frequency-domain analysis to determine the excitation-response characteristics of systems is enhanced by the introduction of the concept of a network function. *Network functions,* also referred to as *system functions,* are defined as the ratio of the response to the excitation in the frequency domain. Mathematically, if $Y(s)$ and $X(s)$ are the Laplace transforms of the response $y(t)$ and excitation $x(t)$, respectively, then the system function $H(s)$ is

$$H(s) = \frac{Y(s)}{X(s)} \tag{5-40}$$

The result in Eq. (5-40) is based on no initial energy storage in the network; that is, the system is at rest prior to the application of the excitation.

The significance of $H(s)$ is that it embodies the input-output relationship which characterizes the network or system. As such, an $H(s)$ can be defined to indicate a voltage response to either a current or voltage excitation. Alternatively, a current response to either a voltage or current excitation can be used to define the network function. Driving-point impedances and admittances are network functions we encountered in earlier chapters. Here the network function is the ratio of either a voltage to a current or a current to a voltage at a specified terminal pair as seen in Fig. 5-13. Often, however, we are interested in the response measured at one pair of terminals (the output) to an excitation (the input) applied to another pair of terminals. This situation is depicted in Fig. 5-14 for which the network function is usually referred to as a *transfer function.* The input-output nature of the transfer function is, for example, frequently used to indicate the gain of an amplifier as the ratio of the output voltage to the input voltage.

The use of the network function in circuit analysis is demonstrated by considering the linear system schematically represented in Fig. 5-15. The time-domain, shown in Fig. 5-15a, illustrates that the excitation $x(t)$ produces a response $y(t)$. [This is often indicated $x(t) \rightarrow y(t)$.] In Fig. 5-15b the frequency-domain equivalent of Fig. 5-15a is depicted for which $X(s)$ applied to the system $H(s)$ results in the response $Y(s)$.

In the time-domain, a differential equation describes the relationship between $x(t)$ and $y(t)$. The general form of this equation is

$$a_m \frac{d^m x}{dt^m} + a_{m-1} \frac{d^{m-1} x}{dt^{m-1}} + \cdots + a_1 \frac{dx}{dt} + a_o x = b_n \frac{d^n y}{dt^n}$$

$$+ b_{n-1} \frac{d^{n-1} y}{dt^{n-1}} + \cdots + b_1 \frac{dy}{dt} + b_o y \tag{5-41}$$

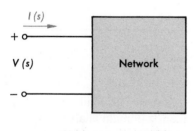

$$Z(s) = \frac{V(s)}{I(s)} \qquad Y(s) = \frac{I(s)}{V(s)}$$

FIGURE 5-13 Representation of driving-point functions.

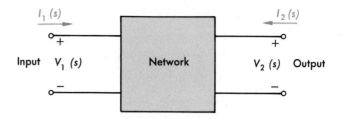

Transfer functions: $\dfrac{V_2\,(s)}{V_1\,(s)} \quad \dfrac{V_2\,(s)}{I_1\,(s)} \quad \dfrac{I_2\,(s)}{V_1\,(s)} \quad \dfrac{I_2\,(s)}{I_1\,(s)}$

FIGURE 5-14 Representation of transfer-type network functions.

The natural response, obtained by the methods in Chap. 4, is

$$y_n(t) = \sum_1^n K_i \epsilon^{s_i t} \tag{5-42}$$

for which K_i are the constants and s_i are the roots of the characteristic equation

$$b_n s^n + b_{n-1} s^{n-1} + \cdots + b_1 s + b_0 = 0 \tag{5-43}$$

The forced response is of the form of $x(t)$ and its unique derivatives.

The frequency-domain solution is obtained by Laplace-transforming the differential equation. Assuming no initial energy storage, the transform of

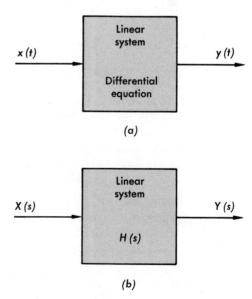

FIGURE 5-15 A linear system in (a) the time-domain and (b) the frequency domain.

Eq. (5-41) becomes

$$(a_m s^m + a_{m-1} s^{m-1} + \cdots + a_1 s + a_0)X(s)$$
$$= (b_n s^n + b_{n-1} s^{n-1} + \cdots + b_1 s + b_0)Y(s) \quad (5\text{-}44)$$

The system function, from Eq. (5-40), can be expressed as

$$H(s) = \frac{Y(s)}{X(s)} = \frac{a_m s^m + a_{m-1} s^{m-1} + \cdots + a_1 s + a_0}{b_n s^n + b_{n-1} s^{n-1} + \cdots + b_1 s + b_0} = \frac{A(s)}{B(s)} \quad (5\text{-}45)$$

Alternatively, $H(s)$ can be used to indicate the action of the system on the excitation that is necessary to produce the response. Combination of Eqs. (5-40) and (5-45) results in

$$Y(s) = H(s)X(s) = \frac{A(s)}{B(s)} X(s) \quad (5\text{-}46)$$

For an excitation $X(s) = \dfrac{P(s)}{Q(s)}$, Eq. (5-46) becomes

$$Y(s) = \frac{A(s)}{B(s)} \frac{P(s)}{Q(s)} \quad (5\text{-}47)$$

To obtain $y(t)$, $Y(s)$ is expanded in partial fractions and the inverse Laplace transform is taken term by term. Thus, Eq. (5-47) is rewritten as

$$Y(s) = \sum_1^n \frac{K_i}{s - s_i} + \sum_1^k \frac{K_j}{s - \alpha_j} \quad (5\text{-}48)$$

The first summation in Eq. (5-48) are the terms in the partial-fraction expansion associated with $B(s)$, in which the values of s_i are the roots of $B(s) = 0$. The second summation represents the terms that are identified with $Q(s)$ in Eq. (5-47). The α_j values are roots of $Q(s) = 0$. The inverse Laplace transform of Eq. (5-48) yields

$$y(t) = \sum_1^n K_i \epsilon^{s_i t} + \sum_1^k K_j \epsilon^{\alpha_j t} \quad (5\text{-}49)$$

In Eq. (5-49), the first summation corresponds to the natural response given in Eq. (5-42) and, as expected, these terms arise from the system function and are independent of the excitation. Consequently, the second summation must be the forced response and, from Eqs. (5-47) and (5-48) can be seen to correspond to terms arising from the excitation $X(s)$.

Example 5-12

The system function for a network is given by

$$H(s) = \frac{2(s + 2)}{(s + 1)(s + 4)}$$

a Determine $y(t)$ for $x(t) = u(t)$.

b Repeat a for $x(t) = \epsilon^{-2t}$.

c Write the differential equation that relates $x(t)$ and $y(t)$.

SOLUTION **a** For $x(t) = u(t)$, $X(s) = \dfrac{1}{s}$. Then, from Eq. (5-46),

$$Y(s) = H(s)X(s) = \frac{2(s + 2)}{(s + 1)(s + 4)} \frac{1}{s}$$

The partial fraction expansion is

$$Y(s) = \frac{K_1}{s + 1} + \frac{K_2}{s + 4} + \frac{K_3}{s}$$

where

$$K_1 = \frac{2(s + 2)}{(s + 4)s}\bigg|_{s=-1} = -\frac{2}{3}$$

$$K_2 = \frac{2(s + 2)}{(s + 1)s}\bigg|_{s=-4} = -\frac{1}{3}$$

and

$$K_3 = \frac{2(s + 2)}{(s + 1)(s + 4)}\bigg|_{s=0} = 1$$

Substitution for K_1, K_2, and K_3 and obtaining the inverse transform gives

$$y(t) = -\tfrac{2}{3}\epsilon^{-t} - \tfrac{1}{3}\epsilon^{-4t} + 1$$

It is noteworthy to observe that K_3 is in actuality $H(s)$ evaluated at the s-value of the excitation ($s = 0$ in this instance). The first two terms are the natural response, the last term is the forced response as these terms arise from $H(s)$ and $X(s)$, respectively.

b For $x(t) = \epsilon^{-2t}$, $X(s) = 1/(s + 2)$ so that

$$Y(s) = H(s)X(s) = \frac{2(s + 2)}{(s + 1)(s + 4)} \times \frac{1}{(s + 2)}$$

$$Y(s) = \frac{K_1}{s + 1} + \frac{K_2}{s + 4} + \frac{K_3}{s + 2}$$

where

$$K_1 = \frac{2(s + 2)}{(s + 4)(s + 2)}\bigg|_{s=-1} = \frac{2}{3}$$

$$K_2 = \frac{2(s + 2)}{(s + 1)(s + 2)}\bigg|_{s=-4} = -\frac{2}{3}$$

$$K_3 = \frac{2(s + 2)}{(s + 1)(s + 4)}\bigg|_{s=-2} = 0$$

From the above,

$$y(t) = \tfrac{2}{3}\epsilon^{-t} - \tfrac{2}{3}\epsilon^{-4t}$$

Note that $y(t)$ contains only the terms associated with the natural response. As $K_3 = 0$, the forced response is also zero. As in part a, $K_3 = H(s)|_{s=-2}$, the s-value of the excitation. This would be evident if the $(s + 2)$ terms in the numerator and denominator were cancelled. Then

$$Y(s) = \frac{2}{(s + 1)(s + 4)} = \frac{2/3}{s + 1} + \frac{-2/3}{s + 4}$$

from which the response $y(t) = \tfrac{2}{3}\epsilon^{-t} - \tfrac{2}{3}\epsilon^{-4t}$ is obtained.

c The differential equation can be obtained by performing the operations which result in deriving Eq. (5-44) from Eq. (5-41) in reverse.

$$Y(s) = H(s)X(s) = \frac{2(s + 2)}{(s + 1)(s + 4)}\, X(s)$$

can be rewritten as

$$(s + 1)(s + 4)Y(s) = 2(s + 2)X(s)$$

or

$$(s^2 + 5s + 4)Y(s) = (2s + 4)X(s)$$

Recognition that multiplication by s in the frequency domain corresponds to differentiation in the time-domain, the differential equation is

$$\frac{d^2y}{dt^2} + 5\frac{dy}{dt} + 4y = 2\frac{dx}{dt} + 4x$$

The results obtained and the methods used in Example 5-12 indicate that both the forced and natural components of $y(t)$ are intimately related to the system function. It is then convenient to identify two properties of $H(s)$ which can be correlated with the response components. For $H(s) = A(s)/B(s)$ [as given in Eq. (5-45)], we note that the solutions of $B(s) = 0$ determine the natural frequencies of the network. When $B(s) = 0$, $H(s)$ tends to infinity. A *pole* is a value of s that causes the network function to become infinite. Thus, the poles of $H(s)$ are the natural frequencies of the network.

Consider a linear system whose excitation $x(t)$ is the unit impulse function $\delta(t)$. In the frequency domain, the response $Y(s) = H(s)X(s)$. As the Laplace transform of $\delta(t)$ is unity, $Y(s) = H(s)$ and

$$y(t) = \mathcal{L}^{-1}[H(s)] = \sum_{1}^{n} K_i\epsilon^{s_it} \tag{5-50}$$

In Eq. (5-50), the values of s_i are the poles of $H(s)$ and the K_i are the coefficients in the partial fraction expansion of $H(s)$. Comparison of Eqs. (5-50) and (5-42) indicates that $\mathcal{L}^{-1}[H(s)]$ yields the natural response of the system.

A *zero* of $H(s)$ is defined as a value of s which causes $H(s) = 0$. For $H(s)$ in the form of Eq. (5-45), one set of values for the zeros occurs when $A(s) = 0$. The physical significance of a zero is that it indicates the complex frequencies of the excitation which cause the forced response to be zero. This case is illustrated in Example 5-12*b*. Practical application of the design of networks whose system function contains one or more zeros at prescribed frequencies is to the class of circuits called electronic filters (introduced in Chap. 10).

For excitations of the form $A_o\epsilon^{s_o t}$, which includes both dc and ac, the frequency-domain response of a network is

$$Y(s) = H(s)\frac{A_o}{s - s_o} \tag{5-51}$$

The partial fraction expansion of $Y(s)$ is expressible as

$$Y(s) = \sum_1^n \frac{K_i}{s - s_i} + \frac{K_o}{s - s_o} \tag{5-52}$$

The term containing the summation in Eq. (5-52) represents the expansion of $H(s)$ at its poles and gives rise to the natural response. The $K_o/(s - s_o)$ term is due to the excitation and leads to the forced component of the response. Evaluation of K_o gives

$$K_o = (s - s_o)Y(s)\Big|_{s=s_o} = A_o H(s_o) \tag{5-53}$$

and results in a forced response

$$y_f(t) = A_o H(s_o)\epsilon^{s_o t} \tag{5-54}$$

In Eqs. (5-53) and (5-54), the significance of $H(s_o)$ is that the ratio of the amplitude of the forced response to the amplitude of the excitation is simply the system function evaluated at the excitation frequency. Alternatively, $H(s_o)$ is the amplitude of the forced response per unit excitation.

For sinusoidal excitation ($s_o = j\omega$), investigation of $\mathbf{H}(j\omega)$ permits the determination of the network response over a range of frequencies without the necessity of evaluating the response due to each excitation frequency separately. One method of analysis which makes use of this concept is the Bode diagram discussed in the succeeding chapter.

Example 5-13

The shaded portion of the circuit shown in Fig. 5-16 is often used to represent a transistor. For this circuit

a Determine the network function $I_o(s)/I_{in}(s)$.

b For $r_\pi = 1,000\ \Omega$, $g_m = 0.2\ \mho$, $C_\pi = 49\ \text{pF}$ and $C_\mu = 1\ \text{pF}$ determine the

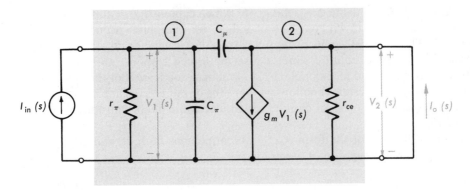

FIGURE 5-16 Circuit for Example 5-13.

response $i_o(t)$ to a unit-step-function excitation. Assume the circuit has no initial energy storage.

c Using the parameter values in part b and for an excitation $\epsilon^{j\omega t}$, determine the value of ω for which $\left|\dfrac{I_o}{I_{in}}(j\omega)\right|$ is unity.

SOLUTION a By use of KCL at ①, and recognizing that $V_2(s) = 0$,

$$-I_{in}(s) + \frac{V_1(s)}{r_\pi} + sC_\pi V_1(s) + sC_\mu V_1(s) = 0$$

from which

$$V_1(s) = \frac{I_{in}(s)}{\dfrac{1}{r_\pi} + s(C_\pi + C_\mu)}$$

At ②, KCL requires

$$I_o(s) = g_m V_1(s) - I_\mu(s) = g_m V_1(s) - sC_\mu V_1(s)$$

Substitution for $V_1(s)$ and forming $I_o(s)/I_{in}(s)$ yields

$$\frac{I_o(s)}{I_{in}(s)} = H(s) = \frac{(g_m - sC_\mu)}{\dfrac{1}{r_\pi} + s(C_\pi + C_\mu)}$$

Alternatively, $H(s)$ can be written as

$$H(s) = \frac{g_m r_\pi (1 - sC_\mu/g_m)}{1 + sr_\pi(C_\pi + C_\mu)} = \frac{-C_\mu(s - g_m/C_\mu)}{(C_\pi + C_\mu)\left[s + \dfrac{1}{r_\pi(C_\pi + C_\mu)}\right]}$$

$H(s)$ has one pole and one zero; the pole is located at

$$s = s_p = -\frac{1}{r_\pi(C_\pi + C_\mu)}$$

and the zero location is*

$$s = s_z = +g_m/C_\mu$$

b For a unit step-function excitation $I_{in}(s) = \dfrac{1}{s}$ and

$$I_o(s) = H(s)\frac{1}{s} = \frac{-C_\mu}{(C_\pi + C_\mu)} \times \frac{s - g_m/C_\mu}{s + \dfrac{1}{r_\pi(C_\pi + C_\mu)}} \frac{1}{s}$$

Substitution of the given parameter values gives

$$I_o(s) = \frac{-\frac{1}{50}(s - 2 \times 10^{11})}{s + 2 \times 10^7} \times \frac{1}{s}$$

Expansion by partial fractions yields

$$I_o(s) = \frac{K_1}{s + 2 \times 10^7} + \frac{K_o}{s}$$

where

$$K_1 = \frac{-(s - 2 \times 10^{11})}{50s}\Bigg|_{s=-2\times10^7} = -200$$

$$K_o = \frac{-(s - 2 \times 10^{11})}{50(s + 2 \times 10^7)}\Bigg|_{s=0} = +200$$

Substitution of K_1 and K_o into $I_o(s)$ and obtaining the inverse transform result in

$$i_o(t) = -200\epsilon^{-2\times10^7t} + 200 = 200(1 - \epsilon^{-2\times10^7t})$$

c For the given parameter values, $H(s)$ from part a becomes

$$H(s) = \frac{200(1 - 5 \times 10^{-12}s)}{1 + s \times 5 \times 10^{-8}}$$

Letting $s = j\omega$ and forming $|H(j\omega)|^2$ yields

$$|H(j\omega)|^2 = \frac{(200)^2(1 + 25 \times 10^{-24}\omega^2)}{1 + 25 \times 10^{-16}\omega^2}$$

By setting $|H(j\omega)|^2 = 1$, the value of ω which satisfies this condition is determined as

$$\omega = 4.00 \times 10^9 \text{ rad/s}$$

The results obtained in this example are typical values that one encounters in practice for the current gain and frequency response of transistors.

PROBLEMS

5-1 By use of Eq. (5-1), verify that $\mathscr{L}[t] = 1/s^2$.

5-2 Repeat Prob. 5-1 to determine: (a) $\mathscr{L}[\sin \omega t]$; (b) $\mathscr{L}[\cos \omega t]$.

5-3 Verify entries 14 and 15 in Table 5-1.

5-4 Evaluate the Laplace transform for $f(t) = \sinh t$.

5-5 Evaluate the Laplace transform for $f(t) = \cosh t$.

5-6 Verify entry 10 in Table 5-1.

5-7 Express the following functions by their partial fraction expansions:

a. $\dfrac{4s}{s^2 + 3s + 2}$

b. $\dfrac{12(s^2 + 4)}{s^3 + 3s^2 + 2s}$

c. $\dfrac{s^2 + 3s + 2}{s^2 + 7s + 12}$

d. $\dfrac{5(s + 2)}{s^3 + 6s^2 + 9s + 4}$

5-8 Express the following functions by their partial fraction expansions:

a. $\dfrac{9(s - 1)}{s^2 + 4s + 3}$

b. $\dfrac{9(s - 1)}{s^3 + 4s^2 + 3s}$

c. $\dfrac{s^2 + 3s + 2}{s^2 + 2s + 2}$

d. $\dfrac{s^2 + 3s + 2}{s^4 + 2s^3 + 2s}$

5-9 Obtain the inverse Laplace transforms of the following functions:

a. $\dfrac{10}{s(s + 2)}$

b. $\dfrac{10(s + 1)}{s(s + 2)}$

c. $\dfrac{10(s + 1)}{s(s + 2)^2}$

d. $\dfrac{10(s + 1)^2}{s(s + 2)}$

5-10 Obtain the inverse Laplace transforms of the following functions:

a $\dfrac{12s}{(s + 2)(s + 4)}$

b $\dfrac{12(s + 1)}{(s + 2)^2(s + 4)}$

c $\dfrac{12s}{(s^2 + 4)(s + 2)}$

d $\dfrac{12(s + 4)}{(s^2 + 4)^2(s + 2)}$

5-11 Repeat Prob. 5-10 for the following functions:

a $\dfrac{12s}{[(s + 2)^2 + 4]}$

b $\dfrac{12s}{[(s + 2)^2 + 4]^2}$

c $\dfrac{12(s + 2)}{s[(s + 2)^2 + 4]}$

d $\dfrac{s^2 - 6s + 25}{s(s^2 + 6s + 25)}$

5-12 Determine the inverse Laplace transform of the following functions:

a $\dfrac{(s + 2)^2}{s(s^2 + 4)}$

b $\dfrac{s[(s + 2)^2 + 12]}{[(s + 1)^2 + 3](s + 2)^2}$

c $\dfrac{(s - 1)}{(s + 1)^2}$

d $\dfrac{s^2 + s + 1}{(s + 1)^3}$

5-13 Use the Laplace transform method to solve for $i(t)$ in the following differential equation for $t \geq 0$.

$$\frac{d^2i}{dt^2} + 4\frac{di}{dt} + 3i = 6u(t)$$

Consider $i(0^+) = [di(0^+)]/dt = 0$.

5-14 Use the Laplace transform method to solve the differential equation for $t \geq 0$. Consider $v(0^+) = 4V$ and $(dv/dt)(0^+) = -4$ V/s.

$$3\frac{d^2v}{dt^2} + 6\frac{dv}{dt} + 2v = 12 \cos 2t$$

5-15 a Solve the differential equation, for $t \geq 0$, when $v(0^+) = 0$ and $(dv/dt)(0^+) = 6$ V/s.

$$\frac{d^2v}{dt^2} + 2\frac{dv}{dt} + 2v = 12 \sin t - 6\epsilon^{-2t}$$

b Identify the natural and forced components of the response.

5-16 a Solve the differential equation, for $t \geq 0$, when $i(0^+) = 10$ A.

$$\frac{di}{dt}(0^+) = -5 \text{ A/s}, \qquad \frac{d^2i}{dt^2}(0^+) = \frac{d^3i}{dt^3}(0^+) = 0$$

$$\frac{d^4i}{dt^4} + 4\frac{d^3i}{dt^3} + 7\frac{d^2i}{dt^2} + 6\frac{di}{dt} + 2i(0^+) = 10\epsilon^{-2t} \cos 2t$$

b Identify the forced and natural response components of the response.

5-17 Use Table 5-2 to determine the Laplace transforms of each of the following:

a $\dfrac{1 - \cos t}{t}$

b $t^2 \epsilon^{-t}$

5-18 Repeat Prob. 5-17 for the following functions:

a $\dfrac{\epsilon^{-2t} \sin 2t}{t}$

b $t \epsilon^{-2t} \sin 2t$

5-19 A particular amplitude-modulated (AM) wave is expressible as $f(t) = \cos 10^4 t \cos 5 \times 10^6 \, t$. Determine $F(s)$.

5-20 Determine the Laplace transform of the waveform in Fig. 5-17.

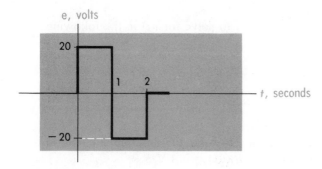

FIGURE 5-17 Waveform for Prob. 5-20.

5-21 Repeat Prob. 5-20 for the trapezoidal waveform in Fig. 5-18.

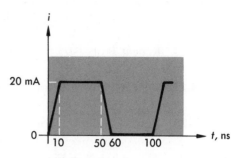

FIGURE 5-18 Waveform for Prob. 5-21.

5-22 Repeat Prob. 5-20 for the half-wave rectified sinusoid shown in Fig. 5-19.

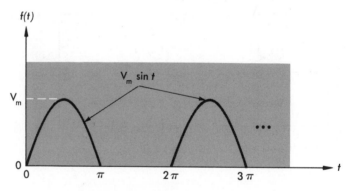

FIGURE 5-19 Half-wave rectified sinusoid for Prob. 5-22.

5-23 Use Table 5-2 to obtain $f(t)$ for the functions

a $F(s) = \dfrac{\epsilon^{-s}}{s + 1}$

b $F(s) = \dfrac{\epsilon^{-2s}}{(s + 2)^2}$

5-24 Repeat Prob. 5-23 for

a $F(s) = \ln \dfrac{s + 3}{s + 4}$

b $F(s) = s \ln \dfrac{s + 3}{s + 4}$

5-25 **a** For each of the functions $F(s)$ in Prob. 5-7, determine the values of $f(0^+)$ and $f(\infty)$ by use of the initial-value and final-value theorems.
b Do these values exist for each of the functions in a?

5-26 Repeat Prob. 5-25 for each of the functions in Prob. 5-8.

5-27 Repeat Prob. 5-25 for each of the functions in Prob. 5-11.

5-28 Repeat Prob. 5-25 for each of the functions given in Prob. 5-12.

5-29 The switch S in Fig. 5-20 is closed at $t = 0$. Determine $v(t)$ for $t \geq 0$ when $C = 0.5$ F using Laplace transform techniques.

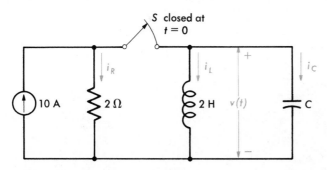

FIGURE 5-20 Circuit for Prob. 5-29.

5-30 The switch S in Fig. 5-21 is closed at $t = 0$. Determine $i(t)$ for $t \geq 0$ when $C = 0.25$ F.

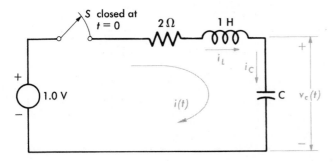

FIGURE 5-21 Circuit for Prob. 5-30.

5-31 The switch S in Fig. 5-22 has been in position 1 for a long time. At $t = 0$ it is instantaneously switched to position 2. Find $v(t)$ for $t \geq 0$ using the Laplace transform.

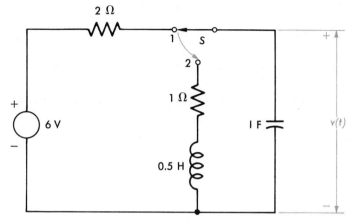

FIGURE 5-22 Circuit for Prob. 5-31

5-32 The excitation $v_s(t)$ in Fig. 5-23 is $10u(t)$. Find $v_2(t)$.

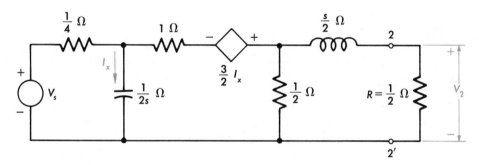

FIGURE 5-23 Circuit for Prob. 5-32.

5-33 Repeat Prob. 5-32 for $v_s(t) = 10\epsilon^{-t}u(t)$.

5-34 Determine $v_o(t)$ in the circuit of Fig. 5-24 for $v(t) = tu(t)$.

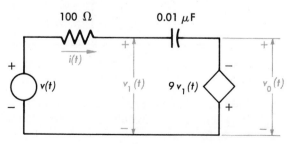

FIGURE 5-24 Circuit for Prob. 5-34.

5-35   Find $i_L(t)$ in the circuit of Fig. 5-25 for $i(t) = t\epsilon^{-0.1t}u(t)$.

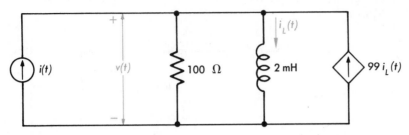

FIGURE 5-25   Circuit for Prob. 5-35.

5-36   The switch S in Fig. 5-26 has been open for a long time. At $t = 0$ it is closed. Find $v(t)$ for $t \geq 0$.

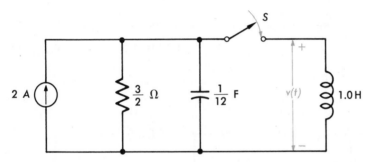

FIGURE 5-26   Circuit for Prob. 5-36.

5-37   In the circuit of Fig. 5-27, $i(t)$ is a pulse of amplitude 50 μA and duration 10 μs. Determine $v_2(t)$ and sketch the result.

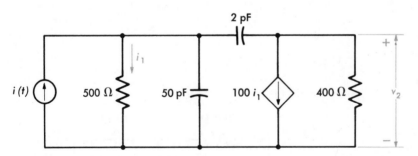

FIGURE 5-27   Circuit for Prob. 5-37.

5-38   a   Repeat Prob. 5-37 if the pulse duration is 10 ns.
b   Compare the result in a with that obtained when the excitation is $0.5 \times 10^{-12} \, \delta(t)$.

5-39 Determine the response $v_o(t)$ of the circuit of Fig. 5-28 to the unit-step voltage excitation.

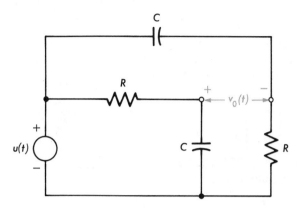

FIGURE 5-28 Circuit for Prob. 5-39.

5-40 The circuit shown in Fig. 5-29 is a representation of a sweep circuit and is used to generate a saw-tooth waveform which provides the horizontal drive (time axis) for a cathode-ray oscilloscope. For the circuit shown in Fig. 5-29, the switch S has been in position 2 for a long time. At $t = 0$, the switch is moved to position 1 for 10 ms. At $t = 10$ ms, the switch is returned to position 2. Determine and sketch the output voltage $v_o(t)$. (*Hint:* It is useful to recall that $\epsilon^{-x} \doteq 1 - x$ for $x \ll 1$.)

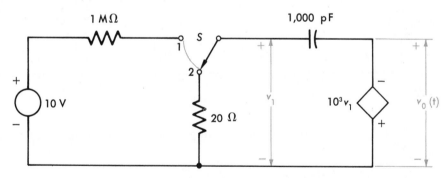

FIGURE 5-29 Sweep circuit representation for Prob. 5-40.

5-41 The circuit shown in Fig. 5-30 is often a basic building block in integrated-circuit filters. A filter is a network used to select one range of frequencies while rejecting all other frequencies. For the circuit in Fig. 5-31, determine:
a The transfer function $V_o(s)/V_s(s)$
b The response $v_o(t)$ for $v_s(t) = u(t)$
c Repeat b for $v_s(t) = \cos tu(t)$

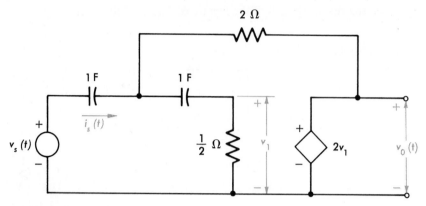

FIGURE 5-30 Circuit for Prob. 5-41.

5-42 Repeat Prob. 5-41 for the filter circuit shown in Fig. 5-31.

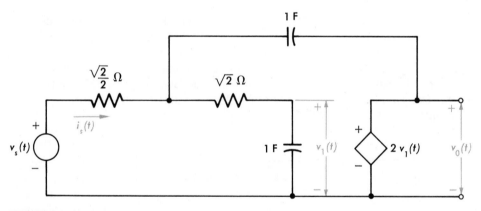

FIGURE 5-31 Filter circuit for Prob. 5-42.

5-43 Determine the driving-point impedance $V_s(s)/I_s(s)$ in the circuit of Fig. 5-31.

5-44 Repeat Prob. 5-43 for the circuit in Fig. 5-30.

5-45 The circuit shown in Fig. 5-32 is an all-electric analog of a dc motor driving a friction-inertia load. The left-hand segment represents the electric input to the motor; the right-hand segment depicts the mechanical output and load. The dependent current source indicates the torque developed (proportional to the electric input) and the voltage $v_2(t)$ is the analog of the motor speed.

a Determine the output speed $v_2(t)$ assuming the switch S is closed at $t = 0$.

b How much time is required for the motor to reach its steady-state speed?

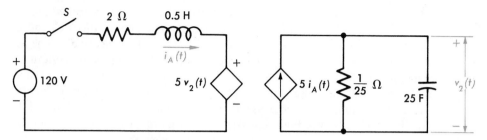

FIGURE 5-32 All-electric-circuit representation of dc motor in Prob. 5-45.

5-46 For the circuit in Fig. 5-32, find $i_A(t)$ assuming the switch S is closed at $t = 0$.

5-47 The switch S in the circuit of Fig. 5-32 and Prob. 5-45 has been closed for a long time. Determine $v_2(t)$ and the time required for the motor to come to rest, for practical purposes, if the switch is opened at $t = 0$.

5-48 The circuit depicted in Fig. 5-33a is the equivalent circuit of a compensated attenuator frequently used in an oscilloscope probe.
a Determine the transfer function $V_2(s)/V_1(s)$.
b For the voltage excitation $v_1(t)$ shown in Fig. 5-33b, find and sketch $v_2(t)$ for $R_1 = 9$ MΩ, $R_2 = 1$ MΩ, $C_1 = 16$ pF, and $C_2 = 54$ pF.
c Repeat b for $R_1 = 9$ MΩ, $R_2 = 1$ MΩ, $C_1 = 2$ pF, and $C_2 = 27$ pF.
d The value of C_1 is changed to 6 pF; repeat b with all remaining element values as given in b.

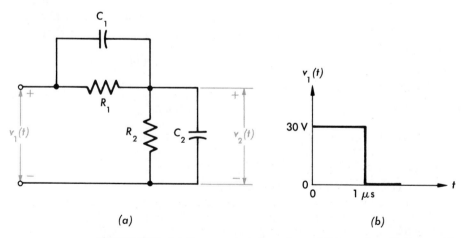

(a)   *(b)*

FIGURE 5-33 (a) Compensated attenuator and (b) waveform for Prob. 5-48.

5-49 Determine the input impedance $[V_1(s)]/[I_{in}(s)]$ in the circuit of Fig. 5-16.

5-50 The circuit of Fig. 5-34 is a "first-order" electrical analog which describes the alcohol content in the blood stream. The capacitances C_1 and C_2 represent the volumes of the alcoholic fluid and blood, respectively. The initial voltage across C_1 indicates the concentration of alcohol in percent in the liquid consumed. The output voltage $v_o(t)$ is the measure in percent of the alcohol concentration in the bloodstream. The resistances R_1 and R_2 reflect the biological impediment to the passage of alcohol into the bloodstream and the alcohol removed by the body's waste processes, respectively. The element values given in Fig. 5-34 are scaled so that the response time in tenths of a second corresponds to real time in hours. The initial voltage and value of C_1 correspond to a "double" of a 100-proof alcoholic beverage.
a Determine the time required for the concentration of alcohol in the bloodstream to be no greater than one-tenth of one percent. Assume the switch S is closed at $t = 0$.
b What is the concentration of alcohol in the bloodstream after one-half hour (real time)? Should you drive your car at this time?

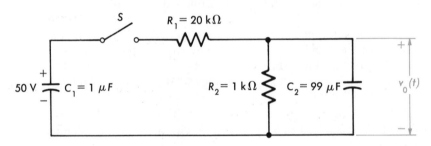

FIGURE 5-34 Circuit for Prob. 5-50.

5-51 Repeat Prob. 5-50 if $C_1 = 2\ \mu F$ and the initial voltage is 12.5 V. This corresponds to consuming 200 ml of wine with 12.5 percent strength.

5-52 The response $v(t)$ of a linear system to a unit-step excitation $i(t)$ is

$$v(t) = (1 - 2e^{-t} + \epsilon^{-2t})\, u(t)$$

Determine the transfer function $H(s) = V(s)/I(s)$.

5-53 The response $y(t)$ of a linear system to a unit-step excitation is

$$y(t) = (2 - 5\epsilon^{-t} + 4\epsilon^{-2t})\, u(t)$$

a Determine the system function.
b At what frequency is the forced response zero?

5-54 The unit impulse response $h(t)$ of a linear system is

$$h(t) = 2.5\epsilon^{-t} \cos(2t - 26.6°)$$

a Determine $H(s)$.
b What is the forced response if the excitation is $2 \cos \sqrt{5}\ t$?

5-55 The response $y(t)$ of a linear system to an excitation $x(t) = \epsilon^{-2t} u(t)$ is

$$y(t) = (t + 1) \epsilon^{-t} u(t)$$

a Determine the poles and zeros of the transfer function.
b What is the form of the natural response of this system to a sinusoidal excitation?
c Determine the response of the system to an excitation $x(t) = \epsilon^{-2(t-2)} u(t - 2)$.

5-56 The shaded portion of the circuit displayed in Fig. 5-35 is the representation of a Wien-bridge oscillator. These oscillators are frequently used in test equipment where variable-frequency sinusoidal signals are needed. For the circuit shown in Fig. 5-36 determine:
a The transfer function $V_o(s)/I(s)$.
b The response $v_o(t)$ when $i(t)$ is the unit-impulse function and $R_1 = R_2 = 10 \text{ k}\Omega$, $C_1 = C_2 = 1{,}000$ pF, and $A = 3$.

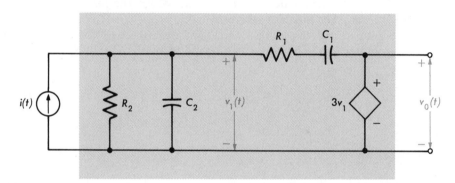

FIGURE 5-35 Wien-bridge circuit for Prob. 5-56.

5-57 The schematic representation of the drive for a phonograph is shown in Fig. 5-36*a* and its electric-circuit analog is depicted in Fig. 5-36*b*. In the circuit of Fig. 5-36*b*, the analogous quantities are:

$\quad i_m(t)$ = torque developed by motor
$\quad G_1, G_2$ = bearing friction
$\quad\quad C_1$ = moment of inertia of motor
$\quad\quad C_2$ = moment of inertia of turntable
$\quad\quad L$ = compliance of flexible coupling
$\quad\quad v_2$ = angular velocity of turntable

The purpose of the flexible coupling is to minimize the effects of the fluctuations in the torque developed by the motor. The numerical values of the components in this problem are selected for numerical convenience so that the effect of the coupling is easily observed. Consider $G_1 = G_2 = 1 \mho$, $C_1 = C_2 = 1$ F, $L = 8$ H.

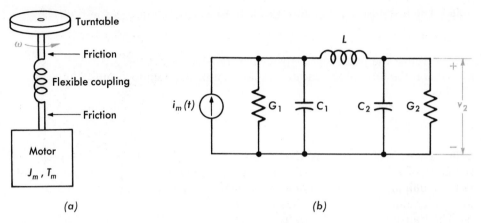

FIGURE 5-36 (a) Turntable and (b) its electric-circuit analog for Prob. 5-57.

a For $i_m(t) = 2u(t)$, determine $v_2(t)$ and sketch the response.
b Repeat *a* for the case where $L = 0$.
c Compare the results in *a* and *b* and describe why the flexible coupling gives better immunity to fluctuations in torque.

REPRESENTATION OF SYSTEMS ☐ **FREQUENCY**
TATION OF SYSTEMS ☐ FREQUENCY **RESPONSE**
ON OF SYSTEMS ☐ FREQUENCY RESPONSE **AND**
FREQUENCY RESPONSE AND **REPRESENTATION**
QUENCY RESPONSE AND REPRESENTATION **OF**
RESPONSE AND REPRESENTATION OF **SYSTEMS**
FREQUENCY RESPONSE AND REPRESENTATION
RESPONSE AND REPRESENTATION OF SYSTEMS
AND REPRESENTATION OF SYSTEMS ☐ FREQUE
REPRESENTATION OF SYSTEMS ☐ FREQUENCY RESPONSE AND RE
OF SYSTEMS ☐ FREQUENCY RESPONSE AND REPRESENTATION OF
SYSTEMS ☐ FREQUENCY RESPONSE AND REPRESENTATION OF SY
FREQUENCY RESPONSE AND REPRESENTATION OF SYSTEMS ☐ FR
RESPONSE AND REPRESENTATION OF SYSTEMS ☐ FREQUENCY RE
AND REPRESENTATION OF SYSTEMS ☐ FREQUENCY RESPONSE AN
REPRESENTATION OF SYSTEMS ☐ FREQUENCY RESPONSE AND RE
OF SYSTEMS ☐ FREQUENCY RESPONSE AND REPRESENTATION OF
SYSTEMS ☐ FREQUENCY RESPONSE AND REPRESENTATION OF SY
FREQUENCY RESPONSE AND REPRESENTATION OF SYSTEMS ☐ FR
RESPONSE AND REPRESENTATION OF SYSTEMS ☐ FREQUENCY RE
AND REPRESENTATION OF SYSTEMS ☐ FREQUENCY RESPONSE AN
REPRESENTATION OF SYSTEMS ☐ FREQUENCY RESPONSE AND RE
OF SYSTEMS ☐ FREQUENCY RESPONSE AND REPRESENTATION OF
SYSTEMS ☐ FREQUENCY RESPONSE AND REPRESENTATION OF SY
FREQUENCY RESPONSE AND REPRESENTATION OF SYSTEMS ☐ FR
RESPONSE AND REPRESENTATION OF SYSTEMS ☐ FREQUENCY RE
AND REPRESENTATION OF SYSTEMS ☐ FREQUENCY RESPONSE AN
REPRESENTATION OF SYSTEMS ☐ FREQUENCY RESPONSE AND RE
OF SYSTEMS ☐ FREQUENCY RESPONSE AND REPRESENTATION OF
SYSTEMS ☐ FREQUENCY RESPONSE AND REPRESENTATION OF SY
FREQUENCY RESPONSE AND REPRESENTATION OF SYSTEMS ☐ FR
RESPONSE AND REPRESENTATION OF SYSTEMS ☐ FREQUENCY RE
AND REPRESENTATION OF SYSTEMS ☐ FREQUENCY RESPONSE AN
REPRESENTATION OF SYSTEMS ☐ FREQUENCY RESPONSE AND RE
OF SYSTEMS ☐ FREQUENCY RESPONSE AND REPRESENTATION OF
SYSTEMS ☐ FREQUENCY RESPONSE AND REPRESENTATION OF SY
FREQUENCY RESPONSE AND REPRESENTATION OF SYSTEMS ☐ FR
RESPONSE AND REPRESENTATION OF SYSTEMS ☐ FREQUENCY RE
AND REPRESENTATION OF SYSTEMS ☐ FREQUENCY RESPONSE AN

system can be considered as a combination of elements and components organized to function as an integral whole. In the context of this book, systems for control, instrumentation, communication, and computation are composed of individual circuits and devices whose appropriate interconnection provides a desired performance. Proper performance depends on the ability of the system to respond in a prescribed manner to a wide range of excitation frequencies. The interaction between constituent parts is also a significant aspect of system operation. Two areas of study, each based on the extension of the concepts and techniques of the previous chapters, evolve from these considerations.

In the first area, network response to sinusoidal signals of variable frequency are investigated. From this study comes an understanding of resonance and frequency response. The second area consists of two-port networks, block diagrams, and signal-flow graphs. These are useful circuit techniques by which constituent parts of systems are represented in terms of their input-output characteristics. It is these subjects with which this chapter deals.

6-1 FREQUENCY RESPONSE OF SIMPLE NETWORKS

One of the most commonly used signals for testing a wide variety of electrical equipment is a sinusoid whose frequency can be varied. The device which produces this signal is called an *oscillator;* it serves as the excitation to the equipment to be tested. The resultant response is a phasor quantity in which both the magnitude and angle are functions of the excitation frequency. For convenience, this response, called the *frequency response,* is often expressed as an output/input ratio; i.e., it is a network function. Graphical display of the frequency-response characteristics of networks is a useful means by which the wide range of frequencies encountered can be accommodated. This section provides an introduction to the techniques of frequency-response determination; further development is provided in subsequent sections of this chapter and elsewhere in the text as needed.

Consider the RC network of Fig. 6-1. The current-source excitation has a constant amplitude I and a varying angular frequency $\omega = 2\pi f$. The network voltage $\mathbf{V}$ has the same frequency as the excitation; however, its amplitude and phase are functions of frequency. The voltage $\mathbf{V}$ is

$$\mathbf{V} = \frac{\mathbf{I}}{\mathbf{Y}} = \frac{\mathbf{I}}{(1/R) + j\omega C} = \frac{IR}{1 + j\omega RC} \tag{6-1}$$

If $\mathbf{I} = I\underline{/0°}$, $\mathbf{V}$, in polar form, becomes

$$\mathbf{V} = \frac{IR}{\sqrt{1 + (\omega RC)^2}} \underline{/-\tan^{-1} \omega RC} = V\underline{/\alpha} \tag{6-2}$$

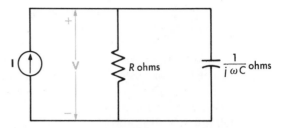

FIGURE 6-1 Parallel *RC* network.

The magnitude and phase of **V** are plotted as a function of ω in Fig. 6-2. The frequency response shown in these curves indicates that the voltage decreases and lags the current more as the frequency is increased. Physically, the curves indicate that the admittance of the parallel *RC* circuit becomes more and more capacitive with increasing frequency. This results in the lagging phase, increasing admittance, and consequently, decreasing voltage.

At

$$\omega_1 = \frac{1}{RC} \tag{6-3}$$

the voltage is reduced to 0.707 of its maximum value and the phase shift is $-45°$. At this frequency, the power delivered to the circuit (V^2/R) is one-half

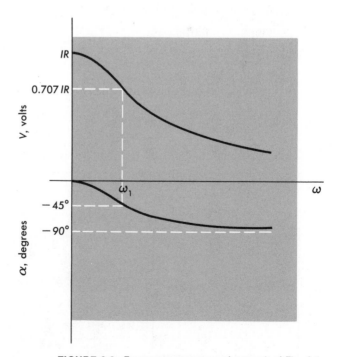

FIGURE 6-2 Frequency response of network of Fig. 6-1.

the maximum power, and as a consequence, ω_1 is often called the *half-power angular frequency*. It is frequently used as a measure of the effectiveness of a circuit in passing or eliminating certain frequencies. In this context, the circuit of Fig. 6-1 passes angular frequencies in the range 0 to ω_1 and rejects angular frequencies in the range ω_1 to infinity. The region of transmitted frequencies is the *passband,* and the range of these frequencies is the *bandwidth.* Beyond the passband, the decreasing response indicates a region of high attenuation, called the *stopband.*

Example 6-1

Plot the network function $\mathbf{V}_L/\mathbf{V}$ as a function of frequency for the network in Fig. 6-3. What are the half-power frequency and passband?

SOLUTION   The network equations are

$$\mathbf{I} = \frac{\mathbf{V}}{\mathbf{Z}} = \frac{\mathbf{V}}{10 + j0.1\omega}$$

and

$$\mathbf{V}_L = j0.1\omega\, \mathbf{I} = \frac{j0.1\omega\, \mathbf{V}}{10 + j0.1\omega} = \frac{j0.01\omega\, \mathbf{V}}{1 + j0.01\omega}$$

Forming the required ratio yields

$$\frac{\mathbf{V}_L}{\mathbf{V}} = \frac{0.01\omega}{\sqrt{1 + (0.01\omega)^2}}\ \underline{/90° - \tan^{-1} 0.01\omega}$$

The results are shown in Fig. 6-4.
The half-power angular frequency is at $\omega = 100$ rad/s, or at a frequency

$$f = \frac{\omega}{2\pi} = \frac{100}{2\pi} = 15.9 \text{ Hz}$$

The passband includes all frequencies above 15.9 Hz.

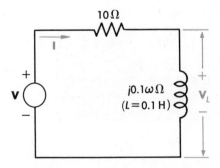

10 Ω

$j0.1\omega\,\Omega$
$(L=0.1\,\text{H})$

FIGURE 6-3 Circuit of Example 6-1.

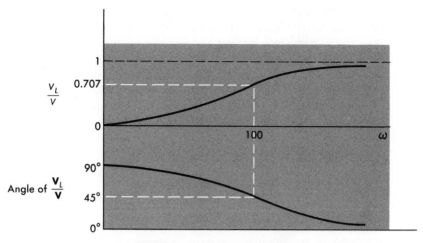

FIGURE 6-4 Frequency response of circuit of Example 6-1.

The frequency responses in Figs. 6-2 and 6-4 display markedly different characteristics. To distinguish various types of network frequency response, it is convenient to classify them. In Fig. 6-2, the passband includes all frequencies below the half-power frequency, while those above the half-power frequency are rejected. This type of response is referred to as a *low-pass* characteristic. The response in Fig. 6-4 is a *high-pass* characteristic as the

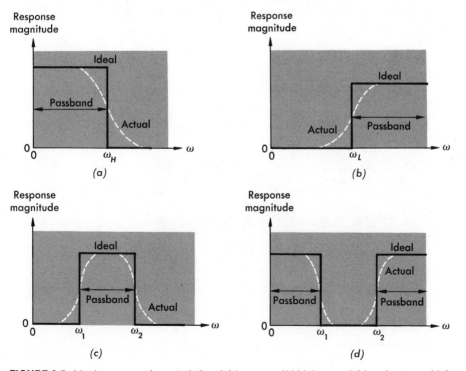

FIGURE 6-5 Ideal response characteristics: (a) low pass, (b) high pass, (c) bandpass, and (d) band reject.

passband includes frequencies beyond the half-power frequency, all others being attenuated. Ideal low- and high-pass responses are illustrated in Figs. 6-5a and b, respectively. In the ideal case, no attenuation occurs in the passband and total attenuation occurs in the stopband. *Bandpass* (Fig. 6-5c) and *band reject* (Fig. 6-5d) are the two remaining classes of frequency response. Resonant circuits, treated in Sec. 6-4, are often used to realize these characteristics. Practical networks, however, can only approximate the ideal characteristics as indicated by the dashed curves in the figure.

6-2 BODE DIAGRAMS

Many electronic circuits must process signals whose frequencies vary widely. Typical examples are audio and video amplifiers in which the signal frequencies range from 50 to 20,000 Hz and from 100 to 4,000,000 Hz, respectively. In addition, the devices used in these systems must be characterized over the range of operating frequencies. Because of the wide variations in signal frequencies, it is often useful to display the frequency-response characteristics of circuits and devices in graphical form. The magnitude and phase of the response are usually both desired. One of the most common forms of these displays is the *Bode diagram*.

In the Bode diagram, the magnitude and phase of the network function $G(j\omega)$ are plotted separately as functions of the frequency variable ω. To accommodate large frequency ranges, a logarithmic scale is used for the frequency variable. The magnitude of the network function G is expressed in

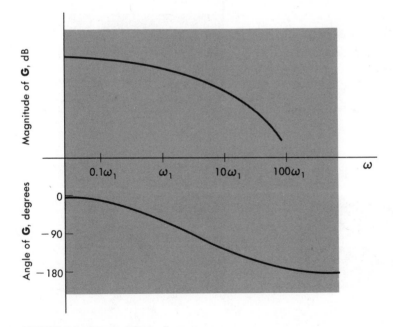

FIGURE 6-6 A typical Bode diagram.

decibels, abbreviated *dB* and defined as

$$G(dB) = 20 \log G \tag{6-4}$$

The two portions of the Bode diagram are thus plotted on semilog paper, and a typical diagram is shown in Fig. 6-6.

Example 6-2

Plot the Bode diagram for the impedance function

$$Z(s) = \frac{100(1 + s/40)}{(1 + s/10)(1 + s/400)}$$

SOLUTION   The function $\mathbf{Z}(j\omega)$ is formed and is

$$\mathbf{Z}(j\omega) = \frac{100(1 + j0.025\omega)}{(1 + j0.1\omega)(1 + j0.0025\omega)}$$

The calculated values of $\mathbf{Z}(j\omega)$ for various values of ω are given in Table 6-1. From the data in Table 6-1, the Bode diagram is drawn and the result depicted in Fig. 6-7.

6-3 ASYMPTOTIC BODE DIAGRAMS

Determining the frequency-response characteristics of a network or system by algebraic manipulation is a moderate chore. In Example 6-2, for example, the work leading to the results presented in Table 6-1 is time-consuming; as the network increases in complexity, the complications in the calculations also increase. For many purposes, an approximate frequency-response char-

TABLE 6-1   IMPEDANCE VALUES FOR EXAMPLE 6-2

ω, RAD/S	MAGNITUDE OF $Z(j\omega)$, OHMS	MAGNITUDE OF $Z(j\omega)$, dB	ANGLE OF $Z(j\omega)$, DEG
1.0	100.0	40.0	−4.4
2.0	98.1	39.8	−8.7
5.0	89.1	39.0	−20.2
10.0	73.3	37.3	−32.4
20.0	49.5	33.9	−39.7
40.0	34.3	30.7	−36.7
80.0	27.2	28.7	−30.8
100.0	26.0	28.3	−30.1
200.0	22.6	27.1	−35.0
400.0	17.8	25.0	−49.3
800.0	11.2	21.0	−65.6
1,600.0	6.1	15.7	−77.0
4,000.0	2.5	7.9	−84.7

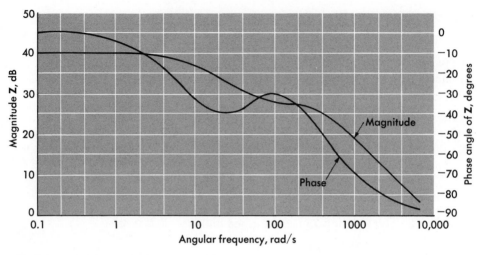

FIGURE 6-7 Bode diagram for Example 6-2.

acteristic is adequate. The nature of the Bode diagram leads to a simply drawn approximate characteristic, called the *asymptotic Bode diagram*.

In general, a network function can be expressed as a quotient of two polynomials in s or $j\omega$. If the network function is put in the form

$$G(s) = K \frac{1 + a_1 s + a_2 s^2 + \cdots + a_m s^m}{1 + b_1 s + b_2 s^2 + \cdots + b_n s^n} \tag{6-5}$$

the numerator and denominator polynomials can be factored and the function represented by

$$G(s) = K \frac{(1 + s/z_1)(1 + s/z_2) \cdots (1 + s/z_m)}{(1 + s/p_1)(1 + s/p_2) \cdots (1 + s/p_n)} \tag{6-6}$$

Note that $-z_1, -z_2$, etc., and $-p_1, -p_2$, etc., are the roots of the numerator and denominator polynomials, respectively, and also that the $-z$ terms are the zeros and the $-p$ terms are the poles of the network function. The frequency-response curve is found by letting s become $j\omega$, giving

$$\mathbf{G}(j\omega) = K \frac{(1 + j\omega/z_1)(1 + j\omega/z_2) \cdots (1 + j\omega/z_m)}{(1 + j\omega/p_1)(1 + j\omega/p_2) \cdots (1 + j\omega/p_n)} \tag{6-7}$$

The value of $\mathbf{G}(j\omega)$ is evidently the product of a constant and a group of terms having the form $(1 + j\omega/\omega_o)$ or $1/(1 + j\omega/\omega_o)$. Each of these terms can be thought of as an individual phasor; the resultant $\mathbf{G}(j\omega)$ has a magnitude which is the product of the magnitudes and an angle which is the sum of the individual angles.

The magnitude-curve portion of the Bode diagram is plotted in decibels and, from Eq. (6-4), is a logarithmic function. Thus the product

$$\left(1 + \frac{j\omega}{z_1}\right)\left(1 + \frac{j\omega}{z_2}\right) \cdots \left(1 + \frac{j\omega}{z_m}\right)$$

becomes the sum

$$\left(1 + \frac{j\omega}{z_1}\right)_{\text{dB}} + \left(1 + \frac{j\omega}{z_2}\right)_{\text{dB}} + \cdots + \left(1 + \frac{j\omega}{z_m}\right)_{\text{dB}}$$

when the individual terms are expressed in dB. As a result, the Bode-diagram magnitude and phase curves can both be considered to be composed of sums produced by the individual factors. The behavior of the $(1 + j\omega/\omega_o)$ and $1/(1 + j\omega/\omega_o)$ terms is then seen to be of importance in the construction of Bode diagrams. The development of their characteristics will show certain simplifying approximations which are useful for the rapid sketching of these diagrams.

Consider the functions

$$G_1(j\omega) = 1 + \frac{j\omega}{\omega_o} \quad \text{and} \quad G_2(j\omega) = \frac{1}{1 + j\omega/\omega_o} \tag{6-8}$$

At low frequencies ($\omega/\omega_o \ll 1$), the magnitude of both functions is approximately

$$G_1(j\omega) = G_2(j\omega) = 1$$

or

$$G_1(j\omega)_{\text{dB}} = G_2(j\omega)_{\text{dB}} = 20 \log 1 = 0 \tag{6-9}$$

At high frequencies ($\omega/\omega_o \gg 1$), the functions become

$$G_1(j\omega) = \frac{\omega}{\omega_o} \quad \text{and} \quad G_2(j\omega) = \frac{1}{\omega/\omega_o}$$

or

$$G_1(j\omega)_{\text{dB}} = 20 \log \frac{\omega}{\omega_o} \quad \text{and} \quad G_2(j\omega)_{\text{dB}} = -20 \log \frac{\omega}{\omega_o} \tag{6-10}$$

The low-frequency magnitudes are seen to be 0 dB (unity magnitude). The high-frequency magnitudes are $G_1 = G_2 = 0$ dB at $\omega/\omega_o = 1$; $G_1 = 20$ dB, $G_2 = -20$ dB at $\omega/\omega_o = 10$; $G_1 = 40$ dB, $G_2 = -40$ dB at $\omega/\omega_o = 100$; etc. The value of G_1 increases (and G_2 decreases) by a factor of 20 dB for each factor of 10 (decade) increase in ω/ω_o. Since factors of 10 are linear increments on the logarithmic frequency scale, Eqs. (6-10) are straight lines on the Bode plots; their slopes are $+20$ dB/decade for G_1 and -20 dB/decade for G_2. Often the slopes of the straight lines are expressed in units of dB/octave, where an octave represents a factor of 2 in frequency. For $\omega/\omega_o = 2$, the values of G_1 and G_2 from Eq. (6-10) are 6 dB and -6 dB, respectively. Thus in one octave, from $\omega/\omega_o = 1$ to $\omega/\omega_o = 2$, G_1 has changed by 6 dB and G_2 by -6 dB. The corresponding slopes are 6 dB/octave and -6 dB/octave. It may be noted that 6 dB/octave and 20 dB/decade define identical slopes.

The angles associated with G_1 and G_2 are

$$\text{Angle } G_1 = \tan^{-1} \frac{\omega}{\omega_o} \quad \text{and} \quad \text{angle } G_2 = -\tan^{-1} \frac{\omega}{\omega_o} \tag{6-11}$$

At $\omega = \omega_o$, the angles of G_1 and G_2 are $+45°$ and $-45°$, respectively. For frequencies where $\omega \geq 10\omega_o$, the angle of G_1 is nearly $90°$, that for the angle of G_2 nearly $-90°$. At low frequencies ($\omega \leq 0.1\omega_o$), both angles are nearly zero. These results lead to the straight-line approximations for the angles of G_1 and G_2 shown in Fig. 6-8. Also depicted in Fig. 6-8 are the straight-line (asymptotic) magnitude characteristics. The dashed curves in the figure indicate the exact magnitude and phase responses. The exact and approximate curves are reasonably close. The greatest error in the asymptotic curve occurs at $\omega = \omega_o$ and is $+3$ dB for G_1 and -3 dB for G_2. One octave away from the corner frequency ($\omega = \omega_o/2$ and $\omega = 2\omega_o$), the error is $+1$ dB for G_1 and -1 dB for G_2. For angular frequencies more than one octave from the break frequency, errors are less than 1 dB and are generally neglected. The maximum error in the phase characteristic occurs at one decade away from the corner and is nearly $6°$. At the break frequency there is zero error; one octave away the error is nearly $5°$. The curves in Fig. 6-8 indicate the algebraic sign of the errors for the angles of G_1 and G_2.

The process of drawing asymptotic Bode plots then becomes one of expressing the function in the form of Eq. (6-7), locating the break frequencies, drawing the component asymptotic curves, and adding these curves to get the resultant.

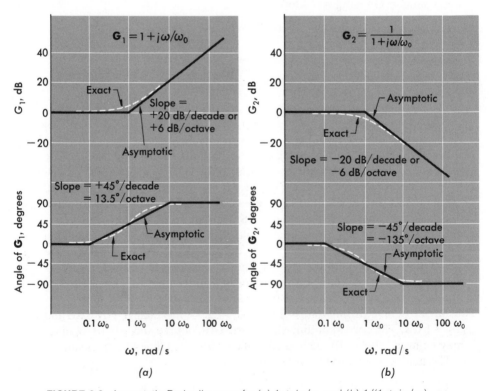

FIGURE 6-8 Asymptotic Bode diagrams for (a) $1 + j\omega/\omega_o$ and (b) $1/(1 + j\omega/\omega_o)$.

Example 6-3

a Sketch the asymptotic Bode diagram for

$$G(s) = \frac{10^4(s + 40)}{s^2 + 410s + 4,000}$$

b Determine the value of $G(j800)$.

SOLUTION **a** The equation for $G(s)$ is factored and put in the form of Eq. (6-7).

$$G(s) = \frac{100(1 + s/40)}{(1 + s/10)(1 + s/400)}$$

or

$$G(j\omega) = \frac{100(1 + j\omega/40)}{(1 + j\omega/10)(1 + j\omega/400)}$$

The break frequencies are 10 and 400 rad/s for the denominator and 40 rad/s for the numerator. The component curves are drawn in Fig. 6-9. Note the constant value of 40 dB = 20 log 100 represents the constant multiplier in $G(j\omega)$. The resultant magnitude and angle characteristics are indicated by the dashed white lines in Fig. 6-9.

b From the resultant curves

$$G(j800) = 22 \text{ dB}$$

so that, from 20 log $G = G(\text{dB})$,

$$G(j800) = \log^{-1} \tfrac{22}{20} = 12.6$$

and

$$\angle G(j800) = -58.5°$$

The function $G(s)$ in this example is identical to $Z(s)$ in Example 6-2. Comparison of the actual values in Fig. 6-7 and Table 6-1 with the curves in Fig. 6-9 indicate good agreement. From Table 6-1, $G(j800) = 21$ dB and $\angle G(j800) = -65.6°$ which compare favorably with the 22 dB and $-58.5°$ obtained from the asymptotic characteristic. An even closer fit to the magnitude curve can be obtained by algebraically adding the error at selected points, one octave away from, and at each corner frequency, to the values given by the asymptotic curves. For example, at $\omega = 40$ rad/s, the error in the factor $(1 + j\omega/40)$ is 3 dB. When added to the 28 dB value at $\omega = 40$ rad/s for the approximate characteristic, a resultant value of 31 dB is obtained. In Table 6-1, the exact value is given as 30.7 dB. Similar corrections to the phase curve also lead to closer fits to actual values.

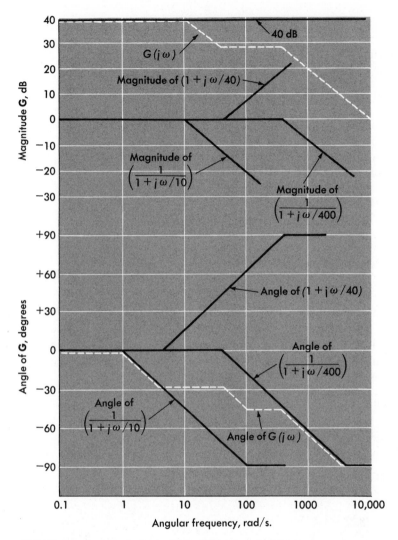

FIGURE 6-9 Asymptotic Bode diagram for Example 6-3.

The frequency response, given by the asymptotic Bode diagram in Fig. 6-9, is determined with far less computation than is the exact characteristic in Fig. 6-7. With little additional effort, incorporating the errors at a few frequencies within the asymptotic plot produces a result of sufficient accuracy for most engineering analyses.

In addition to the techniques described above and illustrated in Example 6-3, two other factors should be considered in drawing asymptotic Bode diagrams. The first of these is the case of a zero or pole at the origin, that is, a case where the limiting value of $G(s)$ as s goes to zero becomes

$$\lim_{s \to 0} G(s) = Ks^n$$

where n can be any positive or negative integer. In this case, the magnitude curve for this factor has a slope of $20n$ dB/decade and goes through K dB at $\omega = 1$. The component angle curve has a constant value of $n90°$.

The second factor which should be considered is the case where some of the zeros or poles occur as complex-conjugate pairs. In this case, a factor pair of the form

$$G_1(s) = \left(1 + \frac{s}{\alpha + j\beta}\right)\left(1 + \frac{s}{\alpha - j\beta}\right)$$

occurs where the poles (or zeros) are $-\alpha \pm j\beta$. The product of the complex-conjugate factors is the real quadratic

$$G_1(s) = 1 + 2\frac{\alpha}{\alpha^2 + \beta^2}s + \frac{s^2}{\alpha^2 + \beta^2} \tag{6-12}$$

The approximate Bode plot of Eq. (6-12) is shown in Fig. 6-10. The magnitude curve is characterized by a 40-dB/decade slope with a break frequency of $\sqrt{\alpha^2 + \beta^2}$, and the angle curve shows a slope of 90°/decade in the range

$$0.1\sqrt{\alpha^2 + \beta^2} \leq \omega \leq 10\sqrt{\alpha^2 + \beta^2}$$

The angle curve is 0° below $0.1\sqrt{\alpha^2 + \beta^2}$, and 180° above $10\sqrt{\alpha^2 + \beta^2}$. If the $G_1(s)$ term appears in the denominator polynomial, the angle and magnitude curves have the opposite sign. The usefulness of the quadratic approxi-

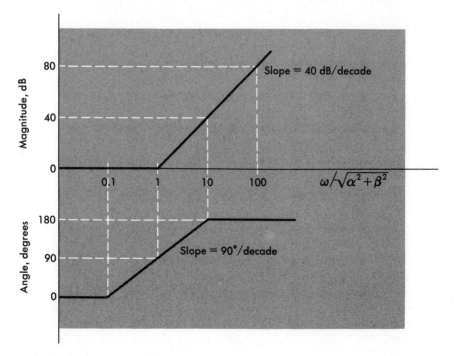

FIGURE 6-10 Asymptotic Bode diagram for the quadratic $[1 + 2\alpha s/(\alpha^2 + \beta^2) + s^2/(\alpha^2 + \beta^2)]$.

mation is limited by the fact that, unlike the first-power terms, the straight-line quadratic representations are not good approximations in the region 1 decade on either side of the break frequency (they are adequate outside this region). Should the doubtful frequency range be of concern, a few points can be calculated to establish the correct curves.

6-4 RESONANCE PHENOMENA

The two preceding sections described techniques to determine network response to a wide range of excitation frequencies. In this section, the response of three circuits to a sinusoidal excitation whose frequency varies over a narrow band is examined. Examples 4-7 and 4-8, in which the natural response of an *RLC* circuit is considered, indicate the possibility of natural frequencies which result in damped oscillations. Interesting effects are produced when these networks are excited with sinusoidal sources having frequencies close to or at these natural frequencies.

The first circuit examined is the *RLC* series circuit in Fig. 6-11. The input admittance of that network is

$$\mathbf{Y}(j\omega) = \frac{1}{R + j\omega L + 1/j\omega C} = \frac{1}{R = j(\omega L - 1/\omega C)} \tag{6-13}$$

Examination of Eq. (6-13) shows that at the frequency

$$\omega_o = \frac{1}{\sqrt{LC}} \tag{6-14}$$

the admittance is

$$\mathbf{Y}(j\omega_o) = \frac{1}{R} = \mathbf{Y}_o \tag{6-15}$$

which is the maximum value Y can have. Also note that at ω_o, $\mathbf{Y}(j\omega_o)$ is a real number. The frequency ω_o at which the impedance or admittance of an *RLC* circuit is real is called a *resonant frequency*. *Series resonance*, as described by Eqs. (6-13) to (6-15), is a condition of maximum admittance and minimum impedance. Thus, for a voltage excitation, the current is a maximum; for a current excitation, the voltage is a minimum.

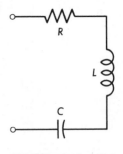

FIGURE 6-11 *RLC* series circuit.

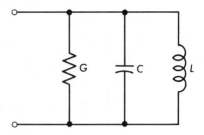

FIGURE 6-12 Parallel *GLC* circuit.

The second circuit to be examined is the parallel *GLC* circuit of Fig. 6-12. Its input impedance is

$$\mathbf{Z}(j\omega) = \frac{1}{G + j\omega C + 1/j\omega L}$$

$$= \frac{1}{G + j(\omega C - 1/\omega L)} \tag{6-16}$$

The maximum impedance of the parallel circuit occurs at

$$\omega_o = \frac{1}{\sqrt{LC}} \tag{6-17}$$

and is

$$\mathbf{Z}(j\omega_o) = \frac{1}{G} = \mathbf{Z}_o \tag{6-18}$$

The *parallel resonant frequency* described by Eq. (6-17) is a condition of maximum impedance or minimum admittance. It produces the opposite effect from the series case. The similarity between Eqs. (6-13) to (6-15) for the series circuit and Eqs. (6-16) to (6-18) for the parallel circuit should be noted. The characteristics of both circuits can be displayed graphically by means of a *universal resonance curve,* which is shown in Fig. 6-13. Derivation of the curve will be based on the series circuit. For this circuit, the ratio of admittance $\mathbf{Y}(j\omega)$ [Eq. (6-13)] to the maximum admittance [Eq. (6-15)] is

$$\frac{\mathbf{Y}}{\mathbf{Y}_o}(j\omega) = \frac{R}{R + j(\omega L - 1/\omega C)}$$

$$= \frac{1}{1 + j(L/R)(\omega - 1/\omega LC)} \tag{6-19}$$

Recognize from Eq. (6-14) that $1/LC = \omega_o^2$. Equation (6-9) can then be expressed as

$$\frac{\mathbf{Y}}{\mathbf{Y}_o}(j\omega) = \frac{1}{1 + j\dfrac{\omega_o L}{R}\left(\dfrac{\omega}{\omega_o} - \dfrac{\omega_o}{\omega}\right)} \tag{6-20}$$

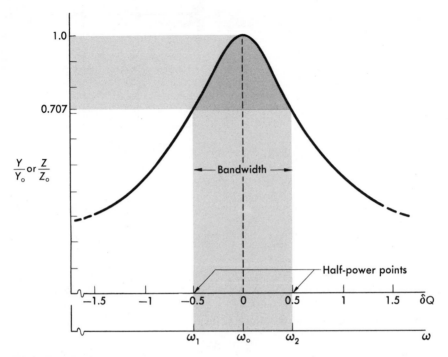

FIGURE 6-13 Universal resonance curve.

Now define a *quality factor* Q_s as the ratio of inductive reactance at resonance to the series resistance, or

$$Q_s = \frac{\omega_o L}{R} \tag{6-21}$$

In terms of Q_s, Eq. (6-20) becomes

$$\frac{\mathbf{Y}}{\mathbf{Y}_o}(j\omega) = \frac{1}{1 + jQ_s\left(\dfrac{\omega}{\omega_0} - \dfrac{\omega_0}{\omega}\right)} \tag{6-22}$$

A family of curves could be drawn by plotting values of the magnitude Y/Y_o as a function of ω/ω_o for several values of Q_s. A more useful curve, however, is found by additional manipulations. This curve is designed to point up the system behavior at or near the resonance frequency, which is usually the area of greatest concern. If the per-unit deviation of the source frequency from the resonant frequency is defined as

$$\delta = \frac{\omega - \omega_0}{\omega_0} \tag{6-23}$$

Eq. (6-22) can be written in terms of δ by substituting for ω

$$\omega = \omega_0(1 + \delta)$$

After some algebraic manipulation, Eq. (6-22) becomes

$$\frac{Y}{Y_o}(j\omega) = \frac{1}{1 + j\delta Q_s \left(\frac{2 + \delta}{1 + \delta}\right)} \tag{6-24}$$

For small frequency deviations around the resonant frequency, that is, for

$$\delta \ll 1$$

Eq. (6-24) reduces to

$$\frac{Y}{Y_o}(j\omega) = \frac{1}{1 + j2\delta Q_s} \qquad \text{near resonance} \tag{6-25}$$

Equation (6-25) is the equation of the universal resonance curve which is plotted in Fig. 6-13. The curve applies equally well to the series circuit for which it was derived and to the parallel circuit of Fig. 6-12. If used for the latter, it gives the values of Z/Z_o when the value of Q is

$$Q_p = \frac{\omega_o C}{G} \tag{6-26}$$

Justification for the parallel case is found by comparing Eq. (6-13) with Eq. (6-16) and by the development leading to Eq. (6-25).

The frequency response displayed in Fig. 6-13 is also useful in representing the relative current as a function of frequency in the series resonant circuit, and the relative voltage in the parallel resonant circuit. If $\mathbf{I}$ is the current in the circuit of Fig. 6-11, then

$$\mathbf{I} = \mathbf{VY}$$

At resonance $\mathbf{Y} = \mathbf{Y}_o$, so that $\mathbf{I}_o$, the current at resonance, equals $\mathbf{VY}_o$. The ratio of $\mathbf{I}/\mathbf{I}_o$ is then simply $\mathbf{Y}/\mathbf{Y}_o$, as the amplitude of the voltage source is constant at all frequencies. Therefore, $\mathbf{I}/\mathbf{I}_o$ is also expressed by the right-hand side of Eq. (6-25). A similar analysis for the voltage $\mathbf{V}$ across the parallel resonant circuit of Fig. 6-12 results in $\mathbf{V}/\mathbf{V}_o = \mathbf{Z}/\mathbf{Z}_o$ as given by Eq. (6-25).

The frequency characteristic displayed in Fig. 6-13 indicates that the response to signals whose angular frequencies are near ω_o are attenuated to a lesser degree than those for frequencies far removed from ω_o. To distinguish the frequency regions of high and low attenuation, it is convenient to use the concept of bandwidth introduced in Sec. 6-1. The bandwidth and half-power points in a resonant circuit can be determined readily from Eq. (6-25). The magnitude of Y/Y_o (or Z/Z_o) is 0.707 when

$$2\delta Q = \pm 1$$

From the definition of δ given in Eq. (6-23), the lower half-power point $(2\delta Q = -1)$ becomes

$$\omega_1 = \omega_o \frac{2Q - 1}{2Q} \tag{6-27}$$

and the upper half-power point ω_2 is

$$\omega_2 = \omega_0 \frac{2Q + 1}{2Q} \tag{6-28}$$

The bandwidth is

$$BW = \omega_2 - \omega_1 = \frac{\omega_0}{Q} \tag{6-29}$$

No subscripts are given for the Q's in Eqs. (6-27) to (6-29) since these equations hold for both the series and the parallel circuits.

Example 6-4

A radio receiver picks up signals having equal amplitudes from two stations. The first station transmits at a frequency of 1,000 kHz, the second at a frequency of 1,020 kHz. It is desired to hear the program on the first station. To minimize the interaction between the signals, the tuning circuit response must attenuate the signal of the second station by four times. A parallel resonant circuit as shown in Fig. 6-12 is to be used to perform the frequency selection.

a Determine the values of G and C, if $L = 50\ \mu H$ (50×10^{-6} H).

b What is the bandwidth of the circuit?

SOLUTION **a** The circuit is to be tuned to the first station so the resonant frequency is 1,000 kHz. The value of C is determined from Eq. (6-17) as

$$2\pi \times 1,000 \times 10^3 = \frac{1}{\sqrt{50 \times 10^{-6}C}}$$

from which

$$C = 5.07 \times 10^{-10} = 507\ \text{pF}$$

The relative response is to be 0.25 at a frequency of 1,020 kHz for an attenuation of four times. Equation (6-25) is used to relate the attenuation to Q. First obtain the magnitude of the relative response as

$$\frac{V}{V_0} = \frac{1}{\sqrt{1 + (2\delta Q)^2}}$$

Solving for δQ gives

$$\frac{1}{4} = \frac{1}{\sqrt{1 + (2\delta Q)^2}}$$

and $\delta Q = 1.94$.

The value of δ, from Eq. (6-23), is

$$\delta = \frac{2\pi \times 1,020 \times 10^3 - 2\pi \times 1,000 \times 10^3}{2\pi \times 1,000 \times 10^3} = 0.02$$

so that

$$Q = \frac{1.94}{0.02} = 97$$

The value of G is obtained by use of Eq. (6-26), from which

$$97 = \frac{2\pi \times 1,000 \times 10^3 \times 507 \times 10^{-12}}{G}$$

and $G = 32.9 \times 10^{-6}$ ℧.

b The bandwidth is, from Eq. (6-29),

$$BW = \frac{2\pi \times 1,000 \times 10^3}{96.8} = 64.9 \times 10^3 \text{ rad/s}$$

or BW = 10.3 kHz.

The values obtained for the bandwidth are typical of those found in commercial AM radios. The 4-to-1 attenuation between adjoining stations is also common in many inexpensive receivers.

The modified parallel circuit, shown in Fig. 6-14, is a third important circuit exhibiting resonance phenomena. Its importance stems from the fact that most inductors contain winding resistance. Usually this resistance is small compared with the inductive reactance at frequencies of interest, making possible assumptions which permit the use of the universal resonance curve with this circuit. The input impedance of the circuit of Fig. 6-14 is

$$\mathbf{Z}(j\omega) = \frac{1}{j\omega C + 1/(R + j\omega L)}$$

$$= \frac{R + j\omega L}{1 + j\omega RC - \omega^2 LC} \qquad (6\text{-}30)$$

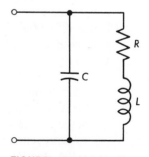

FIGURE 6-14 Modified parallel-resonant circuit.

If $R \ll \omega L$, the numerator of Eq. (6-30) can be approximated by $j\omega L$ and the entire equation written as

$$\mathbf{Z}(j\omega) = \frac{j\omega L}{1 + j\omega RC - \omega^2 LC}$$

$$= \frac{1}{(RC/L) + j(\omega C - 1/\omega L)} \tag{6-31}$$

Equation (6-31) is seen to be of the same form as Eq. (6-16). Thus the conclusions reached for the GLC circuit of Fig. 6-12 apply to the modified circuit of Fig. 6-14. The resonant angular frequency is

$$\omega_o = \frac{1}{\sqrt{LC}} \tag{6-32}$$

the quality factor is

$$Q_m = \frac{\omega_o L}{R} \tag{6-33}$$

and the maximum impedance is

$$Z_{\max} = \frac{L}{RC} = RQ_m^2 \tag{6-34}$$

The universal resonance curve and the conclusions obtained from it also apply to this circuit. It should again be emphasized that these results apply only when $\omega L \gg R$. A generally satisfactory rule of thumb is to use them if $Q \geq 10$.

6-5 TWO-PORT NETWORKS

Many networks can be considered as having two pairs of terminals: a pair of input terminals at which the excitation is usually applied, and a pair of output terminals at which the desired signals are extracted. The use of frequency-response characteristics and transfer functions highlights the importance of input-output relationships in networks and systems. Indeed, as systems are composed of interconnected networks, overall system response is dependent on individual network responses. Just as the Thévenin equivalent is effective in representing the behavior of networks at a pair of terminals, equivalent circuits which focus on input-output characteristics are convenient representations of complex networks.

Networks which contain two pairs of terminals, one input pair and one output pair, are referred to as *two-terminal-pair networks* or *two-port networks*. A *port* refers to a pair of terminals at which energy can be supplied or extracted and at which measurements can be made. It is therefore customary to represent two-port networks as shown in Fig. 6-15, which also indicates the standard convention for positive directions and polarities of the port currents and voltages. In Fig. 6-15, terminals 1-1' represent the input

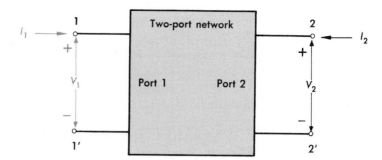

FIGURE 6-15 Representation of a two-port network indicating reference voltage polarities and current directions.

port and terminals 2-2' the output port. Certain two-port and most electronic circuits have the additional property that terminals 1' and 2' are common. Embodied in the definition of a port is that the currents leaving terminals 1' and 2' are exactly equal to the currents entering terminals 1 and 2, respectively. In addition, measurements can be made only at the ports but are not permissible between terminals 1 and 2 or terminals 1' and 2'.

A two-port network can be described by four variables which are the port currents and voltages. Two of the variables may be considered to be the independent variables and two the dependent variables. Because the system behaves linearly, the variables are related by a set of linear equations. These equations relate the port currents and voltages and define a set of *two-port parameters*. There are six combinations by which two of the four variables can be expressed in terms of the remaining two variables. Of the six possible parameter sets, three are used extensively in electronic circuit analysis because of their ease of measurement.

The *two-port admittance parameters,* or *y parameters,* are used to relate the port currents to the port voltages. In the frequency domain, the defining set of equations are

$$I_1 = y_{11}(s)V_1 + y_{12}(s)V_2 \tag{6-35}$$

$$I_2 = y_{21}(s)V_1 + y_{22}(s)V_2 \tag{6-36}$$

The elements $y_{11}(s)$, $y_{12}(s)$, $y_{21}(s)$, and $y_{22}(s)$ have the dimensions of mhos and are called the y parameters. Often the functional dependence of the parameters on the complex-frequency variable s is assumed to be implicit, and the parameters written simply as y_{11}, y_{12}, y_{21}, and y_{22}. Both notations will be used hereafter.

The specific name given each parameter is determined from its volt-ampere relationship. If terminals 2-2' in Fig. 6-15 are short-circuited, the voltage V_2 is constrained to be zero. Under these conditions, Eqs. (6-35) and (6-36) yield

$$y_{11} = \frac{I_1}{V_1}\bigg|_{V_2=0} = \text{short-circuit input admittance} \tag{6-37}$$

$$y_{21} = \frac{I_2}{V_1}\bigg|_{V_2=0} = \text{short-circuit forward transfer admittance} \qquad (6\text{-}38)$$

The term *forward transfer* in Eq. (6-38) indicates the network is being used in its normal fashion with the excitation applied at the input port and the response measured at the output port. If the excitation is applied at port 2 and port 1 is short-circuited, Eqs. (6-35) and (6-36) give

$$y_{12} = \frac{I_1}{V_2}\bigg|_{V_1=0} = \text{short-circuit reverse transfer admittance} \qquad (6\text{-}39)$$

$$y_{22} = \frac{I_2}{V_2}\bigg|_{V_1=0} = \text{short-circuit output admittance} \qquad (6\text{-}40)$$

Reverse transfer indicates excitation applied at the output port and the response measured at the input port.

Another terminology is also used to identify y parameters, particularly when it is employed to describe electronic devices. With this terminology Eqs. (6-35) and (6-36) become

$$I_1 = y_i V_1 + y_r V_2 \qquad (6\text{-}35')$$

$$I_2 = y_f V_1 + y_o V_2 \qquad (6\text{-}36')$$

The subscripts i, r, f, and o indicate that the parameters with which they are associated are the *i*nput, *r*everse transfer, *f*orward transfer, and *o*utput, respectively. The y parameters in Eqs. (6-35') and (6-36') are defined by Eqs. (6-37) to (6-40).

It is often convenient to represent a two-port network by an equivalent circuit which exhibits the same terminal relations as those expressed in the defining equations. Figure 6-16 depicts the *equivalent y-parameter circuit*. In Fig. 6-16, the KCL equation written at node A is

$$-I_1 + (y_{11} + y_{12})V_1 - y_{12}(V_1 - V_2) = 0 \qquad (6\text{-}41)$$

By combining and rearranging the terms, Eq. (6-41) can be shown to be identical to Eq. (6-35). Similarly, the KCL equation written at node B is

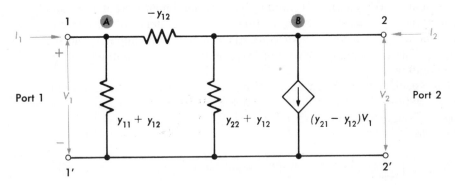

FIGURE 6-16 The y-parameter equivalent circuit.

$$-I_2 + (y_{22} + y_{12})V_2 - y_{12}(V_2 - V_1) + (y_{21} - y_{12})V_1 = 0 \qquad (6\text{-}42)$$

which is identical to Eq. (6-36).

A second set of parameters can be established from the simultaneous solution of Eqs. (6-35) and (6-36) for V_1 and V_2. The results are

$$V_1 = \frac{y_{22}}{y_{11}y_{22} - y_{12}y_{21}} I_1 + \frac{-y_{12}}{y_{11}y_{22} - y_{12}y_{21}} I_2 \qquad (6\text{-}43)$$

$$V_2 = \frac{-y_{21}}{y_{11}y_{22} - y_{12}y_{21}} I_1 + \frac{y_{11}}{y_{11}y_{22} - y_{12}y_{21}} I_2 \qquad (6\text{-}44)$$

In these equations the port voltages V_1 and V_2 are functions of the port currents I_1 and I_2. The general forms of Eqs. (6-43) and (6-44) are

$$V_1 = z_{11}I_1 + z_{12}I_2 = z_iI_1 + z_rI_2 \qquad (6\text{-}45)$$

and

$$V_2 = z_{21}I_1 + z_{22}I_2 = z_fI_1 + z_oI_2 \qquad (6\text{-}46)$$

The parameters z_{11}, z_{12}, z_{21}, and z_{22} are called the *impedance* or *z parameters*. Specific parameters are defined by first open-circuiting port 2, which makes $I_2 = 0$, and exciting port 1, and then repeating the process by open-circuiting port 1 and exciting port 2. The results are

$$z_{11} = z_i = \frac{V_1}{I_1}\bigg|_{I_2=0} = \text{open-circuit input impedance} \qquad (6\text{-}47)$$

$$z_{21} = z_f = \frac{V_2}{I_1}\bigg|_{I_2=0} = \text{open-circuit forward transfer impedance} \qquad (6\text{-}48)$$

$$z_{22} = z_o = \frac{V_2}{I_2}\bigg|_{I_1=0} = \text{open-circuit output impedance} \qquad (6\text{-}49)$$

$$z_{12} = z_r = \frac{V_1}{I_2}\bigg|_{I_1=0} = \text{open-circuit reverse transfer impedance} \qquad (6\text{-}50)$$

The equivalent circuit most often used to represent the z parameters is depicted in Fig. 6-17. The equivalence of this circuit and Eqs. (6-45) and (6-46) can be verified by writing voltage-law equations for the two loops.

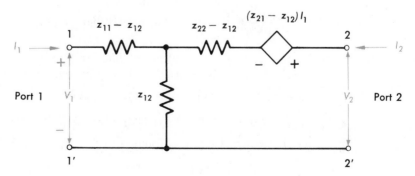

FIGURE 6-17 The z-parameter equivalent circuit.

The relationships between the z and y parameters are given by comparing coefficients of corresponding current terms in Eqs. (6-43) to (6-46).

Example 6-5

Two laboratory tests were run on a two-port network. Reference directions and polarities are given in Fig. 6-15. In test 1, port 1 was excited and port 2 was open-circuited. The results were

$$\mathbf{I}_1 = 10^{-3}\underline{/0°}\text{ A} \qquad \mathbf{V}_1 = 1.41\underline{/45°}\text{ V} \qquad \mathbf{V}_2 = 2.24\underline{/-26.4°}\text{ V}$$

In test 2, port 1 was open-circuited and port 2 was excited. The results were

$$\mathbf{I}_2 = 10^{-3}\underline{/0°}\text{ A} \qquad \mathbf{V}_1 = 1.00\underline{/-90°}\text{ V} \qquad \mathbf{V}_2 = 1.25\underline{/-53°}\text{ V}$$

Both tests were made with a sinusoidal excitation of angular frequency $\omega = 10^6$ rad/s.

a Determine the z parameters of the network at the excitation frequency.

b At the excitation frequency, determine an appropriate equivalent circuit.

SOLUTION **a** From Eqs. (6-47) and (6-48) and the results of test 1,

$$\mathbf{z}_{11}(j10^6) = \frac{1.41\underline{/45°}}{10^{-3}\underline{/0°}} = 1.41 \times 10^3\underline{/45°}$$
$$= 1{,}000 + j1{,}000 \qquad \Omega$$

$$\mathbf{z}_{21}(j10^6) = \frac{2.24\underline{/-26.4°}}{10^{-3}\underline{/0°}} = 2.24 \times 10^3\underline{/-26.4°}$$
$$= 2{,}000 - j1{,}000 \qquad \Omega$$

From Eqs. (6-49) and (6-50) and the results of test 2,

$$\mathbf{z}_{22}(j10^6) = \frac{1.25\underline{/-53°}}{10^{-3}\underline{/0°}} = 1.25 \times 10^3\underline{/-53°}$$
$$= 750 - j1{,}000 \qquad \Omega$$

$$\mathbf{z}_{12}(j10^6) = \frac{1.00\underline{/-90°}}{10^{-3}\underline{/0°}} = 10^3\underline{/-90°} = -j1{,}000 \qquad \Omega$$

b The results of part (a) are used in conjunction with the circuit of Fig. 6-17. The individual impedances used in the equivalent circuit are

$$\mathbf{z}_{11} - \mathbf{z}_{12} = 1{,}000 + j1{,}000 - (-j1{,}000) = 1{,}000 + j2{,}000 \qquad \Omega$$
$$\mathbf{z}_{21} - \mathbf{z}_{12} = 2{,}000 - j1{,}000 - (-j1{,}000) = 2{,}000 \qquad \Omega$$
$$\mathbf{z}_{22} - \mathbf{z}_{12} = 750 - j1{,}000 - (-j1{,}000) = 750 \qquad \Omega$$
$$\mathbf{z}_{12} = -j1{,}000 \qquad \Omega$$

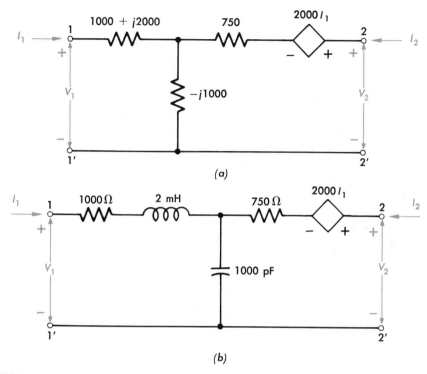

FIGURE 6-18 (a) Equivalent z-parameter circuit for Example 6-5; (b) alternative realization of the circuit shown in a.

The resultant equivalent circuit is shown in Fig. 6-18a.

The results of this example apply only at the angular frequency of the excitation. Should the z parameters be desired at a different angular frequency, either a new set of measurements must be made or the function $z(s)$ must be determined. At the angular frequency of 10^6 rad/s, the impedances indicated in Fig. 6-18a may be identified as resistors, capacitors, or inductors. The elements are each of the form $R + jX$, where R is identified as a resistance and X as either an inductance or capacitance, depending on its sign. Thus $\mathbf{z}_{11} - \mathbf{z}_{12}$ is a resistance of 1,000 Ω in series with an inductance whose value is

$$L = \frac{X}{\omega} = \frac{2,000}{10^6} = 2 \times 10^{-3} \text{ H} = 2 \text{ mH}$$

In similar fashion $\mathbf{z}_{12}$ is a capacitance whose value is

$$C = \frac{1}{\omega X} = \frac{1}{10^6 \times 10^3} = 10^{-9} \text{ F} = 1,000 \text{ pF}$$

and $\mathbf{z}_{22} - \mathbf{z}_{12}$ is simply a resistance of 750 Ω. Based on these element values, the circuit of Fig. 6-18a may be replaced by the circuit shown in Fig. 6-18b.

Again, it should be noted that the circuits of Fig. 6-18 are valid only at $\omega = 10^6$ rad/s.

Example 6-6

The circuit shown in Fig. 6-19 is the equivalent circuit of a field-effect transistor (FET) amplifier stage.

a For the portion of the network contained between terminals 1-1' and 2-2', determine the y parameters.

b For values of $\mu = g_m/g_d \gg 1$, obtain a y-parameter equivalent circuit.

SOLUTION **a** To determine the y parameters, two sets of calculations must be made. The first of these requires exciting port 1 by a voltage source V_1 and short-circuiting port 2. The second requires short-circuiting port 1 and exciting port 2 by a voltage source V_2. These conditions are depicted in Fig. 6-20a and b, respectively. In Fig. 6-20 the voltage source μV_{gs} represents the Thévenin form of the current source $g_m V_{gs}$ in Fig. 6-19.

For the circuit of Fig. 6-20a, $I_1 = 0$. The KVL equation for the output loop is

$$I_2 \frac{1}{g_d} - \mu V_{gs} + I_2 R_K = 0$$

and the constraint equation for the controlled source is

$$V_{gs} = V_1 - I_2 R_K$$

Combination of the two equations results in

$$I_2 \left[(1 + \mu) R_K + \frac{1}{g_d} \right] = \mu V_1$$

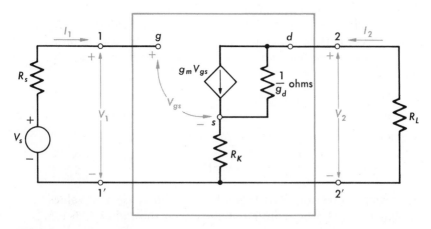

FIGURE 6-19 Circuit for Example 6-6.

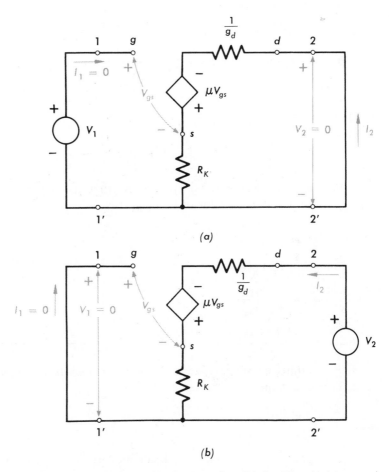

FIGURE 6-20 (a) Circuit to calculate y_{11} and y_{21}; (b) circuit to calculate y_{22} and y_{12} for Example 6-6.

Use of Eqs. (6-37) and (6-38) yields

$$y_{11} = \frac{I_1}{V_1} = 0 \qquad y_{21} = \frac{I_2}{V_1} = \frac{\mu}{R_K(1 + \mu) + 1/g_d}$$

The KVL equation for the output loop in Fig. 6-20b is

$$-V_2 + I_2 \frac{1}{g_d} - \mu V_{gs} + I_2 R_K = 0 \qquad \text{where } V_{gs} = -I_2 R_K$$

Rewriting the voltage-law expression, substituting for V_{gs} and solving for I_2, gives

$$I_2 = \frac{V_2}{(1 + \mu)R_K + 1/g_d}$$

Because of the open circuit between terminals g and s, $I_1 = 0$. Use of Eqs. (6-39) and (6-40) gives

$$y_{22} = \frac{I_2}{V_2} = \frac{1}{(1 + \mu)R_K + 1/g_d} \qquad y_{12} = \frac{I_1}{V_2} = 0$$

b   The equivalent circuit is that given in Fig. 6-16, whose elements are

$$y_{11} + y_{12} = 0 \qquad y_{12} = 0$$

$$y_{21} - y_{12} = y_{21} = \frac{\mu}{(1 + \mu)R_K + 1/g_d}$$

$$y_{22} + y_{12} = y_{22} = \frac{1}{(1 + \mu)R_K + 1/g_d}$$

For the case where $\mu \gg 1$, recalling that $\mu = g_m/g_d$, and dividing numerator and denominator of y_{21} by μ, result in

$$y_{21} = \frac{g_m}{1 + g_m R_K}$$

Similarly, y_{22} becomes

$$y_{22} = \frac{g_d}{1 + g_m R_K}$$

The resultant circuit is shown in Fig. 6-21.

A third set of two-port parameters is called the *hybrid* or *h parameters*. They are defined by the equations

$$V_1 = h_{11}I_1 + h_{12}V_2 = h_i I_1 + h_r V_2 \tag{6-51}$$

$$I_2 = h_{21}I_1 + h_{22}V_2 = h_f I_1 + h_o V_2 \tag{6-52}$$

The specific *h* parameters may be defined by first exciting the input port and short-circuiting the output port ($V_2 = 0$), and then exciting the output port and open-circuiting the input port ($I_1 = 0$). The results are expressed in Eqs. (6-53) to (6-56).

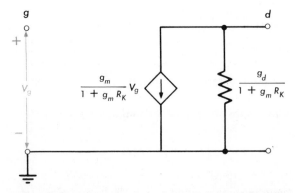

FIGURE 6-21   Equivalent y-parameter circuit of Example 6-6.

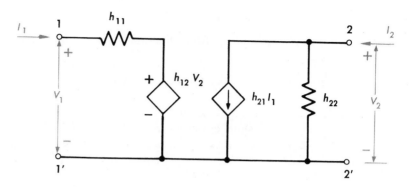

FIGURE 6-22 The *h*-parameter equivalent circuit.

$$h_{11} = h_i = \left.\frac{V_1}{I_1}\right|_{V_2=0} = \text{short-circuit input impedance} \tag{6-53}$$

$$h_{21} = h_f = \left.\frac{I_2}{I_1}\right|_{V_2=0} = \text{short-circuit forward current gain} \tag{6-54}$$

$$h_{22} = h_o = \left.\frac{I_2}{V_2}\right|_{I_1=0} = \text{open-circuit output admittance} \tag{6-55}$$

$$h_{12} = h_r = \left.\frac{V_1}{V_2}\right|_{I_1=0} = \text{open-circuit reverse voltage gain} \tag{6-56}$$

The quantities h_{12} and h_{21} are dimensionless and are each represented by a controlled source in the circuit model shown in Fig. 6-22. The equivalence of the model with Eqs. (6-51) and (6-52) is demonstrated by use of the voltage law for the input loop and the current law at node 2.

Example 6-7

The circuit shown in Fig. 6-23 is the equivalent circuit of a junction-transistor amplifier stage.

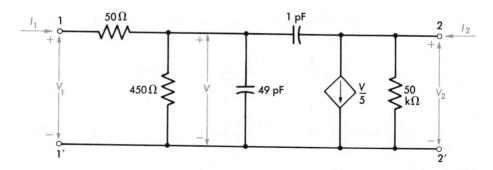

FIGURE 6-23 Circuit for Example 6-7.

a Determine h_{11} and h_{21}.

b Sketch an asymptotic Bode diagram for h_i and h_f for angular frequencies below 10^{10} rad/s.

c Determine the angular frequency at which $|h_f(j\omega)|$ is unity.

SOLUTION **a** From the definitions of Eqs. (6-53) and (6-54), both h_i and h_f are computed by short-circuiting port 2 and exciting port 1 with a current source I_1. The resultant circuits used to determine h_i and h_f are shown in Fig. 6-24a and b, respectively. For the circuit in Fig. 6-24a,

$$V_1 = I_1 \times 50 + I_1 Z$$

where

$$Z = \frac{450 \times 1/s(49 + 1) \times 10^{-12}}{450 + 1/s(40 + 1) \times 10^{-12}} = \frac{450}{1 + s \times 2.25 \times 10^{-8}}$$

Substitution of Z into the equation for V_1 and forming the ratio V_1/I_1 gives

$$\frac{V_1}{I_1} = h_i = 50 + \frac{450}{1 + s \times 2.25 \times 10^{-8}}$$

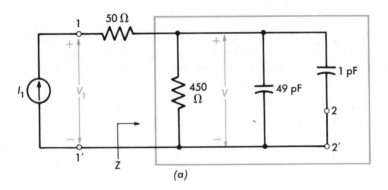

(a)

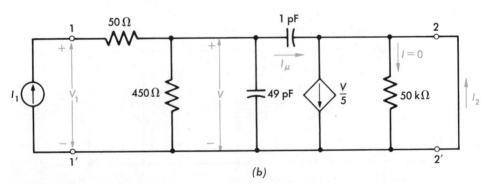

(b)

FIGURE 6-24 (a) Circuit to determine h_i and (b) circuit to determine h_f in Example 6-7.

Rearrangement gives

$$h_i = \frac{500(1 + s/4.44 \times 10^8)}{(1 + s/4.44 \times 10^7)}$$

The circuit of Fig. 6-24b is used to determine h_f. The current I_2, by use of KCL, is

$$I_2 = \frac{V}{5} - I_\mu = \frac{V}{5} - s \times 1 \times 10^{-12} \; V$$

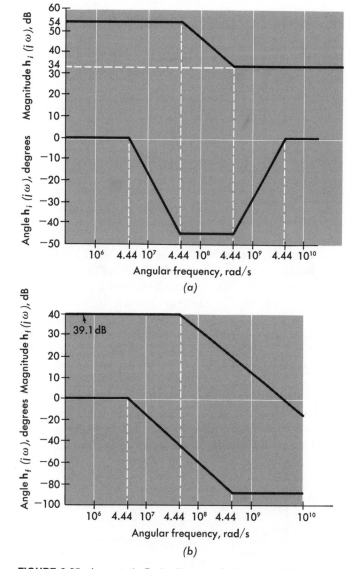

FIGURE 6-25 Asymptotic Bode diagrams for (a) h_i and (b) h_f in Example 6-7.

as the current in the 50-kΩ resistance is zero because $V_2 = 0$. The voltage V is $I_1 Z$ and is given in the calculation for h_i as

$$V = I_1 Z = I_1 \frac{450}{1 + s \times 2.25 \times 10^{-8}}$$

Combination of the equations for V and I_2 and forming the ratio I_2/I_1 yields

$$\frac{I_2}{I_1} = h_f = \frac{90(1 - s/2 \times 10^{11})}{1 + s/4.44 \times 10^7}$$

b   The asymptotic Bode diagrams are plotted in Fig. 6-25.

c   From the Bode diagram, the angular frequency at which $h_f(j\omega) = 0$ dB is 4×10^9 rad/s.

Note, the break frequency at 2×10^{11} rad/s occurs at a frequency far removed from all other critical frequencies of interest. As a result, h_f is often approximated by

$$h_f = \frac{90}{1 + s/4.44 \times 10^7}$$

6-6   BLOCK DIAGRAMS

The determination of the transfer function, and consequently the frequency response, of complex systems is often facilitated by the use of block diagrams. A block diagram is a schematic representation of a system in terms of its component parts in which each component is represented by its transfer function. A typical block used in a block diagram is shown in Fig. 6-26. The function $G(s)$ is the transfer function of the block and represents the transformed or phasor ratio of the output X_2 to the input X_1 and has the same significance as the electric-network transfer function discussed in earlier chapters. A component transfer function may be extremely simple, as in the case of a capacitance, where the ratio of output current $\mathbf{I}$ to the input voltage $\mathbf{E}$ is $j\omega C$. Or it may be moderately complicated as in the function for h_f in Example 6-7.

Two inherent properties of the block diagram are that each block is a unilateral element and that the transfer function of a particular block remains unchanged when other blocks are connected to it. The first property indicates that signal transmission is only in the direction of the arrows, so that $G(s)$ relates the response X_2 to an excitation X_1. The second property indicates that there is no interaction (loading effects) between the various blocks.

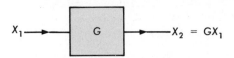

FIGURE 6-26   Elementary block diagram.

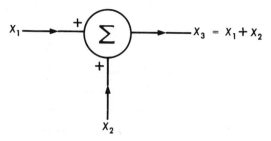

FIGURE 6-27 Symbol for summing point.

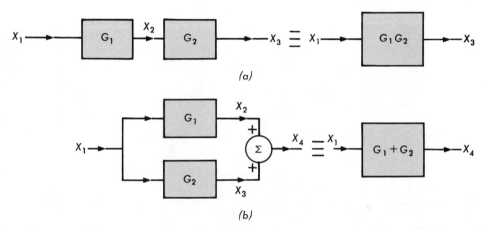

FIGURE 6-28 Block diagram equivalents for (a) cascade structure and (b) parallel-channel struc-ture.

Complex systems are represented by a number of blocks connected to indicate the signal paths which exist and the relationship among the various system components. Figure 6-27 depicts a summing point which is used to show how two or more signals are combined. The polarity markings along-side each arrow specify the polarities of the respective signals. Once the system is represented by appropriate building blocks, the blocks can be combined to reduce the system to a single transfer function or single block. Two useful reductions are illustrated in Fig. 6-28. The cascade structure, Fig. 6-28a, results in a transfer function which is the product of the individ-ual transfer functions. Similarly, the parallel-channel structure of Fig. 6-28b reduces to the sum of the individual transfer functions.

Example 6-8

Determine the transfer function X_o/X_S for the block diagram shown in Fig. 6-29.

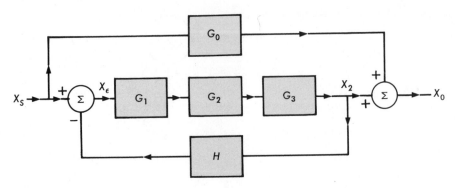

FIGURE 6-29 Block diagram for Example 6-8.

SOLUTION   First combine the three blocks in cascade, which results in $G_A = G_1G_2G_3$ as shown in Fig. 6-30a. The second step is to reduce the loop containing G_A and H.

$$X_2 = G_A X_4 \qquad \text{and} \qquad X_\epsilon = X_S - HX_2$$

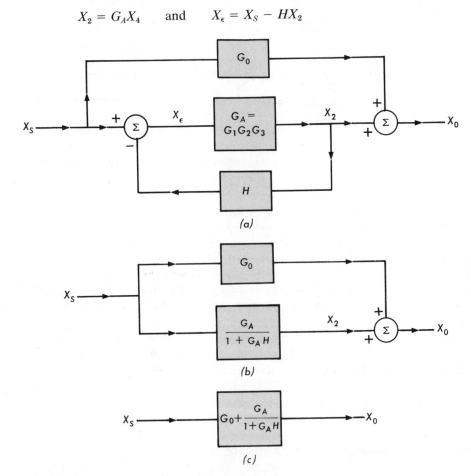

FIGURE 6-30 Reduction of block diagram of Fig. 6-29.

Combination of these equations gives

$$X_2 = \frac{G_A}{1 + G_A H} X_S$$

so that the equivalent block is indicated in Fig. 6-30*b*.

The single block for the system is obtained by adding the transfer functions for the two channels. The result, shown in Fig. 6-30*c*, is

$$\frac{X_o}{X_S} = G_o + \frac{G_A}{1 + G_A H}$$

The block diagram of Fig. 6-29 is that of a class of feedback amplifiers whose transfer function is derived in Example 6-8. Feedback amplifiers are of great importance in electronic circuits and are discussed in Chap. 9.

6-7 SIGNAL-FLOW GRAPHS

Simply stated, a signal-flow graph is a diagrammatic representation of a linear system of equations. As such, a signal-flow graph is often used to schematically describe a system in terms of its constituent parts. The two basic elements in a signal-flow graph are *nodes* and *branches*. A node is used to indicate a variable, a branch to indicate the relationship between a pair of variables. Figure 6-31*a* shows a typical component in the graph. The variables X_1 and X_2 are represented by nodes. The directed arrow is the branch whose branch transmittance G defines the functional relationship $X_2 = GX_1$. The significance of the arrow is that it denotes the unilateral nature of the relation between X_2 and X_1. The arrow, being directed from X_1 to X_2, indicates that X_2 depends on X_1. Thus, in the graph of Fig. 6-31*b*, the dependence of X_3 on X_2 is described by the branch transmittance G_C and that of X_2 on X_3 by the branch transmittance G_D.

In Fig. 6-31*b*, the node variable X_2 is $G_D X_3 + G_A X_1$ and highlights the fact that the value of a node variable is defined only by the branches entering (incident upon) it. Each incident branch contributes to the node value an amount equal to the branch transmittance multiplied by the value of the node variable from which the branch leaves. Nodes having only incident branches are called *sink nodes,* and those having only branches that leave are *source*

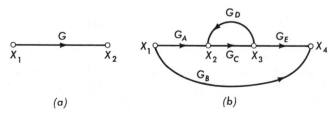

FIGURE 6-31 (a) Signal-flow graph element and (b) a simple flow graph.

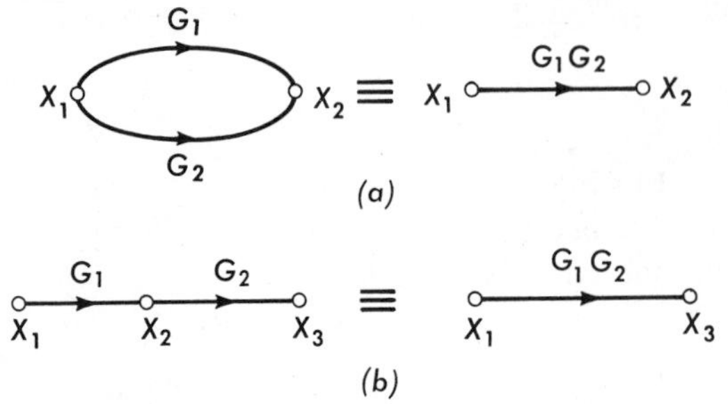

FIGURE 6-32 Signal-flow graph equivalents for (a) parallel configuration and (b) cascade configuration.

nodes. The nodes X_1 and X_4 in Fig. 6-31b are source and sink nodes, respectively.

Because a signal-flow graph describes a set of linear equations, the elements of the graph can be algebraically combined. This process of graph reduction permits the transfer function to be evaluated and, in essence, is a method for solving the set of equations for one of the variables. Two elementary reductions are shown in Fig. 6-32. The parallel-branch configuration in Fig. 6-32a reduces to the sum of branch transmittances and that for the cascade structure is the product of the individual branch transmittances.

Two other commonly encountered configurations are the self-loop and the feedback structure, which are illustrated in Fig. 6-33a and b, respectively. For the self-loop,

$$X_2 = G_1 X_1 + H_1 X_2$$

and, after rearrangement, gives

$$X_2 = \frac{G_1}{1 - H_1} X_1$$

The relation for the equivalent branch in Fig. 6-33a leads to the general rule for eliminating self-loops from a graph. Stated without further proof, this

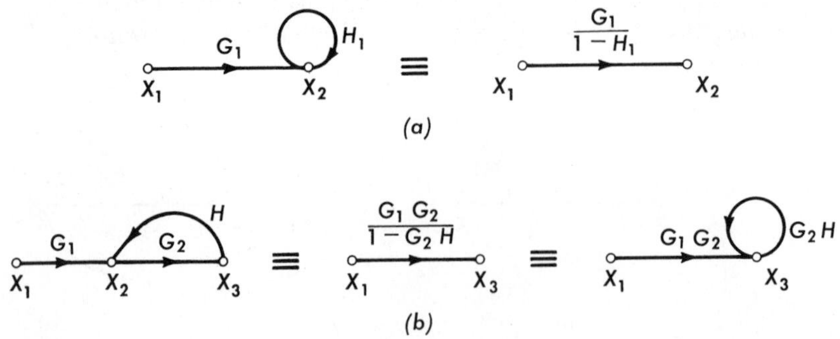

FIGURE 6-33 Signal-flow graph equivalents for (a) self loop and (b) feedback loop.

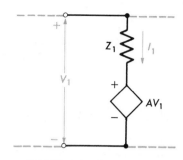

FIGURE 6-34 Circuit segment.

rule is that all branches incident on a node containing a self-loop have their branch transmittances divided by 1 minus the self-loop transmittance.

Figure 6-34a depicts a portion of a circuit and demonstrates how a self-loop can arise in the formulation of network equations. The KVL expression for the circuit segment is

$$V_1 = I_1 Z_1 + A V_1$$

for which the graph in Fig. 6-34b can be drawn. Both using the graph reduction and algebraically solving for V_1 in the KVL expression result in

$$V_1 = \frac{I_1 Z_1}{1 - A}$$

The reduction of the feedback loop in Fig. 6-33b proceeds from

$$X_2 = G_1 X_1 + H_1 X_3$$

and

$$X_3 = G_2 X_2$$

Substituting of X_2 into the equation for X_3 and combining terms yields

$$X_3 = \frac{G_1 G_2}{1 - G_2 H}$$

which is the relation for the equivalent branch in the reduced graph of Fig. 6-33b. Note that the same result is obtained for the flow graph containing the self-loop in Fig. 6-33b.

Example 6-9

The two-port network in Fig. 6-35 is characterized by its y parameters.

a Construct a signal-flow graph for the circuit shown using V_S, V_1, I_1, I_2, and V_2 as nodes.

b Use the flow graph to evaluate the transfer function V_2 / V_s.

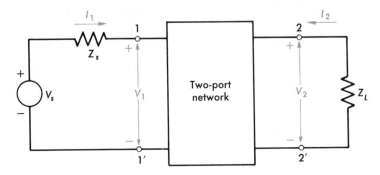

FIGURE 6-35 Circuit for Example 6-9.

SOLUTION **a** The equations which relate the variables must be obtained first. Note, however, that V_s is the excitation and is represented by a source node. No other variable can be indicated by a source node. In general, all other node variables have branches both leaving and entering, except sink nodes. The basic equations which relate I_1, I_2, V_1, and V_2 are the two-port y parameters given in Eqs. (6-35) and (6-36) and restated here:

$$I_1 = y_{11}V_1 + y_{12}V_2$$
$$I_2 = y_{21}V_1 + y_{22}V_2$$

These equations are identified as branches A through D in the signal-flow graph in Fig. 6-36. At port I in Fig. 6-35, the KVL relation is expressible as

$$V_1 = V_s - I_1Z_s$$

and indicated by branches E and F.

The last relation required is the Ohm's law equation at port 2:

$$V_2 = -I_2Z_L$$

which is drawn as branch G. The seven branches constitute one possible signal-flow graph which characterizes the system.

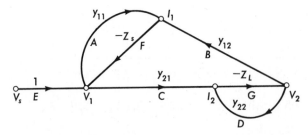

FIGURE 6-36 Signal-flow graph for circuit in Fig. 6-35.

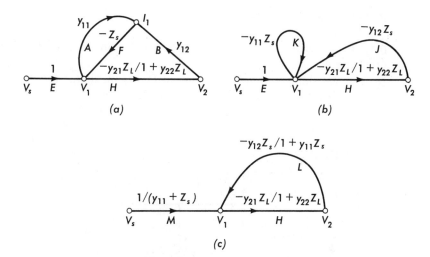

FIGURE 6-37 Reduction of signal-flow graph of Fig. 3-36.

b   The transfer function is obtained by reducing the graph. Branches C, G, and D form a feedback loop. These are replaced by the equivalent branch H (see Fig. 6-33b) whose transmittance is

$$H = \frac{-y_{21}Z_L}{1 + y_{22}Z_L}$$

and shown in Fig. 6-37a. In Fig. 6-37a, a feedback loop is formed by branches B, F, and A. This reduction is depicted in Fig. 6-37b in which the feedback loop is replaced by branch J and self-loop K. The respective branch transmittances are

$$J = -y_{12}Z_s \qquad K = -y_{11}Z_s$$

The flow graph in Fig. 6-37c results when the self-loop is eliminated. The branch transmittances affected by this reduction are E and J whose values are divided by $(1 + y_{11}Z_s)$, the self-loop transmittance. The equivalent branches for J and E are L and M, respectively.

Figure 6-37c indicates that a feedback loop is constituted by branches M, L, and H and their reduction gives

$$\frac{V_2}{V_s} = \frac{\dfrac{1}{1 + y_{11}Z_s} \times \dfrac{-y_{21}Z_L}{1 + y_{22}Z_L}}{1 - \dfrac{-y_{21}Z_L}{1 + y_{22}Z_L} \times \dfrac{-y_{12}Z_s}{1 + y_{11}Z_s}}$$

Clearing fractions and combining terms yields

$$\frac{V_2}{V_s} = \frac{-y_{21}Z_L}{1 + y_{11}Z_s + y_{22}Z_L + Z_sZ_L(y_{11}y_{22} - y_{12}y_{21})}$$

PROBLEMS

6-1 a Determine the transfer function V_2/V_1 in the circuit of Fig. 6-38 in literal terms.
b For the element values indicated in Fig. 6-38, sketch the magnitude and phase of V_2/V_1 as a function of frequency.

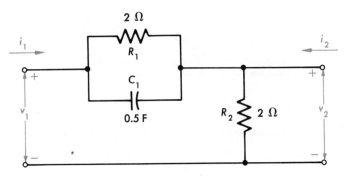

FIGURE 6-38 Circuit for Prob. 6-1.

6-2 Repeat Prob. 6-1 for the circuit depicted in Fig. 6-39.

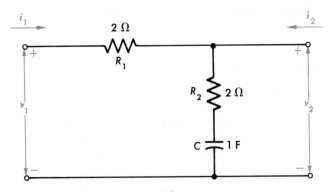

FIGURE 6-39 Circuit for Prob. 6-2.

6-3 For the circuit shown in Fig. 6-40 and using the element values indicated, sketch the magnitude and phase of the system function V_2/V_1 as a function of frequency.

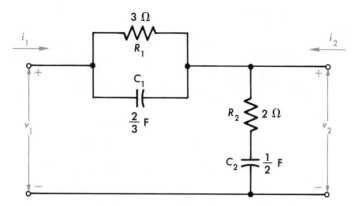

FIGURE 6-40 Circuit for Prob. 6-3.

6-4 The circuit shown in Fig. 6-41 is referred to as a *choke-input filter* and is used in power-supply circuits in which ac is converted to dc. The purpose of the filter is to greatly attenuate the ac component of the input signal v_1 while having little effect on the dc component.
a Determine the transfer function V_2/V_1 and sketch its magnitude and phase as a function of frequency.
b For $v_1(t) = 110\sqrt{2}/\pi + 55\sqrt{2}/2 \cos 377t$ which are the first two terms of the Fourier series for a half-wave rectified sinusoid, determine the output voltage $v_2(t)$ in the steady state.
c Sketch the result obtained in *b*.
d What conclusions can be drawn about the response of this circuit if the higher-order frequency components of the input signal are included?

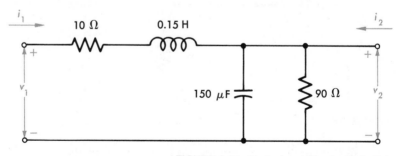

FIGURE 6-41 Choke-input filter for Prob. 6-4.

6-5 The *capacitor-input filter* in Fig. 6-42 serves the same purpose as does the circuit in Fig. 6-41 and Prob. 6-4.
a Determine the system function V_2/V_1 and sketch its magnitude and phase as a function of frequency.
b The first two terms of the Fourier series of a particular full-wave rectified sine wave are

$$v_1(t) = \frac{440\sqrt{2}}{\pi} - \frac{880\sqrt{2}}{3\pi} \cos 200\pi t$$

For this input, determine $v_2(t)$ in the steady state and sketch the result.

c What conclusions can be drawn about the circuit response to an additional term in $v_1(t)$ of value $880\sqrt{2}/15\pi \cos 400\pi t$?

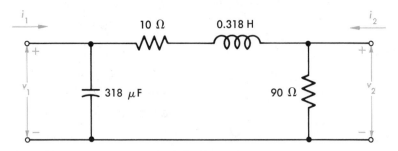

10 Ω   0.318 H

318 μF

90 Ω

FIGURE 6-42 Capacitor-input filter for Prob. 6-5.

6-6 a Determine the transfer function $\mathbf{V}_o/\mathbf{V}_s$ for the circuit in Fig. 6-43.
b Sketch the magnitude and phase of $\mathbf{V}_o/\mathbf{V}_s$ obtained in a.
c Which of the basic frequency-response characteristics does this circuit approximate?

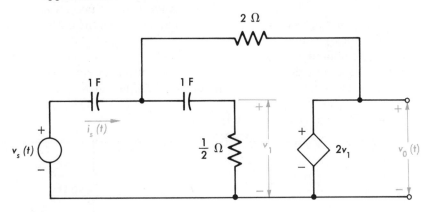

2 Ω

1 F   1 F

$i_s(t)$

$v_s(t)$

$\frac{1}{2}$ Ω

v_1

$2v_1$

$v_0(t)$

FIGURE 6-43 Circuit for Prob. 6-6.

6-7 Sketch the asymptotic Bode diagrams for the following functions:

a $F_1 = \dfrac{200}{(1 + s)(1 + s/10)}$

b $F_2 = \dfrac{200(1 + s/10)}{(1 + s)(1 + s/20)}$

c $F_3 = \dfrac{200s}{(1 + s)(1 + s/10)}$

In each case, smooth the curves to obtain more accurate results and also determine the values of F_1, F_2, and F_3 at $\omega = 5$ rad/s.

6-8 Sketch the asymptotic Bode diagrams for the functions $F_1(s)$, $F_2(s)$, and $F_3(s)$. In each case smooth the curves to obtain more accurate results. Also, determine the values of each function at $\omega = 2$ rad/s.

a $F_1(s) = \dfrac{100}{(1 + s)(1 + s/4)}$

b $F_2(s) = \dfrac{100}{s(1 + s/r)}$

c $F_3(s) = \dfrac{100(1 + s/20)}{s(1 + s/4)}$

6-9 Each of the system functions given is the ratio of an output current I_2 to an input current I_1 in a network. That is, $H(s) = I_2/I_1$.
a Sketch the asymptotic Bode diagram for each function.
b If $i_1(t) = 0.1 \cos 20t$, determine the steady-state value of $i_2(t)$.
c At what angular frequency is the magnitude of $\mathbf{H}(j\omega)$ one-half the magnitude of $\mathbf{H}(j5)$?

$$H_1(s) = \frac{0.5(1 + s/10)}{(1 + s)^2(1 + s/40)} \qquad H_2(s) = \frac{0.5(1 - s)}{(1 + s)(1 + s/10)}$$

$$H_3(s) = \frac{0.5(1 + s)^2}{s(1 + s/10)(1 + s/50)}$$

6-10 Each of the functions $H(s)$ given in this problem is the ratio of an output voltage V_2 to an input voltage V_1 of some network.
a Sketch the asymptotic Bode diagram for each function.
b If $v_1(t) = 10^{-3} \cos 4t$, determine the steady-state response for $v_2(t)$.
c At what angular frequency is $|\mathbf{H}(j\omega)| = 5$?

$$H_1(s) = \frac{32}{(1 + s)(1 + s/8)(1 + s/20)} \qquad H_2(s) = \frac{25(1 - s/8)}{(1 + s)^2(1 + s/8)}$$

$$H_3(s) = \frac{25(1 + s)}{s^2(1 + s/8)}$$

6-11 For each of the following transfer functions
a Sketch the asymptotic Bode diagram.
b Determine the angular frequency at which $|\mathbf{T}(j\omega)| = 0$ dB.
c Find the angular frequency at which $\angle\mathbf{T}(j\omega) = -180°$.

$$T_1(s) = \frac{32}{(1 + s)(1 + s/10)(1 + s/40)} \qquad T_2(s) = \frac{32(1 + s/5)}{(1 + s)(1 + s/10)(1 + s/40)^2}$$

6-12 Repeat Prob. 6-11 for

$$T_1(s) = \frac{20}{(1 + s)(1 + s/10)^2} \qquad T_2(s) = \frac{20(1 + s/8)}{(1 + s)(1 + s/10)^2(1 + s/40)}$$

6-13   **a**   Sketch the asymptotic Bode diagram for V_2/V_1 in the circuit of Fig. 6-38 and Prob. 6-1. Smooth the curves for more accurate results.
b   At what angular frequencies is $\angle V_2/V_1 = 15°$?

6-14   **a**   Sketch the asymptotic Bode diagram for V_2/V_1 in the circuit of Fig. 6-39 and Prob. 6-2.
b   At what angular frequencies is $\angle V_2/V_1 = -15°$?

6-15   **a**   Sketch the asymptotic Bode diagram for V_2/V_1 in the circuit of Fig. 6-40 and Prob. 6-3.
b   Determine the forced response $v_2(t)$ if $v_1(t) = 4 + 2 \cos t$.

6-16   The circuit shown in Fig. 6-44 is the representation of a transistor amplifier stage called an emitter follower. Sketch the asymptotic Bode diagram for V_o/V_s. At what angular frequency is $|V_o/V_s| = -3$ dB?

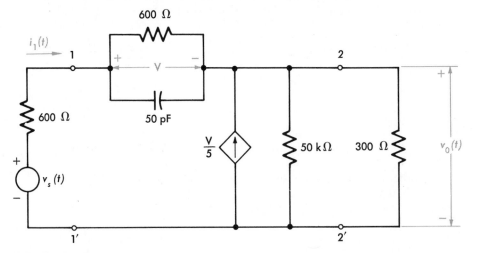

FIGURE 6-44   Circuit for Prob. 6-16.

6-17   In the circuit of Fig. 6-44, sketch the asymptotic Bode diagram for the impedance V_s/I_1.

6-18   The frequency-domain representation of a particular operational amplifier stage is shown in Fig. 6-45.

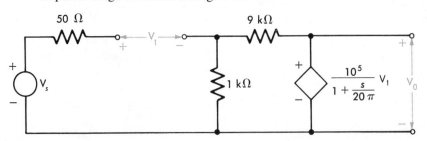

FIGURE 6-45   Circuit for Prob. 6-18.

a Sketch the asymptotic Bode diagram for V_o/V_s.
b Determine the half-power frequency.
c At what frequency is $|V_o/V_s|$ unity?

6-19 A series RLC circuit is excited by a variable-frequency sinusoidal voltage source having an rms amplitude of 2 V. The maximum rms current in the circuit is to be 10 mA at a frequency of 10 kHz. At frequencies between 9.6 kHz and 10.4 kHz the magnitude of the current is to be at least 7.07 mA.
a Determine the values of R, L, and C.
b What is the voltage across the capacitance at resonance?

6-20 An RLC series circuit is to be resonant at $\omega = 10^4$ rad/s and have a bandwidth of 500 rad/s. At resonance a 0.5 V rms input voltage is to produce an rms current of 50 mA.
a Determine the values of R, L, and C.
b What is the voltage across the inductance at resonance?

6-21 A parallel GLC circuit is to be resonant at $\omega = 10^5$ rad/s and have a bandwidth of 2×10^3 rad/s. At resonance, a 10 μA rms input current is to produce a 1.0 V rms output signal.
a Find the values of the circuit elements.
b At what angular frequencies is the output voltage magnitude 0.8 V?

6-22 A parallel GLC circuit is to have a maximum output voltage of 2.0 V rms when excited by a variable-frequency sinusoidal current whose rms amplitude is 5 μA. The circuit is to provide reasonable passage for frequencies between 3.9 and 4.1 kHz.
a Determine the values of the circuit components.
b At what angular frequencies is the output voltage 1.8 V?

6-23 Repeat Prob. 6-22 if the modified parallel resonant circuit of Fig. 6-14 is used in place of the GLC parallel circuit.

6-24 Repeat Prob. 6-21 if the modified parallel resonant circuit of Fig. 6-14 is used in place of the GLC parallel circuit.

6-25 The voltage-transfer function of an electronic amplifier is

$$\frac{V_s}{V_1} = \frac{2 \times 10^4 s}{s^2 + 10^3 s + 10^{10}}$$

Determine the resonant frequency and bandwidth.

6-26 The circuit shown in Fig. 6-46 is the equivalent circuit of a piezoelectric crystal. Determine:
a The impedance $Z_{AB}(s)$
b The resonant frequencies, using the result in a

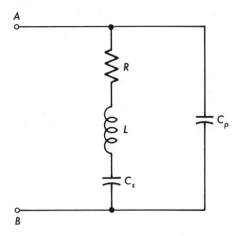

A

B

FIGURE 6-46 Piezoelectric crystal representation for Prob. 6-26.

6-27 The "twin-tee," or "notch," network shown in Fig. 6-47 is frequently used to obtain band-reject characteristics.
a Determine the transfer function V_2/I_1.
b At what angular frequency will the transfer function be zero?

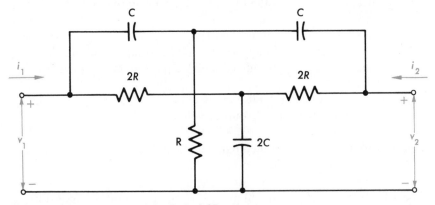

FIGURE 6-47 Twin-tee circuit for Prob. 6-27.

6-28 An amplitude-modulated (AM) signal can be represented as

$$v(t) = (V_s \cos \omega_s t)(V_c \cos \omega_c t)$$

where the term $\cos \omega_s t$ represents the information contained by the carrier ω_c (the frequency of the transmitting station). For $\omega_c = 2\pi \times 10^6$ rad/s, it is desired to pass all frequencies $\omega_s = 2\pi \times 5 \times 10^3$ rad/s. The circuit shown in Fig. 6-48 is to be used to select the desired band of frequencies. For $V_s = 0.1$ V and $V_c = 1.0$ V, it is required that the amplitude of the output signal $v_o(t)$ be at least 1.0 V everywhere in the band. Determine the element values necessary to achieve the response specified.

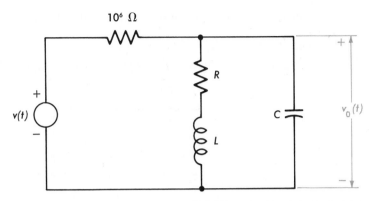

FIGURE 6-48 Circuit for Prob. 6-28.

6-29 Two parallel circuits, resonant at the same frequency, are connected by means of an amplifier as shown in Fig. 6-49. This arrangement is commonly referred to as *synchronous tuning*. Determine:

a The output voltage at the resonant frequency f_o, if $I = 100$ nA

b The bandwidth of the circuit

Sketch the relative response in the range $f_o \pm 2$ BW.

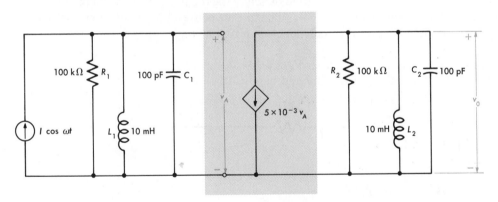

FIGURE 6-49 Double-tuned amplifier for Prob. 6-29.

6-30 The circuit in Fig. 6-49 is used with $C_1 = 95.6$ pF, $R_1 = 230$ kΩ, $C_2 = 104.7$ pF, $R_2 = 210$ kΩ, and $L_1 = L_2 = 10$ mH. As both resonant circuits are tuned to different frequencies, this arrangement is referred to as *stagger tuning*. Determine:

a The output voltage at $\omega = 10^6$ rad/s

b The bandwidth of the circuit

Sketch the relative response in the range 0.95×10^6 rad/s to 1.05×10^6 rad/s.

6-31 The circuit depicted in Fig. 6-50 is used to provide the characteristics of a resonant circuit when technological limitations preclude the use of inductors as is the case in integrated circuits. For the circuit in Fig. 6-50, determine

a The transfer function $V_2(s)/V_1(s)$
b The center frequency and bandwidth, using the results in a

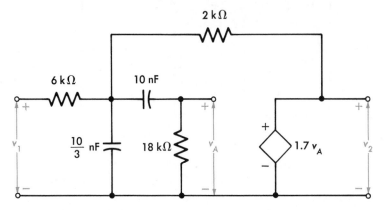

FIGURE 6-50 Active filter circuit for Prob. 6-31.

6-32 An alternative method of achieving the performance of a resonant circuit is depicted in Fig. 6-51. The circuit is useful in applications where the physical size of practical inductors prohibits their use. For the circuit in Fig. 6-51, determine:

a The transfer function V_2/V_1
b The center frequency and bandwidth

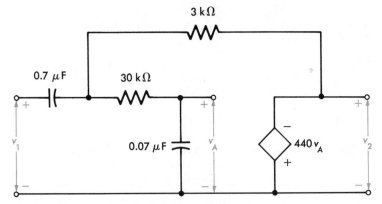

FIGURE 6-51 Circuit for Prob. 6-32.

6-33 The circuit shown in Fig. 6-52 is used to realize the modified parallel resonant circuit of Fig. 6-14. The portion of the circuit enclosed in the shaded rectangle is an ideal representation of an electronic circuit called a *gyrator*. The operation of a gyrator is such that its input impedance at terminals 1-1' is proportional to the admittance connected at terminals 2-2'. As such it is sometimes used in integrated circuits to simulate inductances.

a Show that the impedance seen looking to the right at terminals 1-1′ can be represented by an inductance in series with a resistance.
b For $C_1 = C_2 = 100$ pF, $g = 10^{-3}$ ℧, and $R = 10^5$ Ω, determine the resonant frequency and bandwidth.

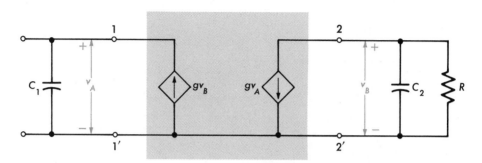

FIGURE 6-52 Gyrator circuit for Prob. 6-33.

6-34 **a** Obtain the z parameters for the circuit shown in Fig. 6-42.
b Determine the transfer function (V_2/V_1) when I_2 is zero.

6-35 **a** Obtain the z parameters for the portion of the network between terminal pairs 1-1′ and 2-2′ in the circuit of Fig. 6-44.
b Determine the transfer function (I_2/I_1) when V_2 is zero.

6-36 **a** Obtain the z parameters for the circuit shown in Fig. 6-50.
b What is the voltage transfer ratio (V_2/V_1) for $I_2 = 0$?

6-37 **a** Determine the y parameters for the circuit shown in Fig. 6-39.
b What is the current transfer function (I_2/I_1) when $V_2 = 0$?

6-38 **a** Determine the y parameters for the circuit shown in Fig. 6-22.
b Evaluate the transfer function (V_2/V_1) when $I_2 = 0$.

6-39 Obtain the y parameters for the circuit shown in Fig. 6-44.

6-40 **a** Determine the h parameters for the circuit shown in Fig. 6-38.
b Obtain the transfer-function (I_2/V_1) when $V_2 = 0$.

6-41 **a** Obtain the h parameters for the circuit shown in Fig. 6-16.
b Determine the transfer function (V_2/V_1) when $I_2 = 0$.

6-42 Obtain the h parameters for the circuit shown in Fig. 6-44.

6-43 The network contained in the shaded area in Fig. 6-53 represents an electronic circuit called a *negative impedance converter*. This circuit is used in some applications in which inductors cannot be employed or where "negative" resistance is beneficial.

a Determine the h parameters for the network defined by terminal pairs 1-1' and 2-2'.

b Determine the impedance seen looking into terminals 1-1' when the resistance R is connected across terminals 2-2'.

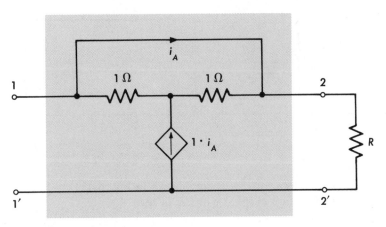

FIGURE 6-53 Negative impedance converter circuit for Prob. 6-43.

6-44 The z parameters of a particular network are given as

$$z_{11} = \frac{2s + 1}{s} \qquad z_{22} = \frac{s + 2}{s} \qquad z_{12} = z_{21} = \frac{1}{s}$$

Determine the component values of a network having the given z parameters. (*Hint:* Use Fig. 6-17 and match the corresponding terms; then identify the elements necessary to synthesize each term.)

6-45 The y parameters of a certain two-port network are:

$$y_{11} = \frac{s^2 + 3s + 1}{s + 2} \qquad y_{22} = \frac{2s + 3}{s + 2} \qquad y_{12} = y_{21} = \frac{1}{s + 2}$$

Synthesize a network and its component values which has the given y parameters. (*Hint:* Match the given parameters with the circuit in Fig. 6-16 and then identify the elements in each term.)

6-46 The block diagram in Fig. 6-54 is the representation of a multiloop control system in which the output can be adjusted to be independent of variations in G_1.

a Determine the transfer function X_o/X_s.

b For $G_2 G_3 H_2 = 1$, evaluate X_o/X_s.

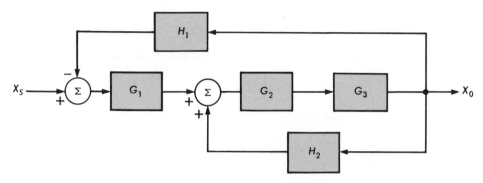

FIGURE 6-54 Block diagram for Prob. 6-46.

6-47 The block diagram in Fig. 6-55 is that of a "nested-loop" feedback system.

a Determine the transfer X_o/X_s.

b If $G_1 = G_2 = G_3 = 10(1 + s/2)$, find H_1, H_2, and H_3 so that

$$\frac{X_o}{X_s} = \frac{8{,}000}{s^3 + 18s^2 + 96s + 160}$$

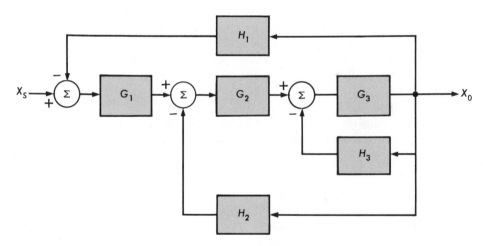

FIGURE 6-55 Block diagram for Prob. 6-47.

6-48 The equations for a two-port network which is excited at port 1 by a voltage source V_1 and in which port 2 is terminated by an impedance Z_L are

$$V_1 = z_{11}I_1 + z_{12}I_2$$

$$0 = z_{21}I_1 + (z_{22} + Z_L)I_2$$

$$V_2 = -I_2Z_L$$

a Construct a block diagram which satisfies the given equations.

b Determine the transfer function V_2/V_1.

6-49 The frequency-domain equations which relate the electric input to the motor V_m to the angular position of the motor shaft Θ are

$$T_m = K_m I_a \qquad V_a = sK_m \qquad V_m = V_a + I_a(R_a + sL_a) \qquad \text{and}$$
$$T_m = Js^2\Theta + Ds\Theta$$

where
I_m = torque developed by the motor
V_a = voltage generated within motor as a result of motion of current-carrying conductors in a magnetic field
I_a = armature current of motor
R_a and L_a = armature resistance and inductance
J = moment of inertia of shaft and load
D = friction of motor and load
K_m = a constant, characteristic of the electromechanical energy conversion in the motor

a Draw a block diagram of the system.
b Determine the transfer function Θ/V_m.

6-50 Variations in component values, such as those caused by tolerances encountered in manufacturing, result in deviations in the desired performance of a system. A principal use of feedback control is to reduce the effect of component variations on system performance. The signal-flow graph in Fig. 6-56 is a representation of one type of multiloop feedback system. For this system,
a Evaluate the transfer function $G = X_o/X_i$.
b Determine the value of $a_1 f_1$ required to make $dG/da_2 = 0$, i.e., the value of $a_1 f_1$ which makes system performance independent of variations in a_2.

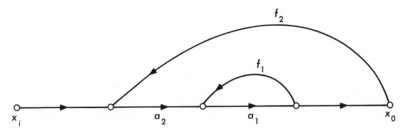

FIGURE 6-56 Signal-flow graph for Prob. 6-50.

6-51 For the signal-flow graph in Fig. 6-57, determine X_o/X_i.

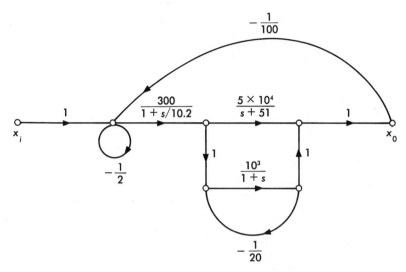

FIGURE 6-57 Signal-flow graph for Prob. 6-51.

6-52 Determine X_o/X_i for the flow graph in Fig. 6-58.

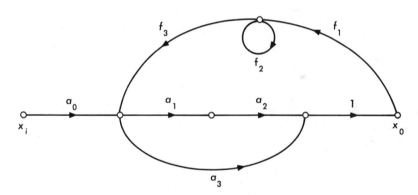

FIGURE 6-58 Signal-flow graph for Prob. 6-52.

6-53 The h parameters of a two-port network are: $h_{11} = 100\ \Omega$, $h_{12} = 0$, $h_{21} = 100$, and $h_{22} = 10^{-4}\ \mho$. Port 1 is excited by a voltage source V_s in series with a 100-Ω resistor so that $V_1 = V_s - 100I_1$. A 10^{-4}-$\mho$ conductance is connected across port 2 so that $I_2 = -10^4\ V_2$.
a Construct a flow graph whose nodes are V_s, V_1, I_1, V_2, and I_2.
b Use the result in a to determine V_2/V_s.

6-54 The mesh equations for a given network are:

$$2I_1 - I_3 = V_1 \qquad 2I_1 + 4I_2 = 0 \qquad -I_1 + 38I_2 + 2I_3 = 0$$

Construct a flow graph having nodes V_1, I_1, I_2, and I_3 and determine I_3/V_1.

NDATIONS OF SEMICONDUCTOR DEVICES □ **THE**
OF SEMICONDUCTOR DEVICES □ THE **PHYSICAL**
UCTOR DEVICES □ THE PHYSICAL **FOUNDATIONS**
CTOR DEVICES □ THE PHYSICAL FOUNDATION **OF**
E PHYSICAL FOUNDATIONS OF **SEMICONDUCTOR**
AL FOUNDATIONS OF SEMICONDUCTOR **DEVICES**
THE PHYSICAL FOUNDATIONS OF SEMICONDUCT
PHYSICAL FOUNDATIONS OF SEMICONDUCTOR D
FOUNDATIONS OF SEMICONDUCTOR DEVICES □
OF SEMICONDUCTOR DEVICES □ THE PHYSICAL FOUNDATION OF S
SEMICONDUCTOR DEVICES □ THE PHYSICAL FOUNDATIONS OF SE
DEVICES □ THE PHYSICAL FOUNDATIONS OF SEMICONDUCTOR DE
THE PHYSICAL FOUNDATIONS OF SEMICONDUCTOR DEVICES □ TH
PHYSICAL FOUNDATIONS OF SEMICONDUCTOR DEVICES □ THE PH
FOUNDATIONS OF SEMICONDUCTOR DEVICES □ THE PHYSICAL FO
OF SEMICONDUCTOR DEVICES □ THE PHYSICAL FOUNDATION OF S
SEMICONDUCTOR DEVICES □ THE PHYSICAL FOUNDATIONS OF SE
DEVICES □ THE PHYSICAL FOUNDATIONS OF SEMICONDUCTOR DE
THE PHYSICAL FOUNDATIONS OF SEMICONDUCTOR DEVICES □ TH
PHYSICAL FOUNDATIONS OF SEMICONDUCTOR DEVICES □ THE PH
FOUNDATIONS OF SEMICONDUCTOR DEVICES □ THE PHYSICAL FO
OF SEMICONDUCTOR DEVICES □ THE PHYSICAL FOUNDATION OF S
SEMICONDUCTOR DEVICES □ THE PHYSICAL FOUNDATIONS OF SE
DEVICES □ THE PHYSICAL FOUNDATIONS OF SEMICONDUCTOR DE
THE PHYSICAL FOUNDATIONS OF SEMICONDUCTOR DEVICES □ TH
PHYSICAL FOUNDATIONS OF SEMICONDUCTOR DEVICES □ THE PH
FOUNDATIONS OF SEMICONDUCTOR DEVICES □ THE PHYSICAL FO
OF SEMICONDUCTOR DEVICES □ THE PHYSICAL FOUNDATION OF S
SEMICONDUCTOR DEVICES □ THE PHYSICAL FOUNDATIONS OF SE
DEVICES □ THE PHYSICAL FOUNDATIONS OF SEMICONDUCTOR DE
THE PHYSICAL FOUNDATIONS OF SEMICONDUCTOR DEVICES □ TH
PHYSICAL FOUNDATIONS OF SEMICONDUCTOR DEVICES □ THE PH
FOUNDATIONS OF SEMICONDUCTOR DEVICES □ THE PHYSICAL FO
OF SEMICONDUCTOR DEVICES □ THE PHYSICAL FOUNDATION OF S
SEMICONDUCTOR DEVICES □ THE PHYSICAL FOUNDATIONS OF SE
DEVICES □ THE PHYSICAL FOUNDATIONS OF SEMICONDUCTOR DE
THE PHYSICAL FOUNDATIONS OF SEMICONDUCTOR DEVICES □ TH
PHYSICAL FOUNDATIONS OF SEMICONDUCTOR DEVICES □ THE PH
FOUNDATIONS OF SEMICONDUCTOR DEVICES □ THE PHYSICAL FO

7

he controlled source and the electronic switch are two of the fundamental electronic circuit components. In the six previous chapters on circuit analysis, the treatment of these elements is predicated on their physical existence. The purpose of this chapter is to present the physical concepts essential to understanding the operation of the devices used to realize controlled sources and switches. Semiconductor devices are dealt with exclusively with major emphasis focused on the characteristics and operation of junction (bipolar) transistors, field-effect transistors, and junction diodes. The electronic properties of semiconductors are used to develop the response of these devices to both constant and time-varying excitations. The response to a constant excitation gives rise to the static characteristics, and the dynamic characteristics are the result of the response to a time-varying excitation. A description of the technologies used to fabricate electronic circuits, in particular, integrated circuits, is also included.

7-1 THE NATURE OF ELECTRONICS

Electronics is the branch of electrical engineering which deals extensively with the transfer and processing of information by means of electromagnetic energy. Electromagnetic energy is used because of its ready availability, the speed and relative ease by which it may be transported, and the usually convenient size of the components involved.

Diverse areas of communication, control, and computation have information transfer as a common characteristic. The information may be simply the output of a thermocouple which is used in both household and industrial temperature-control systems. On the other hand, in a television broadcast system one deals with the audio and visual information of the program originating at the studio. Another example is the information used to instruct a space vehicle to fire its retro-rockets, where the complex signal used is based on data processed by a computer.

Most often, the original information and the information required at the output must be converted to or from electromagnetic energy. The conversion process must be performed by devices and circuits which maintain both the information content and the advantage inherent in using electromagnetic energy. Microphones and the vidicon tubes in television cameras perform the conversion of acoustic and optical information to electrical signals at the studio, while loudspeakers and cathode-ray picture tubes perform the reverse process where the program is observed. A variety of control and computation systems utilize thermocouples, electro-optical devices such as photocells, and tachometers to accomplish the conversion process. Devices such as these are generally referred to as *transducers*.

Once the information is in the form of electromagnetic energy, it must be processed to achieve the system objective. As an example, if the energy of the output of the thermocouple is insufficient to activate the solenoid, an amplifier is used which increases the energy to the required level. Similarly, in order to tune in to a particular television channel, the desired signal must be separated (filtered) from the signals of the other channels present. The filtering process allows only specific frequency components to be transmitted through the system, and the function performed is that of selectivity.

Often, processing consists of generating new information needed to perform certain electronic functions that are essential if the system is to operate effectively. The signals used to synchronize the audio and video portions of the television system and the signals used to set the timing sequence for firing the spacecraft's retro-rockets are examples of the new information which must be generated.

The amount of energy required in the process is, in itself, not the result, but only a consideration to ensure that the desired objective is achieved. Thus, from an electronic viewpoint, a radar system on a small boat and a megawatt system for bouncing signals off the moon have nearly identical characteristics. The difference in the energies required is dictated by the range of the system, not by the electronic process. The electronic process involves the application of signal-transmission properties through devices and their associated circuits.

Two areas of study are evident. One is the physical operation and circuit properties of electronic devices, and the second relates to the circuit functions and systems appropriate to the desired information transfer.

7-2 ELECTRONIC DEVICES

The function of many electronic devices is to provide the characteristics of controlled sources whose circuit properties are utilized in almost every electronic circuit. As an aid to understanding the operation of these devices, it is beneficial to first examine the circuit properties most frequently used.

The first of these properties is illustrated in Fig. 7-1, which depicts a voltage-controlled current source as a circuit element containing four termi-

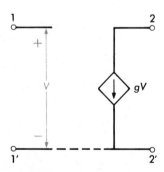

FIGURE 7-1 The schematic representation of an ideal voltage-controlled current source.

nals. The input pair, 1-1', provides the control voltage V, and the output pair, 2-2', exhibits the effect of a current source of strength gV. The parameter g, whose units are expressed in mhos, relates the strength of the source to the control voltage. Because g expresses the relationship between an input quantity and an output quantity, it is usually referred to as the *transconductance* (transfer conductance). Often the terminals 1' and 2' may be common as shown by the dashed line in the figure. Consequently, only three terminals may be required to represent the controlled source as a circuit element. The advantage gained from the use of a three- or four-terminal circuit element is that the control terminals may be connected to components remote from those connected to the source terminals. The controlled source may be thought of as providing "action at a distance" so that various circuit elements may be physically isolated from one another.

A second useful circuit property is that electrical isolation of circuit components is also possible. In Fig. 7-2a a controlled source is shown with excitation signals V_S and I_o impressed at the input and output terminals respectively. The resistors R_S and R_o are the source resistors associated with V_S and I_o, and R_L represents the load resistor. The desired responses are the control voltage V and the load voltage V_L, both of which may be determined by the use of superposition. The components of V and V_L produced by V_S are determined by use of the circuit shown in Fig. 7-2b, and those produced by I_o are determined by use of the circuit in Fig. 7-2c.

Application of Kirchhoff's voltage law to the input circuit of Fig. 7-2b results in

$$- V_S + I_S R_S + V = 0$$

Because terminals 1-1' are open-circuited, I_S is zero, which results in $V = V_S$.

By application of Kirchhoff's current law at the output,

$$gV + \frac{V_L}{R_L} + \frac{V_L}{R_o} = 0$$

As $V = V_S$, V_L becomes

$$V_L = \frac{- gV_S}{1/R_o + 1/R_L}$$

In Fig. 7-2c, the KCL equation at the output is

$$- I_o + \frac{V_L}{R_L} + \frac{V_L}{R_o} = 0$$

from which

$$V_L = \frac{I_o}{1/R_o + 1/R_L}$$

The KVL expression for the input circuit is

$$I_S R_S + V = 0$$

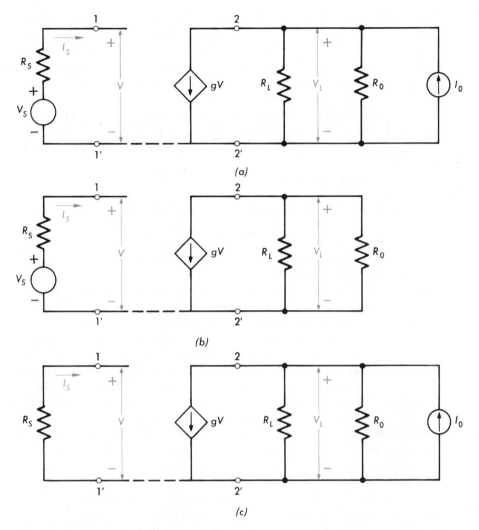

FIGURE 7-2 (a) A voltage-controlled current source with excitations impressed at input and output terminals. (b) Representation of Fig. 7-2a to consider the effect of V_S. (c) Representation of Fig. 7-2a to consider the effect of I_o.

However, I_S is zero because terminals 1-1' are open-circuited, so that V is also zero.

The total responses for V and V_L are the sums of the responses produced by V_S and I_o and are

$$V = V_S \qquad V_L = \frac{I_o - gV_S}{1/R_o + 1/R_L}$$

Examination of these responses indicates that V_L is dependent on both V_S and I_o, while V is dependent on V_S only. The effect of V_S at the input has been transmitted through the controlled source to the output. However, the effect of I_o has not been transmitted through the controlled source from output to input, and I_o is electrically isolated from the input. This result describes the fact that electrical signals may be propagated through a controlled source in only one direction, that being from the control to the source. Circuits and devices which transmit in only one direction are said to be *unilateral*.

The third and most important circuit property is demonstrated by comparing P_L, the power dissipated by the load R_L, with P_S, the power supplied by the source V_S, for the circuit of Fig. 7-3. The KVL equation for the input loop is

$$- V_S + I_S R_S + V = 0$$

Terminals 1-1' are open-circuited, causing I_S to be zero. Consequently, $V = V_S$, and $P_S = I_S V_S = 0$. The value of P_L is given by $I_L^2 R_L$, where I_L is gV_S, so that P_L is equal to $g^2 R_L V_S^2$. As P_L is nonzero and P_S is zero, the additional energy must be provided by the controlled source. The fact that P_L is greater than P_S is significant, for it indicates that the controlled source is capable of increasing the signal energy at one point in a circuit relative to the signal energy at some other point. In addition, the use of small quantities of energy to control large quantities of energy is realizable. This circuit property, often described as power gain, is the basis for electronic amplification. Most electronic systems would either be severely limited in use or would be impractical so as to preclude their use if amplification were not possible.

The additional energy, called the *bias,* is generally supplied to the physical device which provides the controlled-source characteristics. In most electronic devices, direct (constant) energy sources are used to provide the bias energy. The signal energy, in contrast, is supplied by means of a source whose strength varies with time. The increase in signal energy at the output of the controlled source is then obtained from the bias energy supplied to the device.

It is also useful to describe the properties of switches prior to the investigation of how devices perform the switching function. The basic characteristics of a switch can be illustrated using the circuits shown in Fig. 7-4. When

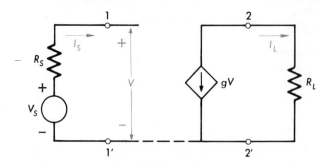

FIGURE 7-3 A voltage-controlled current source with excitation and load.

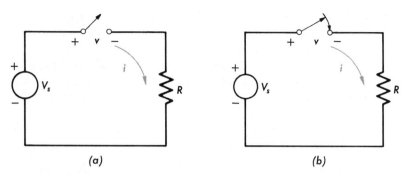

FIGURE 7-4 Circuits to illustrate properties of (a) an open switch and (b) a closed switch.

the switch is open as in Fig. 7-4a, no current exists in the circuit, but by KVL, $v = V_S$. In Fig. 7-4b the closed switch permits current to exist in the circuit, but $v = 0$. The current i becomes V_S/R, its value limited only by V_S and R, elements external to the switch. Thus, the switch is a two-state device; one state being $v \neq 0$, $i = 0$; the second state $v = 0$, $i \neq 0$. For the ideal case described, the power dissipated by the switch is always zero. In practice, however, electronic switches consume a small amount of energy.

In wall switches and on-off switches in electric appliances, mechanical energy is used to control the switch position. Electronic switches are actuated by electronic signals which control the characteristics of the devices used. Most often, as shown in Fig. 7-5, electronic switches consist of three or four terminals. Such devices function as controlled switches as the actuating signal at terminals 1-2 is isolated from terminals 3-4, at which switching action occurs. This allows the expenditure of a small amount of energy to control the switching of larger amounts of energy in the main circuit (terminals 3-4). This property is widely used in control, communication, and computer applications. Because some of the desirable properties in switches are similar to those used for controlled sources, the same devices are generally used to perform both functions.

To understand how various electronic devices provide the characteristics of controlled sources, a knowledge of the physical and electrical phenomena which occur in various materials is required. Similarly, the amount

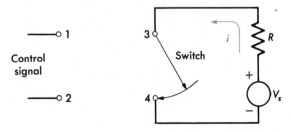

FIGURE 7-5 Representation of a controlled switch.

and form of energy supplied and the response of the devices to this energy must be developed. In addition, the generic properties and terminology associated with most electronic devices need description.

Fundamental to all electronic devices is the use of the controlled flow of charged particles. The information content, converted into electromagnetic energy, forms the signals which are used to control the flow of charges. Basic to most devices are:

1 A source of mobile charged particles

2 A means for controlling the flow of the particles in accordance with the information content

3 A means for gathering the charged particles and ultimately extracting the information to perform a desired function

A portion of the device which displays certain physical properties useful in obtaining one or more of these facets is called an *element* or *electrode*. Most often, the elements are enclosed in a metal, glass, or ceramic case and electrical connections brought through the case to the terminals. Devices in which solids are the medium through which the flow of charged particles occurs (and where control is exerted) are called *solid-state devices*.

Diodes are two-terminal electronic devices. The flow of charge takes place only when an external circuit is provided and energy is supplied to the devices. The control of particle flow is accomplished by means of an external voltage impressed across the diode.

In three-element devices, such as junction transistors and field-effect transistors, two of the elements act as in a diode. When the terminals are connected to external circuits, charge flow may occur. The main source of energy is generally provided by a circuit connecting two of the elements. Control is achieved by use of the external circuit connected to the third element, as well as by the main energy circuit. In silicon-controlled rectifiers as well as other special-purpose devices, more than three elements are used; the additional elements are inserted to obtain desired inherent characteristics.

The understanding of the physical phenomena which give rise to the terminal properties is required for the use of a given device. Emphasis on terminal characteristics is essential, for the terminals represent the only places to which other electrical components can be connected and at which measurements can be made.

In general, both direct (constant) and alternating (time-varying) energy are supplied to the device. Most of the direct energy is supplied by the main power source (the bias) and establishes the particular operating voltage and current levels needed to utilize the device effectively. The alternating energy usually contains the information and produces voltages and currents which reflect the effects of controlled charge flow. Thus, the voltages and currents in the device contain both direct and time-varying components. In this and the succeeding four chapters, it often will be convenient to determine each component separately. The dc, or *static,* behavior is usually described by one or more volt-ampere characteristics which relate the terminal values of voltage and current of device elements. The ac, or *dynamic,* nature is most often represented by a number of circuit elements called an *equivalent circuit* or model and is developed in Chap. 8.

7-3 SEMICONDUCTORS

Semiconductors are solids whose resistivities have values between those of conductors and insulators. In many electronic devices, semiconductors are used to provide the source of mobile charges and constitute the medium through which the charges flow and are controlled. Silicon (Si) and germanium (Ge) are the semiconductors most often used. Both have a valence of 4 and have crystalline structures similar to diamond. As such, strong bonding forces exist between the valence electrons and the atoms in the crystal. In the pure state, with no chemical impurities caused by foreign atoms and with no structural imperfections, these materials are called *intrinsic semiconductors.*

While germanium and silicon crystals are electrically neutral, the application of appropriate forms and quantities of energy may be used to generate mobile charges, called *carriers,* without disturbing the electrical neutrality of the overall system. In intrinsic semiconductors mobile charges are generated by overcoming the bonding forces, usually by means of the thermal energy associated with temperatures greater than 0°K. Two kinds of carriers, holes and electrons, are produced by this process. Once a bond is broken, a valence electron is free to move. The freed electron in turn creates a vacancy of an electron in a bond elsewhere in the crystal. This absence of an electron, called a *hole,* behaves as if it were an independent mobile positive charge. As both electrons and holes are generated in equal numbers, the electrical neutrality of the crystal is maintained. However, both carriers may be influenced by forces and thus may contribute to the charge-flow process.

Because of the strong bonding in intrinsic semiconductors, only a small number of carriers are generated unless large quantities of additional energy are expended. Thus, they are not well-suited for use in devices, for their electrical properties resemble those of insulators.

A very common expedient used to produce large numbers of carriers is to introduce a carefully controlled impurity content into the semiconductor. Semiconductors with added impurities are called *extrinsic semiconductors.* Usually atoms with five valence electrons, such as arsenic (As) and antimony (Sb), or those with three valence electrons, of which gallium (Ga) and boron (B) are typical, are added. Each type of impurity establishes a semiconductor which has a predominance of one kind of carrier.

Mobile carriers are generated through replacement of the host material by impurity atoms at various points in the crystal. The impurity concentrations generally used are in the range of 1 atom of impurity to 10^6 to 10^8 atoms of the host material. Thus, most physical and chemical properties are essentially those of the base semiconductor. Only the electrical properties change markedly, causing the extrinsic material to more closely resemble metals.

To illustrate how carriers are generated, consider two silicon-semiconductor samples in which arsenic impurities have been added to the first, and gallium to the second. When a silicon atom is replaced by an arsenic atom only four of the five valence electrons provided by the arsenic atom are used to form the bonds in the crystal. The remaining valence electron is free to move with the addition of a small amount of energy usually provided by the thermal energy of the crystal. The predominant carriers generated by this means are the negatively charged electrons. Consequently, this type of extrinsic semiconductor is referred to as *n*-type.

In the second sample, a gallium atom replaces a silicon atom in the crystal. The gallium atom can provide only three of the four valence electrons needed to form the bonds. The absence of one electron in the bonding results in the generation of a hole. Motion of the hole occurs when a small amount of energy, usually the thermal energy of the crystal, is added. Extrinsic semiconductors of this type are referred to as *p-type,* since the positively charged holes are the predominant carriers.

Both types of carriers are present in extrinsic semiconductors, for thermal agitation generates a number of electron-hole pairs. However, at normal operating temperatures, the carriers that result from the impurity content are the dominant factor. In extrinsic semiconductors the dominant mobile charge is called the *majority carrier* and the lesser charge the *minority carrier*.

7-4 CHARGE-FLOW PROCESSES

Charge-flow processes refer to the mechanisms which describe the motion of carriers. In extrinsic semiconductors, two such mechanisms exist which are used to effect controlled charge flow. The first of these is the result of the application of electrical forces which influence the motion of the carriers. This forced motion is called *drift* and is the mechanism by which electrical conduction occurs.

In semiconductors, the motion of carriers cannot be described simply. With no external force applied, the thermal energy of the material causes both the electrons and the atoms to vibrate. These vibrations are random in space but are constrained so that the net average velocity of all the particles in a given direction is zero. A current would exist if the average velocity were not zero. The vibrations cause the mobile electrons and atoms in the crystal to interact with one another. Because of the many interactions, we cannot describe the motion of an individual particle but can only speak of the net average motion of the group of particles.

When an external force is applied, usually by means of an electric field, the particles are influenced so that a directed motion is superimposed on the random thermal motion. This results in a net average velocity in the direction of the applied field. Thus, there is a resultant current to which both hole motion and electron motion contribute. Electrons and holes move in opposite directions, but because of their opposite charges, both produce current in the same direction. In extrinsic semiconductors, this current is essentially majority-carrier flow.

The second major flow mechanism is *diffusion,* a force-free process based on a nonuniform distribution of carriers in the system. The process can be illustrated by the classic problem in the kinetic theory of gases. In the system shown in Fig. 7-6, the N_1 gas molecules are distributed uniformly through the volume V_1 and the N_2 molecules are distributed uniformly through V_2. The resultant molecule concentrations are N_1/V_1 and N_2/V_2 and are unequal. If the partition is removed, after some time has elapsed, the gas will be at some new pressure. The gas molecules will have become distributed uniformly throughout the volume (the pressure on each wall will be the same). The redistribution is achieved by a diffusion process based on the unequal concentration in the two sections of the container. The flow of gas molecules is from the region of higher concentration to the region of lower concentration.

The diffusion tendency is present even with the partition in place. However, the rigid nature of the partition constrains the particles to remain in the section in which they exist. Once this constraint is removed, the diffusion process takes place.

In semiconductors, diffusion produces a current without the benefit of an applied field. The flow of charged particles occurs because of the spatial variation of carrier concentration. The resultant electron- and hole-diffusion currents reflect the opposite sign of their charges. Because the direction of motion is toward the region of lower concentration, diffusion is a major factor in describing minority carrier flow.

A third phenomenon occurring in semiconductors, *recombination,* results from the collision of an electron and a hole. The recombination process is essentially the return of an electron to a vacant site in the crystal bonds. The recombination rate is proportional to the carrier concentration; i.e., the

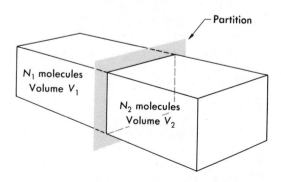

FIGURE 7-6 Rigid partition separating two volumes of an ideal gas.

larger the number of carriers, the more likely is the occurrence of electron-hole recombination. Recombination is important in describing minority-carrier flow because the percentage of the charges affected is significant. As electron-hole pairs are also being generated in the material, it is convenient to refer to the difference between the recombination and generation rates as the *net recombination rate*.

To describe the total flow of carriers in a semiconductor, all three flow phenomena must be included. Each kind of carrier must be treated separately, the number of electrons or holes leaving the sample being accounted for by drift, diffusion, or net recombination. The recombination term is the same for both holes and electrons because one of each kind of carrier is removed in the process. The current in the sample is then the sum of the electron and hole components. These effects are the basis for quantitatively describing the currents and terminal characteristic of devices.

7-5 JUNCTION DIODES

The fundamental building block upon which all semiconductor devices are based is the *p-n* junction. The junction is formed by physically joining *p*-type and *n*-type semiconductors, as shown in Fig. 7-7. In addition, the *p-n* junction is also used as a two-terminal device referred to as a *junction diode* and whose circuit symbol is depicted in Fig. 7-8.

Diodes have several useful circuit properties the most important of which is usually described by means of the *static characteristics,* or *volt-ampere curves*. Figure 7-9 is the static characteristic of the ideal diode. Inspection of the curve indicates that current flows through the device in one direction only so that the diode acts as a unilateral circuit element. The unilateral nature of the device is important in switching, as it provides an ''on-off'' characteristic, and in wave-shaping and rectification, as only appropriate

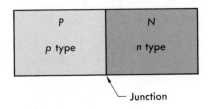

FIGURE 7-7 A *p-n* junction.

FIGURE 7-8 Circuit symbol for a junction diode.

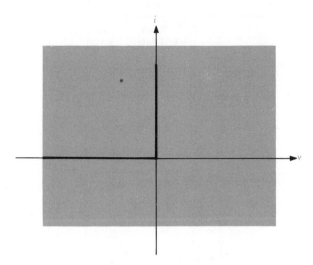

FIGURE 7-9 The volt-ampere characteristic of an ideal diode.

signal polarities may be detected and processed. Note that when v is zero, i is not and vice versa, a condition corresponding to the switch.

The physical operation of the junction is described in terms of the flow processes considered in the previous section. On the assumption that each type of semiconductor has uniform impurity content, there is a greater concentration of holes in the p region than in the n region, and similarly, the electron concentration in the n region is greater than in the p region. The concentration differences establish a gradient across the junction, resulting in a diffusion of carriers; holes diffuse from the p to the n region, electrons from the n to the p region. The result of the diffusion, as shown in Fig. 7-10, is to produce immobile ions of opposite charge on each side of the junction. The region, extending into both the n and p regions, is called the *depletion region* or *space-charge region*. No mobile carriers exist in this region since they have diffused into either the p-type or n-type material.

The immobile ions, or space charge, being of opposite polarity on each side of the junction, establish an electric field. The two effects of the field are the potential barrier formed and the drift current produced. The drift current causes holes to move from the n to the p region and electrons to move from the p to the n region. In equilibrium and with no external circuit, no net current exists; that is, the net number of holes and electrons crossing the junction is each zero. Consequently, the drift and diffusion components of current are equal and oppositely directed.

The potential barrier established across the depletion region acts to prohibit the flow of carriers across the junction without the application of energy from an external source. An external source connected between the n

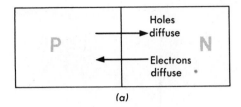

(a)

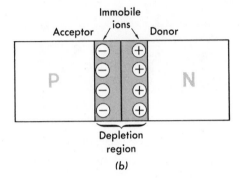

(b)

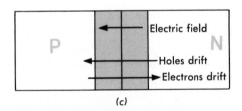

(c)

FIGURE 7-10 A *p-n* junction showing (a) direction of hole and electron diffusion, (b) depletion region, (c) drift of electrons and holes.

and *p* regions, as shown in Fig. 7-11*a*, is used to provide the additional electric energy. The voltage source *V*, called the *bias*, either decreases or increases the potential barrier and controls the flow of carriers across the junction. When *V* = 0, the barrier is unaffected and the circuit behaves as does the open-circuited *p-n* junction. Positive values of *V*, called *forward biasing*, decrease the potential barrier, thereby increasing the number of electrons and holes which diffuse across the junction. The increased diffusion results in a net current, called the *forward current*, from the *p* to the *n* region. As *V* is increased, the current increases rapidly because the barrier is reduced even further. On the other hand, making *V* negative (*reverse biasing*) increases the potential barrier and reduces the number of carriers diffusing across the boundary. A small current, called the *reverse current* or *saturation current*, I_s (essentially the drift component produced by the electric field) exists from the *n* to the *p* region. Increasing the reverse bias does not

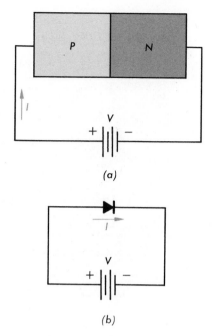

FIGURE 7-11 (a) The biased p-n junction. (b) Diode symbol.

affect the reverse current significantly until breakdown occurs. The magnitude of the saturation current is dependent on the doping levels in the p-type and n-type materials and on the physical size of the junction. Figure 7-11b shows the diagrammatic representation of the p-n junction, or diode, and its external circuit.

The static characteristic of the junction diode, shown in Fig. 7-12, describes the dc behavior of the junction and relates the diode current I and the bias voltage V. These characteristics are expressed analytically in Eq. (7-1), which is often referred to as the *Boltzmann diode equation*.

$$I = I_S(\epsilon^{V/\eta V_T} - 1) \tag{7-1}$$

where η is a property of the semiconductor used (one for G_e and nearly two for S_i), and V_T is the "thermal" voltage given in Eq. (7-2).

$$V_T = \frac{kT}{q} = \frac{T}{11,600} \tag{7-2}$$

In Eq. (7-2), k is Boltzmann's constant and equals 1.38×10^{-23} J/K, q is the magnitude of the electronic charge, and T is the junction temperature in Kelvin. At room temperature ($T = 293$ K), V_T is approximately 0.025 V or 25 mV.

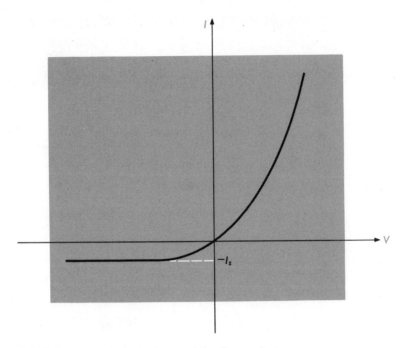

FIGURE 7-12 Volt-ampere characteristic of a *p-n* diode.

Example 7-1

A silicon diode is forward-biased with $V = 0.5$ V at a temperature of 293 K. The diode current is 10 mA. Determine the saturation current for the diode.

SOLUTION   From Eq. (7-1), with $V_T = 25$ mV

$$I_S = \frac{I}{\epsilon^{V/\eta V_T} - 1} = \frac{10 \times 10^{-3}}{\epsilon^{0.5/2 \times 0.025}} = 4.54 \times 10^{-7} = 0.454 \; \mu A$$

The saturation current found in Example 7-1 is significantly smaller than the diode current, even when the forward bias is small. This difference often necessitates using different current scales for forward and reverse bias on the characteristics of Fig. 7-12. Also, when V is greater than zero, $\epsilon^{V/\eta V_T}$ is much greater than unity, so that it is often convenient to approximate the current in the forward-biased case by $I_s \epsilon^{V/\eta V_T}$. Similarly, for V negative, $\epsilon^{V/\eta V_T}$ is much less than unity, so that in reverse bias the diode current is approximately $-I_s$.

Diodes are generally rated by the maximum forward current which can exist and by the maximum instantaneous value of reverse-bias voltage, called the *peak inverse voltage*, which can be applied. The maximum-forward-current rating is based on the ability of the junction to dissipate power in the form of heat. The peak-inverse-voltage rating is a limitation based on the maximum electric field which can exist in the depletion region.

7-6 GRAPHICAL ANALYSIS OF DIODE CIRCUITS

The static characteristic of Fig. 7-12 indicates that the junction diode approximates the ideal diode characteristic of Fig. 7-9. One deviation from the ideal is the nonzero voltage drop across the forward-biased diode. When the diode is used in the circuit of Fig. 7-13, the voltage drop across the diode must be included. The values of current and voltage are obtained from the diode characteristics and the voltage-law constraint

$$V_{BB} = IR + V \tag{7-3}$$

Equation (7-3) is represented graphically by the straight line in Fig. 7-14, and is referred to as the *load line*. The solution is obtained by determining the values of V and I at the intersection Q of the diode characteristic and the load line.

Mathematically, the solution process is the familiar graphical method of solving simultaneous equations. Physically, Q is the only condition satisfying the restrictions imposed by both the diode and the external circuit.

The intersection Q of the two curves is called the *quiescent* or *operating point,* and the values of diode current and voltage at the quiescent point are usually indicated by I_Q and V_Q, respectively. The voltage $V_{BB} - V_Q$ is the voltage which exists across the resistance. By varying the resistance R, the slope of the load line changes and consequently so does Q as indicated in Fig. 7-15a. Similarly, if the supply voltage V_{BB} varies with R held constant, the intercept on the abscissa changes. As depicted in Fig. 7-15b, new load lines parallel to the original line can be drawn, each of which corresponds to a particular value of V_{BB}. Again, the operating point is shifted as each load line intersects the volt-ampere characteristic at a different point.

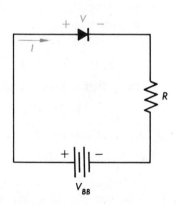

FIGURE 7-13 A biased diode with load resistance.

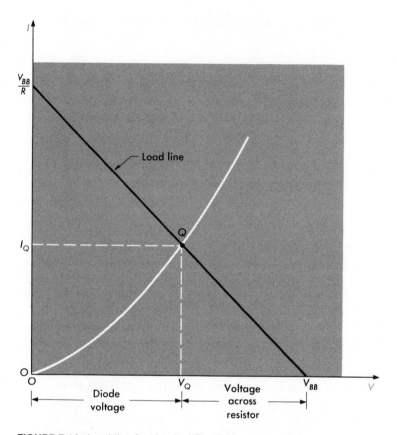

FIGURE 7-14 Load line for circuit of Fig. 7-13.

If, in addition to the constant potential, an alternating or time-varying potential is impressed across the circuit, as shown in the circuit of Fig. 7-16, the dynamic (ac) characteristics of the diode can be obtained. The diode voltage v and current i each contain two components; one is a constant, the other is time-varying. If

$$v_s(t) = \sqrt{2}\, V_s \sin \omega t$$

then the total instantaneous voltage impressed across the circuit is

$$v_B = V_{BB} + \sqrt{2}\, V_s \sin \omega t$$

At each instant of time, the voltage law requires

$$v_B = v + iR \tag{7-4}$$

When ωt is an integral multiple of π, making $\sin \omega t$ equal to zero, v_B equals V_{BB} and Eq. (7-4) becomes Eq. (7-3). Consequently, the load line can be drawn, as shown in Fig. 7-17a and b. At any other instant of time, the value of v_B differs from V_{BB}, and the load line shifts, as it did in the case when the

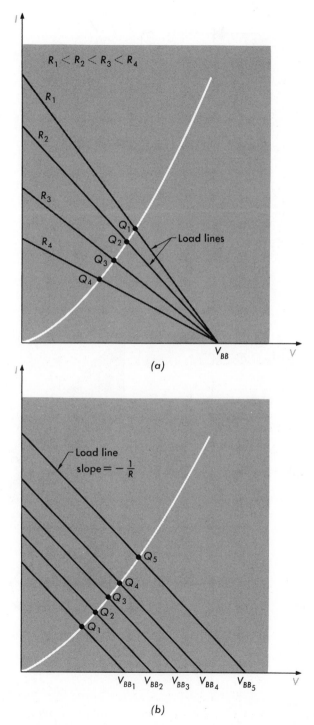

FIGURE 7-15 Load lines for (a) variable resistance and (b) variable supply voltage.

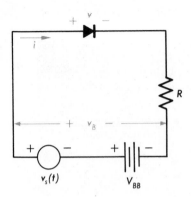

FIGURE 7-16 Diode circuit with dc and ac sources.

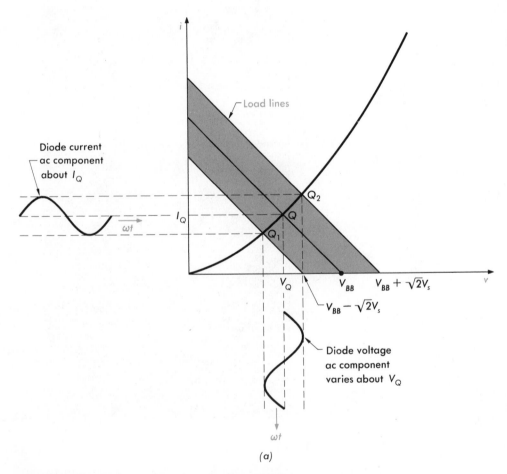

FIGURE 7-17a Small-signal current and voltage waveforms.

supply voltage was varied. As time is a continuous variable, the load-line position corresponds to the particular value of v_B at a given instant. The motion of the load line traces the shaded area of the characteristics, and this area is defined by the maximum and the minimum values of v_B. These values are $V_{BB} + \sqrt{2}\ V_s$ and $V_{BB} - \sqrt{2}\ V_s$ and correspond to values of sin ωt equal to $+1$ and -1, respectively. Figure 7-17a shows the excursion of the load line for values of V_s much less than V_{BB}, while Fig. 7-17b depicts the excursion for values of V_s comparable to those of V_{BB}. The line segment $Q_1 - Q_2$ is the locus of the position of the operating point Q. The waveforms indicate the values of diode voltage and current as functions of time. To plot the waveforms, a point-by-point method must be used.

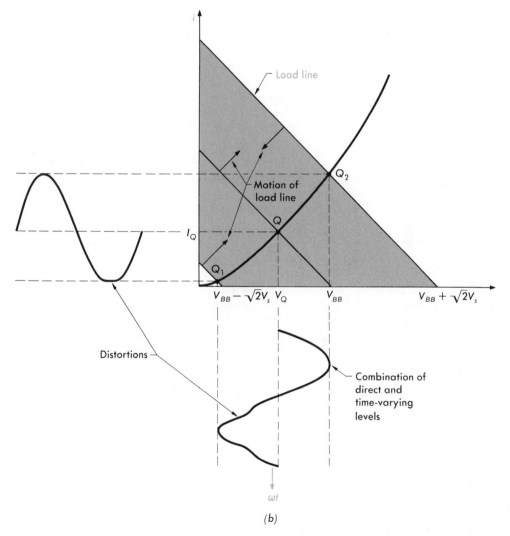

(b)

FIGURE 7-17b Large-signal current and voltage waveforms.

It is evident that when V_s is small compared with V_{BB}, the waveforms (Fig. 7-17a) are sinusoidal and the diode response is the superposition of the alternating wave on the quiescent level. One component represents the response to V_{BB} (the quiescent level), and the second component represents the response to $v_s(t)$, to which it is directly proportional. Under these conditions, the diode is considered to behave linearly, and the segment $Q_1 - Q_2$ is approximated by a straight line. On the other hand, when V_s and V_{BB} are comparable in magnitude, the diode response, indicated by the waveforms in Fig. 7-17b, is distorted and does not represent a simple superposition of the direct and alternating response. The time-varying portion of the response is not directly proportional to $v_s(t)$, and the behavior is nonlinear.

When the diode is reverse-biased, as shown in the circuit of Fig. 7-18, the use of the load line is unnecessary. The current flowing in the circuit is just the diode saturation current I_S. By use of the KVL equation for the circuit,

$$V_{BB} = I_S R + V \tag{7-5}$$

the diode voltage V is determined.

An alternate method for determining the value of I_Q for the forward-biased diode is based on the realization that V_Q is generally small compared with typical values of V_{BB}. As such, the diode characteristic may be represented as shown in Fig. 7-19 which resembles the ideal characteristic. However, the finite voltage drop across the diode is accounted for by V_{on}, which is referred to as the *offset* or *turn-on* or *cutin voltage*. By the use of Eq. (7-3), the value of I_Q for the circuit of Fig. 7-13 is given as

$$I_Q = \frac{V_{BB} - V_{\text{on}}}{R} \tag{7-6}$$

The values of V_{on} encountered in practical junction diodes are based largely on the semiconductor material used. Typical values are 0.6–0.7 V for silicon devices and 0.2–0.3 V for germanium devices.

A closer approximation to the actual diode characteristic than that in Fig. 7-19 is depicted in Fig. 7-20. The straight-line portion of the approxi-

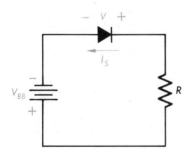

FIGURE 7-18 Circuit with a reverse-biased diode.

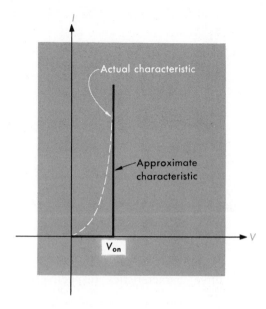

FIGURE 7-19 Alternate representation of a junction diode.

mate characteristic represents a resistance R_f whose value is the reciprocal of the slope. For the diode representation in Fig. 7-20, Eq. (7-6) becomes

$$I_Q = \frac{V_{BB} - V_{on}}{R + R_f} \tag{7-7}$$

where R_f is used to account for the portion of the diode voltage in the region beyond V_{on}. For silicon diodes, typical values of R_f range between 5 and 15 Ω. The lower resistance values are encountered for large values of diode current, and higher values of R_f occur at lower levels of diode current.

The ideal diode characteristic of Fig. 7-9 and the approximations given in Figs. 7-19 and 7-20 can each be used to represent the forward-biased

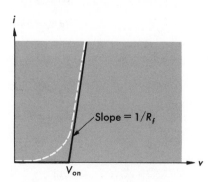

FIGURE 7-20 Diode representation including dynamic resistance.

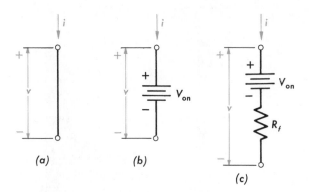

FIGURE 7-21 Diode circuit representations: (a) ideal diode, (b) to include offset voltage, (c) to include offset voltage and dynamic resistance.

diode. The circuits shown in Fig. 7-21a to c are circuit-equivalents of the three characteristics, respectively.

7-7 CHARACTERISTICS OF JUNCTION TRANSISTORS

The *junction* or *bipolar transistor* is a device which can provide the circuit properties of both a controlled source or a switch. As shown in Fig. 7-22a, the junction transistor is a three-element device which is formed of two junctions. Since there are two *p*-type sections, the name *pnp* is given to this type of transistor, which is represented symbolically in Fig. 7-22b. When two

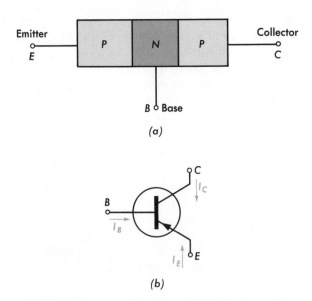

FIGURE 7-22 A *pnp* transistor and its symbolic representation.

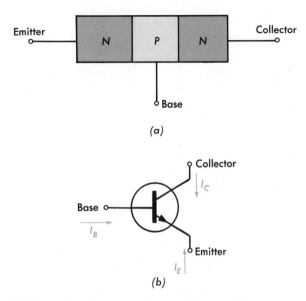

(a)

(b)

FIGURE 7-23 An *npn* transistor and its symbolic representation.

n-type sections are used, an *npn* transistor is formed. Figure 7-23*a* illustrates the *npn* transistor, and Fig. 7-23*b* is its symbolic representation. The three elements are referred to as the *emitter, base,* and *collector.* The emitter acts as the source of mobile carriers, and the collector acts to extract the carriers to perform useful work. The control of the flow of carriers from emitter to collector is accomplished in the base.

Because there are both *pnp* and *npn* transistors, it is convenient to establish the convention, indicated in Figs. 7-22 and 23, that current reference directions are into the transistor. The arrows on the emitter leads in the circuit symbols indicate the actual direction of positive charge flow in the emitter region for each transistor type. The physical mechanisms which govern transistor operation are the same for both *npn* and *pnp* types.

The mode of transistor operation which results in a controlled source is called the *active region.* As shown in Fig. 7-24, the voltages across the junctions in the active region establish forward bias for the emitter-base junction and reverse bias for the collector-base junction. The controlled-source action results from the control of the collector current by the emitter-base junction voltage. Because of the forward emitter-base bias, the emitter injects a large number of carriers into the base region. Most of these carriers diffuse through the base and reach the collector-base junction where the reverse bias on this junction sweeps the carriers into the collector. Because recombination in the base region is kept to a minimum, the collector and emitter currents are nearly equal in magnitude. The difference between emitter and collector currents is accounted for by the base current which

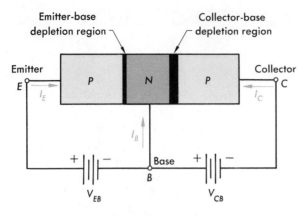

FIGURE 7-24 A *pnp* transistor biased in the active region.

represents the carriers lost by recombination. The emitter-base junction behaves as in the junction diode. A change in the level of forward bias on the emitter-base junction produces a change in the emitter current. Consequently, the collector current also changes by a similar amount indicating the control exerted by the emitter-base voltage.

The performance of the transistor in the active region is described mathematically in terms of the diode action at the emitter-base junction and the controlled source effect at the collector. These relations are given in Eq. (7-8), for $\eta = 1$ and where $\epsilon^{V_{EB}/V_T} \gg 1$.

$$I_E = I_{EO}\epsilon^{V_{EB}/V_T} \tag{7-8a}$$

$$I_C = -\alpha_N I_{EO}\epsilon^{V_{EB}/V_T} = -\alpha_N I_E \tag{7-8b}$$

The quantity I_{EO} is the saturation current of the emitter-base junction and α_N is the *dc common-base short-circuit current gain*. The symbol h_{FB} is also used in place of α_N. The effect of the controlled source is described by α_N, which is defined as

$$h_{FB} = \alpha_N = -\left.\frac{I_C}{I_E}\right|_{V_{CB}=0} \tag{7-9}$$

The relations in Eq. (7-8) represent accurate approximations to the actual equations which describe the operation of the transistor. However, in certain applications the effect of the saturation current I_{CO} associated with the reverse-biased collector-base junction on the total collector current must be included and is given in Eq. (7-10).

$$I_C = -\alpha_N I_E + I_{CO} \tag{7-10}$$

Kirchhoff's current law requires that

$$I_E = -(I_C + I_B) \tag{7-11}$$

Substitution of Eq. (7-8) into Eq. (7-11) results in

$$I_B = -(1 - \alpha_N)I_E \tag{7-12}$$

As it is often desirable to express the collector current in terms of the base current, combination of Eqs. (7-8b) and (7-12) yields

$$I_C = \frac{\alpha_N}{1 - \alpha_N} I_B = h_{FE} I_B \tag{7-13}$$

The quantity h_{FE} is called the *dc common-emitter short-circuit current gain.* For α_N slightly less than 1, h_{FE} is much larger than 1. A value of $\alpha_N = 0.99$ and a corresponding value of $h_{FE} = 99$ are typical. It is then evident that small changes in base current are accompanied by large changes in collector current.

While the results of Eqs. (7-8) and (7-12) apply to both *pnp* and *npn* transistors, the directions of positive charge flow and thus positive current within the transistor differ for the two types. By comparison of the external current conventions indicated in Figs. 7-22 and 23 with the direction of positive carrier flow within the device, it is seen that I_C and I_B are positive quantities and I_E is a negative quantity for *npn* transistors. Similarly, both I_C and I_B are negative quantities in *pnp* transistors while I_E is a positive quantity. The direction of internal transistor currents is one of two major differences between *npn* and *pnp* transistors.

The other major difference is the polarity of the collector-to-base and base-to-emitter potentials. Forward-biasing a junction, as described in Sec. 7-5, requires the p-type region to be at a higher potential than the n-type region, while the inverse is true for reverse-biasing a junction. Thus, both the base-emitter and collector-base voltages are positive for *npn* transistors. For *pnp* transistors the base-emitter and collector-base voltages are negative in the active region.

Because the base current exerts a greater control on the collector current than does the emitter current, the common-emitter configuration, shown in Fig. 7-25 for a *pnp* transistor, is most frequently used. In this con-

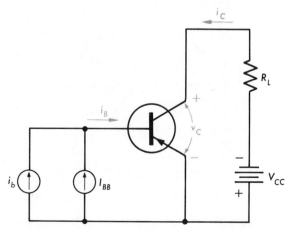

FIGURE 7-25 Common-emitter circuit to indicate voltage and current terminology.

figuration, the emitter is used as the reference node and all other voltages are measured with respect to the emitter.

The common-emitter *collector characteristics* are the volt-ampere curves which relate the collector current I_C and collector-to-emitter voltage V_{CE} for various values of I_B. This family of curves can be obtained either from Eqs. (7-8) or by measurement; they are shown in Fig. 7-26. As indicated in Fig. 7-26, the collector-emitter voltage exerts little control on the collector current, a fact which results from the reverse-biased nature of the collector-base junction and its negligible current component. When the base current is zero, virtually no collector current exists. This is a region of the characteristics referred to as *cutoff* and is a result of both junctions being reverse-biased. At point c on the characteristics, the value of base current is equal to or greater than 15 mA. The collector current does not increase in value if the base current is increased beyond 15 mA. This region, indicated by *ab*, is called *saturation*, which occurs when both junctions are forward-biased. Note that in the saturated region collector current exists at a near-zero value of collector voltage. Similarly, cutoff corresponds to virtually no collector current with large collector voltages. Operation in these regions generally provides the two states in transistor switches.

The maximum-collector-dissipation curve represents a rating determined by the amount of heat that can be dissipated at the collector junction

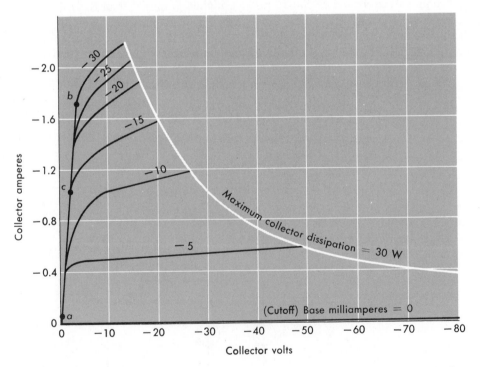

FIGURE 7-26 Common-emitter collector characteristics for a *pnp* transistor (similar to the 2N4999).

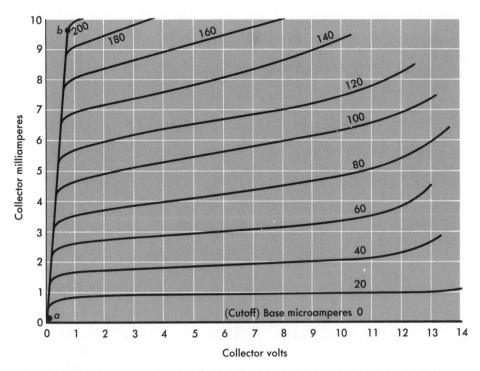

FIGURE 7-27 Common-emitter collector characteristics for an *npn* transistor (similar to the 2N918).

without damage to the transistor. In order not to exceed this rating, the transistor must be operated in a region below and to the left of the dissipation curve. Other ratings usually indicated by the manufacturer are the maximum collector-to-base and emitter-to-base voltages. Both of these ratings apply to reverse-biased junctions and are established by the junction breakdown voltages, similar to that described for the *p-n* diode in Sec. 7-5.

Figure 7-27 displays the common-emitter collector characteristics for an *npn* transistor. Note that the algebraic signs of the quantities differ from the *pnp* case; this is a result of the convention adopted for positive current direction. Figure 7-27 also indicates the cutoff and saturated regions.

The magnitudes of the currents and voltages for the transistors whose characteristics are shown in Figs. 7-26 and 7-27 also differ markedly. These differences are attributed to the fact that the *pnp* transistor is used in high-power applications while the *npn* transistor is used primarily in low-power high-frequency applications.

7-8 GRAPHICAL ANALYSIS OF JUNCTION TRANSISTORS

The static or quiescent conditions in a transistor circuit are established by the main power circuit. The control is provided, usually by means of an ac signal, in the base circuit. As a result, the currents and voltages in the circuit have ac values which are superimposed on the dc operating levels. In order that the symbols representing specific currents and voltages not be too confusing, a standard terminology has been adopted by the Institute of Electrical and Electronic Engineers (IEEE); this is given in Table 7-1.

The system underlying the notation is the following:

1 Lowercase letters v and i designate instantaneous voltage and current, respectively.

2 Uppercase V and I denote effective values of ac components or average values of total quantities.

3 Lowercase subscripts represent only the time-varying, or ac, components of voltages and currents in the transistor.

4 Uppercase subscripts refer to total quantities.

5 Repeated subscripts refer to supply quantities.

6 Currents are taken as positive when entering a terminal from the external circuit.

7 Voltages are taken as positive when measured with respect to the reference element (node). If the reference node is not clearly evident, an additional subscript is used to designate the reference (e.g., the voltage V_{CEQ} is the quiescent value of the collector potential with respect to the emitter).

The circuit of Fig. 7-25 and the characteristics of Fig. 7-28 serve to illustrate the use of the symbols. The line labeled *load line* is a statement of Kirchhoff's voltage law for the collector loop (from collector to emitter through the transistor and from the emitter through the supply and load resistor back to the collector) and may be expressed as

$$v_C + i_C R_L + V_{CC} = 0 \qquad (7\text{-}14)$$

TABLE 7-1 TRANSISTOR VOLTAGE AND CURRENT SYMBOLS

ITEM	SUPPLY	QUIESCENT (STATIC)	AC, OR TIME-VARYING, COMPONENT		TOTAL (DC + AC)	
			INST.	RMS	INST.	AVG.
Collector voltage	V_{CC}	V_{CQ}	v_c	V_c	v_C	V_C
Collector current	. . .	I_{CQ}	i_c	I_c	i_C	I_C
Base voltage	V_{BB}	V_{BQ}	v_b	V_b	v_B	V_B
Base current	I_{BB}	I_{BQ}	i_b	I_b	i_B	I_B
Emitter voltage	V_{EE}	V_{EQ}	v_e	V_e	v_E	V_E
Emitter current	. . .	I_{EQ}	i_e	I_e	i_E	I_E

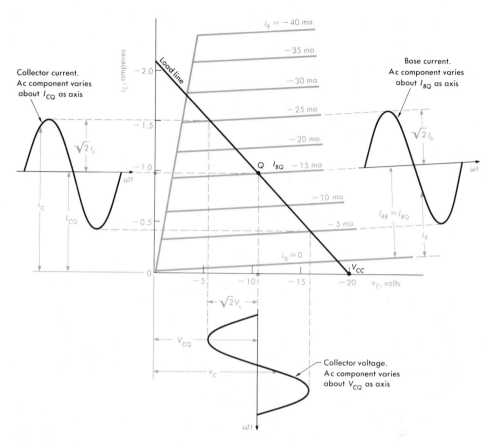

FIGURE 7-28 Sinusoidal variations of collector current i_c, collector voltage v_c, and base current i_b superimposed on the direct values I_{CQ}, V_{CQ}, and I_{BQ}, respectively, for the circuit of Fig. 7-25. The collector characteristics of the *pnp* transistor are idealized.

The base-bias source is I_{BB}, and the controlling signal is $i_b = \sqrt{2} \, I_b \sin \omega t$. The main power source is V_{CC}. This voltage, in conjunction with I_{BB}, is used to establish the operating point Q. Since Kirchhoff's laws must be satisfied at each instant of time, the instantaneous behavior of the circuit is constrained to be along the load line, resulting in the waveforms shown in Fig. 7-28. The values for the currents and voltages indicate that a change in base current of 10 mA (peak value) results in a corresponding change in collector current of approximately 0.6 A, or a current gain of

$$0.6/(10 \times 10^{-3}) = 60$$

The circuit of Fig. 7-25 shows two biasing sources: the collector supply V_{CC} and the base supply I_{BB}. In most practical biasing arrangements, it is desirable to use only one power supply. A simple circuit using an *npn* transistor which contains only one supply is shown in Fig. 7-29a. The terminal

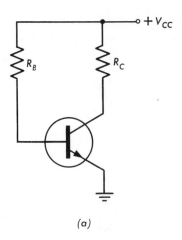

(a)

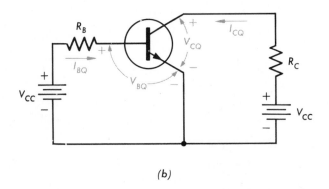

(b)

FIGURE 7-29 Simple one-supply bias circuit.

marked V_{CC} is the positive terminal to which the power supply is connected; the negative terminal is connected to ground. This schematic representation is widely used in practice and is the equivalent of the representation in Fig. 7-29b.

To obtain the operating point, the quiescent value of base current must be determined. The collector current and voltage are constrained by the load line, which satisfies

$$V_C + I_C R_C = V_{CC} \tag{7-15}$$

The voltage law for the base loop (from base to emitter in the transistor and back to the base through the supply and base-bias resistance R_B) is given by

$$V_B + I_B R_B = V_{CC} \tag{7-16}$$

As the emitter-base junction behaves as a junction diode, it is characterized as shown in Fig. 7-19. The value of V_B, approximately 0.7 V for silicon transistors and 0.3 V for germanium transistors, is independent of the external circuit elements R_B and V_{CC}. The base current at the operating point is determined from the solution of Eq. (7-16) for I_B and is

$$I_{BQ} = \frac{V_{CC} - V_B}{R_B} \tag{7-17}$$

Often, V_B is much smaller than V_{CC} so that Eq. (7-17) can be approximated by

$$I_{BQ} \doteq \frac{V_{CC}}{R_B} \tag{7-18}$$

The quiescent value of base current can thus be determined by either Eq. (7-17) or Eq. (7-18). This value in conjunction with the load line determines the operating point. The voltage supply and base-bias resistance fulfill the function of the supply I_{BB} in Fig. 7-25.

Example 7-2

The transistor whose collector characteristics are given in Fig. 7-30 is used in the circuit of Fig. 7-29a. The supply is an 18-V battery. The desired Q point is at a collector voltage of 10 V and a collector current of 16 mA. Determine the values of R_C and R_B needed to establish the operating point.

SOLUTION   The load line is constructed by drawing a straight line through the Q point which intersects the horizontal axis at V_{CC} (see Fig. 7-30). The slope of the load line is $-1/R_C$ and can be determined from the two points. Thus,

$$-\frac{1}{R_C} = \frac{I_{CQ} - 0}{V_{CQ} - V_{CC}} = \frac{16 \times 10^{-3}}{10 - 18} = -2 \times 10^{-3} \quad \mho$$

from which $R_C = 500 \ \Omega$.

The base current at the operating point, obtained from the characteristics, is

$I_{BQ} = 100 \ \mu$A. From Eq. (7-18),

$$R_B = \frac{V_{CC}}{I_B} = \frac{18}{100 \times 10^{-6}} = 180,000 \ \Omega$$

The feasibility of the circuit in Fig. 7-29a is limited by available technology. Practical bias arrangements are discussed in Chap. 8 and are based, in part, on the description of semiconductor technologies at the conclusion of this chapter.

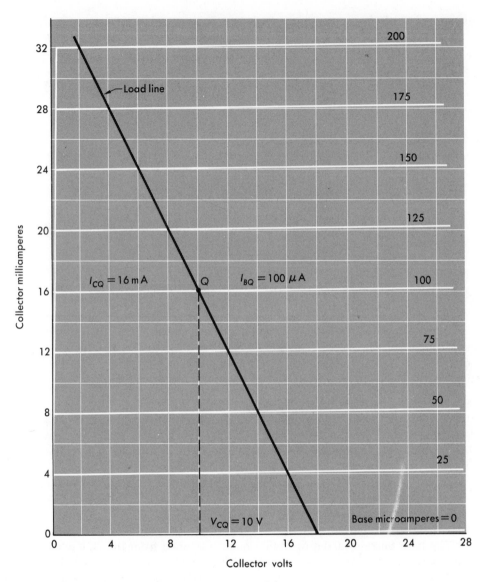

FIGURE 7-30 Collector characteristics for Example 7-2.

7-9 THE CHARACTERISTICS OF FIELD-EFFECT TRANSISTORS

A second important device used to provide the circuit properties of a controlled source is the *junction field-effect transistor,* usually referred to as a *JFET* or simply an *FET*. As shown in Fig. 7-31*a*, the JFET is a three-terminal device consisting of a single junction embedded in a semiconductor sample. Because the base semiconductor forming the channel indicated in Fig. 7-31*a* is *n*-type material, the JFET depicted is referred to as an *n*-

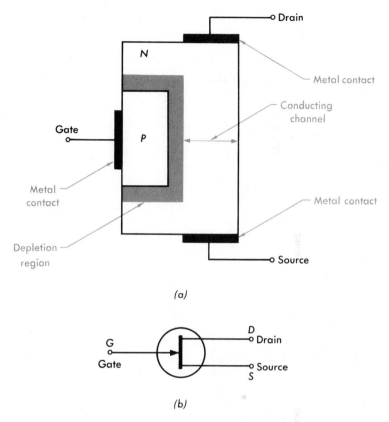

(a)

(b)

FIGURE 7-31 The *n*-channel junction field-effect transistor and its circuit symbol.

channel device whose circuit symbol is given in Fig. 7-31*b*. When the channel is formed of a *p*-type semiconductor, as shown in Fig. 7-32, the device is referred to as a *p*-channel JFET. In both kinds of field-effect transistors, the elements are called the *source, drain,* and *gate.* Their functions are analogous to the emitter, collector, and base in the junction transistor. Thus, the gate provides the means to control the flow of charges between the source and drain.

In normal operation, the junction in the JFET is reverse-biased, and the main power source is connected between drain and source as shown in Fig. 7-33. Both the main power source V_{DD} and the gate supply voltage V_{GG} can establish a reverse-biased junction independently. No gate current flows because of the reverse bias, and all carriers flow from source to drain. The corresponding amount of drain current is dependent on the resistance of the channel and the drain to source voltage V_{DS}. As V_{DS} is increased for a fixed value of V_{GS}, the junction becomes more heavily reverse-biased. Consequently, the depletion region extends further into the conducting channel.

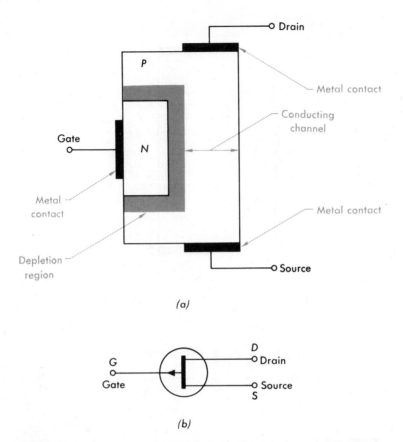

(a)

(b)

FIGURE 7-32 The *p*-channel junction field-effect transistor and its circuit symbol.

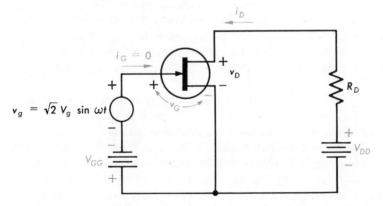

FIGURE 7-33 Common-source circuit to indicate current and voltage terminology.

Further increase of V_{DS} will ultimately block, or *pinch off,* the conducting channel. Once pinch-off is achieved, the drain current I_D remains constant and independent of V_{DS}. By changing the gate to source voltage, one may control where pinch-off occurs and thus control the value of drain current. It is this region beyond pinch-off which is useful for controlled-source operation, as only changes in V_{GS} produce corresponding changes in I_D.

Two regions are indicated by the dashed line in Fig. 7-34, which are the *common-drain output (static) characteristics* of a typical JFET. The region to the left of the dashed line is below pinch-off and is described as the controlled resistive region where I_D is proportional to V_{DS}. The controlled-source region, to the right of the dashed line, is one in which I_D depends on V_{GS} for a given value of V_{DS}.

A second useful characteristic which indicates the strength of the controlled source is the *transfer characteristic* depicted in Fig. 7-35. The transfer characteristic relates the drain current to the degree of negative bias applied between gate and source. The current I_{DSS} is the drain current for the condition that $V_{GS} = 0$ and is useful in mathematically describing the operation of the JFET in the controlled-source region. As shown in Fig. 7-35, a cutoff region exists, indicated by $-V_{PO}$, for which no drain current flows. In this region both V_{GS} and V_{DS} act to eliminate the conducting channel com-

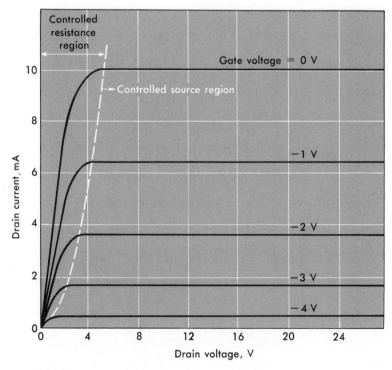

FIGURE 7-34 Common-source drain characteristics of an *n*-channel JFET.

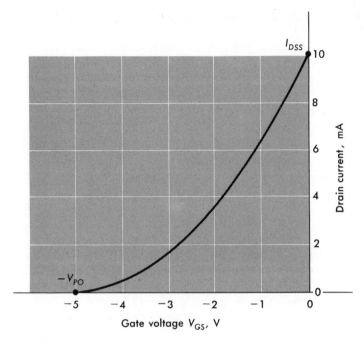

FIGURE 7-35 Transfer characteristic of an *n*-channel JFET.

pletely and permit no current flow. Equation (7-19) is an accurate approximation useful in describing the controlled-source region and is based on the transfer characteristic

$$I_D = I_{DSS} \left(1 + \frac{V_{GS}}{V_{PO}} \right)^2 \tag{7-19}$$

Example 7-3

The following measurements are made for an *n*-channel JFET: For $V_{GS} = -3$ V, $I_D = 1.6$ mA and for $V_{GS} = -4$ V, $I_D = 0.4$ mA. Determine I_{DSS} and V_{PO}.

SOLUTION Each of the conditions given relate two of the quantities in Eq. (7-19). In both cases V_{PO} and I_{DSS} are unknown. The solution utilizes the simultaneous equations which result from each condition. Substitution of each condition into Eq. (7-19) results in

$$1.6 \times 10^{-3} = I_{DSS} \left(1 + \frac{-3}{V_{PO}} \right)^2$$

$$0.4 \times 10^{-3} = I_{DSS} \left(1 + \frac{-4}{V_{PO}} \right)^2$$

Division of the two equations and extracting the square root of both sides of the resulting equation give

$$2 = \frac{1 - 3/V_{PO}}{1 - 4/V_{PO}}$$

from which $V_{PO} = 5$ V.

Substitution of $V_{PO} = 5$ V into the first condition yields

$$1.6 \times 10^{-3} = I_{DSS}(1 - \tfrac{3}{5})^2$$

from which $I_{DSS} = 10$ mA.

Manufacturers' ratings for field-effect transistors are similar to those for junction transistors described in Sec. 7-7. Typically, these include a maximum-power-dissipation rating and maximum gate-to-source and drain-to-source voltages. The maximum voltage ratings are limits to prevent junction breakdown, while the power-dissipation rating is a limit to prevent heating damage of the device.

7-10 GRAPHICAL ANALYSIS OF FIELD-EFFECT TRANSISTORS

The quiescent conditions in a field-effect transistor circuit are established by the main direct-energy supply. Control is usually provided by means of a time-varying signal in the gate circuit. As a result the voltages and currents in the circuit have ac values which are superimposed on the dc operating levels. Table 7-2 lists the standard JFET current and voltage symbols used. The rationale for these symbols is analogous to that for the junction transistor described in Sec. 7-8.

The circuit of Fig. 7-33 and the characteristics of Fig. 7-36 further serve to illustrate the use of the symbols. The load line is the graphical representation of the voltage-law constraint in the drain loop and is given by

$$V_{DD} = i_D R_D + v_D \tag{7-20}$$

TABLE 7-2 JFET VOLTAGE AND CURRENT SYMBOLS

ITEM	SUPPLY	QUIESCENT (STATIC)	AC OR TIME-VARYING, COMPONENT		TOTAL (DC + AC)	
			INST.	RMS	INST.	AVG.
Drain voltage	V_{DD}	V_{DQ}	v_d	V_d	v_D	V_D
Drain current	. . .	I_{DQ}	i_d	I_d	i_D	I_D
Gate voltage	V_{GG}	V_{GQ}	v_g	V_g	v_G	V_G
Gate current	. . .	I_{GQ}	i_g	I_g	i_G	I_G
Source voltage	V_{SS}	V_{SQ}	v_s	V_s	v_S	V_S
Source current	. . .	I_{SQ}	i_s	I_s	i_S	I_S

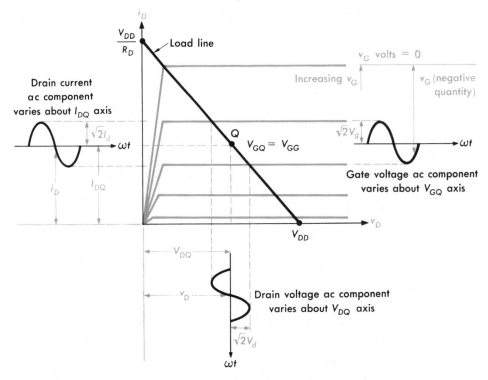

FIGURE 7-36 Sinusoidal variations of drain current i_d, drain voltage v_d, and gate voltage v_g superimposed on the direct values I_{DQ}, V_{DQ}, and V_{GQ}, respectively, for the circuit of Fig. 7-33.

No current flows in the gate loop, so that the gate voltage is

$$v_G = -V_{GG} + \sqrt{2}\ V_g \sin \omega t \qquad (7\text{-}21)$$

The quiescent Q occurs when $\sin \omega t$ is zero. Each of the quantities indicated in Fig. 7-36 is composed of a sinusoidal signal superimposed on the quiescent level.

The biasing arrangement in the circuit of Fig. 7-33 is referred to as *fixed bias* since the quiescent gate voltage is established by the source V_{GG}. *Self-bias,* shown in the circuit of Fig. 7-37, is used most often because the need for the separate gate supply is eliminated. The voltage-law equation for the drain loop is

$$V_{DD} = I_D(R_S + R_D) + V_{DS} \qquad (7\text{-}22)$$

which is plotted as the load line on the characteristics of Fig. 7-38. The voltage-law equation for the gate loop is

$$V_{GS} = -I_D R_S \qquad (7\text{-}23)$$

If the transfer characteristic has been provided by the manufacturer, it may be used to determine the operating point. The straight line *ab* inter-

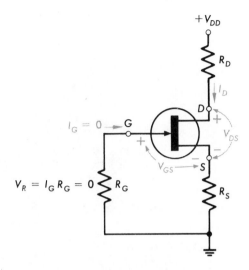

FIGURE 7-37 A self-biased n-channel JFET.

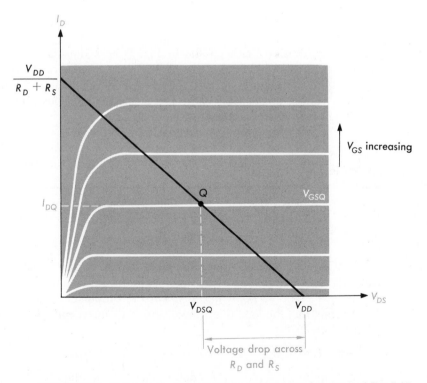

FIGURE 7-38 Volt-ampere characteristics and load line for circuit of Fig. 7-37.

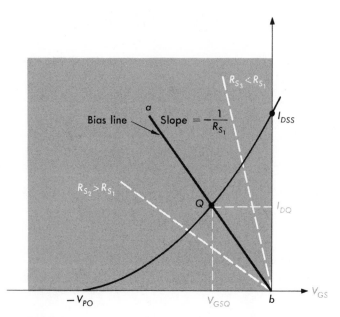

FIGURE 7-39 Transfer characteristic showing bias line for circuit of Fig. 7-37.

secting the transfer characteristic in Fig. 7-39 is the *bias line* and is the graphical representation of Eq. (7-23). For a given value of R_S, the intersection Q is the quiescent point, and the corresponding values of drain current and gate voltage are I_{DQ} and V_{GSQ}. The dashed lines indicate the change in operating point as R_S is varied.

Example 7-4

The characteristics given in Fig. 7-40 are for the JFET used in the circuit of Fig. 7-37. The supply voltage is 36 V, $R_S = 1,000 \ \Omega$, and R_D is 9,000 Ω. Determine the values of V_{GSQ}, I_{DQ}, and V_{DSQ}.

SOLUTION   Plot the bias line, indicated by ab, on the transfer characteristics given in Fig. 7-40a. At the intersection Q, the values of V_{GSQ} and I_{DQ} are $V_{GSQ} = -2.5$ V, $I_{DQ} = 2.5$ mA. On the output characteristics of Fig. 7-40b, construct the load line shown whose V_D intercept is $V_{DD} = 36$ V and whose slope is

$$-1/(R_S + R_D) = -1/10,000 \quad \mho$$

At the quiescent point Q, determined from the bias line, obtain $V_{DSQ} = 11$ V.

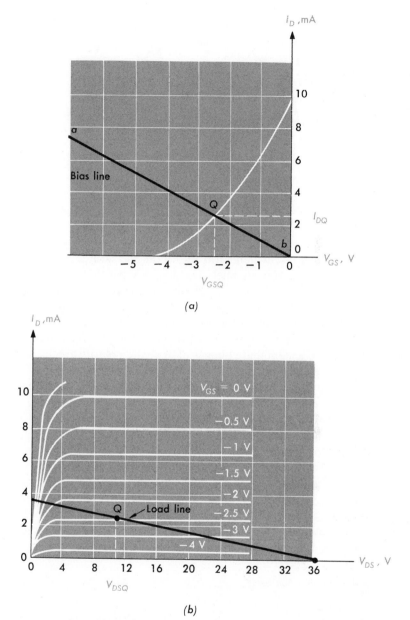

(a)

(b)

FIGURE 7-40 JFET (*a*) transfer characteristic and (*b*) output characteristics for Example 7-4.

7-11 MOSFET CHARACTERISTICS

A third type of semiconductor device, primarily used in switching applications that can also provide controlled-source characteristics, is the *metal-oxide-semiconductor field-effect transistor,* commonly referred to as the *MOSFET.* Two classes of MOS transistors exist; these are the *depletion-*

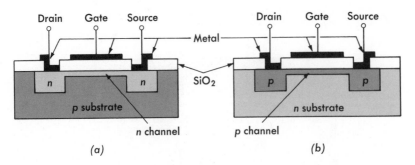

FIGURE 7-41 Depletion-mode MOSFETs: (a) n-channel, (b) p-channel.

mode types shown in Fig. 7-41 and the *enhancement-mode* types displayed in Fig. 7-42.

The *n*-channel depletion-mode MOSFET in Fig. 7-41a consists of two *n*-type regions, which form the drain and source, connected by a narrow *n*-type conducting channel. The gate is separated from the channel by an insulating layer usually made of silicon dioxide (SiO_2). Control is exerted by a potential applied to the gate which depletes the conducting channel of carriers. This action is analogous to pinch-off in the JFET. The operation of the *p*-channel depletion-mode MOSFET of Fig. 7-41b is similar to the *n*-channel operation except that voltages of opposite polarity are used. The volt-ampere characteristics of depletion-mode devices, shown in Fig. 7-43, are similar to those for junction FETs. In both, increasing the magnitude of gate-to-source voltage causes a decrease in drain current. Application of a sufficiently large voltage causes the device to cut off; i.e., no drain current exists.

In the enhancement-mode MOSFETs in Fig. 7-42, the two regions which form the drain and source are separated by the substrate. With no gate voltage applied, the device is cut off and no drain current exists. (Actually, a very small and often negligible leakage current is present as is true in all FET devices.) Application of an appropriate gate voltage induces a small conducting channel to be formed between drain and source so that carrier flow is enhanced. Increases in the magnitude of the gate voltage cause the drain current as indicated in the volt-ampere characteristics in Fig. 7-44.

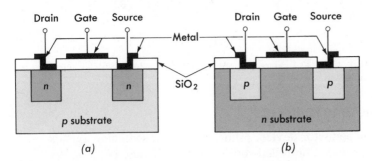

FIGURE 7-42 (a) n-type and (b) p-type enhancement-mode MOSFET structures.

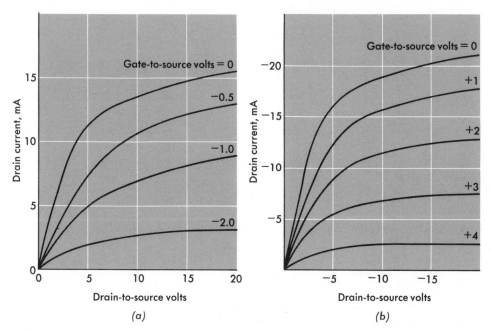

FIGURE 7-43 Volt-ampere characteristics for (a) n-channel and (b) p-channel depletion-mode MOSFETs.

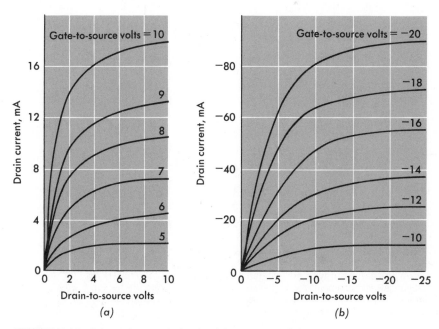

FIGURE 7-44 Output characteristics for (a) n-type and (b) p-type enhancement-mode MOS transistors.

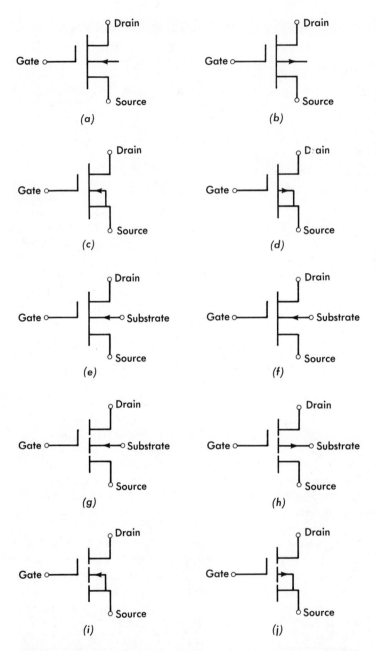

FIGURE 7-45 MOSFET circuit symbols: (a) , (c), and (e), n-channel depletion-mode devices; (b), (d), and (f), p-channel depletion-mode devices; (g) and (i), n-type enhancement-mode devices; (h) and (j) for p-channel enhancement-mode devices. In (c), (d), (i), and (j), the substrate is connected to the source; in (e), (f), (g), and (h), a fourth terminal is connected to the substrate.

Both classes of MOS transistors require that the junctions between the substrate and source and drain be reverse-biased. This is necessitated during operation so that carrier flow between these regions and the substrate is prevented.

In the construction of MOSFET devices, a connection is sometimes made to the substrate so that a four-terminal device results. Often, the substrate is internally connected to the source. These situations are indicated in Fig. 7-45 which displays the standard circuit symbols for MOS transistors.

7-12 SEMICONDUCTOR TECHNOLOGY AND FABRICATION

Three distinct methods or technologies are employed in the fabrication of semiconductor circuits. The first of these is *discrete-component technology* in which each circuit element—transistor, resistor, etc.—is an individual component. Circuit construction is achieved by interconnecting the various components usually by individual wires which are soldered or wrapped to form the connections. Discrete-component fabrication was the original and traditional manner used in circuit construction. Although it is still used for many applications, it is being or has been supplanted by integrated-circuit technologies. The second and third fabrication methods, *monolithic technology* and *hybrid technology,* are two classes of integrated-circuit (IC) construction.

Space, weight, and reliability considerations gave impetus to the development of *integrated circuits.* This area of electronics deals with the manufacture and packaging of entire circuits much in the manner of conventional devices. The advantage of this technique is the reduction in the number of electric connections that must be made (resulting in improved reliability), as well as the evident space and weight reductions. Additionally, there is often an improvement in performance, particularly in high-speed applications such as large-scale computers and data processors. While much of the original development was directed toward data-handling systems and the space program, integrated circuits are now widely used in instrumentation and control systems and communications systems.

Integrated-circuit technology involves the use of only solid-state devices, resistors, and capacitors. The elimination of inductors is necessitated by the fact that typical semiconductors do not exhibit the magnetic properties necessary to realize practical values of inductance.

Monolithic technology is characterized by the use of the same semiconductor crystal, or *chip,* for both the passive components (R and C) and the transistors. Typically, monolithic ICs employ either bipolar devices or MOS devices since fabrication on the same chip is often not feasible. Figure 7-46 illustrates the construction of monolithic resistors and capacitors. The resistor in Fig. 7-46a is formed by the *n*-type material with the metal contacts serving as the resistor terminals. For the simple geometry shown, the resistance R is given by

$$R = \rho \frac{L}{A} \tag{7-24}$$

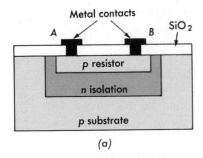

(a)

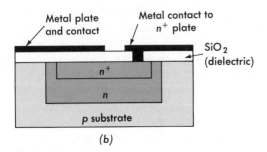

(b)

FIGURE 7-46 Monolithic (a) resistor and (b) capacitor.

where ρ = resistivity in Ω-m
L = length of the resistor in m
A = cross-sectional area in m²

As A is the product of the thickness t and the width W, Eq. (7-24) is often written as

$$R = \frac{\rho}{t} \times \frac{L}{W} \tag{7-25}$$

A common method for indicating the value of resistance and the area on a chip it covers is to express R in ohms per square ($\Omega/\square$). That is, as t is uniform across the resistor area, construction of a resistor of equal length and width results in an R-value equal to the ohms per square.

The capacitor in Fig. 7-46b is a "parallel-plate" capacitor in which the metal contact and n-type layer form the plates. The SiO₂ is the insulator between the plates and is used to increase the capacitance values obtainable.

In both devices, the p-type substrate is used as the foundation upon which the components are built. The p-n junctions that are formed are always reverse-biased so that the components are isolated from other components on the chip.

Bipolar transistors are constructed as illustrated in Fig. 7-47 in which one pnp and two npn transistors are shown. The n^+ symbol represents an n-type region that is more heavily doped than the other n-type regions. The

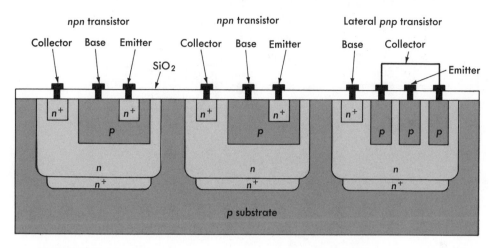

FIGURE 7-47 Monolithic structure containing two *npn* transistors and a lateral *pnp* transistor.

emitter of the *npn* transistor is n^+ so that it is efficient in providing carriers. The region between the collector and substrate, called a *buried layer,* is used to reduce collector series resistance; while the n^+ region below the collector contact serves to improve the electric connection obtained. The *pnp* transistor shown in Fig. 7-47 is referred to as a *lateral pnp* transistor, the term "lateral" referring to the spatial relation of emitter and collector as contrasted with the "vertical" *npn* transistors. The *p*-type region between the devices are isolation islands which keep the various components electrically separated.

The structures shown in Figs. 7-48 and 7-49 are for *n*-channel and *p*-channel MOSFETs, respectively. In each figure both depletion- and enhancement-mode devices are illustrated for which the substrate and source are connected. Generally, MOSFET chips utilize one type of device,

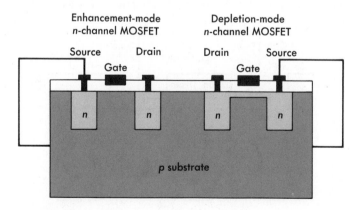

FIGURE 7-48 NMOS structure with one depletion-mode and one enhancement-mode transistor.

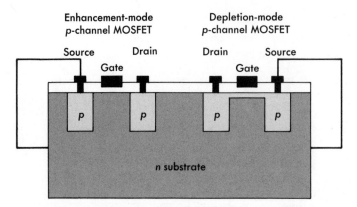

FIGURE 7-49 Depletion-mode and enhancement-mode P-MOS structure.

either *p*-channel or *n*-channel; as such, these chips are often called *P-MOS* and *N-MOS*, respectively. An alternative technology, depicted in Fig. 7-50, uses both *p*-channel and *n*-channel devices to form a compound device. The devices, commonly referred to as *C-MOS* or *complementary MOS*, have the advantage of low power consumption. However, fewer devices can be placed on the chip. MOS technologies are widely used in computer circuits because they have higher packing densities than can be achieved with bipolar devices. Because they respond more quickly, bipolar technologies are used in high-speed applications.

In both MOS and bipolar technology, the fabrication method requires a series of masks, photoetching, and diffusions. The construction of a chip starts with a substrate which is covered with SiO_2 and a photoresistant material, as shown in Fig. 7-51*a*. A mask is placed over the chip and exposed to

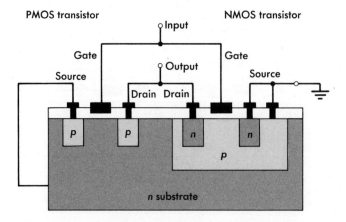

FIGURE 7-50 C-MOS device showing P-MOS and N-MOS enhancement-mode devices and their equivalent terminal connections.

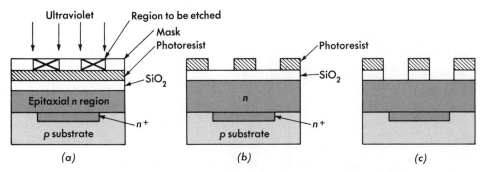

FIGURE 7-51 The photo-etching process for monolithic circuits. (a) The mask is placed over the photoresist and subjected to ultraviolet radiation. (b) The structure after exposure and development. (c) Etching the appropriate regions of SiO₂ to prepare the exposed n regions to diffusion.

ultraviolet radiation, which after development leaves the structure shown in Fig. 7-51b. Etching then removes the SiO₂ exposing the surface of the chip. An n^+ layer is diffused and an epitaxial n-layer grown over the substrate. (Epitaxy refers to the fact that the entire n-layer has a uniform crystal orientation; i.e., it is a single crystal.) The process of masking, photoetching, and diffusion is repeated until the circuit is completed, as illustrated in Fig. 7-52. The last step in the process is metalization, usually using aluminum, to form the connections to and the interconnections between components. When complete the chip is encapsulated and electric connections brought out to the package.

Hybrid technology is a combination of both monolithic and discrete-component technologies. Transistors are fabricated monolithically; while resistors and capacitors are fabricated by the deposition of semiconductor material on a ceramic substrate (thin- or thick-film technology). The interconnection of the various components is made on a printed-circuit (PC) board. Sometimes hybrid circuits employ discrete resistors and capacitors with appropriate interconnections made on a PC-board.

Each technology has unique properties which affect the manner in which they are used in electronic circuit design. In using discrete components, in which all resistors of the same power rating are of the same size and cost, it is usual to minimize the number of transistors. As resistors are cheaper than transistors and since each transistor consumes bias energy, these factors lead to efficient design. A second design rule is that the use of "matched" transistors be avoided because of high cost. A matched pair of transistors refers to two transistors whose characteristics are virtually identical and independent of manufacturing variation. The recent "introduction" of two- or four-transistor packages which contain nearly identical devices is an approach to circumvent the second "rule of thumb."

Because IC chips are small, effective design implies that the components use a minimum amount of the surface area of the chip. As a consequence, large resistors (over 20–30 KΩ) cannot be accommodated. A sec-

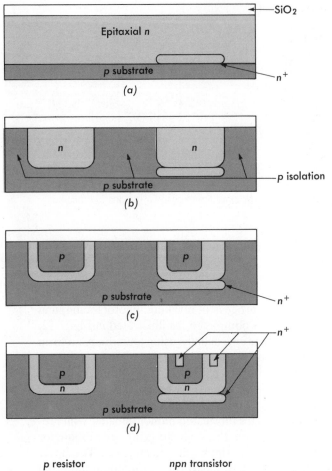

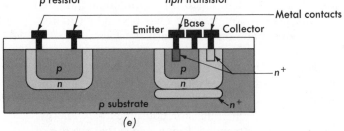

FIGURE 7-52 Steps in the fabrication of an IC: (*a*) the chip after the epitaxial *n* layer is grown. (*b*) First diffusion of *p*-type material for isolation. Each *n* region is now an "island." (*c*) Second diffusion of *p*-type material into the *n*-type regions. These serve as the base of the *npn* transistor and of the resistor. (*d*) Emitter diffusion (n^+) and the n^+ region for collector contact. (*e*) Metalization to connect the terminals and contacts of the devices. Component interconnection (not shown) is also accomplished in this step.

ond design approach is to use transistors in place of resistors because the transistor requires considerably less area to construct than does the resistor. The third important observation useful in design is that monolithic technology can inherently provide virtually identical devices. Indeed, there are technological limitations to the fabrication of transistors of markedly different characteristics on the same chip.

In the succeeding chapters, major attention will be focused on the use of ICs. Discrete-component circuits are also included but with lesser emphasis.

PROBLEMS

7-1 Use the circuit shown in Fig. 7-53 to verify that the following are unilateral and can provide power gain:
a Ideal voltage-controlled voltage source
b Ideal current-controlled voltage source
c Ideal current-controlled current source

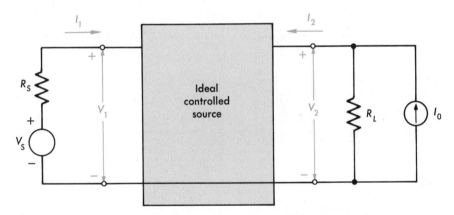

FIGURE 7-53 Circuit for Prob. 7-1.

7-2 With $I_0 = 0$, show that each of the four ideal controlled sources can provide a voltage gain V_2/V_S whose magnitude is greater than unity.

7-3 In the circuit of Fig. 7-54, the resistances R_i and R_0 are the input and output resistances of a practical voltage-controlled current source.
a Determine the transfer function (gain) V_2/I_s.
b Derive an expression for the ratio of the power dissipated by R_L to the power supplied by I_s.
c What conditions must be satisfied if the result obtained in b is to be greater than unity?
d Does the result in c also cause $|V_2/V_1|$ to be greater than 1?
e Is $|I_2/I_s|$ greater than 1 under the conditions of c?

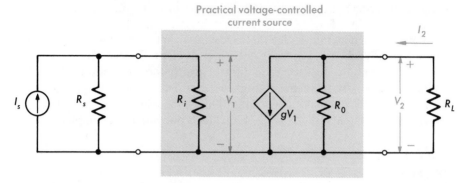

FIGURE 7-54 Circuit for Prob. 7-3.

7-4 Repeat Prob. 7-3 for the current-controlled voltage source in Fig. 7-55.

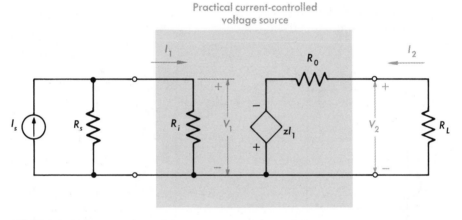

FIGURE 7-55 Circuit for Prob. 7-4.

7-5 The element values in the circuit of Fig. 7-56, employing a practical voltage-controlled voltage source, are $R_s = 300\ \Omega$, $R_i = 10^4\ \Omega$, $A = 20$, and $R_O = 200\ \Omega$.

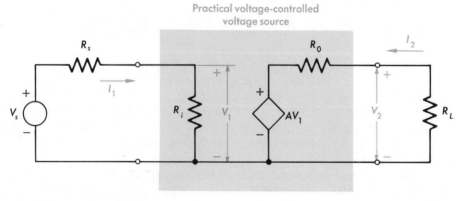

FIGURE 7-56 Circuit for Prob. 7-5.

a Find R_L so that V_2/V_s equals 10.

b Is the power dissipated by R_L in *a* greater than the power provided by V_s? If so, what is the ratio of the two powers?

c Using the value of R_L obtained in *a*, determine $|I_2/I_1|$.

7-6 Repeat Prob. 7-5 for the circuit in Fig. 7-57 for which the element values are $R_s = 600 \ \Omega$, $R_i = 600 \ \Omega$, $R_0 = 20 \ \text{k}\Omega$, and $A = 100$.

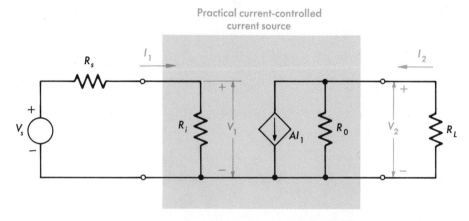

FIGURE 7-57 Circuit for Prob. 7-6.

7-7 The circuit depicted in Fig. 7-58*a* is a conceptual representation of a controlled switch. Control is exerted in the left-hand segment of the circuit and switching action occurs in the right-hand segment. One type of switch remains in its open position when $v_R < 0$ and is closed when $v_R \geq 0$. The source-voltage V_T represents an offset voltage in the device used to realize the switch. The resistances R_{on} and R_{off} are the switch resistances in the closed and open positions, respectively, and reflect the nonideal nature of real-world switching elements. For element values $R_s = 5 \ \text{k}\Omega$, $R = 5 \ \text{k}\Omega$, $R_{on} = 100 \ \Omega$, $R_{off} \to \infty$ (open-circuit), $R_L = 0.9 \ \text{k}\Omega$, $V_{AA} = +5 \ \text{V}$, $V_T = 0$, and $v_s(t)$ given in Fig. 7-58*b*, sketch the output voltage $v_0(t)$ and current $i_L(t)$.

7-8 An alternative method of control of the switching circuit in Fig. 7-58*a* and described in Prob. 7-7 is for the switch to close when $v_R \leq 0$ and remain open for $v_R > 0$. Sketch $v_0(t)$ and $i_L(t)$ for element values $R_s = 10 \ \text{k}\Omega$, $R = 2.5 \ \text{k}\Omega$, $R_{on} = 50 \ \Omega$, $R_L = 200 \ \Omega$, $R_{off} \to \infty$, $V_{AA} = -5 \ \text{V}$, $V_T = 0$, and $v_s(t)$ as given in Fig. 7-58*b*.

7-9 Repeat Prob. 7-7 if V_T is assumed to be 0.6 V.

7-10 Repeat Prob. 7-8 if V_T is assumed to be 0.7 V.

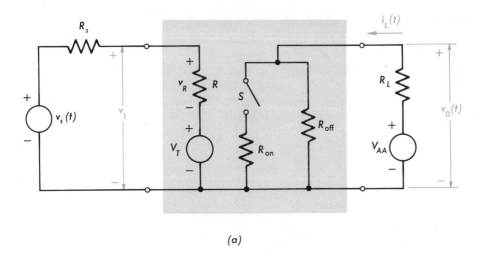

(a)

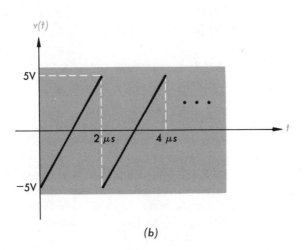

(b)

FIGURE 7-58 (a) Circuit and (b) waveform for Prob. 7-7.

7-11 The element values for the switch described in Prob. 7-7 are $R = 10^5$ Ω, $R_{on} = 20$ Ω, $R_{off} = 100$ kΩ, $V_T = 0.6$ V, $V_{AA} = 15$ V, $R_L = 1$ kΩ, and $v_s(t)$ as given in Fig. 7-58b.
a Sketch $v_0(t)$ and $i_L(t)$.
b Determine the average power dissipated by R_L, and the sum of the average power dissipated by the switch resistances R_{ON} and R_{OFF} during each 2-μs period.

7-12 Repeat Prob. 7-11 for the switch described in Prob. 7-8 whose element values are $R = 200$ kΩ, $R_{on} = 40$ Ω, $R_{off} = 20$ kΩ, $V_T = -0.6$ V, $V_{AA} = 12$ V, $R_L = 0.8$ kΩ, and $v_s(t)$ as given in Fig. 7-58b.

7-13 The volt-ampere relation for a particular junction diode is given in Eq. (7-1) for which $I_S = 0.1\ \mu A$, $V_T = 25$ mV, and $\eta = 2$.
a Sketch the diode characteristic in the range $-0.8 \le V \le 0.8$.
b The diode is used in the circuit of Fig. 7-13 in which $V_{BB} = 10$ V and $R = 100\ \Omega$. Determine the diode current and voltage.

7-14 Repeat Prob. 7-13 for $\eta = 1$ and all other values are as given in Prob. 7-13.

7-15 A diode whose volt-ampere characteristic is given in Fig. 7-59 is used in the circuit of Fig. 7-13. The load resistance is 1 kΩ and V_{BB} is 5 V.
a Determine the current and voltage in the load resistance.
b What is the power dissipated by the diode?
c The load resistance is successively changed to 2, 5, 0.5, and 0.2 kΩ. Determine the load current in each case.

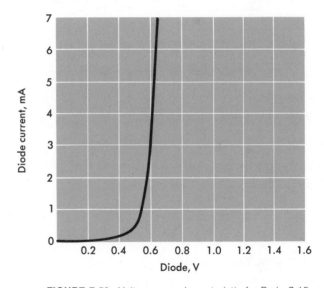

FIGURE 7-59 Volt-ampere characteristic for Prob. 7-15.

7-16 The diode whose volt-ampere characteristic is given in Fig. 7-59 is used in the circuit of Fig. 7-13. The load resistance is 2 kΩ and $V_{BB} = 10$ V.
a Determine the load current and voltage.
b Find the power dissipated by the diode.
c The supply voltage is changed in succession to 2.5, 5.0, 15, and 20 V. For each case determine the load current.

7-17 Two identical junction diodes whose volt-ampere characteristic is described in Prob. 7-13 are connected as shown in Fig. 7-60. The supply volt-

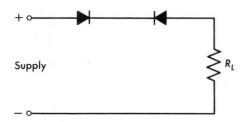

FIGURE 7-60 Circuit for Prob. 7-17.

age is 15 V and the value of $R_L = 100 \text{ k}\Omega$. Find the voltage across each diode and the current in the circuit.

7-18   In the circuit of Fig. 7-61, the junction diode can be represented as shown in Fig. 7-19 with $V_{\text{on}} = 0.6$ V. Determine the current in the diode.

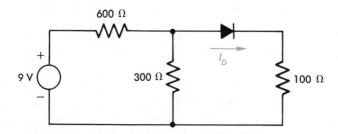

FIGURE 7-61 Circuit for Prob. 7-18.

7-19   Repeat Prob. 7-18.
a   Assume the diode is ideal.
b   Assume the diode is represented as given in Fig. 7-20 with $R_f = 20 \ \Omega$.
c   Compare the results in *a* and *b* with the result in Prob. 7-18.

7-20   The circuit of Fig. 7-13 is used to establish a circuit current of 10 mA with $V_{BB} = 5$ V. Determine the value of R_L if
a   The diode is ideal
b   The diode is represented as in Fig. 7-19 with $V_{\text{on}} = 0.6$ V.
c   The diode is represented as in Fig. 7-20 with $V_{\text{on}} = 0.6$ V and $R_f = 10 \ \Omega$.

7-21   Sketch the output waveform $v_0(t)$ in the circuit of Fig. 7-62*a* for the interval $0 \le t \le 10$ ms.

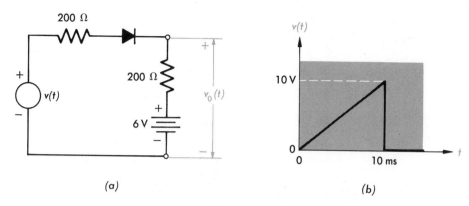

FIGURE 7-62 (a) Circuit and (b) waveform for Prob. 7-21.

7-22 a Repeat Prob. 7-21 assuming V_{on} = 0.5 V.
b Repeat Prob. 7-21 assuming V_{on} = 0.5 V and R_f = 50 Ω.

7-23 Sketch the output waveform $v_o(t)$ in the circuit of Fig. 7-63a for the interval $0 \leq t \leq 5$ ms.

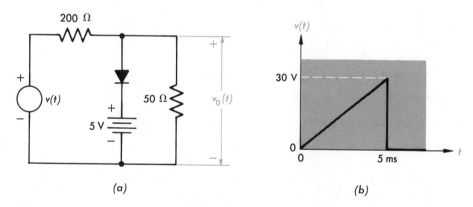

FIGURE 7-63 (a) Circuit and (b) waveform for Prob. 7-23.

7-24 a Repeat Prob. 7-23 assuming V_{on} = 0.6 V.
b Repeat Prob. 7-23 assuming V_{on} = 0.6 V and R_f = 20 Ω.

7-25 Sketch the output voltage $v_o(t)$ for one cycle in the circuit of Fig. 7-64 when
a The diodes are assumed ideal
b The diodes are characterized by V_{on} = 0.5 V
The input voltage is $v(t) = 20 \sin 10^4 t$.

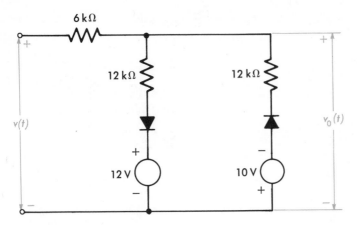

FIGURE 7-64 Circuit for Prob. 7-25.

7-26 The input signal in the circuit of Fig. 7-65 is $v_i(t) = 7.5 \sin 10^5 t$. Sketch the output waveform $v_0(t)$ for one cycle when
a The diodes are assumed ideal
b The diodes are characterized by $V_{on} = 0.5$ V

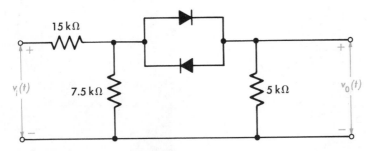

FIGURE 7-65 Circuit for Prob. 7-26.

7-27 The input signal in the circuit of Fig. 7-66 is $v_i(t) = 3 \sin 10^3 t$. Sketch the output waveform $v_0(t)$ for one cycle when
a The diodes are assumed ideal
b The diodes are characterized by $V_{on} = 0.6$ V

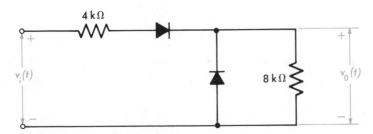

FIGURE 7-66 Circuit for Prob. 7-27.

7-28 Measurements at room temperature made on an *npn* transistor are $I_C = 10.0$ mA, $I_E = -10.1$ mA, $V_{BE} = 0.60$ V, and $V_{CB} = 10.0$ V. Determine
a The values of h_{FE} and α_N
b The value of I_{EO} assuming I_{CO} is negligible

7-29 Room-temperature measurements made on a certain *pnp* transistor are $I_C = 0.50$ mA, $I_B = 10$ μA, $V_{BE} = -0.65$ V, and $V_{CE} = 5$ V. Determine
a The values of h_{FE} and α_N
b The value of I_{EO} assuming I_{CO} is negligible

7-30 The *Darlington compound transistor,* or *Darlington pair,* is depicted in Fig. 7-67. The two-transistor combination is often used as a single three-terminal device. Assume the transistors are identical.
a Show that α_{NC}, the forward short-circuit current gain of the compound transistor is $\alpha_{NC} = 1 - (1 - \alpha_N)^2$.
b Derive a relationship for h_{FE} of the compound in terms of h_{FE} of the individual devices.
c For $\alpha_N = 0.99$, determine α_{NC} and h_{FE} of the Darlington pair.

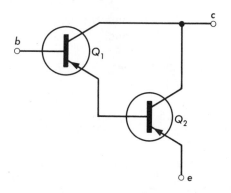

FIGURE 7-67 Darlington compound connection for Prob. 7-30.

7-31 The transistor whose characteristics are given in Fig. 7-27 is used in the circuit of Fig. 7-29. The parameter values are $R_C = 2$ kΩ, $R_B = 235$ kΩ, $V_{CC} = 10$ V, and $V_{BE} = 0.6$ V.
a Determine the values of I_C and V_{CE} at the operating point.
b What is the average power dissipated by the transistor?
c If R_B is varied, what is the minimum value of R_B that just causes saturation? What is the value of I_C at saturation?

7-32 Repeat Prob. 7-31 if $R_C = 1$ kΩ, $R_B = 84$ kΩ, $V_{CC} = 9$ V, and $V_{BE} = 0.6$ V.

7-33 The transistor whose characteristics are displayed in Fig. 7-26 is used in the circuit shown in Fig. 7-68. The relevant parameter values are $R_C = 30$ Ω, $R_B = 6$ kΩ, $V_{CC} = 60$ V, and $V_{BE} = -0.7$ V.

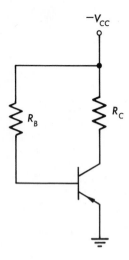

FIGURE 7-68 Circuit for Prob. 7-33.

a Determine the values of I_C and V_{CE} at the quiescent point.
b How much power does the supply V_{CC} provide?

7-34 Repeat Prob. 7-33 when $R_C = 20\ \Omega$, $R_B = 8\ \mathrm{k\Omega}$, $V_{CC} = 50\ \mathrm{V}$, and $V_{BE} = -0.7\ \mathrm{V}$.

7-35 The transistor whose characteristics are shown in Fig. 7-30 is used in the circuit of Fig. 7-29. The supply voltage is 15 V and the desired operating point is $V_{CEQ} = 8\ \mathrm{V}$, $I_{CQ} = 10\ \mathrm{mA}$. The value of $V_{BE} = 0.6\ \mathrm{V}$.
a Determine R_C and R_B.
b A sinusoidal base current $i_b(t) = 5 \times 10^{-5} \sin \omega t$ is applied between base and ground in the circuit described in a. Sketch the waveforms for one cycle of $i_C(t)$ and $v_{CE}(t)$.

7-36 The transistor described in Prob. 7-31 is to be biased at $V_{CEQ} = 5\ \mathrm{V}$ and $I_{CQ} = 3\ \mathrm{mA}$ using the circuit of Fig. 7-29. The values of V_{CC} and V_{BE} are 12 and 0.7 V, respectively.
a Determine the values of R_B and R_C.
b A sinusoidal base current $i_b(t) = 20 \times 10^{-6} \sin \omega t$ is applied between base and ground in the circuit described in a. Sketch the waveforms for $i_C(t)$ and $v_{CE}(t)$ for one cycle.

7-37 A transistor having $h_{FE} = 99$ and $V_{BE} = 0.6\ \mathrm{V}$ is used in the circuit of Fig. 7-69. The element values are $V_{CC} = 10\ \mathrm{V}$, $R_F = 200\ \mathrm{k\Omega}$, and $R_C = 2.7\ \mathrm{k\Omega}$.
a Determine the quiescent values of V_{CE} and I_C.
b If h_{FE} is changed to 199, determine the new operating point.

7-38 In the circuit of Fig. 7-69, the operating point of the transistor is $V_{CEQ} = 5$ V and $I_{CQ} = 5$ mA. The supply voltage is 9 V and the transistor parameters are $V_{BE} = 0.6$ V and $h_{FE} = 100$.
a Determine R_F and R_C.
b Using the values obtained in a, find the new value of I_{CQ} if h_{FE} changes to 50.

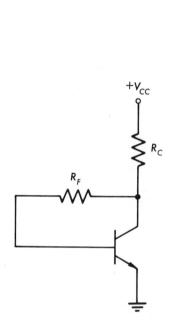

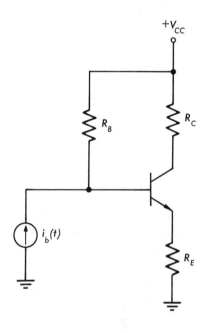

FIGURE 7-69 Circuit for Prob. 7-37.

FIGURE 7-70 Circuit for Prob. 7-39.

7-39 The transistor whose characteristics are shown in Fig. 7-27 is used in the circuit of Fig. 7-70. The parameter values are $V_{CC} = 12$ V, $R_C = 1.25$ kΩ, $R_E = 750$ Ω, $R_B = 100$ kΩ, and $V_{BE} = 0.7$ V.
a For $i_b(t) = 0$, determine V_{CEQ} and I_{CQ}.
b For $i_b(t) = 40 \times 10^{-6} \sin \omega t$, sketch the waveforms of $i_C(t)$ and $v_{CE}(t)$ for one cycle.
c What is the peak amplitude of $i_b(t)$ if the transistor is to remain in the active region at all times?

7-40 The element values in the circuit of Fig. 7-70 are $V_{CC} = 15$ V, $R_C = 1$ kΩ, $R_E = 2$ kΩ, and $R_B = 280$ kΩ. The transistor is characterized by $V_{BE} = 0.6$ V and $h_{FE} = 100$.
a Determine the values of I_{CQ} and V_{CEQ}.
b If h_{FE} becomes 200, what are the new values of I_{CQ} and I_{BQ}?

7-41   The circuit in Fig. 7-70 is to be designed with the transistor biased at $V_{CEQ} = 1.0$ V, $I_{CQ} = 8$ mA. The transistor parameters are $V_{BE} = 0.6$ V and $h_{FE} = 160$ and the supply voltage is 5 V. The circuit is to be designed so that the voltage drops across R_E and R_C are equal. It is convenient to assume $h_{FE} + 1 \doteq h_{FE}$.
a   Determine R_C, R_B, and R_E.
b   If h_{FE} changes to 80, what are the new values of V_{CEQ} and I_{CQ}?

7-42   The transistor whose characteristics are given in Fig. 7-27 is biased using the circuit of Fig. 7-71. The element values are $V_{EE} = 12$ V, $R_C = 1,250$ Ω, $R_E = 750$ Ω, and $R_B = 100$ kΩ. The voltage drop V_{BE} is 0.7 V.
a   Determine the values of V_{CEQ} and I_{CQ}.
b   For the conditions in a, find the value of V_E.
c   Repeat a for $R_E = 1.5$ kΩ.

7-43   A transistor having $h_{FE} = 50$ and $V_{BE} = 0.6$ V is employed in the circuit of Fig. 7-71. The supply voltage is 15 V and $R_E = 1$ kΩ, $R_C = 4$ kΩ and $R_B = 310$ kΩ. Determine the values of
a V_{CEQ} and I_{CQ}
b   The voltage drop from base to ground

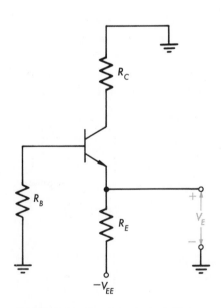

FIGURE 7-71   Circuit for Prob. 7-42.

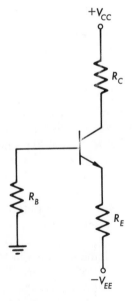

FIGURE 7-72   Circuit for Prob. 7-45.

7-44 The transistor described in Prob. 7-43 is used in the circuit of Fig. 7-71 and is to be biased at $V_{CEQ} = 6$ V, $I_{CQ} = 1$ mA. The supply voltage is 11.2 V and R_C and R_E are equal resistances.
a Determine R_C, R_E, and R_B.
b Using the element values in a, find the new values of V_{CEQ} and I_{CQ} if V_{BE} is changed to 0.7 V.

7-45 The circuit shown in Fig. 7-72 has element values $R_C = 500$ Ω, $R_E = 1$ kΩ and $R_B = 44$ kΩ; the supply voltages are each 15 V. The transistor employed has an $h_{FE} = 100$ and $V_{BE} = 0.6$.
a Determine the values of V_{CEQ} and I_{CQ}.
b Repeat a if h_{FE} becomes 200.
c Repeat a for $h_{FE} = 50$.

7-46 The circuit shown in Fig. 7-73 is used to bias the transistor whose characteristics are $h_{FE} = 100$ and $V_{BE} = 0.7$ V. The element values are $R_1 = 150$ kΩ, $R_2 = 37.5$ kΩ, $R_C = 7$ kΩ, and $R_E = 3$ kΩ.
a Determine the quiescent levels I_{CQ} and V_{CEQ}.
b Repeat a if h_{FE} changes to 200.
c Repeat a for $h_{FE} = 50$.

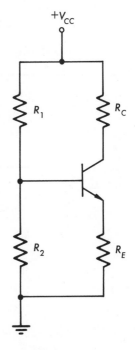

FIGURE 7-73 Circuit for Prob. 7-46.

7-47   The circuit in Fig. 7-73 is designed with $R_1 = 60$ kΩ, $R_2 = 12$ kΩ, $R_E = 10$ kΩ, $R_C = 15$ kΩ, and $V_{CC} = 12$ V. The transistor parameters are $h_{FE} = 50$ and $V_{BE} = 0.6$ V.
a   Determine I_{CQ} and V_{CEQ}.
b   Repeat a for $h_{FE} = 20$.
c   Repeat a for $h_{FE} = 125$.

7-48   The Darlington-pair configuration is biased as shown in Fig. 7-74. The transistors $Q1$ and $Q2$ are identical and are characterized by $h_{FE} = 99$ and $V_{BE} = 0.7$. Determine I_{CQ} and V_{CEQ} for each transistor.

7-49   An n-channel JFET whose characteristics are given in Fig. 7-34 is employed in the circuit of Fig. 7-33. The element values are $V_{DD} = 18$ V, $R_D = 2$ kΩ and $V_{GG} = 2$ V.
a   Determine the values of V_{DQ} and I_{DQ} when $v_g = 0$.
b   For $v_g(t) = 1.0 \sin \omega t$, sketch $v_D(t)$ and $i_D(t)$ for one cycle.

7-50   An n-channel JFET whose transfer characteristic is depicted in Fig. 7-35 is biased using the circuit shown in Fig. 7-37. The circuit-element values are $R_S = 0.5$ kΩ, $R_D = 2.5$ kΩ, $R_G = 200$ kΩ, and $V_{DD} = 18$ V.
a   Find V_{DSQ} and I_{DQ}.
b   Repeat a if R_S changes to 0.6 kΩ.

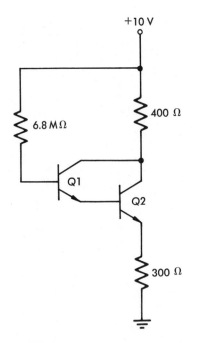

FIGURE 7-74   Circuit for Prob. 7-48.

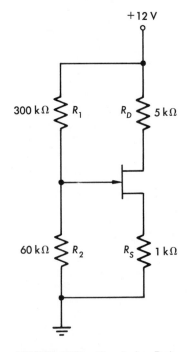

FIGURE 7-75   Circuit for Prob. 7-54.

7-51 The JFET whose characteristics are given in Fig. 7-35 is to be biased at $V_{GS} = -1$ V and $V_{DS} = 5$ V by means of the circuit in Fig. 7-37. Determine R_S and V_{DD} if
a $R_D = 2$ kΩ
b $R_D = 5$ kΩ
The value of $R_G = 500$ kΩ.

7-52 A p-channel JFET is characterized by $V_{PO} = 4$ V and $I_{DSS} = -5$ mA. Using a p-channel version of the circuit in Fig. 7-37, determine the element values necessary to establish $I_{DQ} = -2$ mA and $V_{DSQ} = -4$ V when $V_{DD} = -12$ V.

7-53 **a** Measurements made on an n-channel device indicate that when $V_{GS} = -1$, $I_D = 4$ mA; and when $V_{GS} = -0.5$ V, $I_D = 6.25$ mA. Determine the values of V_{PO} and I_{DSS}.
b The device in a is employed in the circuit of Fig. 7-37 with $V_{DD} = 15$. Determine R_D and R_S so that $I_{DQ} = 4$ mA and $V_{DS} = 4$ V.

7-54 The JFET whose transfer characteristic is shown in Fig. 7-35 is biased by means of the circuit shown in Fig. 7-75. Find the quiescent values V_{DSQ} and I_{DQ}.

7-55 Repeat Prob. 7-54 if the supply voltage is changed to 18 V.

7-56 Describe the process and sketch a cross section of the chip needed to fabricate three npn transistors.

7-57 A monolithic resistor is fabricated from a 0.05-Ω-m n-type semiconductor whose thickness is 10 μm.
a How many ohms per square does this sample contain?
b If the ratio of length to width of the resistor is 10, what is the resistance value?

7-58 The resistivity of the semiconductor material in Prob. 7-57 is changed to 10^{-3} Ω-m. Find the length to width ratio needed for a 10-kΩ resistance.

7-59 Repeat Prob. 7-56 if three p-channel enhancement-type MOSFETs are to be fabricated.

EMENTARY AMPLIFIER STAGES □ **ELEMENTARY**
AMPLIFIER STAGES □ ELEMENTARY **AMPLIFIER**
!ER STAGES □ ELEMENTARY AMPLIFIER **STAGES**
ELEMENTARY AMPLIFIER STAGES □ ELEMENTAF
AMPLIFIER STAGES □ ELEMENTARY AMPLIFIER !
STAGES □ ELEMENTARY AMPLIFIER STAGES □ E
ELEMENTARY AMPLIFIER STAGES □ ELEMENTAF
AMPLIFIER STAGES □ ELEMENTARY AMPLIFIER !
STAGES □ ELEMENTARY AMPLIFIER STAGES □ E
ELEMENTARY AMPLIFIER STAGES □ ELEMENTARY AMPLIFIER STAG
AMPLIFIER STAGES □ ELEMENTARY AMPLIFIER STAGES □ ELEMEN
STAGES □ ELEMENTARY AMPLIFIER STAGES □ ELEMENTARY AMPL
ELEMENTARY AMPLIFIER STAGES □ ELEMENTARY AMPLIFIER STAG
AMPLIFIER STAGES □ ELEMENTARY AMPLIFIER STAGES □ ELEMEN
STAGES □ ELEMENTARY AMPLIFIER STAGES □ ELEMENTARY AMPL
ELEMENTARY AMPLIFIER STAGES □ ELEMENTARY AMPLIFIER STAG
AMPLIFIER STAGES □ ELEMENTARY AMPLIFIER STAGES □ ELEMEN
STAGES □ ELEMENTARY AMPLIFIER STAGES □ ELEMENTARY AMPL
ELEMENTARY AMPLIFIER STAGES □ ELEMENTARY AMPLIFIER STAG
AMPLIFIER STAGES □ ELEMENTARY AMPLIFIER STAGES □ ELEMEN'
STAGES □ ELEMENTARY AMPLIFIER STAGES □ ELEMENTARY AMPL
ELEMENTARY AMPLIFIER STAGES □ ELEMENTARY AMPLIFIER STAG
AMPLIFIER STAGES □ ELEMENTARY AMPLIFIER STAGES □ ELEMEN'
STAGES □ ELEMENTARY AMPLIFIER STAGES □ ELEMENTARY AMPL
ELEMENTARY AMPLIFIER STAGES □ ELEMENTARY AMPLIFIER STAG
AMPLIFIER STAGES □ ELEMENTARY AMPLIFIER STAGES □ ELEMEN'
STAGES □ ELEMENTARY AMPLIFIER STAGES □ ELEMENTARY AMPLI
ELEMENTARY AMPLIFIER STAGES □ ELEMENTARY AMPLIFIER STAG
AMPLIFIER STAGES □ ELEMENTARY AMPLIFIER STAGES □ ELEMEN'
STAGES.□ ELEMENTARY AMPLIFIER STAGES □ ELEMENTARY AMPLI
ELEMENTARY AMPLIFIER STAGES □ ELEMENTARY AMPLIFIER STAG
AMPLIFIER STAGES □ ELEMENTARY AMPLIFIER STAGES □ ELEMEN'
STAGES □ ELEMENTARY AMPLIFIER STAGES □ ELEMENTARY AMPLI
ELEMENTARY AMPLIFIER STAGES □ ELEMENTARY AMPLIFIER STAG
AMPLIFIER STAGES □ ELEMENTARY AMPLIFIER STAGES □ ELEMEN'
STAGES □ ELEMENTARY AMPLIFIER STAGES □ ELEMENTARY AMPLI
ELEMENTARY AMPLIFIER STAGES □ ELEMENTARY AMPLIFIER STAG
AMPLIFIER STAGES □ ELEMENTARY AMPLIFIER STAGES □ ELEMEN'
STAGES □ ELEMENTARY AMPLIFIER STAGES □ ELEMENTARY AMPLI
ELEMENTARY AMPLIFIER STAGES □ ELEMENTARY AMPLIFIER STAG
AMPLIFIER STAGES □ ELEMENTARY AMPLIFIER STAGES □ ELEMEN'
STAGES □ ELEMENTARY AMPLIFIER STAGES □ ELEMENTARY AMPLI

8

he physical operation, volt-ampere characteristics, and fabrication of semiconductor devices were the subject of Chap. 7. This chapter describes the behavior of these devices as circuit elements used in processing electronic signals. The concept of circuit models called equivalent circuits is introduced as a useful method of characterizing the ac, or dynamic, performance of semiconductor devices. Because the ac signals to be processed often have magnitudes smaller than those of the quiescent levels, only small-signal, linear equivalent circuits are developed. These models are then used to describe the dynamic response of elementary amplifier stages. The characteristics of prime importance are gain, frequency response, and impedance levels of the various circuits. Each of the stages selected for investigation can be considered single-stage circuits as either they contain only one elecronic device or their use is similar to that of a single-stage circuit.

The signal-processing capabilities of devices and their associated circuit components depend on the quiescent levels established. Consequently, practical biasing techniques are discussed.

The use of the operational amplifier is introduced in the first section of this chapter. In doing so, we focus attention on the relevant characteristics and analytic techniques used to describe the amplifier stages subsequently discussed.

8-1 BASIC OPERATIONAL AMPLIFIER STAGES

The basic integrated circuit used in the amplification of electronic signals is the *operational amplifier,* colloquially referred to as an *OP-AMP.* The OP-AMP is composed of a number of transistor stages on a single chip and provides the characteristics of a voltage-controlled voltage source. The circuit symbol for the operational amplifier is shown in Fig. 8-1a; its circuit model is shown in Fig. 8-1b. It is assumed that the input to the OP-AMP is open-circuited so that no current enters the OP-AMP. Because the output v_o depends on the difference between two input signals v_1 and v_2, the OP-AMP is an extremely valuable component in measuring small differences such as might arise in the use of strain gauges and thermocouples. Often, in electronic signal processing applications, only one input is applied.

In commercially available operational amplifiers, the value of A, called the *open-loop gain* is large, usually 10^5 or greater. However, the range of frequencies for which this gain is achieved is limited. The asymptotic Bode diagram (see Sec. 6-3) shown in Fig. 8-2 displays the frequency characteristic for the open-loop gain. As indicated in Fig. 8-2, the open-loop gain is constant at a value A_o for frequencies below f_h. For frequencies beyond f_h, the open-loop gain decreases. The frequency f_h is referred to as the open-loop bandwidth as it separates the constant region from the high-frequency band

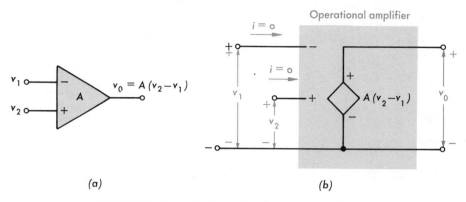

FIGURE 8-1 Operational amplifier (a) symbol and (b) circuit representation.

for which the open-loop gain decreases. (The frequency f_h is also called the half-power frequency as the magnitude of A is 3 dB below its low-frequency value A_o.) The frequency f_{GB}, at which $|A|_{dB} = 0$ or $|A| = 1$, is the *gain-bandwidth product* whose value is $A_o f_h$, which can be determined in the manner illustrated in Example 5-13c. Typical values of f_h and f_{GB} are 10 Hz and 1 MHz, respectively. To overcome this frequency limitation, the circuits shown in Figs. 8-3a and 8-4a are frequently used OP-AMP stages.

The circuit of Fig. 8-3b is the representation of the circuit in Fig. 8-3a in which the model of Fig. 8-1b is substituted for the operational amplifier. The symbol $\perp$ is called *ground* and designates the common or reference node in the circuit. Ground is considered to be at zero potential in much the same fashion as the earth is used as a stationary reference frame in many mechanics problems. At frequencies for which the open-loop gain can be considered constant, the output voltage is

$$v_o = A_o(v_2 - v_1) \tag{8-1}$$

As v_2 is grounded, $v_2 = 0$ so that

$$v_o = -A_o v_1 \tag{8-2}$$

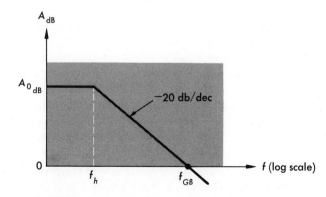

FIGURE 8-2 Asymptotic Bode diagram for open-loop gain.

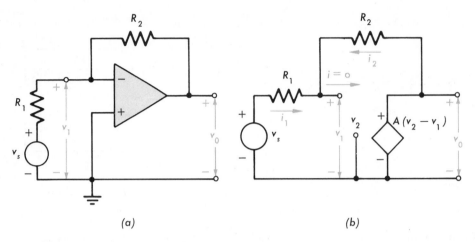

FIGURE 8-3 (a) Basic operational amplifier stage and (b) its circuit model.

The currents i_1 and i_2 are expressible as

$$i_1 = \frac{v_s - v_1}{R_1}$$ (8-3)

and

$$i_2 = \frac{v_o - v_1}{R_2}$$ (8-4)

The KCL expression at node v_1 requires $i_1 = -i_2$ so that a combination of Eqs. (8-3) and (8-4) with a rearrangement of terms gives

$$v_1 = \frac{R_2}{R_1 + R_2} v_s - \frac{R_1}{R_1 + R_2} v_o$$ (8-5)

Substituting Eq. (8-5) into (8-2) and forming the ratio v_o/v_s gives

$$\frac{v_o}{v_s} = G_o = \frac{-A_o R_2}{R_2 + R_1(1 + A_o)}$$ (8-6)

As $A_o \gg 1$ so that $R_1(1 + A_o)$ can be considered much greater than R_2, Eq. (8-6) becomes

$$G_o = -\frac{R_2}{R_1}$$ (8-7)

The quantity $G_o = v_o/v_s$ is called the *voltage gain* or amplification of the stage. The negative sign in Eq. (8-7) indicates a phase reversal; that is, a positive value of v_s produces a negative value of v_o. By making R_2 greater than R_1, however, the magnitude of v_o is larger than v_s, thus achieving voltage amplification. As a consequence of the phase reversal of the output relative to the input, the circuit shown in Fig. 8-3a is called an *inverting OP-AMP stage*, and the input v_1 is called the *inverting input*.

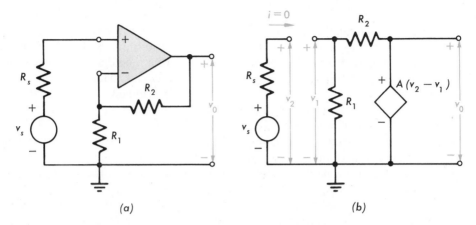

FIGURE 8-4 Operational amplifier stage: (a) circuit diagram, and (b) equivalent representation.

The circuit in Fig. 8-4b is an equivalent representation of the stage depicted in Fig. 8-4a. The output voltage v_o is given by Eq. (8-1). To determine the voltage gain v_o/v_s, v_1 and v_2 must be computed. In Fig. 8-4b, as $i = 0$, v_2 is simply v_s. The voltage v_1 is the output of the voltage divider formed by R_1, R_2, and v_o, and is

$$v_1 = \frac{R_1}{R_1 + R_2} v_o \tag{8-8}$$

Using $v_2 = v_s$, combining Eq. (8-8) with Eq. (8-1) and forming the ratio v_o/v_s yields

$$\frac{v_o}{v_s} = G_o = \frac{A_o(R_1 + R_2)}{R_1(1 + A_o) + R_2} \tag{8-9}$$

For $A_o \gg 1$ so that $R_1(1 + A_o) \gg R_2$, the gain expression in Eq. (8-9) becomes

$$G_o = \frac{R_1 + R_2}{R_1} = 1 + \frac{R_2}{R_1} \tag{8-10}$$

In Eq. (8-10) G_o is positive and there is no phase reversal between output and input. Circuits of the form shown in Fig. 8-4a are known as *noninverting OP-AMP stages*, and the input v_2 is referred to as the *noninverting input*.

The results in Eqs. (8-7) and (8-10) can also be obtained by letting A_o become infinite and evaluating the resultant limits. Because the open-loop gain has such a large value, it is often convenient to assume it is infinite. For this condition the representation illustrated in Fig. 8-1b denotes an ideal operational amplifier. A consequence of an infinite value of A_o is that for the output voltage v_o to remain finite, $v_2 - v_1$ must be zero. For the inverting stage in Fig. 8-3a, since v_2 is zero, the input v_1 must be zero. This situation is often referred to as the input being at *virtual ground*. The implication of an infinite open-loop gain in the circuit of Fig. 8-4a is that $v_1 = v_s$ for $v_2 - v_1$ to be zero.

The expressions in Eqs. (8-7) and (8-10) also indicate that the voltage gain is independent of the operational amplifier parameters, thus making the performance of these stages virtually insensitive to manufacturing variations. Furthermore, the gain of each stage is determined by the ratio R_2/R_1. Fabrication permits resistor ratios to be controlled more precisely than is possible for individual resistances.

Both inverting and noninverting OP-AMP stages are examples of feedback amplifiers of the type described in Chap. 9. In both circuits the resistors R_1 and R_2 form a path from output to input. As a result, the input signal to the operational amplifier is affected by the output voltage. Often, feedback terminology is used to describe the voltage gain G_o of the circuits in Figs. 8-3a and 8-4a; consequently, G_o is referred to as the *closed-loop gain*. The term closed-loop signifies circuit performance when a path exists from output to input; whereas open-loop implies the inherent characteristics of the OP-AMP.

A second property of feedback amplifiers, useful in developing the frequency limitations in operational amplifier stages, is stated as follows: For amplifiers whose frequency response characteristics are shown in Fig. 8-2, the open-loop and closed-loop gain-bandwidth products are equal. This property permits the identification of the closed-loop bandwidth from the expression

$$A_o f_h = |G_o| f_H \qquad \text{or} \qquad A_o \omega_h = |G_o| \omega_H \tag{8-11}$$

where $A_o f_h$ is the open-loop gain-bandwidth product, f_H is the closed-loop bandwidth and $\omega = 2\pi f$. Thus, for an operational amplifier whose gain-bandwidth product is 1 MHz ($A_o = 10^5$, $f_h = 10$ Hz), a stage with a closed-loop gain of 10 has a bandwidth $f_H = 100$ kHz as illustrated in the

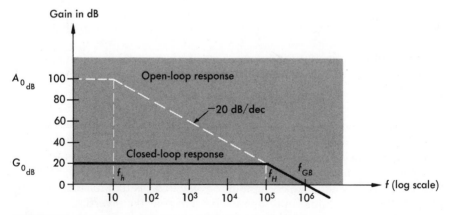

FIGURE 8-5 Frequency response of operational amplifier stage.

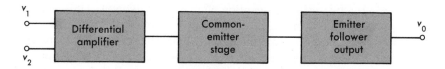

FIGURE 8-6 Block diagram of typical operational amplifier.

asymptotic Bode diagram of Fig. 8-5. The general equation for the characteristic displayed in Fig. 8-5 is

$$G_H(s) = \frac{G_o}{1 + s/\omega_H} \tag{8-12}$$

The discussion just concluded describes the signal-processing characteristics of operational-amplifier stages. To achieve voltage gain, and consequently power gain, the OP-AMP must be biased by a dc source. Terminals at which the dc bias is to be connected are provided on the OP-AMP package with the actual biasing networks internally connected. The manufacturer specifies the permissible range of supply-voltage values and the corresponding OP-AMP characteristics.

Practical OP-AMPs are composed of several amplifier stages, the block diagram of Fig. 8-6 illustrating a typical structure. As indicated in Fig. 8-6, the three building blocks are the input differential amplifier, the common-emitter stage and the emitter-follower output stage. The purpose of the remaining sections of this chapter is to describe the performance characteristics of these and related circuits. As is the case for the OP-AMP stages, the gain and frequency response of these circuits is developed. In addition, appropriate biasing techniques are described.

8-2 INTEGRATED CIRCUIT BIASING

One principal function of semiconductor devices is to physically realize the behavior of controlled sources, particularly as amplifying devices. Ideally, the output of an amplifier should, at a higher level of energy, faithfully reproduce the input signal. As devices themselves have inherent limitations, the circuits in which they are employed cannot exhibit ideal behavior. One set of limitations arises when the range of amplitudes of the input that can be processed effectively (this is referred to as the dynamic range) is considered. Distortion, the degree to which the output does not resemble the input, is one such factor (see Fig. 7-17b). Important among the deviations from the ideal are the limitations on the amplification, or gain, that can be achieved and the frequency range over which amplification exists. These are described later in this chapter by use of equivalent circuits. The parameter values of the equivalent circuits and the dynamic range both depend on the quiescent levels established in the device. In turn, desired dynamic performance is a factor in the selection of an operating point. Thus, the devices must be biased appropriately for signal processing to be accomplished.

Manufacturing tolerances on h_{FE} and the range of temperatures to which the devices are subjected each produce changes in the quiescent levels. While both are important, variations in h_{FE} generally have a more significant effect on biasing; they are treated in this section. The unit-to-unit variation in h_{FE} may be as much as 400 percent for a given transistor type. This wide range of values in h_{FE} does not represent poor manufacturing technique. The manufacturing process controls α_N of the transistor, not $h_{FE} = \alpha_N/(1 - \alpha_N)$. It is the quantity α_N that is related to the structure and properties of the materials employed in fabrication. For a transistor having an h_{FE} range of 50 to 200, the corresponding range of α_N is 0.980 to 0.995, a change of only 1.5 percent.

Two sets of collector characteristics, solid for $h_{FE} = 50$ and dashed for $h_{FE} = 125$ are shown in Fig. 8-7. The effect of h_{FE} on the location of the operating point Q for the circuit of Fig. 8-8 is evident from these curves. For $h_{FE} = 50$, the operating point Q_1 is nearer cutoff than saturation, while the point Q_2 for $h_{FE} = 125$ shows the transistor near saturation. The effect of the shift in operating point on signal processing is as shown in Fig. 8-9 in which the ac collector current i_C for each case is displayed for $i_b(t) = 30 \times 10^{-6}$

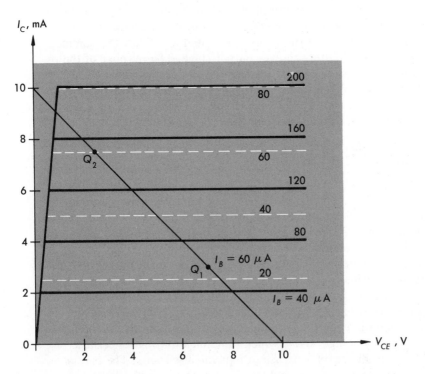

FIGURE 8-7   Collector characteristics showing shift in operating point with changes in h_{fe}.

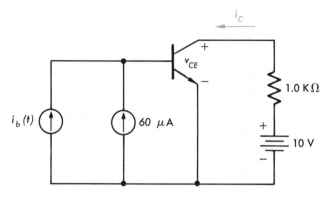

FIGURE 8-8 Transistor stage with constant base-current bias.

sin ωt. When $h_{FE} = 50$, $i_c(t)$ is sinusoidal and virtually distortion-free; whereas, for $h_{FE} = 125$, the collector current is greatly distorted.

The evidence in Figs. 8-7 and 8-9 indicates that for proper circuit performance the variation in the operating point must be controlled; that is, the biasing conditions must be stabilized. To achieve bias stability is to maintain the operating point within some small region over the specified range of h_{FE} values. The import of the previous sentence is that effective bias control occurs when the collector current and, consequently, the emitter current remain essentially constant over the variation in h_{FE}.

The circuit in Fig. 8-10a is typical of the biasing arrangements used in integrated circuits. This and similar circuits are called *current sources* or *current mirrors*. The configuration in Fig. 8-10a is designed to maintain the current I_{C1} at a constant value. The terminal marked $+ V_{CC}$ is used to indicate that the terminal is connected to the positive terminal of the power supply. It is implied that the negative terminal is connected to ground which serves as

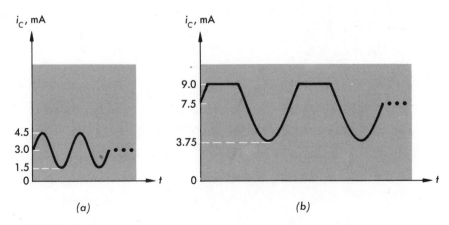

FIGURE 8-9 Effect of changes in h_{FE} on signal: (a) distortion-free and (b) highly distorted.

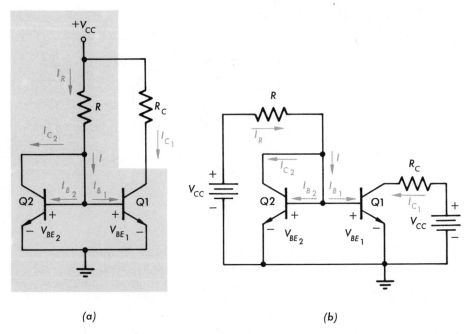

FIGURE 8-10 (a) Current-source biasing and (b) equivalent representation.

the reference node. This notation is standard in the schematic representation of electronic circuits; the equivalent representation is shown in Fig. 8-10b.

The effectiveness of the current-source bias circuit in Fig. 8-10 is predicated on the features of integrated circuit technology. In particular, $Q1$ and $Q2$ are identical transistors and R can be kept within the limits imposed by fabrication. The identical characteristics of $Q1$ and $Q2$ imply that if a transistor having a nominal value of $h_{FE} = 100$ is constructed and h_{FE} of $Q1$ is measured as 150, then for $Q2$, $h_{FE} = 150$. Because $Q1$ and $Q2$ are identical, then, in Fig. 8-10,

$$I_{B1} = I_{B2} = I_B$$

$$I_{C1} = I_{C2} = I_C \tag{8-13}$$

The KVL expression for the loop containing both emitter-base junctions requires that $V_{BE1} = V_{BE2} = V_{BE}$.

The current I_R, called the *reference current*, is determined from the KVL relation for the loop containing V_{CC}, R, and V_{BE}; and is

$$-V_{CC} + I_R R + V_{BE} = 0$$

from which

$$I_R = \frac{V_{CC} - V_{BE}}{R} \tag{8-14}$$

At the junction of the two base terminals, KCL requires

$$I = 2I_B$$

Similarly, the KCL expression involving I_R is

$$-I_R + I_C + I = 0 \tag{8-15}$$

which, after substitution for I and recalling $I_C = h_{FE}I_B$, gives

$$I_R = \frac{h_{FE} + 2}{h_{FE}} I_C = (h_{FE} + 2)I_B \tag{8-16}$$

Combination of Eqs. (8-14) and (8-16) yields

$$I_C = \frac{h_{FE}}{h_{FE} + 2} I_R = \frac{h_{FE}}{h_{FE} + 2} \cdot \frac{V_{CC} - V_{BE}}{R} \tag{8-17}$$

The result given in Eq. (8-17) indicates that I_C remains essentially constant over a wide range of h_{FE} values, thus confirming the constant current nature of the bias network. For $h_{FE} \gg 1$, $h_{FE}/(h_{FE} + 2)$ is virtually unity. As an example, a 3 percent variation in I_C occurs for a range of h_{FE} values from 50 to 200.

Example 8-1

a For the circuit of Fig. 8-10, determine the value of R so that I_C is 1.0 mA for transistors having $h_{FE} = 100$ and $V_{BE} = 0.6$ V. Supply voltage is 10 V.

b Using the result in Part a, determine the percentage change in I_C if

$$h_{FE} = 200$$

c Repeat a for $I_C = 0.1$ mA.

SOLUTION **a** From Eq. (8-17),

$$10^{-3} = \frac{100}{100 + 2} \cdot \frac{10 - 0.6}{R}$$

so that

$$R = 9.22 \text{ k}\Omega$$

b $$I_C = \frac{200}{200 + 2} \cdot \frac{10 - 0.6}{9.22 \times 10^3}$$

$$I_C = 1.009 \text{ mA}$$

Percentage change $= \dfrac{1.009 - 1.0}{1.0} \times 100 = 0.9\%$

c Again, from Eq. (8-17),

$$10^{-4} = \frac{100}{100 + 2} \times \frac{10 - 0.6}{R}$$

which gives

$$R = 92.2 \text{ k}\Omega$$

The results in Example 8-1a and b indicate that the circuit of Fig. 8-10 provides good bias stability with element values well within fabrication capabilities. However, this is not the case in Example 8-1c where fabrication of $R = 92.2 \text{ k}\Omega$ would be extremely difficult and so expensive as to preclude utilization. To overcome this limitation, the asymmetric current source shown in Fig. 8-11 is frequently used to achieve low-value current sources. In the circuit of Fig. 8-11, both transistors are identical. The plus sign at the collector of $Q1$ is used to indicate that the potential at the collector is maintained positive with respect to the base and ensures that $Q1$ is biased in the active region. The function of the resistance R_E is to cause V_{BE2} and V_{BE1} to differ. In this arrangement V_{BE1} is less than V_{BE2} and, consequently, I_{C1} is smaller than I_{C2}. In effect, $Q2$, V_{CC}, and R establish the reference current I_R, and the value of R_E determines how much less than I_R is the current-source current I_{C1}.

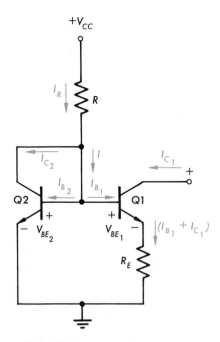

FIGURE 8-11 Current source for small values of current.

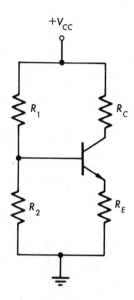

The KVL equation for the emitter-base loop is

$$V_{BE2} = V_{BE1} + (I_{B1} + I_{C1})R_E$$

or

$$V_{BE2} - V_{BE1} = \Delta V_{BE} = (I_{B1} + I_{C1})R_E \qquad (8\text{-}18)$$

As indicated in Eq. (7-8b), the collector current of a transistor depends heavily on the base-emitter voltage. For identical *npn* transistors, I_{C1} and I_{C2} are

$$I_{C1} = \alpha_N I_{E0}\epsilon^{V_{BE1}/V_T} \qquad (8\text{-}19a)$$

$$I_{C2} = \alpha_N I_{E0}\epsilon^{V_{BE2}/V_T} \qquad (8\text{-}19b)$$

from which

$$\frac{I_{C2}}{I_{C1}} = \epsilon^{(V_{BE2}-V_{BE1})/V_T} \qquad (8\text{-}20)$$

Equation (8-20) can be rewritten by taking the natural logarithm of both sides as

$$V_{BE2} - V_{BE1} = \Delta V_{BE} = V_T ln \frac{I_{C2}}{I_{C1}} \qquad (8\text{-}21)$$

Equating ΔV_{BE} in Eqs. (8-18) and (8-21) results in

$$R_E = \frac{V_T}{I_{C1}\left(1 + \dfrac{1}{h_{FE}}\right)} \, ln \, \frac{I_{C2}}{I_{C1}} \qquad (8\text{-}22)$$

The reference current I_R as given in Eq. (8-14) as the KVL expression for the loop containing V_{CC}, $Q2$, and R is the same in both Figs. 8-10 and 8-11. From the KCL expression

$$I_R = I_{C2} + I_{B2} + I_{B1} \tag{8-23}$$

which is rewritten as

$$I_R = I_{C2}\left(1 + \frac{1}{h_{FE}}\right) + \frac{I_{C1}}{h_{FE}} \tag{8-24}$$

As we wish I_{C1} to be smaller than I_{C2}, the term I_{C1}/h_{FE} can be neglected in Eq. (8-23). After combination with Eq. (8-14), Eq. (8-24) yields

$$I_{C2} = \frac{h_{FE}}{h_{FE} + 1} I_R = \frac{h_{FE}}{h_{FE} + 1}\frac{V_{CC} - V_{BE2}}{R} \tag{8-25}$$

The value of I_{C2} is determined from Eq. (8-25) and, as I_{C1} is the specified current value, the resistance R_E is computed from Eq. (8-22). Example 8-2 illustrates the design.

Example 8-2

Determine R_E in the circuit of Fig. 8-11 for $V_{CC} = 10$ V, $R = 9.22$ kΩ, $V_{BE2} = 0.6$ V, $h_{FE} = 100$, and the desired value of $I_{C1} = 0.1$ mA.

SOLUTION   From Eq. (8-25)

$$I_{C2} = \frac{100}{100 + 1}\frac{10 - 0.6}{9,220} = 1.009 \text{ mA}$$

By use of Eq. (8-22)

$$R_E = \frac{0.025}{10^{-4}(1 + \frac{1}{100})} \ln\frac{1.009 \times 10^{-3}}{10^{-4}} = 572 \text{ Ω}$$

Note that although the circuit of Fig. 8-11 uses two resistors and the circuit of Fig. 8-10 contains only one resistor, the total resistance obtained in Example 8-2 is $9,220 + 572 = 9,792$ Ω, while in Example 8-1c the total resistance is $92,200$ Ω. The smaller total resistance occupies far less chip area, and both resistors are easily fabricated. Consequently, the circuit of Fig. 8-11 is of more practical value for use as low-value sources.

As MOS transistors are used predominantly in digital circuits as switching elements, bias considerations are not treated here but are introduced in Chap. 11. Bias stabilization techniques for JFETs are based on the analysis in Sec. 7-10 and are similar to the techniques developed for discrete transistors in the next section. In those ICs which employ FETs, it is often in conjunction with bipolar devices and the current-source biasing techniques are adapted to this situation.

8-3 DISCRETE COMPONENT TRANSISTOR BIASING

Bias stabilization of transistor stages constructed from discrete components is as important a consideration here as it is in integrated circuits. However, as matched-pairs of transistors are costly and since no restriction exists on resistance values, the circuit shown in Fig. 8-12 is quite generally employed. The use of the emitter resistance R_E in conjunction with the base-bias resistors R_1 and R_2 provides effective and practical control of the operating point.

The dc representation of the circuit of Fig. 8-12 is shown in Fig. 8-13a. Replacement of the portion of the circuit to the left of terminals b and g by its Thévenin equivalent results in the circuit of Fig. 8-13b. The values of V_{BB} and R_B are the Thévenin voltage and resistance of the base-bias network and are

$$V_{BB} = \frac{R_2}{R_1 + R_2} V_{CC} \tag{8-26}$$

and

$$R_B = \frac{R_1 R_2}{R_1 + R_2} \tag{8-27}$$

As seen in Sec. 7-8, the operating point can be determined from the voltage-law expressions for the base and collector loops, which are

$$V_{CC} = I_C R_C + V_{CE} + I_C \left(1 + \frac{1}{h_{FE}}\right) R_E \tag{8-28}$$

$$V_{BB} = I_B R_B + V_{BE} + I_B(1 + h_{FE})R_E \tag{8-29}$$

Ideally, it is desired to maintain a fixed collector current over the range of values of h_{FE} encountered for a given transistor. For $h_{FE} \gg 1$, it can be seen in Eq. (8-28) that if I_C is maintained at a constant value, V_{CE} remains constant also. Thus, the operating point is fixed and independent of the variations in h_{FE}. An increase in h_{FE} results in a decrease in I_B, as indicated from examination of Eq. (8-29). Since $I_C = h_{FE}I_B$, the decrease in I_B tends to offset the increase in h_{FE} and helps to maintain I_C at a constant value. A degree of stabilization of the operating point is therefore achieved.

The degree to which the operating point is stabilized depends on the value of R_E used. Solving Eq. (8-29) for I_B, and assuming $h_{FE} \gg 1$, result in

$$I_B = \frac{V_{BB} - V_{BE}}{R_B + h_{FE}R_E} \tag{8-30}$$

As R_E is made larger, the value of I_B becomes more nearly inversely proportional to h_{FE}. For those cases where $h_{FE}R_E \gg R_B$,

$$I_B = \frac{V_{BB} - V_{BE}}{h_{FE}R_E}$$

and the collector current $I_C = h_{FE}I_B$ remains essentially constant.

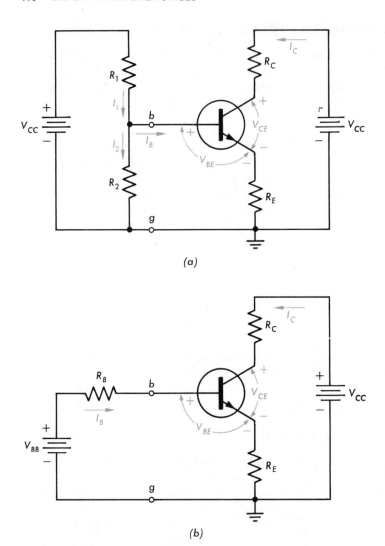

(a)

(b)

FIGURE 8-13 (*a*) Dc equivalent of circuit in Fig. 8-12; (*b*) circuit in *a* with base-bias network replaced by its Thévenin equivalent.

A rule of thumb useful in design is to make $h_{FE}R_E = 10R_B$. For this condition, a 50 to 200 variation in h_{FE} produces a relative change in collector current of nearly 7.5 percent. Selecting R_E so that $h_{FE}R_E > 10R_B$ further reduces the variation in I_C. However, limits on the size of R_E exist, based in part on the a priori specification of V_{CC} and the collector load resistance.

Example 8-3

Element values for the Fig. 8-12 circuit are given as $R_1 = 30$ kΩ, $R_2 = 7.5$ kΩ, $R_C = 3.0$ kΩ, and $R_E = 1.2$ kΩ. The supply voltage is 10 V and $V_{BE} =$

0.6 V. Determine I_{CQ} and V_{CEQ} when

a $h_{FE} = 50$

b $h_{FE} = 150$

SOLUTION **a** From Eqs. (8-26) and (8-27),

$$V_{BB} = \frac{7,500}{30,000 + 7,500} \cdot 10 = 2.0 \text{ V}$$

$$R_B = \frac{30,000 \times 7,500}{30,000 + 7,500} = 6 \text{ k}\Omega$$

The base current is determined from Eq. (8-30) as

$$I_B = \frac{2.0 - 0.6}{6,000 + 50(1,200)} = 21.2 \ \mu\text{A}$$

and

$$I_{CQ} = h_{FE}I_B = 50 \times 21.2 \times 10^{-6} = 1.06 \text{ mA}$$

Substitution of the value of I_{CQ} into Eq. (8-28) results in

$$10 = 1.06 \times 10^{-3} \times 3.0 \times 10^3 + V_{CEQ}$$
$$+ 1.06 \times 10^{-3}(1 + \tfrac{1}{50}) \times 1.2 \times 10^3$$

from which

$$V_{CEQ} = 5.52 \text{ V}$$

b For $h_{FE} = 150$, I_B as given in Eq. (8-30) is

$$I_B = \frac{2.0 - 0.6}{6,000 + 150 \times 1,200} = 7.53 \ \mu\text{A}$$

and

$$I_{CQ} = 150 \times 7.53 \times 10^{-6} = 1.13 \text{ mA}$$

Use of Eq. (8-28) gives

$$10 = 1.13 \times 10^{-3} \times 3 \times 10^3 + V_{CEQ}$$
$$+ 1.13 \times 10^{-3}(1 + \tfrac{1}{150}) \times 1.2 \times 10^3$$

from which

$$V_{CEQ} = 5.24 \text{ V}$$

As is evident from the results in *a* and *b*, the variations in I_{CQ} and V_{CEQ} are small and the bias is stabilized over the range of h_{FE} values encountered.

The circuit analogous to that in Fig. 8-12 and useful for achieving bias stabilization in FETs is depicted in Fig. 8-14a. The dc equivalent of Fig. 8-14a in which the gate-bias network R_1, R_2, and V_{DD} is replaced by its

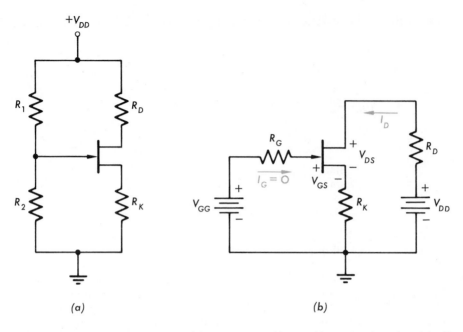

(a)   (b)

FIGURE 8-14 (a) FET bias circuit; (b) circuit in a with gate-bias network replaced by its Thévenin equivalent.

Thévenin equivalent V_{GG} and R_G is shown in Fig. 8-14b. By assuming there is no leakage current in the gate loop ($I_G = 0$), the KVL expression is

$$V_{GG} - I_D R_K = V_{GS} \tag{8-31}$$

Equation (8-31) is the equation for the bias-line which can be plotted on the transfer characteristic as a straight line (see Sec. 7-10). The bias-line of Eq.

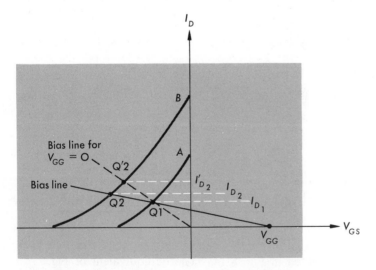

FIGURE 8-15 FET transfer characteristics with bias-lines.

(8-31) is indicated in Fig. 8-15 in which the two transfer characteristics display the unit-to-unit variation encountered for the device. The intersection of the bias-line with the two characteristics shows that the change in I_D is relatively small. The dashed line in Fig. 8-15 is the bias-line, designed to establish the same operating point on characteristic A as that obtained from Eq. (8-31), when $V_{GG} = 0$. Note that the I_D variation for this case is significantly higher and that the slope of the bias-line is correspondingly higher. The higher slope is indicative of the fact that R_K must be decreased to maintain the same operating point when V_{GG} is set to zero. The conclusion drawn is that increased values of R_K improve bias stability. The addition of the gate-bias network, thereby inserting V_{GG} into the gate-source loop, permits larger values of R_K to be utilized without altering quiescent levels.

8-4   JUNCTION-DIODE EQUIVALENT CIRCUITS

The time-varying excitation applied to most electronic circuits contains the information content to be processed. The result of signal processing is generally considered to be the corresponding time-varying component of the response. To relate response and excitation, it is convenient to use an *equivalent circuit,* or *circuit model,* of the devices employed. An equivalent circuit is one which, when appropriately connected, will have the same response to a given excitation as does the actual device. There are two classes of models, referred to as *small-signal,* or incremental, circuits, and *large-signal* equivalent circuits. Small-signal models are used when the magnitude of the excitation is such that the devices can be considered to behave linearly. The fundamental property of linear operation is that superposition applies. Thus, the dc response is proportional solely to the dc excitation and the ac response proportional only to the ac excitation. When superposition is not applicable, nonlinear-device behavior is encountered and large-signal models are appropriate. (Such models are treated in Chap. 11.) These situations are depicted in Fig. 7-17a and b.

   The circuit in Fig. 8-16 is useful in developing the circuit model for a junction diode. The response of this circuit to a sinusoidal excitation is displayed in Fig. 7-17 in which large- and small-signal effects are illustrated. In Fig. 7-17a, the small-signal case, the diode characteristic traversed by the

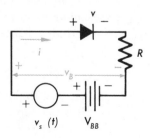

FIGURE 8-16 Diode circuit with ac and dc sources.

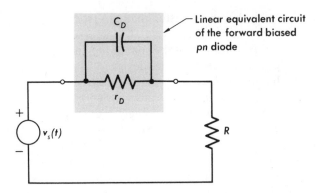

FIGURE 8-17 Small-signal model of junction-diode circuit for forward bias.

signal between Q_1 and Q_2 is easily approximated by a straight line. This is not the case for the large-signal response in Fig. 7-17*b*.

A consequence of small-signal conditions is that the diode can be represented by a linear equivalent circuit, i.e., by standard circuit components such as the resistances, capacitances, and controlled sources discussed in Chaps. 1 to 6. Because small-signal conditions occur so often (so that control is achieved with a small expenditure of energy), linear incremental models are used extensively. For the circuit of Fig. 8-16, with the junction diode used under small-signal conditions, the circuit of Fig. 8-17 can be used to determine the response to $v_s(t)$. The diode model is valid only in the forward-biased direction and $v_s(t)$ is assumed small compared with the bias levels. The resistance r_D, called the *small-signal* or *incremental forward resistance*, relates the change in diode voltage with respect to the change in diode current about a given bias level. Mathematically, this is expressed as

$$r_D = \frac{dv}{di} \qquad (8\text{-}32)$$

evaluated at the quiescent point.

The capacitance element C_D, called the *diffusion capacitance*, is an incremental effect which results from the charge-storage properties of the junction. When the bias level is increased (more heavily forward-biased), a greater number of carriers diffuse across the junction. Thus, more charge is present in the vicinity of the junction, and the change in charge dq is proportional to the change in voltage dv, so that

$$C_D = \frac{dq}{dv} \qquad (8\text{-}33)$$

evaluated at the quiescent point.

It is the capacitor in the model which accounts for the variation in the diode response as the frequency of the excitation is varied. Often it is convenient to divide the small-signal diode behavior into two frequency ranges,

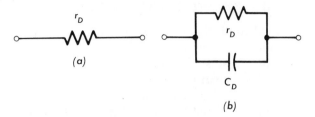

FIGURE 8-18 Junction diode models for (a) low frequency and (b) high frequency.

low and high, to distinguish when C_D affects the response. The circuits in Fig. 8-18a and b are the low- and high-frequency equivalent circuits of the forward-biased diode. Which is used depends on the frequency of the excitation, $v_s(t) = \sqrt{2}\, V_s \sin \omega t$. Values of ω for which the reactance of C_D is much larger than r_D, so that their parallel combination may be approximated by r_D, permit the use of the model of Fig. 8-18a. The high-frequency model is used when values of ω do not permit the reactance of C_D to be neglected when the parallel combination of r_D and C_D is evaluated.

In the reverse-biased condition, because there is negligible diffusion, C_D can be neglected and the model shown in Fig. 8-19 used. The capacitance C_T is the *junction*, or *space-charge*, or *transition* capacitance. The capacitance C_T is also defined as in Eq. (8-33) where the value of dq taken is the change in the charge stored in the depletion region. The value of dv is the change in the junction voltage which produces the change dq. The resistance r_R is the incremental resistance of the reverse-biased diode and is defined as in Eq. (8-32) with the constant bias taken as the quiescent value of direct voltage across the reverse-biased diode. The values of r_R encountered in practice are quite large so that at high frequencies r_R is usually much larger than the reactance of C_T. As such, the equivalent circuit at high frequencies often only includes the capacitance C_T.

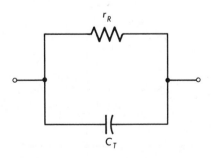

FIGURE 8-19 Small-signal model for a reverse-biased junction diode.

8-5 COMMON-EMITTER EQUIVALENT CIRCUITS

When the voltages and currents may be treated as the superposition of two individual responses, one of which is caused by the dc supply and the other by the ac signal, the transistor is considered to behave as a linear device. This type of response is shown in Figs. 7-28 and 8-9a. To determine the ac response, a small-signal equivalent circuit is used. Linear operation requires the transistor to be biased in its normal mode (the active region) at each instant of time, and requires a reverse-biased collector-base junction and a forward-biased emitter-base junction. The transistor model must therefore reflect the effects of both junctions and their interaction.

The equivalent circuit of Fig. 8-20 is referred to as the *hybrid-π* and is used to represent the transistor in the common-emitter (CE) mode. The controlled source $g_m v_\pi$ reflects the interaction between the forward- and reverse-biased junctions and represents the control exerted by v_π on i_c. The elements r_π and C_π indicate the effect of the forward-biased emitter-base junction. The reverse-biased collector-base junction is indicated by r_μ and C_μ. Both r_x and $r_o = 1/g_o$ reflect two additional effects attributed to carrier flow in the base region.

The first of the base-region factors, called *base-width modulation*, is caused by changes in collector-base voltage and, consequently, changes in the size of the depletion region. The potential barrier at the junction is established by the formation of the depletion region, whose width is related to the junction potential. By changing the collector-base voltage, the barrier potential is increased or decreased, and as a result, the width of the depletion region is varied. The effective base width (the distance between emitter-base and collector-base depletion layers) also varies and is dependent upon collector-base voltage. As a consequence of base-width modulation, the number of carriers which diffuse through the base and are ultimately collected depends on the voltage at the collector-base junction. The value of r_o is in large measure due to base-width modulation.

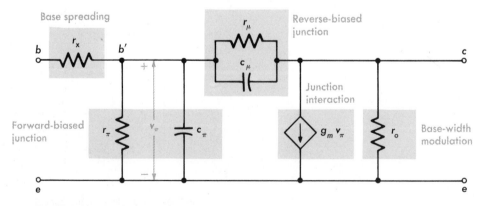

FIGURE 8-20 Hybrid-π equivalent circuit.

The second effect, known as *base spreading,* is attributed to the flow of majority carriers in the base region. As some of the minority carriers injected into the base from the emitter recombine with the majority carriers already present in the base (holes injected from the *p*-type emitter recombine with electrons present in the *n*-type base), additional majority carriers must be injected into the base region (additional electrons must be added). If this action did not take place, the base region would act as a sink for charged particles; this would violate conservation of charge. The element r_x is used to account for base spreading.

The equivalent circuit is used to represent only what occurs within the device. Externally connected elements must be added to the terminals in accordance with the complete circuit diagram. The following examples are used to illustrate the use of the equivalent circuit.

Example 8-4

Draw the small-signal (ac) equivalent circuit for the discrete-component transistor stage shown in Fig. 8-21.

SOLUTION   The equivalent circuit is used to relate only the ac components of current and voltage in the circuit which result from the signal excitation v_s'. The direct voltage and current levels produced by the supply V_{CC} are not included in the representation because the total response is the superposi-

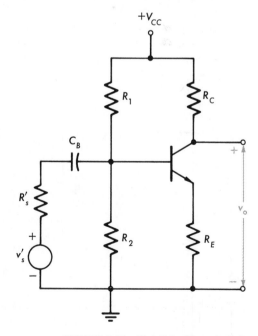

FIGURE 8-21 Circuit for Example 8-4.

tion of the dc and ac components (as indicated in Fig. 7-28 and Table 7-1). The equivalent circuit therefore includes only those elements which influence the values of ac components. The general method involves two steps which are:

1 Replace the transistor by its small-signal equivalent circuit.

2 Connect those components which are external to the transistor and which affect the ac values to the appropriate terminals as indicated in the circuit diagram.

The transistor is replaced by the circuit of Fig. 8-20. All the external circuit elements except V_{CC} influence the ac components. Since V_{CC} is a direct-level source, there is no alternating potential across its terminals. Thus, in the ac model, V_{CC} is represented by an element having a zero voltage drop across it. A zero voltage drop requires that a short circuit be used to replace V_{CC} in the model. To complete the representation:

1 C_B, R'_s, and v'_s are connected in series between base and ground.

2 R_1 is connected between base and ground as V_{CC} is an ac short circuit.

3 R_2 is connected between base and ground.

4 R_E is connected between emitter and ground.

5 R_C is connected between collector and ground as V_{CC} is again represented by a short circuit.

The final small-signal equivalent is shown in Fig. 8-22.

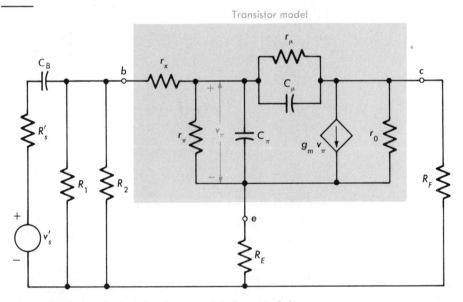

FIGURE 8-22 Equivalent circuit for circuit in Example 8-4.

Example 8-5

Draw the ac equivalent circuit for the circuit of Fig. 8-23.

SOLUTION   Replace the transistor by the circuit of Fig. 8-20; connect the collector to ground. The emitter is connected to ground through R_E; the base circuit is identical to that used in Example 8-4. The resultant circuit is shown in Fig. 8-24. This circuit represents the common-collector model with its various externally connected elements. Often it is convenient to redraw the circuit of Fig. 8-24 as shown in Fig. 8-25, indicating that the collector is the common terminal. In addition, the resistance R_E' is the parallel combination of R_E and the resistance r_o. As typical values of R_E and r_o are 500 and 50,000 Ω, respectively, their parallel combination is most nearly $R_E = 500 \ \Omega$. It is therefore quite common to neglect the effect of r_o in the model.

Note that the transistor model for the *pnp* transistor in Fig. 8-23 is the same as that used for the *npn* transistor in Example 8-4. As small-signal models indicate changes about quiescent levels, these models are identical for both *npn* and *pnp* transistors.

The presence of r_μ and C_μ in the common-emitter hybrid-π model forms a path from the output (collector) to the input (base). Consequently, the transistor no longer behaves as a unilateral device. Because its resistance has a large numerical value, r_μ generally has negligible effect on the response. Therefore, r_μ is almost always omitted from the equivalent circuit and will be in the remainder of this text, except when specifically identified. In frequency ranges where the reactance of C_μ is large and has little effect on the

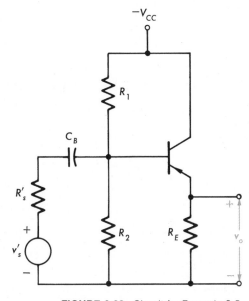

FIGURE 8-23 Circuit for Example 8-5.

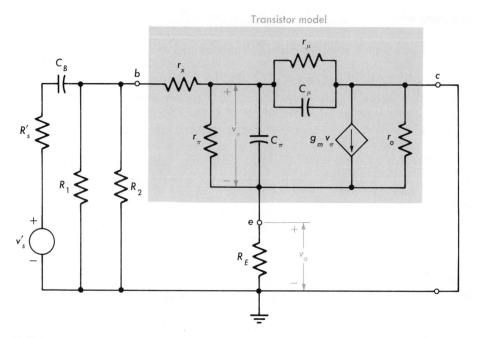

FIGURE 8-24 Equivalent circuit for Example 8-5.

response, C_μ may be treated as an open-circuit. At these frequencies, transistor behavior can be considered unilateral.

At higher frequencies, the reactance of C_μ is decreased requiring its inclusion in the model. However, because the effect of C_μ is more pronounced in the base circuit than in the collector circuit, the high-frequency hybrid-π model can be simplified. In the circuit of Fig. 8-26, the element R_L

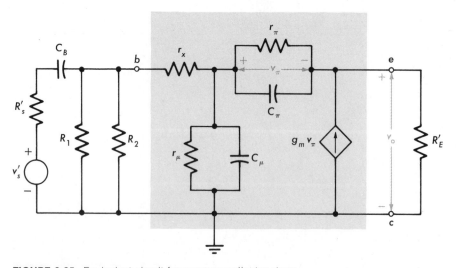

FIGURE 8-25 Equivalent circuit for common-collector stage.

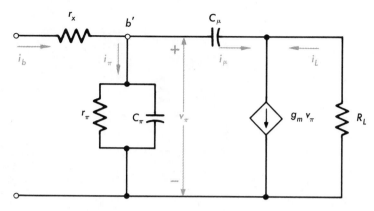

FIGURE 8-26 Hybrid-π circuit with load.

represents the parallel combination of R'_L and r_o. Kirchhoff's current law at node c requires that

$$i_L = g_m v_\pi - i_\mu$$

The current i_μ is usually of the same order of magnitude as the current i_π and cannot be neglected in determining the base current i_b. To account for this component of current, the effect of C_μ is usually reflected into the base circuit and results in a unilateral model. To reflect C_μ to the base circuit, an equivalent admittance Y_R which carries the current I_π is placed between b' and e as indicated in the frequency-domain circuit of Fig. 8-27a.

In Fig. 8-27a, it is evident that

$$I_\mu = Y_R V_\pi \qquad \text{or} \qquad Y_R = \frac{I_\mu}{V_\pi} \tag{8-34}$$

The section of the circuit of Fig. 8-26 to be replaced by the admittance Y_R is shown in Fig. 8-27b; in that circuit,

$$I_\mu = sC_\mu(V_\pi - V_{ce}) \tag{8-35}$$

Also, if I_μ is considered small compared with I_L,

$$V_{ce} = -I_L R_L = -g_m V_\pi R_L \tag{8-36}$$

By combination of Eqs. (8-35) and (8-36),

$$I_\mu = sC_\mu(1 + g_m R_L)V_\pi \tag{8-37}$$

and from Eq. (8-34),

$$Y_R = sC_\mu(1 + g_m R_L) \qquad \text{and} \qquad C_R = C_\mu(1 + g_m R_L) \tag{8-38}$$

For a resistive load, the reflected admittance Y_R is just that of a capacitor whose value is $C_\mu(1 + g_m R_L)$. Thus, at high frequencies, the effect of C_μ is significant and becomes one of the factors which limit the operating fre-

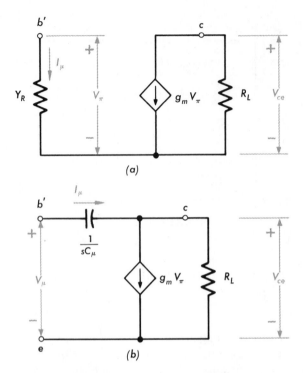

(a)

(b)

FIGURE 8-27 Circuits showing the reflection of C_μ into the input of the transistor. These circuits indicate the portion of the hybrid π between terminals b' and c.

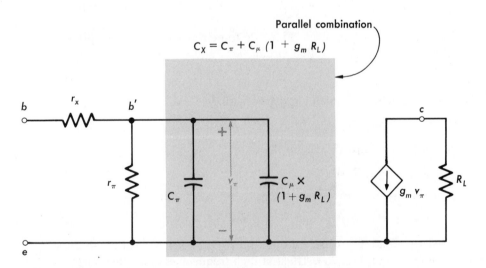

Parallel combination

$$C_X = C_\pi + C_\mu (1 + g_m R_L)$$

FIGURE 8-28 The unilateral hybrid-π equivalent circuit.

quency of the transistor. The *unilateral hybrid* π of Fig. 8-28 is the result of reflecting the effect of C_μ to the input. It will be used almost exclusively to represent the common-emitter configuration in the next three chapters.

8-6 EQUIVALENT CIRCUITS FOR FIELD-EFFECT TRANSISTORS

In practice, JFETs are often employed to act as linear devices. While MOS transistors are usually operated in switching modes, some MOSFET devices have been used as amplifier stages. The models developed in this section apply equally to junction and MOS devices of both n- and p-channel types. When small-signal conditions apply, the ac response is determined by replacing the FET by a linear equivalent circuit.

In the common-source equivalent circuit, depicted in Fig. 8-29a, the voltage-controlled current source $g_m v_{gs}$ is used to represent the control ex-

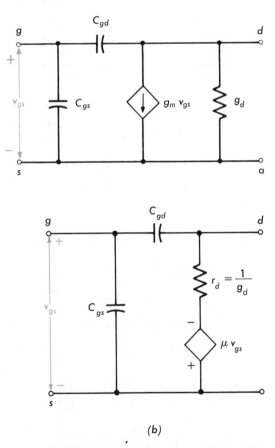

(b)

FIGURE 8-29 (*a*) Equivalent circuit for FET; (*b*) circuit in *a* with current source converted to voltage-source representation.

erted by the gate on the current in the drain-to-source channel. The element g_d is the incremental conductance of the channel measured at constant gate potential. Mathematically, g_m and g_d are defined in a manner similar to that for r_D in the junction diode and are given by

$$g_d = \left.\frac{\partial i_D}{\partial v_{DS}}\right|_{v_{GS}=\text{const}} \qquad g_m = \left.\frac{\partial i_D}{\partial v_{GS}}\right|_{v_{DS}=\text{const}} \tag{8-39}$$

evaluated at the operating point. Partial derivatives must be used in Eq. (8-39) because the drain current is dependent on both gate and drain voltages. The effect of the reverse-biased junction or insulated gate is almost always negligible, so that the gate current can be considered zero. It is often convenient to change the current source to a voltage source by means of a source conversion. The resultant circuit is depicted in Fig. 8-29b. The quantity μ is called the *amplification factor* and is given by

$$\mu = \frac{g_m}{g_d} \tag{8-40}$$

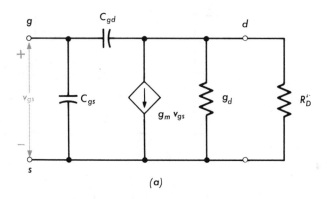

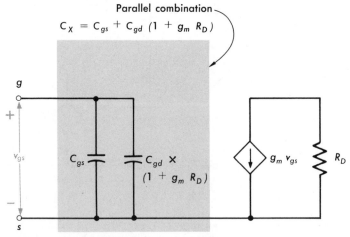

FIGURE 8-30 (a) Field-effect transistor with resistive load; (b) the unilateral representation of the circuit shown in a.

The effect of the capacitance of the reverse-biased diode in JFETs must be included at high frequencies. As both the gate-to-source and drain-to-source voltages can reverse-bias the junction, the capacitive effect is present in both the gate-to-source and gate-to-drain circuits. The capacitors C_{gs} and C_{gd} shown in Fig. 8-29 are used to represent the effect of junction capacitance. These capacitors are also present in MOS transistors and indicate the capacitance that exists between the metal, oxide, and drain and source regions.

The presence of C_{gd} in Fig. 8-29 forms a path from the output (drain) to the input (gate). Consequently, the FET no longer exhibits unilateral behavior. When a load resistance R_D' is connected between drain and source as indicated in Fig. 8-30a, the effect of C_{gd} can be reflected into the gate circuit as shown in Fig. 8-30b. This procedure follows that used to form the unilateral hybrid π described in the previous section. The reflected admittance is a capacitance when the load is resistive, and the value of the total capacitance C_x is

$$C_x = C_{gs} + C_{gd}(1 + g_m R_D) \tag{8-41}$$

where R_D is the parallel combination of R_D' and $1/g_d$.

Example 8-6

The circuit shown in Fig. 8-31 is referred to as the common-drain configuration. Draw the equivalent circuit for this configuration.

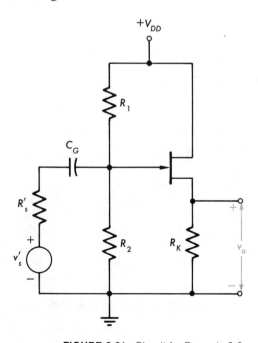

FIGURE 8-31 Circuit for Example 8-6.

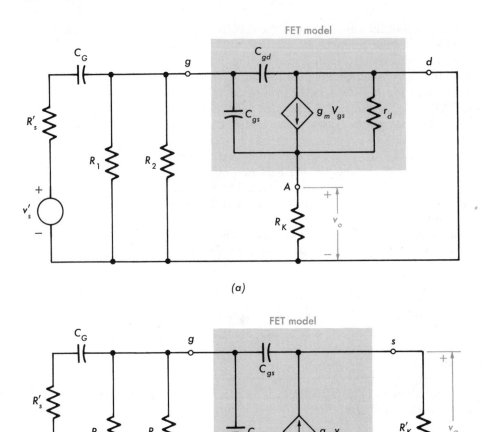

(a)

(b)

FIGURE 8-32   (a) Equivalent circuit for Example 8-6; (b) equivalent circuit for a common-drain amplifier stage.

SOLUTION   Replace FET by the circuit of Fig. 8-29a and connect the drain to ground. The source is connected to ground through R_K and the gate-to-ground circuit contains the parallel branches R_1, R_2 and v_s' in series with R_s' and C_G. Figure 8-32a is the resultant circuit which is usually redrawn for convenience as shown in Fig. 8-32b.

8-7   DETERMINATION OF DEVICE PARAMETERS

Determination of the parameters in the junction-transistor and field-effect-transistor equivalent circuits of Figs. 8-20 and 8-29a by measurement, either directly or indirectly, requires the use of two-port network parameters (Sec. 6-5) and Bode diagrams (Secs. 6-2 and 6-3). The short-circuit admittance (y) and the hybrid (h) parameters are used most commonly. The values of the various y or h parameters are measured easily and can be related to the elements in Figs. 8-20 and 8-29 by the usual methods of circuit theory.

As the transistor may be connected in any one of three possible configurations, an additional subscript representing the common terminal is added to the two-port parameter. For example, h_{fe} is the common-emitter forward current gain (h_{21} in Sec. 6-5).

The following data usually appear on manufacturers' specifications:

h_{ie} the short-circuit input impedance, grounded emitter
h_{fe} the low-frequency common-emitter forward current gain
ω_T the angular frequency at which the magnitude of h_{fe} is unity
C_{ob} the output capacitance, grounded base
C_{case} the capacitance of the leads and transistor package

These parameters are each measured at some specified quiescent level. In IC transistors, C_{case} includes the capacitance between the device and substrate.

At frequencies for which the reactance of C_π and C_μ are negligible so that they can be considered as open-circuits, h_{ie} and h_{fe} are determined as

$$h_{ie} = \left. \frac{v_{be}}{i_b} \right|_{v_{ce}=0} = r_x + r_\pi \tag{8-42}$$

and

$$h_{fe} = \left. \frac{i_c}{i_b} \right|_{v_{ce}=0} = g_m r_\pi = \beta_o \tag{8-43}$$

The symbol β is sometimes used for the common-emitter current gain, and the subscript o is used to denote the low-frequency value.

The quantities g_m and r_π are related to the small-signal diode resistance r_e of the forward-biased emitter junction. Recall that the collector and emitter currents are each approximately h_{fe} times the base current, so that r_π and g_m become

$$r_\pi = h_{fe} r_e \quad \text{and} \quad g_m = \frac{1}{r_e} \tag{8-44}$$

It can be shown that the value of r_e is related to the bias current I_{EQ} by $r_e = V_T/I_{EQ}$ or $r_e = 1/40 I_{EQ}$ at room temperature.

Values of the capacitive elements are usually obtained from high-frequency data, with C_π obtained from the measurement of ω_T, the angular frequency at which $|h_{fe}(j\omega)| = 1$, and C_μ from measurement of y_{ob}, the common-base short-circuit output admittance (y_{22}). (The variation of $|h_{fe}|$ with frequency is derived in Example 5-13.) Based on the measurement of ω_T and y_{ob}, the values of C_π and C_μ, stated without derivation, are

$$C_\pi = \frac{1}{r_e \omega_T} - C_\mu \tag{8-45}$$

$$C_\mu = C_{ob} - C_{case} \tag{8-46}$$

The derivation of Eqs. (8-42), (8-43), and (8-45) are left to the student as an exercise.

Example 8-7

The following data are given for a transistor operating at $I_{EQ} = 1$ mA and $V_{CQ} = 5$ V at room temperature: $\omega_T = 2 \times 10^9$ rad/s, $C_{ob} = 1.5$ pF, $h_{fe} = 100$, $h_{ie} = 2,550$ Ω, and $C_{case} = 0.5$ pF.
Determine the hybrid-π parameters.

SOLUTION   The value of r_e is obtained from the bias current as

$$r_e = \frac{1}{40 \times 1 \times 10^{-3}} = 25 \ \Omega$$

By use of Eq. (8-44),

$$r_\pi = 100 \times 25 = 2,500 \ \Omega \qquad \text{and} \qquad g_m = \tfrac{1}{25} = 0.04 \ \mho$$

The value of r_x is obtained from Eq. (8-42) as

$$2,550 = r_x + 2,500 \qquad \text{or} \qquad r_x = 50 \ \Omega$$

From Eq. (8-46),

$$C_\mu = 1.5 - 0.5 = 1 \text{ pF}$$

and by use of Eq. (8-45),

$$C_\pi = \frac{1}{25 \times 2 \times 10^9} - 1 \times 10^{-12} = 19 \times 10^{-12} = 19 \text{ pF}$$

In similar fashion, the elements in the FET model of Fig. 8-29 can also be determined. The manufacturer usually supplies the y parameters of the device. The additional subscript s indicates that the FET is operated in the common-source mode. Equations (8-47) to (8-50) relate the elements in the model to the y parameters and are stated without derivation.

$$g_m = \text{low-frequency value of } y_{fs} \tag{8-47}$$

$$g_d = \text{low-frequency value of } y_{os} \tag{8-48}$$

$$C_{gd} = C_{rss} \tag{8-49}$$

$$C_{gs} = C_{iss} - C_{rss} \tag{8-50}$$

where C_{rss} is the capacitive part of y_{rs} and C_{iss} is the capacitive part of y_{is}.

It should be noted that the data appearing on the manufacturers' specifications are obtained from measurements made at given quiescent currents and voltages. Each of these parameters will change as the quiescent conditions are altered.

8-8 THE COMMON-EMITTER STAGE

The *differential amplifier* shown in Fig. 8-33 is the basic stage used in integrated circuits. The amplifier stage employed most frequently in discrete-component circuits is depicted in Fig. 8-34. Each circuit employs the common-emitter configuration, the differences in structure reflecting the technology used.

In the differential amplifier, IC fabrication permits identical transistors $Q1$ and $Q2$ and resistors R_C to be made, thus making the circuit symmetrical. The behavior of each section is then identical. The signal sources v_1 and v_2 produce the ac responses superposed on the quiescent levels established by the supply voltages V_{CC} and V_{EE}. When $v_1 = v_2$, the signal current in R_E, i_{e1} plus i_{e2}, is twice the emitter current that exists in one stage and produces a voltage drop across R_E of $2i_{e1}R_E$. Consequently, the ac equivalent of each stage in Fig. 8-33 can be represented as shown in Fig. 8-35a.

The signal current in R_E is zero when $v_1 = -v_2$, as v_1 causes i_{e1} to increase by the same amount that $-v_2$ causes i_{e2} to decrease. As a result, the

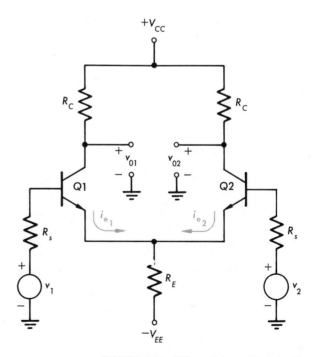

FIGURE 8-33 Differential-amplifier circuit.

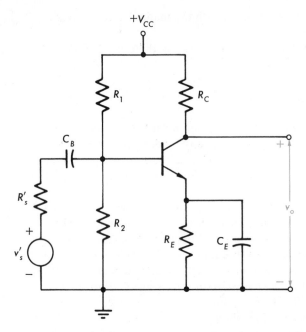

FIGURE 8-34 Discrete-component transistor amplifier stage.

alternating potential across R_E is zero and the voltage from emitter to ground remains constant. Since constant voltages are represented by short circuits in ac models, the circuit displayed in Fig. 8-35b is the ac equivalent of each stage in Fig. 8-33. It is important to note that the representations in Fig. 8-35 are ac equivalents implying that the transistors shown are to be represented by their small-signal equivalent circuits.

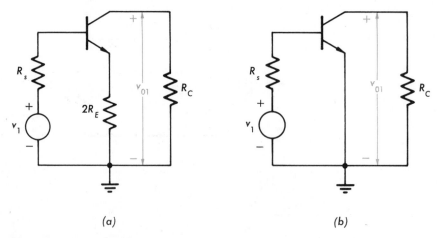

(a)   (b)

FIGURE 8-35 Ac-equivalent representations of differential amplifier; (a) for in-phase input signals and (b) for out-of-phase input signals.

In the discrete-component stage of Fig. 8-34, the signal to be processed is represented by v'_s and R'_s and the load resistor is R_C. The supply voltage V_{CC} and resistors R_1, R_2, and R_E establish the operating point. The capacitor C_B, called a *coupling capacitor,* is used to connect, or couple, the signal source to the transistor and isolate v'_s from the dc bias. Similarly, the *bypass capacitor* is used so that R_E will affect only the dc operation of the circuit. Both C_B and C_E have large values (ideally, they are infinite) so that their reactances are sufficiently small so they can be considered short-circuits. The purpose of the transistor is to provide the controlled source.

At signal frequencies for which C_B and C_E provide ideal coupling and bypass functions, the ac equivalent of the circuit in Fig. 8-34 is given in Fig. 8-36. In Eq. (8-27) in Sec. 8-3, R_1 in parallel with R_2 has been identified as R_B. This is used to obtain the Thévenin equivalent of the portion of the circuit in Fig. 8-36 to the left of terminals $b - g$ as shown in Fig. 8-37. Use of the Thévenin equivalent for the indicated portion of the circuit of Fig. 8-36 converts this circuit to the one shown in Fig. 8-35b.

Often it is desirable not to bypass R_E by omitting C_E. Under this condition, the ac model for the circuit shown in Fig. 8-34 can be represented as shown in Fig. 8-35a when the resistance $2R_E$ is replaced by the value of R_E in the circuit of Fig. 8-34.

The preceding discussion serves to illustrate that the ac performance of both integrated and discrete component circuits can be described by common-emitter stages or by stages with unbypassed emitter resistances. The analysis of these circuits is the subject of the remainder of this and the succeeding section. A more detailed treatment of differential amplifiers is presented in Sec. 8-12.

Analysis of the circuit of Fig. 8-35b is accomplished by representing the transistor by its unilateral hybrid-π equivalent by the method described in Sec. 8-5 (see Fig. 8-28). The frequency domain equivalent of the stage in Fig.

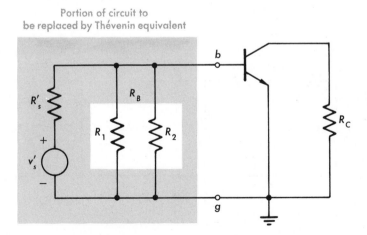

FIGURE 8-36 Ac representation of circuit in Fig. 8-34.

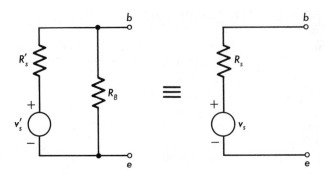

FIGURE 8-37 Thévenin equivalent of signal source and base-bias resistors.

8-35b is depicted in Fig. 8-38. For the circuit in Fig. 8-38 it is convenient to identify the parallel combination of R_C and r_o by R_L. Then,

$$V_o(s) = -g_m R_L V_\pi \tag{8-51}$$

The value of V_π is obtained from the base loop as

$$V_\pi = \frac{z_\pi}{R_s + r_x + z_\pi} V_s \tag{8-52}$$

where z_π is the parallel combination of r_π and $1/sC_x$ and is

$$z_\pi = \frac{r_\pi}{1 + sr_\pi C_x} \tag{8-53}$$

Combination of Eqs. (8-51), (8-52) and (8-53) and formation of the transfer function $V_o(s)/V_s$, referred to as the *voltage gain,* yield

$$\frac{V_o(s)}{V_s} = G_H(s) = \frac{-g_m r_\pi R_L}{R_s + r_x + r_\pi + sC_x r_\pi (R_s + r_x)} \tag{8-54}$$

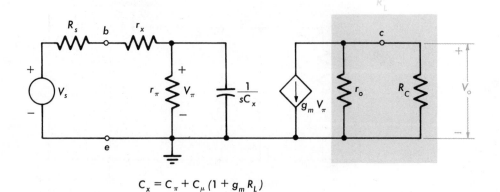

$$C_x = C_\pi + C_\mu (1 + g_m R_L)$$

FIGURE 8-38 Equivalent circuit for common-emitter amplifier stage.

It is often convenient to rewrite Eq. (8-54) by factoring $(R_s + r_x + r_\pi)$ from the denominator as

$$G_H(s) = \frac{-g_m r_\pi R_L/(R_s + r_x + r_\pi)}{1 + sC_x \dfrac{r_\pi(R_s + r_x)}{R_s + r_x + r_\pi}} \qquad (8\text{-}55)$$

The general form of Eq. (8-55) is

$$G_H(s) = \frac{G_o}{1 + sC_x R_x} = \frac{G_o}{1 + s/\omega_H} \qquad (8\text{-}56)$$

in which, after recalling $g_m r_\pi = h_{fe}$,

$$G_o = \frac{-g_m r_\pi R_L}{R_s + r_x + r_\pi} = \frac{-h_{fe} R_L}{R_s + r_x + r_\pi} \qquad (8\text{-}57)$$

$$R_x = \frac{r_\pi(R_s + r_x)}{R_s + r_x + r_\pi} \qquad \text{and} \qquad \omega_H = \frac{1}{R_x C_x} \qquad (8\text{-}58)$$

In Eq. (8-58), R_x can be identified as the parallel combination of r_π and $(R_s + r_x)$. Physically, R_x is the resistance seen by the capacitor C_x in the circuit of Fig. 8-38. The negative sign associated with G_o in Eq. (8-57) is simply interpreted as a phase reversal. (See Fig. 7-28 and compare the sinusoids for i_b and v_{ce}.)

The Bode diagram (see Sec. 6-3) for the transfer function in Eq. (8-56) is displayed in Fig. 8-39 and depicts the decrease in gain with increasing fre-

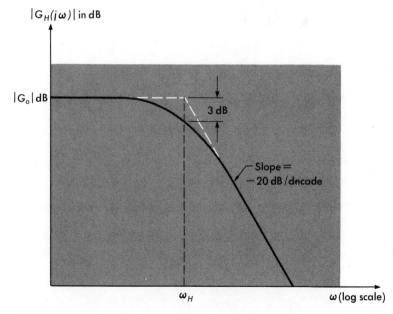

FIGURE 8-39 Bode diagram of high-frequency response of common-emitter stage.

quency. For frequencies below ω_H, the gain is essentially constant at a value G_o. In this range of frequencies, called *midband* (or low frequencies), the reactance of C_x is sufficiently large so that it is neglected. At these frequencies C_x (and consequently C_π and C_μ) is considered an open-circuit and, as such, is omitted from the equivalent circuit of Fig. 8-38. The frequency range for which the effects of C_x must be included is called the *high-frequency range* and is the rationale for referring to the gain in Eq. (8-56) as G_H.

When $s = j\omega_H$, the gain becomes $|G_H(j\omega)| = 0.707 |G_o|$. As the power in the load may be expressed as $|V_o(j\omega)|^2/R_L$, the output power at the corner frequency ω_H is $(0.707 V_o)^2/R_L$. This value is one-half the value of the midband power output, and ω_H is called the *upper half-power angular frequency* of the circuit. Colloquially, ω_H is also referred to as the *upper 3-dB-down frequency*.

It is often important to know the input impedance Z_{in} of the amplifier, which is the impedance seen looking to the right of terminals b and e in Fig. 8-38. In the midband frequency range Z_{in} is purely resistive and is simply r_x in series with r_π and is given by

$$Z_{in} = r_x + r_\pi \tag{8-59}$$

The input impedance at high frequencies must include the effect of the capacitor C_x, as seen in the equivalent circuit of Fig. 8-38. The value of Z_{in} is simply r_x in series with the parallel combination of r_π and $1/sC_x$. The result can be written as

$$Z_{in}(s) = r_x + \frac{r_\pi}{1 + sr_\pi C_x} = (r_x + r_\pi)\frac{1 + sC_x[r_x r_\pi/(r_x + r_\pi)]}{1 + sC_x r_\pi} \tag{8-60}$$

Example 8-8

The amplifier stage shown in Fig. 8-35b is designed with $R_C = 500 \ \Omega$ and $R_s = 50 \ \Omega$. The hybrid-π parameters for the transistor are: $r_x = 50 \ \Omega$, $r_\pi = 1,000 \ \Omega$, $g_m = 0.1 \ \mho$, $r_o = 100 \ k\Omega$, $C_\pi = 50 \ pF$, and $C_\mu = 1 \ pF$.

a   Determine the midband voltage gain and the upper half-power angular frequency.

b   Sketch the asymptotic Bode diagram for the input impedance of the amplifier.

SOLUTION   **a**   The equivalent circuit that is applicable is given in Fig. 8-38 for which

$$R_L = \frac{R_C r_o}{R_C + r_o} = \frac{500 \times 10^5}{500 + 10^5} \doteq 500 \ \Omega$$

The midband gain is found from Eq. (8-57) to be

$$G_o = \frac{-0.1 \times 1{,}000 \times 500}{50 + 50 + 1{,}000} = -45.5$$

To determine ω_H, the value of C_x must be found.

$$
\begin{aligned}
C_x &= C_\pi + C_\mu(1 + g_m R_L) \\
&= 50 \times 10^{-12} + 10^{-12}(1 + 0.1 \times 500) = 101 \times 10^{-12}
\end{aligned}
$$

The values of R_x and ω_H are given in Eq. (8-58) and are

$$R_x = \frac{1{,}000(50 + 50)}{50 + 50 + 1{,}000} = 90.9 \; \Omega$$

and

$$\omega_H = \frac{1}{90.9 \times 101 \times 10^{-12}} = 1.09 \times 10^8 \text{ rad/s}$$

or

$$f_H = \frac{\omega_H}{2\pi} = 17.3 \times 10^6 \text{ Hz} = 17.3 \text{ MHz}$$

b The input impedance at high frequencies is given by Eq. (8-60) as

$$Z_{\text{in}}(s) = (50 + 1{,}000) \frac{1 + s \times 101 \times 10^{-12}[50 \times 1{,}000/(50 + 1{,}000]}{1 + s \times 101 \times 10^{-12} \times 1{,}000}$$

so that

$$Z_{\text{in}}(s) = 1{,}050 \frac{1 + s/2.09 \times 10^8}{1 + s/9.90 \times 10^6}$$

The resultant asymptotic Bode diagram is shown in Fig. 8-40. As indicated in Fig. 8-40, the input impedance for $\omega < 9.90 \times 10^6$ rad/s is 60.4 dB or 1,050 Ω, which corresponds to the midband impedance $r_x + r_\pi =$

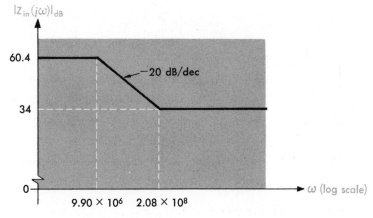

FIGURE 8-40 Asymptotic Bode diagram for Example 8-8.

$50 + 1,000 = 1,050$ Ω. It is to be noted that for $\omega > 2.08 \times 10^8$ rad/s. $|\mathbf{Z}_{\text{in}}(j\omega)|_{\text{dB}} = 34$ dB which corresponds to a value of 50 Ω. By use of Eq. (8-60), it can be shown that this value is just r_x. The method described in this example represents one method for determining r_x by laboratory measurements.

In the discrete-component amplifier stage of Fig. 8-34, the midband gain G_o must be modified to include the effect of the base-bias resistors R_1 and R_2. This is achieved by relating the Thévenin equivalent source used in Fig. 8-38 to the signal source v'_s in Fig. 8-36. The result is given in Eq. (8-61).

$$G_o = \frac{-g_m r_\pi R_L}{R_s + r_x + r_\pi} \frac{R_B}{R_B + R'_s} \tag{8-61}$$

Generally, $R_B \gg R'_s$ so that the values of G_o obtained from Eqs. (8-61) and (8-57) are identical.

The low-frequency response of the amplifier stage in Fig. 8-34 is limited by the finite values of C_B and C_E. In this *low-frequency range*, the reactances of C_B and C_E become sufficiently large so they can no longer be considered short circuits. Figure 8-41 depicts the frequency-domain equivalent circuit of the stage in Fig. 8-34 that is valid at low frequencies. In Fig. 8-41, R_B is the parallel combination of the base-bias resistors R_1 and R_2. Furthermore, because of the frequency range considered, the capacitor C_x is considered an open-circuit.

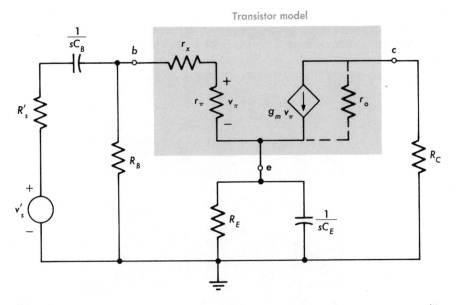

FIGURE 8-41   Low-frequency equivalent circuit of discrete-component common-emitter stage.

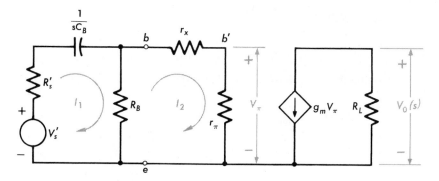

FIGURE 8-42 Low-frequency model of common-emitter stage, assuming emitter by-passing is ideal.

In determining the low-frequency gain of the amplifier, it is convenient to consider the effects of C_B and C_E separately. This procedure is useful in design since it allows performance criteria to be established which are based on each component explicitly. Thus, if the effect of C_B on low-frequency performance is desired, C_E is considered as an ideal bypass capacitor and the equivalent circuit of Fig. 8-41 becomes that shown in Fig. 8-42.

The output voltage in the circuit of Fig. 8-43 is given by Eq. (8-51). To find $V_\pi(s)$, select loop currents in the two windows as indicated. Solving for

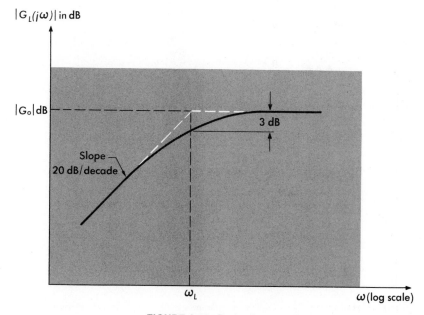

FIGURE 8-43 Bode diagram of low-frequency response.

$V_\pi(s)$ results in

$$V_\pi(s) = \frac{R_B r_\pi V_s'}{(R_B + R_s')(r_x + r_\pi) + R_B R_s' + (R_B + r_x + r_\pi)/sC_B} \qquad (8\text{-}62)$$

Combining Eqs. (8-62) and (8-51) and identifying G_o given in Eq. (8-61) yields

$$G_L(s) = \frac{G_o}{1 + \omega_L/s} \qquad (8\text{-}63)$$

where

$$\omega_L = \frac{1}{R_s' + R_B(r_x + r_\pi)/(R_B + r_x + r_\pi)} \cdot \frac{1}{C_B} \qquad (8\text{-}64)$$

The term R_s' and $R_B(r_x + r_\pi)/(R_B + r_x + r_\pi)$ in Eq. (8-64) can be identified as the series combination of R_s' and a resistance which is R_B in parallel with r_x and r_π in series. Inspection of Fig. 8-42 indicates that this is the resistance seen by the capacitor C_B. Figure 8-43 is the Bode diagram for the low-frequency response, assuming that R_E is ideally bypassed. The angular frequency ω_L corresponds to the *lower half-power frequency* or *lower 3-dB-down frequency*.

To determine the effect of the bypass capacitor C_E on low-frequency performance, the circuit of Fig. 8-41 is used, with C_B considered to be the ideal coupling capacitor (short-circuited), as shown in Fig. 8-44. The portion of the circuit to the left of terminals b and e is replaced by its Thévenin equivalent (as shown in Fig. 8-36), and $R_C \ll r_o$ so that the effect of r_o is neglected. The impedance Z_E is the parallel combination of R_E and C_E and is

$$Z_E = \frac{R_E}{1 + sC_E R_E} \qquad (8\text{-}65)$$

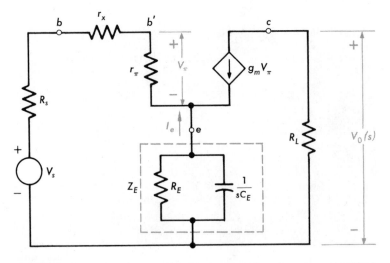

FIGURE 8-44 Low-frequency model of common-emitter stage, assuming ideal base-loop coupling.

The output voltage is given by Eq. (8-51), and the value of $V_{b'e}$ is

$$V_\pi = I_b r_\pi \tag{8-66}$$

The current law at node e requires that

$$I_e = -(I_b + g_m V_\pi) \tag{8-67}$$

while the voltage law for the base-emitter loop requires that

$$V_s = I_b(r_\pi + r_x + R_s) - Z_E I_e \tag{8-68}$$

Combination of Eqs. (8-65) through (8-68) and identification of G_o result in

$$G_L(s) = \frac{G_o(1 + \omega_E/s)}{1 + \omega_L/s} \tag{8-69}$$

where

$$\omega_L = \frac{R_E + (r_\pi + r_x + R_s)/(1 + h_{fe})}{R_E(r_\pi + r_x + R_s)/(1 + h_{fe})} \frac{1}{C_E} \tag{8-70}$$

and

$$\omega_E = \frac{1}{R_E C_E}$$

The term multiplying C_E in Eq. (8-70) can be identified as the resistance seen by C_E which is R_E in parallel with the base-loop resistance divided by $(1 + h_{fe})$. As ω_E is generally much less than ω_L, it has negligible effect on the value of the low-frequency half-power point. Consequently, Eq. (8-69) is approximated by Eq. (8-63). The Bode diagram for Eq. (8-63) is that shown in Fig. 8-43. Because C_B and C_E each affect the low-frequency response, the total low-frequency performance must include their combined effects. Methods for determination of the total low-frequency response are discussed in Chap. 10.

The performance of the amplifier is limited at both high and low frequencies. To describe the frequency range over which the amplifier has useful gain, it is customary to use the concept of *bandwidth* (BW), which is defined as the frequency difference between the upper and lower half-power frequencies. The amplifier bandwidth is

$$\text{BW} = \omega_H - \omega_L \quad \text{rad/s}$$

and is indicated in Fig. 8-45, which is the Bode diagram for the stage over the useful frequency range. Many amplifier circuits are designed to operate from very low frequencies (in the order of 10 to 1,000 Hz) to relatively high frequencies (in the order of 1 to 100 MHz). Such circuits are called *broadband* or *video amplifiers*, and the bandwidth is essentially the upper half-power frequency. It is also noted that in integrated circuit stages where no coupling or bypass capacitors are used, the bandwidth is just the upper half-power frequency.

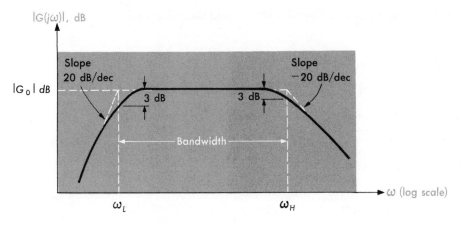

$|G(j\omega)|$, dB

Slope 20 dB/dec

Slope −20 dB/dec

$|G_0|$ dB

3 dB   3 dB

Bandwidth

ω_L ω_H

ω (log scale)

FIGURE 8-45 Bode diagram of amplifier frequency response.

Example 8-9

A discrete-component amplifier stage of the type shown in Fig. 8-34 is designed with element values $R_1 = 60$ kΩ, $R_2 = 12$ kΩ, $R_E = 1$ kΩ, $C_B = 1$ μF, $C_E = 100$ μF, $R'_s = 50$ Ω, and other element values and transistor parameters are those given in Example 8-8. Find

a The lower half-power frequency when C_E acts as an ideal bypass;

b The lower half-power frequency when C_B is a perfect coupling capacitor.

SOLUTION   The low-frequency model is given in Fig. 8-41 for which

$$R_B = \frac{R_1 R_2}{R_1 + R_2} = \frac{60 \times 10^3 \times 12 \times 10^3}{60 \times 10^3 + 12 \times 10^3} = 10 \text{ k}\Omega$$

a When C_E is a perfect bypass, Eq. (8-64) is used to determine the value of ω_L.

$$\omega_L = \frac{1}{50 + 10^4(50 + 1{,}000)/(10^4 + 50 + 1{,}000)} \times \frac{1}{10^{-6}} = 1.00 \times 10^3 \text{ rad/s}$$

$$f_L = \frac{\omega_L}{2\pi} = \frac{1.00 \times 10^3}{2\pi} = 159 \text{ Hz}$$

b With C_B acting as an ideal coupling capacitor, the lower half-power frequency is given in Eq. (8-70). The value of R_s needed in Eq. (8-58) is

$$R_s = \frac{R'_s R_B}{R'_s + R_B} = \frac{50 \times 10^4}{50 + 10^4} = 49.8 \text{ }\Omega$$

The value of R_s is nearly identical with the value of R'_s and using $R_s = 50$ Ω introduces negligible error. Recalling that $h_{fe} = g_m r_\pi = 0.1 \times 1{,}000 = 100$, the value of ω_L becomes

$$\omega_L = \frac{1,000 + (1,000 + 50 + 50)/(1 + 100)}{1,000(1,000 + 50 + 50)/(1 + 100)} \cdot \frac{1}{100 \times 10^{-6}} = 928 \text{ rad/s}$$

and

$$f_L = \frac{\omega_L}{2\pi} = \frac{928}{2\pi} = 148 \text{ Hz}$$

The value of ω_E in Eq. (8-70) is

$$\omega_E = \frac{1}{1,000 \times 100 \times 10^{-6}} = 10 \text{ rad/s}$$

which is considerably less than the ω_L obtained. Therefore, it is reasonable to neglect ω_E in determining the lower-3-dB frequency.

8-9 STAGES CONTAINING EMITTER RESISTORS

It is often desirable and sometimes necessary (see Fig. 8-35a) to utilize amplifier stages in which the emitter resistance affects the response in all frequency ranges. The ac representation of such an amplifier is depicted in Fig. 8-46; the terminals of interest are the base, collector and ground. The emitter node is not an important terminal for measuring the circuit responses to signals. The circuit shown in Fig. 8-47 is the small-signal equivalent circuit for the stage in Fig. 8-46. The transistor and emitter resistance have been replaced by a structure similar to the unilateral hybrid-π of Fig. 8-28. The principal difference between the two circuits is the value of the elements. Table 8-1 compares the element values for the transistor models shown in Figs. 8-28 and 8-47.

As indicated in Table 8-1, the effect of the emitter resistance is to scale those elements connected to the emitter by a factor $(1 + g_m R_E)$. The element corresponding to r_o in the hybrid-π model has been omitted as its value

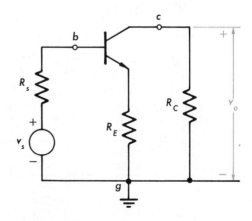

FIGURE 8-46 Ac equivalent of a common-emitter stage containing an emitter resistance.

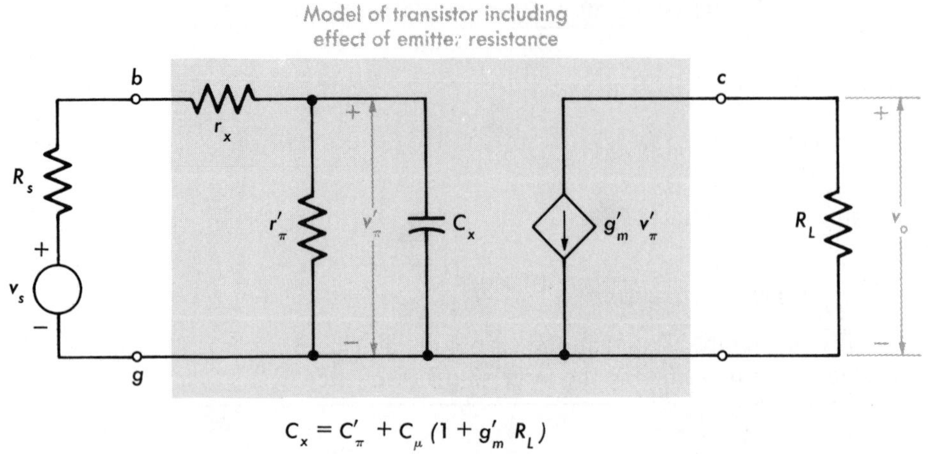

Model of transistor including
effect of emitter resistance

$$C_x = C'_\pi + C_\mu (1 + g'_m R_L)$$

FIGURE 8-47 Equivalent circuit of amplifier stage in Fig. 8-46.

is significantly larger than r_o. As such, its parallel combination with a load resistance R_L is invariably R_L.

Because the circuit in Fig. 8-47 has the same structure as a stage with no emitter resistance, the results for the gain, impedance, and frequency response of the common-emitter stage obtained in Sec. 8-9 are directly applicable upon substitution of the element values in Table 8-1. These are summarized in Eqs. (8-71) to (8-73). The midband gain is, upon recalling $h_{fe} = g_m r_\pi$,

$$G_o = \frac{-g_m r_\pi R_L}{R_s + r_x + r_\pi (1 + g_m R_E)} = \frac{-h_{fe} R_L}{R_s + r_x + r_\pi + h_{fe} R_E} \tag{8-71}$$

The upper half-power frequency becomes

$$\omega_H = \frac{1}{R_x C_x}$$

where

$$R_x = \frac{r_\pi (1 + g_m R_E)(r_x + R_s)}{R_s + r_x + r_\pi (1 + g_m R_E)} \tag{8-72}$$

TABLE 8-1 CORRESPONDENCE OF ELEMENTS IN THE HYBRID-Π MODELS OF CIRCUITS, WITH AND WITHOUT EMITTER RESISTORS

CIRCUITS WITHOUT EMITTER RESISTANCE	CIRCUITS WITH EMITTER RESISTANCE
r_x	r_x
r_π	$r'_\pi = r_\pi (1 + g_m R_E)$
C_π	$C'_\pi = C_\pi / (1 + g_m R_E)$
C_μ	C_μ
g_m	$g'_m = g_m / (1 + g_m R_E)$

The midband input impedance is expressed as

$$Z_{in} = r_x + r_\pi(1 + g_m R_E) = r_x + r_\pi + h_{fe}R_E \qquad (8\text{-}73)$$

For stages which employ coupling capacitors, the low-frequency analysis parallels that in Sec. 8-9. This analysis is part of Prob. 8-54.

Example 8-10

An amplifier stage whose circuit is shown in Fig. 8-46 is designed with $R_C = 1,000 \ \Omega$, $R_E = 45 \ \Omega$, and $R_s = 450 \ \Omega$. The transistor parameters are: $r_x = 50 \ \Omega$, $g_m = 0.2 \ \text{U}$, $h_{fe} = 100$, $C_\pi = 100 \ \text{pF}$, and $C_\mu = 1 \ \text{pF}$. The value of r_o is large and can be neglected.

a Determine the midband gain and upper half-power frequency.

b Repeat a for $R_E = 0$.

c The maximum value of h_{fe} is 200 because of unit-to-unit variation. Determine the percentage difference in midband gain for $h_{fe} = 200$ compared with the results obtained in a and b.

SOLUTION   **a**   The value of r_π is given by

$$r_\pi = \frac{h_{fe}}{g_m} = \frac{100}{0.2} = 500 \ \Omega$$

From Eq. (8-71),

$$G_o = \frac{-100 \times 1,000}{450 + 50 + 500 + 100 \times 45} = -18.2$$

Use of Table 8-1 gives

$$C_\pi' = \frac{100 \times 10^{-12}}{(1 + 0.2 \times 45)} = 10 \ \text{pF}$$

and

$$g_m' = \frac{0.2}{(1 + 0.2 \times 45)} = 0.02 \ \text{U}$$

so that

$$C_x = C_\pi' + C_\mu(1 + g_m' R_L)$$
$$= 10 \times 10^{-12} + 10^{-12}(1 + 0.02 \times 1,000) = 31 \ \text{pF}$$

The upper-half power frequency is given in Eq. (8-72) so that

$$R_x = \frac{500(1 + 0.2 \times 45)(50 + 450)}{450 + 50 + 500(1 + 0.2 \times 45)} = 455 \ \Omega$$

and

$$\omega_H = \frac{1}{455 \times 31 \times 10^{-12}} = 7.09 \times 10^7 \text{ rad/s}$$

b For $R_E = 0$, Eqs. (8-57) and (8-58) are used to obtain the value of G_o and ω_H, respectively. These are

$$G_o = \frac{-0.2 \times 500 \times 1,000}{450 + 50 + 500} = -100$$

$$\begin{aligned} C_x &= C_\pi + C_\mu(1 + g_m R_L) \\ &= 100 \times 10^{-12} + 10^{-12}(1 + 0.2 \times 1,000) = 301 \text{ pF} \end{aligned}$$

$$R_x = \frac{500(450 + 50)}{450 + 50 + 500} = 250 \ \Omega$$

$$\omega_H = \frac{1}{250 \times 301 \times 10^{-12}} = 1.33 \times 10^7 \text{ rad/s}$$

c For $h_{fe} = 200$, $r_\pi = \frac{200}{0.2} = 1,000 \ \Omega$. The gain is given in Eq. (8-71) for $R_E = 45 \ \Omega$ as

$$G_o = \frac{-200 \times 1,000}{450 + 50 + 1,000 + 200 \times 45} = -19.0$$

The percentage change is

$$\text{Relative change} = \frac{19.0 - 18.2}{18.2} \times 100 = 4.4 \text{ percent}$$

For $R_E = 0$, the corresponding results are

$$G_o = \frac{-200 \times 1,000}{450 + 50 + 1,000} = -133$$

$$\text{Relative change} = \frac{133 - 100}{100} = 33 \text{ percent}$$

The results of Example 8-10 indicate that the use of the emitter resistance decreases the gain and increases the upper-3-dB frequency. The disadvantage of decreased gain is often offset by the advantage of increased bandwidth. A more significant advantage is observed in the results of part *c*, in which the stage containing the emitter resistance exhibited better gain control over the range of h_{fe} variation. From Eq. (8-71), when $h_{fe}R_E$ is much larger than $R_s + r_x + r_\pi$, the gain is approximately R_L/R_E and is virtually independent of transistor parameters. This independence is quite desirable as manufacturing tolerances often permit wide variations in the values of h_{fe}.

8-10 THE COMMON-SOURCE AMPLIFIER STAGE

The analysis of FET amplifiers in the common-source (CS) configuration parallels that for the common-emitter stages of the two previous sections.

The basic stage for an *n*-channel depletion-type MOSFET is shown in Fig. 8-48 in which the role of each component is analogous to that for bipolar stages. The equivalent circuit for the stage is depicted in Fig. 8-49a assuming C_G and C_K act as perfect coupling and bypass capacitors. The FET is replaced by its unilateral model developed in Sec. 8-6. As identified in Fig. 8-48a, the parallel combinations of R_1 and R_2 and R_D and r_d are identified as R_G and R_L, respectively. Upon obtaining the Thévenin equivalent of the portion of the circuit identified in Fig. 8-49a, the model shown in Fig. 8-49b results.

For the circuit of Fig. 8-49b, the output voltage V_o is

$$V_o = -g_m R_L V_{gs} \tag{8-74}$$

and

$$V_{gs} = \frac{1/sC_x}{R_s + 1/sC_x} V_s = \frac{1}{1 + sC_x R_s} V_s \tag{8-75}$$

Combining Eqs. (8-74) and (8-75) and forming the ratio V_o/V_s give

$$\frac{V_o}{V_s} = G_H(s) = \frac{-g_m R_L}{1 + sR_s C_x} = \frac{G_o}{1 + s/\omega_H} \tag{8-76}$$

As is indicated in Eq. (8-76), the gain decreases with increasing frequency, the decrease being attributable to the effects of the internal capacitors of the device. (Recall that C_x is composed of C_{gs} and C_{gd}.) At low and midband frequencies C_x is considered open-circuited. The Bode diagram for the gain in Eq. (8-76) is that given in Fig. 8-39 in Sec. 8-8.

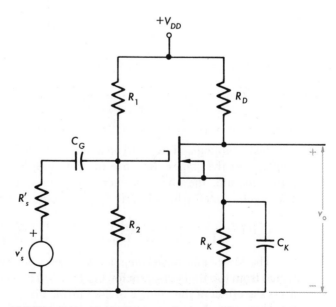

FIGURE 8-48 Field-effect transistor amplifier stage (common-source).

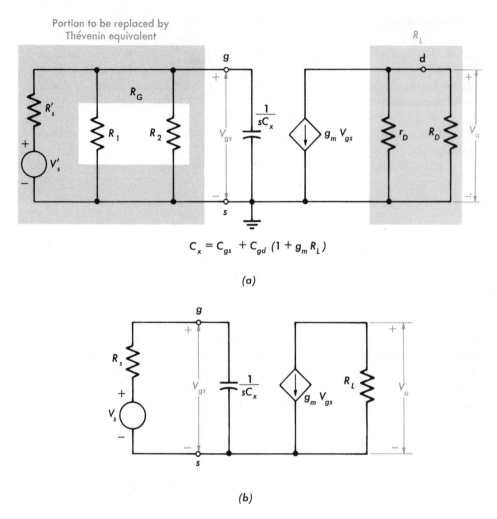

$$C_x = C_{gs} + C_{gd}(1 + g_m R_L)$$

(a)

(b)

FIGURE 8-49 (a) Equivalent circuit of a common-source amplifier stage; (b) the circuit in a with signal-source and gate-bias resistors replaced by their Thévenin equivalent.

The coupling and bypass capacitors C_G and C_K affect the low-frequency response of the stages. Because the gate loop and source loop are electrically separate, the effect of each capacitor is determined independently. The result, stated without derivation, is given in Eq. (8-77) as

$$G_L(s) = \frac{G_o(1 + 1/sR_KC_K)}{[1 + 1/sC_G(R_G + R_s')][1 + (1 + g_mR_K)/sR_KC_K]} \tag{8-77}$$

The lower half-power angular frequency for the stage is generally obtained from the Bode diagram for Eq. (8-77). The value of G_o in Eq. (8-77) reflects the effect of the voltage divider formed by R_s' and R_G used in obtaining the Thévenin equivalent. Thus,

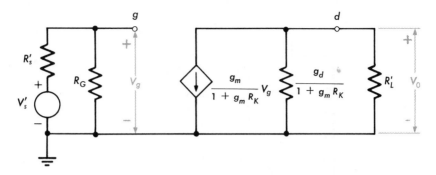

FIGURE 8-50 Midband equivalent circuit of a common-source amplifier stage containing a source resistance.

$$G_o = \frac{V_o}{V_s'} = -g_m R_L \frac{R_G}{R_s' + R_G} \tag{8-78}$$

The value of G_o given in Eq. (8-78) is also used in Eq. (8-76) to obtain the transfer function V_o/V_s'.

For reasons similar to those described in Example 8-10, it is often advantageous to use circuits with source resistors. It is then convenient to use the model developed in Example 6-6 and shown in Fig. 6-20 to represent the FET and source resistance. The midband representation of the circuit of Fig. 8-48, with $C_K = 0$, becomes that depicted in Fig. 8-50. The output voltage V_o is

$$V_o = -\frac{g_m R_L}{1 + g_m R_K} V_g \tag{8-79}$$

where R_L is the parallel combination of R_L' and the conductance $g_d/(1 + g_m R_K)$. For $R_G \gg R_s' V_g$ is simply V_s; the midband gain becomes

$$G_o = \frac{-g_m R_L}{1 + g_m R_K} \tag{8-80}$$

The effect of the unbypassed resistor is to decrease the gain. However, the bandwidth increases in the manner described in Example 8-10. In addition, in Eq. (8-80), if $g_m R_K \gg 1$, G_o becomes $-R_L/R_K$ and the gain is virtually independent of the FET parameters.

8-11 EMITTER-FOLLOWER AND SOURCE-FOLLOWER AMPLIFIERS

A type of amplifier circuit which is valuable for special purposes is shown in Fig. 8-51. It is called an *emitter follower* (or common-collector amplifier). The output voltage is taken across the emitter resistor R_E.

The midband equivalent circuit for the emitter follower is given in Fig. 8-52, with the source and base-bias resistors replaced by their Thévenin

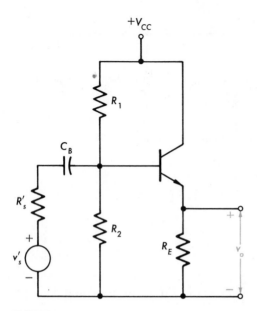

FIGURE 8-51 Emitter-follower, or common-collector, amplifier stage.

equivalent [given by Eq. (8-21) and shown in Fig. 8-37]. The KVL equation for the base loop is

$$V_s = I_b(R_s + r_x + r_\pi + R_E) + g_m V_\pi R_E \tag{8-81}$$

Also

$$V_{b'e} = I_b r_\pi, \quad g_m r_\pi = h_{fe}$$

and

$$V_o = R_E(I_b + g_m V_\pi) \tag{8-82}$$

The voltage gain is found by combining Eqs. (8-81) and (8-82) and is

$$G_o = \frac{V_o}{V_s} = \frac{(h_{fe} + 1)R_E}{R_s + r_x + r_\pi + (h_{fe} + 1)R_E} \tag{8-83}$$

In general, $(h_{fe} + 1)R_E$ is much larger than $R_s + r_x + r_\pi$, and the gain approaches unity.

One of the valuable performance features of the emitter follower is its high input impedance (see Sec. 8-9), making it useful as an initial stage in many applications. A second feature, particularly useful in an output stage, is that the emitter follower constitutes a voltage source with a relatively low internal resistance. The output impedance is the Thévenin equivalent impedance between terminals e and g shown in Fig. 8-52. The open-circuit voltage in Fig. 8-52 is V_o, which is given in Eq. (8-82). The short-circuit current is $I_b + g_m V_\pi$, with R_E set equal to zero, or

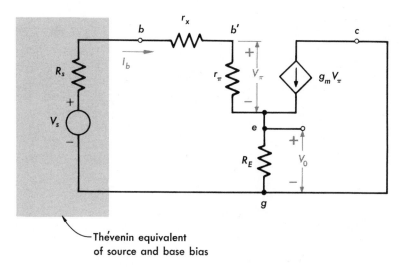

Thévenin equivalent
of source and base bias

FIGURE 8-52 Midband equivalent of emitter-follower stage.

$$I_{sc} = \frac{V_s(1 + g_m r_\pi)}{R_s + r_x + r_\pi}$$

$$= \frac{V_s(1 + h_{fe})}{R_s + r_x + r_\pi}$$

The Thévenin resistance R_o is the ratio of the open-circuit voltage to the short-circuit current and is given by

$$R_o = \frac{R_E[(R_s + r_x + r_\pi)/(h_{fe} + 1)]}{R_E + (R_s + r_x + r_\pi)/(h_{fe} + 1)} \tag{8-84}$$

The value of R_o in Eq. (8-84) is seen to be the parallel combination of R_E and $(R_s + r_x + r_\pi)/(h_{fe} + 1)$ and is much smaller than the value obtained for the same transistor in a conventional stage. Matching or approximate matching with a load resistor for maximum power transfer can often be obtained. The resistance $R_s + r_x + r_\pi$ is the total series resistance in the base portion of the circuit. Its effect, when reflected into the emitter, is reduced by h_{fe}. This result is the inverse of that obtained in Sec. 8-9 for reflecting the effect of an emitter resistance into the base. The emitter follower therefore acts as an impedance converter and is capable of coupling a high-impedance source to a low-impedance load, over a wide range of frequencies. This action is similar to some of the functions performed by a transformer (see Chap. 13).

The impedance-conversion property of the emitter follower is often used to provide electrical isolation between two sections of a system. In the system of Fig. 8-53a, section A is connected to the input of the emitter follower and section B is connected to the output. The input impedance Z_B of section B is reflected by h_{fe} across the output of section A as shown in Fig. 8-53b. The effective impedance is approximately $h_{fe}Z_B$ and is made large compared with the output impedance Z_A of section A. The large effective

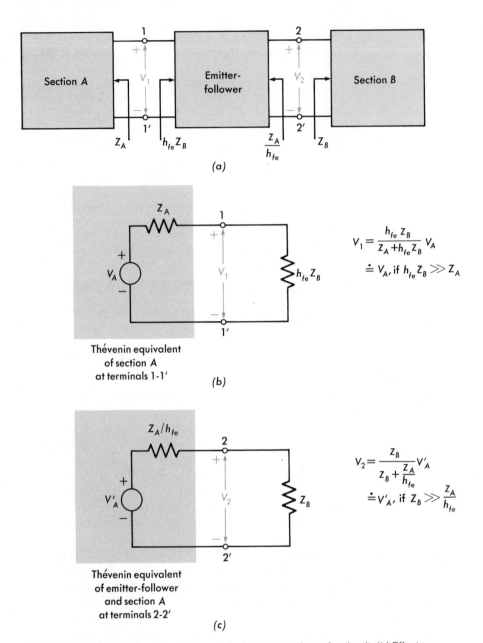

FIGURE 8-53 (a) Use of emitter-follower to isolate two sections of a circuit. (b) Effect on section B on output of section A. (c) Effect of section A on input of section B.

impedance therefore does not alter the electrical properties of section A. Similarly, as shown in Fig. 8-53c, Z_A is reflected by $1/h_{fe}$, into the input circuit of section B. The effect of Z_A is small and does not affect the electrical properties of section B. This isolation property is sometimes referred to as *buffering*, and the emitter follower is often called a *buffer amplifier*.

Example 8-11

A JFET is used in the circuit depicted in Fig. 8-54 and is called a source follower or common-drain amplifier. The equivalent circuit for the stage is that shown in Fig. 8-55. The circuit parameters are $\mu = 200$, $g_d = 10^{-5}$ ℧, and $R_K = 5{,}000$ Ω. For midband frequencies, determine

a The gain V_o/V_s

b The output impedance

SOLUTION **a** The KVL equations for the drain and gate loops, respectively, of the circuit of Fig. 8-55 are

$$\mu V_{gs} = I_d \left(R_K + \frac{1}{g_d} \right) \qquad \text{and} \qquad V_{gs} = V_s - I_d R_K$$

The output voltage is

$$V_o = I_d R_K$$

Combination of the above equations and substitution of values give

$$G_o = \frac{V_o}{V_s} = \frac{\mu R_K}{(1 + \mu)R_K + 1/g_d} = 0.905$$

b The output impedance R_o is found from the ratio of the open-circuit voltage V_{oc} to the short-circuit current I_{sc}. The open-circuit voltage is given by

$$V_{oc} = V_o = \frac{\mu R_K}{(1 + \mu)R_K + 1/g_d} V_s$$

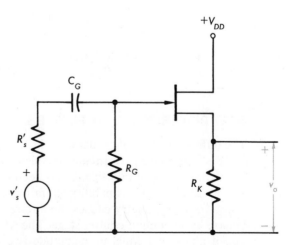

FIGURE 8-54 Source-follower, or common-drain, amplifier stage.

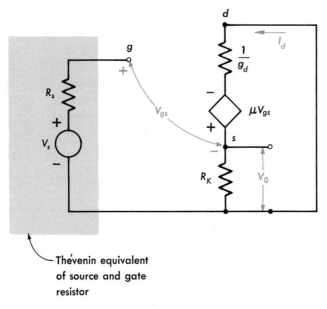

Thévenin equivalent
of source and gate
resistor

FIGURE 8-55 Midband equivalent of circuit in Fig. 8-54.

The short-circuit current is the value of I_d when $R_K = 0$ and is

$$I_{sc} = I_d = g_d\mu V_s$$

Then

$$R_o = \frac{V_{oc}}{I_{sc}} = \frac{R_K}{(1 + \mu)R_K g_d + 1} = \frac{[1/(1 + \mu)g_d]R_K}{[1/(1 + \mu)g_d] + R_K}$$

Based on the values given,

$$R_o = \frac{5 \times 10^3}{(1 + 200) \times 5 \times 10^3 \times 10^{-5} + 1} = 452 \ \Omega$$

Both the emitter follower and the source follower are examples of negative-feedback amplifiers, considered in general terms in Chap. 9.

8-12 DIFFERENTIAL AMPLIFIERS

The differential amplifier, depicted in Fig. 8-56, first introduced in Sec. 8-8, is the stage whose performance enables an operational amplifier to respond to the difference of two signals. As seen in Fig. 8-56, the circuit is symmetric, a result of the capability of fabricating identical transistors and resistors. This symmetry causes the differential amplifier, when $v_1 = -v_2$, to behave as a common-emitter stage. The situation for which v_1 and v_2 are 180° out of phase and equal in magnitude is called the *differential mode (DM)*. The *common mode (CM)* describes circuit operation when v_1 and v_2 are

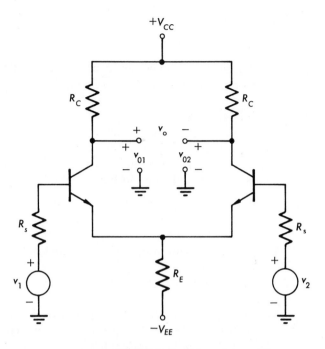

FIGURE 8-56 Differential-amplifier stage.

equal. As shown in Fig. 8-35*b*, circuit behavior is that of a common-emitter stage which contains an emitter resistance of value $2R_E$.

At midband frequencies, the *differential-mode gain* A_{DM} is given by Eq. (8-57), with R_L the parallel combination of R_C and r_o, as

$$A_{DM} = \frac{v_{o1}}{v_{DM}} = \frac{v_{o2}}{v_{DM}} = \frac{-h_{fe}R_L}{R_s + r_x + r_\pi} \tag{8-85}$$

where v_{DM} is the differential-mode signal ($v_1 = -v_2 = v_{DM}$). The output voltages at each of the collectors with respect to ground are then

$$v_{o1} = A_{DM}v_{DM} \quad \text{and} \quad v_{o2} = -A_{DM}v_{DM} \tag{8-86}$$

The voltages v_{o1} and v_{o2} are 180° out of phase, as are their input signals. The voltage v_o, measured between the two collector terminals, is

$$v_o = v_{o1} - v_{o2} = 2A_{DM}v_{DM} \tag{8-87}$$

Similarly, at midband frequencies, Eq. (8-71) is used to determine the *common-mode gain* A_{CM} as

$$A_{CM} = \frac{v_{o1}}{v_{CM}} = \frac{v_{o2}}{v_{CM}} = \frac{-h_{fe}R_L}{R_s + r_x + r_\pi + 2h_{fe}R_E} \tag{8-88}$$

where v_{CM} is the common-mode signal ($v_1 = v_2 = v_{CM}$). The output voltage between each collector and ground is identical because of circuit symmetry;

they are given in Eq. (8-89) as

$$v_{o1} = v_{o2} = A_{CM}v_{CM} \qquad (8\text{-}89)$$

As a result, the voltage measured between collectors is identically zero for common-mode signals. A consequence of the previous statement and Eq. (8-87) is that only differential-mode signals are amplified when the output signal is taken as v_o.

In practice the input signals v_1 and v_2 are arbitrary so that they are neither common-mode nor differential-mode signals. Arbitrary input signals can, however, be decomposed into their differential and common-mode components as indicated in Eqs. (8-90) and (8-91).

$$v_1 = v_{CM} + v_{DM} \qquad (8\text{-}90)$$

$$v_2 = v_{CM} - v_{DM} \qquad (8\text{-}91)$$

Simultaneous solution of Eqs. (8-90) and (8-91) for v_{CM} and v_{DM} gives

$$v_{CM} = \frac{v_1 + v_2}{2} \qquad v_{DM} = \frac{v_1 - v_2}{2} \qquad (8\text{-}92)$$

The significance of Eq. (8-92) is that for arbitrary input signals v_1 and v_2 the circuit of Fig. 8-56 can be represented as depicted in Fig. 8-57. In the circuit of Fig. 8-57, v_1 and v_2 have each been replaced by two signal sources as given

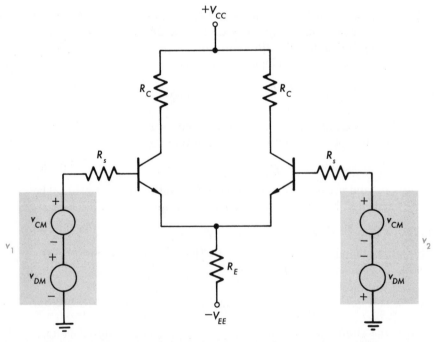

FIGURE 8-57 Differential-amplifier stage showing input signals decomposed into their common-mode and differential-mode signals.

in Eqs. (8-90) and (8-91), respectively. The output voltage v_{o1} can now be determined by the principle of superposition by first evaluating the response component to the differential inputs and then evaluating the common-mode response component. Utilization of Eqs. (8-86) and (8-89) allows the result to be expressed as

$$v_{o1} = A_{DM}v_{DM} + A_{CM}v_{CM} = A_{DM}\left(v_{DM} + \frac{A_{CM}}{A_{DM}}\, v_{CM} \right) \tag{8-93}$$

Substitution of Eq. (8-92) into Eq. (8-93) yields

$$v_{o1} = \frac{A_{DM}}{2}\, (v_1 - v_2) + \frac{A_{CM}}{2}\, (v_1 + v_2)$$

$$= \frac{A_{DM}}{2} \left[(v_1 - v_2) + \frac{A_{CM}}{A_{DM}}\, (v_1 + v_2) \right] \tag{8-94}$$

It is often desirable to use only one output terminal rather than to utilize the two possible output terminals while retaining the characteristic of the output proportional to the difference of the input signals. Examination of Eq. (8-94) indicates that this is accomplished, for practical purposes, when the common-mode output component is negligible when compared with the differential-mode response. Ideally, if A_{CM} were zero, only the differential-mode component would be present in v_{o1}. A measure of the effectiveness of the differential amplifier in this respect is the *common-mode rejection ratio,* designated *CMRR,* and defined as

$$\text{CMRR} = \frac{A_{DM}}{A_{CM}} \tag{8-95}$$

Substitution of Eq. (8-83) into (8-81) gives

$$v_{o1} = A_{DM} \left(v_{DM} + \frac{v_{CM}}{\text{CMRR}} \right) \tag{8-96}$$

which indicates that the CMRR must be very large for v_{o1} to approach $A_{DM}v_{DM} = A_{DM}(v_1 - v_2)/2$. Note that the ideal case is achieved only when the CMRR becomes infinite.

In Sec. 8-1, the operational amplifier stages are described when only one input signal is applied. Assuming an infinite CMRR and using v_{o1} as the output, when v_1 is applied with v_2 being zero, the differential-mode signal is simply $v_1/2$. By use of Eq. (8-96), v_{o1} becomes $A_{DM}v_1/2$. Since A_{DM} indicates an inherent phase reversal between input and output, this situation corresponds to the inverting-mode operation in the OP-AMP. In analogous fashion, when v_1 is zero and v_2 is applied, v_{DM} is $-v_2/2$ and v_{o1} is $-A_{DM}v_2/2$. The output is in phase with the input, and this leads to the noninverting OP-AMP stage.

The CMRR is related to the circuit parameters by substituting Eqs. (8-85) and (8-88) into Eq. (8-95) which yields

$$\text{CMRR} = \frac{R_s + r_x + r_\pi + 2h_{fe}R_E}{R_s + r_x + r_\pi} = 1 + \frac{2h_{fe}R_E}{R_s + r_x + r_\pi} \tag{8-97}$$

As is the usual case, the common-mode rejection ratio is assumed to be much greater than 1, so that Eq. (8-97) becomes

$$\text{CMRR} \doteq \frac{2h_{fe}R_E}{R_s + r_x + r_\pi} \tag{8-98}$$

It is evident in Eq. (8-98) that large values of the CMRR can be obtained only when R_E is large. Both because of limits on the value of R_E and for effective biasing, differential stages are biased by current sources of the type described in Sec. 8-2. The circuit shown in Fig. 8-58a is a differential-amplifier stage with current-source biasing. Figure 8-58b is the representation of Fig. 8-58a and depicts the current source I and its internal resistance R. The value of I must be sufficient to provide the dc current for both transistors. The value of R is the incremental resistance seen looking into terminals AB in Fig. 8-58a and is essentially the output resistance r_o of transistor $Q3$. As values of r_o can be several hundred thousand ohms, common-mode rejection ratios equal to or greater than 80 dB (10^4) can be achieved in practical circuits.

Example 8-12

A differential amplifier of the type shown in Fig. 8-58 is designed with the following circuit values: $R_C = 10$ kΩ, $R_s = 1$ kΩ, and $I = 400$ μA. The transistor parameters are: $r_x = 25\,\Omega$, $h_{fe} = 200$, $r_\pi = 25$ kΩ, and $r_o = 250$ kΩ.

a Determine the CMRR.

b For $v_1 = 0.1 \cos \omega t$ and $v_2 = 0.1 \cos \omega t - 0.005 \cos 2\omega t$, determine the output v_{o1}. Assume the value of ω permits midband analysis.

SOLUTION **a** By use of Eq. (8-98), with R_E taken as the r_o of the transistor in the current source,

$$\text{CMRR} = \frac{2 \times 200 \times 2.5 \times 10^5}{10^3 + 25 + 25 \times 10^3} = 3.84 \times 10^3 = 71.7 \text{ dB}$$

b The value of R_L is

$$R_L = \frac{10^4 \times 250 \times 10^3}{10^4 + 250 \times 10^3} = 9.62 \text{ k}\Omega$$

From Eq. (8-85)

$$A_{DM} = \frac{-200 \times 9.62 \times 10^3}{10^3 + 25 + 25 \times 10^3} = -73.9$$

The common-mode and differential-mode signals are, by use of Eq. (8-92),

$$v_{CM} = \tfrac{1}{2}(0.1 \cos \omega t + 0.1 \cos \omega t - 0.005 \cos 2\omega t)$$
$$= 0.1 \cos \omega t - 0.0025 \cos 2\omega t$$

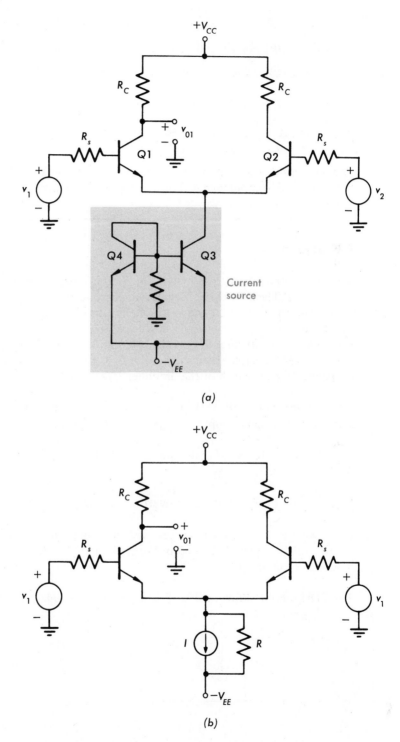

FIGURE 8-58 (a) Differential amplifier with current-source biasing; (b) equivalent representation of circuit in a.

and

$$v_{DM} = \tfrac{1}{2}(0.1 \cos \omega t - 0.1 \cos \omega t + 0.005 \cos 2\omega t) = 0.0025 \cos 2\omega t$$

The output voltage v_{o1} is given in Eq. (8-96). Substitution of values gives

$$v_{o1} = -73.9 \left(0.0025 \cos 2\omega t + \frac{0.1 \cos \omega t - 0.0025 \cos 2\omega t}{3.84 \times 10^3}\right)$$

$$= -0.185 \cos 2\omega t - 1.92 \times 10^{-3} \cos \omega t + 4.81 \times 10^{-5} \cos 2\omega t$$

$$= -0.185 \cos 2\omega t - 1.92 \times 10^{-3} \cos \omega t$$

The amplitude of the differential-mode output signal is nearly 100 times the common-mode output signal. For many practical purposes the output is the differential-mode response.

PROBLEMS

8-1   The amplifier stage shown in Fig. 8-3a is designed with $R_1 = 5$ kΩ and $R_2 = 10$ kΩ. The OP-AMP characteristics are $A_o = 2 \times 10^5$ and $f_h = 10$ Hz. Determine the gain and bandwidth of the amplifier stage.

8-2   The circuit in Fig. 8-4a is designed with $R_1 = 5$ kΩ and $R_2 = 10$ kΩ. The OP-AMP characteristics are $A_o = 3 \times 10^5$ and $f_h = 6$ Hz. Determine the gain and bandwidth of the amplifier stage.

8-3   An inverting OP-AMP stage is to be designed to produce an output signal of 5 V when an input voltage of 500 mV is applied. The bandwidth of the stage is to be 100 kHz.
a   Determine the gain-bandwidth product of the OP-AMP.
b   If $R_1 = 2$ kΩ, find R_2.

8-4   Repeat Prob. 8-3 if a noninverting OP-AMP is to be employed.

8-5   Determine the impedance seen by the signal source as a function of frequency for the stage described in Prob. 8-1 and sketch an asymptotic Bode diagram of the result.

8-6   The circuit shown in Fig. 8-59 is called a *voltage-follower*.

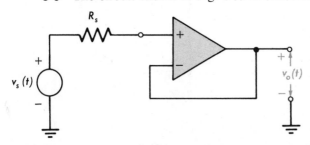

FIGURE 8-59   Voltage-follower circuit for Prob. 8-6.

a Determine the gain and bandwidth of the stage.

b If the OP-AMP characteristics are $A_o = 10^5$ and $f_h = 10$ Hz, and $R_s = 10$ kΩ, what is the output voltage $v_0(t)$ when $v_s(t) = 20 \cos 5 \times 10^5 t$?

8-7 The circuit shown in Fig. 8-60 is often used as an *instrumentation amplifier*, particularly in those instances in which small differences between signals must be measured.

a Derive an expression for v_o in terms of the input signals v_A and v_B and the four resistors. Assume the OP-AMP is ideal ($A_o \to \infty$).

b For $R_1 = R_2 = R_3 = R_4 = R$, determine v_o.

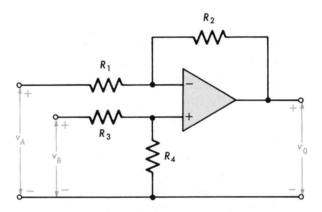

FIGURE 8-60 Instrumentation amplifier for Prob. 8-7.

8-8 The amplifier stage shown in Fig. 8-60 is also used as a *subtractor*. For $R_1 = R_3 = 10$ kΩ, find R_2 and R_4 so that $v_o = 2v_A - 3v_B$. Assume the OP-AMP is ideal.

8-9 The Wheatstone-bridge portion of the circuit in Fig. 8-61 represents a strain gauge in which the change in resistance ΔR is caused by the forces ap-

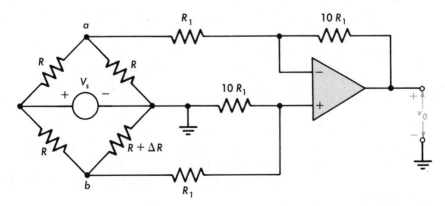

FIGURE 8-61 Strain gauge for Prob. 8-9.

plied to the structural member under test. Assuming the resistance R_1 is much greater than R,

a Derive an expression for v_o in terms of R, ΔR, and V.

b For $R = 120\ \Omega$, $V_S = 1.35$ V, and $R_1 = 10$ kΩ, what is the minimum value of R that can be measured if the smallest output voltage that can be detected is 100 μV?

c Using the result obtained in b, what is the minimum strain that can be measured if $\Delta R/R = \frac{1}{2}\ \Delta l/l$, where l and Δl are the length and elongation of the test member, respectively?

8-10 The transistor parameters in the circuit of Fig. 8-10a are $V_{BE} = 0.6$ V and $h_{FE} = 100$. The supply voltage is 9 V and $R = 10$ kΩ.

a Determine the value of I_C.

b If h_{FE} becomes 50, what is the percentage change in I_C?

c Repeat b for $h_{FE} = 200$.

8-11 The current mirror in Fig. 8-10a is designed with $V_{CC} = 5$ V and $R = 1$ kΩ. The transistors are identical and have $h_{FE} = 150$ and $V_{BE} = 0.7$ V.

a Determine I_C.

b What are the minimum and maximum values of h_{FE} if the change in I_C is to be less than or equal to 1 percent of the value obtained in a.

8-12 A current mirror is to be designed to provide 0.5 mA using a 10-V supply. The transistors used have $V_{BE} = 0.6$ V and $h_{FE} = 125$.

a Determine the value of R.

b Assuming all other parameter values remain constant and the change in V_{BE} is 2.2 mV/°C, what change in temperature is permissible if I_C is to remain within 1 percent of its nominal design value?

8-13 **a** Determine the value of R needed in the circuit of Fig. 8-10a to provide a current of 1 mA using a 15-V supply and the transistors described in Prob. 8-11.

b Assuming all other parameter values are unchanged and V_{BE} changes by 2.2 mV/°C, find the percentage change in I_C for a 50°C change in temperature.

8-14 **a** Determine I_C for the current mirror shown in Fig. 8-62. The transistors used are described in Prob. 8-12.

b What are the minimum and maximum values of I_C if h_{FE} varies between 40 and 300?

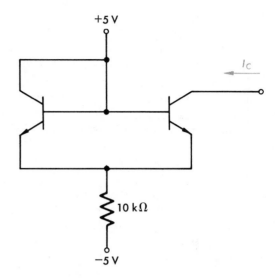

FIGURE 8-62 Current mirror for Prob. 8-14.

8-15 The circuit in Fig. 8-63 is that of a three-transistor current source. The transistors are identical. Derive an expression for I_C in terms of V_{CC}, R, V_{BE}, and h_{FE}.

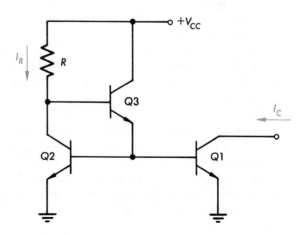

FIGURE 8-63 Circuit for Prob. 8-15.

8-16 Repeat Prob. 8-15 for the Wilson current source in Fig. 8-64.

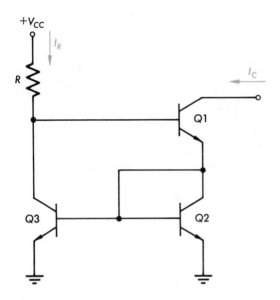

FIGURE 8-64 Wilson current source for Prob. 8-16.

8-17 The circuit parameters in the circuit of Fig. 8-64 are $V_{CC} = 12$ V, $R = 5$ kΩ, $V_{BE} = 0.7$ V and the range of h_{FE} values is between 75 and 250. Determine the range of values for I_C.

8-18 Repeat Prob. 8-17 for the circuit in Fig. 8-63.

8-19 Determine the value of R_E if the current source in Prob. 8-12 is to be converted to that depicted in Fig. 8-11 and used to provide 50 μA.

8-20 Transistors having $h_{FE} = 200$ and $V_{BE} = 0.7$ V are used in the circuit of Fig. 8-11. The supply voltage is 15 V and I_{C1} is to be 75 μA. Find the values of R and R_E if $I_{C1} = I_R/10$.

8-21 The circuit in Fig. 8-11 is designed having the following values: $V_{CC} = 15$ V, $R = 10R_E$, $V_{BE} = 0.6$ V, and $h_{FE} = 100$. Operation is at room temperature.
a Determine the ratio I_{C2}/I_{C1}. (*Hint:* The result is most easily obtained by plotting the graphs of each side of the transcendental equation.)
b For $R = 30$ kΩ, find I_{C1}.

8-22 The circuit in Fig. 8-12 is designed with $R_1 = 36$ kΩ, $R_2 = 12$ kΩ, $R_C = 2$ kΩ, $R_E = 1$ kΩ, and $V_{CC} = 12$ V. The transistor has $V_{BE} = 0.6$ V and $h_{FE} = 50$. Determine:
a The values of I_{CQ} and V_{CEQ}.
b Repeat *a* for $h_{FE} = 200$.

8-23 The bias network in Fig. 8-12 is designed with $R_1 = 72$ kΩ, $R_2 = 18$ kΩ, $R_C = 6$ kΩ, $R_E = 2$ kΩ and $V_{CC} = 15$ V. The transistor parameters are $h_{FE} = 75$ and $V_{BE} = 0.7$ V.
a Determine I_{CQ} and V_{CEQ}.
b What are the maximum and minimum values of h_{FE} permitted if the collector current is to remain within 5 percent of the value in a?

8-24 The effect of variations in V_{BE} on the quiescent point of the circuit described in Prob. 8-23 is to be investigated. Assuming all other parameter values remain constant and the value of V_{BE} changes by -2.2 mV/°C,
a Determine the new values of I_{CQ} and V_{CEQ} that correspond to a ± 50°C change in temperature.
b Repeat a for $h_{FE} = 250$.

8-25 A *pnp* transistor stage of the type shown in Fig. 8-12 is designed having $R_1 = 60$ kΩ, $R_2 = 4.5$ kΩ, $R_C = R_E = 600$, and $V_{CC} = -9$ V. Over the operating temperature range, unit-to-unit variations for the transistor-type employed indicate that when $h_{FE} = 60$, $V_{BE} = 0.75$; and when $h_{FE} = 300$, $V_{BE} = 0.45$. Find the minimum and maximum values of I_{CQ} and V_{CEQ}.

8-26 The circuit shown in Fig. 8-65 is a differential amplifier biased by a current source. All four transistors are identical and have $V_{BE} = 0.6$ V and $h_{FE} = 100$. For $R_C = 5$ kΩ and $V_{BB} = 3$ V, measurements indicate that $V_C = 10$ V and $V_{CE4} = 2.3$ V. Determine the values of R, R_B and V_{CE2}. For the purposes of calculation, assume that $h_{FE} + 2 \doteq h_{FE}$.

8-27 The circuit in Fig. 8-65 is designed with $R = 12$ kΩ and $R_B = 10$ kΩ. The four identical transistors have $V_{BE} = 0.7$ V and $h_{FE} = 100$. Determine:
a The value of R_C needed to make $V_C = 7.5$ V.
b The value of V_{BB} which makes $V_E = 1.5$ V.
For the purposes of calculation, assume $h_{FE} = 2 \doteq h_{FE}$.

8-28 An *n*-channel JFET is biased using the circuit of Fig. 8-14a in which $R_1 = 600$ kΩ, $R_2 = 278$ kΩ, $R_D = R_K = 1$ kΩ and $V_{DD} = 30$ V. The transfer characteristic of the JFET is the typical characteristic given in Fig. 8-66.
a Determine the values of I_{DQ} and V_{DSQ}.
b If the JFET characteristic is the maximum characteristic, what are the new values of I_{DQ} and V_{DSQ}?

8-29 The JFET whose typical characteristic is given in Fig. 8-66 is used in the circuit of Fig. 8-14a. The element values are $R_1 = 1$ MΩ, $R_2 = 250$ kΩ, $R_K = 2$ kΩ, and $V_{DD} = 28$ V. Determine:
a The value of I_{DQ}.
b The value of R_L required to make $V_{DSQ} = 8$ V.
c The minimum and maximum values of I_{DQ} corresponding to the minimum and maximum transfer characteristics in Fig. 8-66?

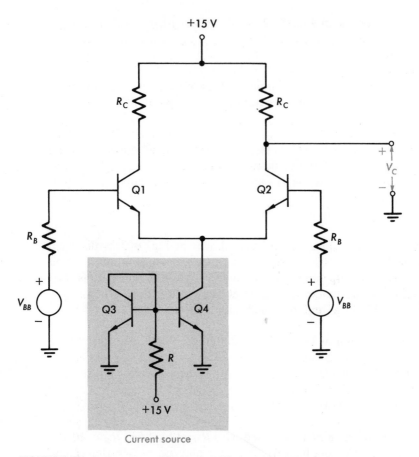

Current source

FIGURE 8-65 Current-source biasing for differential amplifier in Prob. 8-26.

8-30 The manufacturer of a p-channel JFET provides the following data:

	MINIMUM	MAXIMUM
V_{PO}	5	6
I_{DSS}	−2.5 mA	−4.5 mA

A circuit analogous to that of Fig. 8-14a for p-channel devices is to be designed so that I_{DQ} lies between −1.6 and −2.0 mA. The supply voltage is −30 V and $R_G \geq 100$ kΩ.
a Find R_K, R_1, and R_2.
b If $R_L = 10$ kΩ, what are the maximum and minimum values of V_{DSQ}?

8-31 The JFET whose characteristics are given in Fig. 8-66 is to be biased so that the drain current lies between 6 and 7 mA over the entire unit-to-unit variation indicated.

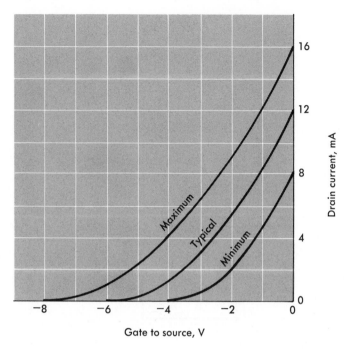

FIGURE 8-66 JFET transfer characteristics for Prob. 8-28.

a For a supply voltage of 50 V and with $R_G \geq 100$ kΩ, determine R_1, R_2, and R_K.
b Using the results obtained in a, find the value of R_D which makes $V_{DSQ} = 10$ V for the typical device.
c Determine the range of values of V_{DSQ}.

8-32 The typical JFET described by the characteristics in Fig. 8-66 is to be biased at $I_{DQ} = 4$ mA.
a Determine R_K if the maximum variation in I_{DQ} is to be 10 percent of its typical value.
b What is the ratio of R_1/R_2 needed to satisfy the conditions in a?

8-33 The JFET described in Prob. 8-30 is used in the circuit of Fig. 8-67. The parameter values are $R_K = 1$ kΩ, $R_G = 150$ kΩ, $V_{GG} = 1.5$ V, $R_C = 10$ kΩ, and $V_{DD} = 40$ V.
a Determine the maximum value of V_s for which the gate junction remains reverse-biased.
b What is the peak-to-peak drain current under the conditions given in a?

8-34 In the circuit of Fig. 8-67, $R_G = 200$ kΩ, $R_K = 3$ kΩ, $V_{GG} = 3$ V, $R_C = 12$ kΩ, and $V_{DD} = 30$ V. The JFET characteristics are described in

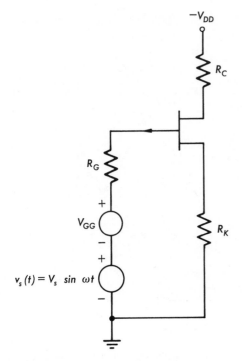

FIGURE 8-67 Circuit for Prob. 8-33.

Prob. 8-30. The input-signal amplitude $V_s = 1.0$ V. Sketch the voltage across R_C as a function of time for the extremes in unit-to-unit variations.

8-35 In the circuit of Fig. 8-68, the voltage source V_{DD} establishes a diode voltage of 0.3 V when $v_s(t) = 0$. The small-signal diode parameters are $r_D = 50 \Omega$ and $C_D = 500$ pF. Determine the ac component of the load current for $v_s(t) = 0.05 \cos 10^6 t$.

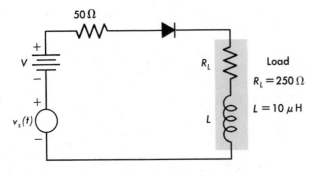

FIGURE 8-68 Circuit for Prob. 8-35.

8-36 The small-signal diode parameters are $r_D = 40\ \Omega$, $C_D = 750\ \Omega$, $r_R = 1\ M\Omega$, and $C_T = 2\ pF$. Determine the ac component of load voltage in Fig. 8-69 for $i_s(t) = 10^{-5} \cos 10^6 t$ when
a The source $V_{DD} > 0$
b The source $V_{DD} < 0$

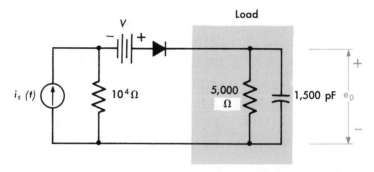

FIGURE 8-69 Circuit for Prob. 8-36.

8-37 a Draw the small-signal equivalent circuit for the amplifier stage shown in Fig. 8-70.
b What modifications to the circuit in *a* result when the frequency of operation permits the reactances of C_B, C_C, and C_E to be considered zero?

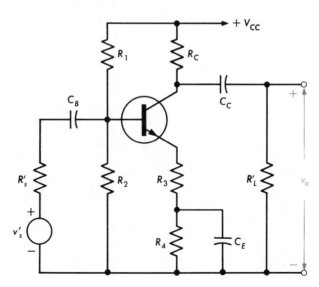

FIGURE 8-70 Circuit for Prob. 8-37.

8-38 a Draw a small-signal equivalent circuit for the amplifier stage shown in Fig. 8-71.

b Repeat *a* when the values of C_B and C_C are considered infinite.

c Repeat *b* if the capacitances C_π and C_μ are zero.

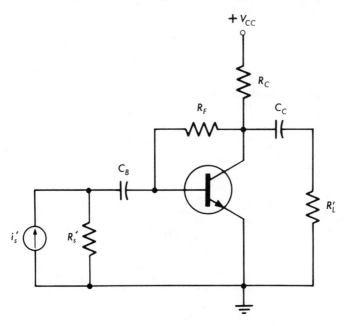

FIGURE 8-71 Circuit for Prob. 8-38.

8-39 a Draw the small-signal equivalent circuit for the amplifier stage in Fig. 8-72.

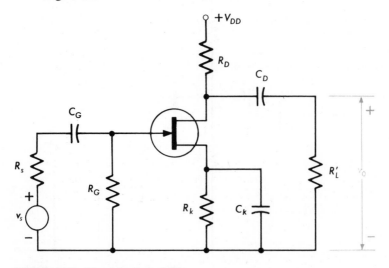

FIGURE 8-72 Circuit for Prob. 8-39.

b Repeat *a* for signal frequencies at which the reactances of C_G, C_K, and C_D are sufficiently small so they may be neglected.

8-40 a Repeat Prob. 8-39*a* for the grounded-gate stage in Fig. 8-73.
b At signal frequencies which permit the reactances of C_{gd} and C_{gs} to be treated as open circuits, repeat *a*.

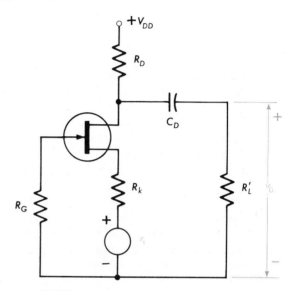

FIGURE 8-73 Grounded-gate stage for Prob. 8-40.

8-41 Measurements made at room temperature on an *npn* transistor biased at $I_{CQ} = 5$ mA and $V_{CEQ} = 10$ V are $h_{ie} = 500$ Ω, $h_{fe} = 80$, $f_T = 600$ MHz, $C_{ob} = 2$ pF, and $g_o = 5$ μ℧. The case and lead capacitances are each 1 pF. Determine the hybrid-π parameters of the transistor.

8-42 A *pnp* transistor is biased at $V_{CEQ} = 5$ V and $I_{CQ} = 2.5$ mA. Measurements at room temperature indicate that $C_{ob} = 1.5$ pF, $g_o = 10$ μ℧, $h_{fe} = 100$, and $r_x = 50$ Ω and the case capacitance is 0.5 pF. In addition, $|h_{fe}(j\omega)| = 10$ at $\omega = 10^9$ rad/s.
a Determine the hybrid-π parameters.
b If a collector resistance of 1 kΩ is used, obtain the unilateral hybrid π.

8-43 The short-circuit input impedance of an *npn* transistor, biased at $I_{CQ} = 10$ mA and $V_{CEQ} = 7.5$ V is 100 at a frequency $f = 50$ MHz. The data provided by the manufacturer is $C_{ob} = 1.5$pF, the case capacitance is 0.5 pF, $r_x = 25$ Ω and $h_{fe} = 200$. Determine:
a The value of ω_T
b The frequency at which $|h_{fe}(j\omega)| = 0.707 h_{fe}$
c The unilateral hybrid-π model if the short circuit is replaced by a 500-Ω resistance. Neglect g_o.

8-44 The manufacturer provides the following data for a p-channel JFET: $C_{rss} = 6$ pF, $C_{iss} = 12$ pF, and the low-frequency values of y_{fs} and y_{os} are 2 and 30 μU, respectively.
a Determine the JFET equivalent circuit.
b Obtain the unilateral model if $R_D = 10$ kΩ.

8-45 The n-channel JFET described by the typical transfer characteristic given in Fig. 8-66. Determine the value of g_m in the small-signal model at a bias current $I_{DQ} = 4.0$ mA.

8-46 An amplifier stage of the type shown in Fig. 8-34 is designed having $R_1 = 180$ kΩ, $R_2 = 60$ kΩ, $R_E = 1$ kΩ, $R_C = 1.5$ kΩ, $C_B = 2.0$ μF, and C_E is sufficiently large so that its reactance is negligible. The transistor parameters are $r_x = 25$ Ω, $r_\pi = 500$ Ω, $g_m = 0.2$ U, $g_o = 20$ μU, $C_\pi = 100$ pF, and $C_\mu = 1.0$ pF. The source resistance is 300 Ω. Determine:
a The midband gain
b The upper and lower half-power frequencies

8-47 The transistor described in Prob. 8-46 is used in an amplifier stage as shown in Fig. 8-34. The element values are $R_1 = 120$ kΩ, $R_2 = 30$ kΩ, $R_E = 1.5$ kΩ, $R_C = 500$ Ω, and $C_E = 100$ μF. The reactance of C_B is sufficiently small at signal frequencies so that it can be neglected. Determine G_0, ω_H, and ω_L of the stage.

8-48 A transistor stage of the type shown in Fig. 8-35b is designed with $R_C = 5$ kΩ and $R_s = 1$ kΩ. The transistor parameters are $h_{fe} = 120$, $g_m = 0.4$ U, $r_x = 0$, $r_o \to \infty$, $C_\mu = 0.5$ pF, and $C_\pi = 50$ pF. Determine the gain and bandwidth of the stage.

8-49 The input signal v_s to the amplifier stage in Prob. 8-48 is a step function of amplitude 5 mV. Determine $v_{01}(t)$.

8-50 The input signal to an amplifier whose transfer function is given by Eq. (8-56) is a unit-step function.
a Determine the times at which the amplifier response is $0.9G_o$ and $0.1G_o$.
b Find the rise time of the amplifier stage, defined as the difference between the response times in a.

8-51 An amplifier stage is to be designed using the transistor described in Prob. 8-46 to provide a 2-V output signal when an input signal of 80 mV is applied. The source resistance is 50 Ω, and the base-bias resistors are sufficiently large so their effect is negligible.
a Determine the value of R_C required.
b What is the value of the upper-half power frequency?
c If the emitter resistance is 2.0 kΩ, what value of C_E is required for $f_L = 50$ Hz? Assume $C_B \to \infty$.

8-52 A common-emitter amplifier stage is designed using the transistor given in Prob. 8-48. The base-bias resistances are $R_1 = 150$ kΩ and $R_2 = 30$ kΩ, and the source resistance is 300 Ω.
a What value of R_C is needed for $\omega_H = 2 \times 10^7$ rad/s?
b Find G_O using the result in a.
c Assuming the emitter resistance is perfectly bypassed, determine C_B so that $\omega_L = 10^3$ rad/s.

8-53 The small-signal transistor parameters given in Prob. 8-48 are specified at an operating point $I_{CQ} = 10$ mA and $V_{CEQ} = 5$ V. The dc behavior of the transistor is described by $V_{BE} = 0.6$ V and $h_{FE} = 120$. A supply voltage of 15 V is available. An amplifier-stage of the type shown in Fig. 8-34 is to be designed which provides a 500-mV output signal for an input signal of 5 mV fed through a 300-Ω resistance. The amplifier response is to be no less than 3 dB below its midband value at all frequencies between 100 Hz and 6 MHz.
a Determine the values of R_1, R_2, R_C, and R_E.
b Find C_B assuming $C_E \to \infty$ needed to achieve the low-frequency response.
c Find C_E assuming $C_B \to \infty$ needed to satisfy the low-frequency specification.
d What is the value of f_H?

8-54 Repeat Prob. 8-46a, b, and d when the emitter resistance is unbypassed ($C_E = 0$).

8-55 An emitter resistance $R_E = 250$ Ω is added to the amplifier stage in Prob. 8-48.
a Determine the values of G_O and ω_H.
b If h_{fe} changes to 200, find the new value of G_O.
c What is the input resistance to the amplifier stage?

8-56 The unit-to-unit variation in h_{fe} of the transistor described in Prob. 8-48 is $60 \leq h_{fe} \leq 240$. The amplifier-stage depicted in Fig. 8-46 is to be designed with a midband gain that is at least 10 and no greater than 11 over the range of h_{fe} values. The source resistance is 50 Ω. Determine:
a R_C and R_E
b The range of values of ω_H
[*Hint:* Form $(G_{MAX} - G_{MIN})/G_{MIN}$ as a function of the circuit elements and $(h_{fe,max} - h_{fe,min})/h_{fe,min}$. Assume I_C is constant.]

8-57 The transistor whose parameters are given in Prob. 8-48 is used in a common-base amplifier stage, the model for which is shown in Fig. 8-74. Find G_O and ω_H.

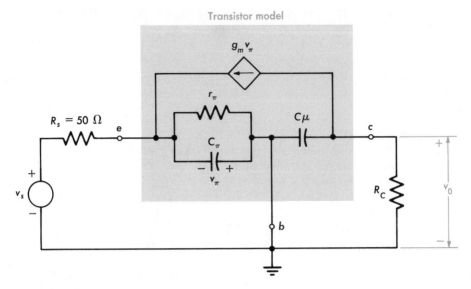

FIGURE 8-74 Circuit for Prob. 8-57.

8-58 The discrete-component amplifier stage in Fig. 8-70 is used when both the dc operating point and midband gain must be stabilized because of variations in h_{FE} and h_{fe}. The transistor parameters are given in Prob. 8-46; the element values are $R_s' = 600\ \Omega$, $R_1 = 300\ \text{k}\Omega$, $R_2 = 75\ \text{k}\Omega$, $R_C = 2\ \text{k}\Omega$, $R_3 = 50\ \Omega$, and $R_L' = 2\ \text{k}\Omega$.

a Determine the midband gain and input impedance.

b If h_{fe} is doubled, what percentage change in the midband gain occurs?

c Repeat b if h_{fe} is halved.

d Find the value of f_H.

e Assuming C_B and C_C to be perfect coupling capacitors, find the value of C_E needed to make $f_L = 100$ Hz.

f Assuming C_B and C_E provide perfect coupling and bypassing, find the value of C_C needed to make $f_L = 100$ Hz.

8-59 The FET amplifier stage shown in Fig. 8-48 is designed having the following element values: $R_1 = 300$ kΩ, $R_2 = 150$ kΩ, $R_D = 10$ kΩ, $R_K = 2$ kΩ, $C_G = 1\ \mu$F, and $C_K = 100\ \mu$F. The FET parameters are $g_m = 5$ m$\mho$, $r_d = 50$ kΩ, $C_{gs} = 10$pF and $C_{gd} = 5$pF. The signal source has an amplitude of 50 mV rms and is fed through a 600-Ω source resistance.

a Determine the peak-to-peak output voltage at midband frequencies.

b Find the upper half-power frequency.

c Assuming C_K acts as a perfect bypass, determine the lower half-power frequency.

8-60 The amplifier stage depicted in Fig. 8-48 is designed using an FET whose parameters are $g_m = 2\ \text{m}\mho$, $r_d = 20$ kΩ, $C_{gs} = 8$ pF, and $C_{gd} = 4$ pF. The circuit elements are $R_K = 1$ kΩ, $R_1 = 160$ kΩ, and $R_2 = 40$ kΩ. The source resistance is 300 Ω.

a Determine R_D if the midband gain is to be 20.
b Assuming C_K acts as a perfect bypass, determine C_G to give a lower half-power frequency of 200 Hz.
c Find the upper half-power frequency.

8-61 The FET whose parameters are given in Prob. 8-59 is used in an amplifier stage of the type shown in Fig. 8-48. The gate-bias resistors are sufficiently large so their effect is negligible and C_K and C_C provide ideal bypass and coupling, respectively. The input signal is supplied through a 1-kΩ source resistance. Determine the value of R_D needed to achieve a midband gain of 40 and the corresponding upper half-power frequency.

8-62 Derive the expressions for the midband gain and input impedance for the circuit shown in Fig. 8-73.

8-63 The circuit in Fig. 8-75 is a MOSFET amplifier stage for which the biasing arrangement is not shown and can be assumed to have negligible effect on the circuit performance. The device parameters are $g_m = 4\,\text{m}\mho$. $r_d = 25\,\text{k}\Omega$, $C_{gs} = 5\,\text{pF}$, and $C_{gd} = 5\,\text{pF}$. The input-signal amplitude is 100 mV peak to peak and is fed through a 600-Ω source resistance. For $R_L = 15\,\text{k}\Omega$, determine:
a The value of R_K needed for a midband gain of 10.
b The upper half-power frequency using the result in a.
c The change in gain and bandwidth of g_m is decreased by 50 percent.

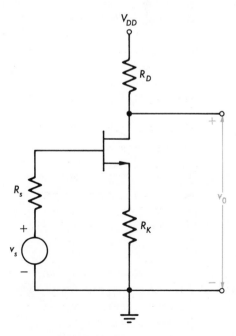

FIGURE 8-75 Circuit for Prob. 8-63.

8-64 a Determine the midband gain and the bandwidth of the amplifier-stage in Prob. 8-62 when $R_K = 0$.
b Repeat a if g_m is decreased by 50 percent.

8-65 The amplifier stage in Prob. 8-59 is operated with $C_K = 0$, that is, R_K is unbypassed.
a Determine the midband gain and upper half-power frequency.
b Repeat a if g_m is increased by 20 percent.
c Repeat a and b when $R_K = 0$.

8-66 The circuit of Fig. 8-51 is designed with a transistor whose parameters are $r_x = 0$, $g_m = 0.1$ ℧, $h_{fe} = 150$, $g_o = 0$, $f_T = 796$ MHz, $C_{ob} = 2.0$ pF, and $C_{case} = 1.0$ pF. The element values are $R_1 = 300$ kΩ, $R_2 = 60$ kΩ, $R_E = 2$ kΩ, and $R'_s = 5$ kΩ. Determine:
a The midband gain and the input and output impedances.
b The value of C_B which produces a value of $f_L = 100$ Hz.

8-67 The transistor whose parameters are given in Prob. 8-66 is used in the emitter-follower circuit shown in Fig. 8-76. Assume that the input signal has both constant and time-varying components such that the constant component establishes the appropriate bias conditions in the circuit. For $R_s = 10$ kΩ,
a Determine R_E so that the output impedance is no greater than 75 Ω.
b What are the gain and input impedance at midband frequencies?
c Find the upper half-power frequency.

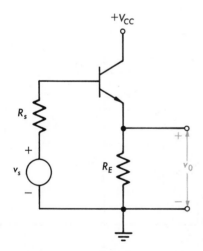

FIGURE 8-76   Circuit for Prob. 8-67.

8-68 Figure 8-77 is the circuit of an emitter-follower which employs a Darlington-pair configuration. The transistors used are identical and their parameters are given in Prob. 8-66. For $R_s = 10$ kΩ and $R_E = 1$ kΩ, determine the midband values of gain and input and output impedances.

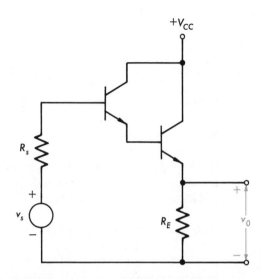

FIGURE **8-77** Emitter-follower stage using a Darlington compound transistor for Prob. 8-68.

8-69 The source-follower stage depicted in Fig. 8-54 is designed with element values: $R_s' = 5$ kΩ, $R_G = 500$ kΩ, $R_K = 5$ kΩ, and $C_G = 2$ μF. Determine:
a The midband gain and output impedance.
b The lower half-power frequency.

8-70 Determine the upper half-power frequency of the source-follower stage described in Prob. 8-69.

8-71 In the differential amplifier of Fig. 8-56, the parameters of the transistor used are given in Prob. 8-66. The element values are $R_C = 10$ kΩ, $R_E = 5$ kΩ, and $R_s = 1$ kΩ. Determine the common-mode gain, the differential-mode gain, and the CMRR.

8-72 The differential amplifier in Fig. 8-58 is designed with $R_C = 50$ kΩ, $R_s = 10$ kΩ, and $R = 100$ kΩ. The transistor parameters are $r_\pi = 50$ kΩ, $r_x = 0$, $h_{fe} = 2 \times 10^3$, and $r_o = 400$ kΩ. Determine:
a A_{DM}, A_{CM}, and the CMRR.
b The common-mode input impedance.

8-73 The transistor whose parameters are described in Prob. 8-72 is used in the circuit of Fig. 8-58. The source resistance is 5 kΩ.
a Determine the value of R needed to give a CMRR = 80 dB.
b The common-mode input impedance.

8-74 The following small-signal measurements are made on the differential amplifier of Fig. 8-56: $v_1 = 20$ mV, $v_2 = 0$, $v_{o1} = -4$ V; $v_1 = v_2 = 20$ mV, $v_{o1} = -4$ mV.
a Determine the CMRR.
b Find v_{o1} when $v_1 = 0$ and $v_2 = 20$ mV.

8-75 The CMRR and differential-mode gain of a differential amplifier are 60 and 80 dB, respectively.
a Determine $v_{o2}(t)$ for input signals $v_1(t) = 3 + 10^{-2} \cos 10^4 t$ and $v_2(t) = 3 + 1.01 \cos 10^4 t$.
b If the transistor whose parameters are given in Prob. 8-66 is used, find the values of R_C and R_E needed in the circuit of Fig. 8-56 to realize the given CMRR and A_{DM}. Assume the signal source is fed through a 2-kΩ resistance.

8-76 The differential amplifier is often used in instrumentation systems to detect small differences between signals.
a If $v_1(t) = 2(1 + 10^{-2} \cos \omega t)$ and $v_2(t) = 2 + 2.10 \times 10^{-2} \cos \omega t$, determine the value of the CMRR if the common-mode output is to be no greater than 1 percent of the differential-mode output.
b The circuit of Fig. 8-58 is used to realize the differential stage and employs the transistor whose parameters are given in Prob. 8-72, find the value of R necessary to achieve the CMRR. The source resistance is 5 kΩ.
c What value of R_C is needed to produce an output voltage of 2 V?

8-77 Assuming the transistors are identical in all respects, determine the upper-half power frequency in both the common- and differential-modes for the amplifier in Prob. 8-71.

8-78 Repeat Prob. 8-77 for the amplifier stage in Prob. 8-72. The additional transistor parameters are $C_\pi = 200$ pF and $C_\mu = 2$ pF.

MULTISTAGE AMPLIFIERS

9

ERS □ MULTISTAGE AMPLIFIERS □ MULTISTAGE
STAGE AMPLIFIERS □ MULTISTAGE AMPLIFIERS
MULTISTAGE AMPLIFIERS □ MULTISTAGE AMPL
AMPLIFIERS □ MULTISTAGE AMPLIFIERS □ MUL
MULTISTAGE AMPLIFIERS □ MULTISTAGE AMPL
AMPLIFIERS □ MULTISTAGE AMPLIFIERS □ MUL
MULTISTAGE AMPLIFIERS □ MULTISTAGE AMPL
AMPLIFIERS □ MULTISTAGE AMPLIFIERS □ MUL
MULTISTAGE AMPLIFIERS □ MULTISTAGE AMPL
AMPLIFIERS □ MULTISTAGE AMPLIFIERS □ MULTISTAGE AMPLIFIEI
MULTISTAGE AMPLIFIERS □ MULTISTAGE AMPLIFIERS □ MULTISTA
AMPLIFIERS □ MULTISTAGE AMPLIFIERS □ MULTISTAGE AMPLIFIEI
MULTISTAGE AMPLIFIERS □ MULTISTAGE AMPLIFIERS □ MULTISTA
AMPLIFIERS □ MULTISTAGE AMPLIFIERS □ MULTISTAGE AMPLIFIEI
MULTISTAGE AMPLIFIERS □ MULTISTAGE AMPLIFIERS □ MULTISTA
AMPLIFIERS □ MULTISTAGE AMPLIFIERS □ MULTISTAGE AMPLIFIEI
MULTISTAGE AMPLIFIERS □ MULTISTAGE AMPLIFIERS □ MULTISTA
AMPLIFIERS □ MULTISTAGE AMPLIFIERS □ MULTISTAGE AMPLIFIEI
MULTISTAGE AMPLIFIERS □ MULTISTAGE AMPLIFIERS □ MULTISTA
AMPLIFIERS □ MULTISTAGE AMPLIFIERS □ MULTISTAGE AMPLIFIEI
MULTISTAGE AMPLIFIERS □ MULTISTAGE AMPLIFIERS □ MULTISTA
AMPLIFIERS □ MULTISTAGE AMPLIFIERS □ MULTISTAGE AMPLIFIEI
MULTISTAGE AMPLIFIERS □ MULTISTAGE AMPLIFIERS □ MULTISTA
AMPLIFIERS □ MULTISTAGE AMPLIFIERS □ MULTISTAGE AMPLIFIEI
MULTISTAGE AMPLIFIERS □ MULTISTAGE AMPLIFIERS □ MULTISTA
AMPLIFIERS □ MULTISTAGE AMPLIFIERS □ MULTISTAGE AMPLIFIEI
MULTISTAGE AMPLIFIERS □ MULTISTAGE AMPLIFIERS □ MULTISTA
AMPLIFIERS □ MULTISTAGE AMPLIFIERS □ MULTISTAGE AMPLIFIEI
MULTISTAGE AMPLIFIERS □ MULTISTAGE AMPLIFIERS □ MULTISTA
AMPLIFIERS □ MULTISTAGE AMPLIFIERS □ MULTISTAGE AMPLIFIEI
MULTISTAGE AMPLIFIERS □ MULTISTAGE AMPLIFIERS □ MULTISTA
AMPLIFIERS □ MULTISTAGE AMPLIFIERS □ MULTISTAGE AMPLIFIEI
MULTISTAGE AMPLIFIERS □ MULTISTAGE AMPLIFIERS □ MULTISTA
AMPLIFIERS □ MULTISTAGE AMPLIFIERS □ MULTISTAGE AMPLIFIEI
MULTISTAGE AMPLIFIERS □ MULTISTAGE AMPLIFIERS □ MULTISTA
AMPLIFIERS □ MULTISTAGE AMPLIFIERS □ MULTISTAGE AMPLIFIEI
MULTISTAGE AMPLIFIERS □ MULTISTAGE AMPLIFIERS □ MULTISTA
AMPLIFIERS □ MULTISTAGE AMPLIFIERS □ MULTISTAGE AMPLIFIEI
MULTISTAGE AMPLIFIERS □ MULTISTAGE AMPLIFIERS □ MULTISTA

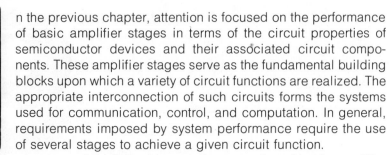

n the previous chapter, attention is focused on the performance of basic amplifier stages in terms of the circuit properties of semiconductor devices and their associated circuit components. These amplifier stages serve as the fundamental building blocks upon which a variety of circuit functions are realized. The appropriate interconnection of such circuits forms the systems used for communication, control, and computation. In general, requirements imposed by system performance require the use of several stages to achieve a given circuit function.

In this chapter and the following one, the properties of basic amplifier stages are used to evaluate the performance of several frequently used multistage circuits. As in the analysis of single-stage amplifiers, multistage amplifier behavior is described by the gain, frequency response, and impedance levels of the circuit. Feedback concepts are introduced as a means to control these and other circuit characteristics. In the last section of the chapter, the operational amplifier is viewed as the multistage circuit it is. This viewpoint serves to bring together many of the concepts introduced in this and the two previous chapters.

9-1 CASCADED OPERATIONAL AMPLIFIER STAGES

The essential purpose of electronic amplifiers is to increase the amplitude and power of a signal so that either useful work can be done or information processing can be realized more easily. The output-signal power is greater than the input-signal power; the additional power is supplied by the bias supply. The amplifier action is then one of energy conversion in which the bias power is converted to signal power within the device.

Most amplifiers consist of several stages because the amplification required over the bandwidth cannot be achieved by a basic amplifier stage of the type described in Chap. 8. Figure 9-1 is the circuit of a three-stage amplifier which is composed of three operational amplifier stages connected in a *cascade configuration*. This type of configuration consists of a number of stages connected so that the output of one stage becomes the input to the succeeding stage. The first stage in the cascade is usually referred to as the *input* or *initial stage;* the last stage is called the *output stage.* All of the other stages are referred to as *intermediate* or *interior stages.*

For the purpose of the ensuing discussion, it is assumed that the operational amplifiers are identical and have zero output impedance and infinite input impedance as indicated in the model of Fig. 8-1b. In the circuit of Fig. 9-1, the output voltage v_1 of the first stage is the input voltage to the second stage. Similarly, the output voltage v_2 of the second stage is the input voltage to the third stage. In order to focus attention on the cascading process, it is desirable not only to obtain a numerical result for the overall amplifier gain v_o/v_s but also to relate the overall gain to the gains of the individual stages which comprise the amplifier. This is accomplished by obtaining the Thévenin equivalent of the circuit to the left of terminals 1 and 1$'$. This por-

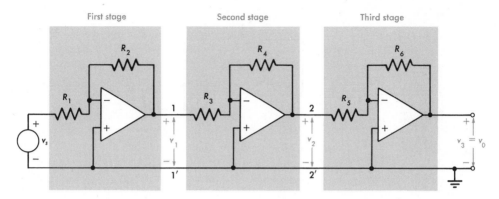

FIGURE 9-1 Three-stage cascade amplifier.

tion of the amplifier is the unloaded first stage, i.e., the first stage without the connection to the second stage. This Thévenin equivalent source becomes the input to the second stage. Repetition of this process for each of the remaining stages allows the output voltage to be related to the input voltage by means of the individual-stage responses.

The unloaded first stage is depicted in Fig. 9-2. Comparison of Fig. 9-2 with Fig. 8-3a shows the former to be the operational-amplifier stage discussed in Sec. 8-1. Therefore, the first-stage closed-loop gain is given by Eqs. (8-6) and (8-7). The open-circuit voltage v_{10} is then the gain of the stage multiplied by the source voltage, or

$$v_{10} = \frac{-A_o R_2}{R_2 + R_1(1 + A_o)}\, v_s = G_{o1} v_s \tag{9-1}$$

For very large values of $A_o (A_o \to \infty)$, Eq. (9-1) becomes

$$v_{10} = -\frac{R_2}{R_1}\, v_s \tag{9-2}$$

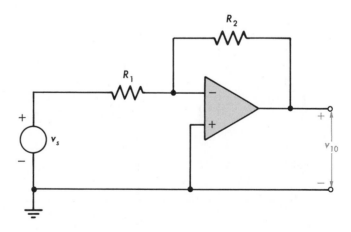

FIGURE 9-2 Unloaded first stage of amplifier in Fig. 9-1.

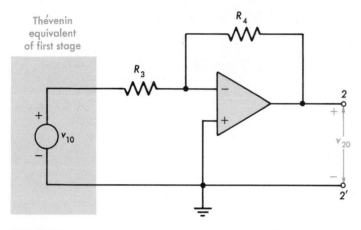

FIGURE 9-3 Second-stage representation of circuit in Fig. 9-1.

Because the operational amplifier is assumed to have zero output resistance, the Thévenin resistance of the unloaded first stage is also zero.

The Thévenin equivalent of the first stage is the input to the second stage, as shown in Fig. 9-3. Comparison of Fig. 9-3 with Fig. 9-2 shows that the circuits are identical if v_{10}, R_3, and R_4 are identified with v_s, R_1, and R_2, respectively. Thus, the Thévenin voltage of the unloaded second stage v_{20} is given by Eqs. (9-1) and (9-2) as

$$v_{20} = \frac{-A_o R_4}{R_4 + R_3(1 + A_o)} \, v_{10} = G_{20} v_{10} \tag{9-3}$$

and, for large values of A_o,

$$v_{20} = -\frac{R_4}{R_3} \, v_{10} \tag{9-4}$$

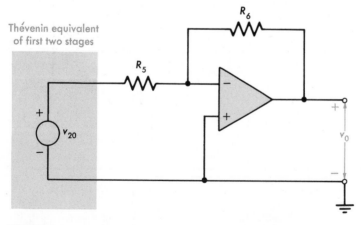

FIGURE 9-4 Representation of third stage of cascade amplifier.

As indicated for the first stage, the Thévenin resistance is zero by virtue of the zero value for the output resistance of the OP-AMP.

As depicted in Fig. 9-4, the Thévenin equivalent of the first two stages forms the input to the third stage. Identification of v_{20}, R_5, and R_6 with v_s, R_1, and R_2, respectively, shows the circuits of Figs. 9-4 and 9-2 to be identical. By following the procedure outlined for the first two stages, the output voltage v_o is expressed as

$$v_o = \frac{-A_o R_6}{R_6 + R_5(1 + A_o)} v_{20} = G_{o3} v_{20} \tag{9-5}$$

As A_o approaches infinity,

$$v_o = -\frac{R_6}{R_5} v_{20} \tag{9-6}$$

Combination of Eqs. (9-1), (9-3), and (9-5) shows the overall amplifier gain G_{vo} to be

$$G_{vo} = \frac{v_o}{v_s} = G_{o1} G_{o2} G_{o3} \tag{9-7}$$

The subscript o in Eq. (9-7) is used to signify a midband value; that is, G_{vo} is the midband gain of the cascade structure. Equation (9-7) indicates that the overall gain of the amplifier is the product of the gains of the individual stages. This statement is general and can be extended to an n-stage cascade structure.

In Figs. 9-1 and 9-2, the signal source is depicted as an ideal voltage source. However, the equivalent series resistance of a practical source can be included in the value of R_1. Similarly, if the Thévenin resistance of each stage is not zero, caused by a nonzero output resistance of the operational amplifier, its effect can be included in the series resistance of the following stage. For typical values of A_o, it can be shown that the Thévenin resistance of the stage is virtually zero, even when the OP-AMP has a nonzero output resistance (see Prob. 2-22). Consequently, it is assumed that the OP-AMP output resistance is zero.

The frequency response of the cascaded amplifier can be determined in a manner similar to the midband gain by considering that the open-loop gain of each OP-AMP is given as shown in Fig. 8-2. In Sec. 8-1, the closed-loop gain of an operational amplifier stage is shown to preserve the gain-bandwidth product. Therefore, from Eq. (8-12) and Fig. 8-5, the closed-loop gain variation with frequency of each stage can be expressed as

$$G_H(s) = \frac{G_o}{1 + \dfrac{s}{\omega_H}} \tag{9-8}$$

The overall gain as a function of frequency $G_{vH}(s)$ can then be written in the form of Eq. (9-7) as

$$G_{vH}(s) = G_{H1}(s) G_{H2}(s) G_{H3}(s) \tag{9-9}$$

Upon substitution of Eq. (9-8) for the responses of the respective stages, Eq. (9-9) becomes

$$G_{vH}(s) = \frac{G_{o1}G_{o2}G_{o3}}{\left(1 + \dfrac{s}{\omega_{H1}}\right)\left(1 + \dfrac{s}{\omega_{H2}}\right)\left(1 + \dfrac{s}{\omega_{H3}}\right)}$$

$$= \frac{G_{vo}}{\left(1 + \dfrac{s}{\omega_{H1}}\right)\left(1 + \dfrac{s}{\omega_{H2}}\right)\left(1 + \dfrac{s}{\omega_{H3}}\right)} \qquad (9\text{-}10)$$

in which ω_{H1}, ω_{H2}, and ω_{H3} are the closed-loop bandwidths of stages one, two, and three, respectively. The terms G_{o1}, G_{o2}, and G_{o3} are the midband gains of the individual stages and are given in Eqs. (9-1), (9-3), and (9-5).

To determine the upper half-power angular frequency ω_{Ho} and, consequently, the bandwidth, a Bode diagram is constructed from Eq. (9-10). A typical response is illustrated in Fig. 9-5 in which it is assumed that $\omega_{H1} < \omega_{H2} < \omega_{H3}$.

Alternatively, Eq. (9-10) can, by performing the indicated multiplications, be rewritten as

$$G_{vH}(s) = \frac{G_{vo}}{1 + a_1 s + a_2 s^2 + a_3 s^3} \qquad (9\text{-}11)$$

where

$$a_1 = \frac{1}{\omega_{H1}} + \frac{1}{\omega_{H2}} + \frac{1}{\omega_{H3}} \qquad (9\text{-}12a)$$

$$a_2 = \frac{1}{\omega_{H1}\omega_{H2}} + \frac{1}{\omega_{H1}\omega_{H3}} + \frac{1}{\omega_{H2}\omega_{H3}} \qquad a_3 = \frac{1}{\omega_{H1}\omega_{H2}\omega_{H3}} \qquad (9\text{-}12b)$$

It is often desirable and sometimes necessary to approximate the value of ω_{Ho} without obtaining the entire frequency response. The first-order approximation of Eq. (9-11) is simply

$$G_{vH}(s) = \frac{G_{vo}}{1 + a_1 s} \qquad (9\text{-}13)$$

from which ω_{Ho} becomes

$$\omega_{Ho} = \frac{1}{a_1} \qquad (9\text{-}14)$$

The result expressed in Eq. (9-14) is often referred to as the *dominant-pole approximation* and allows ω_{Ho} to be computed without construction of the Bode diagram.

The validity of the dominant-pole approximation is justified by considering Fig. 9-5. As ω_{H1} is smaller than either ω_{H2} or ω_{H3}, the resultant bandwidth indicated in Fig. 9-5 is very nearly ω_{H1}. In Eq. (9-12), when ω_{H1} is the smallest of the three terms, the value of a_1 is approximately $1/\omega_{H1}$, and, from Eq. (9-14), the same overall bandwidth is obtained. The two methods just described are illustrated in Example 9-1.

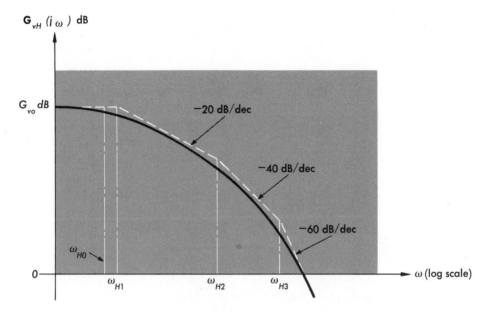

FIGURE 9-5 Bode diagram for cascade amplifier in Fig. 9-1.

Example 9-1

The cascade amplifier depicted in Fig. 9-1 is designed with the following values: $R_1 = 0.5$ kΩ, $R_2 = 20$ kΩ, $R_3 = 2.0$ kΩ, $R_4 = 10$ kΩ, $R_5 = 5.0$ kΩ, and $R_6 = 10$ kΩ. The operational amplifiers are identical and have a mid-band open-loop gain of 10^5 and a gain-bandwidth product of 10^7 rad/s.

a Determine the midband gain.

b Use the Bode diagram to determine the bandwidth of the cascaded amplifier.

c Repeat b using the dominant-pole approximation.

SOLUTION **a** The overall midband gain is given by Eq. (9-7). Since $A_o = 10^5$, Eqs. (9-2), (9-4), and (9-6) can be used to determine the gains of the individual stages. Thus,

$$G_{o1} = -\frac{20 \times 10^3}{500} = -40$$

$$G_{o2} = -\frac{10^4}{2 \times 10^3} = -5$$

$$G_{o3} = -\frac{10^4}{5 \times 10^3} = -2$$

and

$$G_{vo} = (-40)(-5)(-2) = -400$$

b The frequency response of the amplifier is given by Eq. (9-10) in which the terms ω_{H1}, ω_{H2}, and ω_{H3} are related to the gain-bandwidth product of the OP-AMP and the midband closed-loop gain by Eq. (8-11). Therefore,

$$\omega_{H1} = \frac{A_o \omega_h}{|G_{o1}|} = \frac{10^7}{40} = 2.5 \times 10^5 \text{ rad/s}$$

$$\omega_{H2} = \frac{10^7}{5.0} = 2.0 \times 10^6 \text{ rad/s}$$

and

$$\omega_{H3} = \frac{10^7}{2.0} = 5.0 \times 10^6 \text{ rad/s}$$

which permits the gain to be expressed as

$$G_{vH}(s) = \frac{-400}{\left(1 + \dfrac{s}{2.5 \times 10^5}\right)\left(1 + \dfrac{s}{2.0 \times 10^6}\right)\left(1 + \dfrac{s}{5 \times 10^6}\right)}$$

The Bode diagram for the response is shown in Fig. 9-6, from which the bandwidth is, to two significant figures,

$$\omega_{Ho} = 2.4 \times 10^5 \text{ rad/s}$$

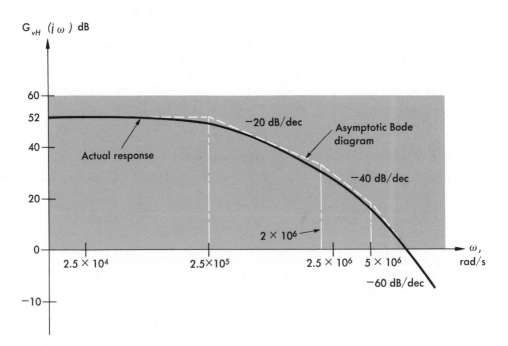

FIGURE 9-6 Bode diagram for Example 9-1.

c For the dominant-pole approximation, a_1 is given by Eq. (9-12) as

$$a_1 = \frac{1}{2.5 \times 10^5} + \frac{1}{2.0 \times 10^6} + \frac{1}{5.0 \times 10^6} = 4.7 \times 10^{-6} \text{ s}$$

and

$$\omega_{Ho} = \frac{1}{a_1} = \frac{1}{4.7 \times 10^{-6}} = 2.1 \times 10^5 \text{ rad/s}$$

The efficacy of the dominant-pole approximation is illustrated in the results obtained in Example 9-1 as the two results only differ by about 12 percent. However, the dominant-pole approximation produces the value of ω_{Ho} with a minimum of computation; furthermore, the approximate value is always less than the actual bandwidth. The student is encouraged to use any of the standard computer programs, or develop one of her or his own, to obtain the Bode diagram. The dominant-pole approximation is then useful as a point of reference to check the results.

The results obtained in Example 9-1 serve to underscore the rationale for the cascading process. The overall midband gain of the amplifier is -400 and its bandwidth is 2.4×10^5 rad/s. The resultant gain-bandwidth product is $400 \times 2.4 \times 10^5 = 9.6 \times 10^7$, which exceeds that for the employed OP-AMP by nearly 10 times. Indeed, this result could be made more dramatic. Increasing the gain of each stage by a factor of 10 makes the midband gain -4×10^5 while the overall bandwidth decreases to 2.4×10^4 rad/s. The gain-bandwidth product for the cascade amplifier is then nearly 10^4 times as great as that for the operational amplifiers used. In essence, the increase in gain-bandwidth product for cascaded structures arises from the fact that the gain increases multiplicatively and the bandwidth shrinkage is "additive." These observations are evident in Eqs. (9-9) and (9-12).

9-2 CASCADED TRANSISTOR STAGES AT MIDBAND FREQUENCIES

The analysis of cascaded transistor-amplifier stages, both for bipolar and field effect devices, parallels the development in the previous section. For expedience, the two-stage, discrete-component, bipolar transistor amplifier, shown in Fig. 9-7, is considered. By this means, both high- and low-frequency responses can be investigated. The results obtained are directly applicable to integrated circuit amplifiers. Those components not present in ICs, such as the coupling and bypass capacitors, and base-bias resistors, can be eliminated from the equations which describe system performance.

The small-signal equivalent circuit, applicable at midband frequencies, for the amplifier in Fig. 9-7 is given in Fig. 9-8. It is assumed that the transistors Q_1 and Q_2 used in the circuit of Fig. 9-7 are identical and thus the elements in their models have the same numeric values. The effective base-bias

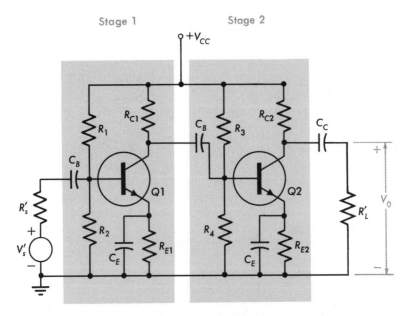

FIGURE 9-7 Bipolar transistor cascade amplifier.

resistors R_{B1} and R_{B2} are the parallel combinations of the resistors R_1 and R_2, and R_3 and R_4, respectively. The circuit of Fig. 9-8 not only applies to the circuit of Fig. 9-7, but is used to represent the more general case where portions of the emitter resistances are unbypassed. In the latter instance, the transistor and unbypassed emitter resistance are replaced by the equivalent circuit of Fig. 8-41, and analyzed as described in Sec. 8-9.

In the circuit of Fig. 9-8, the output voltage V_1 of the first stage is the input voltage to the second stage. The overall gain of the cascade structure is determined by first obtaining the Thévenin equivalent of the unloaded first stage at terminals 1 and 1', as depicted in Fig. 9-9. In Fig. 9-9 the effect of the base-bias resistor R_{B1}, the signal source V_s', and source resistance R_s' are replaced by their equivalent source as illustrated in Fig. 8-31. Comparison of

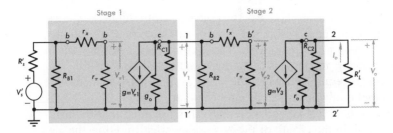

FIGURE 9-8 Midband equivalent circuit of amplifier in Fig. 9-7.

Thévenin equvalent
of signal source and
base-bias resistors

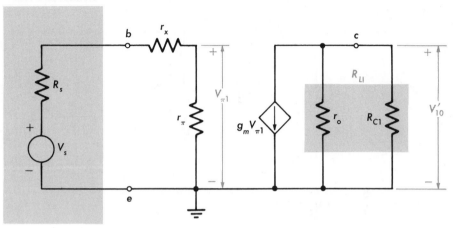

FIGURE 9-9 Unloaded first stage of circuit in Fig. 9-7.

Fig. 9-9 with Fig. 8-32 at midband frequencies (when C_x can be treated as an open circuit) indicates that they are identical. The single-stage circuit in Fig. 8-32 is discussed in Sec. 8-7 and, therefore, the gain of the unloaded first stage is given by Eq. (8-45). The open-circuit voltage V'_{10} is then the gain of the stage multiplied by the source voltage V_s, or

$$V'_{10} = \frac{-h_{fe}R_{L1}}{R_s + r_x + r_\pi} V_s \qquad (9\text{-}15)$$

The effect of the base-bias network is included as given in Eq. (8-49) so that

$$V'_{10} = \frac{-h_{fe}R_{L1}}{R_s + r_x + r_\pi} \frac{R_{B1}}{R_{B1} + R'_s} = G_{01}V'_s. \qquad (9\text{-}16)$$

The Thévenin resistance is the ratio of the open-circuit voltage V'_{10} to the short-circuit current I_{sc}. When R_{L1} is short-circuited, the current that exists between terminals 1 and 1' is the short-circuit current and is equal to $-g_m V_{\pi1}$, or

$$I_{sc} = \frac{-h_{fe}}{R_s + r_x + r_\pi} \frac{R_{B1}}{R_{B1} + R'_s} V'_s \qquad (9\text{-}17)$$

Division of Eq. (9-16) by Eq. (9-17) results in the Thévenin resistance

$$R_{o1} = \frac{V_{10}}{I_{sc}} = R_{L1} \qquad (9\text{-}18)$$

The first stage, represented by its Thévenin equivalent, is the input to the second stage as shown in Fig. 9-10. Replacing V'_{10}, R_{L1}, and R_{B2} by its

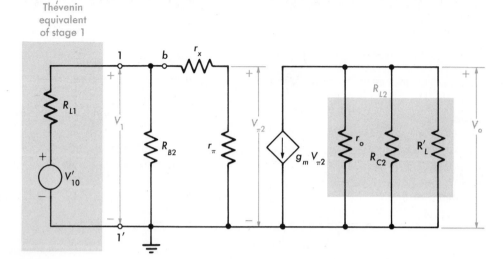

FIGURE 9-10 Second stage of cascade amplifier in Fig. 9-7.

equivalent source V_{10} and R_{s2}, as indicated in Fig. 8-31, results in the circuit depicted in Fig. 9-11. Comparison of Fig. 9-11 with Fig. 9-9 shows that they are identical if V_{10}, R_{s2}, and R_{L2} are identified with V_s, R_s, and R_{L1}, respectively. Thus, the output voltage V_o is found in Eq. (9-15); inclusion of the effect of the base-bias resistor R_{B2} is as given in Eq. (9-16). These results are

$$V_o = \frac{-h_{fe}R_{L2}}{R_{s2} + r_x + r_\pi} V_{o1} \tag{9-19}$$

$$V_o = \frac{-h_{fe}R_{L2}}{R_{s2} + r_x + r_\pi} \frac{R_{B2}}{R_{B2} + R_{L1}} V_{o1}' = G_{o2}V_{o1}' \tag{9-20}$$

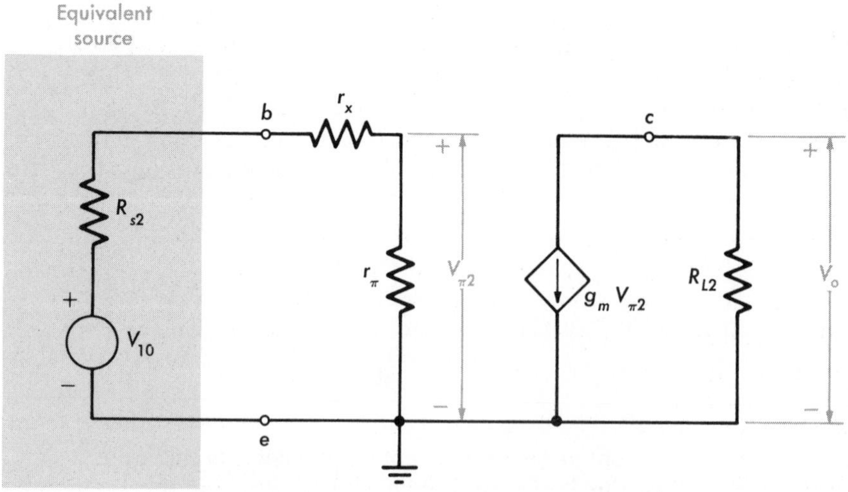

FIGURE 9-11 Equivalent representation of circuit in Fig. 9-10.

The combination of Eqs. (9-16) and (9-20) shows the overall gain of the amplifier to be

$$G_{vo} = \frac{V_o}{V_s'} = G_{o1}G_{o2} \tag{9-21}$$

The result in Eq. (9-21) can be extended to an n-stage amplifier by using the procedure indicated for the first two stages on the succeeding amplifier stages.

As shown in Figs. 9-10 and 9-11, the gain and output voltage of each stage in the cascade reflect the loading effects on both the input and output terminals. The effects of both preceding and succeeding stage impedances influence the gain of a given stage. Note specifically that the effective source resistance of a stage includes the collector load resistance of the previous stage. In fact, for R_{C1} much less than r_o in Fig. 9-9, the value of R_{o1} given by Eq. (9-18) is just R_{C1}. In addition, the effective load resistance of any stage includes the collector resistance of that stage and the input resistance to the following stage, as indicated in Fig. 9-10.

The circuit in Fig. 9-12 is a cascaded amplifier consisting of two common-source JFET stages. The analysis of this circuit directly parallels that for the bipolar transistor stages in this section. The results so obtained, stated without derivation, can be verified by the student.

The overall gain of the amplifier at midband frequencies is given in Eq. (9-21) in which

$$G_{o1} = -g_m R_{L1} \frac{R_{G1}}{R_s' + R_{G1}} \tag{9-22}$$

$$G_{o2} = -g_m R_{L2} \frac{R_{G2}}{R_{L1} + R_{G2}} \tag{9-23}$$

In Eqs. (9-22) and (9-23), the load resistances R_{L1} and R_{L2} are each the parallel combination of the drain resistance of the stage and the output resistance r_d of the FET. Similarly, R_{G1} and R_{G2} are the parallel combination of the gate-bias resistors R_1 and R_2, and R_3 and R_4, respectively. Both Eqs. (9-22) and (9-23) are of the form given in Sec. 8-10 and Eq. (8-78).

Example 9-2

The two-stage cascaded amplifier shown in Fig. 9-13 is driven by a current source I_s in parallel with its source resistance $R_s' = 600 \ \Omega$. The transistors are identical and have parameter values $r_x = 50 \ \Omega$, $r_\pi = 400 \ \Omega$, $g_m = 0.3 \ \mho$, and r_o is sufficiently large so that its effect is negligible.

a Determine the current gain I_o/I_s of the amplifier.

b Determine the values of the effective load and source resistances for each stage.

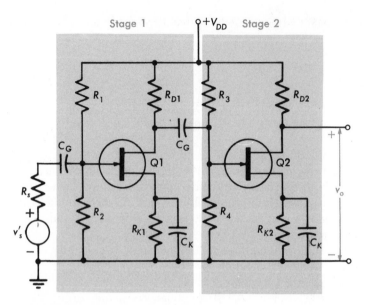

FIGURE 9-12 Two-stage JFET amplifier.

SOLUTION **a** First convert the driving source to its voltage-source equivalent so that $V_s' = I_s R_s'$ appears in series with R_s' at the input to the first stage. The values of R_{B1} and R_{B2} are identical, each being the parallel combination of the 60 kΩ and 12 kΩ resistances, which is 10 kΩ. The amplifier is now in the form of Fig. 9-7 for which Fig. 9-8 is its equivalent circuit. As indicated in the model, the output current I_o is

$$I_o = -\frac{V_o}{R_L'} \quad \text{or} \quad -V_o = I_o R_L'$$

The source $V_s' = I_s R_s'$ and the voltage gain G_v is

$$G_v = \frac{V_o}{V_s'} = \frac{-I_o R_L'}{I_s R_s'} = -\frac{R_L'}{R_s'} = -\frac{R_L'}{R_s'} G_i$$

where $G_i = I_o/I_s$, the current gain.

This result indicates the relation between voltage and current gains and allows one to be calculated from the other.

From the parameter values given

$$h_{fe} = g_m r_\pi = 0.3 \times 400 = 120$$

and

$$R_s = \frac{R_B r_s'}{R_B + R_s'} = \frac{10^4 \times 600}{10^4 + 600} = 566 \ \Omega$$

which permit the first-stage gain to be computed from Eq. (9-16) as

$$G_{o1} = \frac{-120 \times 900}{566 + 50 + 400} \frac{10^4}{10^4 + 600} = -100.3$$

Similarly, from Eq. (9-20), with $R_{L2} = (900 \times 900)/(900 + 900) = 450\ \Omega$ and $R_{s2} = (10^4 \times 900)/(10^4 + 900) = 826\ \Omega$,

$$G_{o2} = \frac{-120 \times 450}{826 + 50 + 400} \frac{10^4}{10^4 + 900} = -38.8$$

The overall voltage gain is given in Eq. (9-21) and results in

$$G_{vo} = (-100.3)(-38.8) = 3{,}892$$

The current gain is then

$$3.892 = -\frac{900}{600}G_i$$

and

$$G_i = -2{,}595$$

It should be noted that if the effect of R_B is neglected in both $R_B/(R_B + R'_s)$ and $R_B R'_s/(R_B + R'_s)$ so that $R_s = 600\ \Omega$, $R_{s2} = 900\ \Omega$, and the resultant voltage gain $G_{vo} = 4{,}120$, an error of about 5 percent. Often, to simplify calculations this approximation is employed as it shall be in **b** of this example.

b   The effective load on each stage is simply the collector resistance in parallel with the input resistance to the following stage. The input resistance of the second stage is $r_x + r_\pi = 50 + 400 = 450\ \Omega$. The effective load on the first stage is then

$$R'_{L1} = \frac{900 \times 450}{900 + 450} = 300\ \Omega$$

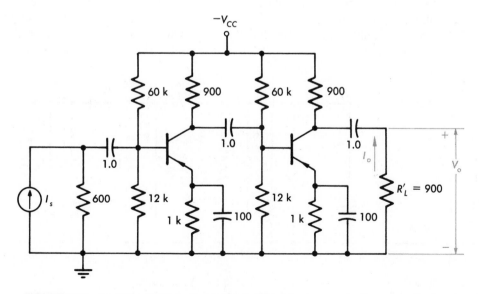

FIGURE 9-13   Amplifier circuit for Example 9-2. Element values are in ohms and microfarads.

The effective load resistance of the second stage is the parallel combination of the collector resistance and the load resistance and is

$$R_{L2} = \frac{900 \times 900}{900 + 900} = 450 \ \Omega$$

The effective source resistance is R'_s for the first stage. For all other stages in the cascade, the effective source resistance is the collector resistance of the previous stage. Thus, for the first stage, the source resistance is 600 Ω and for the second stage it is 900 Ω.

It is important to note that while each stage in the cascade has the same components and thus the stages appear to be identical, they are not. The interaction between stages, which is reflected in both the differing values of source and load resistances, results in a different performance from each stage.

9-3 FREQUENCY RESPONSE OF CASCADED AMPLIFIER STAGES

The determination of the frequency-response characteristics of cascaded amplifier stages is developed in a manner similar to the previous section. Thus, the frequency response of the cascade is obtained from the individual amplifier stages of which it is comprised. As in the case of the single-stage amplifier, it is convenient to determine the high- and low-frequency behavior separately.

The measure of high-frequency amplifier performance is the half-power angular frequency ω_{Ho} at which the magnitude of the gain is 3 dB below its midband value. To obtain the value of ω_{Ho} exactly is both difficult and cumbersome because of the interaction between stages. In the midband frequency range, the various sections of the amplifier are isolated electrically from one another, as is indicated in Fig. 9-8. However, as shown in the high-frequency equivalent circuit of Fig. 9-14, the capacitor C_μ couples

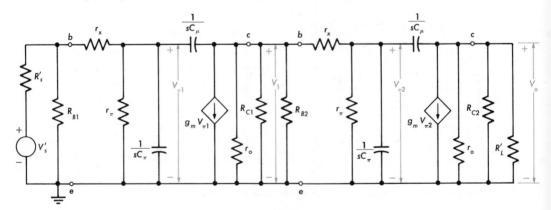

FIGURE 9-14 High-frequency equivalent circuit of two cascaded stages.

stage 1 to stage 2. There is then no isolation between stages. The technique of Sec. 8-5 in which C_μ is reflected into the input circuit assumes a resistive load in the collector of the particular stage. However, as indicated in the circuit of Fig. 9-14, the effective load on stage 1 is the parallel combination of R_{L1}, R_{B1}, and the input impedance of the second stage. At high frequencies, the input impedance contains the reactances of C_μ and C_π and is obviously not resistive. Since the load on the third stage is resistive, the unilateral model of Sec. 8-5 can be used.

Because of the capacitive interaction between stages, exact determination of the high-frequency response requires the solution of the node-voltage equations for the circuit. For the two-stage amplifier of Fig. 9-14, this process involves the solution of the four simultaneous equations that result. The process is both tedious and time-consuming. Computer-aided analysis programs, such as SPICE and ECAP, are invaluable aids where accurate determination of the response is required. For most engineering purposes, sufficiently accurate results are obtained by neglecting the effects of the capacitive coupling, using the unilateral model, and approximating the frequency response by use of the dominant-pole approximation described in Sec. 9-1. Each stage can then be considered separately and these results combined to approximate the value of ω_{Ho}.

As described in Sec. 9-1 and given in Eqs. (9-12)–(9-14), the first-order approximation to the high-frequency response is expressed as

$$G_{vH}(s) = \frac{G_{vo}}{1 + a_1 s} = \frac{G_{vo}}{1 + \dfrac{s}{\omega_{Ho}}} \tag{9-24}$$

where G_{vo} and its constituent parts are given in Eqs. (9-21), (9-16), and (9-20). The value of a_1 is the sum of the reciprocals of the half-power angular frequencies of the individual stages. Thus, for the two-stage amplifier whose model is given in Fig. 9-14

$$a_1 = \frac{1}{\omega_{H1}} + \frac{1}{\omega_{H2}} \tag{9-25}$$

The high-frequency model of the first stage is shown in Fig. 9-15a; its unilateral equivalent is shown in Fig. 9-15b. It is noted that the effective load resistance of the first stage includes the input resistance to the second stage as described in Example 9-2. The circuit depicted in Fig. 9-15b is identical to the circuit of Fig. 8-32 when V_s', R_s', and R_{B1} are replaced by their Thévenin equivalent V_s in series with R_s. The value of ω_{H1} is given by Eq. (8-46), which is restated in Eq. (9-26) as

$$\omega_{H1} = \frac{1}{R_{x1}C_{x2}} \quad \text{and} \quad R_{x1} = \frac{r_\pi(R_s + r_x)}{R_s + r_x + r_\pi} \tag{9-26}$$

The equivalent capacitance in the unilateral model is

$$C_{x1} = C_\pi + C_\mu(1 + g_m R_{L1}') \tag{9-27}$$

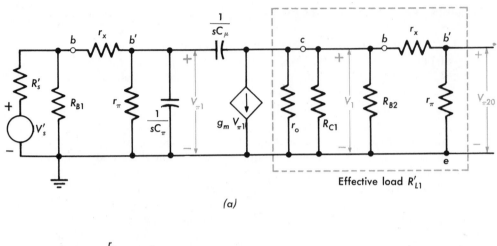

(a)

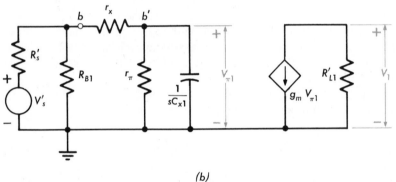

(b)

FIGURE 9-15 (a) High-frequency model and (b) its unilateral equivalent of the first stage of amplifier in Fig. 9-14.

The second-stage equivalent circuit, displayed in Figs. 9-10 and 9-11 and valid at midband frequencies, indicates it is identical in structure to the model of the first stage. Similar conditions apply at high frequencies provided the identifications of R_{s2} and R_{L2} with R_s and R'_{L1}, respectively, are made. Thus, ω_{H2}, R_{x2}, and C_{x2} are given by the same relations expressed in Eqs. (9-26) and (9-27). These are

$$\omega_{H2} = \frac{1}{R_{x2}C_{x2}} \quad \text{and} \quad R_{x2} = \frac{r_\pi(R_{s2} + r_x)}{R_s + r_x + r_\pi} \tag{9-28}$$

$$C_{x2} = C_\mu + C_\pi(1 + g_m R_{L2}) \tag{9-29}$$

Upon evaluation of ω_{H1} and ω_{H2}, Eqs. (9-24) and (9-25) are used to obtain the upper half-power frequency for the cascade. It is important to restate that the effective source and load resistances described in Example 9-2 must be used in conjunction with Eqs. (9-26)–(9-29).

Example 9-3

The transistors in the cascaded amplifier described in Example 9-2 have capacitive parameter values $C_\mu = 1.0$ pF and $C_\pi = 100$ pF. Determine the upper half-power frequency of the cascade.

SOLUTION   To determine the value of ω_{Ho} the upper half-power angular frequency of each stage must be evaluated. Because both R'_s and R'_L are resistive, the 3-dB frequencies obtained for both the voltage gain and the current gain are identical. The values of ω_{H1} and ω_{H2} are obtained by the use of Eqs. (9-26)–(9-29). For the first stage, as given in Example 9-2, $R_s = 600\ \Omega$ and $R'_{L1} = 300\ \Omega$. Thus,

$$R_{x1} = \frac{400(600 + 50)}{600 + 50 + 400} = 248\ \Omega$$

$$C_{x1} = 100 \times 10^{-12} + 10^{-12}(1 + 0.3 \times 300) = 191 \times 10^{-12}\ \text{F}$$

so that

$$\omega_{H1} = \frac{1}{248 \times 191 \times 10^{-12}} = 21.1 \times 10^6\ \text{rad/s}$$

For the second stage, $R_{s2} = R_{C1} = 900\ \Omega$ and $R_{L2} = 450\ \Omega$. Then

$$R_{x2} = \frac{400(900 + 50)}{900 + 50 + 400} = 281\ \Omega$$

$$C_{x2} = 100 \times 10^{-12} + 10^{-12}(1 + 0.3 \times 450) = 236 \times 10^{-12}\ \text{F}$$

which results in

$$\omega_{H2} = \frac{1}{281 \times 236 \times 10^{-12}} = 15.1 \times 10^6\ \text{rad/s}$$

The value of a_1, from Eq. (9-25), is

$$a_1 = \frac{1}{21.1 \times 10^6} + \frac{1}{15.1 \times 10^6} = 1.14 \times 10^{-7}\ \text{s}$$

By use of Eq. (9-24),

$$\omega_{Ho} = \frac{1}{1.14 \times 10^{-7}} = 8.80 \times 10^6\ \text{rad/s}$$

and

$$f_{Ho} = \frac{\omega_{Ho}}{2\pi} = 1.40 \times 10^6\ \text{Hz} = 1.40\ \text{MHz}$$

Computer analysis of the amplifier yields a value of $\omega_{Ho} = 9.71 \times 10^6$ rad/s and so indicates a reasonable correlation between the dominant-pole method and the actual value.

Field-effect transistors have high-frequency models which are similar in structure to those for bipolar transistors. As indicated in Sec. 8-10, the high-frequency response of FET-amplifier stages is determined from the unilateral model shown in Fig. 8-43. Consequently, the analysis for FET-amplifier stages parallels that for junction transistors. The overall gain and frequency response are expressed as given in Eqs. (9-24) and (9-25), in which the constituent parts of G_{vo} are expressed in Eqs. (9-22) and (9-23) for a two-stage amplifier. The expressions for ω_{H1} and ω_{H2} each have the form of Eq. (8-76), in which R_s and C_x are evaluated by use of the effective source and load resistances associated with each stage. The results are summarized in Eqs. (9-30) and (9-31) for the two-stage amplifier depicted in Fig. 9-12.

$$\omega_{H1} = \frac{1}{R_s C_{x1}} \qquad C_{x1} = C_{gs} + C_{gd}(1 + g_m R'_{L1}) \tag{9-30a}$$

$$\omega_{H2} = \frac{1}{R_{s2} C_{x2}} \qquad C_{x2} = C_{gs} + C_{gd}(1 + g_m R_{L2}) \tag{9-30b}$$

It is noted that in Eq. (9-30b), R_{s2} is the effective source resistance and is composed of R_{L1} in parallel with the input resistance of the second stage R_{G2}. Similarly, R'_{L1} in Eq. (9-30a) is the effective load on the stage and equals R_{s2}.

The results in Eqs. (9-25) to (9-30) can easily be extended to the n-stage cascade. The value of a_1 can be expressed as

$$a_1 = \sum_1^n \frac{1}{\omega_{Hi}} = \sum_1^n R_{xi} C_{xi} \tag{9-31}$$

where ω_{Hi} is the 3-dB-frequency for the ith stage. The values of R_{xi} and C_{xi} for each stage are given in Eqs. (9-25) and (9-26) for junction transistor stages and in Eq. (9-30a) for FET stages.

The low-frequency response is affected by the coupling and bypass capacitors associated with each stage. The measure of low-frequency amplifier performance is the lower half-power angular frequency ω_{LO} at which the magnitude of the gain is 3 dB below its midband value. As in the high-frequency case, the stages are not completely isolated from one another at low frequencies. In addition, there is interaction between the emitter bypass and base-coupling networks within each stage. It is therefore convenient to approximate the value of ω_{LO} by a dominant-pole approximation and leave exact response characteristics to computer analysis.

The first-order dominant-pole approximation valid at low frequencies is

$$G_{LO}(s) = \frac{G_{vo}}{1 + \dfrac{b_1}{s}} = \frac{G_{vo}}{1 + \dfrac{\omega_{LO}}{s}} \tag{9-32}$$

where, for a two-stage amplifier,

$$\omega_{LO} = \omega_{L1} + \omega_{L2} \tag{9-33}$$

As both coupling and bypass capacitors affect the low-frequency response of a stage, both ω_{L1} and ω_{L2} are of the form

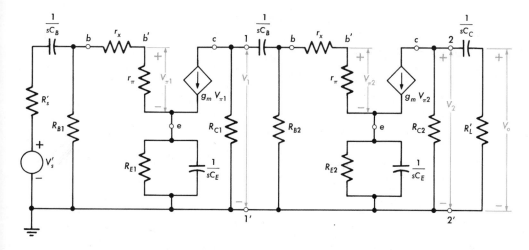

FIGURE 9-16 Low-frequency equivalent of first stage of amplifier in Fig. 9-8.

$$\omega_{Li} = \omega_{Bi} + \omega_{Ei} \tag{9-34}$$

where ω_{Bi} is the 3-dB-down frequency for the base-coupling capacitor of the stage assuming ideal emitter bypassing and ω_{Ei} is the corresponding angular frequency for the bypass capacitor when ideal base coupling is assumed. Therefore, for a two-stage amplifier, Eq. (9-33) contains four terms.

The low-frequency model of the first stage (the portion to the left of terminals 1-1′ in Fig. 9-8) is shown in Fig. 9-16. Comparison of Fig. 9-16 with Fig. 9-16, recalling r_o is assumed negligible, shows the two circuits to be identical. Thus, the low-frequency response of the stage is given by Eqs. (8-64) and (8-70), and

$$\omega_{B1} = \frac{R_{B1} + r_x + r_\pi}{R_s'R_{B1} + (R_s' + R_{B1})(r_x + r_\pi)} \frac{1}{C_B} \tag{9-35}$$

and

$$\omega_{E1} = \frac{R_E + (r_\pi + r_x + R_s)/(1 + h_{fe})}{R_E(r_\pi + r_x + R_s)/(1 + h_{fe})} \frac{1}{C_E} \tag{9-36}$$

The circuit of Fig. 9-16 can also be used to represent the low-frequency characteristic of the sound stage if V_s' is replaced by the open-circuit voltage of the preceding stage and R_s' is replaced by the output resistance of the preceding stage. The output resistance for each stage is the collector resistance of the stage, as described in Sec. 9-2. Equations (9-35) and (9-36) can be used to evaluate the response of the second stage provided the aforementioned identifications are made.

The final coupling capacitor C_C is used to isolate the load resistance R_L' from the collector of the second transistor. The circuit of Fig. 9-17 is used to

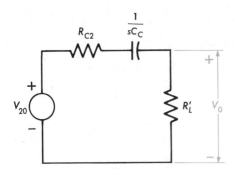

FIGURE 9-17 Portion of circuit used to determine effect of load coupling capacitor.

determine the low-frequency response of this portion of the cascade. The output voltage V_o is given by

$$V_o = \frac{R'_L}{R_{C2} + R'_L + 1/sC_C} V_{2o}$$

$$= \frac{R'_L}{R_{C2} + R'_L} \frac{1}{1 + 1/s[C_C(R'_L + R_{C3})]} V_{2o} \quad (9\text{-}37)$$

The term $1/C_C(R'_L + R_{C2})$ in Eq. (9-37) may be identified as an angular frequency ω_{LL}.

The overall low-frequency response can then be approximated by use of Eqs. (9-32), (9-33), and (9-37). For the two-stage amplifier of Fig. 9-8 this result is

$$\omega_{L0} = \omega_{B1} + \omega_{E1} + \omega_{B2} + \omega_{E2} + \omega_{LL} \quad (9\text{-}38)$$

Example 9-4

The transistor amplifier described in Example 9-2 uses coupling capacitors whose values are each 1.0 μF and bypass capacitors of 100 μF each. Determine the lower half-power frequency.

SOLUTION   Again, as in Example 9-3, the low-frequency responses for the current and voltage gains are identical because both source and load are resistive. The base-coupling capacitors effects on the response are obtained from Eq. (9-35); the contribution of the emitter-bypass capacitors are given in Eq. (9-36). The effect of the load-coupling capacitor is treated by means of Eq. (9-37). Thus, for the first stage, and recalling from Example 9-2 that $h_{fe} = 100$,

$$\omega_{B1} = \frac{10^4 + 50 + 400}{600 \times 10^4 + (600 + 10^4)(50 + 400)} \frac{1}{10^{-6}} = 970 \text{ rad/s}$$

$$\omega_{E1} = \frac{10^3 + (400 + 50 + 566)/(1 + 100)}{10^3(400 + 50 + 566)/(1 + 100)} \frac{1}{10^{-4}} = 1{,}004 \text{ rad/s}$$

For the second stage, the equivalent of R_s' is $R_{Cl} = 900\ \Omega$, and $R_s = (R_{C1}R_{B2})/(R_{C1} + R_{B2})$ equals $826\ \Omega$. Then,

$$\omega_{B2} = \frac{10^4 + 50 + 400}{900 \times 10^4 + (900 + 10^4)(50 + 400)}\ \frac{1}{10^{-6}} = 752\ \text{rad/s}$$

$$\omega_{E2} = \frac{10^3 + (400 + 50 + 826)/(1 + 100)}{10^3(400 + 50 + 826)/(1 + 100)}\ \frac{1}{10^{-4}} = 802\ \text{rad/s}$$

For the load-coupling capacitor

$$\omega_{LL} = \frac{1}{C_C(R_L' + R_{C2})} = \frac{1}{10^{-6}(900 + 900)} = 556\ \text{rad/s}$$

The overall response is given in Eq. (9-38) and results in

$$\omega_{L0} = 970 + 1{,}004 + 752 + 802 + 556 = 4{,}084\ \text{rad/s}$$

and $f_{L0} = 650\ \text{Hz}$

The results indicate that the effect of the emitter-bypass capacitors is as significant as that of the base-coupling capacitors despite the fact that the value of C_E is 100 times that for C_B. This fact underscores one difficulty of the use of capacitors at low frequencies. To lower ω_{L0} by a factor of 20 so that the lower audio frequencies are amplified requires an increase in the size of each of the capacitors by a factor of 20. Thus, the value of C_E becomes $2{,}000\ \mu\text{F}$, which is physically large compared to other components and often costly.

The total response of the amplifier can now be determined by combination of the results of Examples 9-2 to 9-4. In addition, the techniques described for the two-stage cascade are directly extended to amplifiers comprising more than two stages.

The low-frequency response of FET stages is determined in the same manner as for bipolar stages. The overall response is approximated by Eqs. (9-32) and (9-33); the individual terms in Eq. (9-33) reflect the effects of the gate-coupling and source-bypass capacitors. The single-stage, low-frequency response is given in Eq. (8-77) and restated as Eqs. (9-39) and (9-40).

$$\omega_G = \frac{1}{(R_G + R_s')C_G} \tag{9-39}$$

$$\omega_K = \frac{(1 + g_m R_K)}{R_K C_K} \tag{9-40}$$

9-4 BANDWIDTH IMPROVEMENT TECHNIQUES

In most electronic systems, the amount of information that can be processed is restricted by the bandwidth of the system. The standard television system has a bandwidth requirement of nearly 6 MHz to ensure proper reception of both the video and the audio information. Any reduction of the bandwidth would seriously limit the quality of the picture. Similarly, in order to transmit efficiently and economically large numbers of telephone messages on a single communications link, or channel, large bandwidths are required. Wide band channels are also essential in high-speed data-processing applications. The need for large bandwidths is also present in many other control, instrumentation, and communications systems.

The previous sections describe the frequency limitations of amplifier stages. It is often desirable to extend the frequency range over which the amplifier is usable, both at high frequencies and at low frequencies. There are many techniques used to compensate for poor frequency response. Each method utilizes at least one additional component not present in the basic stage to counteract the inherent frequency limitations.

Shunt peaking is a technique, illustrated by the circuits of Figs. 9-18*a* and 9-18*b*, used to improve the high-frequency performance of the stage. The inductor L in series with the load R_L is used to counteract the effect of device capacitance. In the uncompensated common-emitter stage described in Sec. 8-8, the collector current, and consequently the output voltage, decreases with frequency because of the shunting effect of the capacitance C_x. The value of inductance in the compensated stage of Fig. 9-18*a* is selected so that its reactance in the midband range is negligible compared with R_L. At high frequencies, the inductive reactance is no longer negligible, and the effective impedance of the load $\mathbf{Z}_L = R_L + j\omega L$ increases. The magnitude of the output voltage V_o is given as

$$V_o = I_c Z_L$$

The increase of Z_L with frequency tends to offset the decrease in I_c and helps

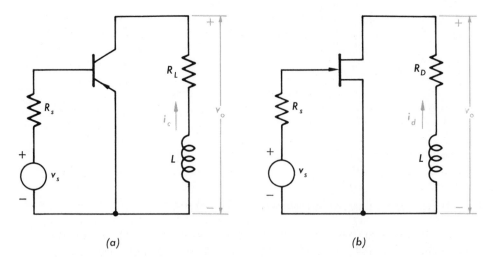

(a)   (b)

FIGURE 9-18 Shunt-peaking in (a) bipolar and (b) JFET amplifier stages.

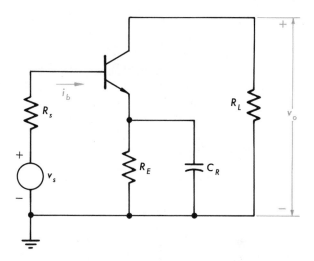

FIGURE 9-19 Transistor stage with emitter regeneration.

to maintain the output voltage at a constant level. The bandwidth obtained by this method can be nearly twice that of the uncompensated stage.

For the common-source stage of Fig. 9-18b, the function of the inductor is to compensate for the effects of the shunt capacitance in much the same fashion as it does in the junction-transistor stage. As the current through R_L decreases because of the shunting effect of the equivalent capacitance, the increased reactance of the inductance L increases the effective load impedance. The increase in load impedance and decrease in load current tend to offset one another, thus maintaining the output voltage at a more nearly constant level.

Another method employed for high-frequency compensation is called *emitter regeneration,* the circuit for which is shown in Fig. 9-19. This method is particularly useful in microelectronic circuits, where inductors cannot be used. The improvement is achieved by the use of the capacitance C_R, whose value is selected so that at midband frequencies its reactance is much larger than the value of R_E. As the frequency increases, the reactance of C_R decreases, making the effective emitter impedance smaller. The decrease in emitter impedance increases the base current and consequently tends to increase the collector current and output voltage. The decrease in emitter impedance thus offsets the decrease in the voltage V_π, helping to maintain the output voltage constant over a wider frequency range. This technique is also used in common-source stages in an almost identical manner.

The circuit of Fig. 9-20 is commonly used to improve the low-frequency performance of an amplifier stage. The parallel combination of R_F and C_F in series with the load is used to counteract the effect of the coupling capacitor C_B. The value of C_F is selected so that at midband its reactance is much smaller than R_F and R_L. The amplifier load is then approximately R_L. As the

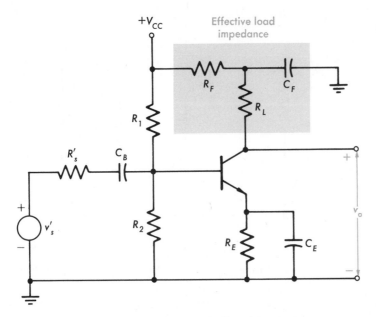

FIGURE 9-20 Low-frequency compensated transistor stage.

frequency is decreased, the reactance of C_F increases, thereby increasing the effective load impedance. The effect of C_B at low frequencies is to reduce the base current and consequently the collector current. By increasing the effective load impedance, the decrease in collector current is offset and the output voltage remains essentially constant over a wider band of frequencies. The bandwidth improvement is achieved at the expense of increasing the power supplied to the circuit. To maintain the same quiescent point in the compensated stage as in the uncompensated stage, the collector supply must be increased by $I_{CQ}R_F$, the direct voltage drop across the compensating network. The increase in power dissipation and the value of supply voltage limit the amount of improvement that can be realized. Often, the conflicting constraints of improved low-frequency response and minimum power dissipation result in design compromises. In a given application, these compromises are based on the relative importance of the two factors.

While other, more complex methods are also used, the manner by which improvement is obtained is similar to the methods described here.

9-5 COMPOUND AMPLIFIER STAGES

There are several widely used circuit configurations which enhance the performance of amplifiers. One such class of circuits is referred to as *compound amplifiers* or *compound devices*. Three types are described in this section: the *Darlington compound,* the *cascode amplifier,* and the

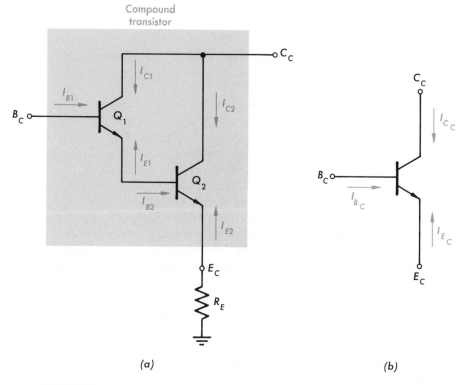

FIGURE 9-21   (a) Darlington compound transistor and (b) its equivalent representation.

complementary-symmetry configuration; each type affords certain advantages which lead to their use, primarily in integrated circuits.

The Darlington compound transistor, displayed in Fig. 9-21a, is designed to yield large equivalent values for both h_{FE} and h_{fe}. The transistor shown in Fig. 9-21b is the equivalent of the compound transistor in Fig. 9-21a. In Fig. 9-21a, assuming identical transistors, the currents in transistor $Q1$ are

$$I_{E1} = -(1 + h_{FE})I_{B1}$$

and

$$I_{C1} = h_{FE}I_{B1}$$

For transistor $Q2$, $I_{B2} = -I_{E1}$ and

$$I_{E2} = -(h_{FE} + 1)I_{B2} = -(h_{FE} + 1)^2 I_{B1}$$

$$I_{C2} = h_{FE}I_{B2} = h_{FE}(h_{FE} + 1)I_{B1}$$

For the compound transistor, the three terminal currents are given as

$$I_B = I_{B1}, \qquad I_E = -(h_{FE} + 1)^2 I_{B1} \qquad\qquad (9\text{-}41)$$

and

$$I_{C_C} = I_{C1} + I_{C2} = h_{FE}I_{B1} + h_{FE}(h_{FE} + 1)I_{B1} \tag{9-42}$$

The equivalent short-circuit current gain h_{FE_C} of the compound is obtained from Eqs. (9-41) and (9-42) as

$$h_{FE_C} = \frac{I_C}{I_B} = h_{FE}(h_{FE} + 2) \tag{9-43}$$

For a value of $h_{FE} = 100$, a corresponding value of $h_{FE} = 10,200$, a larger value than can be obtained from a single transistor. A result identical to that given in Eq. (9-43) is obtained for h_{fe_C} in the small-signal ac case.

Two advantages of the Darlington compound are in the low value of base current and high-input resistance that result. For $I_{C_C} = 1$ mA the corresponding value of I_{B_C} is in the order of 100 nA. Recall that the input resistance for a stage containing an emitter resistance [see Eq. (8-73)] is, for large values of h_{fe}, approximately $h_{fe}R_E$. Therefore, for $h_{fe} = 100$ and $R_E = 100\ \Omega$, a value of R_{in} of 1 MΩ is obtained for the compound transistor, a value realized only with $R_E = 10$ kΩ for a single transistor. The two principal disadvantages are the decrease in high-frequency performance and the increase in the effective turn-on voltage, which is $2V_{BE}$ for the Darlington configuration.

The *cascode amplifier stage* consists of a common-emitter stage driving a common-base amplifier as shown in Fig. 9-22. The principal advantage of this circuit is that gain can be provided to a large-value load-resistance R_L over a wide range of frequencies. The input resistance R_{in} of the common-base stage serves as the load resistance for the common-emitter input stage. As R_{in} is low, the effect of the term $C_\mu(1 + g_mR_L)$ in the common-emitter equivalent circuit is reduced. This results in improved high-frequency performance for the stage. The large load resistance can be

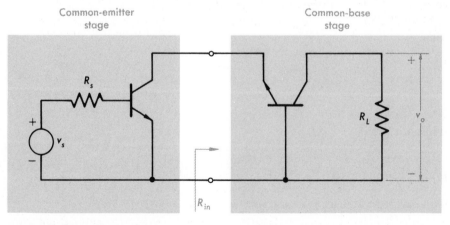

FIGURE 9-22 Cascode amplifier.

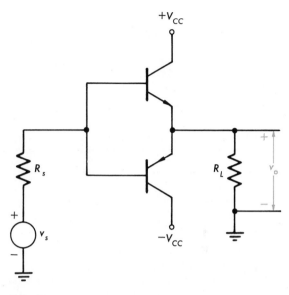

FIGURE 9-23 Complementary-symmetry emitter-follower stage.

accommodated because the output resistance of the common-base stage is larger than that of a common-emitter stage.

The *complementary-symmetry configuration,* depicted in Fig. 9-23, contains both an *npn* and a *pnp* transistor. In Fig. 9-23, the output terminal is at the junction of the emitters of the two transistors so that the circuit is usually referred to as a *complementary-symmetry emitter-follower.* Both transistors are usually biased near cutoff so that when v_s is a positive-going signal, the *pnp* transistor is cut off. The *npn* transistor is in its active region and provides the output signal in the load resistor R_L. For negative-going input signals, the roles of the two transistors are reversed so that the *pnp* transistor provides the output signal. Thus, positive signals are amplified by the *npn* device, negative ones by the *pnp* transistor. The principal virtue of this arrangement, often called "push-pull," is that the dynamic range of the circuit is essentially doubled. That is, the maximum input signal that can be accommodated is larger than can be achieved by either device.

9-6 CHARACTERISTICS AND REPRESENTATIONS OF AMPLIFIERS

In the previous sections of this chapter, the characteristics of multistage amplifiers are described. The results obtained indicate that most types of amplifiers have similar gain and frequency-response properties independent of the devices used. As cascaded-amplifier stages are often used as integral parts of other electronic circuits (such as those discussed in the subsequent sections of this chapter and in Chaps. 10 and 11), it is convenient to repre-

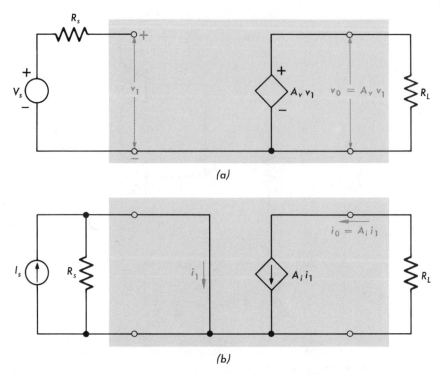

FIGURE 9-24   Ideal (a) voltage amplifier and (b) current amplifier.

sent the cascaded amplifier by an equivalent network. This equivalent must have the same properties at both the input and the output terminals as does the cascaded amplifier and serves to represent it in the analysis and design of more complex electronic circuits.

The development of this type of functional representation makes use of the concept of an ideal amplifier, which has the property that the output is a constant multiple of the input over all frequencies and whose gain is unaffected by either source- or load-impedance levels. Figure 9-24 depicts the ideal voltage and current amplifiers. For the voltage amplifier (Fig. 9-24a), the input impedance is infinite, having the effect of applying the entire driving voltage V_s to the amplifier without having a voltage drop across R_s. At the output, the zero-impedance level results in the entire output voltage $A_V V_s$ appearing across the load. In addition, in the ideal OP-AMP, the gain A_v is assumed infinite. The current amplifier of Fig. 9-24b has zero input impedance, so that the full signal current I_s is applied to the amplifier. The infinite output impedance ensures that the entire amplified current is delivered to the load.

Practical amplifiers can only approximate the ideal because they possess finite, nonzero input and output impedances and inherent frequency limitations of the type described in previous sections. Any amplifier can be

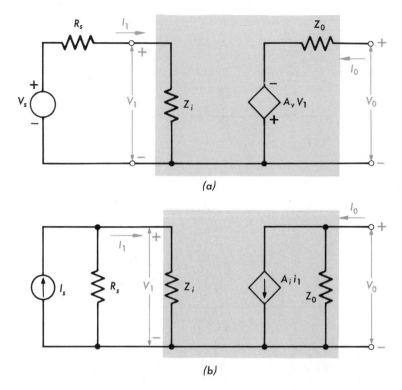

FIGURE 9-25   Representation of practical voltage and current amplifiers.

represented by an ideal amplifier together with additional circuit parameters representing the nonideal behavior, as shown in Fig. 9-25.

Cascaded-amplifier stages can also be represented by the models of Fig. 9-25. Between the driving source and the load, the various stages are used to amplify either the voltage which appears across the input or the current which enters the amplifier input. In Example 9-2, it is shown that the voltage gain and current gain are related. Thus, a practical amplifier can be represented by either the nonideal voltage amplifier or the nonideal current amplifier. This fact can be verified by performing a source conversion on the circuit of Fig. 9-25b. The source conversion results in a voltage source $I_s R_s$ in series with R_s at the input to the amplifier. The voltage V_1 is $I_1 Z_i$, so that the controlled-current source $A_i I_1$ can be expressed as $A_i V_1 / Z_i$. This current source in parallel with Z_o can be converted to a voltage source $-A_i Z_o V_1 / Z_i$, whose positive terminal is the same as is indicated for V_o, in series with Z_o. By this procedure, the current amplifier now has the form of the voltage amplifier in Fig. 9-25a. Amplifier stages may also be represented by voltage-controlled current sources and current-controlled voltage sources as shown in Fig. 9-26. Both controlled-source parameters can be related to the voltage and current amplifiers of Fig. 9-25 as indicated in Fig. 9-26.

Because transistor and FET amplifiers have similar characteristics,

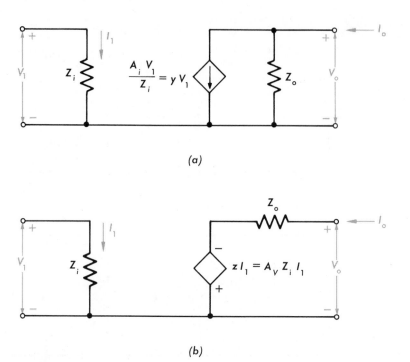

(a)

(b)

FIGURE 9-26 Representation of practical (a) voltage-controlled current source and (b) current-controlled voltage source.

both can be represented by the same type of nonideal model. Only the specific parameter values may be different so as to reflect the differences among the various devices. The essential procedure for obtaining the models of Figs. 9-25 and 9-26 from the actual circuits described earlier in this chapter is similar to that for determining the two-port parameters described in Sec. 6-5. Both transistor and FET amplifiers are assumed to be unilateral; i.e., transmission occurs from input to output only. Thus, to represent the two-stage transistor amplifier of Fig. 9-7 by the equivalent circuit of Fig. 9-25a, only the input impedance and the Thévenin equivalent at the output need be determined. This type of representation can be used in each of the frequency ranges of interest.

Each of the four amplifier types in Figs. 9-25 and 9-26 are often, for convenience, symbolically represented as shown in Fig. 9-27. The subscripts used on conjunction with the gain A refer to the type of amplifier used. Thus, A_v represents a voltage amplifier and A_i a current amplifier. The voltage-controlled current amplifier, or transconductance amplifier, is indicated by A_t; and A_z is used for the transimpedance amplifier (current-controlled voltage amplifier). When the input is applied to the positive terminal, the effect is to reverse the voltage polarity or current directions in the models in Figs. 9-25 and 9-26. To denote an OP-AMP, for which the same symbol is used, no

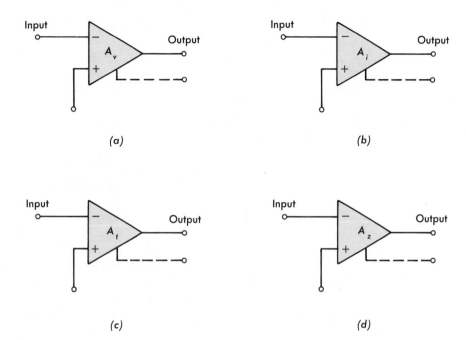

FIGURE 9-27 Symbols for (a) voltage amplifier, (b) current amplifier, (c) transconductance amplifier, and (d) transimpedance amplifier.

indication for A is used. These symbols will be used in this sense in the remainder of this text.

9-7 SOME GENERAL PROPERTIES OF FEEDBACK AMPLIFIERS

In the analysis of amplifier circuits up to this point, the operation of the devices has always been considered to be perfectly linear. Increasing the signal levels, such as is required to drive speakers and many other amplifier loads, may cause the devices to display certain nonlinear characteristics. Distortion is the result of the nonlinear behavior and alters the signal being processed. Other variations in the system performance are caused by deviations in device parameters, of which the change in h_{fe} of a transistor is typical. Changes in the gain of a stage because of device variations may also have an effect on succeeding stages, with a resultant change in the overall system gain.

To control the effects of system variations and reduce distortion and, in general, to have amplifier performance more nearly approach the ideal, *feedback* is frequently employed. Feedback is the process whereby a portion of the output signal is returned to the input and forms part of the signal used to drive the amplifier. When the returned signal acts in opposition to the applied signal, *negative feedback* exists. On the other hand, when the signal returned reinforces the applied signal, *positive feedback* results. In this section, only negative feedback is considered.

Several instances of feedback are encountered in Chap. 8, among them the inverting and noninverting OP-AMP stages and the common-emitter stage containing an emitter resistance. In Sec. 8-1 it is seen that in both basic operational-amplifier stages the use of feedback causes the gain to decrease and the bandwidth to increase so that the gain-bandwidth product remains constant. Furthermore, the gain of the OP-AMP stages is virtually insensitive to changes in the operational-amplifier characteristics. The use of an emitter resistance in a common-emitter stage (see Sec. 8-9) helped to control the gain variation caused by changes in h_{fe}. Again the feedback decreased gain and increased bandwidth.

The signal-flow graph of an ideal feedback amplifier, shown in Fig. 9-28, consists of two main components: the amplifier gain $a(s)$ which constitutes the forward path, and the feedback network whose transfer function is $f(s)$. The structure of the flow-graph in Fig. 9-28 depicts the five major elements of which a feedback system are comprised. These are:

1 the input signal X_s,

2 the output signal X_o,

3 the measure of the output $f(s)$ returned to the input,

4 comparison of the fed back signal and the input resulting in the signal X_c, and

5 the input signal X_c' to the amplifier which produces the output.

The indicated signals X_s, X_c, X_c', and X_o will either be voltages or currents for particular feedback amplifiers such as those discussed in Sec. 9-8.

The various signal quantities can be related by the equations

$$X_c = X_s + f(s)X_0 \tag{9-44}$$

$$X_o = a(s)X_c' \tag{9-45}$$

$$X_c' = X_c \tag{9-46}$$

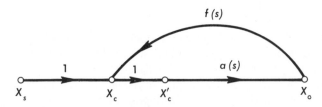

FIGURE 9-28 Signal-flow graph for ideal feedback system.

The gain of the feedback amplifier $G_F(s)$, called the *closed-loop gain*, is found by combining Eqs. (9-44) to (9-46) and is

$$G_F(s) = \frac{X_o}{X_s} = \frac{a(s)}{1 - a(s)f(s)} \tag{9-47}$$

It is of note that $a(s)$, the *open-loop gain*, is the value of $G_F(s)$ when $f(s)$ is zero, a condition corresponding to removing the feedback path. The quantity $a(s)f(s)$ in Eq. (9-47) is, as indicated in Fig. 9-28, a measure of what portion of the amplifier signal X_c' returns and contributes to the comparison signal X_c. As the flow-graph indicates the path from X_c' to X_c is through both the amplifier and feedback path, the quantity $a(s)f(s)$ is referred to as the *loop gain*. The negative of the loop gain, $a(s)f(s)$, is called the *return ratio* $T(s)$. The significance of the return ratio (or loop gain) is that many of the closed-loop performance characteristics can be determined from the open-loop behavior of the amplifier.

In Sec. 8-1, the gain of the inverting operational-amplifier stage, given in Eq. (8-6) at midband, is

$$G_o = \frac{-A_o R_2}{R_2 + R_1(1 + A_o)}$$

which, after multiplication and division by $-R_2$ and factoring $R_1 + R_2$ from the denominator, yields

$$G_o = \frac{-A_o R_2/(R_1 + R_2)}{1 - \dfrac{A_o R_2}{R_1 + R_2} \dfrac{R_1}{R_2}} \tag{9-48}$$

Comparison of Eq. (9-48) with Eq. (9-47) gives

$$a_o = -\frac{A_o R_2}{R_1 + R_2} \quad \text{and} \quad f_o = \frac{R_1}{R_2} \tag{9-49}$$

where a_o and f_o are the midband values of $a(s)$ and $f(s)$. The identifications in Eq. (9-49) reinforce the concept that the basic OP-AMP stages are feedback amplifiers.

An essential aspect of feedback-amplifier performance is that of comparison, achieved by using the signal ($X_c = X_c'$) to drive the amplifier. The amplifier processes the signal X_c', which is composed of both the signal X_s to be amplified and the measure of the output. When the amplifier output X_o is exactly the desired value, X_c' is its prescribed value and maintains the level of X_o. Should X_o increase because of parameter changes in $a(s)$, the value of fX_o increases, causing X_c to decrease. The reduction in the signal processed by the amplifier tends to counteract the increase in $a(s)$, helping to maintain X_o at its original level. The measure of how well the feedback amplifier maintains its desired response despite changes in gain is called the *sensitivity*, defined as

$$S_a^{G_F} = \frac{\Delta G_F/G_F}{\Delta a/a} \tag{9-50}$$

The symbol $S_a^{G_F}$ is called the *sensitivity function* and is referred to as the sensitivity of G_F with respect to a. For a transistor amplifier, the variation of a with h_{fe} can be determined, so that the sensitivity function can be used to relate changes of G_F and h_{fe}.

In Eq. (9-47), if $a(s)$ changes to $a(s) + \Delta a$, $G_F(s)$ becomes $G_F(s) + \Delta G_F$. If the value of ΔG_F is determined and is substituted into Eq. (9-50), the sensitivity can be shown to be

$$S_a^{G_F} = \frac{1}{1 - a(s)f(s) - \Delta a(s)f(s)} \tag{9-51}$$

If the quantity $a(s)f(s)$ is made large, then $S_a^{G_F}$ is small and indicates that the relative change in $G_F(s)$ is significantly smaller than that in $a(s)$.

Equations (9-47) and (9-51) can be rewritten by identifying $-a(s)f(s)$ as the return ratio $T(s)$ and are

$$G_F(s) = -\frac{1}{f(s)} \frac{T(s)}{1 + T(s)} \tag{9-52}$$

$$S_a^{G_F} = \frac{1}{1 + T(s) + \Delta T(s)} \tag{9-53}$$

For convenience, the quantities G_{FO}, T_o, a_o, and f_o are used to denote the midband values of the corresponding functions.

The essence of feedback to control performance is embodied in Eq. (9-52). At midband, for $T_o \gg 1$, Eq. (9-52) becomes

$$G_{FO} = -\frac{1}{f_o} \tag{9-54}$$

The result in Eq. (9-54) indicates that the closed-loop gain is virtually insensitive to changes in open-loop performance. For the inverting OP-AMP stage, substitution of the identification for f_o in Eq. (9-49) into Eq. (9-54) yields $G_{FO} = -R_2/R_1$. This is the result obtained in Sec. 8-1.

The introduction of the feedback loop is not without disadvantage, for the gain of the amplifier is reduced. Thus, to realize a given value of gain and still have the advantage of feedback, more stages must be used than are required in a nonfeedback cascaded amplifier having the same gain.

Example 9-5

What must be the values of a_o and f_o if an amplifier is to have a closed-loop gain at midband of 100 which must not drop by more than 1 percent for a 50 percent reduction in a_o?

SOLUTION The value of T is determined from the constraint on the gain variation. From Eq. (9-50),

$$S_{a_o}^{G_F} = \frac{-1/100}{-\frac{1}{2}a_o/a_o} = \frac{1}{50}$$

The value of T, from Eq. (9-53), is

$$\frac{1}{50} = \frac{1}{1 + T_o - \frac{1}{2}T_o}$$

so that $T_o = 98$. The value of f_o is determined from the closed-loop gain requirement, Eq. (9-52),

$$f_o = \frac{-1}{G_{Fo}} \frac{T_o}{1 + T_o} = \frac{1}{100} \frac{98}{1 + 98} = \frac{98}{9,900} = -0.0099$$

and a_o is obtained from T and f_o as

$$a_o = \frac{98}{98/9,900} = 9,900$$

To check the results, the value of G_{Fo} is determined for $a_o = 9,900$ and $a_o = 4,950$. From Eq. (9-47),

$$G_{Fo} = \frac{9,900}{1 + 9,900 \times 98/9,900} = 100$$

for $a_o = 9,900$, and

$$G_{Fo} = \frac{4,950}{1 + 4,950 \times 98/9,900} = 99.0$$

for $a_o = 4,950$. These values satisfy the design criterion.

9-8 ANALYSIS OF TYPICAL FEEDBACK-AMPLIFIER STRUCTURES

There are four basic feedback-amplifier configurations for which the signal-flow graph of Fig. 9-28 is an approximate, and often valid, representation. Each structure consists of an amplifier, usually multistage, whose performance is to be controlled, and a feedback network used to provide the measure of the output and the signal necessary for comparison. The nonideal amplifier representations described in Sec. 9-6 indicate that such amplifiers can be viewed as two-port networks; that is, there is one pair of terminals for the input and a second terminal-pair for the output. Similarly, the feedback network is a two-port network. Each of the four basic feedback configurations consists of the interconnection of these two-port networks. It is the manner of interconnection from which each structure derives its name.

The four feedback-amplifier configurations are illustrated in Figs. 9-29 and 9-30. The *shunt-feedback amplifier* in Fig. 9-29a indicates that the feedback network and amplifier are connected in parallel at both input and output. The difference signal is derived from the comparison of currents at the input node which is amplified to provide the output voltage. The feedback network measures the output voltage and provides the current required for comparison. Figure 9-29b is the representation of *series-feedback amplifier*,

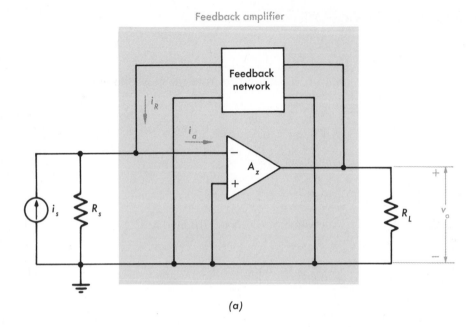

(a)

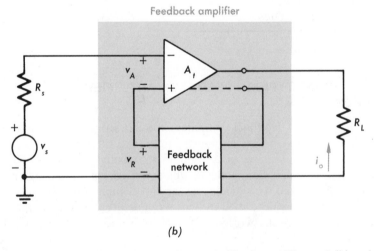

(b)

FIGURE 9-29 Circuit representations of (*a*) shunt-feedback amplifier and (*b*) series-feedback amplifier.

used to control variations in output current. The fed-back, or returned, current establishes the voltage needed for comparison. The *shunt-series feedback amplifier* of Fig. 9-30*a* uses current comparison at the input to control the output current. The circuit of Fig. 9-30*b*, a *series-shunt feedback amplifier,* makes use of voltage comparison to control the output voltage.

Each of the feedback amplifiers in Figs. 9-29 and 9-30, while structurally

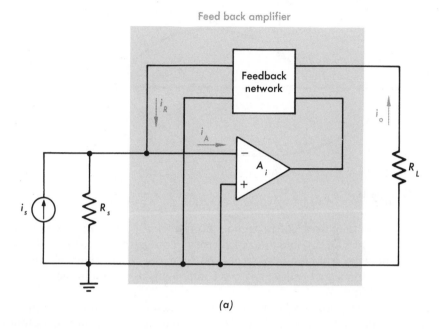

(a)

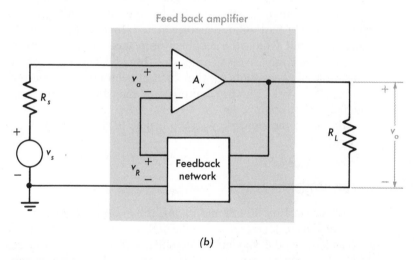

(b)

FIGURE 9-30 Circuit representations of (a) shunt-series amplifier and (b) series-shunt amplifier.

different, performs the same function, that of controlling the output. In each circuit the amplifier provides the output and the feedback network establishes the signal required for comparison. Consequently, the analysis of each configuration proceeds in an identical manner. However, in the signal-flow graph of Fig. 9-28, the feedback network is assumed unilateral; that is, signal transmission exists only from the output to the point of comparison. Prac-

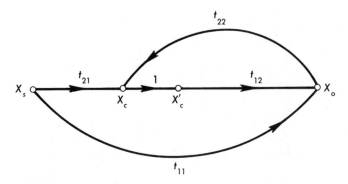

FIGURE 9-31 General signal-flow graph for a feedback amplifier.

tical amplifiers utilize feedback networks in most instances that are comprised of a combination of resistors so a path exists through the feedback network from input to output. As such, the output signal as well as the comparison signal consists of two components as shown in the signal-flow graph in Fig. 9-31.

The flow-graph in Fig. 9-31 characterizes each of the four basic feedback configurations. The various signals may be identified as follows: X_s is the input signal, X_o is the output signal, X_c is the comparison signal, and X_c' is the amplifier input. These signals may be either voltages or currents depending on the particular structure. The transmittance t_{12} is the gain of the amplifier about which the feedback is placed. The measure of the output returned for comparison is indicated by t_{22}. The feed-forward or direct transmission t_{11} accounts for the path between input and output through the feedback network. The transmittance t_{21} measures the fraction of the input signal used for comparison.

Based on the signal-flow graph in Fig. 9-31, the following relations can be written:

$$X_o = t_{11}X_s + t_{12}X_c' \qquad (9\text{-}55)$$

$$X_c = t_{21}X_s + t_{22}X_o \qquad (9\text{-}56)$$

Based on Eqs. (9-55) and (9-56), the method of analysis used to determine the closed-loop gain is similar to both two-port network analysis (Sec. 6-5) and the nodal and mesh methods with controlled sources (Sec. 2-6). First, the t parameters are defined assuming X_c' is an independent source. The second step is to include the constraint equation $X_c' = X_c$ for the dependent source, a relationship evident in the flow-graph. Imposition of the constraint equation permits Eqs. (9-55) and (9-56) to be solved for the closed-loop gain X_o/X_s. The impact of this approach in determining the t-parameters is that one can assume that no connection exists between X_c and X_c', thereby causing the loop to open. Addition of the constraint equation is equivalent to forming the closed-loop system under investigation. This procedure is followed in the ensuing analysis.

From Eq. (9-55), in the manner of the definition of two-port parameters,

$$t_{11} = \left.\frac{X_o}{X_s}\right|_{X_c'=0} \quad \text{and} \quad t_{12} = \left.\frac{X_o}{X_c'}\right|_{X_s=0} \tag{9-57}$$

Similarly, from Eq. (9-56),

$$t_{21} = \left.\frac{X_c}{X_s}\right|_{X_o=0} \quad \text{and} \quad t_{22} = \left.\frac{X_c}{X_o}\right|_{X_s=0} \tag{9-58}$$

By setting $X_c' = 0$ in the definition of t_{11}, the transmission through the amplifier is eliminated. This condition is often referred to as the *dead system* because the amplifier gain, in effect, is zero. The quantity t_{11} is often referred to as the *direct transmission gain* and denoted by G_{DT}.

Application of the constraint equation $X_c = X_c'$ in Eqs. (9-55) and (9-56) and solving for the closed-loop gain gives

$$G_F = \frac{X_o}{X_s} = \frac{t_{11} + t_{12}t_{21}}{1 - t_{12}t_{22}} \tag{9-59}$$

Equation (9-59) can be rewritten in a form similar to Eq. (9-52) as

$$G_F = \frac{G_{DT}}{1 + T} + K\frac{T}{1 + T} \tag{9-60}$$

where

$$T = -t_{12}t_{22}$$

$$K = -\frac{t_{21}}{t_{22}} \quad \text{and} \quad G_{DT} = t_{11} \tag{9-62}$$

In Fig. 9-31, the loop-gain can be identified as $t_{12}t_{22}$, the path through the amplifier and feedback network to the point of comparison. The quantity K, called the gain constant, in Eq. (9-62) is the value of G_F when T becomes infinitely large and corresponds to the amplifier gain becoming infinite. Note that for $t_{11} = 0$, Eq. (9-60) reduces to Eq. (9-52) with K identified as $-1/f(s)$.

The method of analysis described in the development of Eqs. (9-55) to (9-62) is illustrated in the example which follows.

Example 9-6

For the shunt-feedback amplifier represented in Fig. 9-32 determine the t parameters, the return ratio, and the gain constant.

SOLUTION   First, the identification of the variables in the circuit of Fig. 9-32 is made with the signal variables in Fig. 9-31. These are:

$$X_s = I_s \quad X_c = I_1$$

$$X_o = V_o \quad X_c' = I_1'$$

Note that identification of the source zI_1' serves to make that an independent source.

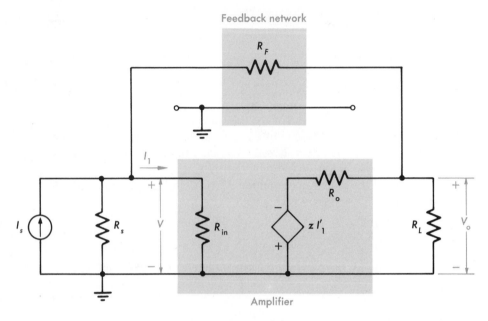

FIGURE 9-32 Shunt-feedback amplifier for Example 9-6.

In the definitions in Eq. (9-57) making $X_s = I_s = 0$ has the effect of open-circuiting the current source and causing $I_1' = X_c' = 0$ makes the source zI_1' a short circuit. The circuit representations in Figs. 9-33a and 9-34a show these conditions for the circuit in Fig. 9-32. In Fig. 9-33b, I_S in parallel with R_S' has been converted to its voltage source equivalent and R_A is parallel combination R_o and R_L. In Fig. 9-33b, the output voltage is simply that for the voltage divider formed by the source I_sR_s, R_s, R_F, and R_A. Therefore,

$$t_{11} = \left.\frac{V_o}{I_s}\right|_{I'=0} = \frac{R_s'R_A}{R_s' + R_F + R_A}$$

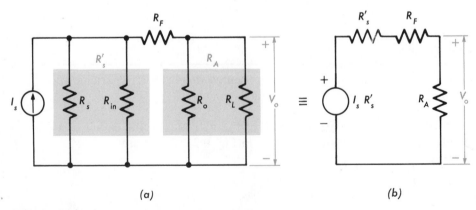

(a)   (b)

FIGURE 9-33 (a) Representation of circuit in 9-32 with $I_1' = 0$ and (b) the circuit in a with the current source converted to a voltage source.

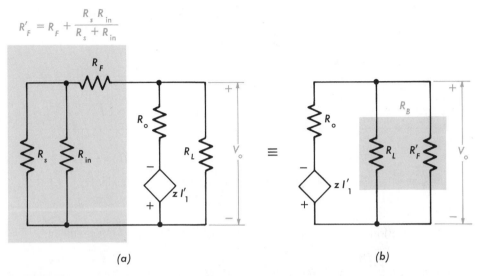

FIGURE 9-34 (a) Representation of circuit in Fig. 9-32 with $I_s = 0$; (b) the circuit in a with the resistances combined.

When the identifications indicated are made, the circuit of Fig. 9-34a can be redrawn as shown in Fig. 9-34b. Again, by use of the voltage-divider technique,

$$t_{12} = \frac{V_o}{I_1'}\bigg|_{I_1 = 0} = \frac{-zR_B}{R_o + R_B}$$

The definitions in Eq. (9-58) when applied to the circuit in Fig. 9-32 lead to the circuits shown in Figs. 9-35a and b. Note that when $V_o = 0$, R_F is connected to ground. Similarly, in Fig. 9-35b only that portion of the circuit which contributes to the comparison need be considered. For the circuit in Fig. 9-35a, the KCL expression is

$$I_s = \frac{V}{R_s} + \frac{V}{R_F} + \frac{V}{R_{in}}$$

The current I_1 is V/R_{in} and combination with the KCL relation gives

$$t_{21} = \frac{I_1}{I_s}\bigg|_{V_o = 0} = \frac{R_s R_F}{R_s R_F + R_s R_{in} + R_F R_{in}}$$

Similarly, for the circuit of Fig. 9-35b, the KCL expression is

$$-\frac{V_o - V}{R_F} + \frac{V}{R_{in}} + \frac{V}{R_s} = 0$$

which, when combined with $I_1 = V/R_{in}$, yields

$$t_{22} = \frac{I_1}{V_o}\bigg|_{I_o = 0} = \frac{R_s}{R_s R_F + R_s R_{in} + R_F R_{in}}$$

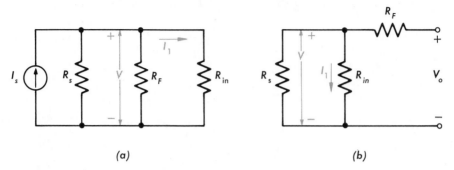

FIGURE 9-35 The circuit portions of Fig. 9-32 used to determine (a) t_{21} and (b) t_{22}.

Substitution of the t parameters into Eqs. (9-61) and (9-62) result in

$$T = \frac{zR_B}{R_o + R_B} \frac{R_s R_F}{R_s R_F + R_s R_{in} + R_F R_{in}}$$

$$K = -R_F$$

The expression developed for t_{11} in Example 9-6 indicates that the feed-forward is a result of both the feedback resistance and the output resistance of the amplifier (R_o is contained in R_A). For the case when $R_o = 0$, as in the operational amplifier, t_{11} becomes zero.

The choice of which feedback configuration is utilized is often dictated by the input and output impedance levels of the amplifier. To demonstrate the effect of feedback on impedance, the input impedance of the circuit in Fig. 9-32 is considered. Based on the identifications given in Example 9-6, Eqs. (9-55) and (9-56) become

$$V_o = t_{11}I_s + t_{12}I_1' \tag{9-63}$$

$$I_1 = t_{21}I_s + t_{22}V_o \tag{9-64}$$

For closed-loop conditions ($I_1 = I_1'$), Eqs. (9-63) and (9-64) can be solved for I_1 in terms of I_s as

$$I_1 = \frac{t_{21} + t_{11}t_{22}}{1 - t_{12}t_{22}} I_s \tag{9-65}$$

In Fig. 9-32, the voltage $V = I_1 R_{in}$, and so by multiplying both sides of Eq. (9-65) by R_{in} and forming the ratio V/I_s one obtains

$$\frac{V}{I_s} = Z_{in} = \frac{R_{in}(t_{21} + t_{11}t_{22})}{1 - t_{12}t_{22}} \tag{9-66}$$

The value of Z_{in} is the resistance seen by the source I_s. The expression in the numerator can be identified as the dead system input impedance because making the system "dead" is equivalent to setting $t_{12} = 0$. The quantity

$1 - t_{12}t_{22}$ in Eq. (9-66) is recognized as $1 + T$ for the amplifier. Thus, Eq. (9-66) indicates that the input impedance of the shunt-feedback amplifier decreases, a result expected when circuits are parallel-connected. The effect of the feedback, however, is significant and is attributed to the $1 + T$ factor.

The input and output impedance in each of the basic feedback configurations is either increased or decreased by the factor $1 + T$. The general form of Eq. (9-66), known as *Blackman's impedance relation*, is

$$Z_{ii} = Z_{ii}^o \frac{1 + T_{sc}}{1 + T_{oc}} \qquad (9-67)$$

where Z_{ii} is the impedance seen at terminals *i-i* (the driving-point impedance), Z_{ii}^o is the dead system value of Z_{ii}, and T_{oc} and T_{sc} refer to the values of T when terminals *i-i* are open-circuited and short-circuited, respectively. For the four configurations shown in Figs. 9-29 and 9-30, either T_{oc} or T_{sc} is zero, illustrating the increase or decrease in impedance level.

Practical circuits used to realize the other feedback-amplifier configurations have gain, bandwidth, and sensitivity properties similar to that of the shunt-feedback amplifier. However, because of differences in the nature of the comparison, variations in performance do exist. Table 9-1 lists the properties of feedback amplifiers for the different configurations and allows for ease in comparison. The various quantities are based on the signal-flow graph in Fig. 9-31.

9-9 STABILITY IN FEEDBACK AMPLIFIERS

The circuits previously discussed employ negative feedback, in which the returned signal is subtracted from the applied signal. The storage elements in the circuit, however, may introduce a sufficient amount of phase shift at some frequency so that the returned signal is in phase with the driving signal. Under these conditions, the amplifier must process a signal which is the sum of the two, causing the output to become large. The cycle repeats itself, and the output is increased without bound until the devices saturate or cut off. This unbounded output results in a condition called *instability*. By defini-

TABLE 9-1 PROPERTIES OF FEEDBACK-AMPLIFIER CONFIGURATIONS

PROPERTY	CONFIGURATION			
	SHUNT	SERIES	SHUNT-SERIES	SERIES-SHUNT
Comparison signal X_c	Current	Voltage	Current	Voltage
Output X_o	Voltage	Current	Current	Voltage
Input connection	Parallel	Series	Parallel	Series
Output connection	Parallel	Series	Series	Parallel
Input impedance	Decreased	Increased	Decreased	Increased
Output impedance	Decreased	Increased	Increased	Decreased

tion, a system is *stable* if its response to an excitation which decays to zero with time also decays to zero.

The consequence of this definition is that all the poles of the transfer function must have negative real parts; i.e., they must lie in the left half of the s plane. In Chaps. 4 and 5, it is shown that the natural response of a system can be obtained from the poles of the system transfer function. Each pole s_p results in a natural-response term of the form $\epsilon^{s_p t}$. The exponential, for $t \geq 0$, will decay if the real part of s_p is negative. If the real part of s_p is positive, the exponential increases without bound. For the cases when the real part is zero, the exponential either has a constant value or varies sinusoidally with time, depending on whether $s_p = 0$ or $s_p = j\omega$. This condition also results in an unstable system because the natural response does not die away with time. Thus, only for poles which have negative real parts is the system stable. If the poles lie in the right half plane or on the j axis, the system is unstable. In unstable systems, the natural frequencies dominate the response; the system is said to *oscillate*. Systems which oscillate, while extremely undesirable in amplifier circuits, are the basis for the wave-form generating circuits described in Chaps. 10 and 11.

To determine if a system is stable, it is necessary to determine either the pole locations or sufficient information about the existence of poles in the right half plane and on the j axis. For systems having many poles, determining all the pole locations is very cumbersome and not particularly useful. The Bode diagram (Sec. 6-3) provides a simple means for determining if any poles exist in the right half plane and thus gives the desired information regarding system stability. The following discussion, based on a particular example, will both illustrate the method and constitute a justification of the results.

The poles of the closed-loop transfer function [Eq. (9-52)] are given by the values of s which satisfy the expression

$$1 - a(s)f(s) = 1 + T(s) = 0 \tag{9-68}$$

The poles are also those values of s which satisfy the equation $T(s) = -1$, enabling us to determine closed-loop stability from the open-loop performance. The open-loop response for simple feedback systems is usually stable, and its performance can be measured and a Bode diagram constructed.

For the purposes of this discussion, consider a system for which $T(s) = K/(s + 1)^3$. The pole locations of the closed-loop system are at $(s + 1)^3 + K = 0$.

The Bode diagram for $\mathbf{T}(j\omega)$ is shown in Fig. 9-36 for the cases where $K = 1, 4, 8$, and 16. As seen in Fig. 9-36, the phase curve is independent of K. An important frequency, called the *phase crossover* ω_ϕ, is the frequency at which the phase of $\mathbf{T}(j\omega)$ is $-180°$. For the example being considered, $\omega_\phi = \sqrt{3}$ rad/s.

A second frequency of importance is the frequency at which $|\mathbf{T}(j\omega)| = 1$, or $|\mathbf{T}(j\omega)|$ is 0 db. It is referred to as *gain-crossover* frequency ω_G. As is evident in Fig. 9-36, ω_G depends on the value of K. As K increases,

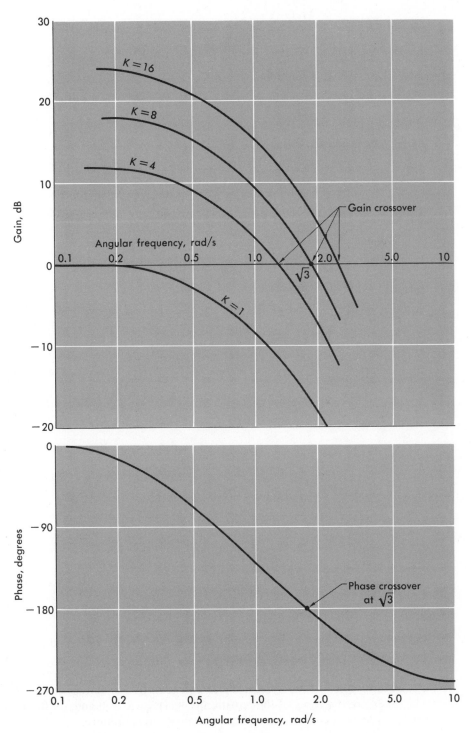

FIGURE 9-36 Bode diagram for amplifier with $T(s) = K/(s + 1)^3$.

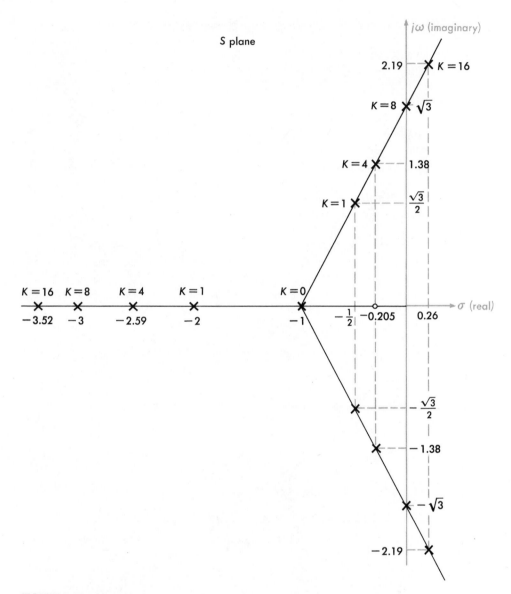

FIGURE 9-37 Closed-loop pole locations in s plane for $T(s) = K/(s + 1)^3$.

the value of ω_G also increases. For $K = 8$, $\omega_G = \sqrt{3}$ rad/s and gain and phase crossovers occur simultaneously.

The pole-zero, or s-plane, diagram showing the position of the closed-loop poles of the system under consideration is shown in Fig. 9-37. The poles are the roots of the equation $(s + 1)^3 + K = 0$ and are found for the different values of K for which Bode diagrams are plotted. In Fig. 9-37, all the poles lie in the left half plane for values of $K < 8$ and two poles are in the right half plane for $K > 8$. When $K = 8$, two of the poles lie on the $j\omega$ axis at $s = \pm j\sqrt{3}$. Thus, from the s-plane diagram, it is evident that the system is unstable for values of $K \geq 8$.

The results of the s-plane diagram can be correlated with those of the Bode diagram. For the stable systems ($K < 8$), it is seen that gain crossover occurs at a lower frequency than phase crossover. Similarly, for $K \geq 8$, gain crossover occurs at frequencies equal to or greater than phase crossover. Thus, for single-loop feedback systems for which the open-loop response is stable, a rule of thumb, often applicable, for stability of the closed-loop system based on the Bode diagram is as follows: *For a system to be stable, gain crossover must occur prior to phase crossover.*

The general criterion for stability is based on the introduction of the quantity called the *phase margin,* PM, defined as

$$\text{PM} = \angle\mathbf{T}(j\omega_G) + 180° \tag{9-69}$$

The phase margin, as given in Eq. (9-69), is a measure of how much the phase of $\mathbf{T}(j\omega_G)$ differs from 180°. As seen in Fig. 9-36, for K > 8, the $\angle\mathbf{T}(j\omega_G)$ is always above $-180°$ while for K < 8 the $\angle\mathbf{T}(j\omega_G)$ is below $-180°$. The two situations described in the previous sentence correspond to phase margins which are positive and negative, respectively. *The general criterion for the closed-loop system to be stable is that the phase margin be positive.* The validity of this criterion presupposes a stable open-loop system.

Another quantity useful in stability analysis is the *gain margin* GM, which is a measure of $T(j\omega_\phi)$ relative to zero dB. Mathematically,

$$\text{GM} = -T(j\omega_\phi)|_{\text{dB}} \tag{9-70}$$

which in Fig. 9-36 is positive for stable systems and negative for unstable ones.

Example 9-7

The return ratio of a certain feedback amplifier is given by the expression

$$T(s) = \frac{100(1 + s/40)}{(1 + s)(1 + s/10)(1 + s/20)}$$

Determine whether the closed-loop amplifier is stable.

SOLUTION The condition for stability is that the phase margin be positive. At gain crossover $|\mathbf{T}(j\omega_G)| = 1$ (0 dB), and at phase crossover the angle of $\mathbf{T}(j\omega_\phi) = -180°$. These quantities are readily determined from the Bode diagram shown in Fig. 9-38. The dashed curve in Fig. 9-38 represents the asymptotic Bode diagram and the solid curve the exact Bode diagram. From the curves, it is determined that

$$\omega_G = 26.1 \text{ rad/s} \quad \text{and} \quad \text{PM} = 3.75°$$

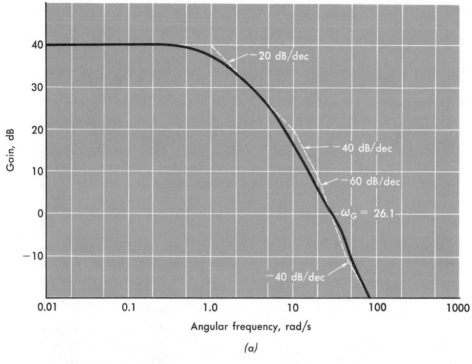

(a)

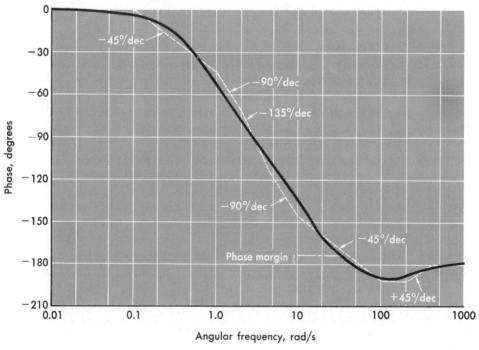

(b)

FIGURE 9-38 Bode diagram for Example 9-7.

As the phase margin is positive the closed-loop amplifier is stable. However, the phase margin is so small that such a system would be impractical.

Feedback amplifiers are used to provide the benefits of sensitivity improvement and reduction in distortion. To realize the desired improvements, a value of T may be required which will cause the system to be unstable. Compensating networks are often employed to ensure stability while permitting an otherwise unstable value of T. Compensation would also be employed to render the system described in Example 9-7 practical.

9-10 THE OPERATIONAL AMPLIFIER

The characteristics of the ideal operational amplifier, first introduced in Sec. 8-1, are infinite input resistance, zero output resistance, and a gain A_o sufficiently large so it can be considered infinite. Practical OP-AMPs, for which the block diagram in Fig. 9-39 is a typical structure, are designed to approximate these characteristics over a range of frequencies usually specified by the gain-bandwidth product. Each of the component circuits in Fig. 9-39 is primarily responsible for realizing one of the characteristics. The interconnection of these circuits, therefore, provides the desired overall performance.

The initial stage is the differential amplifier (see Sec. 8-12) whose function allows for both inverting and noninverting inputs and provides the high-input resistance required. In addition, because integrated circuits are direct-coupled, the dc input, or leakage, current must be small to minimize the interaction of the bias network and signal source. A pictorial representation of the differential amplifier is shown in Fig. 9-40. The current source, often of the type described in Sec. 8-2, both biases the circuit and provides a high value of common-mode rejection ratio. The equivalent resistance of the current source helps ensure that the input resistance is large.

By biasing the devices at low levels of current, leakage current is kept to a minimum. The Darlington compound transistor, shown in Fig. 9-21, is sometimes used for this purpose as is the super-beta transistor. A *super-beta transistor* is one which is fabricated to give values of h_{FE} in the order of 2,000 so that $I_B = I_C/h_{FE}$ can be kept small.

Although the major function of the differential stage is to interface the device with the input signals, it is desirable that the differential-mode gain A_{DM} be as large as is practical. As given in Eq. (8-73), A_{DM} is increased by increasing the value of R_L. The *active load* shown in Fig. 9-41 serves in place of the load resistor and overcomes fabrication limitations on resistance values. The circuit in Fig. 9-41 is seen to be the current source in Fig. 8-58a that

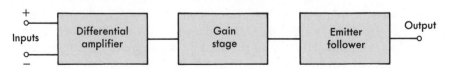

FIGURE 9-39 Block diagram of operational amplifier circuit.

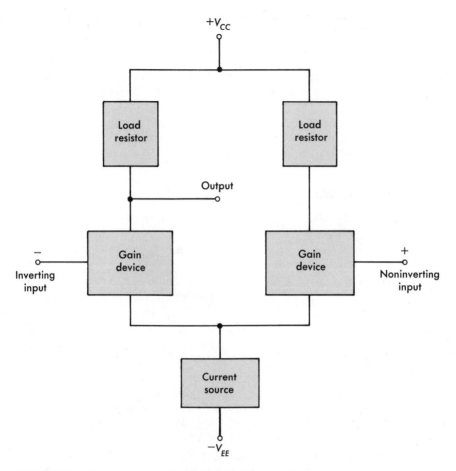

FIGURE 9-40 Pictorial representation of differential amplifier.

utilizes lateral *pnp* transistors. The effective load resistance is then the output resistance r_o of the transistor. To provide the gain over the required frequency range in stages with large load resistances, the cascode circuit of Fig. 9-22 is commonly used as the gain device.

The output stage is the interface between the operational amplifier and the load it drives. To achieve the low-output resistance required, emitter-follower stages described in Sec. 8-10 are generally employed. For improvement in both dynamic range and the maximum signal that can be processed, use is often made of the complementary emitter-follower depicted in Fig. 9-23.

The principal function of the interior stage is to provide the gain and shape the frequency response, as shown in Fig. 8-5, of the operational amplifier. However, as the OP-AMP is almost exclusively used in a feedback configuration, the OP-AMP stages must be stable under closed-loop conditions. A typical asymptotic Bode diagram, in both magnitude and phase, for a

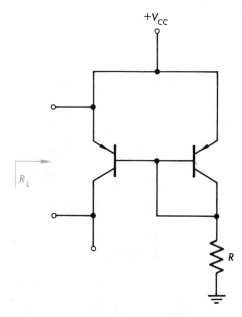

FIGURE 9-41 Active load.

three-stage amplifier is displayed in Fig. 9-42; the characteristic indicated by the dashed line corresponds to the desired operational-amplifier frequency response. When used as the amplifier in a feedback system, the characteristics in Fig. 9-42 are the return ratios as functions of frequency. The solid-curve for the three-stage amplifier results in an unstable closed-loop system, whereas the amplifier characteristic given by the dashed curve leads to a closed-loop response that is stable. To convert the solid characteristic in Fig. 9-42 to the dashed one, compensation is used to deliberately narrow-band the amplifier. This is usually accomplished by using a capacitive feedback loop around the interior stage as depicted in Fig. 9-43a. The equivalent circuit for the stage shown in Fig. 9-43a is given in Fig. 9-43b, in which the amplifier output resistance has been combined with R'_L to form R_L. The circuit in Fig. 9-43b is identical in structure to that shown in Figs. 8-26 and 8-27, in which the capacitor C_μ was reflected into the input stage. The value of the resultant equivalent input capacitance is $C(1 + yR_L)$ and illustrates the multiplication effect of the amplifier. Known as the *Miller effect,* the increased capacitance is a result of the shunt feedback used in the circuit of Fig. 9-43a. As shunt feedback decreases the impedance level, the capacitance must increase (and by a factor $1 + T$). In a common-emitter stage, the use of C increases the value of C_x and, consequently, reduces the bandwidth.

The Miller effect is often used in *IC* circuits to realize capacitance values whose sizes preclude fabrication. The effect is also used in timing circuits described in Chap. 11.

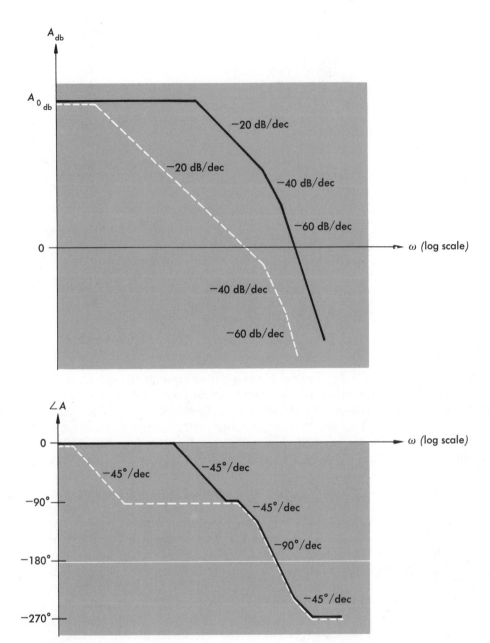

FIGURE 9-42 Typical asymptotic Bode diagram for a three-stage feedback amplifier.

PROBLEMS

9-1 The amplifier depicted in Fig. 9-1 is designed with component values $R_1 = R_3 = R_5 = 5$ kΩ, $R_2 = 50$ kΩ and $R_4 = R_6 = 10$ kΩ. The operational amplifiers are identical and have $A_o = 10^5$ and a gain-bandwidth product of

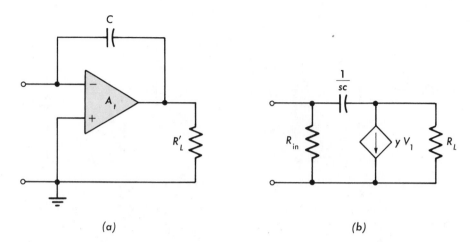

FIGURE 9-43   (a) Representation of capacitive compensation and (b) its equivalent circuit.

2 MHz. Determine the overall gain of the cascade and the bandwidth using both the Bode diagram and dominant-pole approximation.

9-2   Repeat Prob. 9-1 for $R_6 = 6$ kΩ and all other parameters are those given in Prob. 9-1.

9-3   The circuit shown in Fig. 9-44 is designed using identical OP-AMPs whose open-loop gain and bandwidth are 2×10^5 and 10 Hz, respectively. The element values are $R_s = 10$ kΩ, $R_1 = 20$ kΩ, $R_2 = 300$ kΩ, $R_3 = 10$ kΩ, and $R_4 = 90$ kΩ. Determine the midband gain and bandwidth of the cascade. Compare the results obtained from the Bode diagram and the dominant-pole approximation.

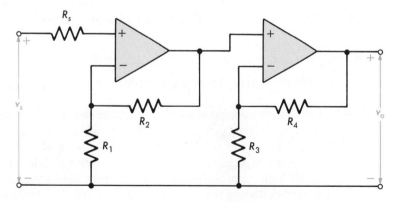

FIGURE 9-44   Circuit for Prob. 9-3.

9-4   The amplifier stage containing R_3 and R_4 in Fig. 9-44 is replaced by an inverting OP-AMP stage. Repeat Prob. 9-3 assuming the same element values and operational amplifiers are used as given in Prob. 9-3.

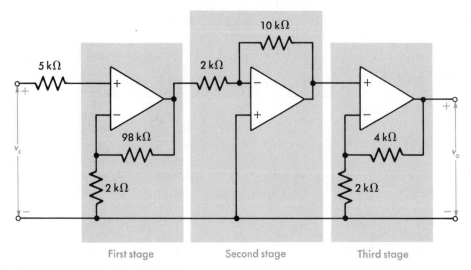

FIGURE 9-45 Circuit for Prob. 9-5.

9-5 Determine the midband gain and bandwidth of the three-stage cascade shown in Fig. 9-45. The operational amplifier characteristics are given in Prob. 9-1.

9-6 Repeat Prob. 9-5 if the first and second stages in Fig. 9-45 are interchanged.

9-7 Repeat Prob. 9-5 if the second and third stages in Fig. 9-45 are interchanged.

9-8 The gain of an operational amplifier stage can be expressed as

$$G_H(s) = \frac{G_o}{1 + s/\omega_H}$$

Show that the bandwidth of n cascaded stages is $\omega_H\sqrt{2^{1/n} - 1}$.

9-9 Often in the design of a cascaded amplifier it is desirable to place the stage with the highest gain (and lowest bandwidth) closest to the input. This configuration can be useful in reducing the effects of the noise (unwanted and often random signals) that is always present with the signal to be processed. With this approach to design, the first stage in the cascade is used to set the bandwidth and the additional stages provide the remainder of the required gain. The import of the previous statement is the implication that the bandwidths of the succeeding stages are sufficiently larger than the initial stage bandwidth. As the designer, you have available OP-AMPs whose open-loop gain is 126 dB and whose gain-bandwidth product is 2 MHz. Design an amplifier whose bandwidth is 5 kHz and which provides a 2-V rms output signal for a 25-μV rms input signal fed through a 5-kΩ resistance. It is desired to have the output out of phase with the input.

9-10 Repeat Prob. 9-9 when the output and input signals are to be in phase.

9-11 Design an amplifier which provides 60 dB of gain and has a bandwidth of 40 kHz. The source resistance is 2 kΩ and the output is to be in phase with the input. The operational amplifier characteristics are $A_o = 10^5$ and $f_h = 10$ Hz.

9-12 Repeat Prob. 9-11 if the output is to be 180° out of phase with the input.

9-13 Manufacturing reliability and technological feasibility often dictate that the ratio of the largest resistance to the smallest resistance used in a circuit be equal to or less than 10.
a Using this restriction, find the maximum midband gain that can be realized by a three-stage amplifier if the output is to be 180° out of phase with the input.
b Use this restriction and assume the OP-AMP characteristics are those given in Prob. 9-11 to find the bandwidth of the amplifier obtained in *a*.

9-14 Repeat Prob. 9-13 if the amplifier output and input signals are to be in phase.

9-15 Repeat Prob. 9-13 employing only inverting OP-AMP stages.

9-16 The amplifier shown in Fig. 9-7 is designed using identical transistors whose parameters are $r_x = 60\,\Omega, r_\pi = 240\,\Omega, g_m = 0.4\mho, C_\pi = 100\,\text{pF}, C_\mu = 2\text{pF}$, and $r_o = 100$ kΩ. The element values are $R_{C1} = R_{C2} = 900\ \Omega$, $R'_s = R'_L = 300\ \Omega$, $R_{E1} = R_{E2} = 1.2$ kΩ and the base-bias resistances R_1, R_2, R_3, and R_4 are sufficiently large so their effect can be considered negligible. Determine
a The midband gain of the cascade
b The upper half-power frequency

9-17 **a** Assuming that C_E acts as a perfect bypass, determine the lower half-power frequency of the amplifier in Prob. 9-16 when $C_B = 2\ \mu$F.
b Assuming C_B acts as a perfect coupling capacitor, determine ω_{LO} for the amplifier in Prob. 9-16 when $C_E = 100\ \mu$F.

9-18 The amplifier in Fig. 9-7 is designed using identical transistors whose parameters are $r_\pi = 600\ \Omega$, $r_x = 0$, $g_m = 0.2\ \mho$, $r_o \to \infty$, $C_\pi = 80$ pF, and $C_\mu = 1$ pF. The circuit-parameters are $R'_s = 600\ \Omega$, $R_{C1} = 1.2$ kΩ, $R_{E1} = 1.2$ kΩ, $R_{C2} = 1.5$ kΩ, $R_{E2} = 0.9$ kΩ, $R'_L = 300\ \Omega$, $R_1 = 18$ kΩ, $R_2 = $

11 kΩ, $R_3 = 56$ kΩ, $R_4 = 27$ kΩ, $C_E = 300$ μF, and $C_B = 5$ μF.
Determine
a   The midband gain
b   The upper half-power frequency of the cascade

9-19   a   Determine the lower half-power frequency of the cascaded amplifier in Prob. 9-19.
b   To what value must C_B be changed if the lower half-power frequency is
to be determined, for all practical purposes, by C_E?

9-20   A three-stage amplifier is constructed using the transistor described in
Prob. 9-16. Assuming that coupling and bypass are ideal and that the
base-bias resistances have negligible effect,
a   Determine the midband gain when $R_{C1} = R_{C2} = R_{C3} = 1.2$ kΩ, $R'_s =$
50 Ω, and $R'_L = 600$ Ω
b   Find the upper half-power frequency
c   Find the power gain of the amplifier

9-21   Biasing arrangements are not indicated in the circuit of Fig. 9-46 and
can be assumed to have negligible effect on amplifier performance. The transistors employed have the parameters given in Prob. 9-18 and the parameter
values are

$$R_{C1} = 2 \text{ k}\Omega, \ R_{E1} = 50 \ \Omega, \ R_{C2} = 500 \ \Omega, \text{ and } R_{E2} = 100 \ \Omega.$$

a   Determine the gain and bandwidth of the cascade.
b   If r_π and consequently h_{fe} increase by 50 percent, repeat a.

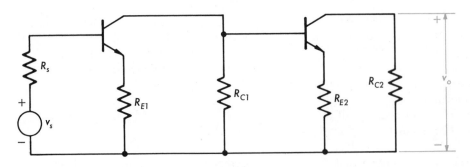

FIGURE 9-46   Circuit for Prob. 9-21.

9-22   The amplifier in Fig. 9-46 is to be designed using the transistor whose
parameters are given in Prob. 9-18. The resistances R_{E1} and R_{C1} are to be
equal to R_{E2} and R_{C2}, respectively. The source resistance R_s is 500 Ω.
a   Determine R_{C1} if the midband gain of the cascade is to be 5,000 when
$R_{E1} = 100$ Ω.
b   Find ω_{HO} for the conditions in a.

9-23 The transistor in Prob. 9-18 is used in the circuit of Fig. 9-18 for which $R_s = 500\ \Omega$ and $R_{C1} = R_{C2} = 2\ \text{k}\Omega$.
a Determine $R_{E1} = R_{E2}$ for an overall midband gain of 5,000.
b Find ω_{HO} for the conditions in a.

9-24 To make the performance of an amplifier relatively insensitive to variations in the signal-source resistance R_s, the configuration in Fig. 9-47 is frequently employed. The input stage is an emitter-follower which acts as a buffer stage. The transistors used are identical and have parameter values: $r_x = 0$, $r_\pi = 1.5\ \text{k}\Omega$, $h_{fe} = 150$, $r_o \to \infty$, $C_\pi = 50\ \text{pF}$, and $C_\mu = 1.0\ \text{pF}$.
a Determine the midband gain when $R_s = 10\ \text{k}\Omega$.
b Repeat a for $R_s = 5\ \text{k}\Omega$ and $R_s = 20\ \text{k}\Omega$.
(*Note:* Bias arrangements are not shown and can be considered as having negligible effect.)

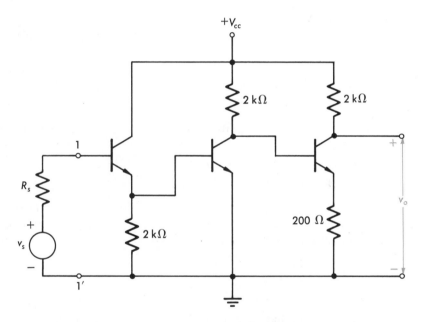

FIGURE 9-47   Circuit for Prob. 9-24.

9-25 Determine the upper half-power frequency of the amplifier in Prob. 9-24 when R_s equals
a 10 kΩ
b 5 kΩ
c 20 kΩ
Assume the frequency response of the emitter-follower is sufficiently high so that it has no effect on the frequency response of the cascade.

9-26 The amplifier shown in Fig. 9-48 employs the transistor described in Prob. 9-24. The purpose of the emitter-follower output stage is to provide a

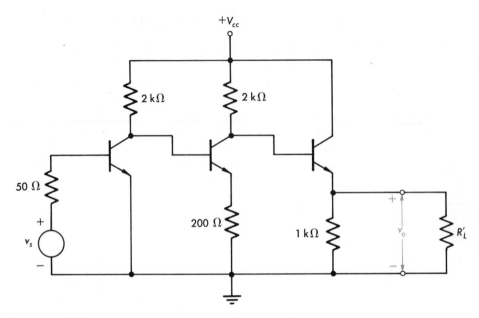

FIGURE 9-48 Circuit for Prob. 9-26.

low output impedance for the cascade and to render the response impervious to changes in R'_L.

a Determine the midband gain for $R'_L = 600\ \Omega$.

b Repeat *a* for $R'_L = 1.2\ \text{k}\Omega$ and $R'_L = 300\ \Omega$.

9-27 Assuming the emitter-follower has negligible effect on the frequency response of the cascade, determine the value of ω_{HO} for the amplifier described in Prob. 9-26 when R'_L equals

a $600\ \Omega$

b $1.2\ \text{k}\Omega$

c $300\ \Omega$

9-28 The amplifier in Fig. 9-12 is designed with identical JFETs whose characteristics are: $g_m = 2m\mho$, $r_d = 100\text{k}\Omega$, $C_{gd} = 5\text{pF}$ and $C_{gs} = 10\text{pF}$. The circuit element values used are: $R_1 = R_3 = 240\ \text{k}\Omega$, $R_2 = R_4 = 80\ \text{k}\Omega$, $R_{D1} = R_{D2} = 10\ \text{k}\Omega$, $R_{K1} = R_{K2} = 5\ \text{k}\Omega$ and $R'_s = 1\ \text{k}\Omega$. Determine:

a The midband gain on the cascade.

b The upper half-power frequency.

9-29 Determine the lower half-power frequency of the amplifier stage in Prob. 9-29 when $C_G = 1.0\ \mu\text{F}$ and $C_K = 500\ \mu\text{F}$.

9-30 The JFET described in Prob. 9-28 is used in the amplifier depicted in Fig. 9-12. The circuit element values are: $R_1 = 300\ \text{k}\Omega$, $R_2 = 100\ \text{k}\Omega$, $R_{D1} =$

25 kΩ, R_{K1} = 2.5 kΩ, R_3 = 240 kΩ, R_4 = 160 kΩ, R_{D2} = 5 kΩ, R_{K2} = 5 kΩ
and the source impedance is 600 Ω.
a Find the input signal amplitude if the output signal has an amplitude of
2V rms at midband frequencies
b At what frequency is the output 4V peak-to-peak?

9-31 For the amplifier in Prob. 9-30,
a Determine the value of C_K which results in a lower half-power frequency
of 159 Hz. Assume C_G acts to provide perfect coupling.
b Determine the value of C_G which results in a lower half-power frequency
of 159 Hz assuming C_K is an ideal bypass.
Using reasonable engineering approximations,
c Find the value of C_G needed to satisfy the conditions in *a*
d Find the value of C_K needed to satisfy the conditions in *b*

9-32 The characteristics of the MOSFETs in Fig. 9-49 are g_m = 2.5 m$\mho$,
r_d = 60 kΩ, C_{gd} = 4 pF, and C_{gs} = 8 pF. Determine the gain and upper half-
power of the cascade. (*Note:* Biasing arrangements are not shown and can
be considered to have negligible effect.)

9-33 The MOSFET described in Prob. 9-32 is used in a three-stage cascade

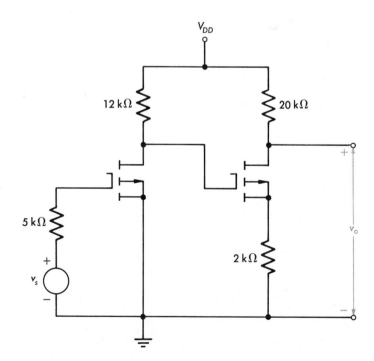

FIGURE 9-49 Circuit for Prob. 9-32.

amplifier for which biasing resistances have negligible effect on performance. The drain resistances of each stage are made equal.
a Determine R_D so that a 5-V output signal results when an input signal of 5
mV is applied through a 2-kΩ resistance. Assume $R_K = 100\ \Omega$ in each stage.
b Evaluate the upper half-power frequency of the cascade using the results
in a.

9-34 The normalized voltage-transfer function of a transistor amplifier
stage employing emitter regeneration can be approximated as

$$G_N(p) = \frac{1 + mp}{1 + p + mp^2}$$

where $\hat{p} = s/\omega_H$ is the frequency variable normalized to the upper half-
power frequency of the uncompensated stage. $G_N = G_H/G_O$ is the transfer
function normalized to the midband gain of the uncompensated stage. m is a
parameter which is directly proportional to the value of the compensation
capacitor C_R. The condition $m = 0$ corresponds to the uncompensated stage
whose normalized gain and upper half-power frequency are both unity.
a For $m = 5/36$, $1/4$, $(\sqrt{2} - 1)$ and $25/36$, sketch a Bode diagram for the
frequency response.
b Determine the upper half-power frequency for each case in a and compare the response to the uncompensated case.

9-35 Determine the response to a unit step of voltage for each value of m
given for the amplifier stage in Prob. 9-34. Compare these results with the
response of the uncompensated stage.

9-36 The transistors in the cascode amplifier in Fig. 9-22 are identical. Its
parameters are $r_\pi = 5$ kΩ, $h_{fe} = 200$, $r_x = 0$, $r_o = 500$ kΩ, $C_\pi = 20$ pF, and
$C_\mu = 0.5$ pF. The load resistance is 500 kΩ and the signal is fed into the
amplifier through a 1.25-kΩ resistance.
a Determine the midband gain of the cascode amplifier.
b What is the upper half-power frequency?

9-37 Cascode amplifiers are often fabricated using "super-beta" transistors, i.e., transistors with extremely high values of h_{fe}. Repeat Prob. 9-36
if all parameters are as given in Prob. 9-36 but $h_{fe} = 2,000$.

9-38 The small-signal parameters of the npn and pnp transistors in Fig.
9-23 are identical. Assuming $r_x = 0$ and $r_o \to \infty$, derive an expression for the
midband input impedance.

9-39 The npn and pnp transistors in Fig. 9-23 have identical small-signal
parameters which are given in Prob. 9-36. The circuit is driven by a signal
source having an internal resistance of 5 kΩ.
a Determine the minimum value of R_L for which the output voltage is 95
percent of the signal voltage at midband frequencies.

b   What is the midband output impedance of the amplifier?

9-40   At midband frequencies represent the amplifier in Prob. 9-21 as a non-ideal current-controlled current source.

9-41   In the midband range, represent the amplifier in Prob. 9-16 as a non-ideal voltage-controlled current source.

9-42   A feedback amplifier as in Fig. 9-28 is designed using

$$a(s) = \frac{-10^5}{1 + s/10}$$

a   Find the value of f which makes the midband value of the closed-loop gain 100.
b   What is the closed-loop bandwidth?
c   If the midband value of $a(s)$ decreases by 20 percent, what change results in the closed-loop gain?

9-43   The midband, closed-loop gain of the feedback amplifier represented in Fig. 9-28 is to be 50 and cannot deviate by more than 2 percent. The gain of the amplifier $a(s)$ can be expected to change by 100 percent.
a   Determine the midband values of a and f.
b   What must the bandwidth of $a(s)$ be if the closed-loop bandwidth is to be 10^5 rad/s.

9-44   Two identical stages are cascaded and used to realize the branch $a(s)$ in Fig. 9-28. If each amplifier stage has a gain-bandwidth product of 10^6 rad/s, design a feedback amplifier which satisfies the midband performance given in Prob. 9-43. What is the bandwidth of this feedback amplifier?

9-45   For $a(s) = a_o/(1 + s/\omega_h)$ and $f(s) = f_o$, show that the open-loop and closed-loop gain-bandwidth products are equal.

9-46   Determine the midband values of G_{DT}, T, and K in Eq. (9-60) for a noninverting OP-AMP stage.

9-47   **a**   Determine the loop-gain, direct transmission gain and gain constant at midband frequencies for the stage depicted in Fig. 9-50. Assume $r_x = 0$ and $r_o \to \infty$ in the small-signal model of the transistor.
b   Evaluate the closed-loop gain using the results in a and compare with the gain of the stage developed in Chap. 8.

9-48   An operational amplifier having an open-loop gain of 10^5 and a gain-bandwidth product of 2 MHz is used in an inverting stage for which $R_1 = 5$ kΩ and $R_2 = 20$ kΩ. Determine
a   The loop gain and gain constant at midband frequencies
b   At what frequency the magnitude of the loop-gain is equal to 100

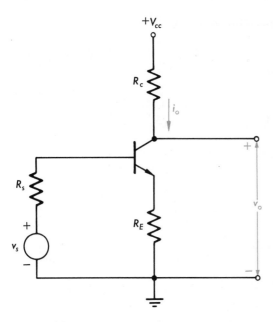

FIGURE 9-50 Circuit for Prob. 9-47.

9-49 Repeat Prob. 9-48 if the OP-AMP and R_1 and R_2 are used in a noninverting stage.

9-50 The circuit in Fig. 9-51 is a representation of a series-shunt amplifier. The element values are $R_i = 1\ \text{k}\Omega$, $R_o = 5\ \text{k}\Omega$, $A_v = 10^3$, $R_2 = 50\ \text{k}\Omega$, $R_1 = 2\ \text{k}\Omega$, $R_s = 500\ \Omega$, and $R_L = 600\ \Omega$. Determine G_{DT}, T, and K.

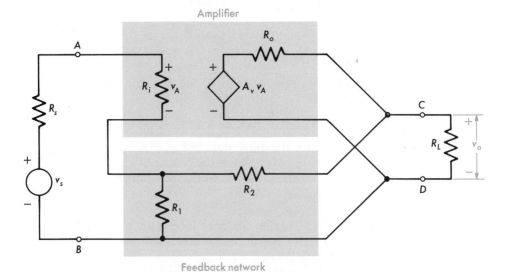

FIGURE 9-51 Series-shunt amplifier for Prob. 9-50.

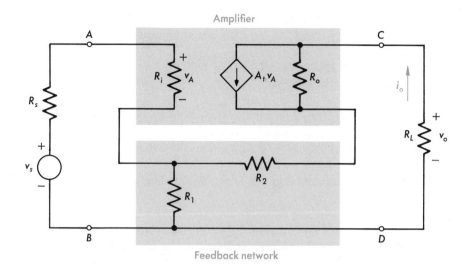

FIGURE 9-52 Series amplifier for Prob. 9-51.

9-51 The circuit in Fig. 9-52 is a series-feedback amplifier. The element values are $R_i = 500\ \Omega$, $R_o = 20\ \text{k}\Omega$, $R_2 = 50\ \text{k}\Omega$, $R_1 = 1\ \text{k}\Omega$, $R_s = 600\ \Omega$, $R_L = 2\ \text{k}\Omega$, and $A_t = 100\ \Omega$. Determine T, K, and G_{DT} assuming the output is i_o.

9-52 The element values in the shunt-series amplifier depicted in Fig. 9-53 are $R_i = 5\ \text{k}\Omega$, $R_o = 500\ \Omega$, $A_z = 100\ \text{k}\Omega$, $R_2 = 10\ \text{k}\Omega$, $R_1 = 1\ \text{k}\Omega$, $R_s = 50\ \Omega$, and $R_L = 2\ \text{k}\Omega$. Determine T, K, and G_{DT}.

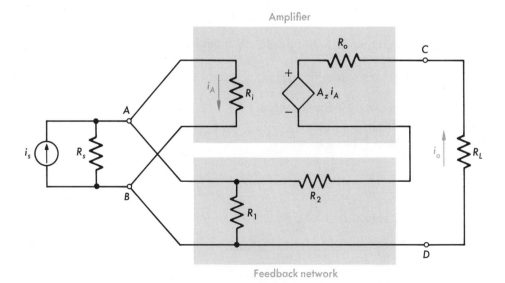

FIGURE 9-53 Shunt-series amplifier for Prob. 9-52.

9-53   The element values in the circuit of Fig. 9-32 are $R_i = 2\,k\Omega$, $R_o = 5\,k\Omega$, $R_F = 10\,k\Omega$, $z = 20\,k\Omega$, $R_L = 1\,k\Omega$, and $R_s = 1\,k\Omega$. Determine T, K, and G_{DT}.

9-54   Use Blackman's relation to find the input impedance Z_{AB} for the amplifier described in Prob. 9-50.

9-55   Repeat Prob. 9-54 for the circuit described in Prob. 9-51.

9-56   Repeat Prob. 9-54 for the circuit described in Prob. 9-52.

9-57   Determine the input and output impedances for the shunt amplifier in Prob. 9-53.

9-58   What is the reduction in the "dead-system" output impedance Z_{CD} that results from the use of feedback in the circuit of Fig. 9-50 and Prob. 9-50.

9-59   Determine the output impedance Z_{CD} for the circuit in Prob. 9-51.

9-60   Repeat Prob. 9-59 for the circuit in Prob. 9-52.

9-61   An inverting OP-AMP stage is designed with $R_1 = 5\,k\Omega$, $R_2 = 20\,k\Omega$ and an operational amplifier whose open-loop gain and bandwidth are 2×10^5 and 10 Hz, respectively. Determine $T(s)$ and the gain crossover frequency for the stage.

9-62   Repeat Prob. 9-61 if the same elements are used in a noninverting stage.

9-63   The return ratio of a feedback amplifier is given by

$$T(s) = \frac{40}{(1 + s)(1 + s/4)(1 + s/40)}$$

Is the closed-loop amplifier stable?

9-64   Repeat Prob. 9-63 when

$$T(s) = \frac{40(1 + s/8)}{(1 + s)(1 + s/4)(1 + s/40)^2}$$

9-65   The return ratio of a feedback amplifier is

$$T(s) = \frac{T_o}{(1 + s)(1 + s/10)(1 + s/50)}$$

a   Determine the largest value of T_o for which the amplifier is stable.

b What value of T_o results in a phase margin of $45°$?

9-66 The return ratio of a feedback amplifier is

$$T(s) = \frac{T_o(1 + s/\sqrt{10} \times 10^6)}{(1 + s/10^5)(1 + s/10^6)(1 + s/10^7)^2}$$

a Determine the largest value of T_o for which the amplifier is stable.
b What value of T_o results in a gain margin of 10 dB?

9-67 The asymptotic Bode diagram in Fig. 9-54 is the open-loop gain of an operational amplifier as a function of frequency. This OP-AMP is used in an inverting stage for which $R_1 = R_2 = 10$ kΩ. Determine
a The midband loop-gain
b The gain margin
c The phase margin

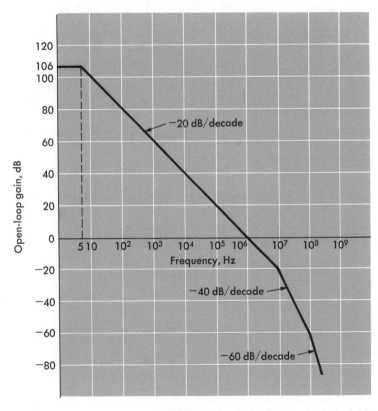

FIGURE 9-54 Bode diagram for Prob. 9-67.

9-68 The open-loop bandwidth of an operational amplifier is improved when pole-zero compensation is used. The resultant open-loop gain charac-

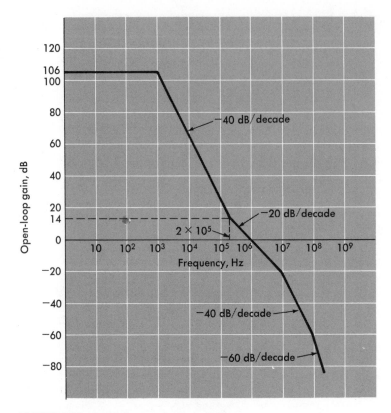

FIGURE 9-55 Bode diagram for Prob. 9-68.

teristic is displayed in Fig. 9-55. This OP-AMP is used in an inverting stage for which $R_1 = R_2 = 10$ kΩ. Determine

a The midband value of the loop gain
b The bandwidth of the loop gain
c The gain and phase margins
d The closed-loop bandwidth

10

ANALOG SIGNAL PROCESSING

iverse areas of application, among them communication, control, computation, and instrumentation systems, have as their essential feature the processing of information. Virtually all these systems require amplification of the electrical waveforms present. In addition, many other circuit functions are necessary to provide the requisite signal processing so these systems can function appropriately. The amplifier circuits described in the preceding two chapters form the basis for several classes of electronic circuits. In this chapter, particular emphasis is placed on sinusoidal oscillator circuits, frequency-selective networks, and circuits which perform analog computation.

10-1 FREQUENCY-DOMAIN SIGNAL PROCESSING

Continuous signals and *discrete signals,* two commonly referred to classes of waveforms, are described on the basis of their behavior as functions of time. Continuous signals are displayed in Fig. 10-1, and the waveforms shown in Fig. 10-2 are discrete signals. As illustrated in Fig. 10-1, continuous signals are described by time functions which are defined for all values of t; i.e., that is t is a continuous variable. The discrete signals exist only at specific instances of time. As such, their functional description is valid only for the discrete-time intervals. Commercial broadcast systems, analog computers, and a variety of control and instrumentation systems process continuous signals. Digital computers, pulsed-communication systems such as modern telephone and radar, and microprocessor-based control systems utilize discrete signals.

Another feature of the waveforms in Figs. 10-1 and 10-2 can be discerned if we consider the signal in Fig. 10-1a to be a voltage representing a physical quantity (perhaps it can be the output voltage of a phonograph cartridge). The discrete signal in Fig. 10-2a has the same amplitude at times $t = 0$, T_1, T_2, and T_3 as does the continuous signal in Fig. 10-1a. Both voltage waveforms have a one-to-one correspondence in time and amplitude with the physical quantity represented.

In the context of this discussion, the sequence of pulses in each time interval in Fig. 10-2b is a numeric, or digital, representation of the corresponding voltage samples shown in Fig. 10-2a. The waveforms in Fig. 10-1b may be the signal, called a clock, which sets the timing sequence used in the generation of the pulses in Fig. 10-2. Neither the amplitude or time of the signals in Figs. 10-1b and 10-2b correspond to the physical quantity v_1. Essentially, these are signals in which the information is contained by the presence or absence of a pulse during a given time interval.

The description of the waveforms in the preceding discussion leads to the classification of systems being analog or digital. The information pro-

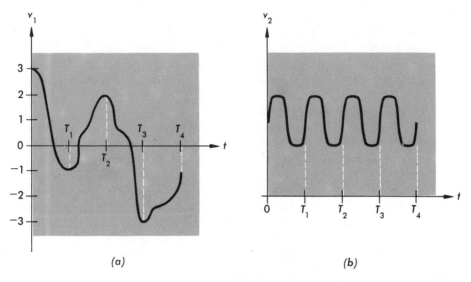

FIGURE 10-1 Continuous signals.

cessed in *analog systems* is contained in the time function which defines the signal. *Digital systems,* as the name implies, process digits, i.e., *pulse trains,* in which the information is carried in the pulse sequence rather than the amplitude-time characterization of the pulses. This chapter deals with analog systems, in particular, those in which continuous signals are processed. Systems which process pulses, notably digital systems, are treated in Chaps. 11 and 12.

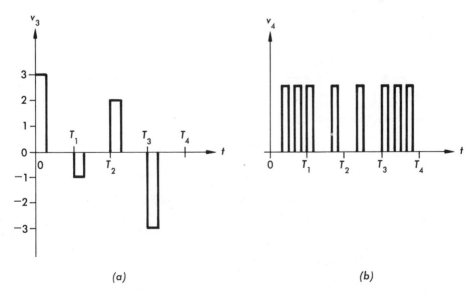

FIGURE 10-2 Discrete signals.

In most analog systems, the signals containing the information to be processed may be represented by a number of sinusoids (see Sec. 3-12). The performance of analog systems is then conveniently described in terms of the steady-state response to each of the sinusoidal excitations. As described in Chaps. 3 to 6, this type of analysis is best performed in the frequency domain. Hence, analog systems are often thought of as performing signal processing in the frequency domain.

To perform signal processing, a number of circuit functions other than amplifiers must be used. The pictorial representation of a commercial AM radio system shown in Fig. 10-3 is useful in indicating several of the required circuit functions needed to process the information. The primary purpose of the system is to transfer the audio information at the transmitting end to the receiving end. The first step in the process is to convert the acoustic energy into an electrical signal. The conversion is effected by a transducer, usually a microphone. As the output of the transducer is a low-level signal, amplification is necessary. Radio-frequency (RF) signals (signals whose frequencies are greater than 500 kHz) are much more readily propagated through the atmosphere than are audio frequencies (20 to 20,000 Hz). Therefore, the audio information is frequency translated to radio frequencies. The frequency translation is achieved by a process called *modulation*.

At the receiver, the process of extracting the information is nearly the reverse of transmission. The received signal is weak and must be amplified. In addition, because many signals (stations) are present at the receiving antenna, the desired signal must be identified and extracted. This function is

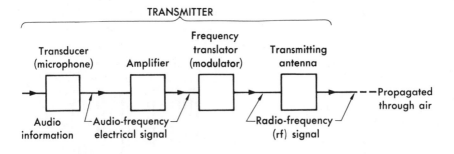

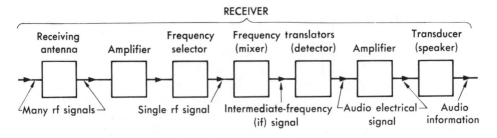

FIGURE 10-3   Pictorial representation of a commercial AM radio system.

referred to as *frequency selection*. Practical considerations dictate two frequency translations, the mixer and the detector, before the desired audio signal is extracted. The final transducer, usually a loudspeaker, reconverts the electrical signal into the audible acoustic wave.

Based on the commercial broadcast system, three basic circuit functions are needed to perform frequency-domain signal processing. These are frequency generation, frequency selection, and frequency translation. The transducers are energy-conversion devices and their functions are accomplished by use of the properties of appropriate materials rather than by circuit techniques.

10-2 SINUSOIDAL OSCILLATORS

As discussed in Sec. 9-9, unstable feedback amplifiers oscillate at the natural frequencies (poles) of the system. When a pair of poles of the closed-loop gain lies on the j axis, the natural response of the amplifier is a sinusoid. Feedback circuits which are designed to have poles on the j axis are called *sinusoidal oscillators*. Among their desirable properties are the ease and flexibility with which they can be tuned over a large range of frequencies and their compact size. Because of these advantages, sinusoidal oscillators are used extensively in test instruments and communications receivers to process signals and in dielectric and induction heating units to provide the required energy.

There are many types of sinusoidal oscillators. The specific properties of each are dependent on the circuit configuration used. For convenience, oscillators are classified by the manner in which they are tuned. Two common classes are resistance-capacitance (RC) oscillators and inductance-capacitance (LC) oscillators. Each is, however, essentially a feedback amplifier whose open-loop gain is selected so that the closed-loop transfer function has a pair of conjugate closed-loop poles on the j axis. In terms of the Bode-diagram analysis of Sec. 9-9, the condition necessary for achieving sinusoidal oscillations is that gain crossover and phase crossover occur simultaneously. At gain crossover, the magnitude of $T(j\omega)$ is unity; and at phase crossover, the angle of $T(j\omega)$ is $-180°$. Thus, if both gain and phase crossovers occur at the same frequency, it is necessary that

$$T(j\omega) = GH(j\omega) = -1 + j0 \tag{10-1}$$

The frequency of oscillation is the value of ω which satisfies Eq. (10-1). The value of open-loop gain necessary to obtain oscillation is also determined from Eq. (10-1). The relation expressed in Eq. (10-1) is known as the *Barkhausen criterion*.

When feedback circuits are used as oscillators, no signal source is required. The ac energy extracted at the output is obtained from the dc energy required to bias the devices used. Practically, a small disturbance is required to start oscillation; this is supplied by extraneous signals such as power-supply ripple and transients which result when the system is turned on.

In the analysis and design of oscillator circuits, it is necessary to relate the frequency of oscillation to the elements in the circuit. Furthermore, the gain of the amplifier must be adjusted to satisfy Eq. (10-1) to ensure that oscillation occurs at the desired frequency.

10-3 *RC*-TUNABLE OSCILLATORS

Oscillators which use *RC* tuning are particularly useful at low frequencies and in integrated circuits, where the use of inductors is impractical. The *phase-shift* and *Wien-bridge oscillators* are commonly used *RC*-tunable-oscillator circuits; typical representations using ideal voltage amplifiers are shown in Figs. 10-4 and 10-7, respectively.

In the phase-shift oscillator of Fig. 10-4, the values of A_v, R, and C are the same for each stage. The *RC* networks are used to provide the $-180°$ phase shift around the feedback loop, and the amplifier stages provide the gain necessary to produce oscillation. The oscillator circuit is a shunt-feedback amplifier (Fig. 9-29a), and the $a(s)$ and $f(s)$ networks are identified in Fig. 10-4.

The forward gain $a(s)$ is determined by finding the ratio V_o/V_1. The voltage V_2 is

$$V_2 = -A_v V_1 \tag{10-2}$$

and

$$V_3 = \frac{1/sC}{R + 1/sC} V_2 = \frac{1}{1 + sRC} V_2 \tag{10-3}$$

Similarly, V_4, V_5, V_6, and V_o can be determined by equations similar to Eqs. (10-2) and (10-3). The resulting value of $a(s)$ can be shown to be

$$a(s) = \frac{-A_v^3}{(1 + sRC)^3} \tag{10-4}$$

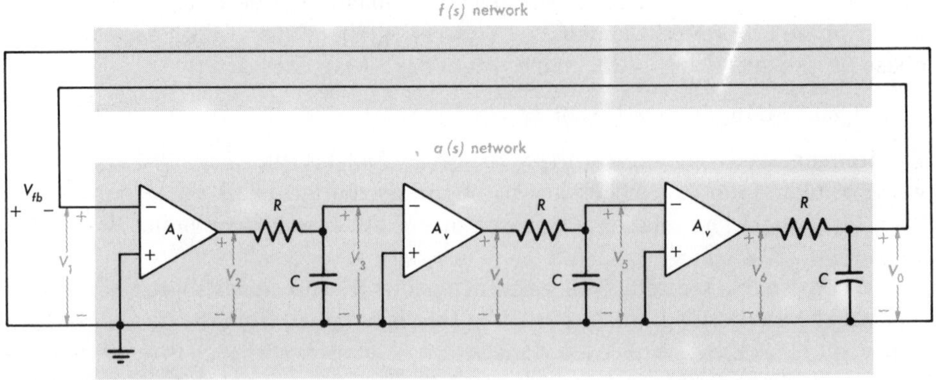

FIGURE 10-4 Phase-shift oscillator.

The value of V_{fb} is minus V_o, so that $f(s) = -1$. The open-loop gain is then

$$T(s) = a(s)f(s) = \frac{A_v^3}{(1 + sRC)^3}$$

or

$$\mathbf{T}(j\omega) = \frac{A_v^3}{(1 + j\omega RC)^3} = \frac{A_v^3}{(1 + \omega^2 R^2 C^2)^{3/2}} \underline{/-3 \tan^{-1} \omega RC} \qquad (10\text{-}5)$$

The frequency of oscillation is found by applying the criterion of Eq. (10-1) to Eq. (10-5). Thus,

$$\angle \mathbf{T}(j\omega_o) = -180° = -3 \tan^{-1} \omega_o RC$$

from which ω_o is

$$\omega_o = \frac{\sqrt{3}}{RC} \qquad (10\text{-}6)$$

The magnitude criterion in Eq. (10-1) gives the value of A_v required. Thus,

$$\frac{A_v^3}{(\sqrt{1 + 3})^3} = 1$$

from which $A_v^3 = 8$ and $A_v = 2$. The results for the phase-shift oscillator of Fig. 10-4 are consistent with those for the feedback amplifier discussed in Sec. 9-9.

Two practical circuits for phase shift oscillators are shown in Figs. 10-5 and 10-6. In the FET circuit shown in Fig. 10-5, the RC-interstage networks provide the required phase shift and the gain is provided by the FETs. The

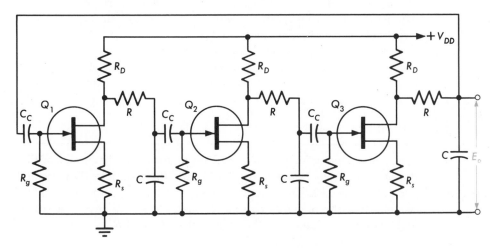

FIGURE 10-5 FET phase-shift oscillator.

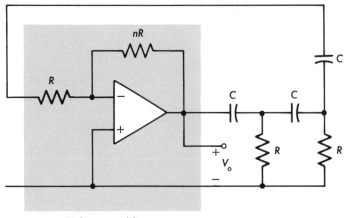

Voltage amplifier

FIGURE 10-6 Phase-shift oscillator using one OP-AMP.

source resistance R_K provides both bias and gain stability. The frequency of oscillation is such that the effects of the coupling capacitors C_C and internal device capacitances C_{gs} and C_{gd} are negligible. The oscillators are tuned (i.e., their frequencies are adjusted to the desired value) by adjusting the values of either R or C. In most practical circuits, the capacitor C is usually made variable and the value of R is selected so that it is much larger than the drain resistance R_D. Under these conditions, adjusting the phase-shift networks does not affect the gain of the stage.

The circuit in Fig. 10-6 uses only one operational amplifier to achieve the gain necessary for oscillation. The RC-feedback network provides the needed phase shift and adjustment of the values of C are used to vary the frequency. The positions of R and C in the phase-shift network are reversed from those shown in Fig. 10-4. This is done so that the resistor R at the OP-AMP input in Fig. 10-6 can be used for both the phase-shift newtork and to control the gain of the amplifier. As the circuits in Figs. 10-4 and 10-6 differ, so do the conditions necessary to sustain oscillation. In addition, phase-shift oscillators are often constructed using the circuit of Fig. 10-4, in which the positions of the resistor and capacitor are interchanged.

Example 10-1

It is desired to design a 5-kHz phase-shift oscillator using the circuit of Fig. 10-5. Biasing conditions establish $R_s = 1\ k\Omega$ and R_G is sufficiently large and can be neglected. The FET parameters are: $g_m = 5 \times 10^{-3}\ \mho$ and $r_d = 20\ k\Omega$. It is assumed that the effect of C_{gd} and C_{gs} and C_C are all negligible at the operating frequency. Determine the values of R, C, and R_D.

SOLUTION   Based on $\mathbf{T}(j\omega) = -1$, to sustain oscillation, the gain of each stage must be 2. From Eq. 8-80, using the concept of the equivalent FET presented in Sec. 8-10, the gain of one stage is

$$-2 = \frac{-5 \times 10^{-3}R_L}{1 + 5 \times 10^{-3} \times 10^3}$$

or

$$R_L = 2,400 \; \Omega$$

As R_L is the parallel combination of R_D and $r_d(1 + g_mR_K)$,

$$2,400 = \frac{R_D \times 20 \times 10^3(1 + 5 \times 10^{-3} \times 10^3)}{R_D + 20 \times 10^3(1 + 5 \times 10^{-3} + 10^3)}$$

from which $R_D = 2,400 \; \Omega$. To prevent the phase-shift network from loading the stage, the value of R is made much larger than the value of R_D. Usually, the value of R is at least 10 times that of R_D, and a good rule of thumb is to select $R = 50R_D$, if possible. Thus, R is selected as $50 \times 2, 400$, or $120 \; k\Omega$. The value of C is determined from Eq. (10-6) as

$$2\pi \times 5 \times 10^3 = \frac{\sqrt{3}}{1.2 \times 10^5C}$$

or $C = 459 \; pF$

The *Wien-bridge oscillator* of Fig. 10-7 is widely used in laboratory oscillators for audio frequencies. The bridgelike structure formed by R_1C_1 and

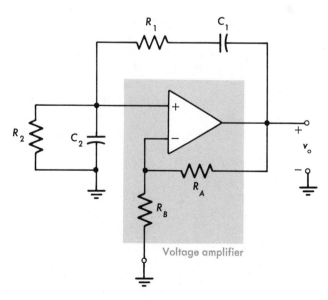

FIGURE 10-7 Wien-bridge oscillator.

R_2C_2 provides both tuning and the phase shift necessary to produce oscillations. Tuning is usually accomplished by simultaneously adjusting all the elements in both phase-shift networks. The major advantages of the Wien-bridge oscillator are a fairly wide tuning range (a few Hz to 100 kHz is typical) and a high degree of frequency stability. The analysis is left to the student as an exercise (see Prob. 10-7).

10-4 *LC*-TUNABLE OSCILLATORS

Two circuits commonly used as radio-frequency oscillators are the *Hartley oscillator* (Fig. 10-8a) and the *Colpitts oscillator* (Fig. 10-8b). Either oscillator can be represented by the equivalent circuit shown in Fig. 10-8c, and each is essentially a series-feedback amplifier of the type shown in Fig. 9-29b. However, both circuits may also employ shunt feedback as shown in the typical practical realizations of Fig. 10-9. The *RF* choke in Fig. 10-9 is an inductance whose reactance is very large at the frequency of oscillation and serves to isolate the bias supply V_{CC} from the high-frequency signal generated.

In the equivalent circuit of Fig. 10-8c, the impedance Z_1 provides the returned signal and is analogous to an unbypassed emitter resistor used in a single stage. The three reactive elements provide the 180° phase shift necessary for oscillation. The current amplification A_i is selected to sustain oscillation in accordance with the criterion of Eq. (10-1). Tuning is accomplished by varying either L or C.

The analysis of both the Hartley and Colpitts oscillators can be accomplished by determination of the conditions for which $\mathbf{T}(j\omega) = -1$. However, in this case, the analysis is facilitated by writing the mesh equations for the circuit of Fig. 10-8c and allowing the circuit determinant Δ to become zero. The circuit determinant is sometimes used to signify the denominator determinant formed in the use of Cramer's rule to solve simultaneous linear equations (see Chap. 2). Letting $\Delta(j\omega) = 0$ corresponds to the condition $\mathbf{T}(j\omega) + 1 = 0$ and results in the conditions necessary for the circuit to have poles on the j axis. The circuit in Fig. 10-10 is that of Fig. 10-8c, for which the current source $A_i\mathbf{I}_1$ and the impedance $\mathbf{Z}_2$ have been converted to their voltage-source equivalent. The mesh equations are

$$\mathbf{I}_1(R_{in} + \mathbf{Z}_1) - \mathbf{I}_2\mathbf{Z}_1 = 0 \tag{10-7}$$

$$-\mathbf{I}_1(\mathbf{Z}_1 - A_i\mathbf{Z}_2) + \mathbf{I}_2(\mathbf{Z}_1 + \mathbf{Z}_2 + \mathbf{Z}_3) = 0 \tag{10-8}$$

The value of the determinant Δ is

$$\Delta = \mathbf{Z}_1[R_{in} + \mathbf{Z}_2(1 + A_i) + \mathbf{Z}_3] + R_{in}(\mathbf{Z}_2 + \mathbf{Z}_3) \tag{10-9}$$

For the Hartley oscillator $\mathbf{Z}_1 = j\omega L_1$, $\mathbf{Z}_2 = j\omega L_2$, and $\mathbf{Z}_3 = -j/\omega C$. Substituting these values into Eq. (10-9) and letting both the real and imaginary

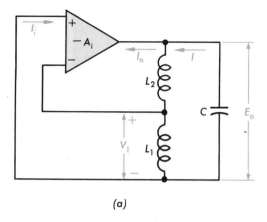

(a)

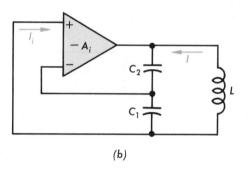

(b)

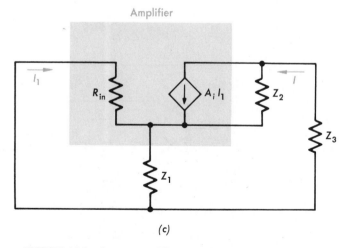

(c)

FIGURE 10-8 Current-amplifier representation of *LC*-tunable oscillators; (*a*) Hartley oscillator; (*b*) Colpitts oscillator; (*c*) equivalent circuit.

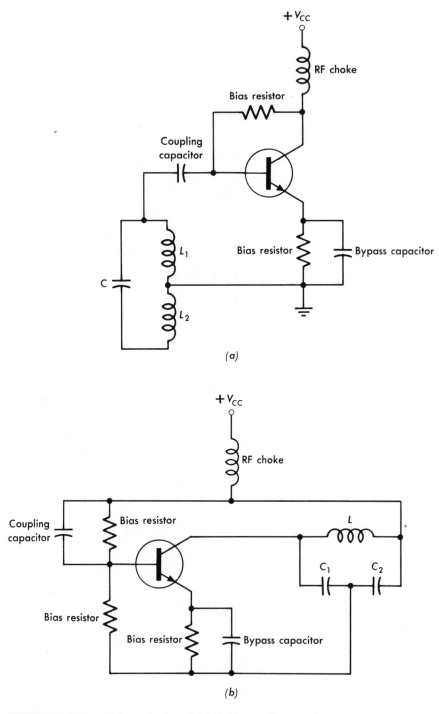

FIGURE 10-9 Transistor realization of (a) Hartley oscillator and (b) Colpitts oscillator.

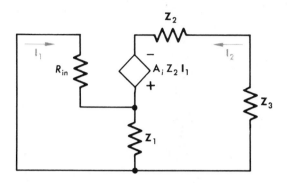

FIGURE 10-10 Circuit of Fig. 10-8c with voltage-source equivalent.

parts of $\Delta(j\omega)$ be zero result in the conditions necessary to sustain oscillation given in Eq. (10-10).

$$\omega_o = \frac{1}{\sqrt{(L_1 + L_2)C}} \qquad A_i = \frac{L_1}{L_2} \tag{10-10}$$

The analysis for the Colpitts oscillator is identical and results in

$$\omega_o = \frac{1}{\sqrt{C_1 C_2 L/(C_1 + C_2)}} \qquad A_i = \frac{C_2}{C_1} \tag{10-11}$$

It is often desirable to generate sinusoidal waveforms whose frequencies are constant over extended time periods. (Stability requirements in the order of a few parts per million are common.) *Crystal oscillators*, of which the circuit of Fig. 10-11 is typical, are used extensively as single-frequency oscillators meeting this requirement. Crystals which exhibit the *piezoelectric effect*, a phenomenon in which an impressed electrical signal produces a mechanical vibration in the material, are used to control the frequency of oscillation. The crystal has its own natural resonant circuit of the form shown in Fig. 10-12. If the crystal in Fig. 10-11 is replaced by the equiv-

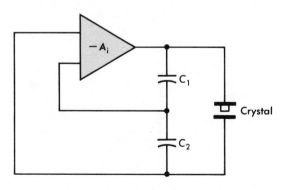

FIGURE 10-11 Crystal oscillator.

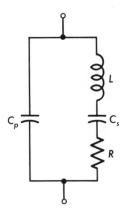

FIGURE 10-12 Equivalent circuit of crystal.

alent circuit of Fig. 10-12, the crystal oscillator is seen to be similar to the Colpitts oscillator of Fig. 10-8b.

In many control and instrumentation applications, oscillators are needed in which the frequency of oscillation can be controlled automatically. One such application is the *automatic frequency control* (AFC) required in the tuning section of an FM receiver. The *voltage-controlled oscillator* (VCO) is generally used for this purpose. The principle of operation is schematically shown in Fig. 10-13, in which the MOSFET circuit is used to replace the capacitor in the phase-shift network of the Hartley oscillator of Fig. 10-8a. In Fig. 10-13, the MOSFET $Q2$ serves as the active load-resistance R_L of the stage containing $Q1$. As described in Sec. 9-10, the capacitance seen at terminals A and B is the value of the feedback capacitance C enhanced by the Miller effect. Thus, $C_{AB} = C(1 + g_m R_L)$. By biasing $Q2$ in the voltage-controlled-resistance region (see Fig. 7-34), the bias control is used to vary the effective value of R_L. Consequently, the bias control varies

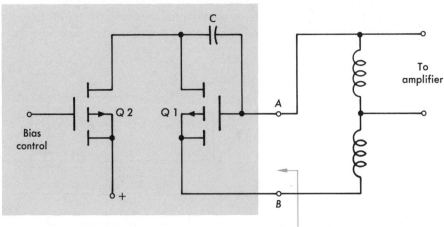

Equivalent capacitor

FIGURE 10-13 Schematic representation of voltage-controlled oscillator (VCO).

the value of C_{AB} in the phase-shift network, thereby producing a change in the frequency of oscillation.

This technique can be used with other oscillator configurations with similar results. In addition, the method for obtaining a voltage-variable capacitor is often used in *voltage-to-frequency converters*. The output of voltage-to-frequency converters is a form of frequency modulation (FM) and is widely used in electronic music synthesizers. Often, a signal to be processed varies slowly with time as is the case where the signal represents variations in mass flow or temperature in a manufacturing process. Such signals contain very low frequency components and voltage-to-frequency conversion translates the information to a higher range of frequencies. These signals are processed more conveniently at high frequencies. Consequently, control of the manufacturing operation is more readily achieved.

10-5 FREQUENCY-SELECTIVE AMPLIFIERS

The process of frequency selection is used to separate a band of frequencies (the desired channel or station) from the entire band of frequencies received. Because signal levels are small, and must be amplified, most circuits used to provide frequency selection also provide amplification. The various types of frequency characteristics, originally displayed in Fig. 6-5 and repeated as

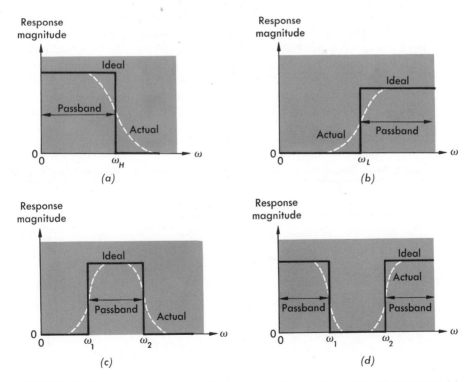

FIGURE 10-14 Ideal frequency-selective characteristics; (a) low-pass; (b) bandpass; (c) high-pass; (d) band-elimination.

Fig. 10-14, are conveniently classified in terms of their use. The ideal *low-pass amplifier,* whose characteristic is depicted in Fig. 10-14a, is used to provide a constant gain A from zero frequency (dc) to some angular frequency ω_H, and zero gain for frequencies greater than ω_H. The broadband amplifier stages described in Chap. 9 are used to approximate this response. The characteristic of Fig. 10-14b is that of a *bandpass amplifier,* whose use is to provide gain between two angular frequencies ω_1 and ω_2. The *high-pass amplifier* characteristic of Fig. 10-14c is used to provide gain above some angular frequency ω_L. To reject a given band of frequencies between ω_1 and ω_2 while amplifying all others, one uses a *band-elimination,* or *band-reject, amplifier* whose characteristic is displayed in Fig. 10-14d.

In most applications, frequency-selective amplifiers are constructed by cascading a number of individual stages. As described in Chap. 9, the performance of cascaded stages is conveniently described by first considering the characteristics of the individual stages.

The transfer function $G(s)$ of a generalized frequency-selective amplifier-stage is

$$G(s) = \frac{a_2 s^2 + a_1 s + a_o}{b_2 s^2 + b_1 s + b_o} \tag{10-12}$$

The expression for $G(s)$ in Eq. (10-12) is called a *biquadratic function,* or simply a *biquad,* because both numerator and denominator are quadratics in the frequency variable s. All four characteristics in Fig. 10-14 can be expressed in the form of Eq. (10-12) by appropriate adjustment of the coefficient values. For the bandpass case in Fig. 10-14c, the function is zero at both $s = j\omega$ equal to zero and infinity. In order that $G(0)$ in Eq. (10-12) be zero, the coefficient a_o must be zero. Similarly, if $G(j\omega)$ is to tend towards zero as ω approaches infinity, it is necessary that a_2 also be zero. Similar analysis for the three remaining cases leads to the results given in Table 10-1.

TABLE 10-1 TRANSFER FUNCTIONS FOR FREQUENCY-SELECTIVE AMPLIFIER STAGES

TYPE OF CHARACTERISTIC	FORM OF TRANSFER FUNCTION	
Low-pass	$\dfrac{K}{s^2 + \dfrac{\omega_o}{Q} s + \omega_o^2}$	$\dfrac{K(s + z)}{s^2 + \dfrac{\omega_o}{Q} s + \omega_o^2}$
High-pass	$\dfrac{Ks^2}{s^2 + \dfrac{\omega_o}{Q} s + \omega_o^2}$	$\dfrac{Ks(s + z)}{s^2 + \dfrac{\omega_o}{Q} s + \omega_o^2}$
Bandpass	$\dfrac{Ks}{s^2 + \dfrac{\omega_o}{Q} s + \omega_o^2}$	
Band-elimination	$\dfrac{K(s^2 + \omega_r^2)}{s^2 + \dfrac{\omega_o}{Q} + \omega_o^2}$	

The transfer functions listed in Table 10-1 are in a standard form suitable for determining the response in the frequency range of interest.

In the two subsequent sections of this chapter, circuits capable of realizing transfer functions of the type expressed in Table 10-1 are described.

10-6 TUNED AMPLIFIERS

One type of bandpass amplifier, called a *narrowband* amplifier, is used to select a small portion of the frequency spectrum; in such cases, ω_1 is very nearly the value of ω_2. A simple and common method of achieving a narrowband characteristic is the use of the resonant circuits described in Sec. 6-4. This is illustrated in the OP-AMP stage in Fig. 10-15 and the tuned-drain circuit of Fig. 10-16. In the operational-amplifier circuit, the parallel resonant circuit is between the noninverting input and ground so that the OP-AMP stage amplifies the output signal v_2 of the resonant circuit. In the circuit of Fig. 10-16, the parallel resonant circuit is the load driven by the common-source stage. The operation of this amplifier is based on the assumption that the bias resistor R_G and coupling and bypass capacitors C_G, C_D, and C_K have negligible effect on the response at frequencies in the vicinity of resonance.

The equivalent circuit in Fig. 10-17 is the approximate representation of the circuit in Fig. 10-16 valid at frequencies near resonance. For convenience G_{eq} is identified as the parallel combination of G_L', G_D, and $1/r_d$; C_{eq} is the parallel combination of C and the FET output capacitance. With these identifications, the output portion of the model in Fig. 10-17 is seen to be the parallel of resonant circuit of Fig. 6-12. As indicated in Eq. (6-16), the transfer function of the parallel resonant circuit, when s is substituted for $j\omega$, is that given in Table 10-1 for the bandpass case.

The resonant angular frequency ω_o of the circuit is

$$\omega_o = \frac{1}{\sqrt{LC_{eq}}} \tag{10-13}$$

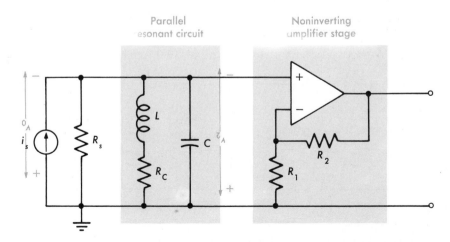

FIGURE 10-15 Tuned operational-amplifier stage.

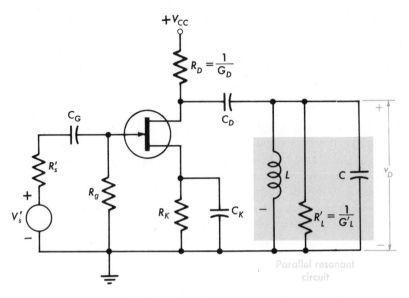

FIGURE 10-16 Tuned-drain JFET stage.

At resonance, the output voltage V_o is

$$V_o = - \frac{g_m V_g}{G_{eq}} \qquad (10\text{-}14)$$

since the total susceptance of the parallel combination of L and C is zero. The voltage V_g is equal to V_s, so that the gain at resonance G_o becomes

$$G_o = \frac{V_o}{V_s} = \frac{-g_m}{G_{eq}} \qquad (10\text{-}15)$$

The value of $1/Q_p = G_{eq}/\omega_o C_{eq}$ [from Eq. (6-26)] is a measure of the bandwidth (BW) of the amplifier. The bandwidth is the frequency difference between the half-power frequencies (ω_1 and ω_2 in the ideal characteristic of

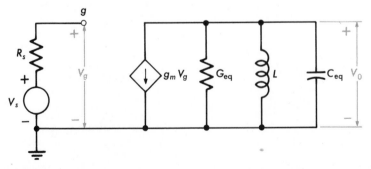

FIGURE 10-17 Approximate equivalent circuit of amplifier stage in Fig. 10-16, valid at frequencies near resonance.

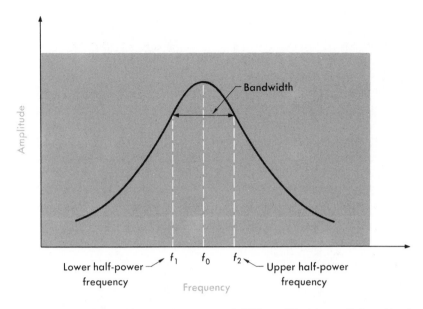

FIGURE 10-18 Frequency-response of JFET-amplifier stage with tuned load.

Fig. 10-14c). It is expressed in Eq. (6-29) and is given again here for convenience:

$$\text{BW} = \frac{\omega_o}{Q_p} \qquad (10\text{-}16)$$

As seen in Eq. (10-16), the higher the value of Q_p, the smaller the bandwidth. The frequency response of the amplifier is that given by the universal resonance curve (described in Sec. 6-4) and is shown in Fig. 10-18.

Example 10-2

Certain amplifiers in standard broadcast receivers are tuned to 456 kHz and are required to have a bandwidth of 20 kHz. They are called *intermediate-frequency (IF) amplifiers* (see Fig. 10-3). An FET whose parameters are $C_x = 50$ pF, $g_m = 5 \times 10^{-3}$ ℧, $C_o = 10$ pF, and $r_d = 50$ kΩ is used. The amplifier is to have a gain of 100 at resonance, and the load resistance is 40 kΩ.

Determine the element values of the parallel resonant circuit that must be used.

SOLUTION The circuit model is shown in Fig. 10-17. The parameters of the tuned circuit are the inductance L, the coil conductance G_L', and the capacitance C. From Eq. (10-15),

$$100 = \frac{5 \times 10^{-3}}{G_{\text{eq}}} \qquad \text{and} \qquad G_{\text{eq}} = 5 \times 10^{-5} \text{ ℧}$$

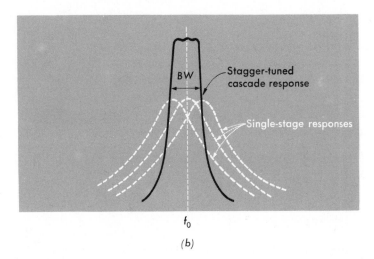

FIGURE 10-19 (a) Synchronous-tuned and (b) stagger-tuned responses.

As G_{eq} is the parallel combination of G_D, G'_L, and $1/r_d$,

$$G'_L = G_{eq} - G_D - \frac{1}{r_d} = 5 \times 10^{-5} - \frac{1}{4 \times 10^4} - \frac{1}{5 \times 10^4} = 5 \times 10^{-6} \,\mho$$

The bandwidth is given by Eq. (10-16), so that

$$2\pi(20 \times 10^3) = \frac{2\pi(456 \times 10^3)}{Q_p} \quad \text{and} \quad Q_p = 22.8$$

The value of C_{eq} can be determined from Eq. (6-26) and is given by

$$22.8 = \frac{2\pi(456 \times 10^3)C_{eq}}{5 \times 10^{-6}} \quad \text{from which} \quad C_{eq} = 79.5 \text{ pF}$$

This capacitance is the parallel combination of C and C_o, so that

$$C = C_{eq} - C_o = 79.5 - 10 = 69.5 \text{ pF}$$

From Eq. (10-13),

$$2\pi(456 \times 10^3) = \frac{1}{\sqrt{L \times 69.5 \times 10^{-12}}} \quad \text{and} \quad L = 1.75 \text{ mH}$$

The element values of the tuned circuit that must be specified are

$$L = 1.75 \text{ mH} \qquad R'_L = 200 \text{ k}\Omega \qquad C = 79.5 \text{ pF}$$

Tuned-amplifier stages can be cascaded in the manner described in Chap. 9. When each stage in the cascade is tuned to the same frequency, the cascade is said to be *synchronously tuned*. *Stagger tuning* exists when each stage in the cascade is tuned to a slightly different frequency. The frequency responses of each are displayed in Fig. 10-19. For synchronous tuning, the overall gain increases and the bandwidth decreases. As indicated in Fig. 10-19*a*, the response of the cascade has a steeper slope and narrower peak than does the single-stage circuit. The stagger-tuned response of Fig. 10-19*b* indicates increased gain, a broader peak, and steeper slope than does the single-stage circuit. The use of stagger tuning permits closer approximation of the ideal bandpass characteristic of Fig. 10-14*c* than does synchronous tuning. The disadvantages of stagger tuning when compared to synchronous tuning are that tuning must be more precise, and, since each tuned circuit is different, the amplifier may be more costly.

10-7 ACTIVE-*RC* FILTERS

Limitations in technology require that integrated circuits be fabricated without the use of inductors. In many applications where discrete components are employed, the physical size and component quality render inductors impractical in circuit design. To realize the frequency-selective characteristics in Fig. 10-14 the combined use of resistance-capacitance (*RC*) networks and amplifiers in a feedback configuration are utilized. Circuits designed to realize specific frequency-response characteristics are called *filters;* those which employ only resistors, capacitors, and amplifiers are referred to as *RC-active filters* or simply active filters. Each type of active-filter structure is used to realize one of the biquadratic functions given in Table 10-1. Higher-order filters result when such stages are cascaded.

The structure of one commonly utilized class of *RC*-active filters is shown in Fig. 10-20. The basic amplifier in Fig. 10-20 is assumed to provide constant gain over the entire frequency spectrum; in practice, all that is required is that the bandwidth of the amplifier be large compared to the

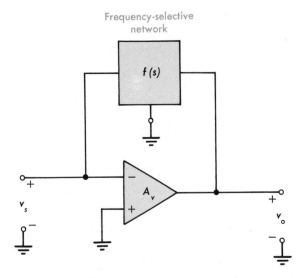

FIGURE 10-20 Typical structure of active-*RC* filter.

operating frequencies of the filter. The feedback network $f(s)$, composed only of resistors and capacitors, is used to control the frequency response of the circuit.

The circuit in Fig. 10-21 is a narrowband amplifier and is used to replace the tuned amplifiers of Figs. 10-15 and 10-16. The amplifier A_v is considered to be an ideal voltage amplifier (usually realized by operational-amplifier

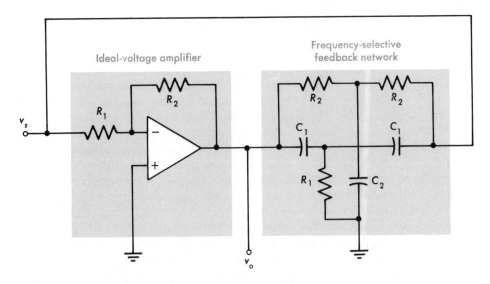

FIGURE 10-21 Narrowband active-*RC* stage.

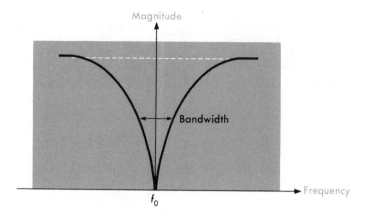

FIGURE 10-22 Notch characteristic of twin-tee network.

stage), and the feedback circuit $f(s)$ is called a *twin-tee* or *parallel-tee net-work*.

The twin-tee network is designed to act as a band-elimination circuit whose frequency response is depicted in Fig. 10-22. At the angular frequency ω_o, there is no transmission through the $f(s)$ network and consequently no feedback. Thus, the output voltage is $A_v v_s$. At frequencies different from ω_o there is transmission through $f(s)$, and negative feedback exists which reduces the output voltage. Because more feedback is present the further the frequency is from ω_o, the overall response of the amplifier becomes that shown in Fig. 10-23.

All bandpass circuits employed in analog systems need not be of the narrowband variety. Some instrument landing systems, in which the Doppler effect is used to determine aircraft velocity, often require bandpass filters whose passband is 600 Hz to 6 kHz. Filters of this type exhibit a *broad-*

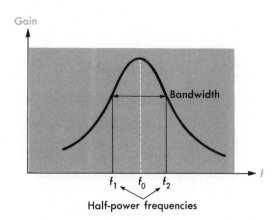

FIGURE 10-23 Frequency response of twin-tee active filter.

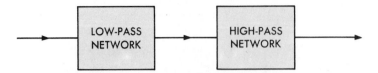

FIGURE 10-24 High-pass and low-pass cascade structure.

band bandpass response as the band of frequencies selected is such that ω_2 is substantially larger than ω_1 (see Fig. 10-14c).

Broadband bandpass amplifiers are often constructed, as shown in Fig. 10-24, by cascading a low-pass circuit with a high-pass circuit. The response of each circuit is shown in Fig. 10-25a. The angular frequency ω_H in the low-pass circuit is designed to be greater than ω_L for the high-pass section. The composite frequency response is that given in Fig. 10-25b and indicates that the high-pass section is used to eliminate angular frequencies below ω_L while the low-pass circuit rejects frequencies above ω_H.

The combination of high-pass and low-pass networks can also be used to obtain broadband, band-reject characteristics as shown in Fig. 10-26. For the low-pass and high-pass responses shown in Fig. 10-25a, the frequency response of the system in Fig. 10-26 is displayed in Fig. 10-27. In the configu-

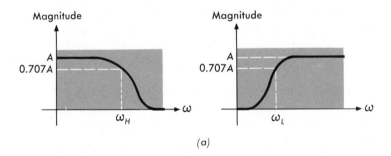

(a)

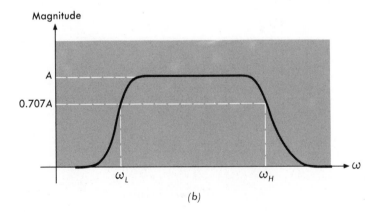

(b)

FIGURE 10-25 (a) Low-pass and high-pass responses and (b) composite characteristic.

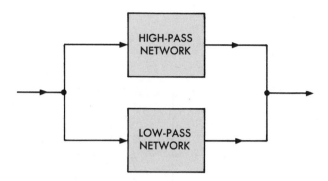

FIGURE 10-26 Parallel-channel high-pass and low-pass networks to realize band-reject characteristics.

ration in Fig. 10-26, if ω_L is less than ω_H, angular frequencies between ω_L and ω_H are eliminated as both the high-pass and low-pass networks act to reject these frequencies. For frequencies below ω_H, the low-pass network provides the output; frequencies above ω_L are transmitted by the high-pass network.

Figure 10-28*a* depicts one type of realization of a low-pass network, and the high-pass realization is that given in Fig. 10-28*b*. Comparison of the two circuits indicates that they have the same structure but that the position of the resistors and capacitors is interchanged. In the low-pass circuit, as the frequency increases, the reactances of the capacitors decrease, causing V_1 to decrease. As $V_o = A_v V_1$, it, too, decreases and results in high-frequency rejection. For the high-pass amplifier, the series capacitors C_1 and C_2 behave in a manner similar to the coupling capacitors in an amplifier stage. At low frequencies the reactances of C_1 and C_2 become appreciable and reduce the value of V_1. Consequently, $V_o = A_c V_1$ is also reduced, and low-frequency signals are rejected.

One version of a popular class of active filters which realize biquadratic functions is shown in Fig. 10-29 and is referred to as a *universal* or *state-*

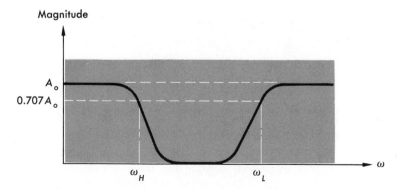

FIGURE 10-27 Frequency response of circuit in Fig. 10-26.

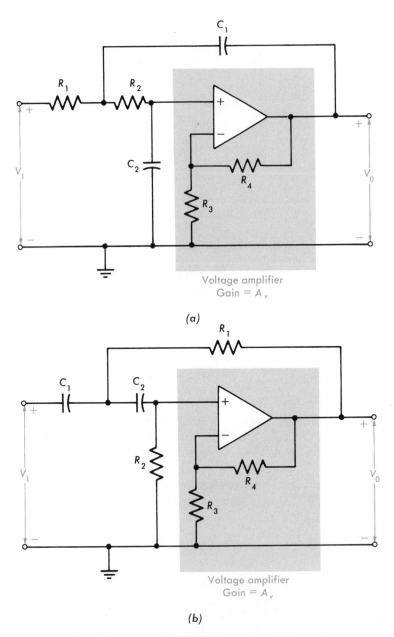

FIGURE 10-28 (a) Low-pass and (b) high-pass biquadratic sections.

variable filter. These filters are commercially available in a single package and, at a minimum, contain the three operational amplifiers and the two capacitors. All of some of the resistors are added externally to control the response characteristics. In the circuit of Fig. 10-29 the voltage v_1 serves the input; the terminals indicating v_2, v_3, and v_4 are accessible output terminals.

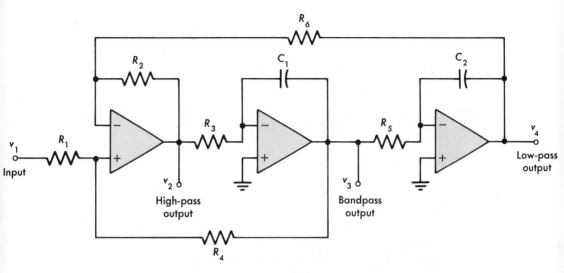

FIGURE 10-29 Universal or state-variable filter.

The circuit is called a *universal filter* because all possible frequency-selective characteristics are available. In essence, the availability of low-pass, bandpass and high-pass outputs is based on the control of the frequency response of the feedback networks. When v_4 is the output, the overall feedback is resistive and therefore frequency insensitive. Circuit behavior is similar to the feedback amplifiers described in Chap. 9 which have low-pass responses. Selection of either v_2 or v_3 as the output voltage makes the feedback-loop frequency-dependent and performance is similar to the notch-network filter shown in Fig. 10-21. As a result, bandpass and high-pass responses are available at v_3 and v_2, respectively. Band-elimination filters are constructed of the form in Fig. 10-26 by summing the low-pass and high-pass output signals.

There are many other circuit configurations used as active-*RC* filters. However, the principle of operation in each is the same; the basic amplifier is used to provide the gain while the *RC* network and feedback are used to control the frequency response.

10-8 MODULATION AND DETECTION

Many of the techniques of electrical control, computation, measurement, and communication involve processing signals which are the output of devices such as microphones, television cameras, punched-card readers, and tachometers. These signals are almost always time-varying quantities whose amplitude and frequency contain the desired information. It often happens that the signal is inherently not in the best form or in the best frequency range to be processed. One example is the commercial AM broadcast

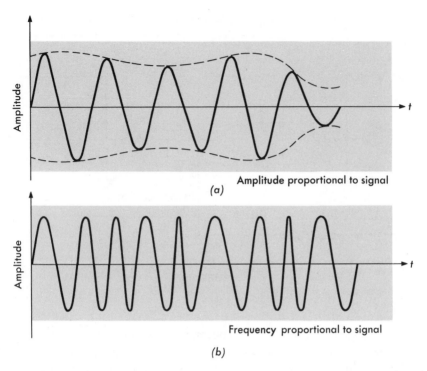

Amplitude proportional to signal

(a)

Frequency proportional to signal

(b)

FIGURE 10-30 Modulated waveforms: (a) amplitude modulation (AM); (b) frequency modulation (FM).

system described in Sec. 10-1. A second example is one type of position control where a slowly varying unidirectional signal is used to cause rotation in an ac motor. Because the signal size is small, amplification, best handled by an ac amplifier, is required. Under these circumstances, the signal can be caused to modify, or modulate, the amplitude, frequency, or other characteristic of a higher-frequency signal. The higher-frequency signal can then be processed in the desired manner, and the desired information can be extracted by demodulation. Both modulation and demodulation, therefore, can be considered as *frequency-translation* mechanisms.

As technically defined by the Institute of Electrical and Electronics Engineers, *modulation* is "the process whereby the amplitude (or other characteristic) of a wave is varied as a function of the instantaneous value of another wave. The first wave, which is usually a single-frequency wave, is called the *carrier wave;* the second is called the *modulating wave*." *Demodulation* or *detection* is "the process whereby a wave resulting from modulation is so operated upon that a wave is obtained having substantially the characteristics of the original modulating wave." Modulation and demodulation are thus seen to be reverse processes.

Examples of modulated waveforms are shown in Fig. 10-30. *Amplitude*

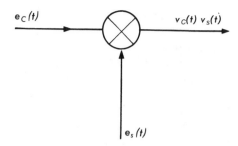

FIGURE 10-31 Ideal multiplier symbol.

modulation (*AM*), is illustrated by the waveform in Fig. 10-30a, and the waveform in Fig. 10-30b depicts *frequency modulation* (*FM*). In these cases, the amplitude or frequency of a sinusoidal wave (the carrier) is varied in accordance with the magnitude of the modulating signal. In order that the modulated carrier contain substantially all the information present in the modulating signal, the frequency of the carrier should be much higher than the frequencies present in the modulating signal. Ideally, a wave whose instantaneous values are proportional to those of the modulating signal is obtained after demodulation.

The basic operation needed to build modulators is the multiplication of two signals. The ideal multiplier is symbolically represented in Fig. 10-31. The voltage $v_c(t) = \sqrt{2}\,V_c \cos \omega_c t$ is the carrier and is generated by means of an oscillator. The modulating signal $v_s(t)$, usually the amplified output of a transducer, has the form $V_s(1 + m \cos \omega_s t)$. This voltage consists of both the direct level voltage V_s and the sinusoidal signal $m V_s \cos \omega_s t$ one expects at the output of an amplifier. The multiplier output $v_o(t)$ is

$$V_o(t) = v_c(t)v_s(t) = \sqrt{2}\,V_c \cos \omega_c t\, V_s(1 + m \cos \omega_s t) \tag{10-17}$$

Expansion of Eq. (10-17) and use of the trigonometric identity

$$\cos x \cos y = \tfrac{1}{2}[\cos (x + y) + \cos (x - y)] \tag{10-18}$$

allows $v_o(t)$ to be written as

$$v_o(t) = \sqrt{2}\,V_c V_s \cos \omega_c t \;+ \frac{\sqrt{2}\,m V_c V_s}{2}$$

$$[\cos (\omega_c + \omega_s)t + \cos (\omega_c - \omega_s)t] \tag{10-19}$$

The form of Eq. (10-19) is that for a standard AM signal shown in Fig. 10-30a and contains three frequency components. These are the carrier ω_c and the sum and difference frequencies $\omega_c + \omega_s$ and $\omega_c - \omega_s$, usually called the upper and lower *sidebands*. For signal frequencies very much smaller than the carrier frequency, $\omega_c + \omega_s$ and $\omega_c - \omega_s$ are both nearly ω_c. The information content of the signal is now contained in a narrow band of frequencies centered about the carrier frequency. Such signals are amplified and distinguished from each other by means of the narrowband amplifiers described in the two previous sections. In addition, this type of modulator is an integral component in frequency modulators.

Frequency translation from high- to lower-frequency signals is also accomplished by using circuits which multiply two signals. Such circuits are often called *mixers* or *down converters*. In AM receivers (see Fig. 10-3) mixers are used to obtain signals at the intermediate frequency (IF), which are then amplified as described in Example 10-2.

10-9 CIRCUITS FOR ANALOG COMPUTATION

The analog computer is an electronic system whose primary function is to obtain the solutions of a system of differential equations. The technique employed consists of interconnecting a number of electronic circuits, each of which performs a specific mathematical operation, and whose overall behavior is governed by the differential equations to be solved. The solutions are then the corresponding voltage and current responses in the electronic system. For example, if knowledge of the velocity of a mass in a mechanical system is required, the force-current analog in Table 3-2 can be used to convert the mechanical system into an electric circuit. The velocity becomes the voltage across the capacitor which represents the mass; measurement of the voltage response is the desired solution. Generally, the operations needed to solve differential equations are integration, addition, and multiplication.

Often, analog computation circuits are incorporated within measurement and control systems. In addition to the mathematical operations useful

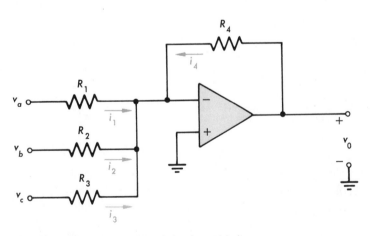

FIGURE 10-32 Summing-circuit (analog adder).

to solve differential equations, circuits for which the output is the logarithm of the input are often used because of the conveniences they afford. (The student need only refer to the Bode diagram to discern several advantages of logarithms.)

The operational amplifier is the basic element used to realize the several mathematical operations needed for analog computation. As described in Sec. 8-1, both the inverting and noninverting OP-AMP stages perform the operation of multiplication by a constant. The basic OP-AMP stage can also be converted to do addition as shown in the *adder* or *summer* shown in Fig. 10-32. Each of the input voltages v_a, v_b, and v_c in Fig. 10-32 is measured with respect to ground. Recall that inverting input is at virtual ground and that no current enters the OP-AMP so KCL at the inverting input gives

$$\frac{v_a}{R_1} + \frac{v_b}{R_2} + \frac{v_c}{R_3} + \frac{v_o}{R_4} = 0$$

Solving for v_o yields

$$v_o = -R_4 \left(\frac{v_a}{R_1} + \frac{v_b}{R_2} + \frac{v_c}{R_3} \right) \qquad (10\text{-}20)$$

In Eq. (10-20), if the values of all resistors are equal, v_o is simply the negative of the sum of the individual voltages. Differing resistance values permit a weighted sum to be obtained.

Subtraction is performed using the inherent differential nature of the operational amplifier. The circuit shown in Fig. 10-33 provides an output which is proportional to the difference of the two inputs. The analysis is left for the student as an exercise (see Prob. 10-48).

When the resistor R_2 is replaced by a capacitor C, the inverting

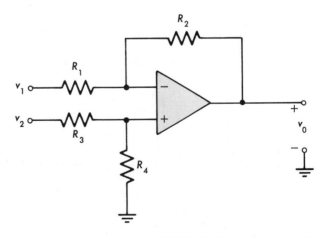

FIGURE 10-33   Subtractor circuit.

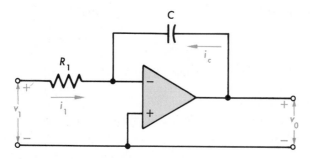

FIGURE 10-34 Integrator.

OP-AMP stage becomes an *integrator*, shown in Fig. 10-34. Use of the virtual-ground concept results in the KCL expression

$$\frac{v_1}{R_1} + \frac{C\,dv_o}{dt} = 0$$

from which

$$\frac{dv_0}{dt} = \frac{v_1}{R_1 C}$$

and

$$v_o = \frac{1}{R_1 C} \int v_1 dt \tag{10-21}$$

The circuit depicted in Fig. 10-35 is the basic configuration for a *logarithmic amplifier*. As seen in Fig. 10-35, the diode replaces the resistor R_2 in the inverting OP-AMP stage. The operation of the circuit is based on the exponential volt-ampere characteristic of the diode which, from Eq. (7-1) and $e^{v/\eta V_T} \gg 1$, is

$$i_d = I_s e^{v/\eta V_T}$$

Again, by use of the virtual-ground concept i equals i_d and the relationship

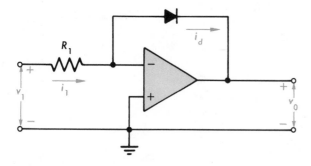

FIGURE 10-35 Logarithmic amplifier.

between the voltages is expressed as

$$\frac{v_1}{R_1} = I_s e^{v_0/\eta V_T} \tag{10-22}$$

By taking the logarithm of both sides and solving for v_0 gives

$$v_0 = -\eta V_T(\ln v_1/R_1 - \ln I_s) \tag{10-23}$$

The result in Eq. (10-23) shows that for instances where the diode saturation current I_s is negligible, the output voltage v_0 is proportional to the logarithm of the input.

The multiplication of two-time functions is accomplished by modifying the current-source bias in a differential amplifier. In the circuit of Fig. 10-36, the current source I_{EE} is the usual ac bias supply in the differential amplifier (see Sec. 8-12). The two signals to be multiplied are v_1 and i_s and, as indicated in Fig. 10-36, the current source modulates the bias current of the amplifier. In Sec. 8-7, it is shown that the value of g_m in the hybrid $-\pi$ model of the transistor is proportional to the emitter current. Similarly, the collector current i_c proportional to gm v_{DM}, where v_{DM} is the differential-mode signal, equals $v_1/2$. Thus, using $g_m = i_E/\eta V_T$,

$$i_c = g_m v_1 = \frac{i_E v_1}{2\eta V_T} \tag{10-24}$$

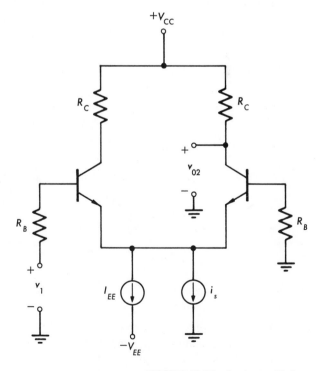

FIGURE 10-36 Analog multiplier.

Since $i_E = (I_{EE} + i_s)/2$ for any one stage, Eq. (10-24) becomes

$$i_c = \frac{I_{EE} v_1}{4\eta V_T} + \frac{i_s v_1}{4\eta V_T} \tag{10-25}$$

The first term in Eq. (10-25) is the usual expression for the collector current under constant bias current conditions. The second term indicates the multiplication of the two signals. *Analog multipliers* which utilize circuits of the type shown in Fig. 10-36 are commercially available as a single-chip unit. Analog multipliers can also be used as the basis for modulator circuits described in Sec. 10-8.

PROBLEMS

10-1 The oscillator in Fig. 10-4 is designed to provide a variable-frequency sinusoidal signal. The resistance R can assume one of two values: 50 or 500 kΩ. The capacitance can be varied continuously from 0.01 to 0.1 μF. Determine the range of frequencies of oscillation.

10-2 The voltage amplifiers in Fig. 10-4 are realized using operational amplifiers whose gain-bandwidth product is 2 MHz. The bandwidth of the voltage amplifier should be at least 10 times the oscillation frequency so the amplifier does not load the RC phase-shift network. Under these conditions,
a Determine the maximum oscillation frequency.
b Determine the value of C needed to obtain the maximum frequency when $R = 10$ kΩ.

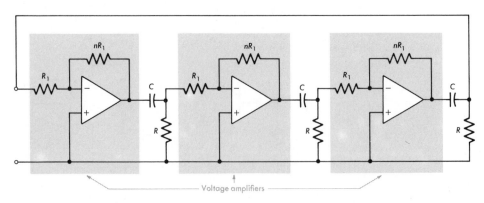

FIGURE 10-37 Circuit for Prob. 10-3.

10-3 The circuit depicted in Fig. 10-37 is an alternative realization of the phase-shift oscillator in Fig. 10-4.
a Determine the frequency of oscillation and the gain of the voltage amplifiers necessary to sustain oscillations.
b Repeat Prob. 10-1 for the circuit in Fig. 10-37.

10-4   An alternate type of phase-shift oscillator is shown in Fig. 10-6.
a   Determine the frequency of oscillation.
b   What value of n is needed to sustain oscillation?
c   If the bandwidth of the voltage amplifier is to be at least 10 times the frequency of oscillation, what is the maximum frequency of oscillation permissible? Express the answer as a fraction of the gain-bandwidth product of the OP-AMP.

10-5   In the circuit of Fig. 10-6, the value of C is 0.1 μF. Find the values of R and n necessary to sustain oscillation at 5 kHz.

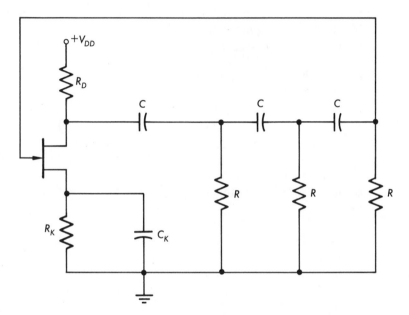

FIGURE 10-38   Circuit for Prob. 10-6.

10-6   The circuit in Fig. 10-38 uses a JFET amplifier stage to realize the voltage amplifier in Fig. 10-6. The bypass capacitor C_K behaves ideally at the frequency of oscillation. The JFET parameters are $g_m = 5$ m℧ and $r_d = 50$ kΩ. The capacitances C_{ds} and C_{gs} have negligible effect at the operating frequency. Determine R_D and C, for $R = 100$ kΩ, necessary to sustain oscillation at 10 kHz.

10-7   Determine the frequency of oscillation and required gain for the Wien-bridge oscillator of Fig. 10-7.

10-8   In the design of a particular variable-frequency Wien-bridge oscillator, the values of R_1 and C_2 are adjustable according to the relations $0.1 \le R_1/R_2 \le 10$ and $0.1 \le C_2/C_1 \le 10$.
a   Find the minimum and maximum values of the oscillation frequency for $R_2 = 10$ kΩ and $C_1 = 0.1$ μF.

b What must be the minimum gain-bandwidth product of the voltage amplifier if the amplifier frequency response is to have a negligible effect on oscillator performance?

10-9 a Design the circuit in Fig. 10-7 to oscillate at 2 kHz. Choose $R_1 = R_2$ and $C_1 = C_2$ with the minimum resistance allowable equal to 1 kΩ.
b If the technology used to fabricate the circuit allows the resistors and capacitors to be matched within 1 percent of nominal values, what is the range of oscillation frequency?

10-10 A Hartley oscillator, shown in Fig. 10-8a, is designed with $L_1 = 2$ mH and $L_2 = 20$ μH and a variable capacitance.
a Determine the range of capacitance values if the frequency of oscillation is varied between 950 and 2,050 kHz.
b If a bipolar transistor is used to realize A_i, what must be its minimum value of h_{fe}?

10-11 A Colpitts oscillator, shown in Fig. 10-8b, is designed with $C_1 = 100$ pF and $C_2 = 7,500$ pF. The inductance is variable.
a Determine the range of inductance values if the frequency of oscillation is to vary between 950 and 2,050 kHz.
b If a bipolar transistor is used to realize A_i, what must be its minimum value of h_{fe}?

10-12 An alternate realization of a Hartley oscillator is depicted in Fig. 10-39 in which $\mathbf{Z}_1 = j\omega L_1$, $\mathbf{Z}_2 = j\omega L_2$ and $\mathbf{Z}_3 = -j/\omega C$. The voltage amplifier has infinite input resistance and an output resistance R_o. Determine the value of A_v needed to sustain oscillation and the oscillation frequency.

10-13 The circuit shown in Fig. 10-39 can be used as a Colpitts oscillator when $\mathbf{Z}_1 = -j/\omega C_1$, $\mathbf{Z}_2 = -j/\omega C_2$ and $\mathbf{Z}_3 = j\omega L$. The voltage amplifier has an output resistance R_o and an infinite input resistance. Determine the frequency of oscillation and the value of A_v needed to sustain oscillation.

10-14 In this problem, the effect of practical elements on oscillator performance is examined. Consider the circuit in Fig. 10-8b in which $C_1 = 500$ pF and $C_2 = 0.05$ μF and the 20-μH inductance contains series resistance $R_L = 4$ Ω. Determine the frequency of oscillation and the value of A_i needed to sustain oscillation.

10-15 The effect of the series resistance associated with practical inductors on the performance of the circuit in Fig. 10-8a is to be investigated. For $C = 100$ pF, $L_1 = 99$ μH in series with 10 Ω, and $L_2 = 1$ μH in series with 1 Ω, find the frequency of oscillation and the value of A_i needed to sustain oscillation.

10-16 The tuned-amplifier stage in Fig. 10-15 is designed having element values: $R_1 = 10$ kΩ, $R_2 = 110$ kΩ, $C = 1,250$ pF, $L = 5$ mH, $R_C = 80$ Ω, and $R_s = 50$ kΩ. Determine:
a The resonant frequency and bandwidth of the stage.
b The rms output voltage when the rms signal current is 10 μA.

10-17 The circuit in Fig. 10-15 is to be resonant at 10 kHz and have a bandwidth of 1 kHz. The maximum output signal required is 10-V rms when the circuit is excited by a sinusoidal current source whose rms amplitude is 5 μA and which has a source resistance of 100 kΩ. The closed-loop gain of the operational amplifier is 10.
a Determine L, C, and R_c.
b At what frequency is the rms output voltage 5 V?

10-18 The JFET stage shown in Fig. 10-16 is designed with the following element values: $L = 10$ mH, $C = 90$ pF, $R_L' = 80$ kΩ, and $R_D = 30$ kΩ. The capacitors C_G and C_K have negligibly small reactances at the frequencies of interest and R_G is sufficiently large so as to have negligible effect. The JFET parameters are: $g_m = 3$ m$\mho$, $C_{ds} = 10$ pF, $C_{gs} = 10$ pF, $C_{gd} = 5$ pF, and $r_d = 60$ kΩ.
a Determine the resonant frequency and bandwidth.
b If the maximum rms output signal is 5 V, what is the largest value of input-signal amplitude permissible?

10-19 The JFET whose parameters are given in Prob. 10-18 is used in the circuit shown in Fig. 10-16. The effects of C_G, C_K, and R_G are negligible at the signal frequencies and $L = 60$ μH and $R_D = 20$ kΩ.
a Determine the value of R_L' and the range of values for C if the resonant frequency is to vary between 500 and 1,600 kHz. The bandwidth is to be 10 kHz and independent of the resonant frequency.
b At resonance, find the gain of the circuit.

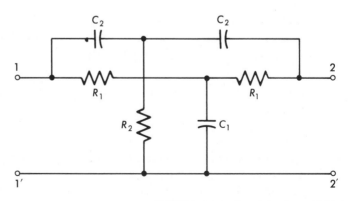

FIGURE 10-39 Circuit for Prob. 10-20.

10-20 The feedback network in the frequency-selective amplifier of Fig. 10-20 is shown in Fig. 10-39.

a For $f(s) = -y_{12}(s)$, show that when $R_1C_1 = 4R_2C_2$, $f(s)$ has zeros at

$$s = \pm j\omega_o = \pm j\frac{\sqrt{8R_2/R_1}}{R_1C_1}$$

b Determine the transfer function (V_o/V_s) when $a(s) = A_o$.
c Show that the function obtained in b is a bandpass function and identify ω_o and Q.

10-21 **a** Show that the transfer function of the circuit in Fig. 10-28a is given by

$$\frac{V_o}{V_1} = \frac{A_v}{1 + s[R_2C_2 + R_1C_2 + R_1C_1(1 - A_v)] + s^2R_1R_2C_1C_2}$$

b Identify the values of ω_o and Q for the result obtained in a.

10-22 One approach to the design of low-pass filter sections of the type shown in Fig. 10-28a is to make $R_1 = R_2 = R$ and $C_1 = C_2 = C$. This choice leads to certain fabrication economies as the number of components of different values required is minimized.

a Using the approach described, determine the values of R and A_v needed to realize $\omega_o = 4\pi \times 10^3$ rad/s and $Q = \sqrt{2}/2$ when $C = 5{,}000$ pF.
b Select the values of R_3 and R_4 so that only one different resistance value is used.
c What is the output voltage when the input voltage is 1.0 V?
d Find the magnitude of the output voltage when $v_1(t) = \sqrt{2} \cos 4\pi \times 10^3 t$.
e Repeat d for $v_1(t) = \sqrt{2} \cos 8\pi \times 10^3 t$.

10-23 A low-pass filter section using an ideal inverting amplifier stage and originally proposed by Sallen and Key is displayed in Fig. 10-40.

a Show that the transfer function V_o/V_1 is given by

$$\frac{V_o}{V_1} = \frac{-A_vR_3}{R_3 + R_1(1 + A_v) + s[R_3(R_1C_1 + R_1C_2 + R_2C_2)] + s^2R_1R_2R_3C_1C_2}$$

b Identify ω_o and Q in the result in a.

10-24 **a** Design the circuit of Fig. 10-40 and described in Prob. 10-23 to realize $\omega_o = 8 \times 10^3$ rad/s and $Q = 1$. Select $C_1 = C_2 = 0.01$ μF and $R_1 = R_2 = R_3 = R$.
b Determine the value of V_o for $\omega = 0$, $\omega = \omega_o$, and $\omega = 2\omega_o$.

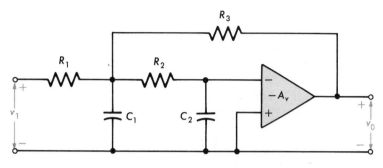

FIGURE 10-40 Circuit for Prob. 10-23.

10-25 Show that the transfer function of the circuit in Fig. 10-28b is given by

$$\frac{V_o}{V_1} = \frac{A_v R_1 R_2 C_1 C_2 s^2}{1 + s[R_1 C_1 + R_1 C_2 + R_2 C_2(1 - A_v)] + s^2 R_1 C_1 R_2 C_2}$$

10-26 A common design approach for the circuit in Fig. 10-28b is to use a unity-gain amplifier stage. In addition, the ratio of the resistances and the ratio of the capacitances are to be equal. That is, R_1/R_2 and C_1/C_2 are equal. Design the circuit for $\omega_o = 2\pi \times 10^3$ and $Q_o = 1/\sqrt{2}$ when the largest resistance permissible is 50 kΩ.

10-27 The circuit in Fig. 10-41 is a Sallen and Key high-pass filter section which employs an inverting amplifier-stage. Show that the transfer function is given by

$$\frac{V_o}{V_1} = \frac{-A_v R_1 R_2 C_1 C_2 s^2}{1 + s[R_2 C_2 + R_1(C_1 + C_2 + C_3)] + s^2 R_1 R_2 C_3[C_1 + C_3(1 + A_v)]}$$

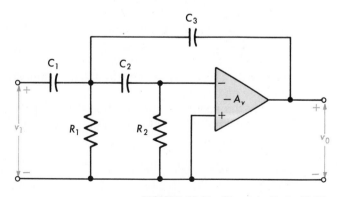

FIGURE 10-41 Circuit for Prob. 10-27.

10-28 The circuit in Fig. 10-41 is to be designed for $f_o = 10$ kHz and $Q = \sqrt{3}/3$. The values of $R = R_1 = R_2$ and $C = C_1 = C_2 = C_3$ are restricted so that $R \leq 30$ kΩ and $C \leq 1,000$ pF. Determine the range of permissible values for R and C and the value of A_v which satisfy the design specifications.

10-29 The circuit depicted in Fig. 10-42 is a Sallen and Key bandpass circuit. Show that the transfer function can be expressed as

$$\frac{V_o}{V_1} = \frac{sA_vR_2C_2}{(1 + R_1/R_3) + s[R_1C_1 + R_2C_2 + R_1C_2 + R_1R_2C_2(1 - A_v)/R_3] + s^2R_1R_2C_1C_2}$$

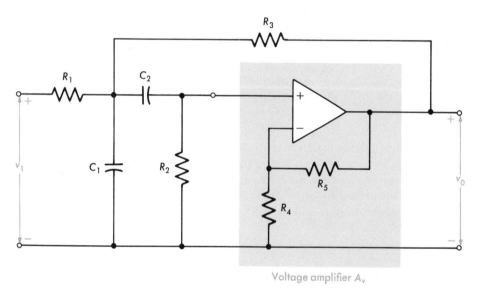

FIGURE 10-42 Sallen and Key bandpass circuit for Prob. 10-29.

10-30 a The circuit shown in Fig. 10-42 whose transfer function is given in Prob. 10-29 is to be designed for $\omega_o = \pi \times 10^4$ rad/s and $Q = 10$. One approach that is readily fabricated is to set the gain A_v equal to two and to make both capacitors equal. Using this approach, determine R_1, R_2, and R_3 for $C = 10,000$ pF.
(*Note:* The resistance values are not unique. It is convenient to define $a = R_1/R_3$ and $b = R_1/R_2$ and to select components which keep a and b between 1 and 10.)
b Using the results in a, determine the gain at the angular frequency ω_o.
c Select R_4 and R_5 so that the number of different resistance values required is minimum.

10-31 The purpose of this problem is to demonstrate why the schematic arrangement in Fig. 10-24 cannot be used to realize narrow-band bandpass circuits. In Fig. 10-24, consider both the low-pass and high-pass sections to be ideal with $\omega_L = 9,900$ rad/s and $\omega_H = 10,100$ rad/s.

a Find the bandwidth of the circuit.
b Determine the range of values for the center frequency and bandwidth if both ω_H and ω_L can be controlled to within 1 percent of their nominal values.

10-32 The circuit arrangement in Fig. 10-26 is designed with ideal low-pass and high-pass sections for which $\omega_L = 10,000$ rad/s and $\omega_H = 9,900$ rad/s.
a Sketch the overall response and determine the bandwidth and center frequency.
b Both ω_L and ω_H can be controlled to within one percent of their nominal values. Determine the range of values of the bandwidth and center frequency that can result.
c Comment on the practical effectiveness of the circuit in Fig. 10-26 in light of the results in *b*.

10-33 The OP-AMPs in the circuit of Fig. 10-29 are ideal. Show that the transfer function (V_4/V_1) is a low-pass characteristic.

10-34 Determine the transfer function (V_3/V_1) for the circuit of Fig. 10-29 and show that it is a bandpass characteristic. Assume the OP-AMPs are ideal.

10-35 Show that the transfer function (V_2/V_1) for the circuit of Fig. 10-29 is a high-pass characteristic. Assume ideal OP-AMPs.

10-36 The circuit shown in Fig. 10-43 is an alternative version of the biquadratic section in Fig. 10-29 in which v_4 is the output and either v_1, v_2, or v_3 is the input. Show that low-pass, bandpass, and high-pass character-

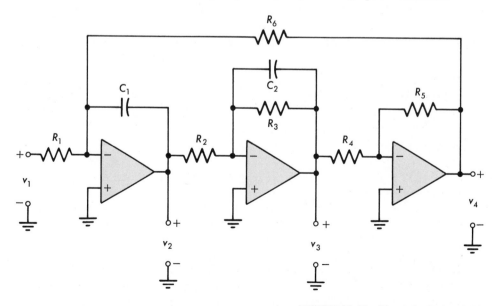

FIGURE 10-43 Circuit for Prob. 10-36.

istics are obtained for input signals applied at v_1, v_2, and v_3, respectively. Assume ideal OP-AMPs.

10-37   The circuit in Fig. 10-43 is often used with the location of the elements R_5 and C_1 being interchanged. Show that the outputs at v_2, v_3, and v_4 result in high-pass, bandpass, and low-pass characteristics, respectively, when v_1 is the applied signal. Assume the OP-AMPs are ideal.

10-38   Most practical systems require more than one filter section be used to achieve the desired frequency-response characteristics. Consequently, filter sections are cascaded. The purpose of this problem is to illustrate the differences in response that result when stagger-tuning and synchronous tuning are employed.
a   Consider the cascade of two identical filter sections for which $\omega_o = 10^4$ rad/s and $Q = 12.88$. Plot the magnitude response in the frequency range $9.5 \times 10^3 \leq \omega \leq 10.5 \times 10^3$ rad/s. Assume the gain is adjusted to make the maximum output unity.
b   Repeat a when two different sections are cascaded, one of which is characterized by $\omega_o = 10,177$ rad/s and $Q = 28.78$, the second by $\omega_o = 9,823$ rad/s and $Q = 27.78$.
c   Which circuit arrangement more closely approximates the ideal bandpass characteristic?
(*Note:* Both circuits are adjusted to provide the same overall bandwidth. To obtain the overall response it may be useful to use the universal resonance curve.)

10-39   An ideal multiplier is used to obtain an AM signal. The carrier is $\sqrt{2}\, V_c \cos 2\pi f_c t$ and the modulating signal is $\sqrt{2}\, V_s \cos 2\pi f_s t$. The values of f_c and f_s used in standard AM broadcast systems are: $540 \leq f_c \leq 1600$ kHz and $20 \leq f_s \leq 5,000$ Hz. Determine the output signal and the range of frequencies present in the output.

10-40   The system depicted in Fig. 10-44 is a functional representation of a *mixer* and the intermediate-frequency (IF) amplifier used in most AM radio receivers. If the frequency range of the RF signal is between 540 and 1,600 kHz, what must the corresponding range of local oscillator frequency be?

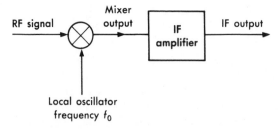

FIGURE 10-44   Mixer representation for Prob. 10-40.

Two possible ranges exist; comment briefly on the practical feasibility of each range.

10-41 In certain instrumentation systems it is required to generate a voltage whose amplitude is proportional to the frequency of the applied signal. The circuit shown in Fig. 10-45 is a simple circuit which can achieve this result over a specified range of frequencies.
a Determine the transfer function (V_o/V_s).
b Under what conditions can the result in *a* be used to achieve an output directly proportional to the applied frequency?
c In terms of the time constant, what is the highest frequency that can be applied so that the output deviates by no more than one percent from a linear function of frequency?

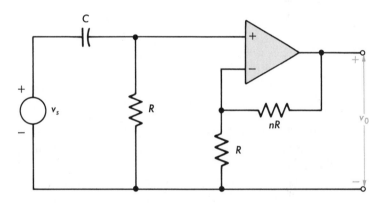

FIGURE 10-45 Frequency-voltage converter for Prob. 10-41.

10-42 The block diagram in Fig. 10-46 uses the frequency-to-voltage converter shown in Fig. 10-45 and described in Prob. 10-41. The schematic in Fig. 10-46 is to be employed as an instrumentation system for the measurement of force, displacement or pressure. The VFO is a Hartley oscillator whose frequency of oscillation is changed by mechanically varying the tuning capacitor. The change in capacitance is directly proportional to the force to be measured. The crystal oscillator is used to generate a fixed-frequency sinusoid so that when no force is applied both the VFO and crystal oscillator generate the same frequency. The low-pass filter is adjusted to pass only the difference frequency component of the multiplier output. The frequency-to-voltage converter provides an output voltage proportional to the difference frequency and thus allows an output dependent on the force applied to be displayed. A specific system is designed with the following characteristics: The VFO has a nominal frequency of oscillation of 500 kHz for a capacitance of 4,000 pF. The maximum force to be measured causes a change in capacitance of 80 pF. The low-pass filter can be consid-

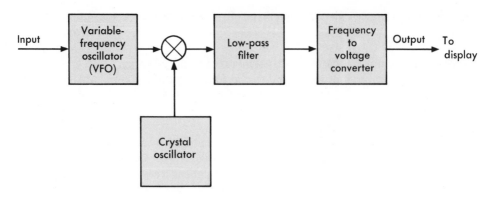

FIGURE 10-46 Block diagram for Prob. 10-42.

ered ideal and has a passband of 5.1 kHz and provides a 7-V-rms output signal throughout the passband.
a For $R = 4.5k$, $C = 1,000$ pF, find n so that the maximum applied force produces a 10-V-rms output.
b The minimum output that can be accurately detected is 10 mV. Find the change in capacitance that corresponds to the minimum output signal.

10-43 The system shown in Fig. 10-46 and described in Prob. 10-42 is to be used to measure the change in concentration of a particular dielectric material in a mixture. The mixture forms the dielectric in the tuning capacitor of the VFO. When the concentration of this material is 30 percent the VFO oscillates at the same frequency as does the crystal oscillator. Using the system characteristics and component values given in Prob. 10-42,
a Determine the maximum change in concentration that can be detected if the output is not to exceed 10-V-rms. The value of n is 9.
b Sketch the calibration curves of output voltage vs. concentration and output voltage vs. change in concentration.
c Comment on the usefulness of this system.

10-44 The two input signals to an ideal multiplier are $v_1(t) = \cos 10^9 t$ and $v_2(t) = 100 \cos 2 \times 10^9 t$. The output of the multiplier is fed through a filter which passes only the angular frequency $\omega = 10^9$ rad/s. Show that the amplitude of the multiplier output is greater than the input amplitude at $\omega = 10^9$ rad/s. (This feature is often the basis of a high-frequency amplifier called a *parametric amplifier*.)

10-45 **a** Determine the output v_o as a function of v_1, v_2, and v_3 for $R_1 = 2.5$ kΩ, $R_2 = 5$ kΩ, and $R_3 = R_4 = 10$ kΩ in the circuit of Fig. 10-32.
b The resistance values in a are binary-weighted. Show that if v_1, v_2, and v_3 are independently either zero or one, the magnitude of the output voltage is the decimal integers zero through seven.

10-46 In the circuit of Fig. 10-32, $R_1 = 1$ kΩ, $R_2 = 10$ kΩ, and $R_3 = R_4 = 100$ kΩ. The voltages v_1, v_2, and v_3 can be independently set to be 0, 1.0, 2.0, 3.0, . . . , 9.0 mV. Show that the magnitude of the output voltage, in millivolts, can assume all integral values between 0 and 999.

10-47 The circuits of Fig. 10-32 can be augmented to handle more than three inputs by the addition of branches which contain resistances and signal sources similar to $v_1 - R_1$, $v_2 - R_2$, etc. Using this arrangement, we wish to design a summing circuit whose output in volts is the class average on a quiz. The class size is 25 and all grades are integers between one and ten. The maximum output voltage is 10 V and the minimum resistance that can be used is 1 kΩ.
a Design the circuit.
b Check your design with the following grade distribution:

Number of students	0	1	0	2	1	4	7	4	3	3
Quiz grade	1	2	3	4	5	6	7	8	9	10

10-48 The circuit in Fig. 10-33 is to be designed to provide an output function $v_0 = 5v_1 - 3v_2$. The minimum resistance that can be used is 1 kΩ and the maximum output voltage is to be limited to 15 V.
a Determine the circuit elements.
b If the maximum values of v_1 and v_2 are equal, determine this maximum value. Assume all inputs are always positive.
c Repeat b if the inputs can be positive or negative.

10-49 The circuit in Fig. 10-33 is to be designed with $R_1 = 5$ kΩ, and $R_2 = R_3 = R_4 = 10$ kΩ.
a Determine the output voltage for $v_1 = 2.50 + 1.5 \cos \omega t$ and $v_2 = 2.51 + 1.5 \cos \omega t - 0.5 \cos 2\omega t$.
b The OP-AMP used has $A_o = 10^5$ and $\omega_h = 20$ rad/s. Repeat a for $\omega = 10^4$ rad/s and $\omega = 5 \times 10^5$ rad/s.

10-50 The element values used in the circuit of Fig. 10-34 are $R_1 = 10$ kΩ and $C = 1,000$ pF.
a Assuming the operational amplifier is ideal, determine the output voltage v_o when the input signal is a 1.0 V positive pulse whose duration is 100 μs.
b Repeat a if the OP-AMP characteristics are those given in Prob. 10-49b.

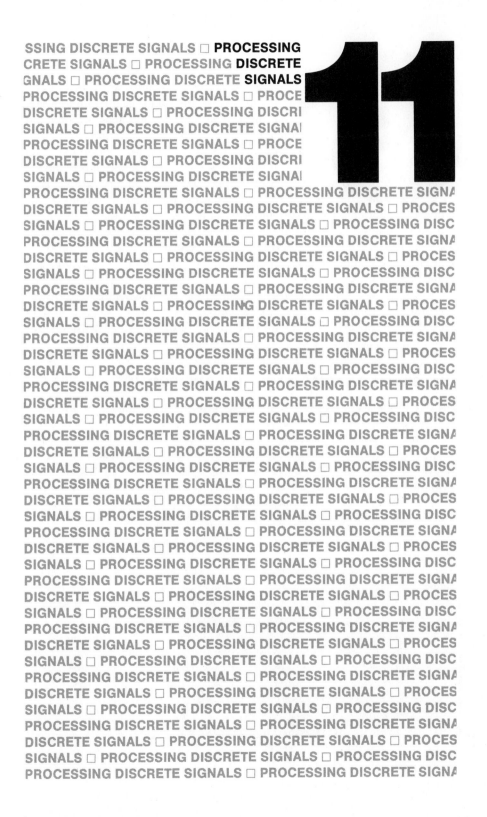

ystems which utilize discrete signals often offer the advantages of economy in time and power consumption, accuracy, and reliability. Invariably, discrete signals are sequences of pulses in which the information is contained in the pulse characteristics and the relation between one pulse and others in the sequence during a specified time interval. Consequently, time-domain signal processing is used; the circuit functions required parallel those employed in frequency-domain processing described in the preceding chapter. These circuits include those which establish timing and generate the necessary waveforms, and those which perform time selection and time translation. System aspects of digital computation are treated in Chap. 12.

Because semiconductor devices are used extensively as switches in these circuits, their characteristics as switching elements are discussed. In addition, the conversion of continuous to discrete and discrete to continuous signals is described.

11-1 SIGNAL PROCESSING IN THE TIME DOMAIN

The quantities used to describe the behavior of most physical systems, such as pressure, temperature, and velocity, are generally continuous (analog) functions. Usually, transducers, some of which are briefly described in Chap. 17, convert these quantities to continuous electrical signals. As described in Sec. 10-1, two classes of discrete signals exist. In one class, information is contained in the characteristics of the pulse; in the other, the presence or absence of a pulse at a specific time signifies the information content. The signals in Figs. 11-1 and 11-2 illustrate two methods in which

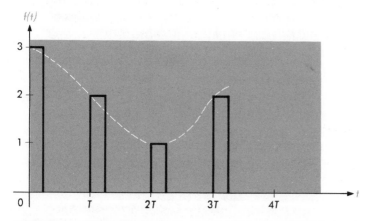

FIGURE 11-1 Sampled-data or pulse-amplitude-modulated (PAM) signal.

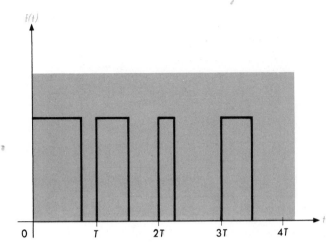

FIGURE 11-2 Pulse-width-modulated (PWM) signal.

information content resides in the pulse characteristics. The first of these, illustrated in Fig. 11-1, is often called a sampled-data signal. The dashed portion of the waveform in Fig. 11-1 is the continuous electrical signal; the solid pulses represent the sampled-data signal obtained by sampling the continuous signal at periodic intervals of time. The pulses have uniform but short duration, and their amplitudes are the values of the original signal at the various sampling times. The resultant pulse sequence, or pulse train, is then useful in representing the original signal. This technique is also used to obtain a modulated signal, called *pulse-amplitude modulation* (*PAM*), and corresponds to amplitude modulation in analog systems. As shown in Fig. 11-2, a pulse train can be generated in which the duration of each pulse is proportional to the amplitude of the continuous signal. This method is referred to as *pulse-width modulation* (*PWM*) and is similar in concept to the frequency-modulated continuous signal. An important similarity of PAM and PWM signals is that in each case a characteristic of an individual pulse directly corresponds to the instantaneous value of the analog signal represented.

The second method essentially converts the physical quantity into a numeric code as depicted in Fig. 11-3. The sequence of pulses in each time interval is used to represent the numeric value of the amplitude of the corresponding pulse in Fig. 11-1. Thus, the pulse train in the first time interval in Fig. 11-3 corresponds to the first sampled-data pulse; each successive pulse train corresponds to each consecutive pulse in Fig. 11-1. The individual pulses in Fig. 11-3 are of both uniform duration and amplitude, and no single pulse contains information regarding the amplitude of the original signal at any instant of time. Usually, generating a numeric signal from a continuous one is called *analog-to-digital* conversion and is often indicated by *A/D*. In communication systems the conversion to numeric data and subsequent transmission of a modulated carrier is called *pulse-code modulation* (*PCM*).

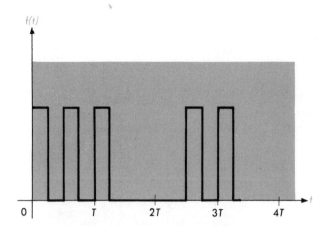

FIGURE 11-3 Numeric or pulse-code-modulated (PCM) signal.

The signal shown in Fig. 11-3 represents a number and need not have physical significance. The characteristics of the pulse train that represent the number 10 are independent of whether 10 signifies the voltage in a circuit or the cost of an item recorded on a charge statement. The systems which utilize this kind of signal process numeric data to perform their functions, and, as such, are called *digital systems*. Often, the popular context of the term digital systems refers to those systems whose operation involves discrete signals. Arithmetic and logical operations are the only functions performed in processing numeric data. The arithmetic functions combine the data in a prescribed fashion, logic functions are used to ensure that the correct data is used and to establish the proper sequence of arithmetic operations that need be performed. Once the data is processed, the numeric signals can be used to generate continuous signals by means of *digital-to-analog* conversion (*D/A*).

Many of the same circuits are used in the processing of both classes of discrete signals. The establishment of appropriate timing sequences, including time translation and selection, and the generation of pulses, square waves, and triangular waveforms are among the processing functions that are common. In this chapter, the behavior of these circuits, including device characteristics that are utilized, is described.

11-2 WAVESHAPING CIRCUITS AND COMPARATORS

A class of diode circuits which process signals by changing their waveforms is often used in electronic systems. The unilateral nature of the diode is used either to maintain certain voltage levels or to control the path through which the signals may be transmitted.

One typical circuit, shown in Fig. 11-4, is used to convert the sinusoidal input waveform to the nearly trapezoidal output waveform shown. When the source-voltage amplitude v_s is zero, both diodes D_1 and D_2 are reverse-

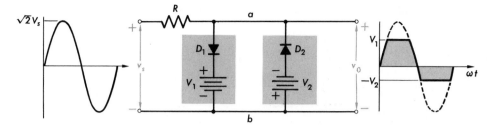

FIGURE 11-4 Diode clipper circuit.

biased and therefore do not conduct, resulting in an output voltage v_o of zero. Both D_1 and D_2 remain open-circuited until the signal amplitude reaches the value V_1, at which time D_1 becomes forward-biased and conducts. With the voltage drop across D_1 neglected, the voltage across nodes ab, and consequently the output voltage, becomes V_1. The output voltage is unchanged until the signal level drops below V_1, again reverse-biasing D_1. During the negative half cycle, diode D_2 begins to conduct when the signal amplitude exceeds V_2, and thus the output voltage becomes V_2. This level is maintained until the value of v_s drops below V_2, which reverse-biases D_2. The process repeats itself during each cycle. The output waveform is a clipped sinusoid, and this type of circuit is often referred to as a *diode clipper*.

In Chaps. 1, 3, and 4, simple circuits using resistors, inductors, and capacitors are shown to have waveshaping properties. When used in conjunction with diodes, the resulting circuits provide a great deal of variety in the waveforms that can be obtained. As an example, consider the circuit of Fig. 11-5a, which can be used to convert the rectangular-wave input (Fig. 11-5b) to the approximately triangular waveform (Fig. 11-5c) at the output. When the signal amplitude V_s is positive, diode D, being reverse-biased, is open-circuited. Consequently, the capacitor C charges through the path shown in the equivalent circuit of Fig. 11-6a. If the circuit time constant is larger than the duration of the positive portion of the wave, the capacitor charging is incomplete and the peak positive output voltage V_o is less than V_s (Fig. 11-5c). During the negative portion of the cycle, the capacitor discharges through the path indicated in Fig. 11-6b until its voltage, which is also the diode voltage, is zero. When the capacitor voltage becomes zero, diode D conducts, maintaining the output at 0 volts. The output is maintained at 0 volts until the cycle is repeated. The operation of the diode in this circuit is often referred to as *clamping*.

The operation of the diodes in Figs. 11-4 and 11-5 can be considered as performing comparison. The diodes change state whenever the voltage across them changes from positive to negative, or vice versa; that is, the diodes switch to and from conduction and cutoff. In effect, the diodes in Fig. 11-4 act to compare the reference voltages V_1 and V_2 with the input signal and in so doing control the output level.

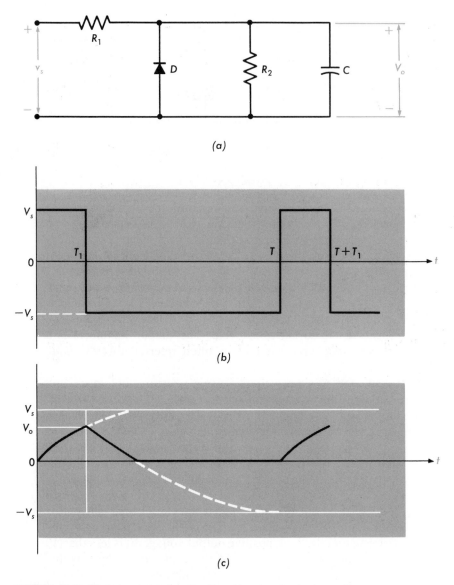

FIGURE 11-5 Diode waveshaping circuit and input and output waveforms.

Specially constructed diodes, called *Zener diodes* and whose circuit symbol and volt-ampere characteristics are depicted in Fig. 11-7, act in much the same manner as the diode-reference voltage combinations in Fig. 11-4. Zener diodes exhibit a stable breakdown voltage in reverse bias, as indicated in Fig. 11-7b, and have properties similar to a junction diode when forward-biased. For reverse-biased levels below the Zener voltage V_z, the diode is essentially an open circuit. However, when reverse-biased, the voltage across the Zener diode can never exceed V_z. For example, in the circuit

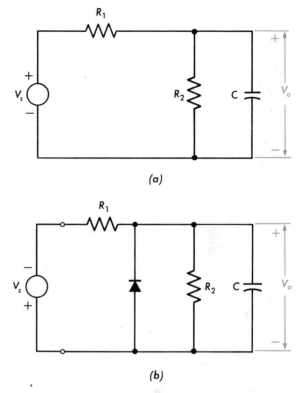

FIGURE 11-6 Equivalent circuits for circuit of Fig. 11-5*a*.

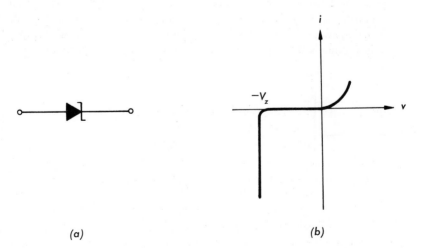

FIGURE 11-7 Zener diode (*a*) circuit symbol and (*b*) volt-ampere characteristic.

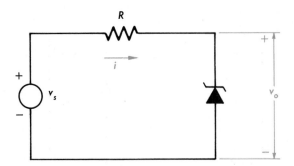

FIGURE 11-8 Zener diode circuit.

of Fig. 11-8, when the input voltage v_s is less than the Zener voltage V_z, the diode is open-circuited, no current exists in the circuit, and the output voltage v_o equals v_s. For the case where v_s is greater than V_z, the output voltage is V_z. The resistor voltage is $v_s - V_z$ and produces a current i equal to $(v_s - V_z)/R$. Zener diodes are often used in regulator circuits, which are treated in Chap. 17, as well as in comparison circuits in signal processing.

The combination of an operational amplifier and Zener diodes is used to form a commercially available circuit called a *comparator*. The function performed by a comparator is to provide one of two output levels depending on the relative magnitudes of the signals being compared. In the comparator circuit displayed in Fig. 11-9, the operational amplifier is used to compare the two voltages v_1 and v_2, one of which is generally a reference voltage. The back-to-back Zener diodes are used to establish the two output levels. Except during the transition between output levels, one Zener diode is forward-biased and the other is at its breakdown voltage. The gain provided by the OP-AMP permits the comparison of low-level signals.

The use of positive feedback in the comparator circuit shown in Fig. 11-10, known as a *Schmitt trigger,* improves both the sensitivity and transition speed. The positive feedback increases the gain of the OP-AMP stage and reduces the comparison voltage necessary to cause a transition between

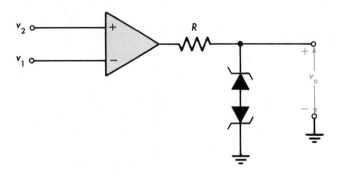

FIGURE 11-9 Comparator circuit.

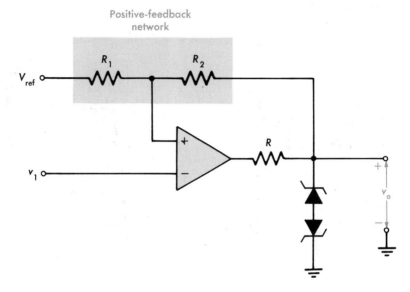

FIGURE 11-10 Schmitt-trigger circuit.

output levels. The smaller comparison voltage is accompanied by an increase in the transition speed.

Comparators are widely used in signal detection and are often components in commercial integrated circuits.

11-3 BIPOLAR AND MOS TRANSISTOR SWITCHES

The diodes in the waveshaping circuits discussed in the previous section behave as switching elements as an on-off behavior is exhibited. Similarly, comparator circuits provide a two-level output signal and the two output levels can be thought of as on and off conditions. The operation can be considered to be like a controlled switch; one level of input produces the "on" condition; the second input level results in the "off" condition. In addition, individual pulses are described by the two amplitude levels which exist and, therefore, can be generated by a switching process. Often, the devices which act as switches to generate and process discrete signals operate alternately in a condition at or near cutoff and a condition at or near saturation. Consequently, the combination of these devices and their associated circuit elements is considered to switch between cutoff and saturation.

The bipolar transistor circuit of Fig. 11-11a can be used as a controlled electronic switch. The bias at the base-emitter junction is controlled by the signal applied. When V_s is negative, the junction is reverse-biased and the transistor cut off, so that, ideally, there is no collector current. Actually, there is a small leakage current. Making V_s more negative has no effect on the collector current because the emitter-base junction remains reverse-biased. Ideally, since both junctions are now reverse-biased, no current

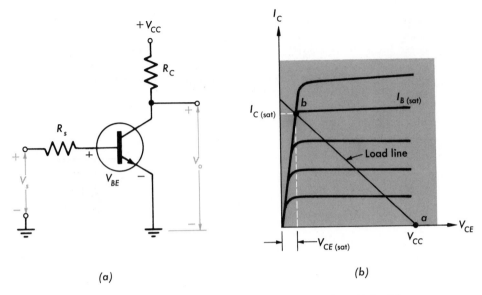

FIGURE 11-11   (a) Transistor switch and its (b) output characteristics with load line.

exists anywhere in the circuit. This condition is shown by point a on the characteristics displayed in Fig. 11-11b.

When the signal V_s (Fig. 11-11a) is positive and greater than the turn-on voltage, the base-emitter junction is forward-biased and there is a collector current. If V_s is sufficiently large, the transistor will become saturated and the collector current will be limited essentially by V_{CC} and R_C. Ideally, V_{CE} is zero in saturation; practically, there is a small voltage drop, called the *saturation voltage* $V_{CE(sat)}$. Increasing V_s further will have no effect on this value. The saturated condition is indicated by point b in Fig. 11-11b. The base current $I_{B(sat)}$ is the minimum value necessary to cause saturation. Note that the operating point for the circuit remains at a even if an increase in V_s increases the value of I_B.

In the course of switching from the saturated mode to the cutoff mode, or vice versa, the transistor operation traverses the active or normal mode, the models for which are discussed in Chap. 8. Three calculations are required to describe the performance of the switch. Two of these are used to determine the voltages and currents which exist in the saturated and cutoff circuits and involve dc analysis. The third calculation describes the transient behavior of the circuit and is used to determine the switching speed which is limited by the transistor capacitances and charge-storage effects.

The transistor switch of Fig. 11-11a is either open or closed and is thus a two-state circuit. (A *state* is a condition of fixed current and voltage levels in a circuit.) The transistor described is switched from one state (cutoff) to the second state (saturation) only by changing the value of V_s, which is often called the *trigger signal* or, simply, the *trigger*. The switch has two stable states that can be maintained indefinitely provided V_s does not change. The concept of a controlled switch is manifested in the control of the collector current (output current) by the trigger signal applied to the base. The transistor may be thought of as a three-terminal switch; the main current path is the output circuit containing the collector and emitter, while the base circuit exhibits the control.

Example 11-1

The transistor switch shown in Fig. 11-11a is designed to operate between cutoff and saturation with the signal depicted in Fig. 11-12a applied. The circuit is to operate with a supply voltage of 5 V with a collector resistance of 0.5 kΩ. The transistor, whose characteristics are given in Fig. 11-12b, has a turn-on voltage V_{on} of 0.6 V.

a Determine the maximum value of R_s which causes saturation.

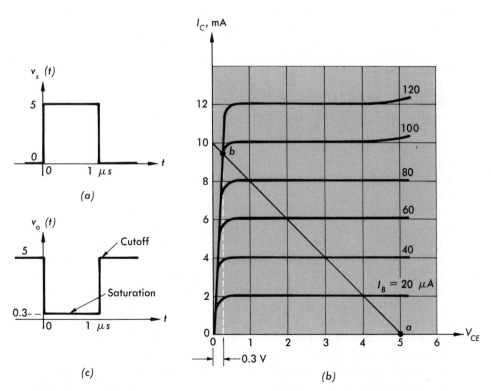

FIGURE 11-12 (a) Input signal waveform; (b) output voltage waveform; (c) transistor output characteristics and load line for switching circuit in Example 11-1.

b Sketch the output voltage as a function of time neglecting switching transients.

SOLUTION **a** The load line is shown on the characteristics in Fig. 11-12b on which saturation is indicated by point b. The minimum base current for saturation is, by interpolation, found to be 95 μA. In the circuit of Fig. 11-11a, the KVL expression for the base-emitter loop is

$$-V_s + R_s I_{B(\text{sat})} + V_{BE} = 0$$

Substitution of values gives

$$-5 + R_s \cdot 95 \times 10^{-6} + 0.6 = 0$$

from which

$$R_s = 46.3 \text{ k}\Omega$$

b For $t < 0$, the input signal $V_s(t)$ is zero so that the transistor is cut off (point a on Fig. 11-12b) and the output level is 5 V as depicted in Fig. 11-12c. During the interval between $t = 0$ and 1 μs, the transistor is switched into a saturated condition for which the output voltage decreases to $V_{CE(\text{sat})} = 0.3$ V. The transistor is returned to cutoff for $t > 1$ μs and the output level again becomes 5 V.

The circuits shown in Figs. 11-13a and 11-14a depict MOS switches using n-channel depletion and enhancement-mode devices, respectively. The behavior of both circuits parallels that for the bipolar switch shown in Fig. 11-11a. In Fig. 11-13b, the load line drawn on the depletion-mode MOS characteristics indicates that in cutoff the current is essentially zero and corresponds to a gate-to-source voltage V_{GS} equal to or less than pinch-off. Saturation, identified at point b, occurs for V_{GS} equal to zero and the output voltage is V_{ON}. Similarly, in Fig. 11-14b, the cutoff condition occurs for $V_{GS} = 0$ and the output voltage is essentially V_{DD}. Application of a positive gate-to-source voltage of sufficient magnitude saturates the circuit and results in an output voltage of V_{ON}. The voltage V_{ON} is a threshold voltage and is analogous to $V_{CE(\text{sat})}$ in the bipolar switch. Note that in both n-channel MOS switches, a positive-going input signal is required to produce a transition from cutoff to saturation. However, for the depletion-mode transistor, the input signal remains negative, whereas the applied signal is positive in the enhancement-mode circuit.

To eliminate the large chip area required by the resistor in the circuits of Figs. 11-13a and 11-14a, MOS switches are generally implemented with active loads as shown in Fig. 11-15. The load transistor $Q2$ is always maintained in its constant-current region (active region); the transistor $Q1$ acts as the switch controlled by the input signal. The advantage of this realization is that only MOS transistors must be fabricated and isolation between adjacent devices and circuits is unnecessary.

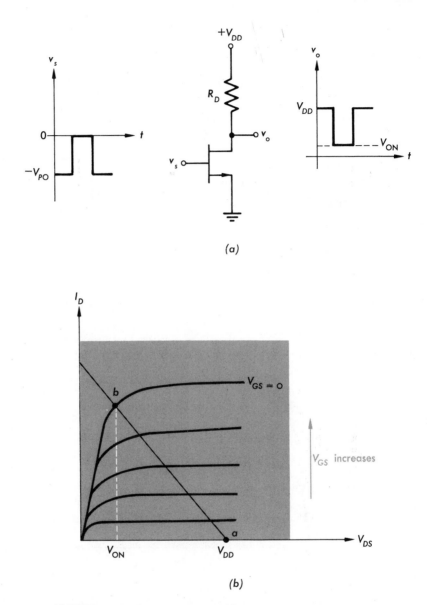

(a)

(b)

FIGURE 11-13 Depletion-mode MOS transistor switch and characteristics.

11-4 TIME-BASE GENERATORS

Many electronic systems require signal waveforms which vary linearly with time. This type of signal, shown in Fig. 11-16, is called a *sawtooth* and is generated by a class of circuits called *sweep circuits,* or *time-base generators.* A major application of sweep circuits is in systems which use cathode-ray-tube displays. The horizontal axis is used as the time axis, or time base, so the

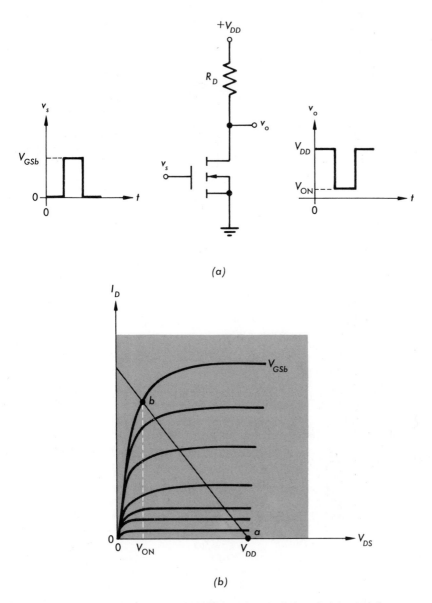

FIGURE 11-14 Enhancement-mode MOS transistor switch and characteristics.

display presented shows the amplitude-time relationship of a signal. The output of the sweep circuits is used to produce an electron beam whose horizontal deflection is proportional to time.

The model of Fig. 11-17 is the basic representation of a sawtooth voltage generator. The switch K periodically changes its position from point 1 to point 2. (Although a mechanically operated switch is implied in the represen-

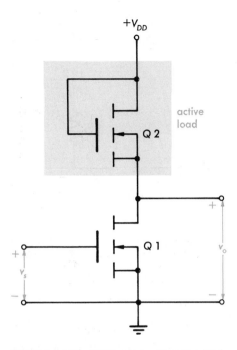

FIGURE 11-15 MOS switch with active load.

tation of Fig. 11-17, the switching action is actually performed electronically by means of a separate circuit not shown.) This switching action causes the capacitor C to be alternately charged and discharged. The resulting waveform of output voltage v_o is shown in Fig. 11-18. The time T_1 is called the *sweep time,* and T_2 is the *retrace* or *return time.* The return time is needed in display systems to return the beam to its original starting point before the next cycle begins.

With the switch in position 1 (Fig. 11-17), the capacitor tends to charge to the supply voltage V, called the *aiming potential.* Before the capacitor can charge completely, the switch is put in position 2, and the capacitor dis-

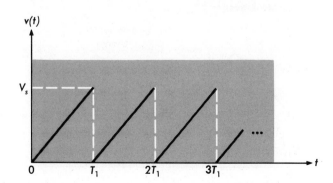

FIGURE 11-16 Ideal sawtooth waveform for several cycles.

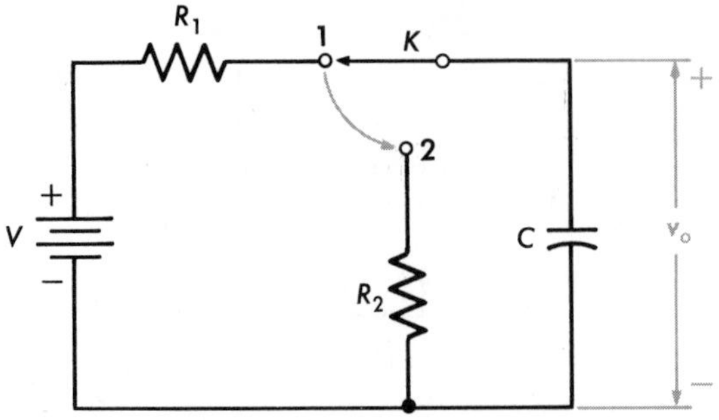

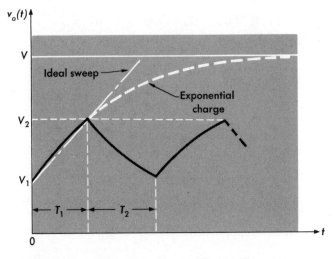

FIGURE 11-17 Representation of a sawtooth generator.

charges until the switch returns to position 1, at which time the cycle is repeated. The output voltage $v_o(t)$ is given by

$$v_o(t) = V - (V - V_1)\epsilon^{-t/R_1C} \qquad \text{for } 0 \le t \le T_1 \tag{11-1}$$

and

$$v_o(t) = V_2\epsilon^{-(t-T_1)/R_2C} \qquad \text{for } T_1 \le t \le T_2 \tag{11-2}$$

When $t = T_1$, $v_o(T_1) = V_2$, so that

$$T_1 = R_1C \ln \frac{V - V_1}{V - V_2} \tag{11-3}$$

and at $t = T_2$, $v_o(T_2) = V_1$, so that

$$T_2 = R_2C \ln \frac{V_2}{V_1} \tag{11-4}$$

FIGURE 11-18 Waveforms for circuit of Fig. 11-17.

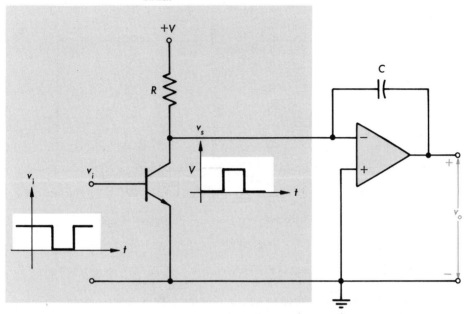

FIGURE 11-19 Miller sweep circuit.

The sweep amplitude V_S is given by

$$V_S = V_2 - V_1 \tag{11-5}$$

The sweep linearity is a measure of how close to the ideal the exponential waveform comes. From the waveforms in Fig. 11-18, it is seen that for small values of V_2, the straight-line and exponential charges are nearly identical. In general, the sweep voltage should be a small fraction of the aiming potential if good linearity is to be achieved.

The circuit of Fig. 11-19 is the basic schematic of a widely used sawtooth generator called the Miller sweep. The amplifier acts to increase the aiming potential; thus, linearity is improved and the output amplitude increased. It is of note that the circuit of Fig. 11-19 and the integrator of Fig. 10-34 are virtually identical. As the integral of a step function is a ramp, it is evident that the Miller sweep should provide a sawtooth output.

11-5 MULTIVIBRATORS

A class of two-state circuits useful for generating pulses and square waves is called *multivibrators*. These circuits usually consist of a pair of amplifiers coupled to each other in a positive-feedback arrangement, as shown in Fig. 11-20. Each amplifier is composed of a basic stage of the types described in Chap. 8, which can be switched from one state to another by means of an

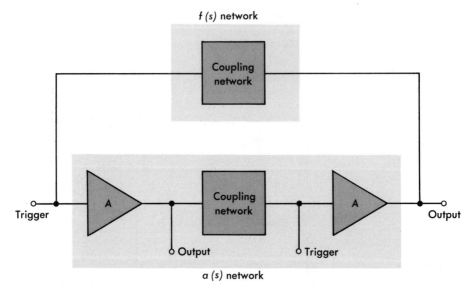

FIGURE 11-20 Basic configuration of multivibrator.

appropriate trigger applied to either amplifier. Thus, each amplifier acts as a switch and each has an available output.

The operation of the circuit is such that when one amplifier is cut off (OFF), the positive-feedback loop maintains the other amplifier in a conducting, or ON, state. When a trigger causes one amplifier to change state, the coupling networks act to change the state of the second amplifier. The outputs are opposite in sense, one indicating a transition from an OFF to an ON state and the other the opposite transition.

As there are two possible types of states, it is convenient to classify multivibrator circuits in terms of the number of stable states each possesses. The *bistable multivibrator* requires the application of two triggers to return the circuit to its original state. The first trigger causes the ON transistor to be cut off, and the second trigger causes a transition back to the conducting state. Because two triggers are required, bistable circuits are sometimes called *flip-flops*. A single trigger applied to a *monostable,* or *one-shot, multivibrator* cuts off the normally ON transistor, causing the cutoff transistor to conduct. This new state is a quasi-stable one which ultimately causes a transition back to the original state. The *astable,* or *free-running, multivibrator* contains two quasi-stable states, and oscillation occurs with the amplifiers continually switching from one state to the other. The circuit is self-excited and requires no external trigger.

In some instances, the transistor in the conducting state operates in the saturated region. Often, however, it is undesirable for it to operate in saturation because of the inherent limitations on switching speed; the conducting state is then restricted to the active region. Consequently, multivibrators are referred to as *saturating* or *nonsaturating* circuits. Only saturating circuits

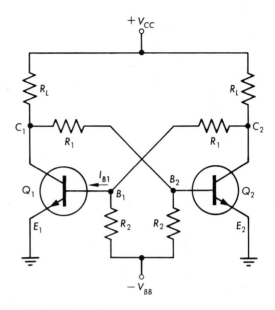

FIGURE 11-21 Two-supply bistable multivibrator.

are discussed in the following sections; however, the principles of operation of the nonsaturating circuit are similar.

The circuit of Fig. 11-21 is that of a typical two-supply bistable multivibrator. Each amplifier stage consists of a transistor Q and load resistor R_L. The value of R_L is selected to achieve the desired collector current in saturation. The supply voltages V_{BB} and V_{CC} are used to establish the appropriate bias conditions on each transistor. The resistors R_1 and R_2 form the coupling networks whose function is to transmit the change in state of one transistor to the other. The values of R_1 and R_2 are selected so that they are much greater than R_L and consequently have no loading effects on the amplifier circuit.

To describe the operation of the circuit first assume that Q_2 is cut off and Q_1 is conducting under saturated conditions. As a result, the voltages V_{B1} between point B_1 and ground (points E_1 and E_2) and V_{C1} between point C_1 and ground are both small. The base-bias voltage V_{B2} on Q_2 is then

$$V_{B2} = -V_{BB} \frac{R_1}{R_1 + R_2} \qquad (11\text{-}6)$$

which, being negative, keeps Q_2 in its cutoff state. Since R_L is much smaller than R_1 and the collector current of Q_2 is assumed zero, the voltage V_{C2} is $+V_{CC}$. The base current I_{B1} is then

$$I_{B1} = \frac{V_{CC} - V_{B1}}{R_1} - \frac{V_{BB} + V_{B1}}{R_2} \qquad (11\text{-}7)$$

and, if V_{B1} can be considered negligibly small,

$$I_{B1} = \frac{V_{CC}}{R_1} - \frac{V_{BB}}{R_2} \qquad (11\text{-}8)$$

If Q_1 is to keep conducting, the value of I_{B1} must be sufficiently positive to ensure continued collector current; that is,

$$I_{B1} \geq \frac{I_{C1}}{h_{FE}} \qquad (11\text{-}9)$$

or, since $I_{C1} = (V_{CC} - V_{CE(\text{sat})})/R_L$

$$I_{B1} \geq \frac{V_{CC} - V_{CE(\text{sat})}}{h_{FE}R_L} \qquad (11\text{-}10)$$

The values of V_{CC}, V_{BB}, R_1, and R_2 are chosen so that Eq. (11-10) is satisfied, and consequently a stable state exists with Q_2 cut off and Q_1 in saturation. From the symmetry of the circuit, it is evident that a second stable state exists with Q_1 cut off and Q_2 in saturation.

The circuit can be switched from one state to another by applying either a negative-voltage pulse to the base of Q_1 or a positive pulse to the base of Q_2. A negative voltage applied to the base of Q_1 will cause it to stop conducting, and the voltage V_{C1} will increase from essentially zero to V_{CC}. Transistor Q_2 will then have a base current given by Eq. (11-7) and will start conducting, at which time the circuit will assume its new stable state.

Two points should be noted about the switching process: First, the pulse used for switching need only be applied for a time sufficient for the changeover; it then can be removed. Second, if a negative pulse is used to effect switching (as described above), a positive pulse is required for the next switching, a negative pulse for the one following, etc. Idealized output waveforms for the bistable multivibrator are shown in Fig. 11-22 as solid curves. The dotted waveforms are those of an actual circuit; deviations from the ideal result from switching transients. There are several other circuit configurations that are used to realize flip-flops. The properties and utilization of these circuits as digital computer elements is treated in Sec. 12-3.

The monostable-multivibrator circuit of Fig. 11-23 is designed for operation with Q_1 normally conducting and Q_2 cut off. Operation in this stable state is ensured if the biasing resistor R is selected to allow sufficient base current in Q_1 for operation in the saturated region. Transistor Q_2 is kept in cutoff by the R_1, R_2, and V_{BB} biasing circuit. In the stable state with Q_2 cut off, V_{C2} is approximately V_{CC}; therefore, since the voltage V_{B1} is negligibly small when Q_1 is conducting, the capacitor voltage V_C is also approximately V_{CC}.

The circuit can be triggered by applying a positive-voltage pulse to the base of Q_2, causing it to conduct. The collector voltage V_{C2} will then drop to about zero, the voltage V_{B1} on the base of Q_1 will be $-V_C$, and Q_1 will stop conducting. (The biasing network for Q_2 is designed to ensure continued conduction so long as Q_1 is cut off.) The capacitor voltage V_C, which now

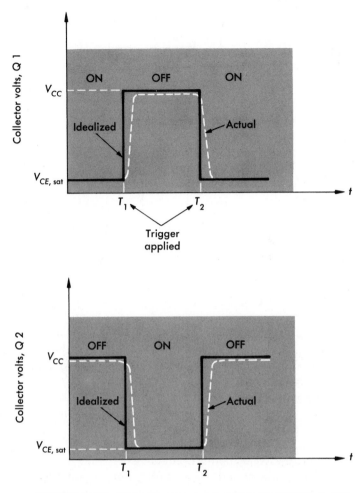

FIGURE 11-22 Collector voltage waveforms for bistable multivibrator.

acts as the base supply for Q_1, will not remain constant, however. It will decrease in magnitude, discharging through R, and will then tend to charge up to V_{CC} in the reverse direction. As soon as V_{B1} equals the turn-on voltage, however, Q_1 will again conduct, causing Q_2 to cut off and the circuit to return to its stable state. Typical waveforms of the base and collector voltages of Q_1 and Q_2 for the ideal (solid) and practical (dotted) cases are shown in Fig. 11-24. The duration T of the quasi-stable state can be shown to be approximately

$$T = RC \ln 2 \tag{11-11}$$

The design of monostable circuits involves setting the duration T and establishing the bias conditions required to maintain the stable state. The value of the bias resistance R is selected so that the base current of the normally

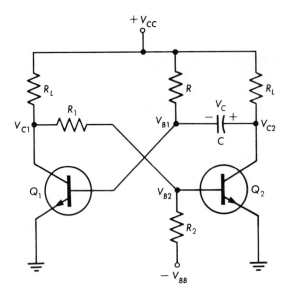

FIGURE 11-23 Monostable multivibrator.

conducting transistor is sufficient to maintain it in saturation. Recognition of these constraints guides the selection of all the circuit components.

The astable circuit of Fig. 11-25 has two quasi-stable states, and the analysis is similar to that for the quasi-stable section of the monostable multivibrator. The time interval during which each stage is cut off is given by Eq. (11-11). Typical output waveforms for one transistor are shown in Fig. 11-26.

The astable multivibrator shown in Fig. 11-25 produces a symmetric square wave as both RC coupling networks are identical. By making the two time constants different, an asymmetric output waveform is obtained. In either case, the periodic pulse train is useful in establishing the timing sequences necessary for synchronization of the operations performed in a system. When utilized in this manner, the free-running multivibrator acts as a *clock*, the output waveform referred to as the *clock pulses*.

As indicated in their output and input waveforms, both bistable and monostable multivibrators can be considered as time-translation and time-selection circuits. Commercially available integrated circuits, called *timers*, are employed in many control and instrumentation applications as a basic building block which provides monostable and astable operation. The IC chip usually contains a flip-flop and two or three comparators and is used in conjunction with external RC elements. The appropriate interconnection of the resistors and capacitors, generally one or two of each kind of element, results in the desired type of operation. Selection of component values is determined by the required output pulse durations. In addition, with the external RC components, the timers can be used to generate several pulse-modulated signals.

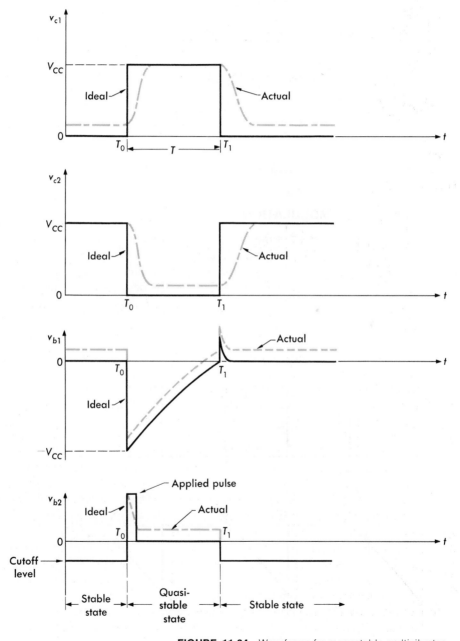

FIGURE 11-24 Waveforms for monostable multivibrator.

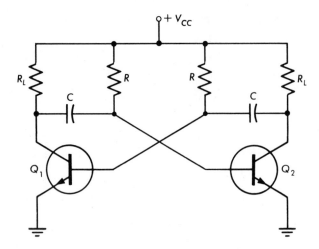

FIGURE 11-25 Astable multivibrator.

11-6 PULSE MODULATION

The circuits described in the preceding sections of this chapter are also utilized to generate the pulse-modulated signals shown in Figs. 11-1 and 11-2. Both types of modulation require that the continuous signal be sampled. Conceptually, the circuit shown in Fig. 11-27a performs the sampling function. The switch S is usually open and is periodically closed for a short dura-

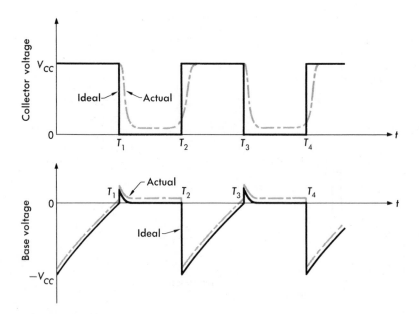

FIGURE 11-26 Waveforms for astable multivibrator.

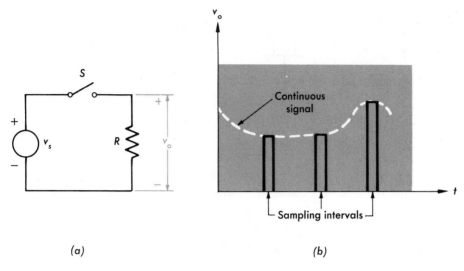

(a)   (b)

FIGURE 11-27   (a) Representation of sampling circuit and (b) output signal.

tion. The output waveform displayed in Fig. 11-27b indicates that during the time interval when S is closed, the amplitudes of the output pulses are just the input signal amplitudes at the sampling times.

The circuit shown in Fig. 11-28 performs the function illustrated in Fig. 11-27 for positive input signals. The clock pulses are obtained from an asymmetric astable multivibrator whose period establishes the sampling rate. The

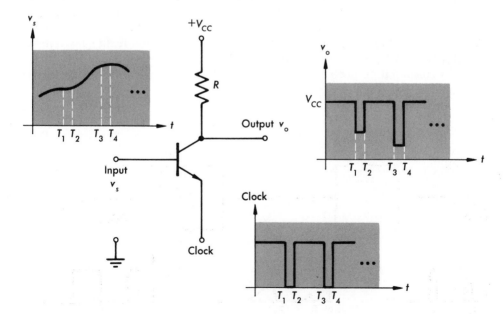

FIGURE 11-28   Transistor sampling circuit and its waveforms.

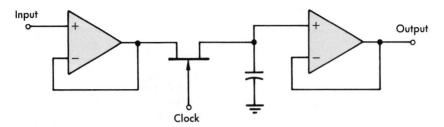

FIGURE 11-29 Sample-and-hold circuit.

clock-pulse amplitude is sufficient to maintain the transistor in cutoff except during the intervals between T_1 and T_2, T_3 and T_4, etc., at which times sampling occurs. At the sampling times, the transistor is in the active region and the output signal is proportional to the input signal.

By replacement of the resistor in Fig. 11-27a by a capacitor one obtains a *sample-and-hold circuit* as the capacitor tends to maintain a constant amplitude throughout the sampling interval. Operation of this type is an important feature of analog-to-digital conversion. A practical sample-and-hold circuit is shown in Fig. 11-29 in which the FET acts as the controlled switch and the unity-gain amplifiers serve as buffers.

Pulse-width-modulated (PWM) signals can be derived from sampled signals by the use of waveshaping circuits and comparators. The underlying concept is similar to the generation of the triangular waveform shown in Fig. 11-5. A pictorial representation of a PWM circuit is depicted in Fig. 11-30. Time constants in the RC waveshaping circuit are adjusted so that the input pulse is converted into a triangular wave. The duration of the triangular wave under these conditions is made proportional to the input amplitude. The comparator acts to convert the triangular signal into rectangular pulses by changing state at the amplitude indicated by the dashed line in Fig. 11-30. The resultant output pulse durations are directly proportional to the amplitudes of the input pulses.

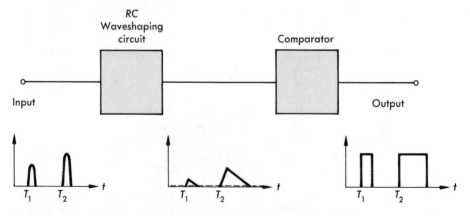

FIGURE 11-30 Pictorial representation of pulse-width-modulation circuit.

11-7 THE BINARY SYSTEM

The pulse sequence during a specific time interval in a digital system represents a number which corresponds to the information to be processed. Both logical and arithmetic operations are required in digital computers and pulse-code-modulated communication systems. The switching circuits used to perform the numerical processing are often the same as those which perform time translation in other pulse-processing systems.

Two-state electronic switches are fast, readily available, cheap, reliable, and easily manufactured in large numbers. Consequently, digital systems operate on a *binary* or *base* (radix) 2 system. The binary system is a number or logic system which contains only two permissible digits or states. Numerically, the binary digits are 1 and 0; in logic systems, the two states are either true and false, or yes and no. Electronically, the states are represented as either ON and OFF, or HI and LO. The HI and LO symbols generally correspond to the voltage or current levels in a switching element or in some trigger pulse. Often, the designations $V(1)$ and $V(0)$ are used to indicate the voltage levels corresponding to the binary digits 1 and 0, respectively. Because two states are available, each *binary digit*, or *bit,* is capable of conveying information. Binary *words* are groups of bits that have collective significance. Often, a group of eight bits is called a *byte.* Both the logic (control) and arithmetic operations are performed in binary in electronic digital systems.

The representation of numbers in binary exactly parallels that used in the decimal (base 10) system. The decimal number 378 is, in reality, simply $300 + 70 + 8$ or $3 \times 10^2 + 7 \times 10^1 + 8 \times 10^0$. Each place in the decimal number represents a power of ten; each digit is the number of times the power of ten is counted. A binary number is composed of a sequence of the digits 1 and 0, each of which multiplies a power of two. The number 101011 is $1 \times 2^5 + 0 \times 2^4 + 1 \times 2^3 + 0 \times 2^2 + 1 \times 2^1 + 1 \times 2^0$ and equals the decimal number 43. Three decimal digits can be used to represent 1,000 different numbers from 0 to 999, the highest number being $10^3 - 1$. Similarly, a 6-bit number can be used to represent 2^6 different values from 0 to $2^6 - 1$. This relationship is generalized and is given in Eq. (11-12).

$$N = 2^n - 1 \tag{11-12}$$

where n is the number of bits and N is the largest decimal number represented by n bits.

A decimal number D can be converted to a binary number B by the following technique:

1 Form two rows of numbers as shown in Table 11-1.

2 Start at the extreme right, divide D by 2 and place the whole-number portion of the quotient D_1 in the first column of the D row.

3 Place the remainder R_1, if any, in the first column of the B row. (The value of R_1 will be 1 or 0 as D is either odd or even.)

TABLE 11-1 ARRAY FOR DECIMAL-TO-BINARY CONVERSION

COLUMN $k + 1$	COLUMN k	COLUMN 2	COLUMN 1	
0	$D_k = D_{k-1}/2$	$D_2 = D_1/2$	$D_1 = D/2$	Decimal No. D (D row)
0	R_k	R_2	R_1	Binary No. B (B row)

4 Divide D_1 by 2 and place quotient D_2 in column two of the D row.

5 Place the remainder R_2, a 0 or 1, in column 2 of the B row.

6 Repeat 4 and 5 until the resultant quotient is zero. The digits in the B row, when read left to right, are the binary representation of the decimal number D.

Example 11-2

Convert the decimal number 73 into binary.

SOLUTION   Set up an array similar to Table 11-1 as shown in Table 11-2.
    The binary representation of 73 is the 7-bit number 1001001. This result is readily checked as

$$1001001 = 1 \times 2^6 + 0 \times 2^5 + 0 \times 2^4 + 1 \times 2^3 + 0 \times 2^2 + 0 \times 2^1 + 1 \times 2^0$$
$$= 64 + 8 + 1 = 73$$

TABLE 11-2 ARRAY FOR EXAMPLE 11-2

8	7	6	5	4	3	2	1	
0	$\frac{1}{2} = 0$	$\frac{2}{2} = 1$	$\frac{4}{2} = 2$	$\frac{9}{2} = 4$	$\frac{18}{2} = 9$	$\frac{36}{2} = 18$	$\frac{73}{2} = 36$	$D = 73$
0	1	0	0	1	0	0	1	B

The extension of binary numbers to fractions, negative numbers, and exponential representation is treated in Sec. 12-2 in relation to binary arithmetic.

11-8 LOGIC GATES

The aspect of control of the data processing in digital systems is of paramount importance. Primarily, the control unit fulfills four functions:

1 To determine which data is used in a specific arithmetic operation.

2 To determine the nature of the arithmetic process used, i.e., addition, subtraction, etc.

3 To determine the sequence of arithmetic operations needed.

4 To determine how the output of an arithmetic operation is to be used. That is, is it to be stored in memory? Is it to be used in another arithmetic operation? etc.

Each of the four functions can be considered as the instructions that must be followed in order to process the data to achieve a desired objective. In general-purpose digital computers, the instructions are introduced by means of the program. Special-purpose computers used in specific applications are often programmed internally. That is, since only certain sequences of operations are needed, the instructions are permanently designed within the computer. The "inboard computer" in a space capsule and microprocessors used in many industrial control systems are typical examples of special-purpose computers. In other types of digital systems, the control function is integrated into the overall system, and external programs may be unnecessary.

In all electronic digital systems, the instructions are converted to electrical signals. To ensure that the desired processing occurs, the set of instructions must be consistent, i.e., they must follow a logical sequence from the given data to the desired result in much the same manner as a student solves a homework problem. Self-consistency is achieved in digital systems by the use of mathematical logic. Because two-state electronic switches are convenient circuits, and for compatibility with the representation of the data, the binary system is also used to obtain logical consistency.

Boolean algebra is the two-state (binary) symbolic logic used in digital systems. A Boolean variable A assumes only one of two permissible values, 1 or 0. Thus, A may be 1 ($A = 1$) or A may be 0 ($A = 0$); if A is not 1, A must be 0. In Boolean algebra, only three basic functions, called AND, OR, and NOT, are required for logical consistency.

A logic gate, or simply a *gate,* is an electronic circuit used to implement a Boolean function. Most often, gates are complex electronic switches which have, as component parts, switches of the type described in Sec. 11-3. One type of logic gate is used to implement each of the basic Boolean functions. Combinations of these gates then serve to implement more complex logical functions.

The first of these is the *AND gate* whose standard symbol is shown in Fig. 11-31. Symbolically, the AND operation is written as

$$Y = A \cdot B = A \text{ AND } B \qquad (11\text{-}13)$$

where A, B, and Y are Boolean variables. Often the "dot" is omitted and the AND operation is written as $Y = AB$. As indicated in Fig. 11-31, A and B are the input variables, and Y is the output variable. The meaning of the AND statement is that $Y = 1$ only when A and B are both 1, and is demonstrated by the waveforms shown in Fig. 11-32. The $V(0)$ and $V(1)$ amplitude levels in Fig. 11-31 correspond to the binary values 0 and 1. An output pulse ($Y = 1$)

Inputs   Output

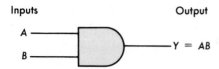

$Y = AB$

FIGURE 11-31 AND gate symbol.

occurs only during the interval of time for which both input pulses exist ($A = B = 1$). At all other times, the output is at the 0 level. Thus, in order to produce an output pulse, the input pulses of an AND gate must be in time coincidence, a property often used in time-selection circuits.

The waveforms shown in Fig. 11-32 indicate the use of *positive logic;* that is, the larger voltage corresponds to the binary 1, the smaller voltage to binary 0. *In negative logic* implementations, depicted in Fig. 11-33, the reverse is true. The actual voltage values have no significance in determining

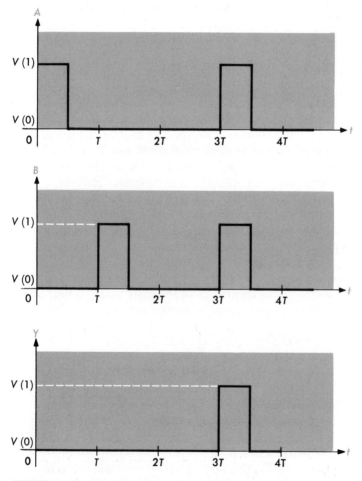

FIGURE 11-32 Waveforms for two-input AND gate (positive-logic).

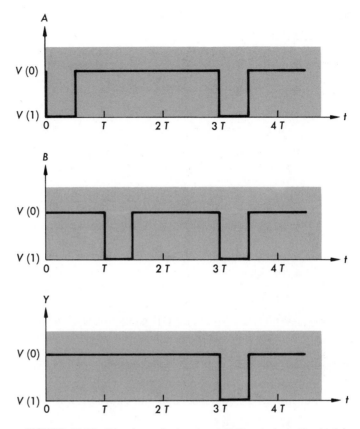

FIGURE 11-33 Waveforms for two-input AND gate (negative-logic).

logic levels. All that is necessary is that the two levels be readily distinguished.

The operation of an AND gate is also demonstrated in Table 11-3, which is called a *truth table*. The truth table is a list of all possible input combinations and the corresponding output that results from each input. Three useful identities for the AND operation, given in Eq. (11-14), are derived from Table 11-3.

$$A \cdot 0 = 0 \quad A \cdot 1 = A \quad A \cdot A = A \tag{11-14}$$

TABLE 11-3 TRUTH TABLE FOR A TWO-INPUT AND GATE

INPUT STATES		OUTPUT STATE
A	B	Y
0	0	0
1	0	0
0	1	0
1	1	1

Inputs   Output

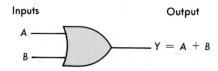

$Y = A + B$

FIGURE 11-34 OR gate symbol.

The symbol in Fig. 11-34 is that of an *OR gate* and is symbolically written as

$$Y = A + B = A \text{ OR } B \tag{11-15}$$

The OR operation is an inclusive statement requiring either $A = 1$ or $B = 1$ or both $A = B = 1$ for Y to be in its 1 state. The truth table for a two-input OR gate is given in Table 11-4, which is demonstrated by the waveforms of Fig. 11-35 for positive logic. As shown in Fig. 11-35, an output pulse occurs in any time interval in which either input pulse is present. By means of Table 11-4, the identities in Eq. (11-16) can be derived.

$$A + 0 = A \quad A + 1 = 1 \quad A + A = A \tag{11-16}$$

While the AND and OR gates are shown for only two-input variables, most gates have several inputs. The basic relations for multiple inputs are extended directly from the two-input case.

The third basic gate is referred to as a *NOT gate* or *INVERTER*. Its symbol for an inverted input is shown in Fig. 11-36*a* and for an inverted output in Fig. 11-36*b*. The Boolean expression for an inverter is

$$Y = \overline{A} = \text{NOT } A \tag{11-17}$$

The function of the NOT gate is logical negation; the bar over the A in Eq. (11-17) is used to indicate negation. Table 11-5 is the truth table for the *INVERTER*. The simple transistor switch of Example 11-1 whose waveform is shown in Fig. 11-11 can be thought of as a NOT gate. Equations (11-18) are identities based on logical negation.

$$A \cdot \overline{A} = 0 \quad A + \overline{A} = 1 \tag{11-18}$$

Most processes in computers involve combinations of logical functions. One may describe rules for the combination of logical functions in a manner similar to the associative, distributive, and commutative properties of ordi-

TABLE 11-4 TRUTH TABLE FOR TWO-INPUT OR GATE

INPUT STATES		OUTPUT STATE
A	B	Y
0	0	0
1	0	1
0	1	1
1	1	1

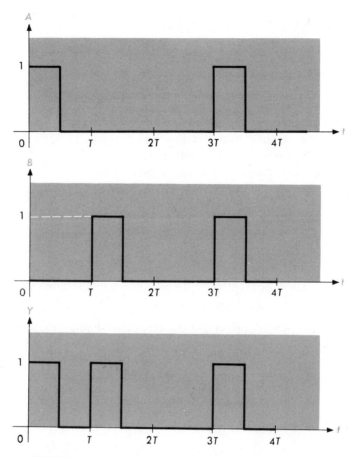

FIGURE 11-35 Waveforms for two-input OR gate (positive-logic).

nary algebraic variables. Several useful theorems, stated without proof, that result from these properties are given in Eqs. (11-19) to (11-22).

$$A \cdot (B + C) = A \cdot B + A \cdot C \tag{11-19}$$

$$A + B \cdot C = (A + B) \cdot (A + C) \tag{11-20}$$

$$\overline{(A + B)} = \overline{A} \cdot \overline{B} \tag{11-21}$$

$$\overline{(A \cdot B)} = \overline{A} + \overline{B} \tag{11-22}$$

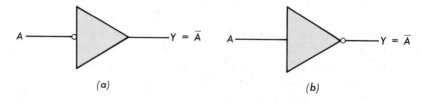

(a)   *(b)*

FIGURE 11-36 NOT gate symbols: *(a)* inverted input; *(b)* inverted output.

TABLE 11-5 INVERTER TRUTH TABLE

INPUT STATE A	OUTPUT STATE Y
0	1
1	0

The results stated in Eqs. (11-21) and (11-22) are known as *DeMorgan's theorems,* which, together with Eqs. (11-19) and (11-20), are useful in reducing complex Boolean logic functions.

Example 11-3

Reduce the logic function $Y = \overline{A}(\overline{B} + C) + A\overline{B}C$ to its simplest form.

SOLUTION By use of Eqs. (11-21) and (11-22), the statement is reduced to

$$\overline{Y} = (A + \overline{B}\overline{C})(\overline{A} + B + \overline{C})$$

From Eq. (11-19) and the identities in Eqs. (11-14), (11-16), and (11-18),

$$\overline{Y} = AB + A\overline{C} + B\overline{C}$$

Again, by the use of DeMorgan's theorems

$$Y = (\overline{A} + \overline{B})(\overline{A} + C)(\overline{B} + C)$$

which, by the identities in Eqs. (11-14), (11-16), (11-18), and (11-19), reduces to

$$Y = \overline{B}(\overline{A} + C) + \overline{A}C$$

The logic function expressed by Y is shown in Fig. 11-37a in which AND, OR, and INVERTER gates are used. The function Y can also be written, upon application of Eqs. (11-21) and (11-22), as

$$Y = (\overline{\overline{Y}}) = \overline{[B(A + \overline{C}) + A\overline{C}]}$$

for which the circuit in Fig. 11-37b is a realization.

Several other gate circuits are widely used because their circuit realizations are simple and compatible with standard technologies. Each of these can be represented by combinations of the three basic gates. The *NAND gate* is an AND gate with an inverted output and may be realized as shown in Fig. 11-38a. The standard NAND gate symbol is given in Fig. 11-38b. The schematic in Fig. 11-39a is that of a *NOR gate,* which is the cascade of an OR gate and an INVERTER. The common NOR gate symbol is depicted in Fig. 11-39b. The *INHIBITOR* is an AND gate in which one input is negated. Figure 11-40a is the realization of the INHIBITOR by an AND gate and a NOT gate, and Fig. 11-40b is its standard symbol.

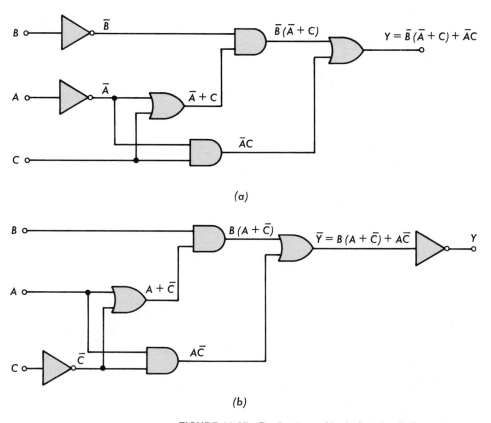

(a)

(b)

FIGURE 11-37 Realizations of logic function in Example 11-3.

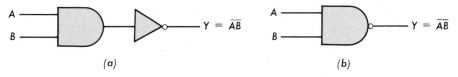

(a)   (b)

FIGURE 11-38 NAND gate (a) realization and (b) symbol.

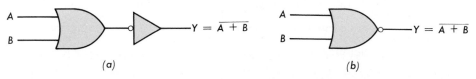

(a)   (b)

FIGURE 11-39 NOR gate (a) realization and (b) symbol.

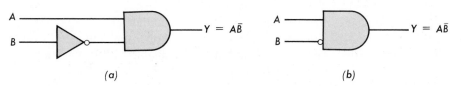

(a)   (b)

FIGURE 11-40 INHIBITOR realization and symbol.

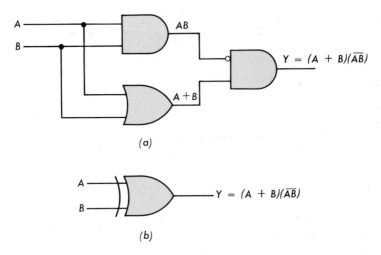

FIGURE 11-41 EXCLUSIVE-OR gate realization and symbol.

The *EXCLUSIVE OR gate,* depicted in Fig. 11-41*a*, is used to realize the condition *A* or *B* but not *A* and *B*. The symbol for the EXCLUSIVE OR gate is shown in Fig. 11-41*b*; it is functionally expressed as

$$Y = A \oplus B = (A + B) \cdot \overline{(A \cdot B)} \tag{11-23}$$

The symbol $\oplus$ is used to denote the EXCLUSIVE OR operation and distinguish it from the OR symbol.

11-9 LOGIC-GATE REALIZATIONS

The variety of integrated circuits available to process digital signals can contain a few to several thousand components on a single chip. One convenient method of digital IC classification, based on the number of components per chip, is as follows:

1 *S*mall-*s*cale *i*ntegration (SSI); contains fewer than 100 components.

2 *M*edium-*s*cale *i*ntegration (MSI); contains 100 to 1,000 components.

3 *L*arge-*s*cale *i*ntegration (LSI); contains 1,000 to 10,000 components.

4 *V*ery *l*arge-*s*cale *i*ntegration (VLSI); contains more than 10,000 components.

Digital computers and other large digital systems utilize mainly LSI and MSI chips to reduce cost and increase reliability and operating efficiency. However, most of the same circuit functions are available as SSI chips and are used as components in numerous applications. At present there are more than a dozen different techniques to implement integrated circuit logic gates.

The ensuing discussion is confined to four of the most widely used types, two of which utilize bipolar technology and two employ MOS technology.

Speed, power consumption, and the number of gates on the chip are three of the more important characteristics by which digital ICs are compared. In general, MOS fabrication achieves a higher density of gates per chip than bipolar devices permit because no isolation islands are needed between adjacent devices (see Sec. 7-12). However, MOS devices operate at inherently slower speeds than bipolar devices. As is the usual case, the engineer's decision, in a given application, is based on a trade-off between desirable characteristics and their accompanying undesirable ones.

MOS integrated circuit chips employ the NAND, NOR, and INVERTER gates as the three basic logic elements. The MOS switch, depicted in Fig. 11-14 and described in Sec. 11-3, is the INVERTER gate. The MOS load transistor, eliminating the need for a resistor and conserving valuable chip area, is also used in the positive-logic NAND and NOR gates shown in Fig. 11-42a and b, respectively. In both circuits, $Q3$ is the load transistor and

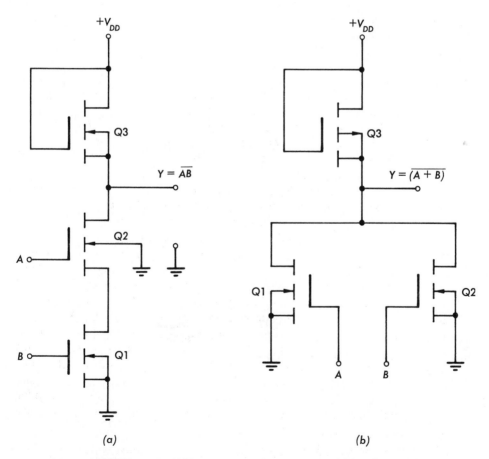

FIGURE 11-42 MOS realizations of positive-logic (a) NAND and (b) NOR gates.

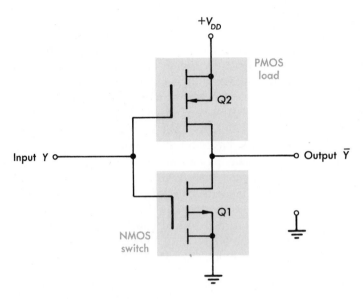

FIGURE 11-43 CMOS INVERTER circuit.

$Q1$ and $Q2$ are the switching transistors. In the NAND gate of Fig. 11-42*a*, transistors $Q1$ and $Q2$ are essentially in series, so that if either input is at $V(0)$, the device is cut off and behavior is similar to the INVERTER of Fig. 11-14. The transistors $Q1$ and $Q2$ are parallel-connected in the NOR gate in Fig. 11-42*b*. Application of $V(1)$ to either input produces a load current and the output Y becomes $V(0)$.

The circuits shown in Fig. 11-42 have a *fan-in* of two; that is, two input signals can be applied. Another important measure of gate performance is the *fan-out*, or number of gates the output is capable of driving.

Complementary MOS (CMOS) gates significantly reduce power consumption (ideally it is zero), and are insensitive to temperature variations. However, this is achieved at the expense of substantial increase in the chip area required for the gate. A typical CMOS INVERTER is shown in Fig. 11-43. Because *p*-channel (PMOS) and *n*-channel (NMOS) devices respond to opposite input polarities, one device is off when either $V(1)$ or $V(0)$ is applied to the input. An input signal $V(0)$ cuts off the NMOS transistor causing the output to be $V_{DD} = V(1)$. The PMOS load transistor is cut off for an input $V(1)$ which, in turn, makes the current in $Q1$ zero. Consequently, the output is at logical 0.

CMOS, NAND and NOR gates can be implemented in a manner similar to the circuits in Fig. 11-42. For the NAND gate the NMOS transistors are series-connected and the PMOS loads are placed in parallel. The reverse connections apply for the NOR gate.

Transistor-transistor logic (TTL or T²L) is the most widely used logic family fabricated with bipolar transistors. The use of multiple-emitter tran-

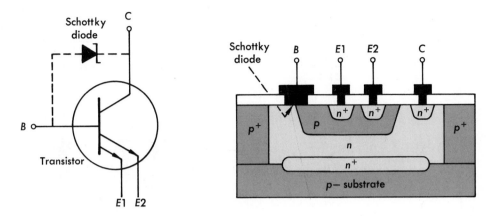

FIGURE 11-44 Multiple-emitter transistor (*a*) symbol and (*b*) its structural representation.

sistors, symbolically and structurally represented in Fig. 11-44, increases gate density and improves switching speed. The NAND gate is the basic logic function implemented and is shown for a fan-in of two in Fig. 11-45. Invariably, TTL chips are operated with $V_{CC} = 5$ V.

The multiple-emitter transistor in Fig. 11-45 can be considered to perform the AND operation. Transistor $Q1$ acts as a buffer between Q and the INVERTER circuit containing $Q2$. The circuit is then analogous to the realization shown in Fig. 11-38 with the INVERTER of the type depicted in Fig. 11-11*a*. When either (both) input signal is $V(0)$, the transistor Q is saturated

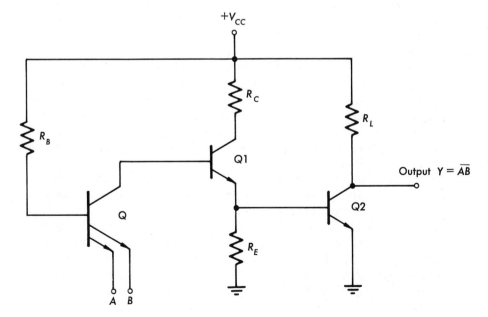

FIGURE 11-45 TTL NAND gate realization.

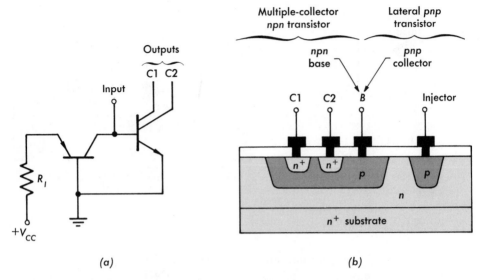

FIGURE 11-46 (a) Integrated-injection logic (I²L) INVERTER circuit and (b) cross section of IC chip.

as both junctions are forward-biased. The resultant collector voltage $V_{CE(sat)}$ at Q is insufficient to cause $Q1$ to conduct. Consequently, the emitter voltage at $Q1$ is zero, $Q2$ is cutoff, and the output is $V(1)$. With both inputs at $V(1)$, the emitter-base junction is reverse-biased. However, the collector-base junction remains forward-biased and Q acts as an *inverted transistor* (the roles of collector and emitter are reversed). The current produced by the inverted transistor saturates $Q1$, which, in turn, saturates $Q2$ and causes an output of $V(0)$.

The transistors in standard TTL gates are saturated when conducting. *Storage time* is the time required for the device to enter the active region from saturation; physically it is the time required to remove the excess carriers in the base that are injected by the collector. The effect of storage time is to increase the time required to switch from one state to the other, thereby reducing the speed of the gate. Often, a *Schottky diode* is used to prevent the transistor from becoming saturated. A Schottky diode is a metal-semiconductor diode whose characteristics are similar to those of a *p-n* junction diode and, as illustrated in Fig. 11-44, fabrication is compatible with TTL technology. Currently, this combination is in widespread use in commercially available TTL chips.

A major LSI logic family based on bipolar technology is integrated injection logic (I²L). Whereas TTL gates make use of multiple-emitter transistors, I²L employs multiple-collector devices, which results in an increased gate density on the chip. The circuit in Fig. 11-46a is the basic I²L INVERTER gate for which a cross section of the chip is displayed in Fig. 11-46b. The multiple-collector transistor Q acts as the INVERTER and the function of the lateral *pnp* transistor $Q1$ is to provide the base-bias current for Q. The injector current level is established by V_{CC} and the resistor R_I

which are external to the chip. When the input is set to $V(0)$, for positive logic, transistor Q is cut off so that the outputs at each of its collectors is at $V(1)$. For this condition the injector current exists in the loop formed by the *pnp* transistor and the input. Application of $V(1)$ at the input is sufficient to saturate Q, producing $V(0)$ at the output. A NAND gate is readily constructed from two INVERTER gates by connecting their respective collectors in parallel.

One advantage of I²L fabrication is that the speed of operation is controlled by the value of injector current. Slower speed is accompanied by low levels of injector current and a subsequent decrease in power dissipation. Similarly, high-speed operation occurs only when the power consumption is increased. As the injector current is controlled by the value of the external resistor R_1, the speed of operation of I²L chips can be controlled externally.

In the following chapter most of the major functional elements which comprise a digital computer can be constructed using any of the logic families described in this section.

PROBLEMS

11-1 The signal source $v_s(t)$ in the circuit of Fig. 11-47 is $15 \sin 10^3 t$. Sketch the output voltage v_o as a function of t for one cycle indicating appropriate amplitude and time values.

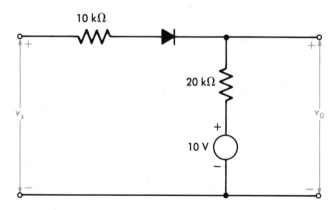

FIGURE 11-47 Circuit for Prob. 11-1.

11-2 Repeat Prob. 11-1 if the diode in Fig. 11-47 is connected in reverse.

11-3 Repeat Prob. 11-1 if the polarity of the 10-V source is reversed.

11-4 The input voltage v_s in the circuit of Fig. 11-48 is $100 \sin 100t$.
a Sketch one cycle of the output voltage $v_o(t)$ assuming that the diodes are ideal.
b Determine the maximum instantaneous diode currents.

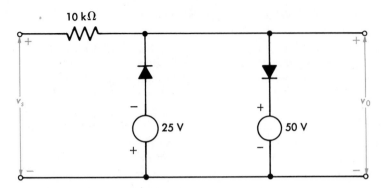

FIGURE 11-48 Circuit for Prob. 11-4.

11-5 Repeat Prob. 11-4 for the circuit in Fig. 11-49.

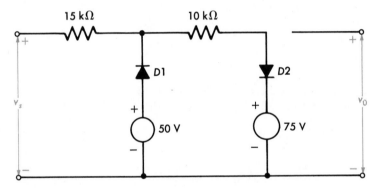

FIGURE 11-49 Circuit for Prob. 11-5.

11-6 The voltage v_s in Prob. 11-4 is applied to the circuit in Fig. 11-50.
a Sketch one cycle of the output voltage assuming the diodes are ideal.
b Determine the maximum current in each of the diodes.

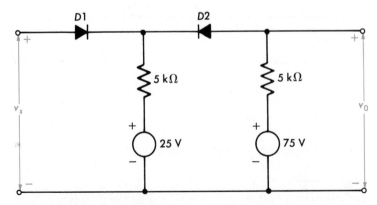

FIGURE 11-50 Circuit for Prob. 11-6.

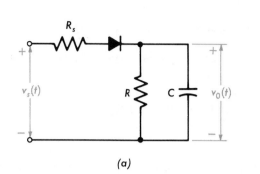

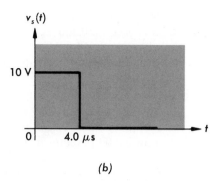

FIGURE 11-51 (a) Circuit and (b) input waveform for Prob. 11-7.

11-7 Sketch the waveform for $v_o(t)$ in the circuit of Fig. 11-51a. The input signal is shown in Fig. 11-51b and the element values are $R_s = 5$ kΩ, $R = 20$ kΩ, and $C = 1,000$ pF. Assume that the diode is ideal.

11-8 Repeat Prob. 11-7 for $R = 1.25$ kΩ and all other element values as given in Prob. 11-7.

11-9 The input signal $v_s(t)$ in the circuit of Fig. 11-52 is a 15-V pulse whose duration is 100 ns.
a Sketch the waveform of $v_o(t)$.
b Describe the effect that halving the input pulse duration has on the output amplitude and duration.

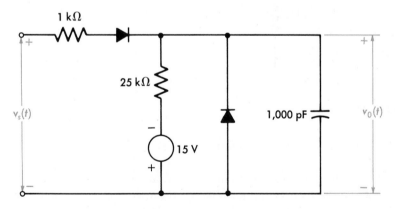

FIGURE 11-52 Circuit for Prob. 11-9.

11-10 A silicon transistor whose characteristics are shown in Fig. 11-12c is used in the switch depicted in Fig. 11-11a. The circuit element values are $R_C = 2$ kΩ, $R_s = 1$ kΩ, and $V_{CC} = 6$ V.
a Determine the minimum value of V_s necessary to cause saturation.
b If the value of h_{FE} of the transistor is halved, what is the new value of V_s required to just cause saturation?

11-11 A transistor switch is designed with $V_{CC} = 10$ V and $R_L = 500 \ \Omega$.
a For $h_{FE} = 50$ and $V_{BE} = 0.6$ V, determine R_s so that the transistor is just barely saturated for $V_s = 10$ V.
b Determine the new values of base and collector currents for $h_{FE} = 100$.

11-12 The transistor whose characteristics are displayed in Fig. 11-12c is to be used as a switch to provide a 5-mA current pulse. The values of V_{CC} and V_s are each 5 V. Determine the values of R_L and R_s needed if the base current is 200 percent of its minimum value for saturation.

11-13 A transistor having an h_{FE} at saturation of 25, $V_{BE} = 0.7$ V, and $V_{CE(\text{sat})} = 0.3$ V is used in the circuit of Fig. 11-53.

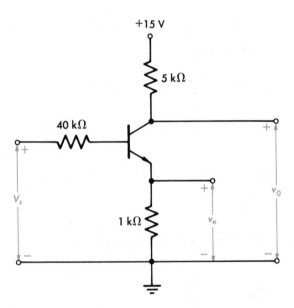

FIGURE 11-53 Circuit for Prob. 11-13.

a Determine the minimum value of V_s necessary to barely saturate the transistor.
b A pulse whose amplitude is the value of V_s determined in a and whose duration is 1 ms is applied. Sketch the waveforms that result for v_o and v_e.

11-14 The transistor described in Prob. 11-13 is used in the circuit of Fig. 11-54. The input signal is a pulse of amplitude -10 V and whose duration is 100 μs.
a Determine the value of R_s required to barely saturate the transistor.
b Sketch the output voltage v_o.

11-15 The switch shown in Fig. 11-13a is designed using the MOSFET

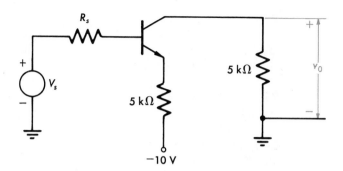

FIGURE 11-54 Circuit for Prob. 11-14.

whose characteristics are given in Fig. 7-43 for $V_{DD} = 15$ V and $R_D = 450$ Ω.
a Determine the input signal amplitude needed to effect switching.
b Sketch the output voltage if the input voltage is a pulse of duration 1 ms and whose amplitude is obtained in *a*.

11-16 The switch shown in Fig. 11-14*a* uses the MOSFET whose characteristics are depicted in Fig. 7-44. The input signal is a pulse whose amplitude is 10 V and whose duration is 0.5 ms. The drain supply is 10 V. Determine the value of R_D and sketch the output waveform.

11-17 Repeat Prob. 11-15 for $R_D = 1.5$ kΩ, all other values remaining as given.

11-18 Repeat Prob. 11-16 for $V_{DD} = 15$ V, all other values remaining as given.

11-19 The element values in the circuit of Fig. 11-17 are $V = 1{,}000$ V, $R_1 = 10$ kΩ, $R_2 = 100$ Ω, and $C = 0.01$ μF. At $t = 0$ the switch is placed in position 1 and remains there for 100 μs. The switch is then placed in position 2 for 3 μs and the cycle is repeated. Determine the sweep amplitude and sketch the output waveform.

11-20 The operational amplifier in the circuit of Fig. 11-19 has a gain of 10^5. The supply voltage V is 10 V, $R = 1$ kΩ, and $C = 100$ pF. The trigger pulse duration is 100 μs. Sketch the output waveform for a 200-μs interval.

11-21 The bistable circuit in Fig. 11-21 is designed with $R_L = 5$ kΩ, $R_1 = 20$ kΩ, and $R_2 = 30$ kΩ and uses supply voltages $V_{CC} = 6$ V and $V_{BB} = 3$ V. The transistors are assumed ideal with $h_{FE} = 24$. Determine:
a The collector current when the transistor conducts.
b The base current in the ON transistor.
c The trigger amplitude necessary to turn the OFF transistor to the ON state.

11-22 In the bistable circuit in Fig. 11-21, the supply voltages are $V_{CC} =$ 5 V and $V_{BB} = 5$ V. The collector current in the ON state is to be 10 mA. Assume ideal transistors with an $h_{FE} = 20$. When cut off, a reverse bias of 1 V exists across the emitter-base junction. Find R_1, R_2, and R_L.

11-23 Repeat Prob. 11-22 if the transistors are characterized by $V_{CE(sat)} =$ 0.3 V and $V_{BE} = 0.6$ V.

11-24 The monostable multivibrator of Fig. 11-23 is designed with element values $R_L = 1$ kΩ, $R_1 = R_2 = 20$ kΩ, $R = 60$ kΩ, and $C = 100$ pF. The supply voltages are 6 V each. The transistors may be assumed ideal with $h_{FE} = 60$. Determine the amplitude and duration of the current pulse in R_L.

11-25 The circuit in Fig. 11-23 is to be designed to have an output pulse of 1-μs duration and 10-mA amplitude. The transistors may be assumed ideal having an $h_{FE} = 20$. When OFF, a reverse-bias of 2 V exists across the emitter-base junction and when ON, the transistors are just barely saturated. Determine R_1, R_2, C, R_L, and R.

11-26 The circuit in Fig. 11-25 is designed to provide a square-wave whose period is 5 μs and whose current amplitude is 5 mA using a supply voltage of 15 V. The transistors are ideal having $h_{FE} = 50$ and are just barely saturated when in the ON state. Find R, R_L, and C.

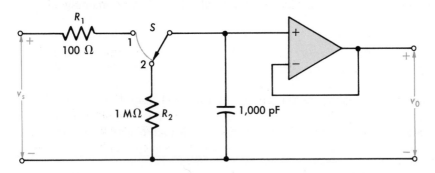

FIGURE 11-55 Circuit for Prob. 11-27.

11-27 The circuit shown in Fig. 11-55 is an equivalent representation of the sample-and-hold circuit depicted in Fig. 11-29. The switch S is clocked to position 1 every 100 μs and remains there for a 1-μs duration. At all other times it is in position 2. The resistance R_1 is used to indicate the combination of the output resistance of the signal-source voltage v_s and the ON resistance of the switch. The OFF resistance of a practical switch is represented by R_2. The signal v_s is generally the output of the input voltage-follower stage (not shown). Sketch the output voltage for one cycle if the input signal is 10 sin $10^3 t$. (*Note:* Only 10 sampling intervals are required.)

11-28 Convert the following decimal numbers to binary numbers: 35, 43, 72, 255, 535, and 999.

11-29 Convert the following decimal numbers to binary numbers: 38, 82, 127, 270, 360, and 1066.

11-30 Convert the following binary numbers to decimal numbers: 1010, 11011, 100110, 110011, 1101101, and 10101010.

11-31 Convert the following binary numbers to decimal numbers: 1101, 10110, 110101, 101100, 1010010, and 11010101.

11-32 Simplify the following Boolean expressions and realize each of them using combinations of AND, OR, and NOT gates.
a $(\overline{A} + B)(A + \overline{C})(\overline{B} + \overline{C})$
b $A\overline{C} + ABC + \overline{AB(\overline{C} + D)}$
c $\overline{AB} + A\overline{B}C + \overline{(A + B + \overline{C})}$

11-33 Repeat Prob. 11-32 for the following functions:
a $\overline{(A\overline{B} + \overline{A}C + BC)}$
b $(AB + CD)(AC + BD)(AD + BC)$
c $(A + \overline{B})(\overline{A} + B + \overline{C})(\overline{A} + \overline{B} + C)$

11-34 Simplify the following Boolean expressions and realize each of them using combinations of AND, OR, and NOT gates:
a $(A + B)\overline{C} + (B + C)\overline{A} + (A + C)\overline{B}$
b $ABC + D(\overline{A} + B) + \overline{C}(ABD)$
c $\overline{(AB + CD)(A\overline{C} + BD)(AD + \overline{BC})}$

11-35 Construct the truth table for a three-input AND gate.

11-36 Construct the truth table for a three-input OR gate.

11-37 Repeat Prob. 11-35 for a three-input NAND gate.

11-38 Repeat Prob. 11-36 for a three-input NOR gate.

11-39 The waveforms displayed in Fig. 11-56 are the inputs to an OR gate. Sketch the output waveform.

11-40 Repeat Prob. 11-39 if the waveforms in Fig. 11-56 are the inputs to an AND gate.

11-41 The inputs to a NAND gate are given in Fig. 11-56. Sketch the output waveform.

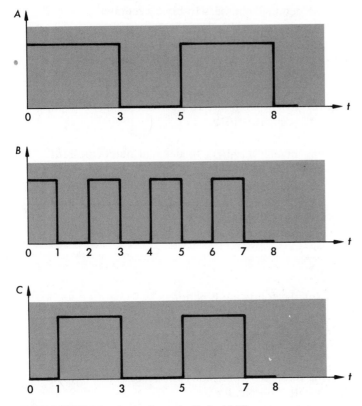

FIGURE 11-56 Input waveforms for Prob. 11-39.

11-42 Repeat Prob. 11-41 if the waveforms in Fig. 11-56 are the inputs to a NOR gate.

11-43 Determine the output function Y for the logic circuit in Fig. 11-57.

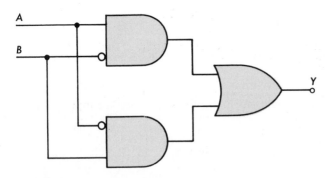

FIGURE 11-57 Logic circuit for Prob. 11-43.

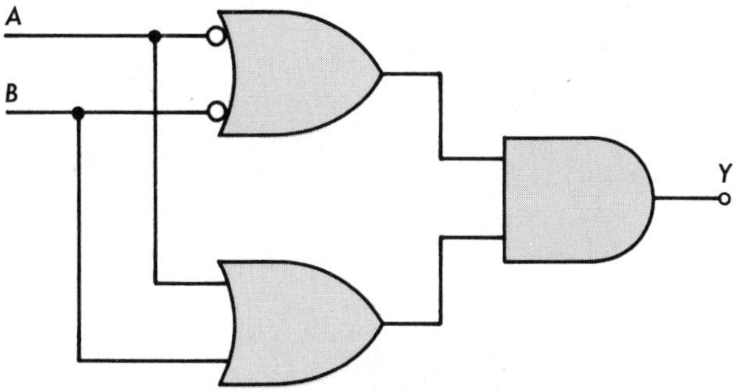

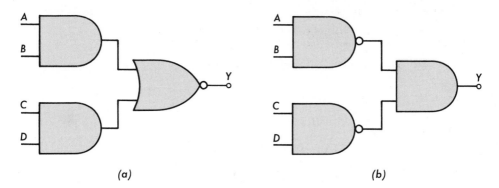

FIGURE 11-58 Circuit for Prob. 11-44.

11-44 Repeat Prob. 11-43 for the logic circuit in Fig. 11-58.

11-45 The logic circuits in Fig. 11-59 are two possible realizations of a commonly fabricated combinatorial logic function called the AND-OR-INVERT gate (AOI). Determine the output function Y for each circuit and show they are equivalent.

(a) *(b)*

FIGURE 11-59 Two realizations of AND-OR-INVERT gate for Prob. 11-45.

11-46 The seven-segment array, shown in Fig. 11-60a, is widely used to form the decimal digits 0 to 9 in LED displays, as indicated in Fig. 11-60b. Develop the logical expression for which the segment Y_1 is turned ON when the inputs are the 4 binary digits used to represent the decimal digits 0 to 9. Indicate one possible logic circuit which realizes the output Y_1.

11-47 Repeat Prob. 11-46 for the segment Y_2.

11-48 Repeat Prob. 11-46 for the segment Y_3.

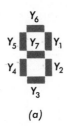

(a)

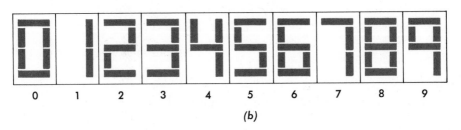

(b)

FIGURE 11-60 *(a)* Seven-segment array and *(b)* its use in forming the decimal digits 0-9.

11-49 Repeat Prob. 11-46 for the segment Y_4.

11-50 Repeat Prob. 11-46 for the segment Y_5.

11-51 Repeat Prob. 11-46 for the segment Y_6.

11-52 Repeat Prob. 11-46 for the segment Y_7.

DIGITAL COMPUTATION □ **DIGITAL**
PUTATION □ DIGITAL **COMPUTATION**
DIGITAL COMPUTATION □ DIGITAL C
COMPUTATION □ DIGITAL COMPUTA
DIGITAL COMPUTATION □ DIGITAL C
COMPUTATION □ DIGITAL COMPUTA
DIGITAL COMPUTATION □ DIGITAL C
COMPUTATION □ DIGITAL COMPUTA
DIGITAL COMPUTATION □ DIGITAL C
COMPUTATION □ DIGITAL COMPUTATION □ DIGITAL COMPUTATIOI
DIGITAL COMPUTATION □ DIGITAL COMPUTATION □ DIGITAL COMF
COMPUTATION □ DIGITAL COMPUTATION □ DIGITAL COMPUTATIOI
DIGITAL COMPUTATION □ DIGITAL COMPUTATION □ DIGITAL COMF
COMPUTATION □ DIGITAL COMPUTATION □ DIGITAL COMPUTATIOI
DIGITAL COMPUTATION □ DIGITAL COMPUTATION □ DIGITAL COMF
COMPUTATION □ DIGITAL COMPUTATION □ DIGITAL COMPUTATIOI
DIGITAL COMPUTATION □ DIGITAL COMPUTATION □ DIGITAL COMF
COMPUTATION □ DIGITAL COMPUTATION □ DIGITAL COMPUTATIOI
DIGITAL COMPUTATION □ DIGITAL COMPUTATION □ DIGITAL COMF
COMPUTATION □ DIGITAL COMPUTATION □ DIGITAL COMPUTATIOI
DIGITAL COMPUTATION □ DIGITAL COMPUTATION □ DIGITAL COMF
COMPUTATION □ DIGITAL COMPUTATION □ DIGITAL COMPUTATIOI
DIGITAL COMPUTATION □ DIGITAL COMPUTATION □ DIGITAL COMF
COMPUTATION □ DIGITAL COMPUTATION □ DIGITAL COMPUTATIOI
DIGITAL COMPUTATION □ DIGITAL COMPUTATION □ DIGITAL COMF
COMPUTATION □ DIGITAL COMPUTATION □ DIGITAL COMPUTATIOI
DIGITAL COMPUTATION □ DIGITAL COMPUTATION □ DIGITAL COMF
COMPUTATION □ DIGITAL COMPUTATION □ DIGITAL COMPUTATIOI
DIGITAL COMPUTATION □ DIGITAL COMPUTATION □ DIGITAL COMF
COMPUTATION □ DIGITAL COMPUTATION □ DIGITAL COMPUTATIOI
DIGITAL COMPUTATION □ DIGITAL COMPUTATION □ DIGITAL COMF
COMPUTATION □ DIGITAL COMPUTATION □ DIGITAL COMPUTATIOI
DIGITAL COMPUTATION □ DIGITAL COMPUTATION □ DIGITAL COMF
COMPUTATION □ DIGITAL COMPUTATION □ DIGITAL COMPUTATIOI
DIGITAL COMPUTATION □ DIGITAL COMPUTATION □ DIGITAL COMF
COMPUTATION □ DIGITAL COMPUTATION □ DIGITAL COMPUTATIOI
DIGITAL COMPUTATION □ DIGITAL COMPUTATION □ DIGITAL COMF
COMPUTATION □ DIGITAL COMPUTATION □ DIGITAL COMPUTATIOI
DIGITAL COMPUTATION □ DIGITAL COMPUTATION □ DIGITAL COMF
COMPUTATION □ DIGITAL COMPUTATION □ DIGITAL COMPUTATIOI
DIGITAL COMPUTATION □ DIGITAL COMPUTATION □ DIGITAL COMF
COMPUTATION □ DIGITAL COMPUTATION □ DIGITAL COMPUTATIOI

12

he digital computer is used, to an ever increasing degree, as an integral component of the systems with which engineers and scientists are involved. Further impetus in this direction results from the introduction of the microprocessor, particularly in control and instrumentation systems. The function of this chapter is to introduce the building blocks which comprise a digital computer and briefly describe their salient features. Included are those elements which are required to implement arithmetic, store data, and control the computation process.

As most physical quantities, such as temperature, pressure, acceleration, etc., are analog quantities analog-to-digital and digital-to-analog converters are treated. These are used to illustrate the interface between the computer and the analog components in the system.

12-1 COMPUTER ARCHITECTURE

Charles Babbage, a mid-nineteenth-century English mathematician, failed in his attempt to construct a mechanical digital computer, called the analytic engine, because the necessary technology was unavailable. However, Babbage's machine contained the five essential elements we identify as comprising the modern, stored-program electronic computer. The five constituent elements are input and output devices (I/O), memory, and control and arithmetic units. A stored-program computer refers to the fact that the operations and instructions are entered and stored in the machine's memory. The programs we write to solve problems are a set of instructions which are translated by the machine into a set of basic operations it can perform.

The function of the input devices is to enter both the external instructions (program) and the data to be processed into the computer memory. As the computer operates on binary digits, or binary-derived number systems, the input devices must enter the information in this form. Typical input devices are punched cards and tapes, magnetic tapes, keyboards, and acoustic couplers (dataphones). In digital control and communication systems, analog-to-digital converters are often used to enter the data into the computer.

Cathode-ray tubes (CRTs), paper printers, and LED (light-emitting diode) displays are among the common output devices. Their function is to present the processed data in a form convenient to the user. When the output of the computer is required to actuate control and communication systems, digital-to-analog converters are often utilized.

The arithmetic unit is responsible for performing the calculations required. Most computation is based on binary addition making the adder a principal arithmetic circuit function. As binary codes are employed for both arithmetic and logical operations, both are usually performed by a single unit called the *arithmetic logic unit* or *ALU*.

Two classes of memory are usually employed to store instructions and data. One class contains information that is permanently stored and which can be "read" from memory when needed. In general, this information cannot be altered once it is entered in memory and, as such, is referred to as *read-only memory* (*ROM*). Dedicated computers, i.e., those designed to perform a specific task, have their programs stored in ROMs. *Read-write memories* permit data to be entered (written) or extracted (read) as required and the information stored can be changed. *Random-access memories* (*RAMs*) and *charge-coupled device* (*CCD*) shift registers are examples of LSI read-write memories. Floppy disks, tapes, and bubble memories are examples of read-write memories which use magnetic materials to store information. In addition to data and instructions, their location (address) in the memory unit must also be stored.

The control unit coordinates the operation of the computer so the program is carried out in the proper sequence and operates on the appropriate data. Two steps are required to accomplish control. First, the control unit *fetches* a numerically encoded instruction contained in memory and interprets or decodes it. Second, the functions performed by each of the constituent parts of the machine are initiated and sequenced so the instruction is *exe-*

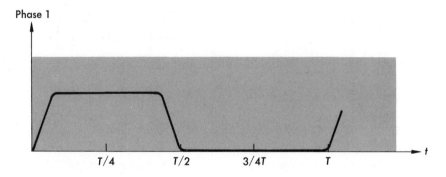

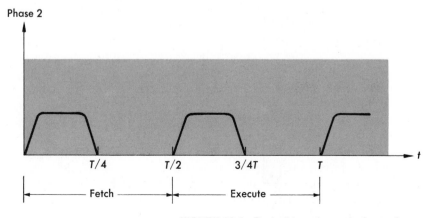

FIGURE 12-1 Typical two-phase clock waveforms.

cuted. Timing, decoding, counting, communicating, and storing information temporarily are all required to effect control.

Generally, a square-wave generator whose period is precisely controlled is used as the clock which establishes the timing in a computer. The amplitude levels of the square wave correspond to logical 1 and logical 0. Often, a two-phase (2ϕ) clock is utilized to divide the fetch and execute steps into two intervals, the first in which interpretation occurs, the second in which the step is performed. Typical two-phase clock waveforms are shown in Fig. 12-1.

Interpreting an address or instruction in memory is performed by a decoder. This action does not occur until a program counter indicates the next operation is ready to be performed.

The control unit must be capable of communicating the interpreted instructions to each of the elements in the computer. In addition, transmission of the data to and from memory and the arithmetic logic unit is to be effected. Usually, control lines and a data bus (an electrical path) are provided for this purpose.

The function of temporary memory, usually a register, is to hold, for a short period of time, data or instructions until the operation is complete. Par-

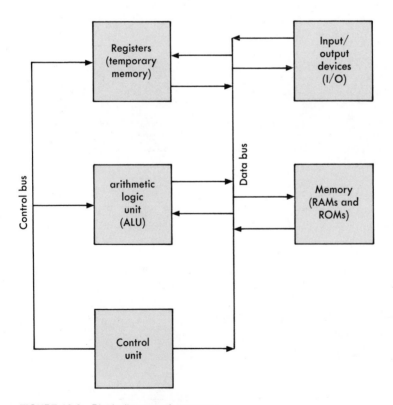

FIGURE 12-2 Block diagram of computer.

tial results are also stored until needed again in a particular calculation or set of calculations. For example, the addition of three quantities, A, B and C, requires the storage of the sum of A and B until C is obtained from memory and the addition completed. An *accumulator* is a special register in which results are stored. The next instruction indicates what is to be done with the result.

The organization of a computer to perform the five basic functions, i.e., the computer architecture, is illustrated in the block diagram of Fig. 12-2. The arrows indicate the communication between the various sections of the computer. The clock generator is not indicated in the figure but is assumed to be present. In the subsequent sections of this chapter, the circuit building blocks used to realize many of the functions in the arithmetic, control, and memory units are described. These circuits are, in turn, comprised of the logic gates described in Chap. 11.

12-2 BINARY ARITHMETIC

The fundamental arithmetic operation performed in a digital computer is binary addition. All other mathematical operations are decomposed into sequences of addition. Familiarity with decimal addition indicates that subtraction, multiplication, and division are all performed using addition. Algorithms to evaluate more complex mathematical functions are methods which enable computation using the arithmetic operations of which the computer is capable.

The rules for binary addition parallel those for decimal arithmetic. The rules for adding bits in binary addition are:

1 0 plus 0 = 0

2 1 plus 0 = 1

3 1 plus 1 = 0 and carry 1

The operation *carry* 1 refers to adding 1 to the next higher binary bit, much as $6 + 4 = 0 +$ carry 1 to the tens column in decimal addition. The following example is useful to illustrate the process of binary addition and its parallel with decimal addition

```
46        = 0101110
29        = 0011101
65          110011
 1  Carry     1100   Carry
75  Sum     0101011
               1      Carry
            0001011
              10      Carry
            1001011   Sum
```

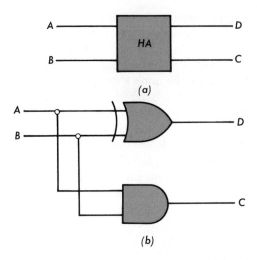

FIGURE 12-3 Half-adder (*a*) symbol and (*b*) realization.

The result of a simple check indicates that the decimal equivalent of the binary sum is 75.

To perform addition, a *binary adder* is used which must keep track of the sum of the bits and also the carry bit, if required. The basic building block in a binary adder is the *half-adder,* whose symbol is given in Fig. 12-3a. The input variables A and B are the digits to be added, and the outputs D and C are the sum bit D and carry bit C, respectively. The schematic of Fig. 12-3b is the circuit implementation of the half-adder by means of an EXCLUSIVE OR gate and an AND gate. The EXCLUSIVE OR gate provides the sum digit in accordance with the previously stated rules. The AND gate provides an output only when both A and B are 1, and thus provides the carry digit. The waveforms in Fig. 12-4 indicate the process of addition. In the time intervals when either A or B is 1, the output of the EXCLUSIVE OR gate D is also 1 but the output of the AND gate C is zero. When both inputs are 1, the output D becomes zero while the output C becomes a 1.

The full binary adder or *full adder* is generally composed of two half-adders and an OR gate as depicted in Fig. 12-5. One half-adder is used to add the incoming bits A_n and B_n. The digit outputs D_n and C_{n-1} (the digit carried from the previous binary column) are the inputs to the second half-adder. The carry bits C_{n1} and C_{n2} from each half-adder are the two inputs to the OR gate, whose output is the carry digit C_n to be added to the next column.

To add two multiple-bit numbers, several full adders of the type shown in Fig. 12-5 are connected to receive their inputs in parallel. Each full adder is used to add a particular column of digits, that is, the 2^1, 2^2, . . . , 2^n column, as illustrated in the 4-bit adder in Fig. 12-6. As the 2^0 is the least significant bit, no carry bit appears at the input of the 2^0 full adder (FA).

The inclusion of negative numbers converts the process of subtraction to one of addition; e.g., $75 - 46 = 75 + (-46)$. In the binary system, nega-

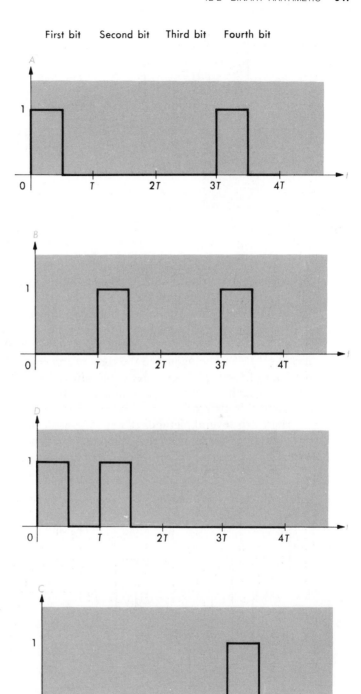

FIGURE 12-4 Waveforms for half-adder for four consecutive binary additions.

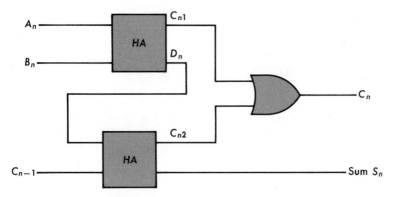

FIGURE 12-5   Full-adder realization.

tive numbers are identified by means of an additional bit called the *sign bit*. To illustrate the use of the sign bit, the positive decimal number 46 is written in binary as $0 \wedge 0101110$, whereas negative 46 is $1 \wedge 0101110$. The sign bit is the digit to the left of the caret ($\wedge$); as indicated the sign bit is zero for positive numbers and is one for negative numbers. The magnitude of the number is represented by the binary digits to the right of the sign bit, with the most significant bit (MSB) being just to the right of the caret. The student should be cognizant of the fact that a caret is not always used to separate the sign bit from the magnitude of the binary number. However, it is a convenience we will use in this chapter.

There are several methods of binary subtraction; the *two's-complement* is commonly used and is readily implemented with standard logic gates. Furthermore, the use of the two's-complement eliminates the difficult-to-

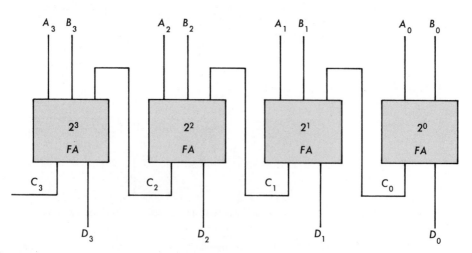

FIGURE 12-6   4-bit adder.

implement process of exchanging or borrowing and lending used in decimal subtraction. For example,

$$
\begin{array}{r}
75 \\
-46 \\
\hline 29
\end{array}
=
\begin{array}{cc}
6 & 15 \\
4 & 6 \\
\hline 2 & 9
\end{array}
\quad \text{(exchange one 10 for ten 1s)}
$$

The *complement* of a binary number is the logical negation (inverter) of each of its bits. Thus, each 1 becomes a 0 and each 0 becomes a 1, a process which results in the *one's-complement*. To form the two's-complement, a 1 is added to the least significant bit of the one's-complement as illustrated in the following:

$$
\begin{array}{ll}
Y = 0 \wedge 110101 & \text{Binary number} \\
\overline{Y} = 1 \wedge 001011 & \text{One's-complement} \\
\underline{\qquad\qquad 1} & \text{Add 1} \\
1 \wedge 001100 & \text{Two's-complement}
\end{array}
$$

Note that the sign bit is included in the complement operation. The two's-complement of the positive number Y has a sign bit equal to 1, indicating a negative number. Negative numbers are often represented in two's-complement as well as the sign-magnitude representation previously described.

To subtract the subtrahend B from the minuend A $(A - B)$, B is first expressed in two's-complement form and the result added to A. The following example illustrates the method.

Example 12-1

Obtain the differences $A - B$ and $B - A$ for $A = 0 \wedge 1001011 =$ decimal 75 and $B = 0 \wedge 0101110 =$ decimal 46.

SOLUTION   To obtain $A - B$, B is expressed in two's-complement form as

$$
\begin{array}{ll}
1 \wedge 1010001 & \text{One's-complement} \\
\underline{\qquad\qquad 1} & \text{Add 1} \\
1 \wedge 1010010 & B \text{ in two's-complement} \\
0 \wedge 1001011 & \text{Add } A \\
& \text{Disregard 1) carry} \\
0 \wedge 0011101 & \text{Difference}
\end{array}
$$

The decimal equivalent of the difference $75 - 46 = 29$ can easily be verified. The carry bit associated with the addition of the sign bit is disregarded as the sign information is contained only in the first digit to the left of the caret.

For the difference $B - A$, the two's-complement of A is formed as

$$
\begin{array}{ll}
1_{\wedge}0110100 & \text{One's-complement of } A \\
\underline{\hspace{1.8em}1} & \text{Add 1} \\
1_{\wedge}0110101 & \text{Two's-complement of } A \\
\underline{0_{\wedge}0101110} & \text{Add } B \\
1_{\wedge}1100011 & \text{Difference}
\end{array}
$$

The sign bit is 1, indicating the result is negative $(46 - 75 = -29)$, with the difference expressed in two's-complement. To verify that the magnitude of the difference is the binary equivalent of 29, the two's-complement is formed.

$$
\begin{array}{ll}
{}_{\wedge}0011100 & \text{One's-complement of difference} \\
\underline{\hspace{1.8em}1} & \text{Add 1} \\
{}_{\wedge}0011101 & \text{Two's-complement of magnitude}
\end{array}
$$

The binary number 0011101 is decimal 29.

The multiplication of one binary digit by another obeys only two rules:

1 $1 \times 0 = 0$

2 $1 \times 1 = 1$

Using these rules, the process for *binary multiplication* is illustrated in a similar manner to decimal multiplication as follows:

$$
\begin{array}{ll}
1010 & \text{Multiplicand} \\
\underline{\times\,1101} & \text{Multiplier} \\
\left.\begin{array}{l}
1010 \\
0000 \\
1010 \\
\underline{1010}
\end{array}\right\} & \text{Partial products} \\
\underline{10000010} & \text{Product}
\end{array}
$$

Most .... Least
significant significant
bits .... bits
(MSB) .... (LSB)

As seen in the example, the process involves copying the multiplicand if the multiplier bit is 1 and entering a row of zeros when the multiplier bit is 0. For each successive multiplier bit (right to left) the partial product term is shifted one bit to the left. The partial products may be summed progressively; i.e., each partial product term can be added to the previous partial product. In the computer, this is more easily done as intermediate results are stored in the accumulator and each successive partial product added to the result stored in the accumulator.

While the method realized can be implemented using standard computer elements it has one serious drawback. The product of two 4-bit numbers is an 8-bit number (a 9-bit number if a sign bit is included). Typically, the accumulator and other registers in the arithmetic logic unit can store only as many bits as are contained in the data. Thus, for 4-bit numbers as shown in the illustration, the accumulator stores only four bits. These are generally the first four bits obtained in the product; for the left-shifting multiplication process, the four least significant bits are stored. A second register can be used in conjunction with the accumulator to store the overflow (the four most significant bits). However, it is preferable to store the most significant bits in the accumulator and to use the auxiliary register to handle the least significant bits. This is effected by means of a right-shift algorithm, which is demonstrated for the multiplication of the same 4-bit binary numbers previously used. The information to the left of the vertical line is stored in the accumulator, that on the right is in the auxiliary register. Registers 1 and 2 contain the multiplicand and multiplier.

1010		Number stored in register 1
x 1101		Number stored in register 2
0000		Initial value in accumulator
1010		Enter multiplicand; first multiplier digit is 1
1010		Add to value in accumulator
0101	0	Shift result one bit to right
0000		Enter zero; second multiplier bit is zero
0101	0	Add
0010	10	Shift result one bit to right
1010		Enter multiplicand; third multiplier bit is 1
1100	10	Add
0110	010	Shift result one bit to the right
1010		Enter multiplicand, fourth multiplier bit is 1
10000	010	Add
1000	0010	Shift result one bit to right; this is the product

The resultant products for both the right-shift and left-shift methods are identical; the only difference is which bits are stored in the accumulator. To implement the multiplication process, registers which can shift a binary number are required. These are discussed in Sec. 12-5.

Binary division is also similar to the long-division method with which we are familiar for decimal numbers. For reasons paralleling those given for binary multiplication, binary division is generally performed using a repetitive subtraction and left-shifting sequence rather than the right-shift method of long division. Again, registers which shift the bits are needed.

In the process of performing arithmetic and logical functions, it is sometimes essential and often convenient to discern whether one binary number is equal to, greater than, or less than some other binary number. The *digital comparator* is used for this purpose; a 1-bit version is depicted in Fig. 12-7. For $A = 1$ and $B = 0$, the output C is 1 as both A and $\overline{B}$ are 1. Similarly, as $\overline{A}$ and B are both zero, D is zero, and the negation of $(A\overline{B} + \overline{A}B)$ causes E to be

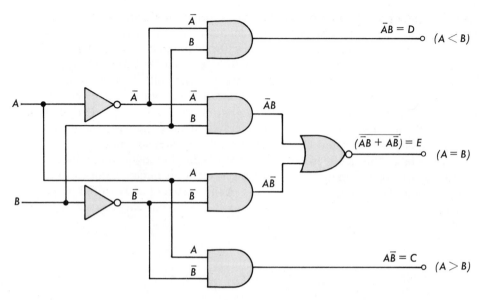

FIGURE 12-7   Digital comparator.

zero. The symmetry of the circuit leads to the conclusion that D is 1 and both C and E are zero when $A = 0$ and $B = 1$. For the case where $A = B = 1, \overline{A}B$ and $A\overline{B}$ are both zero and consequently C and D are zero and E is one. A similar conclusion results when $A = B = 0$. The operation of this circuit is such that whichever output is one gives the desired magnitude comparison.

12-3   BINARY-DERIVED NUMBER REPRESENTATIONS

The binary numbers used to illustrate the arithmetic processes in the preceding section were all integers. Fractional values are included by the introduction of the *binary point,* a method by which a number is divided into positive and negative powers of 2 just as the decimal point partitions a number into positive and negative powers of 10. For example,

$$101.011 = 1 \times 2^2 + 0 \times 2^1 + 1 \times 2^0 + 0 \times 2^{-1} + 1 \times 2^{-2} + 1 \times 2^{-4}$$

Binary point
$$= 4 + 0 + 1 + 0 + \tfrac{1}{4} + \tfrac{1}{8} = \tfrac{43}{8} = 5.375$$

Most arithmetic processes become difficult to implement in the computer when numbers are represented by both integral and fractional values. In *fixed-point representation,* a number is represented as a fractional number multiplied by an appropriate power of 2. As an illustration, the binary number 101.011 is represented as $.101011 \times 2^3$; this is analogous to expressing 5.375 as $.5375 \times 10^1$. All arithmetic operations are performed using the binary fraction and the placement of the binary point is usually remembered by the operator.

Floating-point representation of a binary number permits the storage of both the exponent and fractional value. Thus, 001110101100 is a 12-bit number in which the first four bits represent the exponent and the remaining eight bits are used for the binary fraction. The location of the binary point is always to the left of the most significant bit of the fraction. The number 001110101100 is simply .101011 $\times$ 2^3. In most large computers, arithmetic is performed using floating-point numbers. Typically, a 32-bit (4-byte) number is used in which the first byte contains both the sign bits and exponent; the remaining 24 bits, called the *mantissa,* express the fraction. For example, if the first byte is 10100110, the most significant bit indicates the sign of the mantissa (it is negative), the next bit denotes the sign of the exponent (it is positive), and the remaining six bits give the value of the exponent (38 for the byte given). The range of values that can be expressed is between 2^{-63} and 2^{+63} or between 10^{-19} and 10^{+19}. Similarly, a 24-bit mantissa can be used to express seven decimal places.

The use of 24 bits to express decimal numbers only of the order of 10^7 is often inconvenient. *Octal,* base 8, and *hexidecimal,* base 16, number systems can represent the same decimal number in 8 and 6 digits, respectively. Furthermore, both are binary-derived as octal is represented by three binary digits and hexidecimal by four bits. In an octal system, the digits used are 0 through 7 and the octal number $3425_8 = 3 \times 8^3 + 4 \times 8^2 + 2 \times 8^1 + 5 \times 8^0$. The decimal equivalent is 1,813 and its binary representation is 011 100 010 101, in which each group of three bits corresponds to the octal number.

There are 16 symbols required to represent a hexidecimal number. The first 10 are the decimal digits 0 to 9; the remaining six are the letters A through F, which correspond to the decimal numbers 10 to 15, respectively. The hexidecimal representation of 1,813 (decimal) is 715_{16} and the binary equivalent is written as 0111 0001 0101. As indicated, each group of four bits corresponds to a hexidecimal digit. Similarly, $7D_{16} = 7 \times 16^1 + 13 \times 16^0 = 125$ and is written in binary as 0111 1101.

Binary-coded decimal (BCD) representations are often used in applications in which decimal displays are employed. Many calculators, digital voltmeters, electronic cash registers, and counters are among these applications. Each decimal digit is written as an individual 4-bit word in BCD notation. There are several methods of representing the 10 decimal digits in a 4-bit code which has 16 possible combinations. The natural or 8421 BCD representation utilizes the standard binary representation of decimal digits. The decimal number 259 is written as a 12-bit word as 0010 0101 1001, where each group of four bits corresponds to the respective decimal digit.

Another binary-derived code is the Gray code, which has the virtue of requiring a change of only one bit between consecutive numbers. In the Gray code 000,001,011,010, and 110 represent the decimal digits 0, 1, 2, 3, and 4, respectively. The benefit of changing only one bit at a time is that only one gate must change state to alter the numbers by one unit. This reduces possible error to the least significant bit used whereas in standard binary it can be as great as the most significant bit. A 4-bit Gray code is illustrated in the optical shaft-encoder of Fig. 17-26.

Arithmetic operations can be performed using each of the binary-derived codes by techniques analogous to those described in Sec. 12-2 for standard binary numbers.

12-4 FLIP-FLOPS

The *flip-flop* is the basic computer element used to store one bit of data in the central processor unit. As described in Sec. 11-5, the flip-flop is a bistable circuit having two characteristics important in information storage. The first is that it remains in a given state, 1 or 0, until an input signal (trigger) produces a change in state. By this means, a bit is stored once an input signal establishes a particular state; that bit remains in the flip-flop till such time as a second input signal is applied. The second property is that the two output signals provided are the complements of one another.

The basic flip-flop, or *latch,* whose symbol is shown in Fig. 12-8a, can be realized by use of a pair of cross-coupled NAND gates as depicted in Fig. 12-8b. At least one of the inputs in the circuit of Fig. 12-8b must be at logical 1 to insure operation; $A = B = 0$ is not a permissible logical state. The two conditions for "writing" a bit into the flip-flop are: $A = 1, B = 0$ for which $Y = 0$ and $A = 0$, and $B = 1$, resulting in $Y = 1$. When $A = B = 1$, Y remains unchanged.

The latch in Fig. 12-8b forms the basis of a group of clocked flip-flops used in the computer. As many operations in the computation process are sequential, it is important that the storage circuits are enabled at the proper time. The clocked-R-S flip-flop (*set-re*set) is indicated by the circuit symbol in Fig. 12-9a, in which the designation CK is the clock-pulse input. Basically, this flip-flop is a latch which can only change state when a clock pulse is applied. Therefore, the R-S flip-flop remains in a particular state until a clock pulse enables a transition to occur. The schematic of a clocked R-S flip-flop is shown in Fig. 12-9b, in which the two input NAND gates act to steer the input signals into the latch. The truth table in Fig. 12-9c indicates

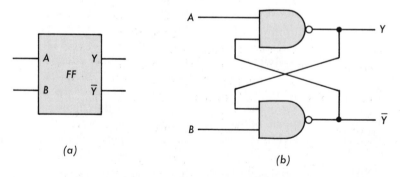

(a)

(b)

FIGURE 12-8 Flip-flop (latch) (a) symbol and (b) NAND gate realization.

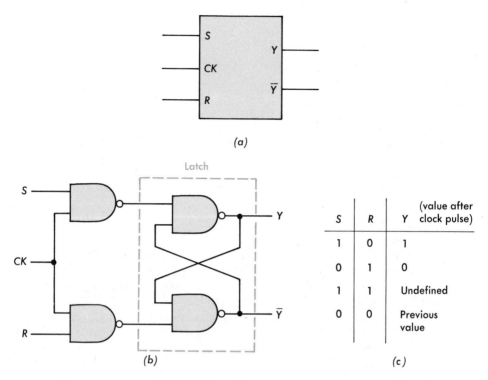

FIGURE 12-9 Clocked *R-S* flip-flop (*a*) symbol and (*b*) realization.

that for $S = R = 1$, the output is undefined and this logical state is not permitted to exist.

The *J-K* flip-flop, symbolically depicted in Fig. 12-10*a*, is a clocked *R-S* circuit with the addition of a third input to the steering NAND gates, as shown in Fig. 12-10*b*. The third input is $\overline{Y}$ for the *J* input and *Y* for the *K* input. This feedback eliminates the undefined state for the *R-S* flip-flop given

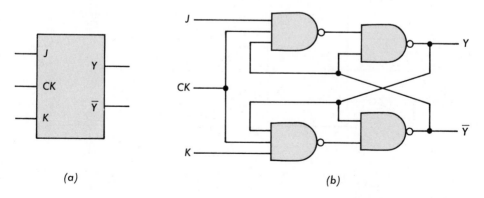

FIGURE 12-10 *J-K* flip-flop symbol and circuit realization.

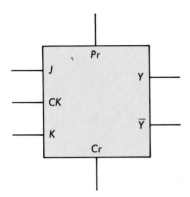

FIGURE 12-11 *J-K* flip-flop with preset and clear inputs.

in the truth table of Fig. 12-9c. The truth table for the *J-K* flip-flop is the same as that for the *R-S* flip-flop except that inputs $J = K = 1$ cause a change in state. That is, if Y is 1 when J and $K = 1$ are applied, Y switches to a 0; if Y is 0, inputs of logical 1 at both J and K produce an output $Y = 1$.

The initial state of the *J-K* flip-flop can be controlled by preset (Pr) and clear (Cr) signals applied to the NAND gates which form the latch. The symbol for the *J-K* flip-flop with preset and clear inputs is shown in Fig. 12-11. The initial state of the output Y is set to one when Pr = 1, Cr = CK = 0; reversing the preset and clear levels sets $Y = 0$.

The *D- and T-type flip-flops* are both derived from the *J-K* flip-flop as shown in Fig. 12-12. The use of the inverter at the K input in Fig. 12-12a converts the *J-K* flip-flop into a delay or *D*-type circuit. The output of the *D*-type circuit when $D = J = \overline{K} = 1$ is the value of Y set during the previous clock pulse application. Consequently, when an input is applied it is held there (delayed) until the next clock pulse and only then is the bit transferred to the output.

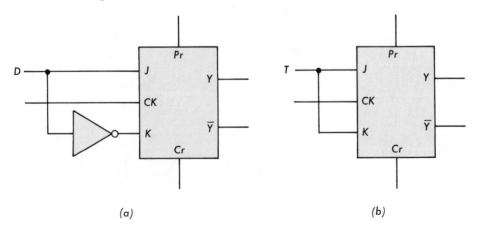

(a)   *(b)*

FIGURE 12-12 (a) *D*-type and (b) *T*-type flip-flops.

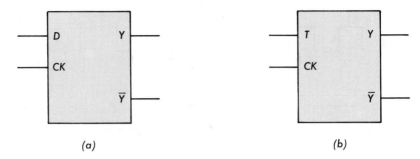

FIGURE 12-13 Symbols for (*a*) *D*-type and (*b*) *T*-type flip-flops.

Connecting a single-input line T to both J and K inputs as indicated in Fig. 12-12*b* converts the *J-K* circuit to a *T*-type flip-flop. This flip-flop acts as a "toggle switch" in that the output changes state each time a clock pulse is applied.

The symbols for *D*- and *T*-type flip-flops are depicted in Fig. 12-13. All four types of flip-flops are used as building blocks for a variety of sequential circuits, some of which are treated in the next section. In addition, SSI and MSI chips which contain several flip-flops are commercially available.

12-5 REGISTERS AND COUNTERS

Each of the clocked flip-flops described in the preceding section is capable of storing one bit of data. It is natural to assume that if N flip-flops are appropriately interconnected an N-bit word can be stored. Circuits constructed in this manner are called *shift registers*. The circuit depicted in Fig. 12-14 is a 4-bit shift register which uses *D*-type flip-flops. As indicated in the timing waveforms of Fig. 12-15, each successive clock pulse transfers, or shifts, the data bit from one flip-flop to the next one. The waveforms in Fig. 12-15 also

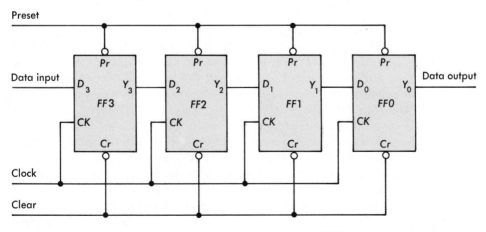

FIGURE 12-14 4-bit shift register.

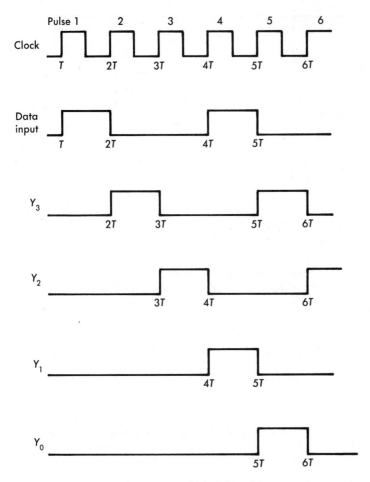

FIGURE 12-15 Timing diagram for 4-bit shift register.

show that the data is serially read into the register and also appears at the output in serial form. As such, the shift register in Fig. 12-14 is referred to as a *serial-in, serial-out* (*SISO*) register. The register in Fig. 12-14 is comprised of *D*-type latches; however, *J-K* or *S-R* flip-flops are also used in shift-register construction.

Parallel-in, serial-out (*PISO*), *serial-in, parallel-out* (*SIPO*), and *parallel-in, parallel-out* (*PIPO*) are other registers which are often used. As their names indicate, such registers are used to read in the input data and read out the output data in the form most convenient for the operations involved. For example, the output of the memory is available in parallel so that reading it out of memory is preferably done in parallel.

The shift register shown in Fig. 12-14 shifts the data stored one bit to the right with each clock pulse. This operation is used in the multiplication algorithm described in Sec. 12-2. Division requires a left-shifting register,

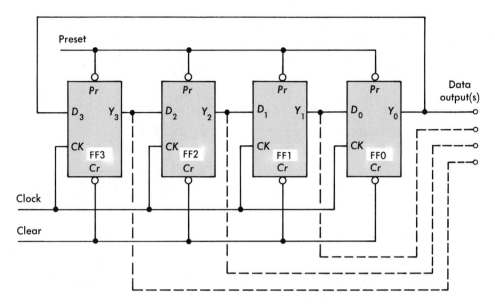

FIGURE 12-16 Ring-counter circuit.

which can be achieved when additional gates are used in conjunction with the basic register. Registers capable of shifting the data left or right are called *bidirectional shift registers*. The *universal register* is an IC component in which additional gates, as well as the register, are on a single chip. The logical arrangement of these gates makes it possible for the universal register to be used in any of the modes described in the three previous paragraphs.

Because the data is shifted for each clock pulse, the shift register can be used as a counter. The circuit in Fig. 12-16, called a *ring counter,* is simply the shift register in Fig. 12-14 for which the output Y_o is connected to the data input. The state of FFO is preset to logical 1 so that each successive clock pulse transfers the count through the register. The flip-flop whose output is 1 indicates the number of input pulses. For the circuit of Fig. 12-14, four pulses are required for Y_o to return to a 1 state. The ring counter can then be used as a 4:1 divider. An N-bit register connected as a ring counter can be used as an $N:1$ divider.

Other counters can be constructed using the interconnection of flip-flops. Figure 12-17 depicts the circuit configuration for a 3-bit *ripple counter,* which utilizes the T-type flip-flop as the basic component. With the input held at logical 1 each pulse applied at the clock input causes a change in state which is propagated, or "rippled," through the register. As seen in the waveforms in Fig. 12-18, the input flip-flop changes state on the falling edge of the clock pulses and each successive transition in the other flip-flops occurs only when the previous flip-flop changes its state from 1 to 0.

The addition of a two-input AND-OR gate between stages converts the ripple counter to an up-down counter. Each AND-OR gate uses the output

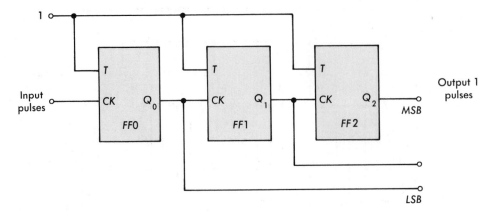

FIGURE 12-17 3-bit ripple counter.

of the preceding flip-flop as one input; the second input is an external pulse which controls whether the count is in the forward (up) or reverse (down) direction.

Each stage in the ripple counter in Fig. 12-17 responds only after the previous stage has completed its transition. In a many-bit counter the time delay through the system, called the *carry-propagation delay,* can be long. The *synchronous counter,* depicted in Fig. 12-19, is often used to increase the speed of the response. The circuit operates synchronously as all *T*-type flip-flops are excited by the input pulses. The AND gates carry forward the transitions of the flip-flops and thereby improve the speed. The circuit shown is referred to as a *series-carry synchronous counter.* Further increases in speed can be obtained using parallel-carry, a configuration for which each AND gate has as its input, the outputs of all previous flip-flops.

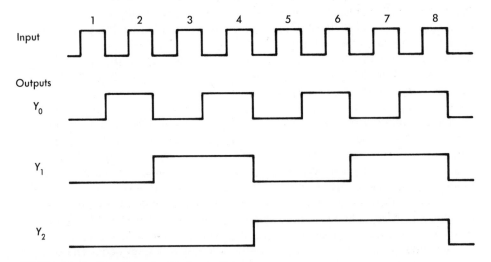

FIGURE 12-18 Waveforms for ripple counter.

Input pulses

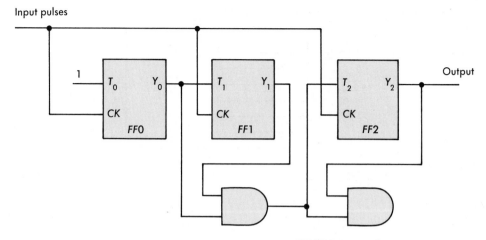

FIGURE 12-19 Synchronous counter.

Counters have many applications other than in computer systems. Among them are digital clocks, frequency counters, industrial process-control timers, and frequency dividers. In addition, with augmentation of additional logic, the counters can be used to count in different number systems, one of the most common being the binary-coded decimal system.

12-6 SEMICONDUCTOR MEMORIES

The storage and retrieval of information is the function served by the memory units in any digital processing system, whether that system is used for computation, control, or communication. Both MOS and bipolar technologies are employed in the fabrication of the read-only and read-write chips contained in most memory units. As described in Sec. 11-9 for logic-gate realizations, MOS memories generally afford higher component density but have lower access speeds than do bipolar memories.

Read-only memories (*ROMs*) are organized with M input lines and N output lines describing a 2^M by N matrix array as shown in Fig. 12-20. A memory cell exists at each location in the array so that there are $2^M \times N$ such cells in the matrix. For each M-bit binary input signal, a single N-bit output signal is defined. As such, a ROM can be considered to be a translator, first decoding an M-bit word and then encoding it as an N-bit word. The circuit shown in Fig. 12-21 is a two-input, four-output decoder. As seen in Fig. 12-21 a two-bit input word describes four states, each of which selects one line in the output.

Alternatively, an M-bit input signal can be thought of as defining 2^M addresses; the content of each address (row) is the N-bit word defined by the states of the N memory cells in that row. The data stored at each address are a function of the hardware used to effect the code conversion from an M-bit to an N-bit word. Consequently, the data are permanently stored and, in general,

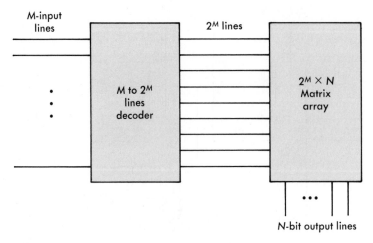

FIGURE 12-20 $M \times N$ bit read-only memory (ROM).

cannot be altered without changing the hardware. The data can be retrieved (read) but not electrically entered (written) and hence, the name *read-only memory*. ROMs are nonvolatile; that is, the information stored is not destroyed when the power to the device is shut off.

The ROM configuration in Fig. 12-20 employs one-dimensional addressing as the M input lines define 2^M N-bit words. Two-dimensional addressing, a form of multiplexing and depicted in Fig. 12-22, decreases the number of decoder gates needed and is generally a more efficient organization. The $L:1$

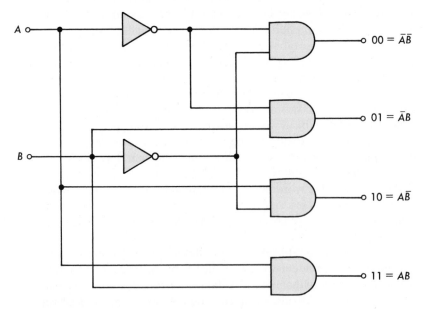

FIGURE 12-21 A two-bit decoder.

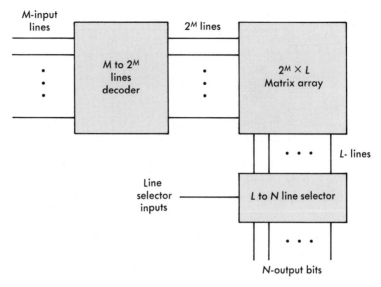

FIGURE 12-22 Two-dimensional addressing.

selectors act as a Y-dimension address decoder in that they select N lines of the matrix. The decoding process allows for L/N different combinations to appear at the output. Therefore, each of the 2^M addresses (X-dimension) define L/N different N-bit words. For example, a 64×64 array which can store 4,096 bits (called a 4-kb ROM) can be organized as shown in Fig. 12-23. The six input lines define the 64 rows and the eight 8:1 line selectors provide an 8-bit output word. The logic actuating the line selectors can select any one of 8 different 8-bit output words for a given input address. The total number of words stored is then $64 \times 8 = 512$. To store 512 8-bit words using one-dimensional addressing requires 9 input lines (2^9) and, therefore, increased complexity in the decoder.

In addition to storing instructions, ROMs are useful elements in automatic test equipment. By storing the values of sin x, for specific values of x, the ROM can be used as a function generator to provide a variety of test signals to a device undergoing test. Comparison of measured results with expected results (also stored in memory) determines whether the device in question meets design specifications.

A *PROM* is a *field-programmable read-only memory* which is constructed with all possible memory values present. Each outcome is connected to a fused link which can be opened by injecting a current into the memory cell. Once the fused link is opened, the program is set and the action of a PROM is that of a standard ROM. The advantage is that the user can program each PROM to fit a specific application and still retain the advantage of using several of the same components in a system. *EPROMs* are *E*rasable *PROMs* in which special construction allows for the link to be opened or reclosed by means of ultraviolet radiation. Consequently, the program can

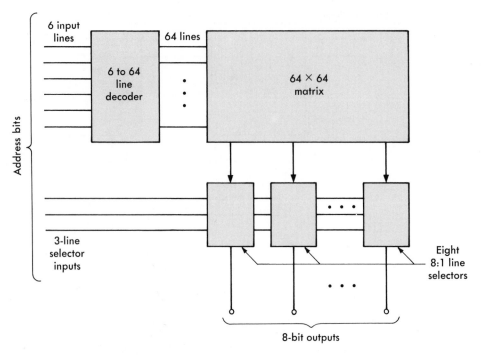

FIGURE 12-23 A 64 × 64 bit array with two-dimensional addressing.

be altered after it is entered. However, the process for changing a program is slow (20 to 30 min) so that program alterations cannot be made during operation.

The *random-access memory (RAM)*, a basic read-write memory, is also organized in a matrix array. As illustrated in Fig. 12-24, two-dimensional addressing is used for the 4-kb RAM whose outputs are 4,096 one-bit words. To insert data into the RAM the write-logic is enabled and the read is inhibited. The reverse is true to read a bit out of the memory by means of the sense amplifier. To store 4,096 eight-bit words, eight arrays of the form shown in Fig. 12-24 are used, each of which is addressed in parallel. Each array, however, has provision for individual data-in and data-out signals.

The addressing of a RAM allows a bit to be entered or retrieved at any location in the memory; i.e., it is randomly accessed. The data stored can be electrically changed; however, RAMs are volatile memories.

The circuits shown in Fig. 12-25 are 1-bit *static* and *dynamic MOSFET memory cells*. The static memory cell in Fig. 12-25a employs a pair of cross-coupled INVERTERs with active loads (see Secs. 11-8 and 11-9). As the INVERTER outputs are complementary both logical 0 and 1 can be entered or retrieved. The transistors labelled $Q1$ are used for X-dimension addressing; Y-dimension addressing and the read-write circuitry is not

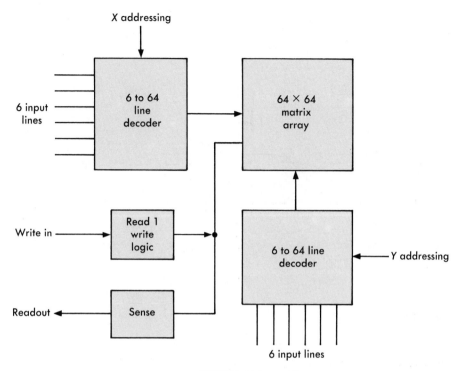

FIGURE 12-24 4-kB random-access memory (RAM).

shown. In the dynamic MOSFET memory cell in Fig. 12-25*b*, logical 1 and 0 are determined by the charge on the capacitor. Because the charge on the capacitor can decrease through leakage, *refresh* circuitry is provided in dynamic MOS memories to maintain the charge. If refresh circuitry is not provided, excessive leakage makes it impossible to distinguish the 1 and 0 levels.

Storage of logical 1 and logical 0 in ROMs can be effected as shown in Fig. 12-26. Opening the connection (the fusable link) stores a 0; keeping the connection stores logical 1.

Charge-coupled devices (CCD) are MOS memories used to construct large shift registers. As depicted in Fig. 12-27 a CCD is a single MOS transistor with a long channel between drain and source. Many gates, very closely spaced, are inserted between draw and source. In each gate, the silicon dioxide and the semiconductor form a capacitor. A signal corresponding to logical 1 applied to the source causes the capacitance at the gate closest to the source to be charged. Clocking the next gate shifts the charge stored to the capacitor formed by the second gate. Repeating this process produces a shift in the logical 1 pulse through the shift register to the output at the drain. Commercial CCD memories (RAMs) are available from 4 to 65 kb.

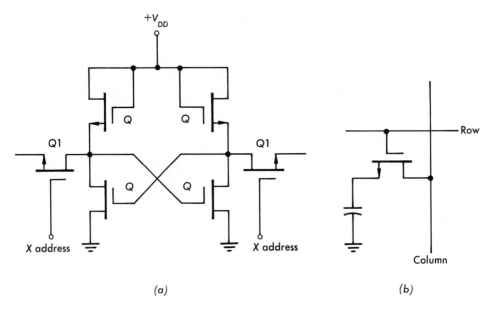

FIGURE 12-25 Static and dynamic 1-bit MOSFET memory cells.

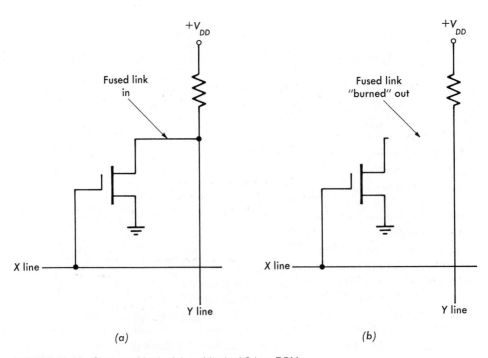

FIGURE 12-26 Storage of logical 1 and logical 0 in a ROM.

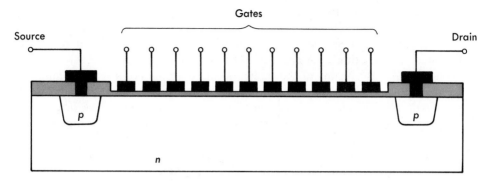

FIGURE 12-27 Charge-coupled device (CCD) shift-register memory.

12-7 MICROPROCESSORS

A *microprocessor* is a single-chip central processing unit (CPU) which contains the arithmetic and control units of a digital computer and their associated registers (see Fig. 12-2). When memory units and the clock generator are incorporated on the same chip, the microprocessor becomes a digital computer. The architecture of a typical microprocessor includes:

1 Clock generator (CK)

2 Control logic

3 Arithmetic logic unit (ALU)

4 Registers

5 Read-only memory (ROM)

6 Random-access memory (RAM)

7 Program-event counter (PC)

8 External control and interrupt logic

9 Input-output ports (I/O)

The functions of these units and some of their circuit implementation is described in the preceding sections of this chapter.

In many microprocessors, the read-only memory is a PROM so that the program stored can be entered by the user. The input-output ports provide access to the central processing unit from a variety of standard devices. In addition, these ports can be used to add additional memory to the system through such devices as floppy disks and off-chip RAMs and ROMs.

The interrupt logic is particularly useful in control applications (see Sec. 17-10). The function of the interrupt routine is to halt the program in process and execute the set of instructions in a specific subroutine. Upon completion of the subroutine, the processor is instructed to complete the program that was interrupted. The interrupt logic performs two tasks: First it halts the program in progress and places in memory the next program step to be executed. At this time, the control unit is instructed to execute the interrupt subroutine. By means of the interrupt and external control logic units, the microprocessor can be used effectively in distributed control. That is, each microprocessor can be programmed to perform a specific task or group of tasks for "local" control. Interconnection of several such microprocessors so they can communicate provides distributed control over a wide range of tasks. This method is often more effective and generally less expensive than centralized control using a large digital computer.

Microprocessors are often used as dedicated computers, one example of which is the digital clock. They also serve as the central processing unit in many common items. Among these are hand-held calculators, automatic fuel-injection systems, electronic music synthesizers, and television games. Industrial applications to "smart" instruments and automatic test equipment are ever-expanding areas in which microprocessors are employed.

12-8   DIGITAL-TO-ANALOG CONVERSION

There are advantages which result, in many applications, from processing digital, rather than analog, signals. However, many physical systems require analog signals if they are to be operated in an effective manner. As a result, it is often necessary to convert the binary output signals of a digital processor into an equivalent analog signal. This transformation is accomplished by means of a *digital-to-analog,* or simply *D/A*, converter.

One of the simplest methods of digital-to-analog conversion is the use of the binary-weighted resistor network depicted in Fig. 12-28. The circuit shown in Fig. 12-28 converts a 4-bit digital signal into a proportional voltage.

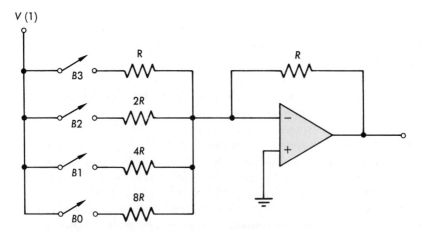

FIGURE 12-28   4-bit digital-to-analog (D/A) converter.

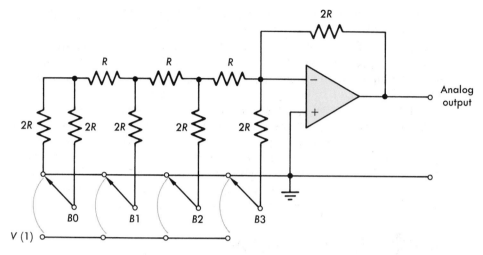

FIGURE 12-29 *R-2R* ladder digital-to-analog converter.

The switches BO-B3 are open when these bits are logical 0 and are closed for logical 1. For each of the $2^4 = 16$ input combinations a discrete output level exists; the magnitude of this output varies in steps of $V(1)/8$ from 0 to $15V(1)/8$. With one or more switches closed, the operational amplifier circuit behaves as an analog summing circuit (see Chap. 10). Increased resolution is obtained when the binary input word contains more bits.

The range of resistor values in the circuit of Fig. 12-28 becomes impractical for binary words of longer than 4 bits. Similarly, the dynamic range of the OP-AMP limits the selection of resistance values. To overcome this limi-

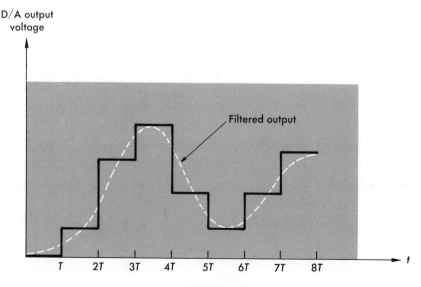

FIGURE 12-30 Analog output of D/A converter.

tation, the *R-2R* ladder network depicted in Fig. 12-29 is used in the D/A converter. Although the circuit in Fig. 12-29 illustrates a 4-bit converter, the circuit configuration can easily be extended to an *N*-bit input word. The analysis of this circuit is left to the student (see Prob. 12-50). Let it suffice at present to indicate that there are 16 discrete output levels corresponding to the 16 binary input values.

To obtain an analog signal which is a function of time, the D/A converter will respond to a periodic sequence of binary inputs. It is as if we are obtaining the analog signal from a sampled signal. The output level remains constant between consecutive binary inputs as shown in Fig. 12-30, often making it necessary to smooth the output by means of a filter. Furthermore, the "waveform" in Fig. 12-30 highlights the fact that a closer approximation to the actual waveform results when more data points are used.

12-9 ANALOG-TO-DIGITAL CONVERTERS

To utilize the advantages of digital processing, the essentially analog signals which arise in most physical systems must be converted into corresponding binary signals. *Analog-to-digital converters (A/D)* are circuits used for this purpose; one type called the *single-counter A/D converter* is depicted in Fig. 12-31.

To convert a time-varying analog signal into a sequence of binary words, it is generally necessary to periodically sample the analog signal. Sample-and-hold circuits described in Sec. 11-6 are often used to generate the sampled-data signal. In the circuit of Fig. 12-31, a binary counter is used to count the number of clock pulses. This count forms the input to the D/A converter, whose output voltage is compared with the analog input by means of a comparator. The comparator output is at $V(1)$ as long as the output of the D/A converter is less than the analog signal level. The AND gate can then transmit clock pulses to the counter. When the D/A output signal is

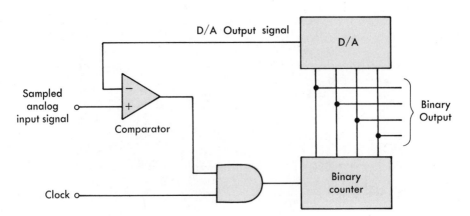

FIGURE 12-31   Single-counter analog-to-digital converter (A/D).

greater than the analog signal, the comparator switches to $V(0)$, consequently inhibiting the AND gate and halting the count. The output of the binary counter is then the desired binary word which represents the sampled analog system.

Several other methods for A/D conversion are employed in IC packages that are available "off the shelf." The *successive-approximation* type is among the fastest as it requires only N clock pulses to generate an N-bit word. The circuit structure is similar to that shown in Fig. 12-31 except the AND gate and counter are replaced by a successive-approximation register. Essentially, this method involves a comparison of one bit of the output word with the analog input for each clock pulse. The most significant bit is set to 1 and compared during the first clock pulse. If the analog input is greater than the D/A output voltage, the most significant bit in the register is set to 1. This bit is set to 0 in the register for the opposite condition. The next most significant bit is set in similar fashion during the next clock pulse and the process continues until the least significant bit is set. The binary output can now be read in parallel from the register. Some commercially available successive-approximation A/D converter chips have provision for serial as well as parallel output data.

PROBLEMS

12-1 Perform the following additions:

100110	10011010	10011000
010101	10110111	01101100

12-2 a Perform the indicated decimal additions in binary.

31	1,018	946
106	474	989

b Express the results in a as an 8-bit number. What is the percentage roundoff error, if any?

12-3 a Perform the indicated additions in binary. Express the result as an 8-bit number.

36	94	50
43	57	125
21	104	100

b If three 8-bit numbers are to be added, what is the number of bits the accumulator must be capable of storing if the entire result is to be stored?
c If the accumulator can store only 8 bits, what is the maximum error?

12-4 The importance of De Morgan's theorems is manifested in the fact that the basic logic operations can be realized by use of INVERTERs and NAND gates only. For the circuits displayed in Fig. 12-32:

a Construct the truth table and identify the type of gate realized.

b Write an expression for the logical output.

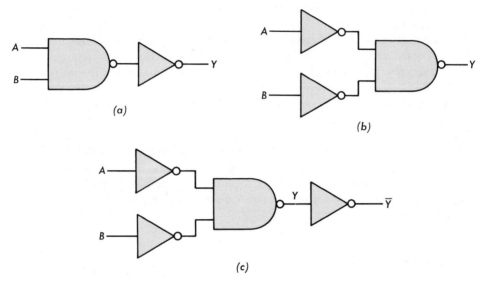

FIGURE 12-32   Circuits for Prob. 12-4.

12-5   Repeat Prob. 12-4 for the NOR and INVERTER realizations shown in Fig. 12-33.

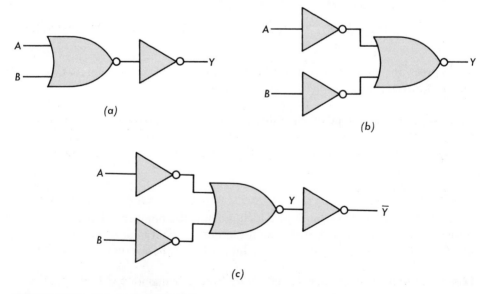

FIGURE 12-33   Circuits for Prob. 12-5.

12-6 Construct a half-adder using only NAND gates and INVERTERs.

12-7 Construct a half-adder using only NOR gates and INVERTERs.

12-8 Write a logic expression for D_n and C_n, the outputs of a full adder, for input digits A_n and B_n and the carry bit C_{n-1}.

12-9 Realize a full adder using NAND and INVERTER gates only.

12-10 Repeat Prob. 12-9 if only NOR and INVERTER gates are used.

12-11 Express the following in two's-complement notation: 28, -28, 109, and -109.

12-12 Express the following in two's-complement notation: 13, -13, 162, and -162.

12-13 Each of the following is expressed in two's-complement notation. Perform the indicated sum and express the result in decimal notation.

$$\begin{array}{ccc} 100110 & 10011010 & 10011000 \\ \underline{010101} & \underline{10110111} & \underline{01101100} \end{array}$$

12-14 **a** Perform the following operations in two's-complement notation.

$$\begin{array}{cccc} -27 & 27 & -27 & 27 \\ \underline{30} & \underline{-30} & \underline{-30} & \underline{30} \end{array}$$

b Express the results in a as decimal numbers.

12-15 **a** Perform the following operations in two's-complement notation.

$$\begin{array}{cccc} 103 & -103 & -103 & 103 \\ \underline{-24} & \underline{24} & \underline{-24} & \underline{24} \end{array}$$

b Express the results in a as decimal numbers.

12-16 A, B, C, and D are the digits in a 4-bit binary number in which A is the sign bit, B is the most significant bit (MSB), and D is the least significant bit (LSB). Show a realization of a logic circuit which converts the 4-bit number to its two's-complement representation.

12-17 **a** Perform the following multiplications using a right-shift algorithm.

$$\begin{array}{ccc} 101101 & 100100 & 11001100 \\ \underline{\times 010011} & \underline{\times 011001} & \underline{\times 10101010} \end{array}$$

b If an 8-bit accumulator is used, what percentage error exists in the results?

12-18 **a** Perform the following multiplications in binary

$$
\begin{array}{ccc}
14 & 57 & 68 \\
\times 13 & \times 21 & \times 39
\end{array}
$$

b If an 8-bit accumulator is available, what is the percentage error in the result?

12-19 Express the following decimal numbers in octal notation: 12, 345, and 6,789.

12-20 Express the following decimal numbers in octal notation: 98, 765, and 4,321.

12-21 Express the following octal numbers by their decimal equivalents: 123, 456, and 7,123.

12-22 Express the following octal numbers by their decimal equivalents: 76, 543, and 2,176.

12-23 Express the following decimal numbers in hexidecimal notation: 97, 864, and 5,321.

12-24 Express the following decimal numbers in hexidecimal notation: 268, 39, and 5,741.

12-25 Convert the following hexidecimal numbers to their decimal equivalents: 6B, 1F4, and C59.

12-26 Convert the following hexidecimal numbers to their decimal equivalents: 2E, 8A3, and D70.

12-27 Express the octal numbers in Prob. 12-22 in hexidecimal notation.

12-28 Convert the hexidecimal numbers in Prob. 12-25 to octal notation.

12-29 Convert the following numbers to binary-coded decimal (BCD) notation: 40, 719, and 6,582.

12-30 Represent the following numbers in BCD notation: 29, 306, and 5,741.

12-31 Express 12.6, 7.49, and 5.08×10^{-3} in
a Fixed-point binary.
b Floating-point binary.

12-32 Express 1.85, 37.6, and 9.02×10^{-2} in

a Fixed-point binary.
b Floating-point binary.

12-33 Show that the circuit in Fig. 12-34 is a NOR-gate realization of a flip-flop.

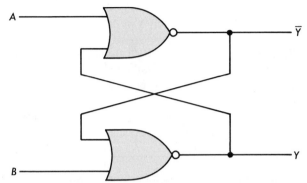

FIGURE 12-34 Circuit for Prob. 12-33.

12-34 **a** What is the number of gates required if the circuit of Fig. 12-7 is realized using NAND and INVERTER gates only?
b Repeat **a** if only NOR and INVERTER gates are used.
c Which realization requires the fewest number of gates?

12-35 Sketch the output waveforms for the register in Fig. 12-14 if J-K flip-flops are used in place of the D-type flip-flops.

12-36 Sketch the waveforms for a 4-bit ripple counter which uses T-type flip-flops.

12-37 Sketch the output waveforms for the synchronous counter in Fig. 12-19.

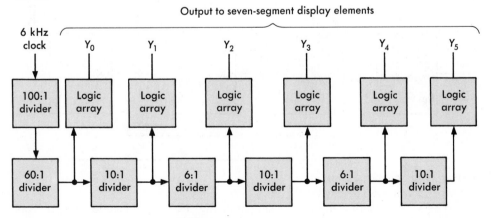

FIGURE 12-35 Representation of a digital clock for Prob. 12-38.

12-38 The schematic representation of a digital clock is shown in Fig. 12-35. Counters are used to realize the various dividers. The blocks indicated as LOGIC ARRAY are logic-gate combinations required to activate the appropriate segments so that the digits can be displayed.

a Verify that the six outputs Y_0 to Y_5 display the number of hours, minutes, and seconds.

b What additional circuitry is needed if the date is also to be displayed?

12-39 a Describe how 10:1 dividers can be used as frequency counters.

b If a 6-digit display is available, what is the smallest change in frequency that can be detected if the maximum frequency is 10 MHz?

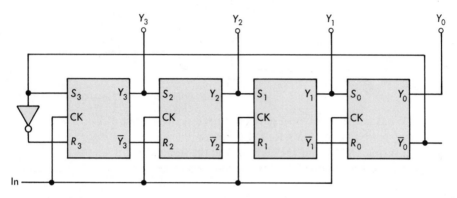

FIGURE 12-36 Johnson counter for Prob. 12-40.

12-40 The circuit in Fig. 12-36 is 3-bit *Johnson, twisted-ring,* or *moebius counter.* Assume all flip-flops are initially zero and that triggering occurs on the negative-going edge of the pulse. Show that this circuit is a 6:1 divider.

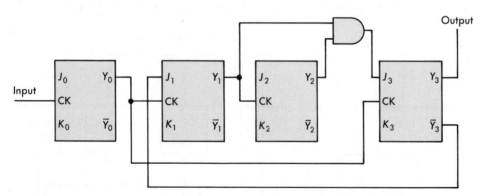

FIGURE 12-37 Circuit for Prob. 12-41.

12-41 Verify that the circuit shown in Fig. 12-37 is a decade counter. Assume the outputs are initially zero and $J_0 = K_0 = K_1 = K_2 = K_3 = 1$.

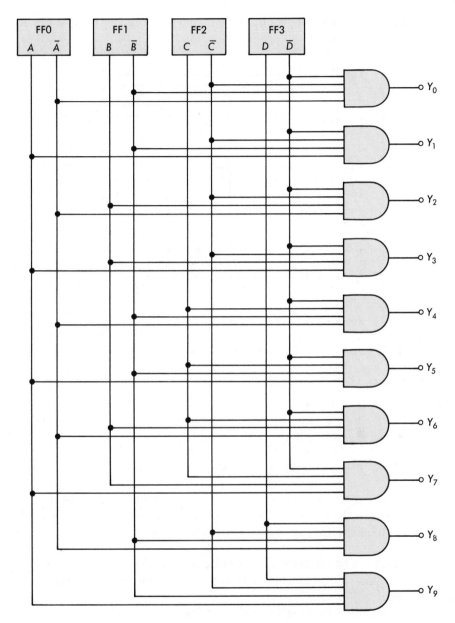

FIGURE 12-38 Circuit for Prob. 12-42.

12-42 Verify that the circuit shown in Fig. 12-38 is a BCD to decimal decoder.

12-43 Develop the necessary logical expressions and construct a circuit which converts a 4-bit binary number to a 4-bit Gray code.

12-44   Show the logic-circuit that can be used to convert a 4-bit Gray code into standard binary.

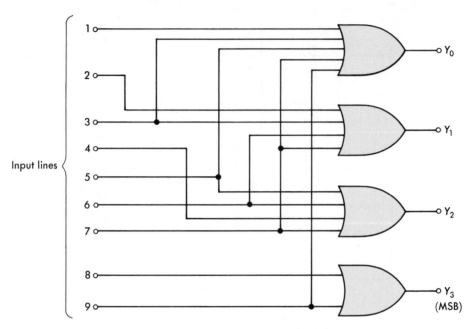

FIGURE 12-39   Circuit for Prob. 12-45.

12-45   Verify that the circuit shown in Fig. 12-39 is a decimal to BCD encoder. Assume that a logical 1 activates each input line.

12-46   Design a decimal to Gray code encoder.

12-47   A 32-kb ROM is used to provide an 8-bit output word.
a   Show the circuit arrangement if one-dimensional addressing is used.
b   Repeat **a** for two-dimensional addressing.

12-48   A 64-kb ROM is to provide a 16-bit output word.
a   Show the organization of the ROM if one-dimensional addressing is used.
b   Repeat **a** for two-dimensional addressing.

12-49   Verify that the circuit in Fig. 12-28 provides an output range from 0 to 9.375 V in increments of 0.375 V. Assume $V(1) = 5$ V.

12-50   For $V(1) = 5$ V, determine the range of the output voltage in the circuit of Fig. 12-29. What is the smallest increment?

12-51   **a**   Design a 6-bit R-$2R$ ladder.

b For $V(1) = 5$ V, what is the maximum output voltage?

c What is the output voltage increment?

d If the output voltage is to indicate increments of 0.1 V, how many bits must be used?

12-52 Describe a system in which the D/A converter in Fig. 12-31 can be employed in a digital voltmeter.

HREE-PHASE CIRCUITS □ **MAGNETIC**
SE CIRCUITS □ MAGNETIC **CIRCUITS,**
AGNETIC CIRCUITS, **TRANSFORMERS,**
TIC CIRCUITS, TRANSFORMERS, **AND**
TRANSFORMERS, AND **THREE-PHASE**
RMERS, AND THREE-PHASE **CIRCUITS**
CIRCUITS □ MAGNETIC CIRCUITS, TR
MAGNETIC CIRCUITS, TRANSFORMEF
CIRCUITS, TRANSFORMERS, AND THF
TRANSFORMERS, AND THREE-PHASE CIRCUITS □ MAGNETIC CIRCU
AND THREE-PHASE CIRCUITS □ MAGNETIC CIRCUITS, TRANSFORMI
THREE-PHASE CIRCUITS □ MAGNETIC CIRCUITS, TRANSFORMERS,
-PHASE CIRCUITS □ MAGNETIC CIRCUITS, TRANSFORMERS, AND T
CIRCUITS □ MAGNETIC CIRCUITS, TRANSFORMERS, AND THREE-PH
MAGNETIC CIRCUITS, TRANSFORMERS, AND THREE-PHASE CIRCUI
CIRCUITS, TRANSFORMERS, AND THREE-PHASE CIRCUITS □ MAGNI
TRANSFORMERS, AND THREE-PHASE CIRCUITS □ MAGNETIC CIRCU
AND THREE-PHASE CIRCUITS □ MAGNETIC CIRCUITS, TRANSFORMI
THREE-PHASE CIRCUITS □ MAGNETIC CIRCUITS, TRANSFORMERS,
-PHASE CIRCUITS □ MAGNETIC CIRCUITS, TRANSFORMERS, AND TI
CIRCUITS □ MAGNETIC CIRCUITS, TRANSFORMERS, AND THREE-PH
MAGNETIC CIRCUITS, TRANSFORMERS, AND THREE-PHASE CIRCUI
CIRCUITS, TRANSFORMERS, AND THREE-PHASE CIRCUITS □ MAGNI
TRANSFORMERS, AND THREE-PHASE CIRCUITS □ MAGNETIC CIRCU
AND THREE-PHASE CIRCUITS □ MAGNETIC CIRCUITS, TRANSFORMI
THREE-PHASE CIRCUITS □ MAGNETIC CIRCUITS, TRANSFORMERS,
-PHASE CIRCUITS □ MAGNETIC CIRCUITS, TRANSFORMERS, AND TI
CIRCUITS □ MAGNETIC CIRCUITS, TRANSFORMERS, AND THREE-PH
MAGNETIC CIRCUITS, TRANSFORMERS, AND THREE-PHASE CIRCUI
CIRCUITS, TRANSFORMERS, AND THREE-PHASE CIRCUITS □ MAGN
TRANSFORMERS, AND THREE-PHASE CIRCUITS □ MAGNETIC CIRCU
AND THREE-PHASE CIRCUITS □ MAGNETIC CIRCUITS, TRANSFORM
THREE-PHASE CIRCUITS □ MAGNETIC CIRCUITS, TRANSFORMERS,
-PHASE CIRCUITS □ MAGNETIC CIRCUITS, TRANSFORMERS, AND TI
CIRCUITS □ MAGNETIC CIRCUITS, TRANSFORMERS, AND THREE-PH
MAGNETIC CIRCUITS, TRANSFORMERS, AND THREE-PHASE CIRCUI
CIRCUITS, TRANSFORMERS, AND THREE-PHASE CIRCUITS □ MAGN
TRANSFORMERS, AND THREE-PHASE CIRCUITS □ MAGNETIC CIRCU
AND THREE-PHASE CIRCUITS □ MAGNETIC CIRCUITS, TRANSFORMI
THREE-PHASE CIRCUITS □ MAGNETIC CIRCUITS, TRANSFORMERS,
-PHASE CIRCUITS □ MAGNETIC CIRCUITS, TRANSFORMERS, AND T
CIRCUITS □ MAGNETIC CIRCUITS, TRANSFORMERS, AND THREE-PH
MAGNETIC CIRCUITS, TRANSFORMERS, AND THREE-PHASE CIRCUI
CIRCUITS, TRANSFORMERS, AND THREE-PHASE CIRCUITS □ MAGN
TRANSFORMERS, AND THREE-PHASE CIRCUITS □ MAGNETIC CIRCU
AND THREE-PHASE CIRCUITS □ MAGNETIC CIRCUITS, TRANSFORM

he remainder of this book is devoted to study of the devices used in the interconversion of electric and mechanical energy and of the control systems associated with them. Attention is also given to the transformer, which, although not an energy-conversion device, is an important auxiliary in the transfer and conversion of electric energy.

Practically all transformers and electric machinery use magnetic material for shaping the magnetic fields which act as the medium for transferring and converting energy. The relationships between the magnetic-field quantities and the electric circuits with which they interact play an important part in describing the operation of electromagnetic devices. The magnetic material determines the size of the equipment, its capability, and the limitations on its performance.

In this chapter, we shall proceed from introductory magnetic-circuit concepts to relatively simple magnetic structures and then to a consideration of transformers. We shall conclude with a treatment of three-phase circuits, an aspect of circuit theory important to the discussion of electric machines and of the bulk transfer of electric energy.

13-1 MAGNETIC EFFECTS OF AN ELECTRIC CURRENT

In the neighborhood of a current-carrying conductor, effects are produced very like those in the vicinity of a permanent magnet: Forces act on magnets, iron, or other current-carrying conductors which are introduced. In addition, if the current is changing in magnitude, voltages are induced in nearby circuits.

Because of the forces which are produced, the neighborhood of the conductor is regarded as possessing a *force field.* It is a region of energy storage possessing the ability to produce forces and do work. This field, called the *magnetic field,* exists simultaneously with the electric field described in Sec. 1-2. The magnetic field can be mapped by drawing lines which, along their entire length, indicate the direction of the force on the north magnetic pole of a permanent magnet (the north tip of a compass needle, for example). This procedure is similar to drawing flow lines to show the direction of water flow in a stream. The map can be made quantitative by drawing many such lines in an area where the force is great and only a few where the force is small. In other words, the *density* of the lines at a particular point is made proportional to the force exerted on the magnetic pole when it is there. In this respect, the lines resemble somewhat the contour lines on a topographical map, contour lines being densest where the slope is greatest.

These lines are called *flux lines.* A simple map is given in Fig. 13-1 showing the magnetic flux surrounding a long, straight conductor carrying current into the paper. The innermost circle is the conductor; it has a plus

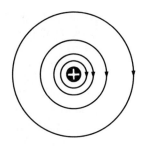

FIGURE 13-1 Flux lines surrounding conductor.

sign in it, representing the tail of an arrow, to denote current into the paper. (Had current been out of the paper, a dot, representing the head of an arrow, would have been used.) Note that the density of the lines is greatest near the conductor, indicating that the force is greatest there. Note also that the flux lines are continuous lines which close on themselves. The direction of the lines is given by the *right-hand rule,* which states that if the conductor is grasped by the right hand, with the thumb extending along it and pointing in the current direction, flux direction will be that in which the fingers wrap around the conductor.

In order to think of a specific number of flux lines (or, as it is commonly expressed, a specific amount of flux), a definite area must be involved for these lines to pass through. The density of the flux lines, or *flux density,* on the other hand, will often change from point to point and must generally be given with specific reference to a particular point. In Fig. 13-1, for example, the flux density decreases as the point moves radially outward from the conductor; for any point on a circle concentric with the conductor, however, the flux density is constant. When the flux ϕ is distributed uniformly over an area A, the flux density at any point in that area is

$$B = \frac{\phi}{A} \tag{13-1}$$

We must also have units of measurement of flux ϕ and flux density B. The unit for flux which has been standardized internationally is the *weber* (abbreviated Wb), named after one of the early physicists who worked on magnetism; the corresponding unit for flux density is the weber per square meter. Both of these units are part of the *rationalized meter-kilogram-second,* or *rationalized MKS, system.* This system is so named because the fundamental units are the meter for length, the kilogram for mass, and the second for time. The electrical units which we have been dealing with (ampere, volt, watt, ohm) are also part of this system. As we shall see in Sec. 13-3, another system of magnetic units is also still in use in English-speaking countries.

A current-carrying conductor will, strictly speaking, set up flux in all space surrounding it. Certain materials, called *ferromagnetic materials* (notably, iron and certain of its alloys with cobalt, tungsten, nickel, aluminum, and other metals), normally are far more receptive to magnetic flux than is air or free space. These materials are used in most devices to concentrate and confine the preponderant portion of the flux within them, thereby greatly increasing the effectiveness for the intended purpose. This action is somewhat like confining water to a pipe rather than letting it flow naturally. We then have a *magnetic circuit*. Magnetic-circuit considerations are closely analogous to those of the electric circuit, with one outstanding exception: The magnetic circuit is usually nonlinear (i.e., the cause-and-effect relation is not a straight-line function). Graphical methods are therefore often used.

13-2 MAGNETIC-CIRCUIT CONCEPTS

Consider a toroidal ring of ferromagnetic material with a coil of wire wound tightly around it (see Fig. 13-2a). When a current is present in the coil, the resulting flux is confined essentially to the ring. Only variations of magnetic quantities around this definite path need be considered. The flux lines will be concentric circles, and the area of the path will be the same at any perpendicular section. If the width of the ring is small compared with its inside or outside diameter, the length of the flux path will be essentially the same along any circle and the flux will be distributed uniformly over the area. Concepts developed from this idea are applicable with sufficient accuracy to widely differing geometries.

The flux-producing ability of the coil in Fig. 13-2a or of a coil on any other magnetic circuit is proportional to the number of turns N and the current I. It is measured by the *magnetomotive force* $\mathscr{F}$ (abbreviated MMF) given by

$$\mathscr{F} = NI \qquad \text{ampere-turns} \tag{13-2}$$

The MMF is the magnetic potential difference tending to force flux around the ring.

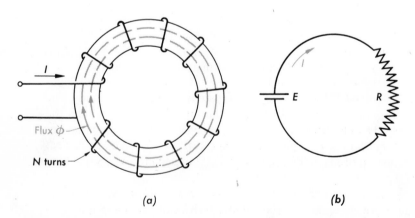

FIGURE 13-2 Toroidal magnetic circuit and the analogous electric circuit.

The resulting flux in the ring, besides being dependent on the MMF, is also a function of the opposition of the iron to carrying flux. This opposition is called the *reluctance* $\mathcal{R}$ of the magnetic circuit. The flux ϕ is then

$$\phi = \frac{\mathcal{F}}{\mathcal{R}} \quad \text{webers} \tag{13-3}$$

As is the case with resistance in the electric circuit, reluctance is directly proportional to length l, inversely proportional to cross-sectional area A, and dependent on the material of the magnetic circuit. With l and A in meters and square meters, respectively,

$$\mathcal{R} = \frac{l}{\mu A} \quad \text{ampere-turns/weber} \tag{13-4}$$

when the flux is constant over the length and uniform over the area. The quantity μ expresses the property of the magnetic material called its *permeability*. Permeability is a measure of the receptiveness of the material to having magnetic flux set up in it. For free space, the permeability μ_o is $4\pi \times 10^{-7}$ in the rationalized MKS system. (The appearance of the factor 4π in this constant is the reason that this system of units is called *rationalized;* by having 4π as a part of this constant, we avoid having it as a factor in several other equations.) The permeability of ferromagnetic material may range up to thousands of times that of free space, indicating why flux tends to concentrate in such material.

Equation (13-3) is sometimes referred to as Ohm's law for the magnetic circuit. It serves to emphasize the mathematical analogy between the magnetic circuit and the dc electric circuit. Analogous quantities in the two circuits are listed in Table 13-1. On this basis, the electric circuit of Fig. 13-2b is the analog of the magnetic circuit of Fig. 13-2a, and the same general thought process is used in analyzing each.

TABLE 13-1 ANALOGOUS QUANTITIES IN MAGNETIC AND ELECTRIC CIRCUITS

Magnetomotive force $\mathcal{F}$	Electromotive force E
Flux ϕ	Current I
Reluctance $\mathcal{R}$	Resistance R
Flux density B	Current density I/A

Example 13-1

Consider a small toroidal core with an inside diameter of 0.050 in and an outside diameter of 0.080 in. The core cross section is rectangular, with a height of 0.025 in. It is made of a material called *ferrite,* having a permeability 1,000 times that of free space. The core flux density is to be 0.15 Wb/m². Determine the core flux and the number of ampere-turns which must be wound on the core to produce this flux density.

SOLUTION   The cross-sectional area of the toroidal core is

$$A = 0.025 \times 0.015 = 0.000375 \text{ in}^2 = 2.42 \times 10^{-7} \text{ m}^2$$

Hence

$$\phi = BA = 0.15 \times 2.42 \times 10^{-7} = 3.63 \times 10^{-8} \text{ Wb}$$

The mean length of the magnetic flux path is that of a circle midway between the inside and outside diameters, or

$$l = \pi(0.065) = 0.204 \text{ in} = 5.18 \times 10^{-3} \text{ m}$$

Then

$$\mathscr{R} = \frac{l}{\mu A} = \frac{5.18 \times 10^{-3}}{(1{,}000)(4\pi \times 10^{-7})(2.42 \times 10^{-7})} = 0.171 \times 10^8 \text{ At/Wb}$$

The MMF required to establish the flux is

$$\mathscr{F} = \phi\mathscr{R} = (3.63 \times 10^{-8})(0.171 \times 10^8) = 0.62 \text{ At}$$

There is inherently some error in this computation because more of the flux will take the shorter paths near the inner circumference of the ring than the longer path near the outer circumference. Magnetic-circuit computations rarely have high precision, however.

When the foregoing concepts are applied to more irregular and complicated magnetic circuits, the circuit is divided into a number of elements so chosen that the flux density is approximately the same at all points in an element. The method is then applied to each element individually. For a number of elements in series, the same total flux must exist in each. The equivalent of the two Kirchhoff laws can be written for the magnetic circuit. The first, analogous to the current law, is that the sum of the flux entering a junction in a magnetic circuit equals the flux leaving the junction. The second, analogous to the voltage law, is that around any closed path in the magnetic circuit, the algebraic sum of the MMFs required to force the flux through the elements must equal the net ampere-turns of excitation.

Magnetic circuits differ from electric circuits in one important respect: The reluctance of a magnetic-circuit branch containing iron or other ferromagnetic material is a function of the flux in that branch. As the flux increases, a larger change in MMF is required to produce the same change in flux. The circuit is said to *saturate*. Saturation usually has started within the normal operating region of the circuit; or, said differently, economic design of the magnetic circuit often requires that the iron reach partial saturation.

We see, then, that both permeability and reluctance may change from one operating condition to another. Hence direct numerical application of the reluctance concept and Eq. (13-3) is rare. Their principal value lies in the guidance they provide to the thought process, largely by virtue of the electric-circuit analogy. For quantitative analysis, graphical methods are generally used because they can easily be adapted to the nonlinearities involved. They are presented in the following sections.

13-3 MAGNETIZATION CURVES

To develop a graphical approach to magnetic-circuit computation, let us first combine Eqs. (13-1) and (13-4) into one compact relation. Thus,

$$B = \frac{\phi}{A} = \frac{\mathscr{F}}{A\mathscr{R}} = \frac{\mathscr{F}}{A(l/\mu A)} = \mu \frac{\mathscr{F}}{l} = \mu H \qquad \text{webers/square meter} \quad (13\text{-}5)$$

where

$$H = \frac{\mathscr{F}}{l} = \frac{NI}{l} \qquad \text{ampere-turns/meter} \tag{13-6}$$

The quantity H is called the *magnetizing force;* it is the MMF per unit length of magnetic circuit. (Note that the length here is that of the magnetic circuit, not that of the coil furnishing the MMF.) For a uniform magnetic circuit, such as that of Fig. 13-2a, the MMF expended per unit length of core is constant around the flux path. The magnetizing force H is therefore also constant along the path.

Equation (13-5) is a basic relation containing the magnetic property of the material in the quantity μ. When we are given the magnetic properties of a specific material by its manufacturer, a curve of flux density B as a function of magnetizing force H ordinarily is included. It is usually called a *BH curve* or *magnetization curve.* A typically shaped curve for iron is sketched in Fig. 13-3. The *BH* curve is dependent on the material only and not upon the dimensions of a specific piece. It is well adapted to graphical analysis. For example, if the ampere-turns NI of the coil and the mean length of the flux path around the toroid of Fig. 13-2a are known, the *BH* curve is entered with the

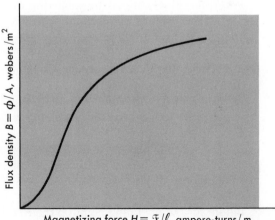

Magnetizing force $H = \mathscr{F}/l$, ampere-turns/m

FIGURE 13-3 Typical *BH* curve.

TABLE 13-2 COMPARISON OF RATIONALIZED MKS AND MIXED ENGLISH SYSTEMS

QUANTITY	SYMBOL	MIXED ENGLISH UNITS	RATIONALIZED MKS UNITS
MMF	$\mathscr{F}$	Ampere-turn	Ampere-turn
Flux	ϕ	Line or kiloline	1 weber = 10^8 lines
Flux density	B	Line/inch² or kiloline/ inch²	1 weber/meter² = 64.5 kilolines/inch²
Magnetizing force	H	Ampere-turn/inch	1 ampere-turn/meter = 0.0254 ampere-turn/inch
Permeability of free space	μ_0	3.19×10^{-3} kiloline/ (ampere-turn)(inch)	$4\pi \times 10^{-7}$ weber/(ampere-turn)(meter)

magnetizing force NI/l on the horizontal axis. The corresponding flux density B is given by the vertical ordinate, and the flux ϕ is BA. The process obviously can be reversed when the flux is known and the MMF of the coil is desired.

In English-speaking countries, engineering computations relating to magnetic circuits may also be carried out in a so-called *mixed English system* of units. The system stems from the habit of using the inch as the unit of length. To a certain extent, also, this system is retained because it tends to avoid very large or very small numbers (as illustrated in Example 13-1, although the situation is somewhat exaggerated there by the small dimensions of the core). Table 13-2 gives a comparison of the mixed English and rationalized mks units. The differences are that inches and square inches are used for length and area in the mixed system and *lines* or *kilolines* (= 1,000 lines) are used as units of flux. The mixed English system is introduced at this point because it is still fairly common to find magnetization curves in these units, so that they may be retained in carrying out computations.

13-4 CHARACTERISTICS OF MAGNETIC MATERIALS

When the MMF on the toroidal ring of Fig. 13-2a is varied from $+\mathscr{F}_1$ to $-\mathscr{F}_1$, the magnetizing force anywhere in the magnetic circuit varies correspondingly from $+H_1$ to $-H_1$. If this cycle is passed through a number of times, the flux density then likewise varies cyclically from $+B_1$ to $-B_1$, but not in a manner which is a single-valued function of H. Instead, the variation of B with H is around the loop $a_1bcdefa_1$ (Fig. 13-4). If the magnetizing force H_1 is removed, an amount of *remanent* or *residual magnetism* equal to Ob remains. To remove it, the magnetizing force must be reversed and made equal to the *coercive force* Oc.

This phenomenon is known as *hysteresis*, and the loop $a_1bcdefa_1$ is called a *hysteresis loop*. If the variation should be smaller, one of the two smaller loops of Fig. 13-4 may be traversed. Because of hysteresis, some uncertainty is unavoidably attached to the result of a magnetic-circuit computation, for the operating point on the BH plot in any given instance depends on the previous magnetic history of the circuit. There may be a wide difference, for example, between the results of ascending to a specified value of

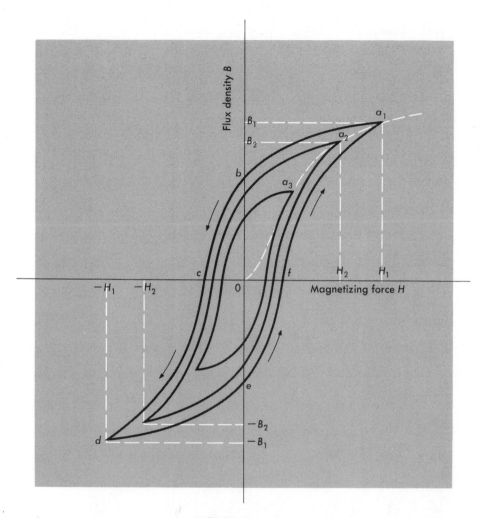

FIGURE 13-4 Hysteresis loops and magnetization curve.

H and descending to it. To obtain average results, the *normal magnetization curve,* or *BH* curve, representing the characteristics of a particular material is taken as the line $Oa_3a_2a_1$ joining the tips of successively larger hysteresis loops.

Such normal magnetization curves for a few common materials are shown in Fig. 13-5. The units on the coordinates of these curves are in the mixed English system. Also included is a *BH* curve for air, which is simply a plot of the fact that when B is measured in kilolines per square inch,

$$H_{air} = 313B_{air} \qquad \text{ampere-turns/inch} \qquad (13\text{-}7)$$

The magnetic properties of materials depend appreciably on heat treatment and mechanical handling. Consequently, the characteristics of a given sample may differ somewhat from those of another sample of the same mate-

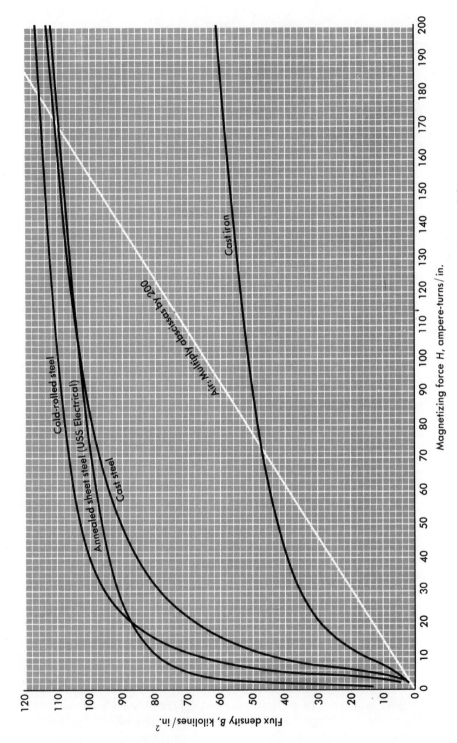

FIGURE 13-5 Normal magnetization curves for common magnetic materials.

rial, an additional source of uncertainty in magnetic calculations. High precision should therefore not be attempted in these calculations.

Numerous alloys of iron with other metals are used when special magnetic properties are desired. These alloys are generally known by various names. For example, when a permanent magnet is to be constructed, the material should have a relatively high remanent flux density. One of a series of aluminum-nickel-cobalt-iron alloys known as *alnico* may then be the most effective material. When very high permeability is desired, nickel-iron alloys known as *Permalloy* or *Hipernik* may be used. *Perminvar*, a nickel-cobalt-iron alloy, shows great constancy of permeability over a significant range of magnetization. The list of such special alloys is a rather long one.

An important group for electronic and control application is that known as *square-loop magnetic materials*. As shown in typical form in Fig. 13-6a and in idealized form in Fig. 13-6b, the materials have a nearly rectangular hysteresis loop. The remanent flux density B_r is nearly equal to the maximum density B_m. The sides of the loop have steep slopes, so that a small change in H causes a large change in B. The core, in effect, becomes substantially a *bistable* magnetic element because it can be in one of two states: fully saturated either in one direction or in the other. Application of the full MMF in a specific direction is required to change it from one state to the other.

Square-loop materials are of two general types. One type, the *ferrites*, is sometimes referred to as a magnetic ceramic. They are formed at high tem-

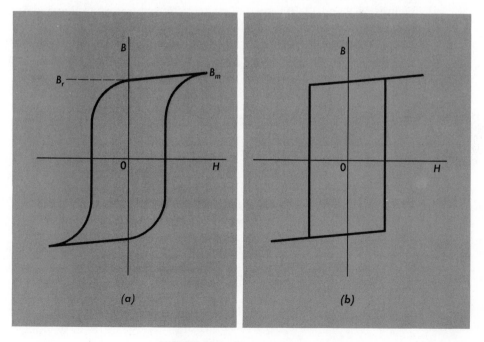

FIGURE 13-6 Characteristics of square-loop magnetic materials.

peratures from mixtures of pulverized iron oxide and other metallic oxides, such as manganese. The second type is metallic and is composed of high-permeability alloys like Permalloy. Square-loop materials are used in switching circuits, as storage elements in computers, and in special types of transformers in electronic circuits.

13-5 PRACTICAL MAGNETIC CIRCUITS

A representative group of magnetic circuits commonly encountered is presented in Figs. 13-7 to 13-10. In all these circuits, the larger amount of flux is confined to the intended path, but a small amount must inevitably leak through the surrounding air. This stray flux is called *leakage flux.* Computations concerned with the main magnetic circuit are usually made with the effect of leakage flux included empirically or neglected entirely. Separate studies of leakage must frequently be made, however, especially for ac machines and transformers, because their performance may be profoundly affected by it.

Figure 13-7 shows a core-type transformer. The magnetic path forms a simple series circuit, and the total MMF is the sum of that for the four parts *ab, bc, cd,* and *da* of the path. A shell-type transformer is shown in Fig. 13-8. The center leg *da* contains the entire main flux, and the paths *abcd* and *ab'c'd* in parallel complete the circuit, each carrying half the flux. Since the same MMF provides for both parallel branches, the total MMF is that for the path *dabcd* or *dab'c'd* alone.

Figures 13-9 and 13-10 show the magnetic structures of a relay and of a magnetic clutch, respectively. In both cases, an air element is an integral part of the magnetic circuit. The flux will naturally fringe out somewhat when it reaches the air gap and will occupy a somewhat greater area than in

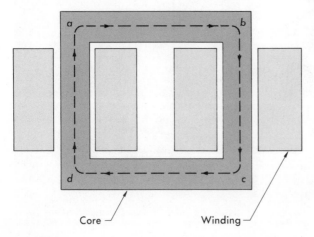

FIGURE 13-7 Magnetic circuit for core-type transformer.

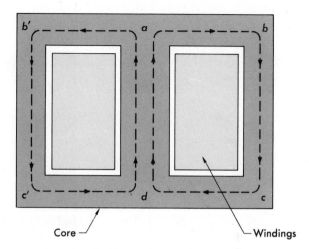

FIGURE 13-8 Magnetic circuit for shell-type transformer.

the adjacent iron. Correction for such *fringing* in short air gaps is made by adding the gap length to each of the two dimensions making up its area.

The parts of the magnetic circuits of Figs. 13-7 to 13-10 subjected to alternating MMF are constructed of laminations punched from thin sheets. As a result of the small amount of space unavoidably present between successive laminations, together with the coating of insulating varnish frequently used on the laminations, the effective magnetic cross-sectional area is less than the overall area of the stack. The effective area equals the overall area times a *stacking factor,* which depends on the lamination thickness and is about 0.90 for 0.014-in (29-gauge) laminations.

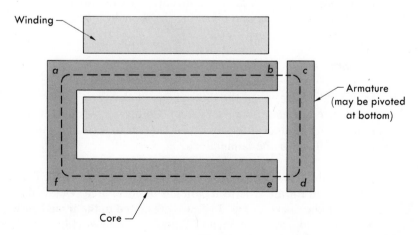

FIGURE 13-9 Magnetic structure for a relay.

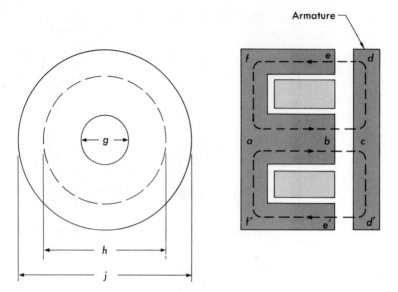

FIGURE 13-10   Structure of a magnetic clutch.

13-6 MAGNETIC CIRCUITS WITH DC EXCITATION

The two general problems arising are to find the excitation required to produce a specified flux or flux density and to find the flux or flux density resulting from a specified excitation. The methods of analysis will be illustrated by specific examples, the solutions being in terms of the quantities
given in Fig. 13-5. Such is the common procedure, instead of use of the reluctance concept and solutions in terms of the permeability μ, because it is
more direct. The reluctance concept, with its electric-circuit analogy, is
useful in formulating and checking the thought process, however.

Example 13-2

In Fig. 13-11 are given the dimensions of the core and armature of a relay
constructed of electrical sheet steel. Compute the number of ampere-turns
required to product a flux of 5,000 lines in this circuit. Assume a stacking
factor of 0.90 for the laminations.

SOLUTION This is a series circuit in which the total MMF is the sum of that
for iron and for air. The iron path has uniform area. Detailed computations,
based on the curve in Fig. 13-5, are presented in the table below. The length
of the path in iron is the mean length, *abcd* plus *ef* (Fig. 13-11). The air-gap
area is corrected for fringing.

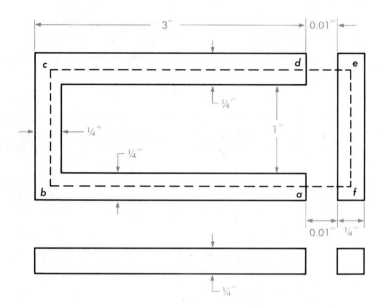

FIGURE 13-11 Magnetic circuit for Examples 13-2 and 13-3.

PART	AREA, IN²	LENGTH, IN	*B*, KILOLINE/ IN²	*H*, AT/IN	*NI*, AT
Iron	(0.25)(0.25)(0.90) = 0.0563	8.5	89	24	(24)(8.5) = 204
Air	(0.25 + 0.01)(0.25 + 0.01) = 0.0676	0.02	74	23,200	(23,200)(0.02) = 464

Total MMF = 204 + 464 = 670 At

Example 13-3

Compute the flux resulting from an MMF of 1,000 At on the magnetic circuit for Example 13-2.

SOLUTION This is the converse problem to that of Example 13-2. Since the division of the total ampere-turns between the iron and the air, and hence the ampere-turns per inch for each portion, is unknown, the magnetization curve cannot be entered directly. The solution therefore must be based on assuming trial values of flux or flux density and checking each trial by the method of Example 13-2 until one is found requiring the specified excitation.

 Completely random choice of values for the first trial can be avoided by examining the results of two extreme assumptions: (1) that all the MMF is used in the air gap, and (2) that all the MMF is used in the iron. Both assumptions will yield flux densities which are too high. A value somewhat smaller than the lower of the two may then be selected for the first trial.

Thus, if all the MMF were used in the air gap, then, from Eq. (13-7),

$$B_{air} = \frac{1,000}{(0.02)(313)} = 160 \text{ kilolines/in}^2$$

If all the MMF were used in the iron, the magnetizing force would be 1,000/8.5 or 118 At/in, corresponding to a density of 105 kilolines/in² in the iron (see Fig. 13-5). Let us therefore try a density of 100 kilolines/in² in the iron.

Computational details are summarized in the table at the end of this solution. Since the first trial requires too high an excitation, a second trial at 95 kilolines/in² is made. By interpolation between the two trials, the final iron density is

$$95 + (100 - 95)\frac{1,000 - 870}{1,160 - 870} = 97 \text{ kilolines/in}^2$$

and the total flux is

$$\phi = (97,000)(0.0563) = 5,500 \text{ lines}$$

The result of this interpolation is checked in the table. The check shows a total NI of 980 At, 2 percent lower than the specified value of 1,000. Because of the uncertainties mentioned earlier, agreement within about 5 percent between the computed and the specified MMF may be regarded as sufficiently close.

TRIAL	B_{iron} (ASSUMED)	H_{iron} (FIG. 13-5)	$(NI)_{iron}$ $(8.5H_{iron})$	B_{air} $\left(B_{iron}\frac{0.0563}{0.0676}\right)$	H_{air} [EQ. (13-7)]	$(Ni)_{air}$ $(0.02H_{air})$	TOTAL NI $[(NI)_{iron} + (NI)_{air}]$
First	100	75	638	83.3	26,100	522	1,160
Second	95	44	374	79.2	24,800	496	870
Check	97	56	476	81.6	25,200	504	980

The foregoing two examples are based on simple series magnetic circuits. Analysis of series-parallel circuits may be arithmetically more cumbersome but utilizes the same approach based on the magnetic-circuit form of Kirchhoff's laws.

Example 13-4

The magnetic circuit of Fig. 13-12 has a cast-steel core whose dimensions are given in the table below. The area of the air gap bc has been corrected for fringing.

PORTION	ab	bc	cd	ad	dea
Mean length, in	10	0.01	10	8	15
Area, in²	4	4.04	4	2	6

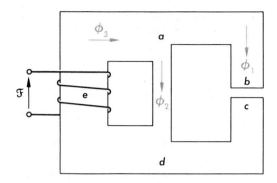

FIGURE 13-12 Magnetic circuit for Example 13-4.

Determine the coil MMF required to establish an air-gap flux of 200 kilo-lines.

SOLUTION The analysis is based upon summing the MMF or magnetic-potential drops around the two loops, together with summing the fluxes at junction a or d.

The flux density in legs ab and cd, which are identical, is $\frac{200}{4}$ = 50 kilo-lines/in². That in gap bc is $\frac{200}{4.04}$ = 49.5 kilolines/in². By use of Fig. 13-5 for the two legs and Eq. (13-7) for the gap, the MMF across ad becomes

$$\mathscr{F}_{ad} = 12(10 + 10) + 313(49.5)0.01 = 395 \text{ At}$$

Leg ad, being 8 in long, is accordingly subjected to a magnetizing force H of $\frac{395}{8}$ = 49.4 At/in. The corresponding flux density, from Fig. 13-5, is 90 kilolines/in², so that

$$\phi_2 = (90)(2) = 180 \text{ kilolines}$$

At junction a, ϕ_3 must be the sum of ϕ_1 and ϕ_2, or

$$\phi_3 = 200 + 180 = 380 \text{ kilolines}$$

The flux density in leg dea is then $\frac{380}{6}$ = 63.5 kilolines/in². From Fig. 13-5, the corresponding magnetizing force is 18 At/in, so that

$$\mathscr{F}_{dea} = 18(15) = 270 \text{ At}$$

The required MMF for the coil is that for dea plus that for ad, or

$$\mathscr{F} = 270 + 395 = 665 \text{ At}$$

13-7 MAGNETICALLY INDUCED VOLTAGES; SELF-INDUCTANCE

A very important effect of a magnetic field on an electric circuit is that when the flux linking the circuit changes, a voltage is induced. Electromagnetic induction of voltage is basic to the operation of transformers, motors, and generators. The effect is described by Faraday's law, which states that the voltage induced in an N-turn coil when the flux ϕ linking or threading through it changes at the rate $d\phi/dt$ is

$$e = N \frac{d\phi}{dt} \qquad \text{volts} \tag{13-8}$$

with ϕ in webers and time t in seconds. The direction of the induced voltage, in accordance with Lenz's law, is such as to produce a current opposing the flux change.

This voltage has not entered the discussions of magnetic circuits with dc excitation because they have been considered to be in the steady state with constant current, MMF, and flux. The induced voltage is then zero. It will play an important role in circuits with ac excitation, however, and in the discussions of all types of machines in later chapters.

The induced voltage appears in any circuit which may be linked by the changing flux, including the circuit giving rise to the flux. Thus, current in a circuit produces a magnetic field linking the circuit and permeating the medium around it. Growth, decay, or any other change of current with time causes corresponding changes of the magnetic flux and induces a voltage in the circuit. From Eq. (13-8), this voltage of self-induction is

$$e = N \frac{d\phi}{dt} = N \frac{d\phi}{di} \frac{di}{dt}$$

$$= L \frac{di}{dt} \qquad \text{volts} \tag{13-9}$$

where

$$L = N \frac{d\phi}{di} \qquad \text{henrys} \tag{13-10}$$

The quantity L is the *self-inductance*, or simply the *inductance*, of the circuit and is in henrys when ϕ is in webers. Self-inductance is introduced as a circuit parameter in Sec. 1-8, where Eq. (13-9) appears as Eq. (1-17). Whenever an inductance term appears in circuit theory, it represents the reaction of the magnetic field upon the circuit.

The inductance concept is of greatest usefulness when the flux is directly proportional to the current, which implies that the permeability of the medium comprising the flux path is constant. Practically speaking, this statement means that no significant part of the path of the flux concerned is in saturated iron. Under these circumstances, Eq. (13-10) reduces to

$$L = \frac{N\phi}{i} \qquad \text{henrys} \tag{13-11}$$

With ϕ and i directly proportional, inductance is a constant independent of current and dependent only on the geometry of the circuit element and the permeability of the magnetic medium. For many configurations, evaluation is difficult because of the complexity of the magnetic field. From Eq. (13-11), inductance is seen to be proportional to N^2, for ϕ itself is proportional to N.

Example 13-5

The operating coil of the relay in Example 13-2 has 10,000 turns and a resistance of 3,000 Ω. It is to be operated from a 230-V dc source. To cause the relay armature to close requires a current (the *pickup current*) sufficient to produce the 670 At found in Example 13-2.

a Compute the approximate inductance of the coil.

b Because this relay is intended for fast operation, it is desired to check approximately the required operating time. Compute the time required for the current to reach the pickup value when a direct voltage of 230 V is suddenly impressed.

SOLUTION **a** Inductance will be defined in accordance with Eq. (13-11). The current required for 670 At is

$$i = \frac{670}{10,000} = 0.067 \text{ A}$$

Then

$$L = \frac{N\phi}{i} = \frac{(10,000)(5,000 \times 10^{-8})}{0.067} = 7.5 \text{ H}$$

b The KVL equation for the coil is

$$7.5\frac{di}{dt} + 3,000i = 230$$

Using the techniques of Chap. 4 to solve for the current gives

$$i = \frac{230}{3,000}(1 - \epsilon^{-3,000t/7.5})$$

If $i(t_1)$ is to be 0.067 A,

$$0.067 = \frac{230}{3,000}(1 - \epsilon^{-3.000t/7.5})$$

yielding $t_1 = 0.0052$ s.

13-8 MAGNETIC CIRCUITS WITH AC EXCITATION

The magnetic circuits of transformers (Figs. 13-7 and 13-8), ac machines, and many other electromagnetic devices are excited from ac rather than dc sources. With dc excitation, the steady-state current is determined by the impressed voltage and the resistance of the circuit, the inductance entering only into transient processes such as that of Example 13-5. The flux in the magnetic circuit then adjusts itself in accordance with this value of current so that the relationship imposed by the magnetization curve is satisfied. With ac excitation, however, inductance enters into the steady-state performance as well; the result for most magnetic circuits, although not for all, is that, to a close approximation, the flux is determined by the impressed voltage and frequency, and the magnetizing current must adjust itself in accordance with this flux so that the relationship imposed by the magnetization curve is satisfied.

Except where preservation of linear relationships is of great importance, economic utilization of material dictates that the normal working flux density in a magnetic circuit be beyond the linear portion of the magnetization curve for the overall circuit. Hence accurate analyses cannot be predicated on constant self-inductance. Equivalent circuits containing parameters that do remain substantially constant are used instead, as will be shown in subsequent sections. The reactive effect of the time-varying flux on the exciting circuit can readily be shown from Faraday's law, Eq. (13-8).

Consider the N-turn iron-core coil of Fig. 13-13. A magnetic flux ϕ is produced by an exciting current i. Let the flux vary sinusoidally with time t, as in

$$\phi = \phi_m \sin 2\pi ft \tag{13-12}$$

ϕ_m being the maximum value in webers and f the frequency. The induced voltage in accordance with Faraday's law [Eq. (13-8)] is

$$e = N\frac{d\phi}{dt} = 2\pi fN\phi_m \cos 2\pi ft \tag{13-13}$$

and its effective value is

$$E = \frac{2\pi}{\sqrt{2}}fN\phi_m = 4.44fN\phi_m \qquad \text{volts} \tag{13-14}$$

The polarity of the voltage must, in accordance with Lenz's law, oppose the flux change and therefore is as shown in Fig. 13-13 when the flux is increasing. Since the current produces the flux, the two may be considered in phase. From Eq. (13-13), the induced voltage leads the flux, and hence the current, by 90°. The induced voltage and the coil resistance drop oppose the impressed voltage.

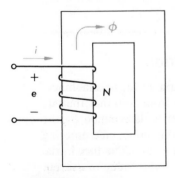

FIGURE 13-13 Magnetic circuit with ac excitation.

The resistance drop does not exceed a few percent of the impressed voltage in ac machines, most transformers, and many other ac electromagnetic devices. To a close approximation, resistance drop may be neglected, and the impressed and induced voltages may be considered equal. The flux ϕ_m is then determined by the impressed voltage in accordance with Eq. (13-14), even if maintenance of this flux requires a magnetizing current far in excess of rated current for the device.

Example 13-6

In order to limit the magnetizing current to a reasonable value, the maximum flux density of a 60-Hz core-type transformer is not to exceed 70 kilolines/in². There are 200 turns on the 2,300-V primary winding.

a What is the minimum value of the gross cross-sectional area of the annealed sheet-steel core? Assume a stacking factor of 0.90.

b If the mean length of the core is 100 in, what is the peak value of the magnetizing current in the 2,300-V winding?

SOLUTION **a**

$$\phi_m = \frac{E}{4.44fN}$$

$$= \frac{2,300}{(4.44)(60)(200)} = 0.0431 \text{ Wb} = 4,310 \text{ kilolines}$$

Hence

$$\text{Area} = \frac{4,310}{(70)(0.90)} = 68.4 \text{ in}^2$$

b This flux density requires a magnetizing force of 5.5 At/in (see Fig. 13-5). Peak magnetizing current is then

$$\frac{(5.5)(100)}{200} = 2.8 \text{ A}$$

13-9 HYSTERESIS AND EDDY-CURRENT LOSSES

When magnetic circuits are subjected to time-varying flux densities, there are two causes of power loss in the form of heat in the iron core. These losses are significant in determining the heating, rating, and efficiency of machines, transformers, and ac-operated magnetic devices.

The first loss is associated with the phenomenon of hysteresis, discussed in Sec. 13-4, and is an expression of the fact that when ferromagnetic material is involved, not all the energy of the magnetic field is returned to the circuit when the MMF is removed. It is known as *hysteresis loss*. When the flux varies cyclically from $+B_1$ to $-B_1$ (Fig. 13-4) at the frequency f, the hysteresis loss per unit volume of core material may be shown to be proportional to the area of the hysteresis loop and to the number of loops traversed per second. On the basis of experimental studies, the hysteresis loss in a given core may be approximated empirically by the expression

$$P_h = k_h f B_m^n \qquad \text{watts} \tag{13-15}$$

k_h being a characteristic constant of the core and B_m the maximum flux density, while n, called the *Steinmetz exponent*, may vary from about 1.5 to over 2.0 and is often taken as 1.6.

The second loss arises from the fact that the core itself is composed of conducting material, so that the voltages induced in it by the changing flux produce circulating currents in the iron. These are called *eddy currents* and are accompanied by an i^2r loss in the core, called the *eddy-current loss*. Since the eddy currents depend upon the rate of change of flux as well as the resistance of the path, it is reasonable to expect this loss to vary as the square of both the maximum flux density and the frequency. Eddy-current loss is given by the expression

$$P_e = k_e f^2 B_m^2 \qquad \text{watts} \tag{13-16}$$

when the flux varies sinusoidally, k_e being a characteristic constant of the core. To increase the core resistance and thereby minimize eddy currents, magnetic cores subject to alternating fluxes are assembled from thin sheets with an insulating layer (surface oxide or varnish) between successive laminations; k_e varies as the square of the thickness of these laminations. Eddy currents also have a magnetic effect, tending to make the flux density at the center lower than at the surface. This *screening effect* is negligible in properly laminated cores at power frequencies but may be of great importance at higher frequencies.

Hysteresis and eddy-current losses, taken together, are known as *core loss*. Core loss is present in dc machines as well as in ac machines, for the rotor iron of a rotating dc machine contains flux which varies cyclically in both magnitude and direction. Many machines are operated at constant voltage and constant frequency or speed and, as a result, have substantially constant core losses regardless of load. The operating voltage and frequency of a power transformer, for example, are normally constant; by virtue of Eq. (13-14), the maximum flux and flux density are therefore constant, and the core loss is the same whether the transformer is loaded or not.

13-10 INTRODUCTION TO TRANSFORMERS

Essentially, a transformer consists of two or more windings interlinked by a common or mutual magnetic field. If one of these windings, the *primary*, is connected to an alternating-voltage source, an alternating flux will be produced whose amplitude will depend on the primary voltage and number of turns. This flux is described in Sec. 13-8, and its magnitude is that denoted by ϕ_m in Eq. (13-14) when N is the number of primary turns. The mutual flux will link the other winding, the *secondary,* in which it will induce a voltage whose value will depend on the number of secondary turns and will be characterized also by Eq. (13-14) when N is the number of secondary turns and ϕ_m refers to the mutual flux. When the numbers of primary and secondary turns are properly proportioned, almost any desired voltage ratio, or *ratio of transformation,* can be obtained. Alternating-voltage levels can thus readily be changed by means of transformers; as we shall see, changes in both current and impedance levels are also involved. There are, of course, no reasons why a third winding (*tertiary*) or additional windings cannot be introduced to interconnect a variety of voltage levels.

Transformer action requires only the existence of an alternating flux linking both the windings. Such action will be obtained if an air core is used, giving rise to an *air-core transformer.* It will be obtained much more effectively, however, with a core of iron or other ferromagnetic material. Most of the flux is then confined to a definite path, which links both windings and has a much higher permeability than air. Such a transformer is an *iron-core transformer.* The majority of transformers are of this type, the principal exceptions being air-core transformers for use at high frequencies beyond the audio range.

To reduce the losses caused by eddy currents in the core, the magnetic circuit usually consists of stacks of thin laminations; two common types of construction are shown in Figs. 13-7 and 13-8. Silicon-steel laminations 0.014 in thick are generally used for transformers operating at frequencies below a few hundred hertz. Silicon steel has the desirable properties of low cost, low core loss, and high permeability at high flux densities (65 to 90 kilolines/in²). The cores of small transformers used in communications circuits at high frequencies and low energy levels may be made of compressed powdered ferromagnetic alloys such as one of the ferrites.

The transformer is one of the principal reasons for the widespread use of ac power systems, for it makes possible electric generation at the most economical generator voltage, power transfer at the most economical transmission voltage, and power utilization at the most suitable voltage for the particular utilization device. When reference is made to the windings of power transformers, the terms *high-voltage winding* and *low-voltage winding* are usually used. Either winding, of course, is capable of acting as primary or secondary.

The transformer is also widely used in low-power electronic and control circuits. There it performs such functions as matching the impedances of a source and its load for maximum power transfer, insulating one circuit from another, or isolating direct current while maintaining ac continuity between two circuits.

13-11 THE IDEAL TRANSFORMER

The most important aspects of transformer action can be brought out by idealizing the transformer. Figure 13-14 shows schematically a transformer having two windings with N_1 and N_2 turns, respectively, on a common mag-

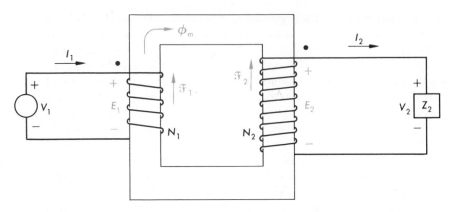

FIGURE 13-14 Ideal transformer.

netic circuit. Assume that (1) all the flux is confined to the core and links both windings, (2) winding resistances are negligible, (3) core losses are negligible, and (4) the permeability of the core is so high that only a negligible MMF is required to establish the flux. These properties of the ideal transformer are closely approached but never actually attained in real transformers.

Let a sinusoidal voltage with rms value V_1 and frequency f be applied to the primary winding. In accordance with Kirchhoff's voltage law, a voltage E_1 must be induced in the winding equal and in opposition to V_1. From Eq. (13-14), the maximum core flux is

$$\phi_m = \frac{V_1}{4.44fN_1} \qquad \text{webers} \tag{13-17}$$

This core flux also links the secondary winding and produces an induced emf E_2 and an equal secondary-terminal voltage V_2 given by

$$V_2 = E_2 = 4.44fN_2\phi_m \qquad \text{volts} \tag{13-18}$$

which, upon substitution of Eq. (13-17), becomes

$$V_2 = \frac{N_2}{N_1}V_1 \qquad \text{and} \qquad \frac{V_1}{V_2} = \frac{N_1}{N_2} \tag{13-19}$$

In an ideal transformer, then, the ratio of the primary to the secondary voltage is the ratio of the primary to the secondary turns.

The secondary induced voltage is produced by the same flux as the primary induced voltage and is either in phase with or 180° out of phase with the primary voltage, depending on the manner in which the coils are wound on the core. In order to be specific in this regard, dots are placed near the transformer terminals in diagrams such as that of Fig. 13-14. The dots indicate that the induced voltage rises in the two windings are in phase when both are defined as rises from the unmarked to the dotted terminal. Thus in Fig. 13-14, the voltages E_1 and E_2 are in phase.

Now let an impedance Z_2 be connected across the secondary terminals. A current

$$I_2 = \frac{V_2}{Z_2} \tag{13-20}$$

will then be present in the secondary winding. The current I_2 will produce an MMF

$$\mathscr{F}_2 = N_2 I_2 \tag{13-21}$$

in the direction shown in Fig. 13-14. This secondary MMF $\mathscr{F}_2$ is in opposition to the core flux, and unless it is counteracted by a primary MMF, the core flux will be radically changed and voltage balance required by Kirchhoff's law in the primary and achieved through Eq. (13-17) will be disturbed. Hence a compensating primary MMF $\mathscr{F}_1$ and current I_1 must be called into being such that

$$\mathscr{F}_1 = N_1 I_1 = \mathscr{F}_2 = N_2 I_2 \tag{13-22}$$

or

$$I_1 = \frac{N_2}{N_1} I_2 \quad \text{and} \quad \frac{I_1}{I_2} = \frac{N_2}{N_1} \tag{13-23}$$

The currents in the primary and secondary of an ideal transformer are then in the inverse ratio of the turns.

In the ideal transformer, all losses have been neglected. From conservation of energy, then, the voltampere input and output must be identical. The same conclusion can be reached by combining Eqs. (13-19) and (13-23):

$$V_2 I_2 = \left(\frac{N_2}{N_1} V_1 \right)\left(\frac{N_1}{N_2} I_1 \right) = V_1 I_1 \tag{13-24}$$

Next examine the effect in the primary circuit of the secondary impedance Z_2. Evidently, from Eqs. (13-19) and (13-23)

$$\frac{V_1}{I_1} = \left(\frac{N_1}{N_2} \right)^2 \frac{V_2}{I_2} \tag{13-25}$$

Consequently, as far as its effect is concerned, the impedance Z_2 in the secondary may be replaced by an equivalent impedance Z_1 in the primary, provided that

$$Z_1 = \left(\frac{N_1}{N_2} \right)^2 Z_2 \tag{13-26}$$

Thus, the three circuits of Fig. 13-15 are indistinguishable as far as their steady-state performance viewed from terminals a and b is concerned. Transferring an impedance from one side of a transformer to the other in this fashion is called *referring the impedance* to the other side. In a similar manner, voltages and currents may be *referred* to one side or the other by

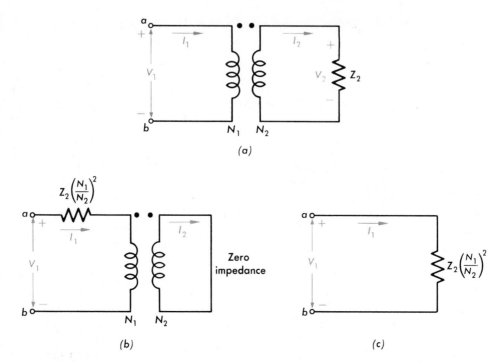

FIGURE 13-15   Three circuits which give identical performance at terminals a and b when the transformer is ideal.

using Eqs. (13-19) and (13-23) to evaluate the equivalent voltage and current on that side.

To sum up, *in an ideal transformer, voltages are transformed in the direct ratio of turns, currents in the inverse ratio, and impedances (and, of course, resistances and reactances individually) in the direct ratio squared, and power and voltamperes are unchanged.*

Example 13-7

A transformer with a 10:1 turns ratio and rated 50 kVA, 2,400/240 V, 60 Hz is used to step down the voltage of a distribution system. The low-tension voltage is to be kept constant at 240 V.

a   What load impedance connected to the low-tension side will cause the transformer to be fully loaded?

b   What is the value of this load impedance referred to the high-tension side?

c   What are the values of the load current referred to the low-tension and high-tension sides?

SOLUTION **a** The current for 50 kVA at 240 V is

$$\frac{50,000}{240} = 208 \text{ A}$$

The impedance of the load is then

$$\frac{240}{208} = 1.15 \ \Omega$$

b Referred to the high-tension side, this impedance is

$$1.15(10)^2 = 115 \ \Omega$$

c As in part (**a**), the low-tension current is 208 A. Referred to the high-tension side, it becomes

$$\frac{208}{10} = 20.8 \text{ A}$$

Example 13-8

An audio-frequency transformer is used to couple a 50-Ω resistive load to an electronic source which can be represented by a constant voltage of 5 V in series with an internal resistance of 2,000 Ω. The transformer can be considered ideal.

a Determine the primary-to-secondary turns ratio required to ensure maximum power transfer by matching the load and source impedances (i.e., by causing the 50-Ω secondary impedance to be 2,000 Ω when it is referred to the primary).

b What will be the current, voltage, and power in the load under these conditions?

SOLUTION **a** From Eq. (13-26),

$$2,000 = \left(\frac{N_1}{N_2}\right)^2 50$$

or

$$\frac{N_1}{N_2} = \sqrt{40} = 6.32$$

b Viewed on the primary side of the transformer, the circuit will consist of the 5-V source, the 2,000-Ω internal resistance, and the referred 2,000-Ω load resistance, all in series. The primary current is

$$I_1 = \frac{5}{2,000 + 2,000} = 0.00125 \text{ A} = 1.25 \text{ mA}$$

Then

$$I_2 = 1.25 \sqrt{40} = 7.90 \text{ mA}$$

$$V_2 = I_2 R_{\text{load}} = (0.00790)(50) = 0.395 \text{ V}$$

and

$$P_2 = I_2^2 R_{\text{load}} = (0.00790)^2(50) = 0.0031 \text{ W} = 3.1 \text{ mW}$$

We can refer to the voltage and current results in several ways. We can say that the actual secondary or load voltage is 0.395 V or that it is 0.395 $\sqrt{40}$ = 2.5 V referred to the primary. Similarly, we can say that the true load current is 7.90 mA or that the load current referred to the primary is 1.25 mA.

Note that we might also have solved part (*b*) with all quantities referred to the secondary. We would then have a series circuit with a 5/$\sqrt{40}$ = 0.79-V source, a 50-Ω internal resistance, and a 50-Ω load resistance. The results, of course, would be the same.

13-12 THE TRANSFORMER AS A CIRCUIT ELEMENT

In discussing the ideal transformer, we have brought out the main reasons for inserting transformers in circuits. Well-designed practical transformers accomplish very close to these same results, but there are inevitably some side effects caused by departures from the four idealizing assumptions. When the suitability of a proposed transformer for a specific application is to be judged, these side effects must be examined.

A transformer coupling a source having internal voltage E_G and resistance r_G with a load resistance r_L is shown schematically in Fig. 13-16*a*. The same circuit with the transformer completely represented by an ideal transformer plus circuit elements is given in Fig. 13-17*a*. We shall now see how

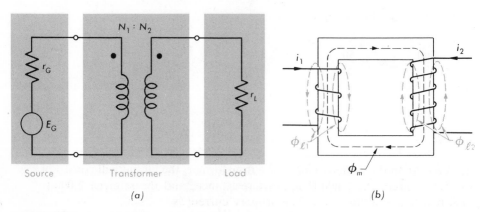

FIGURE 13-16 (*a*) Transformer coupling a source to a load. (*b*) Component fluxes in the transformer.

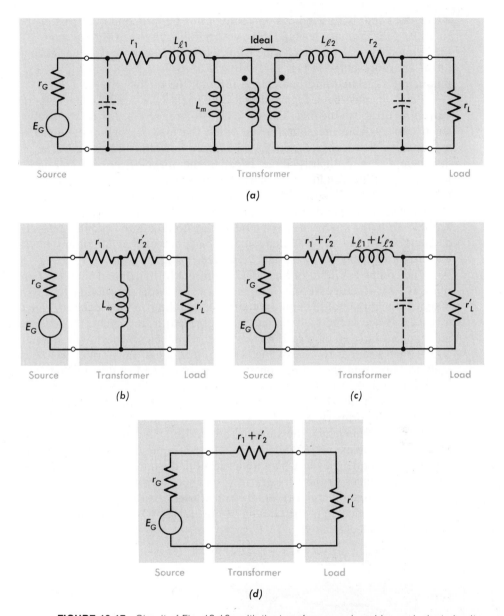

FIGURE 13-17 Circuit of Fig. 13-16a with the transformer replaced by equivalent circuits.

these circuit elements take account of the differences between actual and ideal transformers.

First, the windings of the real transformer obviously have resistance; these are represented by the primary and secondary resistances r_1 and r_2. Second, not all the flux in the core links both windings. As indicated in Fig. 13-16b, the core flux can be divided into three parts: ϕ_m, the mutual flux, which does link both windings and constitutes the preponderant part of the total; ϕ_{l1}, the primary leakage flux, which links only the primary; and ϕ_{l2}, the secondary leakage flux, which links only the secondary. The leakage fluxes are represented in Fig. 13-17a by the *leakage inductances* L_{l1} and L_{l2}, flux being translated into inductance through the medium in Eq. (13-10).

Third, it must be recognized that a physical core requires magnetizing current to establish the flux. Thus, the transformer in Fig. 13-16a will draw a small current from the source even when the load is open-circuited. The *magnetizing inductance* L_m in Fig. 13-17a provides for this current. The inductance L_m is ordinarily large because the magnetizing current under normal operating conditions usually does not exceed a few percent of the full-load current. When core losses are to be included, a resistance is included in series or parallel with L_m; often these losses are ignored, however, as they are here. Lastly, it may be necessary to take into account stray capacitances, which can be an unavoidable part, however small, of any physical circuit element. They are shown dotted in Fig. 13-17a.

The complete circuit of Fig. 13-17a is rather formidable to analyze. Fortunately, satisfactory results can be obtained with simplifying approximations, the nature of the approximations depending on the frequency range involved. Since inductive reactance decreases with frequency, the leakage inductances have negligible effect at low frequencies. The low-frequency circuit then reduces to that of Fig. 13-17b. The ideal transformer has been omitted here, and the primes on the constants r_2' and r_L' denote that they have been referred to the primary through use of Eq. (13-26). Simply remember that the ideal transformer is there and that all quantities have been referred to the primary.

At high frequencies, the shunting effect of L_m becomes negligible, and the circuit becomes that of Fig. 13-17c. There is generally also an intermediate band of frequencies in which none of the inductances is important, and the circuit reduces to the network of resistances shown in Fig. 13-17d. To simplify the situation still further, the transformer resistances r_1 and r_2 or r_2' can often be neglected in the circuits of Fig. 13-17. When transformer resistances are neglected in Fig. 13-17d, we are, of course, back to the ideal transformer.

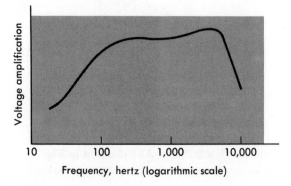

FIGURE 13-18 Frequency-response curve for iron-core transformer.

When transformers are used in electronic and control circuits, their frequency-response characteristics are often a matter of importance. Figure 13-18 gives an example of the response of an iron-core audio-frequency transformer in the circuit of Fig. 13-16a, with the source representing the output of an amplifier. The drop-off at the low-frequency end is caused by the drain imposed by the magnetizing inductance L_m (Fig. 13-17b). Note that there is a significant midband-frequency range where the response is substantially flat. Here the behavior of the transformer is close to ideal (Fig. 13-17d). The drop-off at the high-frequency end is caused by leakage inductance (13-17c). The rise just before this drop-off is caused by the distributed capacitance shown dotted in Fig. 13-17c. The normal performance of a transformer in a power-supply circuit or system is determined at the single supply frequency (e.g., 60 Hz). The equivalent circuit of Fig. 13-17c is sufficiently accurate, and the transformer resistances can often be neglected.

Example 13-9

The audio transformer of Example 13-8 has the following constants: $L_{l1} = 5$ mH, $L_{l2} = 0.125$ mH, L_m (measured on primary side) $= 0.632$ H, $N_1/N_2 = \sqrt{40}$. The winding resistances, core losses, and stray capacitance can be ignored. The transformer is still used in the circuit of Example 13-8. Determine the secondary-terminal voltage under the following conditions:

a At 15,000 Hz, omitting the magnetizing inductance

b At 100 Hz, neglecting leakage inductances

c At 5,000 Hz, omitting all inductances

SOLUTION **a** The high-frequency circuit of Fig. 13-17c, with numerical values pertinent to this example, is reproduced in Fig. 13-19a. The load resistance referred to the primary is $r_L' = 50(\sqrt{40})^2 = 2{,}000$ Ω. The referred secondary leakage inductance is $L_{l2}' = 0.125(\sqrt{40})^2 = 5$ mH.

The total leakage reactance at 15,000 Hz is

$$2\pi(15{,}000)(10 \times 10^{-3}) = 942 \ \Omega$$

The rms primary current in Fig. 13-19a is

$$I_1 = \frac{5}{\sqrt{(4{,}000)^2 + (942)^2}} = 0.001216 \ \text{A}$$

Then

$$V_1 = 0.001216\sqrt{(2{,}000)^2 + (942)^2} = 2.69 \ \text{V}$$

$$V_2' = 0.001216 \times 2{,}000 = 2.43 \ \text{V}$$

$$V_2 = \frac{N_2}{N_1} V_2' = \frac{2.43}{\sqrt{40}} = 0.384 \ \text{V}$$

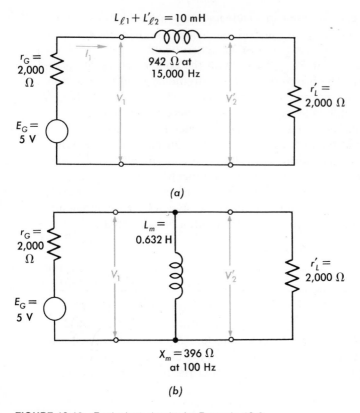

FIGURE 13-19 Equivalent circuits for Example 13-9.

b The low-frequency circuit of Fig. 13-17*b* is reproduced in Fig. 13-19*b*. The equivalent impedance of L_m and r'_L in parallel is

$$\frac{(0 + j396)(2,000 + j0)}{(0 + j396) + (2,000 + j0)} = 75.4 + j380 \qquad \Omega$$

With $\mathbf{E}_G$ as reference phasor, the primary current is

$$\mathbf{I}_1 = \frac{5 + j0}{2,000 + 75.4 + j380} = 0.00237\underline{/-10.4°}\text{ A}$$

Then

$$\begin{aligned}
\mathbf{V}'_2 = \mathbf{V}_1 &= \mathbf{E}_G - \mathbf{I}_1 r_G \\
&= 5 - (0.00237\underline{/-10.4°})(2,000) = 0.921\underline{/68.3°}
\end{aligned}$$

and

$$V_2 = \frac{N_2}{N_1}V'_2 = \frac{0.921}{\sqrt{40}} = 0.146\text{ V}$$

c Since winding resistances are ignored, the transformer is ideal in the midband. From Example 13-8, $V_2 = 0.395$ V.

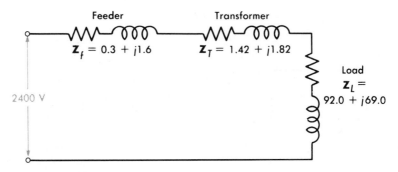

FIGURE 13-20 Equivalent circuit for transformer of Example 13-10.

Example 13-10

The 50-kVA 2,400/240-V 60-Hz transformer of Example 13-7 has a leakage impedance of $0.72 + j0.92 \ \Omega$ in the high-tension winding and $0.0070 + j0.0090 \ \Omega$ in the low-tension winding (both at 60 Hz). The magnetizing current is negligibly low. The transformer is used to step down the voltage at the load end of a feeder whose impedance is $0.3 + j1.6 \ \Omega$. The impedance of the load on the low-tension side of the transformer is $0.920 + j0.690 \ \Omega$.

With 2,400 V at the sending end of the feeder, find the voltage at the load.

SOLUTION The complete circuit with all quantities referred to the high-tension side is given in Fig. 13-20, the transformer circuit being that of Fig. 13-17c. The total impedance is

$$\mathbf{Z}_f + \mathbf{Z}_T + \mathbf{Z}_L = 93.7 + j72.4 = 118\underline{/37.7°} \ \Omega$$

Hence

$$I = \frac{2,400}{118} = 20.4 \ \text{A}$$

The load voltage is then

$$V_L = IZ_L = 20.4\sqrt{(92.0)^2 + (69.0)^2} = 2,340 \ \text{V}$$

referred to the high-tension side. The actual voltage at the load is 2,340/10, or 234 V.

13-13 PULSE TRANSFORMERS

Many of the circuits found in applications such as radar, television, and digital computers are called *pulse* or *digital circuits* because the voltage and current waveforms are pulses. The transformers used in such circuits are

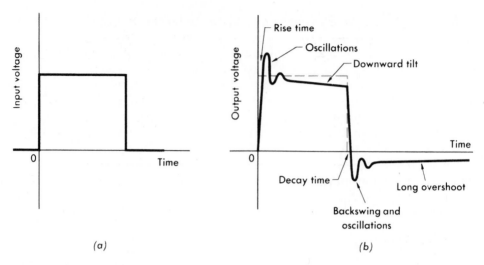

FIGURE 13-21   (a) Square-wave input voltage-to-pulse transformer and (b) the corresponding output-voltage waveform.

pulse transformers. They are inserted for the same general reasons that they appear in the more conventional electronic circuits—to change the amplitude of a pulse, to couple successive stages of pulse amplifiers, to change impedance levels, to isolate direct current from a circuit element, etc.

In fulfilling these requirements, it is important that the transformer reproduce the input pulse as faithfully as possible at its secondary terminals. Figure 13-21a shows a square-wave input pulse. The pulse width will usually range from a fraction of a microsecond to about 20 μs, and a relatively long time will elapse before the pulse repeats. To determine the output waveform requires a transient rather than a steady-state examination. A typical result is that shown in Fig. 13-21b.

The response to the leading edge of the pulse is determined by the high-frequency equivalent circuit of Fig. 13-17c (including the stray capacitance). Because of the presence of the leakage inductance, an appreciable time is required for the output voltage to build up to the desired value. This is called the *rise time*. Because of the stray capacitance, there will usually be oscillations resulting in successive overshooting and undershooting of the desired value. Leakage inductance is kept to a minimum for the shortest rise time.

The response to the flat-top portion of the input pulse is determined by the low-frequency equivalent circuit of Fig. 13-17b. The output voltage cannot remain flat, for that would be the equivalent of transmitting steady direct current through the transformer. Instead, the waveform shows a downward *tilt*, or drop-off of voltage (see Fig. 13-21b). In time, of course, the voltage would become zero, but the duration of the pulse is short compared with this time. The tilt of the pulse top is kept within allowable limits by having high magnetizing inductance, that is, by constructing the core of high-permeability material.

When the input voltage is removed at the termination of the pulse, there is an appreciable *decay time* before the secondary voltage reaches zero. There is also a significant *backswing* associated with a damped oscillation and a long-duration negative overshoot of the voltage. In effect, during this period, the magnetizing inductance is discharging the energy of the decaying magnetic field through the stray capacitance and circuit resistances. The resultant waveform is that of a parallel *RLC* circuit.

Pulse transformers are of a small physical size and have relatively few turns in order to minimize leakage inductance. The cores are constructed of ferrites or of wound strips of high-permeability alloys such as Hipersil (a special high-permeability silicon steel) or Permalloy. Since the time interval between pulses is long compared with the pulse duration, the load-carrying duty on the transformer is light. As a result, a very small transformer can handle surprisingly high pulse-power levels.

13-14 THREE-PHASE VOLTAGES, CURRENTS, AND POWER

Most of the generation, transmission, and heavy-power utilization of electric energy involves polyphase systems, i.e., systems in which several sources equal in magnitude but differing in phase from each other are available. Because it possesses definite economic and operating advantages, the three-phase system is by far the most common. A three-phase source is one which has available three equal voltages which are 120° out of phase with one another. As we shall point out in Sec. 14-5, all three voltages are usually generated in the same machine. A three-phase load is one which can utilize the output of a three-phase source. Three voltage sources forming a three-phase system are shown in Fig. 13-22a; a phasor diagram of these voltages is shown in Fig. 13-22b. Note that the phasor sum of the three-voltages is zero.

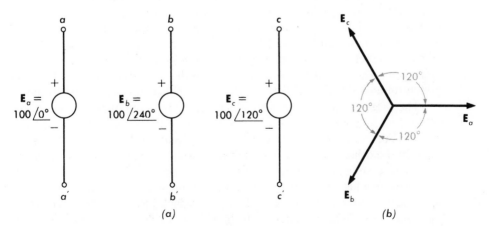

FIGURE 13-22 (a) Three-phase source; (b) phasor diagram.

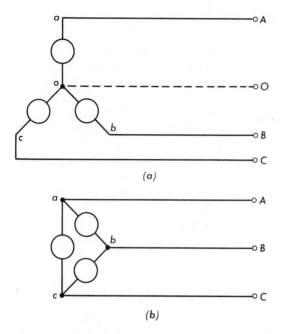

FIGURE 13-23 Three-phase connections: (*a*) Y connection; (*b*) Δ connection.

There are two possibilities for the utilization of voltages generated in this manner: The six terminals a, a', b, b', c, and c' of the winding may be connected to three independent single-phase systems, or the three phases of the winding may be interconnected and used to supply a three-phase system. The latter procedure is the one adopted almost universally. The three phases of the winding can be interconnected in two possible ways, as shown in Fig. 13-23: Terminals a', b', and c' may be joined to form the neutral o, yielding a *Y connection*, or terminals a and b', b and c', and c and a' may be joined individually, yielding a Δ *connection.* In the Y connection, a *neutral conductor,* shown dotted in Fig. 13-23*a*, may or may not be brought out. If a neutral conductor exists, the system is a *four-wire three-phase system;* if not, it is a *three-wire three-phase system.* In the Δ connection (Fig. 13-23*b*), no neutral exists and only a three-wire three-phase system can be formed.

The three phase voltages, Fig. 13-22, are equal and phase-displaced by 120 electrical degrees, a general characteristic of a *balanced three-phase system.* Furthermore, the impedance in any one phase is equal to that in either of the other two phases, so that the resulting phase currents are equal and phase-displaced from each other by 120 electrical degrees. Likewise, equal power and equal reactive power flow in each phase. An *unbalanced three-phase system,* on the other hand, may lack any or all of these equalities and 120° displacements. It is important to note that *only balanced systems are treated in this book and that none of the methods developed or*

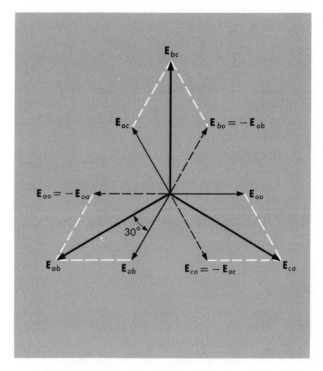

FIGURE 13-24 Voltage phasor diagram for Y connection.

conclusions reached applies to unbalanced systems. The great majority of practical problems are concerned with balanced systems. Many industrial loads are three-phase loads and therefore inherently balanced, and in supplying single-phase loads from a three-phase source, definite efforts are made to keep the three-phase system balanced by assigning approximately equal single-phase loads to each of the three phases.

When the three phases of Fig. 13-22a are Y connected, as in Fig. 13-23a, the phasor diagram of voltages is that of Fig. 13-24. The *phase order* or *phase sequence* in Fig. 13-24 is *abc;* that is, the voltage of phase *a* reaches its maximum 120° before that of phase *b*. The use of *double-subscript notation* in Fig. 13-24 greatly simplifies the task of drawing the complete diagram. The subscripts indicate the points between which the voltage exists, and the order of subscripts indicates the direction in which the voltage rise is taken.[1] Thus, $\mathbf{E}_{ao} = -\mathbf{E}_{oa}$.

The three phase voltages are E_{oa}, E_{ob}, and E_{oc}. They are also called *line-to-neutral voltages*. The three voltages E_{ab}, E_{bc}, and E_{ca}, called *line*

---

[1] The use of voltage rises is widespread in the power industry. However, note that the rise from *o* to *a* is identical to the drop from *a* to *o*. This observation allows the phasor diagram in Fig. 13-24 to be used to relate the voltage drops in a three-phase system.

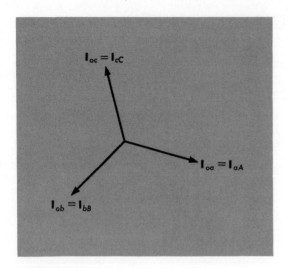

FIGURE 13-25 Current phasor diagram for Y connection.

voltages or, more specifically, *line-to-line voltages,* are also important. By Kirchhoff's voltage law, the line voltage $\mathbf{E}_{ab}$ is

$$\mathbf{E}_{ab} = \mathbf{E}_{ao} + \mathbf{E}_{ob} = -\mathbf{E}_{oa} + \mathbf{E}_{ob}$$
$$= \sqrt{3}\ \mathbf{E}_{ob}\underline{/-30°} \tag{13-27}$$

as shown in Fig. 13-24. Similarly,

$$\mathbf{E}_{bc} = \sqrt{3}\ \mathbf{E}_{oc}\underline{/-30°} \tag{13-28}$$

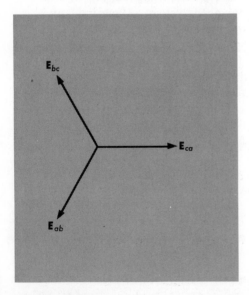

FIGURE 13-26 Voltage phasor diagram for Δ connection.

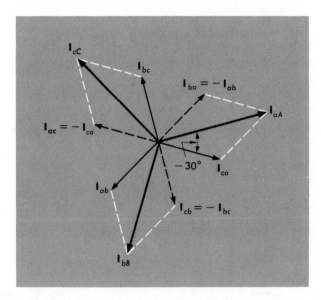

FIGURE 13-27 Current phasor diagram for Δ connection.

and

$$\mathbf{E}_{ca} = \sqrt{3}\ \mathbf{E}_{oa}\underline{/-30°} \tag{13-29}$$

Stated in words, these equations show that, *for a Y connection, the line voltage is $\sqrt{3}$ times the phase voltage, or the line-to-line voltage is $\sqrt{3}$ times the line-to-neutral voltage.*

The corresponding current phasors for the Y connection of Fig. 13-23a are given in Fig. 13-25. Obviously, *for a Y connection, the line currents and phase currents are equal.*

When the three phases in Fig. 13-22a are Δ-connected, as in Fig. 13-23b, the phasor diagram of voltages is that of Fig. 13-26. Obviously, *for a Δ connection, the line voltages and phase voltages are equal.*

The corresponding phasor diagram of currents is given in Fig. 13-27. The three phase currents are I_{ab}, I_{bc}, and I_{ca}, the order of subscripts indicating the current directions. By Kirchhoff's current law, the line current $\mathbf{I}_{aA}$ is

$$\begin{aligned}\mathbf{I}_{aA} &= \mathbf{I}_{ba} + \mathbf{I}_{ca} = -\mathbf{I}_{ab} + \mathbf{I}_{ca}\\ &= \sqrt{3}\ \mathbf{I}_{ca}\underline{/30°}\end{aligned} \tag{13-30}$$

as shown in Fig. 13-27. Similarly,

$$\mathbf{I}_{bB} = \sqrt{3}\ \mathbf{I}_{ab}\underline{/30°} \tag{13-31}$$

and

$$\mathbf{I}_{cC} = \sqrt{3}\ \mathbf{I}_{bc}\underline{/30°} \tag{13-32}$$

Stated in words, Eqs. (13-30) to (13-32) show that *for a Δ connection, the line*

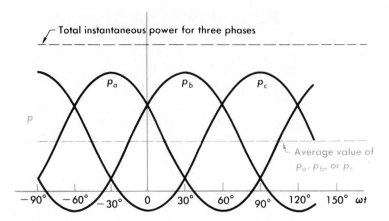

FIGURE 13-28 Instantaneous power in a three-phase system.

current is $\sqrt{3}$ *times the phase current.* Evidently, the relations between phase and line currents of a Δ connection are similar to those between phase and line voltages of a Y connection.

For both Y- and Δ-connected systems, the instantaneous power per phase can be determined in exactly the same manner as for a single-phase system. With the time origin taken at the maximum positive point of the phase-*a* current wave, the instantaneous power for phase *a* is given by Eq. (5-27), where *E* and *I* are the phase voltage and current, respectively. Similarly, the instantaneous powers for phases *b* and *c* may be obtained from Eq. (5-27) when the appropriate 120° phase displacements are inserted in the voltage and current terms. Instantaneous powers for the three phases are plotted in Fig. 13-28, together with the total instantaneous power, which is the sum of the three individual waves. *Notice that the total instantaneous power for a three-phase system is constant and is equal to three times the average power per phase.*

In general, it can be shown that the total instantaneous power for any balanced polyphase system is constant. This is one of the outstanding advantages of polyphase systems. It is of particular advantage in the operation of polyphase motors, for example, for it means that the shaft power output is constant and that torque pulsations, with the consequent tendency toward vibration, do not result from pulsations inherent in the supply system.

On the basis of single-phase considerations, the average power per phase P_p for either a Y- or Δ-connected system is

$$P_p = E_p I_p \cos \theta = I_p^2 R_p \tag{13-33}$$

where E_p, I_p, and R_p are the voltage, current, and resistance, respectively, all per phase. The total three-phase power *P* is

$$P = 3P_p \tag{13-34}$$

Similarly, for reactive power per phase Q_p and total three-phase reactive power Q,

$$Q_p = E_p I_p \sin \theta = I_p^2 X_p \tag{13-35}$$

and

$$Q = 3Q_p \tag{13-36}$$

The voltamperes per phase $(VA)_p$ and total three-phase voltamperes VA are

$$(VA)_p = E_p I_p = E_p^2 / Z_p \tag{13-37}$$

and

$$VA = 3(VA)_p \tag{13-38}$$

In Eqs. (13-33) and (13-35), θ is the angle between phase voltage and phase current. As in the single-phase case, it is given by

$$\theta = \tan^{-1} \frac{X_p}{R_p} = \cos^{-1} \frac{R_p}{Z_p} = \sin^{-1} \frac{X_p}{Z_p} \tag{13-39}$$

The power factor of a balanced three-phase system is therefore equal to that of any one phase.

13-15   Y- and Δ-CONNECTED CIRCUITS

Three specific examples will be given to illustrate the computational details of Y- and Δ-connected circuits. Explanatory remarks which are generally applicable are incorporated in the solutions.

Example 13-11

In Fig. 13-29 is shown a 60-Hz transmission system consisting of a line having the impedance $Z_l = 0.05 + j0.20$ Ω, at the receiving end of which is a load of equivalent impedance $Z_L = 10.0 + j3.00$ Ω. The impedance of the return conductor should be considered zero.

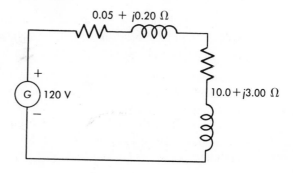

FIGURE 13-29   Circuit for Example 13-11a.

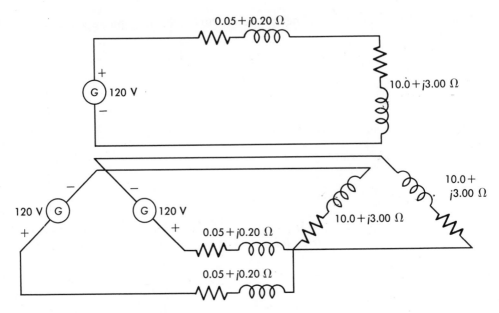

FIGURE 13-30 Circuit for Example 13-11*b*.

a Compute (1) the line current I; (2) the load voltage E_L; (3) the power, reactive power, and voltamperes taken by the load; and (4) the power and reactive-power loss in the line.

Suppose now that three such identical systems are to be constructed to supply three such identical loads. Instead of drawing the diagrams one below the other, let them be drawn in the fashion shown in Fig. 13-30, which is, of course, the same electrically.

b Give, for Fig. 13-30, (1) the current in each line; (2) the voltage at each load; (3) the power, reactive power, and voltamperes taken by each load; (4) the power and reactive-power loss in each of the three transmission systems; (5) the total power, reactive power, and voltamperes taken by the loads; and (6) the total power and reactive-power loss in the three transmission systems.

Next consider that the three return conductors are combined into one and that the phase relationship of the voltage sources is such that a balanced four-wire three-phase system results, as in Fig. 13-31.

c Give, for Fig. 13-31, (1) the line current; (2) the load voltage, both line-to-line and line-to-neutral; (3) the power, reactive power, and voltamperes taken by each phase of the load; (4) the power and reactive-power loss in each line; (5) the total three-phase power, reactive power, and voltamperes taken by the load; and (6) the total power and reactive-power loss in the lines.

d In Fig. 13-31, what is the current in the combined return or neutral conductor?

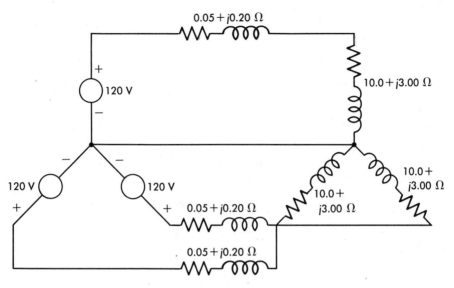

FIGURE 13-31   Circuit for Example 13-11c, d, and e.

e   May this conductor be dispensed with in Fig. 13-31 if desired? (Assume now that this neutral conductor is omitted. This results in the three-wire three-phase system of Fig. 13-32.)

f   Repeat part c for Fig. 13-32.

g   On the basis of the results of this example, outline briefly the method of reducing a balanced three-phase Y-connected circuit problem to its equiva-

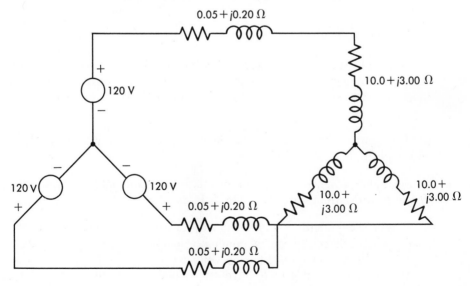

FIGURE 13-32   Circuit for Example 13-11f.

lent single-phase problem. Be careful to distinguish between the use of line-to-line and line-to-neutral voltages.

SOLUTION   **a**

(1) $$I = \frac{120}{\sqrt{(0.05 + 10.0)^2 + (0.20 + 3.00)^2}} = 11.4 \text{ A}$$

(2) $$E_L = IZ_L = 11.4\sqrt{(10.0)^2 + (3.00)^2} = 119 \text{ V}$$

(3) $$P_L = I^2R_L = (11.4)^2(10.0) = 1,300 \text{ W}$$

$$Q_L = I^2X_L = (11.4)^2(3.00) = 390 \text{ var}$$

$$(\text{VA})_L = I^2Z_L = (11.4)^2\sqrt{(10.0)^2 + (3.00)^2} = 1,360 \text{ VA}$$

(4) $$P_l = I^2R_l = (11.4)^2(0.05) = 6.5 \text{ W}$$

$$Q_l = I^2X_l = (11.4)^2(0.20) = 26 \text{ var}$$

b Parts (1) through (4) obviously have the same values as in *a*.

(5)   Total power $= 3P_L = 3(1,300) = 3,900$ W

Total reactive power $= 3Q_L = 3(390) = 1,170$ var

Total VA $= 3(\text{VA})_L = 3(1,360) = 4,080$ VA

(6)   Total power loss $= 3P_l = 3(6.5) = 19.5$ W

Total reactive-power loss $= 3Q_l = 3(26) = 78$ var

c The results obtained in *b* are unaffected by this change. The voltage in *b*2 and *a*2 is now the line-to-neutral voltage. The line-to-line voltage is

$$\sqrt{3} \ 119 = 206 \text{ V}$$

d By Kirchhoff's current law, the neutral current is the phasor sum of the three line currents. These line currents are equal and phase-displaced 120°. Since the phasor sum of three equal phasors 120° apart is zero, the neutral current is zero.

e The neutral current being zero, the neutral conductor may be dispensed with if desired.

f Since the presence or absence of the neutral conductor does not affect conditions, the values are the same as in *c*.

g A neutral conductor may be assumed, regardless of whether one is physically present. Since the neutral conductor in a balanced three-phase circuit carries no current and hence has no voltage drop across it, the neutral conductor should be considered to have zero impedance. Then one phase of the

Y, together with the neutral conductor, may be removed for study. Since this phase is uprooted at the neutral, *line-to-neutral voltages must be used.* This procedure yields the equivalent single-phase circuit, in which all quantities correspond to those in one phase of the three-phase circuit. Conditions in the other two phases being the same (except for the 120° phase displacements in the currents and voltages), there is no need for investigating them individually. Line currents in the three-phase system are the same as in the single-phase circuit, and total three-phase power, reactive power, and volt-amperes are 3 times the corresponding quantities in the single-phase circuit. If line-to-line voltages are desired, they must be obtained by multiplying voltages in the single-phase circuit by $\sqrt{3}$.

Example 13-12

Three impedances of value $\mathbf{Z}_p = 4.00 + j3.00 = 5.00\underline{/36.9°}$ Ω are connected in Y, as shown in Fig. 13-33. For balanced line-to-line voltages of 208 V, find the line current, the power factor, and the total power, reactive power, and voltamperes.

SOLUTION   The line-to-neutral voltage across any one phase, such as ao, is

$$E_p = \frac{208}{\sqrt{3}} = 120 \text{ V}$$

Hence

$$I_l = I_p = \frac{E_p}{Z_p} = \frac{120}{5.00} = 24.0 \text{ A}$$

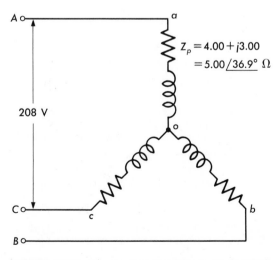

FIGURE 13-33 Y-connected circuit for Example 13-12.

Power factor $= \cos \theta = \cos 36.9° = 0.80$ lagging

$$P = 3P_p = 3I_p^2R_p = 3(24.0)^2(4.00) = 6,910 \text{ W}$$

$$Q = 3Q_p = 3I_p^2X_p = 3(24.0)^2(3.00) = 5,180 \text{ var}$$

$$\text{VA} = 3(\text{VA})_p = 3E_pI_p = 3(120)(24.0) = 8,640 \text{ VA}$$

It should be noted that phases a and c (Fig. 13-33) do not form a simple series circuit. Consequently, the current cannot be found by dividing 208 V by the sum of the phase a and c impedances. To be sure, an equation can be written for voltage between points a and c by Kirchhoff's voltage law, but this must be a phasor equation taking account of the 120° phase displacement between the phase-a and phase-c currents. As a result, the method of thought outlined in Example 13-11g leads to the simplest solution.

Example 13-13

Three impedances of value $\mathbf{Z}_p = 12.00 + j9.00 = 15.00\underline{/36.9°}\ \Omega$ are connected in Δ, as shown in Fig. 13-34. For balanced line-to-line voltages of 208 V, find the line current, the power factor, and the total power, reactive power, and voltamperes.

SOLUTION   The voltage across any one phase, such as ca, is evidently equal to the line-to-line voltage. Consequently,

$$E_p = 208 \text{ V}$$

and

$$I_p = \frac{E_p}{Z_p} = \frac{208}{15.00} = 13.87 \text{ A}$$

Power factor $= \cos \theta = \cos 36.9° = 0.80$ lagging

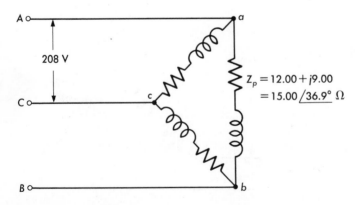

FIGURE 13-34   Δ-connected circuit for Example 13-13.

From Eq. (13-30),

$$I_l = \sqrt{3}\, I_p = \sqrt{3}\, (13.87) = 24.0 \text{ A}$$

Also

$$P = 3P_p = 3I_p^2 R_p = 3(13.87)^2(12.00) = 6{,}910 \text{ W}$$

$$Q = 3Q_p = 3I_p^2 X_p = 3(13.87)^2(9.00) = 5{,}180 \text{ var}$$

and

$$\text{VA} = 3(\text{VA})_p = 3E_p I_p = 3(208)(13.87) = 8{,}640 \text{ VA}$$

It should be noted that phases ab and bc do not form a simple series circuit, nor does the path cba form a simple parallel combination with the direct path through the phase ca. Consequently, the line current cannot be found by dividing 208 V by the equivalent impedance of $\mathbf{Z}_{ca}$ in parallel with $\mathbf{Z}_{ab} + \mathbf{Z}_{bc}$. Kirchhoff's law equations involving quantities in more than one phase can be written, but they must be phasor quantities taking account of the 120° phase displacement between phase currents and between phase voltages. As a result, the method outlined above leads to the simplest solution.

Comparison of the results of Examples 13-12 and 13-13 leads to a valuable and interesting conclusion. It will be noted that the line-to-line voltage, line current, power factor, total power, reactive power, and voltamperes are precisely equal in the two cases; in other words, conditions viewed from the terminals A, B, and C are identical, and one cannot distinguish between the two circuits from their terminal quantities. It will also be seen that the impedance, resistance, and reactance per phase of the Y connection (Fig. 13-33) are exactly one-third of the corresponding values per phase of the Δ connection (Fig. 13-34). Consequently, a balanced Δ connection may be replaced by a balanced Y connection provided that the circuit constants per phase obey the relation

$$\mathbf{Z}_Y = \tfrac{1}{3}\mathbf{Z}_\Delta \qquad\qquad (13\text{-}40)$$

Conversely, a Y connection may be replaced by a Δ connection provided Eq. (13-40) is satisfied. The concept of this Y-Δ equivalence stems from the general Y-Δ transformation and is not the accidental result of a specific numerical case.

Two important corollaries follow from this equivalence. First, a general computational scheme for balanced circuits may be based entirely upon Y-connected circuits or entirely on Δ-connected circuits, whichever one prefers. Since it is frequently more convenient to handle a Y connection, the former scheme is the one usually adopted. Second, in the frequently occurring problems in which the connection is not specified and is not pertinent to the solution, either a Y or a Δ connection may be assumed. Again the Y connection is more commonly selected. In analyzing three-phase motor performance, for example, the actual winding connections need not be known unless the investigation is to include detailed conditions within the coils themselves. The entire analysis may be based on an assumed Y connection.

13-16 ANALYSIS OF BALANCED THREE-PHASE CIRCUITS; SINGLE-LINE DIAGRAMS

By combining the principle of Δ-Y equivalence with the technique revealed by Example 13-11, a simple method of reducing a balanced three-phase-circuit problem to its corresponding single-phase problem may be developed. All the methods of single-phase-circuit analysis thus become available for its solution. The end results of the single-phase analysis are then translated back into three-phase terms to give the final results.

In carrying out this procedure, phasor diagrams need be drawn for but one phase of the Y connection, the diagrams for the other two phases being unnecessary repetition. Furthermore, circuit diagrams may be simplified by drawing only one phase. Examples of such *single-line diagrams* are given in Fig. 13-35, showing two three-phase generators with their associated lines or cables supplying a common substation load. Specific connections of apparatus may be indicated if desired. Thus, Fig. 13-35b shows that G_1 is Y-connected and G_2 is Δ-connected. Impedances are given in ohms per phase.

Evidently the choice of the phasor or voltampere method for the detailed analysis may be made on the basis of convenience. In applying the voltampere method, it is sometimes more convenient to deal with the entire three-phase circuit at once instead of concentrating on one phase. This pos-

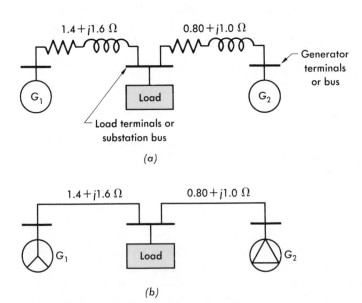

FIGURE 13-35   Examples of single-line circuit diagrams.

sibility arises because simple expressions for three-phase power, reactive power, and voltamperes can be written in terms of line-to-line voltage and line current regardless of whether the circuit is Y- or Δ-connected. Thus, from Eqs. (13-33) and (13-34), three-phase power is

$$P = 3P_p = 3E_pI_p \cos \theta \qquad (13\text{-}41)$$

For a Y connection, $I_p = I_{\text{line}}$ and $E_p = E_{\text{line}}/\sqrt{3}$. For a Δ connection, $I_p = I_{\text{line}}/\sqrt{3}$ and $E_p = E_{\text{line}}$. In either case, Eq. (13-41) becomes

$$P = \sqrt{3} \, E_{\text{line}} I_{\text{line}} \cos \theta \qquad (13\text{-}42)$$

Similarly,

$$Q = \sqrt{3} \, E_{\text{line}} I_{\text{line}} \sin \theta \qquad (13\text{-}43)$$

and

$$\text{VA} = \sqrt{3} \, E_{\text{line}} I_{\text{line}} \qquad (13\text{-}44)$$

The voltampere triangle then holds for either three-phase or single-phase quantities. It should be borne in mind, however, that the power-factor angle θ, given by Eq. (13-39), is the angle between $\mathbf{E}_p$ and $\mathbf{I}_p$ and not that between $\mathbf{E}_{\text{line}}$ and $\mathbf{I}_{\text{line}}$.

The following example illustrates the use of these methods. Line-to-line voltages which depart considerably from standard nominal values appear in this example because it is adapted from a single-phase example previously given. Common nominal line-to-line voltages in this range are 208, 230, 460, 575, and 2,300 V.

Example 13-14

Figure 13-35 is the equivalent circuit of a load supplied from 2 three-phase generating stations over lines having the impedances per phase given on the diagram. The load requires 30 kW at 0.80 power-factor lagging. Generator G_1 operates at a terminal voltage of 797 V line to line and supplies 15 kW at 0.80 power-factor lagging.

Find the load voltage and the terminal voltage and power and reactive power output of G_2.

SOLUTION   The diagram of Fig. 13-35 represents one phase of the three-phase system, the system being considered Y-connected. The voltage at G_1 for one phase is

$$\frac{797}{\sqrt{3}} = 460 \text{ V line to neutral}$$

The power output for one phase of G_1 is

$$\tfrac{15}{3} = 5 \text{ kW at 0.80 power-factor lagging}$$

The power to one phase of the load is

$\frac{30}{3} = 10$ kW at 0.80 power-factor lagging

Conditions within one phase of the system are therefore identical with those of Example 3-13. From the results of Example 3-13, it follows that the desired three-phase quantities are

$E_L = \sqrt{3}\,(432) = 748$ V line to line

$P_2 = 3(5,450) = 16,350$ W

$Q_2 = 3(4,290) = 12,870$ var

and

$E_2 = \sqrt{3}\,(451) = 780$ V line to line

where the notation is the same as in Example 3-13.

ALTERNATIVE SOLUTION   The voltampere method may also be applied directly to a three-phase circuit by using Eqs. (13-42) to (13-44). Note the step-by-step similarity of the following solution to that for Example 3-13.

$$I_1 = \frac{P_1}{\sqrt{3}\,E_1 \cos\theta_1} = \frac{15,000}{\sqrt{3}\,(797)(0.80)} = 13.6 \text{ A}$$

$P_{r1} = P_1 - 3I_1^2R_1 = 15,000 - 3(13.6)^2(1.4) = 14,220$ W

$Q_{r1} = Q_1 - 3I_1^2X_1 = 15,000\tan(\cos^{-1}0.80) - 3(13.6)^2(1.6)$
$= 10,350$ var

The factor 3 appears before $I_1^2R_1$ and $I_1^2X_1$ in the last two equations because the current I_1 exists in all three lines.

$$E_L = \frac{\text{VA}}{\sqrt{3}\,\text{(current)}} = \frac{\sqrt{(14,220)^2 + (10,350)^2}}{\sqrt{3}\,(13.6)}$$
$$= 748 \text{ V line to line}$$

Since the load requires 30,000 W and 30,000 $\tan(\cos^{-1}0.80)$ or 22,500 var,

$P_{r2} = 30,000 - 14,220 = 15,780$ W

and

$Q_{r2} = 22,500 - 10,350 = 12,150$ var

$$I_2 = \frac{\text{VA}}{\sqrt{3}\,\text{(voltage)}} = \frac{\sqrt{(15,780)^2 + (12,150)^2}}{\sqrt{3}\,(748)} = 15.4 \text{ A}$$

$P_2 = P_{r2} + 3I_2^2R_2 = 15,780 + 3(15.4)^2(0.80) = 16,350$ W

$Q_2 = Q_{r2} + 3I_2^2X_2 = 12,150 + 3(15.4)^2(1.0) = 12,870$ var

$$E_2 = \frac{\text{VA}}{\sqrt{3}\,\text{(current)}} = \frac{\sqrt{(16,350)^2 + (12,870)^2}}{\sqrt{3}\,(15.4)} = 780 \text{ V line to line}$$

PROBLEMS

13-1 Draw electric circuits which are analogous to the magnetic circuits of Figs. 13-7 to 13-10.

13-2 Three coils are all acting on the same magnetic circuit. Coil 1 has 4,000 turns and 40-Ω resistance; coil 2, 3,000 turns and 30 Ω; coil 3, 2,000 turns and 20 Ω. Certain combinations of these coils are connected to a 115-V dc supply. Find the resultant MMF on the magnetic circuit under the following conditions:
a All three coils electrically in series: (1) all three aiding magnetically, (2) coil 3 opposing coils 1 and 2 magnetically
b All three coils electrically in parallel: (1) and (2) as in part **a**.

13-3 The toroidal magnetic circuit of Fig. 13-2a is constructed of cast iron (see Fig. 13-5 for magnetization curve). The inside diameter is 5.0 in, and the outside diameter 7.0 in. The ring is circular in cross section and is wound with a 500-turn coil. Determine the core flux in kilolines for the following values of coil current:
a 1.1 A **b** 2.0 A **c** 3.0 A **d** 4.0 A

13-4 For the toroid of Prob. 13-3, find the coil currents required to produce the following values of core flux:
a 10,000 lines **b** 25 kilolines **c** 5×10^{-4} Wb

13-5 A coil of 1,000 turns is wound on a laminated core of annealed sheet steel having a cross section 2 in square and a mean length of 20 in. The stacking factor is 0.90. What coil current is required to produce a core flux of 300 kilolines?

13-6 An air gap 0.10 in long is cut perpendicular to the center line of one leg of the core in Prob. 13-5. To what value must the coil current be increased in order to maintain the same core-flux density? Allowance should be made for fringing at the air gap.

13-7 To what value must the current in Prob. 13-6 be increased if the cut is made at a 45° angle?

13-8 The coil current in the electromagnet of Prob. 13-6 is adjusted to 3.0 amp. What is the total core flux?

13-9 A sectional view of a cylindrical iron-clad plunger magnet is given in Fig. 13-36. The small air gap between the sides of the plunger and the iron shell is uniform and 0.01 in long. Magnetic leakage and fringing of flux at the air gaps can be ignored. The coil has 1,000 turns and carries a direct current of 3.0 A.

Consider that the iron of the magnetic circuit requires only a negligibly

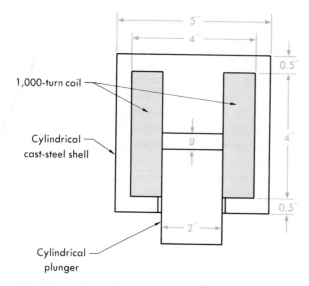

FIGURE 13-36 Plunger magnet for Probs. 13-9 and 13-10.

small MMF. Compute the flux density (in both webers per square meter and kilolines per square inch) at the main air gap g for values of g equal to:
a 0.10 in **b** 0.20 in **c** 0.50 in

13-10 Consider again the plunger magnet of Prob. 13-9. For a series of values of flux, the MMFs required by the iron portion of the magnetic circuit are given below:

Flux, kilolines:	100	150	200	250	260	270	275
MMF, At:	60	95	150	305	425	600	725

Plot curves of flux in webers versus total MMF in ampere-turns for the entire magnetic circuit (the iron, the main air gap, and the gap between the plunger and the shell). These are magnetization curves for the electromagnet. Consider values of g of:
a 0.50 in **b** 0.10 in

13-11 The following changes are made in an electromagnet consisting of an iron core with a single exciting coil. State the effect of these changes on the core-flux density, the coil current, and the coil I^2R loss.
a *DC excitation:* core cross section doubled; coil voltage, resistance, and turns kept constant
b *AC excitation:* same as part **a**
c *DC excitation:* number of turns doubled, coil voltage and resistance kept constant
d *AC excitation:* same as part **c**

e *AC excitation:* frequency halved, coil voltage kept constant
f *AC excitation:* frequency and voltage halved
    Assume operation on the straight part of the magnetization curve in all cases. In ac cases, consider the impressed and induced voltages to be approximately equal, and neglect hysteresis and eddy currents.

13-12 A coil of 200 turns is wound on a laminated core of annealed sheet steel having a cross-sectional area of 2 in² and a mean length of 20 in. The stacking factor is 0.90, and the maximum allowable flux density is 80 kilo-lines/in².
    Compute the effective value of the largest sinusoidal 60-Hz voltage which should be applied to the terminals of this coil. Neglect leakage flux and coil resistance.

13-13 A source which can be represented by a constant voltage of 5 V (rms) in series with an internal resistance of 2,000 Ω is connected to a 50-Ω load resistance through an ideal transformer. Plot the power in milliwatts supplied to the load as a function of the transformer ratio, covering ratios ranging from 0.1 to 10.0.

13-14 An ideal transformer has its primary connected to a voltage source consisting of a 100-V (rms) sinusoidal voltage in series with a 1,600-Ω resistance. A 16-Ω resistance is connected across the secondary. Plot curves of the following quantities as functions of primary-to-secondary turns ratio N_1/N_2, covering the range from 1 to 20.
a Load resistance referred to the primary
b Load power
c Load voltage
d Load current
e Primary current

13-15 An ideal transformer is used to step down the voltage of a 13.8-kV circuit to 2,300 V. It is fully loaded when it delivers 50 kVA.
a Find the turns ratio and the rated currents for each winding.
b Determine the magnitude of the load impedance corresponding to full load. In doing so, refer the magnitude (1) to the high-voltage side, and (2) to the low-voltage side.

13-16 In the circuit of Fig. 13-17a, the numerical values of the circuit quantities are as follows: $E_G = 100$ V, $r_1 = 200$ Ω, $r_2 = 2$ Ω, $r_G = 1,600$ Ω $L_m = 4.95$ H, $r_L = 16$ Ω, $L_{l1} = 50$ mH, and $L_{l2} = 0.5$ mH. The turns ratio is $N_1/N_2 = 10$.
    Draw the equivalent circuits of Fig. 13-17b, c, and d with the appropriate numerical values and with all quantities referred to:
a The primary
b The secondary

13-17 The resistances and leakage reactances of a 10-kVA 60-Hz 2,300/230-V distribution transformer are as follows: $r_1 = 4.20$ Ω, $r_2 = 0.0420$ Ω, $x_{l1} = 5.50$ Ω, and $x_{l2} = 0.0550$ Ω. Subscript 1 denotes the 2,300-V winding, subscript 2 the 230-V winding, and x_{l1} and x_{l2} are the leakage reactances. Each quantity is referred to its own side of the transformer.
a Give the total leakage impedance referred to (1) the high-voltage side, and (2) the low-voltage side.
b Consider the transformer to deliver its rated kilovolt-amperes at 0.80 power factor lagging to a load on the low-tension side with 230 V across the load. Find the high-tension terminal voltage.
c Repeat part **b** for a power factor of 0.80 leading.
d Consider the core loss to be 70 W. Find the efficiency under the conditions of part **b**.
e Will the efficiency be different under part **c**?
f Suppose that the load in part **b** should accidentally become short-circuited. Find the steady-state current in the high-voltage lines, assuming the voltage impressed on the transformer to remain the same as in part **b**.

13-18 An audio-frequency transformer delivers power from a source to a 300-Ω resistive load. The turns ratio of the transformer is chosen so that the value of the resistance load referred to the primary is 7,500 Ω for maximum power transfer. The transformer has primary and secondary resistances of 20 Ω each, primary and secondary leakage inductances of 1.0 mH each, and a magnetizing inductance of 25.0 mH, all referred to the secondary winding. Core losses are negligible.
a Find the transformer turns ratio.
b Draw the equivalent circuit, and label all resistances and reactances with their values referred to the primary winding for a frequency of 15,000 Hz.
c Find the voltage across the 300-Ω load for a frequency of 15,000 Hz. Exciting impedance can be neglected.

13-19 A balanced Y-connected load consisting of three impedances, each $16/\underline{30°}$ Ω, is supplied with the following balanced line-to-neutral voltages: $\mathbf{E}_{an} = 240/\underline{0°}$ V, $\mathbf{E}_{bn} = 240/\underline{240°}$ V, $\mathbf{E}_{cn} = 240/\underline{120°}$ V.
a Find the phasor current in each line.
b Find the line-to-line phasor voltages.
c Find the total active and reactive power supplied to the load.

13-20 Each leg of a balanced Δ-connected load consists of 7 Ω resistance in series with 4 Ω inductive reactance. The line-to-line voltages may be taken as $\mathbf{E}_{ab} = 2,360/\underline{0°}$ V, $\mathbf{E}_{bc} = 2,360/\underline{-120°}$ V, and $\mathbf{E}_{ca} = 2,360/\underline{120°}$ V. Determine:
a The phasor currents $\mathbf{I}_{ab}$, $\mathbf{I}_{bc}$, and $\mathbf{I}_{ca}$
b Each line current and its associated phase angle
c The load power factor
d The impedance per phase of an equivalent Y-connected load which draws the same power at the same power factor

13-21 Three equal impedance units having a resistance of 2 Ω and a capacitive reactance of 1 Ω are connected in Δ across a 230-V three-phase circuit. At the same point, three other equal impedance units each having a resistance of 1.5 Ω and an inductive reactance of 1 Ω are connected in Y across the circuit. Determine:

a The line current
b The total power supplied
c The power factor of the combined circuit

13-22 A 20-kW 0.8-power-factor (lagging) three-phase Y-connected load is supplied from a Δ-connected generator through feeders having a resistance of 1.0 Ω and an inductive reactance of 1.0 Ω each. The line-to-line voltage at the load is 230 V. Compute:

a The phase voltage at the load
b The line-to-line voltage at the generator
c The load current
d The current in each phase of the generator
e The power factor at the generator

13-23 A balanced three-phase load which has a power factor of 75 percent (lagging) draws a current of 200 A per conductor. Power is supplied over a transmission line of impedance $3 + j4$ Ω per phase. The sending-end terminal voltage is 13.8 kV line to line. Determine:

a The load voltage, line to line
b The load power
c The sending-end power factor

13-24 The toroidal magnetic circuit of Fig. 13-2a is constructed of cast steel for which the magnetization curve is given in Fig. 13-5. The inside diameter is 10 cm and the outside diameter is 15 cm. The cross section is circular and wound with 400 turns. Determine the core flux in webers for coil currents equal to (a) 1.5 A, (b) 3.0, (c) 5.0 A.

13-25 The impedances in the transformer equivalent circuit can be determined from two simple tests. The first of these is the short-circuit test in which the primary (high-voltage) side is excited by a voltage that produces rated primary current with the secondary short-circuited. Because the magnetization current is negligibly small compared to rated current, this test indicates copper losses. The quantities measured are the primary voltage, current, and power. In the open-circuit test, primary voltage, current, and power are measured when the primary is excited at rated voltage with the secondary open-circuited. For practical purposes, core losses are obtained from this test.

The following data is obtained for a 60-Hz, 25-kVA, 2,400: 240-V transformer:

Open-circuit test   Short-circuit test

$E_1 = 2,400$ V $E_1 = 50$ V
$I_1 = 0.5$ A $I_1 = 10.4$ A
$P = 200$ W $P = 650$ W

Determine the transformer equivalent circuit referred to the high side.

13-26 To determine the parameters of the transformer equivalent circuit, the short-circuit test is performed as described in Prob. 13-25. The open-circuit test may be performed by exciting the secondary at rated voltage with the primary open-circuited. Measurement of secondary power, current, and voltage can be used to determine core losses.

The following data based on such data are obtained for a 60-Hz, 2,400:240-V, 50-kVA transformer.

Short-circuit test   Open-circuit test

$E_1 = 48$ V $E_2 = 240$ V
$I_1 = 20.8$ A $I_2 = 5.41$ A
$P = 617$ W $P = 186$ W

Determine the equivalent circuit of the transformer, referred to the primary side.

13-27 For the transformer described in Prob. 13-26, determine the efficiency at full load.

13-28 Determine the full-load efficiency of the transformer described in Prob. 13-25.

13-29 Repeat Prob. 13-22 if the load has a power factor of 0.8 leading.

13-30 Repeat Prob. 13-20 when each leg of the Δ-connected load consists of 7-Ω resistance and 4-Ω capacitive reactance.

13-31 Each leg of a balanced Y-connected load consists of an impedance of $24\underline{/60°}$ and is supplied by a balanced three-phase 120-V source. Find
a The phasor current in each line.
b The per-phase active and reactive power supplied to the load.

13-32 The excitation frequency of the three-phase supply in Prob. 13-31 is 50 Hz. Determine the capacitance that must be placed across each leg of the load in order for the effective power factor to be
a Unity.
b 0.9 (lagging).
c 0.9 (leading).

NVERSION □ **ELECTROMECHANICAL**
I □ ELECTROMECHANICAL **ENERGY**
IECHANICAL ENERGY **CONVERSION**
ELECTROMECHANICAL ENERGY CO
ENERGY CONVERSION □ ELECTROI
CONVERSION □ ELECTROMECHANI
ELECTROMECHANICAL ENERGY CO
ENERGY CONVERSION □ ELECTROI
CONVERSION □ ELECTROMECHANI

14

ELECTROMECHANICAL ENERGY CONVERSION □ ELECTROMECHANI
ENERGY CONVERSION □ ELECTROMECHANICAL ENERGY CONVERS
CONVERSION □ ELECTROMECHANICAL ENERGY CONVERSION □ EL
ELECTROMECHANICAL ENERGY CONVERSION □ ELECTROMECHANI
ENERGY CONVERSION □ ELECTROMECHANICAL ENERGY CONVERS
CONVERSION □ ELECTROMECHANICAL ENERGY CONVERSION □ EL
ELECTROMECHANICAL ENERGY CONVERSION □ ELECTROMECHANI
ENERGY CONVERSION □ ELECTROMECHANICAL ENERGY CONVERS
CONVERSION □ ELECTROMECHANICAL ENERGY CONVERSION □ EL
ELECTROMECHANICAL ENERGY CONVERSION □ ELECTROMECHANI
ENERGY CONVERSION □ ELECTROMECHANICAL ENERGY CONVERS
CONVERSION □ ELECTROMECHANICAL ENERGY CONVERSION □ EL
ELECTROMECHANICAL ENERGY CONVERSION □ ELECTROMECHANI
ENERGY CONVERSION □ ELECTROMECHANICAL ENERGY CONVERS
CONVERSION □ ELECTROMECHANICAL ENERGY CONVERSION □ EL
ELECTROMECHANICAL ENERGY CONVERSION □ ELECTROMECHAN
ENERGY CONVERSION □ ELECTROMECHANICAL ENERGY CONVERS
CONVERSION □ ELECTROMECHANICAL ENERGY CONVERSION □ EL
ELECTROMECHANICAL ENERGY CONVERSION □ ELECTROMECHAN
ENERGY CONVERSION □ ELECTROMECHANICAL ENERGY CONVERS
CONVERSION □ ELECTROMECHANICAL ENERGY CONVERSION □ EL
ELECTROMECHANICAL ENERGY CONVERSION □ ELECTROMECHANI
ENERGY CONVERSION □ ELECTROMECHANICAL ENERGY CONVERS
CONVERSION □ ELECTROMECHANICAL ENERGY CONVERSION □ EL
ELECTROMECHANICAL ENERGY CONVERSION □ ELECTROMECHANI
ENERGY CONVERSION □ ELECTROMECHANICAL ENERGY CONVERS
CONVERSION □ ELECTROMECHANICAL ENERGY CONVERSION □ EL
ELECTROMECHANICAL ENERGY CONVERSION □ ELECTROMECHANI
ENERGY CONVERSION □ ELECTROMECHANICAL ENERGY CONVERS
CONVERSION □ ELECTROMECHANICAL ENERGY CONVERSION □ EL
ELECTROMECHANICAL ENERGY CONVERSION □ ELECTROMECHANI

ecause electric energy can be transmitted and controlled simply, reliably, and efficiently, other forms of energy are often converted to and from electrical form. Thus, energy-conversion devices are required at both ends of the typical electric system. Among the most important of these devices are those for the conversion of energy from mechanical to electrical form or from electrical to mechanical form. This introductory chapter deals with the basis of operation of the common electromechanical conversion devices.

14-1 BASIC PRINCIPLES

An electromechanical energy-conversion device, or electromechanical *transducer,* is a link between an electric and a mechanical system. It makes possible the conversion of energy from electrical to mechanical form or from mechanical to electrical form. In a device acting as a *generator,* conversion from mechanical to electric energy takes place. In a device acting as a *motor,* the conversion is from electrical to mechanical form. The conversion process is essentially reversible. Hence generators can be made to act as motors and motors as generators.

The coupling between the electric and mechanical systems is through the medium of the fields of electric currents or charges. Electromechanical energy conversion therefore depends on the existence in nature of phenomena interrelating magnetic and electric fields on the one hand and mechanical force and motion on the other. The three principal phenomena with which we shall be concerned are presented briefly in the remainder of this section.

1 *Generation of voltage.* We have seen from Eq. (13-8) that a voltage is induced in a coil when there is a change in the flux linking the coil. In the transformer, a static device, the voltage is induced by changing flux magnitude. Such a voltage is sometimes called a *transformer voltage.* Voltage can also be induced in the coil by motion of either the coil or the magnetic field relative to each other. It is then called a *motional voltage, speed voltage,* of if the motion is rotary, *rotational voltage.* Both transformer and motional voltages are given by Faraday's law, Eq. (13-8). Electromechanical energy conversion takes place when the change in flux is associated with mechanical motion.

2 *Force on iron.* A mechanical force is exerted on ferromagnetic material, tending to align it with or bring it into the position of the densest part of the magnetic field. When the magnetic field is created by a current-carrying coil, the energy-conversion process is reversible because motion of the material will cause a change in the flux linking the coil and the change of flux linkages will induce a voltage in the coil.

This force is the familiar attraction of a magnet for pieces of iron in its field. Thus, in the magnetic circuits of Figs. 13-9 to 13-11, for example, definite forces are exerted on the iron at the air-iron boundary. If the energy changes associated with a differential displacement of the iron are considered, it can be shown that under certain assumptions, the force acting on two plane parallel iron faces is given by

$$F = \frac{B^2A}{2\mu_0} \quad \text{newtons} \tag{14-1}$$

where the air-gap density B, cross-sectional area A, and permeability μ_0 for free space are in rationalized MKS units. The *newton* is the unit of force in the rationalized MKS system; it is equal to a force of 0.225 lb. Alternatively, if B is in kilolines per square inch and A in square inches, the force is

$$F = 0.0139B^2A \quad \text{pounds} \tag{14-2}$$

where the permeability μ_0 has been included numerically.

This force is the essential operating mechanism of many electromagnetic devices. Among them are lifting magnets, magnetic clutches, magnetic chucks, magnetically operated brakes, contactors (magnetically operated switches), and many types of relays. Magnetically operated valves (often called solenoid-operated valves because *solenoid* is another name for the operating coil) are common elements in many piping systems. In control-system terminology, they are simple *actuators,* a general term denoting devices for converting electric, pneumatic, or hydraulic inputs into mechanical force or torque.

Example 14-1

Evaluate the force acting on the armature in Example 13-2.

SOLUTION   The area of each gap, with correction for fringing, is

$$A = (0.25 + 0.01)(0.25 + 0.01) = 0.0676 \text{ in}^2$$

and the flux density is

$$B = \frac{5,000}{0.0676} \, 10^{-3} = 74 \text{ kilolines/in}^2$$

The total force for the two gaps is

$$F = 0.0139(74)^2(0.0676)(2) = 10.2 \text{ lb}$$

3 *Force on conductor.* A mechanical force is exerted on a current-carrying conductor in a magnetic field and between current-carrying circuits by means of their magnetic fields. The energy-conversion process is reversible because a voltage is induced in a circuit undergoing motion in a magnetic field. This process, in association with that mentioned first above, will be found to be a dominant one in studying the behavior of rotating machines.

14-2 GENERATED VOLTAGE

Faraday's law (Eq. 13-8) summarizes the basic principles underlying the generation of voltage in electric machines. The first requirement is a winding or group of coils in which the desired voltage is to be induced. This winding is called an *armature winding*, and the structure containing it is called an *armature*. The armature of a dc machine is shown being wound in Fig. 14-1. The second need is to create a magnetic field. In the smaller machines, the field may be that of permanent magnets. In the majority of machines, however, the flux will be created by a separate winding, usually called a *field winding*. Finally, rotation must cause continuing changes in the amount of flux linking the armature coils.

The most common means of obtaining this last action is to rotate the armature through the magnetic field or to rotate the magnetic field past the armature winding. The armature of the dc machine, for example, is the rotating member, or *rotor*, and the field structure is the stationary member, or *stator*. The armature windings of most ac machines, on the other hand, are on the stator. Thus, Fig. 14-2 shows a cutaway view of a turbine-driven ac generator. The armature winding is on the stator, and the field winding is on the cylindrical rotor structure.

Both the armature and the field coils are wound on iron cores in order that the flux path through them will be as effective as possible. Because the armature iron is subjected to a varying magnetic flux, eddy currents will be induced in it. To minimize eddy-current loss, the armature iron is built up of thin laminations. Schematic diagrams showing the flux paths in a dc machine and in one type of ac machine are given in Figs. 14-3 and 14-4, respectively. The paths are indicated by dotted lines, of which path *abcda* is typical in both sketches. The field windings in these diagrams are *concentrated*

FIGURE 14-1 Direct current generator or motor armature in process of being wound. One side of each coil is placed in the bottom of a slot; the other side is placed in the top of a slot. (*General Electric Co.*)

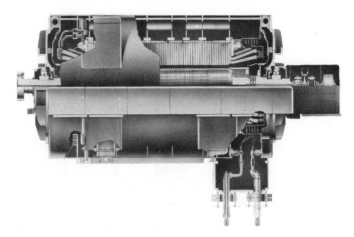

FIGURE 14-2 Section view of a steam-turbine-driven synchronous generator. The ac armature winding is on the stator. The dc field winding is on the cylindrical rotor. (*General Electric Co.*)

windings excited by direct current. Armature windings are almost invariably *distributed windings,* with coils arranged around the entire armature circumference for best use of space and material in the machine.

The radial distribution in the air gap of the flux created by the field winding can be pictured by developing in the descriptive-geometry sense the cylinder forming the armature structure; i.e., the cylinder is cut and laid out flat. The flux distribution for the dc machine of Fig. 14-3 is shown in Fig. 14-5 with the effect of the armature slots on this distribution ignored. Only a two-dimensional plot need be used because the densities are the same for all points in the gap on a line parallel to the shaft of the machine, a statement which neglects the relatively minor effect of fringing flux at the ends of the rotor and stator. The ordinate of this curve at any point gives the flux density in the air gap at that point on the armature periphery. Thus, the flux densities which the armature conductors are sweeping through as the machine rotates can readily be seen. The analogous plot for a typical ac machine will usually be more nearly sinusoidal. For most types of generators and motors in the steady state, the amplitude and shape of the air-gap flux distribution remain constant. Under these circumstances, the only armature-generated voltage results from mechanical motion of either the armature or the field. The induced voltage can then readily be correlated with the flux distribution. Thus, consider any conductor on the armature of Fig. 14-5 to have a linear circumferential velocity v relative to the flux wave and a length l parallel to the shaft. At a particular instant of time, it is located at a point where the ordinate of the flux wave is B. (All quantities are measured in MKS units.) In the time dt the conductor sweeps out an area $lv\,dt$. The associated contribution made by motion of this conductor to the change in flux linking the arma-

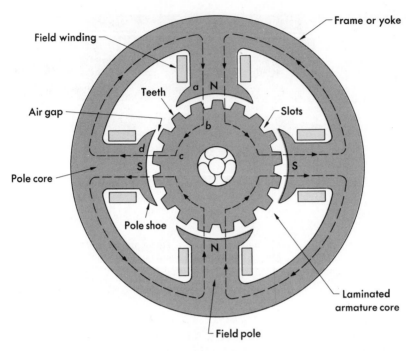

FIGURE 14-3 Diagrammatic sketch of dc machine.

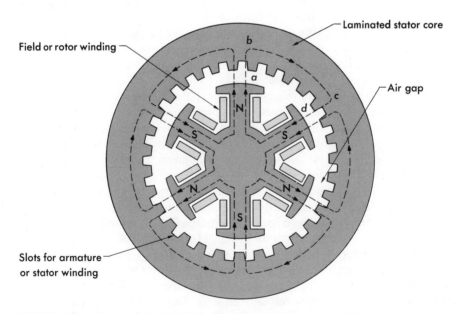

FIGURE 14-4 Diagrammatic sketch of salient-pole synchronous machine.

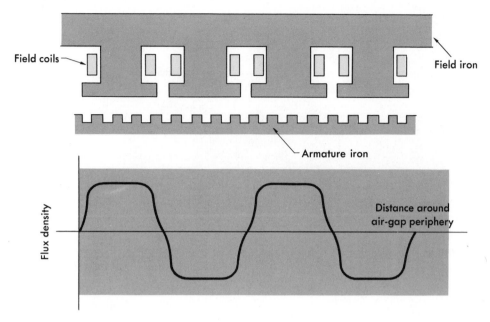

FIGURE 14-5 Developed sketch of dc machine structure showing the air-gap flux distribution.

ture circuit is $Blv \, dt$. In accordance with Eq. (13-8), the contribution of the conductor to the instantaneous armature-generated voltage is

$$e = Blv \qquad \text{volts} \tag{14-3}$$

The instantaneous voltage for the entire armature winding can be computed by adding algebraically the voltages of all series conductors.

Example 14-2

The air-gap flux distribution in a somewhat idealized two-pole machine has the rectangular waveform shown in Fig. 14-6. The flux per pole is 0.01 Wb. The armature is rotating at a speed of 1,200 r/min.

The position of one armature coil at the instant when it is linking maximum flux is shown at ab. Half a revolution later, the coil is at $a'b'$, again linking maximum flux but in the opposite direction. There are four turns in the coil, and during this time it carries a steady current of 10.0 A.

a By direct application of Faraday's induction law, compute the coil voltage during this motion.

b Compute the coil voltage by the Blv equation.

c Compute the electric power associated with this motion.

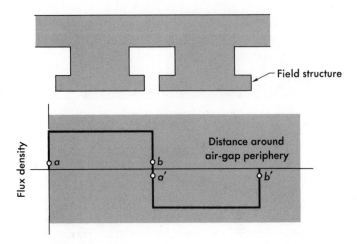

FIGURE 14-6   Armature coil in rectangular flux distribution.

SOLUTION   **a**   Because of the rectangular flux distribution and uniform speed of the armature, the flux linkages with the coil will change at a constant rate during this period. Halfway between ab and $a'b'$, for example, the net linkages will be zero, so that the flux linking the coil changes from the full flux per pole to zero in one-quarter of a revolution. Since one-quarter of a revolution requires

$$\frac{1}{4}\frac{60}{1,200} = 0.0125 \text{ s}$$

the generated voltage is

$$e = N\frac{d\phi}{dt} = 4\frac{0.01}{0.0125} = 3.2 \text{ V}$$

b   When l is the axial length of the armature iron and r the radial distance from the center of the shaft to the air gap, the air-gap flux density is

$$B = \frac{\text{flux per pole}}{\text{area per pole}} = \frac{0.01}{\pi l r}$$

The linear velocity of the conductors is

$$v = \frac{1,200}{60}\,2\pi r = 40\pi r$$

Since there are eight conductors involved in the two coil sides, the generated voltage, from Eq. (14-3), is

$$e = 8\frac{0.01}{\pi l r}\,(l)(40\pi r) = 3.2 \text{ V}$$

c The associated electrical power is

$$ei = (3.2)(10.0) = 32 \text{ W}$$

14-3 ELECTROMAGNETIC TORQUE

Torque considerations in machines may be based on simple formulation of the fact that a mechanical force is exerted on a current-carrying conductor in a magnetic field. Quantitatively, the force on a straight section of conductor of length l meters perpendicular to a magnetic field of density B webers per square meter and carrying the current i amperes is given by

$$f = Bli \quad \text{newtons} \tag{14-4}$$

When r is the radial distance in meters from the center of the rotor shaft to the conductor, the associated torque is

$$T = Blri \quad \text{newton-meters} \tag{14-5}$$

The torque associated with the entire winding is the summation of the torques for the individual conductors or coil sides. Torque produced by such electromagnetic action is called *electromagnetic torque*. Electromagnetic torque, in association with rotation, gives rise to *electromagnetic power,* the torque, power, and speed being interrelated in the same manner as in ordinary mechanics.

From the *Blv, Bli* relations, it follows that generator and motor operation is based on the physical reactions undergone by conductors located in a magnetic field. When there is relative motion between the flux and the conductor, a voltage is generated in the conductor; when the conductor carries current, a force is exerted on it. Generator and motor action go hand in hand in the windings of a rotating machine. Both generators and motors, when operating, have current-carrying conductors in a magnetic field, and the conductors and the flux are traveling at a definite speed relative to each other. Hence both a torque and a voltage of rotation are produced. In fact, within the windings themselves one could not distinguish between generation and motoring as the desired operational function without ascertaining the direction of power flow. Constructionally, generators and motors of the same type differ only in the details necessary for best adaptation of a machine for its intended service; any motor or generator can be used for energy conversion in either direction.

In a generator, the resulting torque is a *counter torque* opposing rotation. It is the torque which the prime mover must overcome, and it is the mechanism through which greater electric power output calls for greater mechanical power input. In a motor, the rotational or speed voltage in the armature acts in opposition to the applied voltage and is termed a *counter EMF*. All these statements are specific, tangible expressions of the reversibility which is characteristic of all electromechanical energy-conversion processes. One interesting application of this reversibility is in the electric braking of motor drives by causing the motor to act as a generator receiving mechanical energy from the moving parts and converting it to electric energy, which is dissipated in a resistor or pumped back into the power line. Thus, an electric locomotive on a downgrade may supply through two stages of conversion some of the energy required by another locomotive going upgrade.

It is naturally to be expected, then, that detailed analyses of generators and motors will be alike in their essentials. Examination of torque production in motors is at the same time examination of counter-torque production in generators; examination of EMF production in generators is at the same time examination of counter-EMF production in motors. These two examinations form the foundation of machinery analysis.

Example 14-3

For the idealized machine of Example 14-2, Fig. 14-6:

a Determine the mechanical torque exerted on the armature by electromagnetic action.

b Compute the mechanical power associated with part **b**, and compare it with the electric power in Example 14-2c.

SOLUTION **a** The torque computed from Eq. (14-5) is

$$T = 8 \frac{0.01}{\pi l r} (lr)(10.0) = \frac{0.8}{\pi} \quad \text{N-m}$$

b When torque and speed are known, power can be obtained from the relation

$$\text{Power} = 2\pi T \times \text{speed in r/s}$$

$$= 2\pi \frac{0.8}{\pi} \frac{1,200}{60} = 32 \text{ W}$$

The results here and in Example 14-2c are naturally equal, since both express the fact that electric and mechanical energy are being interconverted at the rate of 32 W.

Nowhere in this or the previous example has it been stated whether the machine is a generator or a motor. It may be either. If the generated voltage is of such polarity as to oppose the 10.0 A, electric power is absorbed and the machine is acting as a motor. If the generated voltage is of such polarity as to aid the 10 A, the machine is acting as a generator.

14-4 INTERACTION OF MAGNETIC FIELDS

An alternative viewpoint of machine action is to regard it as the result of two component magnetic fields trying to line up so that the center line of a north pole on one machine member is directly opposite the center line of a south

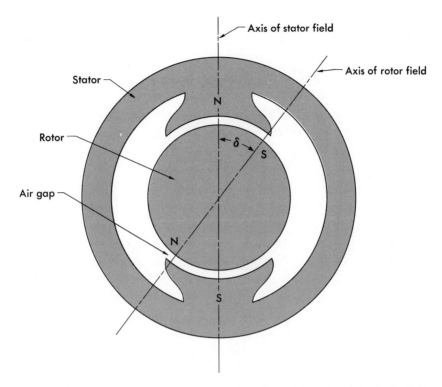

Axis of stator field

Axis of rotor field

Stator

Rotor

Air gap

FIGURE 14-7 Simplified two-pole machine with salient stator poles. Axis of rotor field depends on electrical conditions in rotor windings.

pole on the other member. The component fields are created by the respective windings on the two members. Viewed physically, the process is the same as that by which two bar magnets, pivoted at their centers on the same shaft, line up so that their axes are parallel with unlike poles adjoining.

Currents in the machine windings create magnetic flux in the air gap between the stator and the rotor, the flux path being completed through the stator and the rotor iron, as indicated in Figs. 14-3 and 14-4. This condition corresponds to the appearance of magnetic poles on both stator and rotor, the number of such poles depending on the specific winding design. These poles are shown schematically in Fig. 14-7 for a greatly simplified two-pole machine. The field axes indicated here are the magnetic center lines of the several poles. The stator in Fig. 14-7 is shown with *salient poles* (i.e., poles which stick out from the cylindrical surface), and the rotor with *nonsalient poles* (i.e., a completely cylindrical surface), although practical machines may have nonsalient poles on either the rotor or the stator, or both. Figure 14-8 shows a similar machine with nonsalient poles on both rotor and stator; a practical example of such a machine is the *turbo alternator* (i.e., steam-turbine-driven ac generator) whose cutaway view is shown in Fig. 14-2. The axes of these fields do not necessarily remain fixed in space or with respect to the stator or rotor structure, a fact which is perhaps more easily visualized for the nonsalient structure of Fig. 14-8. In some machines, the axes do re-

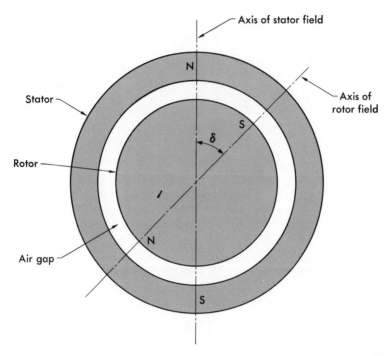

FIGURE 14-8   Simplified two-pole machine with nonsalient poles on rotor and stator. Axes of fields depend on electrical conditions in windings.

main fixed in space; in others, however, they rotate at a uniform angular velocity. In most of the machines which we shall study, the flux per pole remains constant for steady operating conditions.

Torque is produced by interaction of the two component magnetic fields. In Figs. 14-7 and 14-8, for example, the north and south poles on the rotors are attracted by the south and north poles and repelled by the north and south poles, respectively, on the stator, resulting in a counterclockwise rotor torque. Notice that the same magnitude of electromagnetic torque acts on both the rotor and stator structures: The torque on the stator is simply transmitted through the frame of the machine to the foundation, since the stator is not free to rotate.

The magnitude of the torque is proportional to the product of the field strengths. It is also a function of the angle δ between the field axes, varying in a two-pole machine from zero at $\delta = 0°$ to a maximum at $\delta = 90°$, back to zero at $\delta = 180°$, and repeating this half cycle in the negative direction for the range 180 to 360°. For sinusoidal distribution of flux in the air gap (i.e., flux density varying sinusoidally with distance around the air-gap periphery), the torque can be shown to be proportional to the sine of the angle δ; this case is the usual one for ac machines and gives qualitatively correct conclusions for dc machines. The angle δ is commonly referred to as the *torque angle* or *power angle*.

Example 14-4

By using the concept of interaction between magnetic fields, show that electromagnetic torque cannot be obtained by using a four-pole rotor in a two-pole stator.

SOLUTION Consider the simplified motor of Fig. 14-9, with two poles on the stator and four on the rotor. The axes of the rotor field are at an arbitrary angle with the axis of the stator field. The two pairs of poles on the rotor have equal strengths in order to avoid unbalanced radial magnetic pull and the consequent poor bearing operation and tendency to vibrate.

On the N_1N_2 axis, pole N_1 is repelled by pole N and attracted by pole S, causing a counterclockwise torque. Pole N_2 is likewise repelled by N and attracted by S, causing an equal clockwise torque. Hence the net torque is zero. A similar situation exists on the S_1S_2 axis, so that no net electromagnetic torque is produced.

The same result is obtained by examination of any combination of unequal numbers of rotor and stator poles. Thus, the general conclusion is that *all rotating machines must have the same number of poles on the stator and on the rotor*. This conclusion applies equally well to generators, for if no

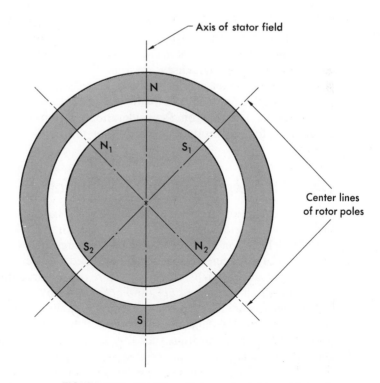

FIGURE 14-9 Machine with two-pole stator and four-pole rotor.

counter torque is produced, there can be no reaction on the mechanical driver and hence no energy conversion.

In motor and generator action, then, the magnetic fields tend to line up, pole to pole. When their complete alignment is prevented by the application of a mechanical torque to the rotor from a source of mechanical energy, mechanical to electric energy conversion takes place: Generator action is obtained. When, on the other hand, their complete alignment is prevented by the need to furnish torque to a mechanical shaft load, electric to mechanical energy conversion takes place: Motor action is obtained.

The qualitative conditions necessary for the production of torque in rotating machines follow readily from this viewpoint. First, it must be borne in mind that the rotor torque should act in the same direction at all times and should preferably be a steady torque. The angle δ between field axes (Figs. 14-7 and 14-8) should therefore remain constant, for a change in δ will cause a change in torque and a linear variation of δ with time will cause periodic variation of torque with reversal of direction each half cycle. Consequently, for the production of a steady, unidirectional torque, *the axis of the magnetic field produced by the rotor winding must remain fixed in space relative to that of the field produced by the stator winding.* Since torque is proportional to the product of the field strengths, *best operation results if the flux per pole does not fluctuate with time.* Otherwise, torque pulsations are produced which may cause undesirable mechanical vibration of the motor and the equipment driven. In some motor types (most single-phase ac motors, for example), the field strengths must unavoidably vary in magnitude sinusoidally with time. In such cases, the time variations of stator and rotor pole strengths must be as nearly in phase as possible in order to produce the greatest possible net torque in the forward direction per unit of current input to the motor. The foregoing considerations pertain equally to motors and generators, for the production of counter torque is essential to generator operation.

14-5 ALTERNATING-CURRENT GENERATORS

With rare exceptions, the armature winding of a *synchronous generator* is on the stator. The field winding is on the rotor, with the field current conducted to it through carbon brushes bearing on slip rings or collector rings. Preliminary ideas of ac generator action can be gained by discussing the elementary ac generator of Fig. 14-10. On the armature of this three-phase two-pole generator are three coils, aa', bb', and cc', whose axes are displaced 120° in space from one another. The field winding is excited by direct current (or, in small machines, the flux may be furnished by a permanent magnet). The rotor is turned at a constant speed by a source of mechanical power connected to its shaft.

The machine is normally so designed that the distribution of flux around the air-gap circumference is a sine wave, as illustrated in Fig. 14-11a. The re-

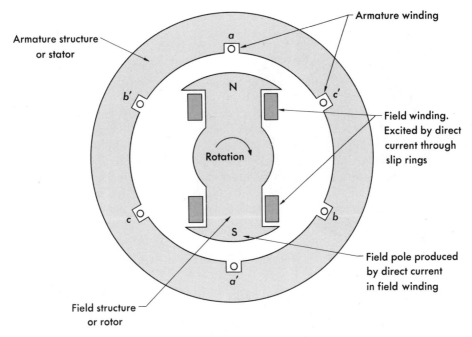

FIGURE 14-10   Elementary three-phase two-pole generator.

sulting voltage in any one armature coil (Fig. 14-11*b*) therefore varies sinusoidally with time. Because of the 120° displacement between coils, the three coil voltages can be represented by a phasor diagram like that of Fig. 13-24. We therefore have a three-phase machine, the most common for this type.

The voltage in any one coil passes through a complete cycle of values for each revolution of the two-pole machine of Fig. 14-10. Its frequency in

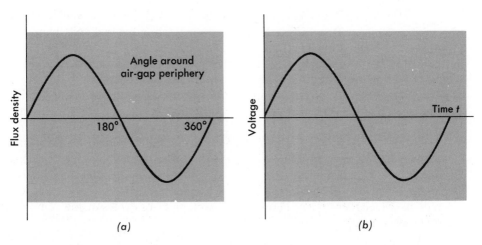

FIGURE 14-11   (*a*) Space distribution of flux density and (*b*) corresponding waveform of the generated voltage.

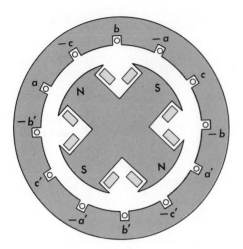

FIGURE 14-12 Elementary four-pole three-phase alternator.

hertz is the same as the speed of the rotor in revolutions per second; i.e., the electrical frequency is synchronized with the mechanical speed. This is the reason that these machines are often called *synchronous machines*. Thus, a two-pole synchronous machine must revolve at 3,600 r/min to produce a 60-Hz voltage. A cutaway view of a large steam-turbine-driven 3,600-r/min 60-Hz synchronous generator is given in Fig. 14-2.

A great many synchronous machines have more than two poles. As a specific example, Fig. 14-12 shows an elementary four-pole three-phase generator. The field coils on the rotor are connected so that the poles are of alternate north and south polarity. There are then two complete wavelengths or cycles in the flux distribution around the complete circumference. One phase of the armature winding now consists of two coils, such as a, $-a$ and a', $-a'$, which are connected in series. Figure 14-13 shows how all the coils are interconnected to form a Y connection. Each individual coil spans one pole or one-half wavelength of flux. The generated voltage in the coil now goes through two complete cycles per revolution of the rotor. The frequency f in hertz then is twice the speed in revolutions per second.

When a machine has more than two poles, it is often convenient to concentrate on a single pair of poles and to recognize that the electrical, magnetic, and mechanical conditions associated with every other pole pair are repetitions of those for the pair under consideration. For this reason, it is convenient to express angles in *electrical degrees* or *electrical radians* rather than in mechanical units. One pair of poles in a p-pole machine or one cycle of flux distribution equals 360 electrical degrees or 2π electrical radians. Since there are $p/2$ complete wavelengths or cycles in one complete revolution, it follows that

$$\theta = \frac{p}{2}\,\theta_m \tag{14-6}$$

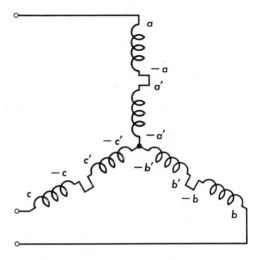

FIGURE 14-13 Schematic diagram for Y connection of winding in Fig. 14-12.

where θ is the angle in electrical units and θ_m is the mechanical angle. The coil voltage of a p-pole machine passes through a complete cycle every time a pair of poles sweeps by, or $p/2$ times each revolution. The frequency of the voltage wave is therefore

$$f = \frac{p}{2} \frac{n}{60} \qquad \text{hertz} \tag{14-7}$$

where n is the mechanical speed in revolutions per minute and $n/60$ is the speed in revolutions per second. The radian frequency ω of the voltage wave is

$$\omega = \frac{p}{2} \omega_m \tag{14-8}$$

where ω_m is the mechanical speed in radians per second.

The constructional reasons for some synchronous generators having salient-pole rotor structures and others having cylindrical rotor structures can be more fully appreciated with the aid of Eq. (14-7). Most power systems in the United States operate at a frequency of 60 Hz. A salient-pole construction is characteristic of hydroelectric generators because hydraulic turbines operate at relatively low speeds, and as a result, a relatively large number of poles are required to provide the desired frequency; the salient-pole construction is better adapted mechanically to this situation. Steam turbines, on the other hand, operate best at relatively high speeds, and turboalternators are commonly two- or four-pole cylindrical-rotor machines.

14-6 COMMUTATOR ACTION IN DC MACHINES

A very elementary two-pole dc generator is shown in Fig. 14-14. The armature winding, consisting of a single coil of N turns, is indicated by the two coil sides a and $-a$ placed at diametrically opposite points on the rotor with the conductors parallel to the shaft of the machine. The rotor is normally turned at a constant speed by a source of mechanical energy connected to the shaft. The air-gap flux distribution usually approximates a flat-topped wave like that of Fig. 14-5.

Although the ultimate purpose is the generation of a direct voltage, the voltage produced in an individual armature coil is an alternating voltage which, for constant machine speed, has the same waveform in time as the flux distribution of Fig. 14-5 has in space. The voltage must therefore be rectified; that is, the alternating voltage must be converted to a direct voltage. Rectification is sometimes provided externally—for example, by means of semiconductor rectifiers (see Sec. 17-1). In a conventional dc machine, rectification is provided mechanically by means of a *commutator,* which is a cylinder formed of copper segments insulated from one another by mica and mounted on, but insulated from, the rotor shaft. Stationary carbon *brushes* held against the commutator surface connect the winding to the armature terminals of the machine.

For the elementary generator, the commutator takes the form shown in Fig. 14-14. For the direction of rotation shown, the commutator at all times connects the coil side which is under the south pole to the positive brush and

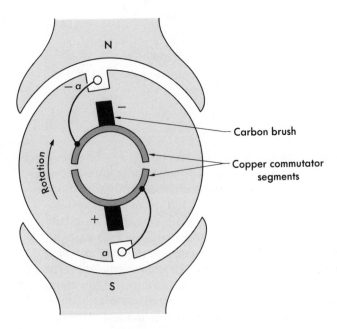

FIGURE 14-14 Elementary dc machine with commutator.

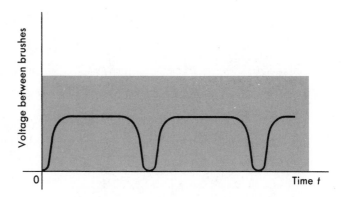

FIGURE 14-15 Waveform of voltage between brushes.

the coil side which is under the north pole to the negative brush. If the direction of rotation reverses, the polarity of the brushes reverses. The commutator provides full-wave rectification for the single armature coil, transforming the voltage waveform between brushes to that of Fig. 14-15 and making available a unidirectional voltage to the external circuit.

The armature winding of Fig. 14-14 is simplified to the point of being unrealistic in the practical sense. A more representative winding is shown diagrammatically in Fig. 14-16. The commutator is represented by the ring of segments in the center. Two stationary brushes are represented by the black rectangles inside the commutator (although the brushes usually contact the outer surface in an actual machine).

The winding in Fig. 14-16 is closed upon itself. The coil sides in the slots are shown in cross section by the small circles with dots and crosses in them, indicating currents toward and away from the reader, respectively. Only slots 1 and 7 are specifically shown. The connections of the coils to the commutator segments are shown by the circular arcs. The end connections at the back of the armature are shown dotted for the two coils in slots 1 and 7, and the connections of these coils to adjacent commutator segments are shown by the heavy arcs. All coils are identical. The back end connections of the other coils have been omitted to avoid complicating the figure. They can easily be traced by remembering that each coil has one side in the top of a slot and the other side in the bottom of the diametrically opposite slot.

In Fig. 14-16a, the brushes are in contact with commutator segments 1 and 7. Current entering the right-hand brush divides equally between two parallel paths through the winding. The first path leads to the inner coil side in slot 1 and finally ends at the brush on segment 7. The second path leads to the outer coil side in slot 6 and also finally ends at the brush on segment 7. The current directions in Fig. 14-16a can readily be verified by tracing these two paths. The effect is identical to that of a coil wrapped around the armature with its magnetic axis vertical, and a clockwise magnetic torque is ex-

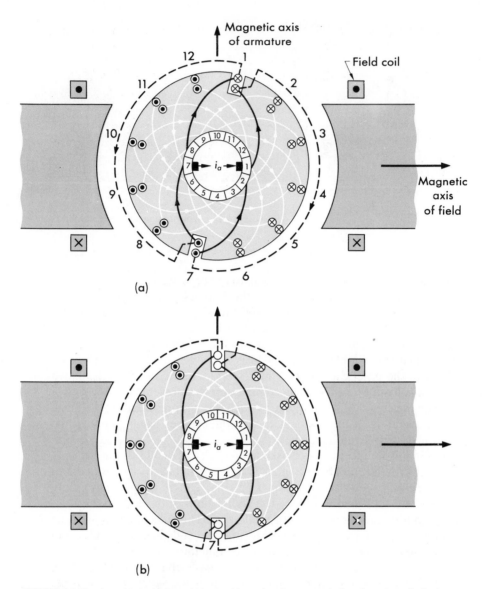

FIGURE 14-16 Armature winding of dc machine, showing current directions in coils for two positions of the armature.

erted on the armature, tending to align its magnetic field with that of the field winding.

Now suppose the machine is acting as a generator driven in the counterclockwise direction by an applied mechanical torque. Figure 14-16b shows the situation after the armature has rotated through the angle subtended by half a commutator segment. The right-hand brush is now in contact with both segments 1 and 2, and the left-hand brush is in contact with both segments 7 and 8. The coils in slots 1 and 7 are now short-circuited by the brushes. The currents in the other coils are shown by the dots and crosses, and they produce a magnetic field whose axis again is vertical.

After further rotation, the brushes will be in contact with segments 2 and 8 and slots 1 and 7 will have rotated into the positions which were previously occupied by slots 12 and 6 in Fig. 14-16*a*. The current directions will be similar to those of Fig. 14-16*a* except that the currents in the coils in slots 1 and 7 will have reversed. The magnetic axis of the armature is still vertical.

During the time when the brushes are simultaneously in contact with two adjacent commutator segments, the coils connected to these segments are temporarily removed from the main circuits through the winding, short-circuited by the brushes, and the currents in them are reversed. Ideally, the current in the coils being commutated should reverse linearly with time. The waveform of the current in any coil as a function of time is then trapezoidal.

Thus we see that the switching action of the commutator makes both the voltage and the current unidirectional in the external circuit even though they change direction in the coils themselves. Likewise, the axis of the armature magnetic field remains fixed, usually with a space displacement of 90° from the stator-field axis.

14-7 DIRECT-CURRENT GENERATORS

The armature winding of a dc generator is on the rotor, with the current conducted from it by means of carbon brushes. The field winding is on the stator. The machine is of the general construction sketched in Fig. 14-3.

As in the alternator, the field winding of a dc generator carries a steady direct current and must therefore be connected to a dc source. Two general possibilities exist: The field winding may be connected to a dc source which is electrically independent of the machine, resulting in a *separately excited generator,* or it may be recognized that the armature itself constitutes a dc source capable of supplying electric power not only to the load on the machine but also to its own field winding, resulting in a *self-excited generator.* In the latter case, residual magnetism must be present in the machine iron to get the self-excitation process started.

The symbolic connection diagram of a separately excited generator is given in Fig. 14-17. The required field current is a very small fraction of the armature current—of the order of 1 to 3 percent in the average generator. A small amount of power in the field circuit of a separately excited generator may control a relatively large amount of power in the armature circuit. The simplest manner of achieving such control is through a rheostat in the field circuit. The rheostat controls the field current and hence the height of the flux-density wave; it follows that control of the generated voltage is thereby obtained. Separately excited generators are commonly used when control over the armature voltage and output is of special importance.

The field windings of self-excited generators can be supplied in three different ways. The field may be connected in series with the armature (Fig.

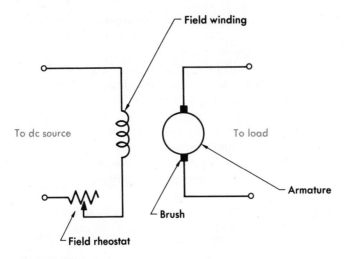

FIGURE 14-17 Separately excited dc generator.

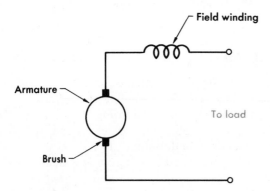

FIGURE 14-18 Series dc generator.

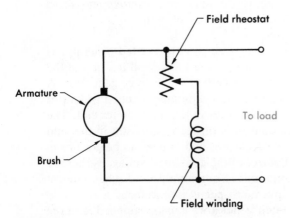

FIGURE 14-19 Shunt dc generator.

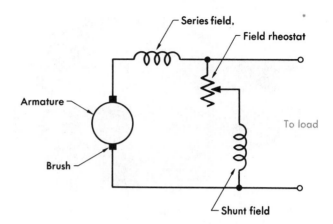

FIGURE 14-20   Compound dc generator.

14-18), resulting in a *series generator*. The field may be connected in parallel with the armature (Fig. 14-19), resulting in a *shunt generator*. Or the field may be in two sections (Fig. 14-20), one of which is connected in series and the other in parallel with the armature, resulting in a *compound generator*. The field current of a series generator is the same as the load current, so that the air-gap flux and hence the voltage vary widely with load. As a consequence, series generators are not very often used. The voltage of shunt generators drops off somewhat with load, but not in a manner which is objectionable for many purposes. Compound generators are normally connected so that the MMF of the series winding aids that of the shunt winding. The advantage is that through the action of the series winding, the flux per pole can increase with load, resulting in a voltage output which is nearly constant or which even rises somewhat as load increases. The shunt winding usually contains many turns of relatively small wire. The series winding, wound on the outside, consists of a few turns of comparatively heavy conductor because it must carry the full armature current of the machine. The voltage of both shunt and compound generators can be controlled over reasonable limits by means of rheostats in the shunt field.

14-8   ELECTRIC MOTORS

The essential parts of electric motors, like those of generators, include two sets of windings wound on or embedded in slots in iron cores. Either or both windings may be excited by alternating or direct currents to form the commonly encountered varieties of practical electric motors. As illustrated in simple fashion by Example 14-4, the rotor and stator must be wound for the same number of poles to produce a successful motor.

When alternating current is supplied to one winding (the armature) and direct current to the other (the field), the motors are known as *synchronous motors* and are the counterparts of the synchronous generators considered

FIGURE 14-21 Cutaway view of high-speed synchronous machine. At the left end is a shaft-mounted dc generator to furnish excitation to the motor field. (*General Electric Co.*)

in Sec. 14-5. As in these generators, the dc winding of synchronous motors is almost invariably placed on the rotor. Since induced voltages as well as torques are associated with both motors and generators, we intuitively expect an induced counter EMF in the armature winding of the motor similar to that in an alternator. Moreover, in order that the counter EMF and the impressed armature voltage may enter into an appropriate Kirchhoff-law balance in the steady state, we expect a relation between operating speed and impressed frequency not unlike that given by Eq. (14-7). In fact, we shall find in Chap. 16 that exactly this relation is obeyed. Synchronous motors therefore operate at an absolutely constant average speed determined by the number of poles and the impressed frequency. Significant departure from this speed, occasionally caused by electrical or mechanical disturbances, results in loss of motor action and shutdown of the machine.

The broad constructional features of general-purpose synchronous motors are much the same as for the synchronous generator sketched diagrammatically in Fig. 14-4. A cutaway view of a three-phase 60-Hz synchronous motor is shown in Fig. 14-21.

A second class of motors, the *dc motor,* arises when direct current is supplied to both rotor and stator windings (i.e., to both the armature and the field). The general appearance of a typical dc motor is shown in Fig. 14-22. Any of the methods of excitation used for generators can also be used for motors. Thus, the connection diagrams of Figs. 14-17 to 14-20 can represent, respectively, the *separately excited, series, shunt,* and *compound motors* when the words "to load" are replaced by "to mains." In a motor, the armature current is in the direction opposite to that in a generator and the electromagnetic torque is in the direction to sustain rotation of the armature. The generated EMF in the armature is smaller than the terminal voltage for a motor, whereas it is larger for generator operation. In both cases, the difference between the generated EMF and terminal voltage is the voltage drop in the armature resistance.

FIGURE 14-22 Cutaway view of typical dc motor. (*Westinghouse Electric Corp.*)

The outstanding advantages of dc motors will ultimately be shown to lie in the variety of performance characteristics offered by the possibilities of shunt, series, and compound excitation and in the relatively high degree of adaptability to control, both manual and automatic. To mention but one simple aspect: As load is added to the motor shaft, the series motor operates at speeds which decrease rapidly, the shunt motor operates at almost constant speed, and the compound motor operates with any degree of droop between these extremes, depending on the relative strengths of series and shunt fields.

The third variation on exciting motor stator and rotor windings is to supply alternating current to both windings. The most common example is the *induction motor,* in which alternating current is supplied directly to the stator and by induction (i.e., transformer action) to the rotor. Although the induction motor is the most common of all motors, the induction machine is seldom used as a generator. Its performance characteristics as a generator are unsatisfactory for most applications. The induction machine is a versatile device, however, and it finds several applications for other than simple motor action. It is, in effect, a transformer with one additional degree of freedom: Its secondary can rotate.

In the induction motor, the stator winding is essentially the same as that in the synchronous machine. The simplified stator windings of Figs. 14-10, 14-12, and 14-13 are therefore representative for three-phase machines. The rotor winding is electrically closed on itself and has currents induced in it by transformer action from the stator winding. In fact, the rotors of many induction motors have no external terminals. A cutaway view of such a motor is given in Fig. 14-23. The rotor winding, called a *squirrel-cage rotor,* con-

FIGURE 14-23 Cutaway view of an induction motor with squirrel-cage rotor. (*Westinghouse Electric Corp.*)

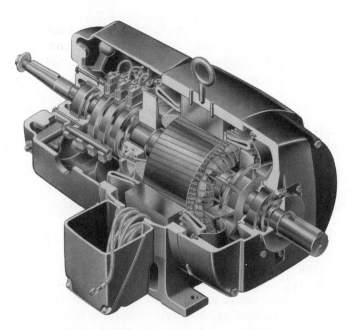

FIGURE 14-24 Cutaway view of three-phase induction motor with a wound rotor and slip rings connected to the three-phase rotor winding. (*General Electric Co.*)

sists of conducting bars embedded in slots in the rotor iron and short-circuited at each end by conducting end rings. Another type of rotor, called a *wound rotor,* is shown in Fig. 14-24. This rotor carries a polyphase winding similar to and wound for the same number of poles as the stator. The terminals of the rotor winding are connected to insulated slip rings mounted on the shaft. Carbon brushes bearing on these rings make the rotor terminals available external to the motor.

The extreme simplicity and ruggedness of the squirrel-cage construction are outstanding advantages of the induction motor. Such a motor runs at a speed somewhat below the synchronous speed as determined by the number of poles and impressed frequency in accordance with Eq. 14-7. Its usual characteristic is that the speed drops off slightly as load is added to its shaft, although variations in the degree of drop may be obtained. A wound rotor is generally more expensive than a cage rotor and is used only for severe starting requirements or when speed control is desired. Both requirements can be fulfilled by inserting external resistance in the wound-rotor circuit.

14-9 LOSSES AND EFFICIENCY

Electromechanical energy conversion is necessarily accompanied by a certain amount of irreversible conversion of energy to heat in the conversion device or machine. These energy losses arise because of circuit resistances (copper losses), because of the existence of alternating or fluctuating magnetic fields (core losses), and from mechanical factors (friction and windage). Although they play essentially no basic part in the energy-conversion process, losses are nevertheless important factors in the practical application of machines.

Treatment of machine losses is important for two reasons: Losses determine the efficiency and hence appreciably influence the operating cost of the machine, and losses determine the heating of the machine and hence fix the rating or power output that can be obtained without deterioration of the insulation because of overheating. Consideration of the machine parts in which such losses can occur will show that the following list includes all the possibilities within the machine itself:

1 *The I^2R losses* in the rotor and stator windings. Resistances should, by convention, be taken at a temperature of 75°C. In synchronous and dc shunt motors (except where variable speed is obtained by control of shunt-field current), the field copper loss is constant because such motors are normally operated at constant field current. In series and induction motors, the losses in both windings vary as the square of the line current.

2 *Core losses,* consisting of eddy-current and hysteresis losses. In synchronous and induction machines, these losses are confined essentially to the stator iron, and in dc machines essentially to the rotor iron, although in both cases a small core loss will be present in the other member because of small flux variations caused by the slots. The iron in the offending member is laminated to reduce the eddy-current loss. In all except series motors, variable-speed shunt motors, and to a lesser degree, compound motors, the air-gap flux and hence the core losses are sensibly constant regardless of load.

3 *Friction and windage losses,* which are constant unless the speed varies appreciably. The sum of the friction and windage and core losses is called the *rotational loss.*

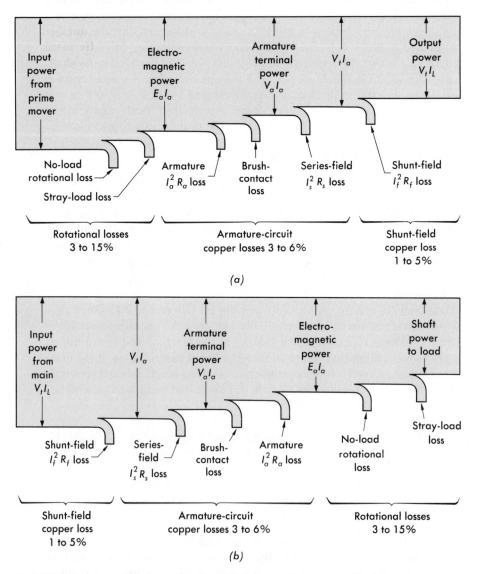

FIGURE 14-25   Power division in (a) a dc generator and (b) a dc motor. V_t is the terminal voltage of the machine, V_a the terminal voltage of the armature, E_a the internally generated EMF in the armature, I_L the line current, I_f the shunt-field current, and I_s the series-field current.

4 *Stray-load losses,* a term designating the additional hysteresis and eddy-current losses arising from any distortion in flux distribution caused by the load currents. These losses are difficult to measure and are usually about 1 percent of the machine output.

As an example, graphical power balances for dc generators and motors having both shunt and series fields are given in Fig. 14-25. Orders of magnitude of losses at full load, expressed in percentage of input, are quoted in Fig. 14-25 for generators in the 1- to 100-kW range and for motors in the 1- to 100-hp range. The smaller percentage losses apply to the larger machines.

Machine efficiency, like that of any energy-transforming device, is

$$\text{Efficiency} = \frac{\text{output}}{\text{input}} \tag{14-9}$$

which can also be written

$$\text{Efficiency} = \frac{\text{input} - \text{losses}}{\text{input}} \tag{14-10}$$

$$= \frac{\text{output}}{\text{output} + \text{losses}} \tag{14-11}$$

Rotating machines in general operate efficiently except at light loads. The full-load efficiency of average motors is in the neighborhood of 74 percent for 1-hp motors, 88 percent for 50-hp, 94 percent for 500-hp, and 97 percent for 5,000-hp. The efficiency of slow-speed motors is usually lower than that of high-speed motors, the total spread being 3 or 4 percent.

Example 14-5

The losses at full load for a certain 10-hp 230-V motor are found to be as follows: rotational loss, 620 W; stator copper loss, 310 W; rotor copper loss, 370 W; stray-load loss, 70 W. It is not stated whether the motor is an induction, synchronous, or dc machine.

Compute the full-load efficiency.

SOLUTION   By substitution in Eq. 14-11,

$$\text{Efficiency} = \frac{(10)(746)}{(10)(746) + 620 + 310 + 370 + 70}$$

$$= 0.845 = 84.5 \text{ percent}$$

Inspection of the foregoing list of losses will show that for qualitative purposes, the losses may be divided into two categories: (1) constant losses, including friction and windage, core losses, and those field copper losses which are sensibly constant; and (2) losses that vary approximately as the square of the load, including copper losses in the windings carrying load current. (The series motor is perhaps the outstanding exception to this classification.) Under these circumstances, it can be shown that maximum efficiency occurs at the load for which the constant and variable losses are equal. This point is, of course, under the control of the designer and is usually placed at the average operating load, about two-thirds to three-quarters of the rated load. The value of maximum efficiency, as well as of full-load efficiency, is a function of the amount of iron and copper used. It is thus the result of economic balance between the incremental cost of losses and that of material. As a result, the efficiency of large machines is higher than that of small ones.

14-10 MACHINE-APPLICATION CONSIDERATIONS

In actual use, both motors and generators must play the role of very willing servants to the loads they are supplying with energy, and insofar as it is practically and economically feasible, the master should have every requirement spontaneously satisfied.

The power or torque supplied by an electric motor to the mechanical equipment driven is determined in large degree by the requirements of the equipment. The motor must satisfy these requirements or quit trying. (Quitting usually takes the form of the motor being disconnected from the line by automatic circuit breakers because of the excessive currents associated with the attempt of the motor to fulfill the requirements.) The operating speed is fixed by the point at which the power or torque that the motor can furnish electromagnetically is equal to the power or torque that the load can absorb mechanically. In Fig. 14-26, for instance, the solid curve is a plot of the operating speed of an induction motor as a function of its mechanical torque output. The dotted curve is a plot of the mechanical torque input required by a fan for various operating speeds. When the fan and motor are coupled, the operating point of the combination is at the intersection of these two curves—where what the motor can give is the same as what the fan can take.

Motor power or torque requirements depend on conditions within the driven equipment. The requirements of some motor loads are satisfied by a speed which remains approximately constant as load varies; an ordinary hydraulic pump is an example. Others, like a phonograph turntable, require absolutely constant speed. Still others require a speed closely coordinated with another speed, the raising of both ends of a vertical bridge, for instance. Some applications, such as cranes and many traction-type drives, inherently demand low speeds and heavy torques at one end of the range and relatively high speeds and light torques at the other—a varying-speed characteristic, in other words. Others may require an adjustable-constant speed (e.g., some machine-tool drives in which the speed of operation may require adjustment over a wide range but must always be carefully predetermined) or an adjustable-varying speed (a crane is again an example). In almost every ap-

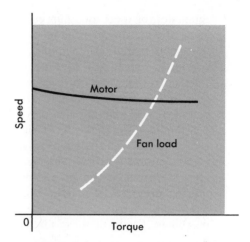

FIGURE 14-26 Superposition of motor and load characteristics.

plication, the torque which the motor is capable of supplying while starting, the maximum torque which it can furnish while running, and the current requirements are items of importance, and not infrequently of determining importance.

Similar remarks can be made for generators. For example, the terminal voltage and power output of a generator are determined by the characteristics of both the generator and its load. Thus, the solid curve of Fig. 14-27 is a plot of the terminal voltage of a dc shunt generator as a function of its electric power output. The dotted curve is a plot of the electric power input required by a load at various impressed voltages. When the load is connected to the generator terminals, the operating point of the combination is at the in-

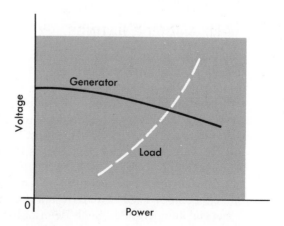

FIGURE 14-27 Superposition of generator and load characteristics.

tersection of these two curves—where what the generator can give is the same as what the load can take. Often, as in the usual central station, the requirement is that terminal voltage shall remain substantially constant over a wide load range. Not infrequently, however, a motor is associated with its own individual generator in order to provide greater flexibility and more precise control of motor output. Then it may be desired that the terminal voltage vary with load in some particular fashion.

Among the features of outstanding importance, therefore, are the torque-speed characteristics of motors and the voltage-load characteristics of generators, together with knowledge of the limits between which these characteristics can be varied and ideas of how such variations may be obtained. Moreover, pertinent economic features are efficiency, power factor, comparative costs, and the effect of losses on the heating and rating of the machines. Study of these features is accordingly the object of machinery analyses.

To a large extent, this study is based on steady-state characteristics, for initial mastering of machine workings is most easily gained thereby. In an increasingly important class of applications in the field of automatic control, however, the emphasis is rather on the dynamic behavior of the complete electromechanical system of which the machine is one component. For example, it may be desired to control the speed or position of a shaft driving a load in accordance with some specified function of time or of some other variable. A typical industrial application is the accurate control of tension in a process involving the winding of long strips of material, such as paper, on a reel. In applications of this kind, the electromechanical transient behavior of the system as a whole is a major consideration; the system should respond accurately and rapidly to the control function, and oscillations should die out quickly. Not only the electrical characteristics but also the mechanical properties of the system, such as stiffness, inertia, and friction, must be considered and indeed may become the predominant factors.

14-11   MACHINE RATINGS

One of the most common and important questions in the application of machines, transformers, and other electrical equipment is: What maximum output can be obtained? The answer depends on various factors, for the machine, while providing this output, must meet definite performance standards. A universal requirement is that the life of the machine shall not be unduly shortened by overheating. The temperature rise resulting from the losses considered in Sec. 14-9 is therefore a major factor in the rating of a machine.

The operating temperature of a machine is closely associated with its life expectancy because deterioration of the insulation is a function of both time and temperature. Such deterioration is a chemical phenomenon involving slow oxidation and brittle hardening and leading to loss of mechanical durability and dielectric strength. A very rough idea of the life-temperature relation can be obtained from the old and very approximate rule of thumb that the time to failure of organic insulation is halved for each 8 to 10°C rise.

Both normal life expectancy and service conditions will vary widely for different classes of electric equipment. Life expectancy, for example, may be a matter of minutes in some military and missile applications, may be 500 to 1,000 h in certain aircraft and electronic equipment, and may range from 10 to 30 y or more in large industrial equipment. Test procedures to evaluate insulation and insulating systems will accordingly vary with the type of equipment. Accelerated life tests on models, called *motorettes*, are commonly used.

For the allowable temperature limits of insulating systems used commercially, the latest standards of the Institute of Electrical and Electronic Engineers (IEEE) and the National Electrical Manufacturers Association (NEMA) should be consulted. The three NEMA insulation-system classes of chief interest for industrial machines are Class B, Class F, and Class H. Class B insulation includes mica, glass fiber, asbestos, and similar materials with suitable bonding substances. Class F insulation also includes mica, glass fiber, and synthetic substances similar to those in Class B, but the system must be capable of withstanding higher temperatures. Class H insulation, intended for still higher temperatures, may consist of materials such as silicone elastomer and combinations including mica, glass fiber, asbestos, etc., with bonding substances such as appropriate silicone resins. Experience and tests showing the material or system to be capable of operation at the recommended temperature form the important classifying criteria.

When the temperature class of the insulation is established, the permissible observable temperature rises for the various parts of industrial-type machines may be found by consulting the appropriate standards. Reasonably detailed distinctions are made with respect to type of machine, method of temperature measurement, machine part involved, whether the machine is enclosed or not, and the type of cooling (air-cooled, fan-cooled, hydrogen-cooled, etc). Distinctions are also made between general-purpose machines and definite- or special-purpose machines. The term *general-purpose motor* refers to one of standard rating "up to 200 hp with standard operating characteristics and mechanical construction for use under usual service conditions without restriction to a particular application or type of application." In contrast a *special-purpose motor* is "designed with either operating characteristics or mechanical construction, or both, for a particular application." For the same class of insulation, the permissible rise of temperature is lower for a general-purpose motor than for a special-purpose motor, largely to allow a greater factor of safety where service conditions are unknown. Partially compensating the lower rise, however, is the fact that general-purpose motors are allowed a service factor of 1.15 when operated at rated voltage; *service factor* is a multiplier which, applied to the rated output, indicates a permissible loading which may be carried continuously under the conditions specified for that service factor.

Examples of allowable temperature rises may be seen from the table below, excerpted from NEMA standards. The table applies to integral-horsepower induction motors, is based on 40°C ambient temperature, and assumes measurement of temperature rise by determining the increase of winding resistances.

	ALLOWABLE TEMPERATURE RISE, °C		
	CLASS B	CLASS F	CLASS H
Motors with 1.15 service factor	90	115	
Motors, 1.00 service factor, encapsulated windings	85	110	
Totally enclosed, fan-cooled motors	80	105	125
Totally enclosed, nonventilated motors	85	110	135

The most common machine rating is the *continuous rating* defining the output (in kilowatts for dc generators, kilovolt-amperes at a specified power factor for ac generators, and horsepower for motors) which can be carried indefinitely without exceeding established limitations. For intermittent, periodic, or varying duty a machine may be given a *short-time rating* defining the load which can be carried for a specified time. Standard periods for short-time ratings are 5, 15, 30, and 60 min. Speeds, voltages, and frequencies are also specified in machine ratings, and provision is made for possible variations in voltage and frequency. Motors, for example, must operate successfully at voltages 10 percent above and below rated voltage and, for ac motors, at frequencies 5 percent above and below rated frequency; the combined variation of voltage and frequency may not exceed 10 percent. Other performance conditions are so established that reasonable short-time overloads may be carried. Thus, the user of a motor may expect to be able to apply for a short time an overload of, say, 25 percent at 90 percent of normal voltage with an ample margin of safety.

The required size of motor for a particular application is easily found when the motor operates continuously at a substantially constant load, since only computations of the power requirements of the driven equipment are involved. When the motor is operated on a more or less repetitive cycle of duty, however, the average heating must be found from the motor losses during the various parts of the cycle and the change in ventilation with motor speed must be included. For estimating the required size of a constant-speed motor, armature heating may be assumed to be determined by the variable losses and thus to depend on the square of the horsepower output. As with the periodically varying currents of Sec. 5-2, the average heating may then be reflected by finding the rms value of the horsepower-time curve and choosing a motor of at least this rating. Formulated, this becomes

$$\text{rms horsepower} = \sqrt{\frac{\Sigma(\text{hp})^2(\text{time})}{\text{running time} + (\text{standstill time}/k)}} \qquad (14\text{-}12)$$

where the constant k accounts for the poorer ventilation at standstill and equals approximately 3 for an open motor. The time for a complete cycle must be short compared with the time for the motor to reach a steady temperature. Special consideration should be given to motors that are frequently started or reversed, for such duties are the equivalent of heavy overloads, especially if a high-inertia load is involved.

Example 14-6

A synchronous motor operates continuously on the following duty cycle: 50 hp for 10 s, 100 hp for 10 s, 150 hp for 6 s, 120 hp for 20 s, idling for 14 s. Compute the size of motor required.

SOLUTION   The constant k does not enter because there is no standstill period.

$$\text{rms hp} = \sqrt{\frac{(50)^2(10) + (100)^2(10) + (150)^2(6) + (120)^2(20) + (0)^2(14)}{10 + 10 + 6 + 20 + 14}}$$

$$= 95.7$$

Here choose a 100-hp motor.

Consideration must also be given to duty cycles having such high torque peaks that motors with continuous ratings chosen on purely thermal bases would be unable to furnish the torques required. It is to such duty cycles that special-purpose motors with short-time ratings are often applied. Short-time-rated motors in general have better torque-producing ability than motors rated to produce the same power output continuously, although they have a lower thermal capacity. In general, the ratio of torque capacity to thermal capacity increases as the period of the short-time rating decreases. A motor with a 150-hp 1-h 50°C rating, for example, may have the torque ability of a 200-hp continuously rated motor; it will be able to carry only about 0.8 of its rated output, or 120 hp, continuously, however. In many cases, it will be the economical solution for a drive requiring a continuous thermal capacity of 120 hp but having torque peaks which require the ability of a 200-hp continuously rated motor.

PROBLEMS

14-1   A sectional view of a cylindrical iron-clad plunger magnet is given in Fig. 13-36. The small air gap between the sides of the plunger and the iron shell is uniform and 0.01 in long. Magnetic leakage and fringing of flux at the air gaps can be ignored. The coil has 1,000 turns and carries a direct current of 3.0 A. Consider that the iron of the magnetic circuit requires only a negligibly small MMF.

Determine the pull (in both newtons and pounds) on the plunger for values of g equal to:

a 0.10 in   **b** 0.20 in   **c** 0.50 in

14-2 For the plunger magnet of Prob. 14-1, plot curves of pull in pounds on the plunger versus coil current for values of g equal to:

a 0.10 in   **b** 0.20 in

Currents should range from 0 to 5.0 A.

14-3 Magnetically operated relays are very commonly used for the automatic control and protection of electric equipment. Consider that the core and armature of the relay shown schematically in Fig. 13-9 are constructed of magnetic material having the same magnetization curve as that for annealed sheet steel. The core has a circular cross section and is 0.5 in in diameter. Its effective magnetic length is 5.0 in. The armature has a rectangular cross section 0.25 by 1.0 in, and its effective magnetic length is 2.0 in. The armature is so supported that the two air gaps are always equal and of uniform length over their areas. Motion of the armature is restrained by a spring whose force opposes the magnetic pull. The operating coil has 1,800 turns.

With the air gaps set at 0.05 in each, what pull must be exerted by the spring if the armature is to pull up when the current in the coil is 1.0 A? Approximate correction for fringing should be made.

14-4 An elementary single-phase two-pole ac generator has a stator winding consisting of a single coil of 100 turns. The rotor produces a sinusoidal space distribution of flux at the stator surface, the peak value of the flux-density wave being 0.80 Wb/m². The inside diameter of the stator is 0.10 m, and its axial length is 0.10 m. The rotor is driven at a speed of 3,600 r/min.

a Taking zero time at the instant when maximum flux density is cutting the coil sides, write the expression for the instantaneous voltage generated in the stator coil as a function of time.

b Sketch the stator, showing how two more coils can be added to form a three-phase generator.

14-5 In the idealized two-pole machine of Fig. 14-6, the ordinate of the rectangular flux-density wave is 40 kilolines/in². The axial length of the armature is 4.5 in, and the armature diameter is 11.0 in. The speed is 1,200 r/min.

Repeat parts **a** and **b** of Example 14-2 for the four-turn coil under these conditions.

14-6 Figure 14-28 shows a two-pole rotor revolving inside a smooth stator which carries a coil of 100 turns. The rotor produces a sinusoidal space distribution of flux at the stator surface, the peak value of the flux-density wave being 0.80 Wb/m² when the current in the rotor is 10 A. The magnetic circuit is linear. The inside diameter of the stator is 0.10 m, and its axial length is 0.10 m. The rotor is driven at a speed of 60 r/s.

Stator coil sides

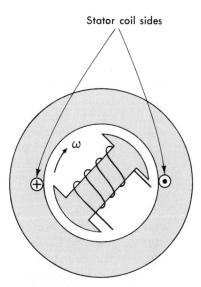

FIGURE 14-28 Elementary generator, Prob. 14-6.

a The rotor is excited by a direct current of 10 A. Taking zero time as the instant when the axis of the rotor is vertical, find the expression for the instantaneous voltage generated in the open-circuited stator coil.

b The rotor is now excited by a 60-Hz sinusoidal alternating current whose rms value is 7.07 A. Consequently, the rotor current reverses every half revolution; it is timed to go through zero whenever the axis of the rotor is vertical. Taking zero time as the instant when the axis of the rotor is vertical, find the expression for the instantaneous voltage generated in the open-circuited stator coil. This scheme is sometimes suggested as a dc generator without a commutator, the thought being that, if alternate half cycles of the alternating current generated in part **a** are reversed by reversal of the polarity of the field (rotor) winding, then a pulsating direct voltage will be generated in the stator. Explain whether this invention will work as described.

14-7 Determine the torque in newton-meters produced by the coil of Prob. 14-5 when its sides are centered under the poles and it is carrying a current of 10.0 A.

14-8 For the dc armature winding of Fig. 14-16, start at the right-hand brush and list in order the conductors in the two parallel paths to the left brush. Do this for:
a Fig. 14-16 *a*
b When the armature has rotated 30° counterclockwise from its position in part **a**

14-9 A small experimental three-phase four-pole alternator has the Y-connected armature winding shown diagrammatically in Fig. 14-13. Each coil (that represented by coil sides a and $-a$, for example) has two turns, and all the turns in any one phase are connected in series. The flux per pole is 25.0 megalines and is sinusoidally distributed in space. The rotor is driven at 1,800 r/min.

a Determine the rms-generated voltage to neutral.

b Determine the rms-generated voltage between lines.

c Consider an *abc* phase order, and take zero time at the instant when the flux linkages with phase a are a maximum. Write a consistent set of time equations for the three phase voltages from terminals a, b, and c to neutral.

14-10 For 60-Hz synchronous motors, make a table showing operating speeds as the number of poles increases from 2 to 20.

14-11 Electric power is to be supplied to a three-phase 25-Hz system from a three-phase 60-Hz system through a motor-generator set consisting of two directly coupled synchronous machines.

a What is the minimum number of poles which the motor can have?

b What is the minimum number of poles which the generator can have?

c At what speed will the set specified in parts a and b operate?

14-12 A manufacturer of airplane equipment requires a source of exactly 800 Hz in the test laboratory. It is proposed to use a motor-generator set with the motor operating from the 230-V 60-Hz line. The generator will be directly coupled to the motor.

a What kind of motor would you use? Why?

b What are the smallest possible numbers of poles that can be used for the motor and the generator?

c What is the speed of the motor-generator set?

14-13 A small high-speed induction motor to be used for textile machines must operate at a speed somewhat under 72,000 r/min. What must be the frequency of the power supply? Use the smallest feasible multiple of 60 Hz.

14-14 A load run on a three-phase 230-V 60-Hz six-pole 3-hp squirrel-cage induction motor yields the following data:

LINE VOLTAGE, V	LINE CURRENT, A	POWER INPUT, W	SPEED, r/min	SHAFT TORQUE, lb-ft
230	4.1	210	1,195	0
230	8.2	2,600	1,100	14.3

a Compute the output and efficiency of the motor at 14.3 lb·ft torque.

b Compute the power factor of the motor at this load and at no load.

14-15 A 230-V 40-hp 1,150-r/min dc shunt motor has a full-load efficiency of 89.9 percent. Motor losses at full load expressed in percentage of power input are as follows: rotational loss, 3.7 percent; armature copper loss, 3.9 percent; field copper loss, 2.5 percent. Stray-load losses can be neglected. Find:

a Armature current at full load
b Field current
c Armature current at half load
d Armature current at no load
e Shaft torque at full load

14-16 The distribution of losses in a 50-hp 230-V dc shunt motor, expressed in percentage of total power input to the motor, is as follows: rotational losses, 3.5 percent; field copper loss, 2.3 percent; armature-circuit copper loss, 3.7 percent. Stray-load losses are to be neglected. The distribution is for full load.

As load on the motor changes, the rotational and field losses remain constant and the armature-circuit copper loss varies as the square of the armature current.

a Obtain points for and plot a curve of armature current as a function of horsepower output. For uniformity, points should be computed for armature currents of 50, 100, 150, 200, and 250 A.
b Compute the no-load armature current and the no-load line current.
c Obtain points for and plot a curve of motor efficiency as a function of horsepower output. Use the points forming the basis of computation in part a.
d Consider that the motor speed is 1,200 r/min at no load and 1,150 r/min at full load and that the speed drops linearly with load. Obtain points for and plot a curve of shaft torque in pound-feet as a function of armature current. Use the armature currents listed in part a.

14-17 Specifications are to be written for a three-phase 60-Hz synchronous motor for a rolling-mill drive in a steel plant. Points on the motor-duty cycle, estimated on the basis of the proposed rolling schedule and previous experience with mills of this type, are given in the following table:

Time, s	0	5	36	39	55	80	
Output, hp	160	1,101	1,470	400	160	160	Repeat cycle

The complete curve can be obtained by joining these points with straight lines.

Specify the continuous horsepower rating of the motor.

14-18 Following are the points on a curve showing the estimated duty cycle for a three-phase 230-V induction motor:

Time	0	40	160	180	240	260	280
Horsepower	0	3	3	1.5	1.5	0	0

The complete curve can be obtained by joining these points with straight lines.

Specify the minimum allowable continuous horsepower rating when the time in the above table is measured

a In seconds
b In minutes

14-19 The core of the relay schematically depicted in Fig. 13-9 has a circular cross section of diameter 1.5 cm and an effective magnetic length of 15 cm. The armature has a rectangular cross section 1 cm by 3 cm and an effective magnetic length of 6 cm. The two air gaps are equal and have uniform length over their areas. Both core and armature are constructed of annealed sheet steel. Motion of the armature is restrained by a spring whose force opposes the magnetic pull. The coil has 1,000 turns and the air gaps are set 0.1 cm each. What is the pull, in newtons, that must be exerted by the spring if the armature is to pull up for a 2.0-A coil current?

14-20 The elementary single-phase ac generator whose parameters are given in Prob. 14-4 is to be used to generate a 25-Hz alternating voltage.
a With all other parameters as given in Prob. 14-4, what is the new value of rotor speed?
b Using the result in **a**, determine the new peak value of flux density if the rms voltage generated is 240 V.

14-21 The electric supply in many naval vessels is a 400-Hz alternating voltage. To provide this source when the vessel is in dry dock for servicing, a motor-generator set is to be used. The generator is directly coupled to the motor which operates from a 240-V, 60-Hz source.
a What kind of motor would you use and why?
b What are the minimum number of poles that can be used in both motor and generator?
c For the conditions in **b**, find the operating speed of the motor-generator set.

14-22 A three-phase, 230-V, 50-Hz squirrel-cage induction motor has the following load characteristics:

LINE VOLTAGE V	LINE CURRENT A	INPUT POWER W	SPEED r/min	SHAFT TORQUE N-m
220	10.4	530	1,780	0
230	31.7	17.2 kW	1,650	86.0

a Compute the output and efficiency of the motor at 14.3 lb·ft torque.
b Compute the power factor of the motor at this load and at no load.

RENT MACHINES □ **DIRECT-CURRENT**
INES □ DIRECT-CURRENT **MACHINES**
MACHINES □ DIRECT-CURRENT MAC
DIRECT-CURRENT MACHINES □ DIRE
-CURRENT MACHINES □ DIRECT-CUR
MACHINES □ DIRECT-CURRENT MAC
DIRECT-CURRENT MACHINES □ DIRE
-CURRENT MACHINES □ DIRECT-CUR
MACHINES □ DIRECT-CURRENT MAC
DIRECT-CURRENT MACHINES □ DIRECT-CURRENT MACHINES □ DIR
-CURRENT MACHINES □ DIRECT-CURRENT MACHINES □ DIRECT-CU
MACHINES □ DIRECT-CURRENT MACHINES □ DIRECT-CURRENT MA
DIRECT-CURRENT MACHINES □ DIRECT-CURRENT MACHINES □ DIR
-CURRENT MACHINES □ DIRECT-CURRENT MACHINES □ DIRECT-CU
MACHINES □ DIRECT-CURRENT MACHINES □ DIRECT-CURRENT MA
DIRECT-CURRENT MACHINES □ DIRECT-CURRENT MACHINES □ DIR
-CURRENT MACHINES □ DIRECT-CURRENT MACHINES □ DIRECT-CU
MACHINES □ DIRECT-CURRENT MACHINES □ DIRECT-CURRENT MA
DIRECT-CURRENT MACHINES □ DIRECT-CURRENT MACHINES □ DIR
-CURRENT MACHINES □ DIRECT-CURRENT MACHINES □ DIRECT-CU
MACHINES □ DIRECT-CURRENT MACHINES □ DIRECT-CURRENT MA
DIRECT-CURRENT MACHINES □ DIRECT-CURRENT MACHINES □ DIR
-CURRENT MACHINES □ DIRECT-CURRENT MACHINES □ DIRECT-CU
MACHINES □ DIRECT-CURRENT MACHINES □ DIRECT-CURRENT MA
DIRECT-CURRENT MACHINES □ DIRECT-CURRENT MACHINES □ DIR
-CURRENT MACHINES □ DIRECT-CURRENT MACHINES □ DIRECT-CU
MACHINES □ DIRECT-CURRENT MACHINES □ DIRECT-CURRENT MA
DIRECT-CURRENT MACHINES □ DIRECT-CURRENT MACHINES □ DIR
-CURRENT MACHINES □ DIRECT-CURRENT MACHINES □ DIRECT-CU
MACHINES □ DIRECT-CURRENT MACHINES □ DIRECT-CURRENT MA
DIRECT-CURRENT MACHINES □ DIRECT-CURRENT MACHINES □ DIR
-CURRENT MACHINES □ DIRECT-CURRENT MACHINES □ DIRECT-CU
MACHINES □ DIRECT-CURRENT MACHINES □ DIRECT-CURRENT MA
DIRECT-CURRENT MACHINES □ DIRECT-CURRENT MACHINES □ DIR
-CURRENT MACHINES □ DIRECT-CURRENT MACHINES □ DIRECT-CU
MACHINES □ DIRECT-CURRENT MACHINES □ DIRECT-CURRENT MA
DIRECT-CURRENT MACHINES □ DIRECT-CURRENT MACHINES □ DIR

15

he outstanding advantages of dc machines arise from the wide variety of operating characteristics which can be obtained. In this chapter, the performance of the machines will be examined quantitatively for both steady-state and dynamic operating conditions. A number of idealizing assumptions will be made —assumptions whose effects would require examination in more thorough studies of critical cases.

15-1 ELECTRIC-CIRCUIT ASPECTS

Rotation of the armature winding in the air-gap magnetic field induces a voltage known as the *armature EMF* and commonly referred to as the *generated EMF* in a generator and the *counter EMF* in a motor. To find the EMF magnitude, consider that the machine has p poles, each pole producing a flux in the air gap of Φ webers. A specific inductor (such as the upper inductor in slot 1 in Fig. 14-16) then cuts $p\Phi$ webers of flux in one complete revolution. If the speed is n revolutions per minute, the inductor cuts this amount of flux $n/60$ times per second. The average rate at which the inductor cuts flux is then

$$p\Phi\,\frac{n}{60} \qquad \text{webers/s, or volts}$$

which is the contribution of each individual inductor to the total armature EMF. The total EMF E_a is the sum of the contributions from all the inductors which are in series between brushes.

In Fig. 14-16 there is a total of 24 inductors. But there are also two parallel paths between brushes, so that only 12 of the inductors are in series between brushes. In a more general winding, there will be Z inductors, a parallel paths, and Z/a inductors in series per path. The armature-generated EMF is then

$$E_a = \frac{Z}{a}\,p\Phi\,\frac{n}{60} \qquad \text{volts} \tag{15-1}$$

It is sometimes convenient to work with angular velocity ω_m in mechanical radians per second rather than speed in revolutions per minute. Since

$$\omega_m = \frac{2\pi n}{60} \tag{15-2}$$

Eq. (15-1) becomes

$$E_a = \frac{Z}{a}\,p\Phi\,\frac{\omega_m}{2\pi} \tag{15-3}$$

or

$$E_a = K_a\Phi\omega_m \tag{15-4}$$

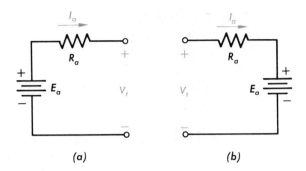

FIGURE 15-1 Circuit representation for armature of (a) dc generator, and (b) dc motor.

The quantity K_a is a constant fixed for any specific machine by the design of the winding. It is given by

$$K_a = \frac{pZ}{2\pi a} \qquad (15\text{-}5)$$

Equation (15-4) shows that the armature EMF is directly proportional to speed or angular velocity and flux per pole. As illustrated in Fig. 15-1a, the armature of a dc generator can be thought of as the *generated EMF* E_a in series with the armature resistance R_a. The armature terminal voltage V_t is smaller than E_a by the armature-resistance drop; that is,

$$V_t = E_a - I_a R_a \qquad (15\text{-}6)$$

where I_a is the armature current in amperes.

As shown in Fig. 15-1b, a similar representation can be used for a motor, except that the current direction must be reversed. The generated voltage E_a is then a *counter EMF* opposing the current. We then have

$$V_t = E_a + I_a R_a \qquad (15\text{-}7)$$

and E_a is smaller than the armature terminal voltage V_t.

Example 15-1

A 230-V shunt motor has an armature resistance of 0.1 Ω. This motor operates on the 230-V mains at a speed of 1,150 r/min and takes an armature current of 100 A. An external resistance of 1.0 Ω is now inserted in series with the armature, the field-rheostat setting remaining unchanged and the load being so adjusted that the armature current remains the same.

Compute the effect of this external resistance on the armature-circuit input, armature EMF, and speed.

SOLUTION Before the resistance is inserted,

Armature-circuit input $= V_t I_a = (230)(100) = 23{,}000$ W

$$E_a = V_t - I_a R_a = 230 - (100)(0.1) = 220 \text{ V}$$

After the resistance is inserted,

Armature-circuit input $= (230)(100) = 23{,}000$ W

$$E_a = 230 - (100)(0.1 + 1.0) = 120 \text{ V}$$

Since the field-rheostat setting is unchanged, I_f and hence Φ are unchanged. Consequently, from Eq. (15-4), a direct proportion exists between E_a and n, and

$$n = \frac{120}{220}(1{,}150) = 627 \text{ r/min}$$

From the circuit representations of Figs. 14-17 to 14-20, it is evident that the internal armature power is

$$P_m = E_a I_a \qquad \text{watts} \tag{15-8}$$

It is the *electromagnetic power* resulting from the energy-conversion process. The electromagnetic power in a generator is greater than the generator output by the amount of the copper losses and less than the mechanical power input by the amount of the rotational losses. In a motor, it is greater than the mechanical shaft output by the amount of the rotational losses and less than the motor electrical input by the amount of the copper losses. These facts have already been illustrated graphically in Fig. 14-25.

The *electromagnetic torque* T_m corresponding to the power P_m at the operating speed n/min is, from the familiar power torque relation of mechanics,

$$T_m = \frac{60}{2\pi n} P_m \qquad \text{newton-meters} \tag{15-9}$$

By use of Eq. (15-8),

$$T_m = \frac{60}{2\pi} \frac{E_a}{n} I_a \qquad \text{newton-meters} \tag{15-10}$$

An alternative form is obtained by substitution of Eq. (15-1); that is,

$$T_m = K_a \Phi I_a \qquad \text{newton-meters} \tag{15-11}$$

where the constant K_a is given by Eq. (15-5).

The torque T_m opposes rotation (i.e., is a counter torque) in a generator but maintains rotation in a motor. The form of expression given in Eq. (15-10) is often convenient for computations. That given in Eq. (15-11), however, is often useful as a basis of qualitative reasoning concerning motor performance. It states that the electromagnetic torque is directly proportional to the field flux and to the armature current.

Example 15-2

Approximate qualitative discussion of machine performance sometimes involves neglect of all machine losses to permit concentration on the principal happenings. Accordingly, consider an idealized dc motor with no armature resistance, no field copper loss, and no rotational losses. The mechanical equipment driven by the motor requires a constant torque regardless of the speed.

a The flux per pole Φ is halved, the terminal voltage V_t remaining constant. What happens to I_a, n, and the mechanical power output?

b Answer part **a** for V_t halved and Φ remaining constant.

SOLUTION **a** From Eq. (15-11), with T_m constant, halving Φ doubles I_a. From Eq. (15-4), with E_a constant because $V_t = E_a$ and is constant, n is doubled. From Eq. (15-8) or (15-9), P_m is doubled.

b From Eq. (15-11), I_a remains the same. From Eq. (15-4), n is halved. From Eq. (15-8) or (15-9), P_m is halved.

Note that so far in our analysis, we have been using capital-letter symbols for quantities such as V_t, I_a, and E_a. The implication, of course, is that we are concerned with steady-state events rather than with the dynamic readjustment processes between successive steady states. We shall continue to be preoccupied with the steady state through the next four sections. When we consider dynamic aspects in Secs. 15-6 and 15-7, certain of the symbols will require modification.

15-2 MAGNETIC-CIRCUIT ASPECTS

One additional relation is essential to the determination of dc machine performance: that between field current or field ampere-turns and armature EMF E_a. It may be obtained from design data by magnetic-circuit computations, Eq. (15-1) being used to replace the flux Φ in the magnetic circuit in terms of the voltage E_a in the electric circuit. Or it may be obtained experimentally by driving the machine at the desired speed as an unloaded generator and recording the series of values of armature voltage corresponding to a series of values of field current. The resulting curve, of which those in Fig. 15-2 are typical, is the *magnetization curve, saturation curve,* or *open-circuit characteristic*.

The speed of a dc machine, of course, need not remain constant. For this reason, magnetization curves at several different speeds are given in Fig. 15-2, any one curve being obtainable from any other by recognizing from Eqs. (15-1) and (15-4) that voltage is directly proportional to speed for a fixed flux or field current. Also, a dc machine may have both a shunt and a

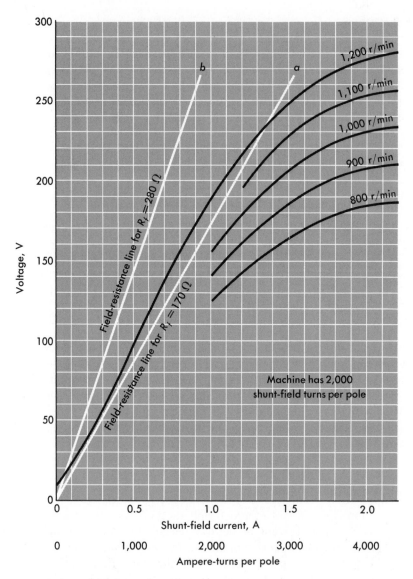

FIGURE 15-2 Magnetization curve for a 230-V 1,200 r/min machine, typical for a 15-kW generator or 15-hp motor.

series winding connected so that their MMFs add. For a compound machine, the magnetization curve is taken with current in the shunt field only. As indicated on Fig. 15-2, the abscissa scale can be changed to ampere-turns per pole by multiplying it by the number of shunt-field turns per pole. The new abscissa may then be interpreted as the net result of all MMFs on the magnetic circuit.

Example 15-3

A 15-kW 230-V dc generator has an armature resistance of 0.1 Ω and the magnetization curve of Fig. 15-2. With the armature current kept constant at 50 A, over what range must the field MMF be varied in order to keep the armature terminal voltage constant at 230 V while the speed is decreased from 1,200 to 1,100 r/min?

SOLUTION   The armature EMF must be kept constant at

$$E_a = V_a + I_aR_a = 230 + (50)(0.1) = 235 \text{ V}$$

At 1,200 r/min, the corresponding field MMF from Fig. 15-2 is 2,800 At/pole. At 1,100 r/min, the MMF is 3,300 At/pole. In terms of excitation from the shunt field alone, the change in field current is from 1.40 to 1.65 A.

It should be borne in mind that any magnetization curve represents the average operating conditions within a hysteresis loop (see Sec. 13-4). The voltage of a generator or speed of a motor for fixed operating conditions will therefore depend on the particular part of the loop being worked on and hence on the magnetic history of the iron circuit. Some degree of uncertainty is thus inevitable in results such as those of Example 15-3.

Many dc machines have auxiliary field windings in addition to the main field. All but the smallest machines have a *commutating field* or *interpole field*. As shown in Fig. 15-3, the interpole is a narrow pole midway between the main poles. Its purpose is to produce a small amount of flux to be cut by the conductors undergoing commutation so as to aid in current reversal without sparking at the brushes.

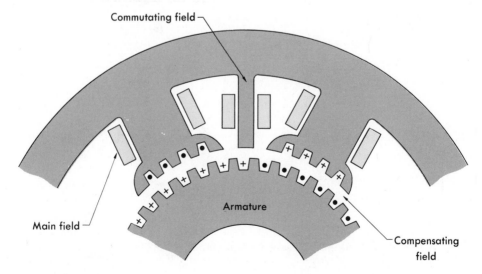

FIGURE 15-3 Section of a dc machine showing commutating and compensating fields.

Sometimes it is also necessary to recognize that the armature winding is an MMF source which may significantly distort the flux distribution created by the main field. This effect, called *armature reaction,* is ignored in later discussions of dc-machine operation. In machines subjected to heavy overloads or wide speed ranges, it may interfere seriously with commutation. It can be neutralized in such cases by means of a *compensating field* or *pole-facing winding.* As shown in Fig. 15-3, this winding is embedded in slots in the pole face and has a polarity opposite to that of the adjoining armature winding. Both the commutating and the compensating windings are connected in series with the armature.

15-3 GENERATOR PERFORMANCE

The four general possibilities for exciting the field windings of dc generators, introduced in Sec. 14-7, are summarized in Fig. 15-4. Superimposed on these diagrams are the Kirchhoff-law relations among line current I_L, armature current I_a, and shunt-field current I_f. From these relations, together with

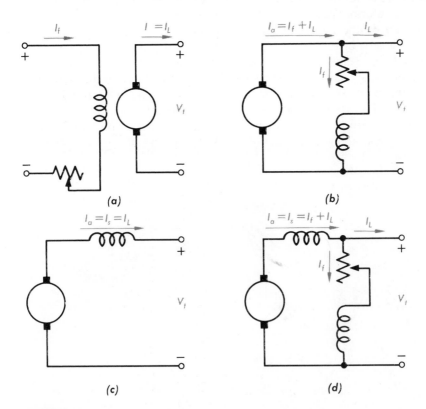

FIGURE 15-4 Types of dc generators, showing relations among line current I_L, armature current I_a, shunt-field current I_f, and series-field current I_s. (*a*) Separately excited generator; (*b*) shunt generator; (*c*) series generator; and (*d*) compound generator.

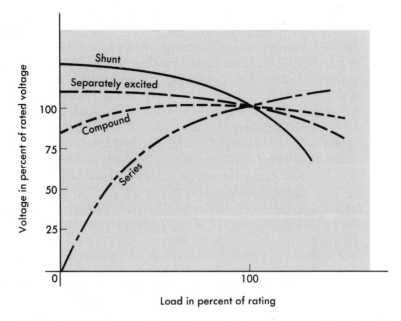

FIGURE 15-5 Direct-current generator load-voltage curves. Generator adjusted for rated voltage at full load.

Eqs. (15-1) and (15-6) and the magnetization curve, some of the more important performance characteristics can be deduced.

These considerations are particularly easy to apply to a separately excited generator, for the field current is furnished from a separate source and is therefore physically independent of armature conditions. At constant field current and constant speed, the terminal voltage in this case drops off somewhat as load increases because of the increasing armature-resistance drop. This decrease is shown in the dashed curve of Fig. 15-5.

In a shunt generator, the field excitation and the terminal voltage are directly related by Ohm's law:

$$V_t = I_f R_f \tag{15-12}$$

A graphical representation of Eq. (15-12), such as line Oa in Fig. 15-2, is called the *field-resistance line*. It is the locus of the terminal-voltage versus shunt-field-current operating point. Thus, line Oa is drawn for $R_f = 170\ \Omega$ and hence passes through the origin and the point (1.0 A, 170 A).

The usefulness of the field-resistance line can be illustrated by first examining the *buildup* of voltage for a shunt generator and then the change in voltage as load is added. Consider the generator to be driven at 1,200 r/min. When the field circuit is closed, the small voltage from residual magnetism (the 8-V intercept of the magnetization curve, Fig. 15-2) causes a small field current. If the flux produced by the resulting ampere-turns adds to the residual flux, progressively greater voltages and field currents are obtained, and the generator is said to build up. If the field ampere-turns oppose the residual magnetism, the shunt-field terminals must be reversed to obtain buildup. Buildup continues until the relations represented by the magnetization curve and field-resistance line are simultaneously satisfied, that is, at their intersection (230 V in Fig. 15-2). Notice that if the field resistance is too high, as shown by line *Ob* for $R_f = 280\ \Omega$, the intersection is at a very low voltage and buildup is not obtained. Notice also that if the field-resistance line is tangent to the lower part of the magnetization curve, corresponding to 190 Ω in Fig. 15-2, the intersection may be anywhere from about 40 to 160 V, resulting in very unstable conditions. This is the *critical field resistance*, beyond which buildup will not be obtained.

The terminal voltage of a shunt generator will decrease as load is added because of the increase in armature-resistance drop and the decrease in field current accompanying the drop in voltage. The drop in voltage with load is shown in Fig. 15-5. Since the field-resistance line is the locus of V_t versus I_f and the magnetization curve is the locus of E_a versus I_f, the vertical distance between the two curves at any value of I_f must be the I_aR_a drop at the load corresponding to that condition. This fact can be utilized to find the terminal voltage at a stated armature current and field resistance.

Example 15-4

A 15-kW shunt generator has an armature resistance of 0.0985 Ω and the magnetization curve of Fig. 15-2. It is driven by a constant-speed 1,200-r/min prime mover. The shunt-field rheostat is adjusted for a no-load terminal voltage of 230 V. Determine the terminal voltage and power output for an armature current of 66 A.

SOLUTION Line *Oa* (Fig. 15-2) passes through the 230-V point on the magnetization curve and hence is the field-resistance line for this case. The I_aR_a drop is $(66)(0.0985) = 6.5$ V. At $I_f = 1.20$ A, line *Oa* and the magnetization curve are separated vertically by 6.5 V. From this point on *Oa*, the terminal voltage is 206 V. Also

$$I_L = I_a - I_f = 66 - 1.2 = 64.8\ \text{A}$$

and

$$\text{Output} = V_tI_L = (206)(64.8) = 13,350\ \text{W}$$

In series generators, the flux varies widely with load because the load current is also the field current. The voltage-load characteristic, shown dotted in Fig. 15-5, resembles the magnetization curve but is somewhat lower because of resistance drop. Because industrial power systems are almost exclusively constant-voltage systems, series generators are seldom used.

Buildup of an unloaded compound generator is the same as for a shunt generator since the series field carries zero or negligible current at no load. When the generator is loaded, both the shunt- and series-field MMFs are present. The series field may be connected either *cumulatively,* so that its MMF adds to that of the shunt field, or *differentially,* so that it opposes. The differential connection is rarely used.

For a compound generator, use of ampere-turns or ampere-turns per pole is convenient for the abscissa of the magnetization curve. This quantity is then simply the sum of the contributions from the shunt and series fields for a cumulative connection. The causes of voltage decrease from no load to full load in a shunt generator may be partially or completely compensated or overcompensated by the addition of the series field. In a *flat-compounded generator,* they are completely compensated, so that the no-load and full-load terminal voltages are equal. If the full-load voltage is lower than the no-load voltage, the generator is *undercompounded;* if higher, *overcompounded.* The compound characteristic of Fig. 15-5 is for an overcompounded generator.

Example 15-5

There are 2,000 turns/pole on the shunt-field winding of the generator in Example 15-4. Determine the number of series-field turns required to flat-compound the generator at 230 V. Ignore the small resistance of these turns.

SOLUTION   Since the no-load excitation is entirely in the shunt winding, $I_f = 1.34$ for a no-load voltage of 230 V. This value must also be that for I_f at full load since the field resistance is unchanged and the terminal voltage is still 230 V; that is, at both no load and full load, the shunt-field MMF is $(2,000)(1.34) = 2,680$ At/pole.

At full load,

$$I_L = \frac{15,000}{230} = 65.2 \text{ A}$$

$$I_a = 65.2 + 1.34 = 66.5 \text{ A}$$

and

$$E_a = 230 + (66.5)(0.0985) = 236.5 \text{ V}$$

From the magnetization curve, this value requires an MMF of 2,820 At/pole. The series field therefore requires

$$\frac{2,820 - 2,680}{66.5} = 2.1 \text{ turns/pole}$$

This value must be rounded up to the nearest half turn. Actually, more turns would usually be wound and the machine adjusted for the desired compounding by diverting some of the armature current from the series winding by placing a low-resistance shunt in parallel with that winding.

15-4 MOTOR PERFORMANCE

The four possibilities for providing dc-machine excitation are repeated once more in Fig. 15-6. Now, however, they are drawn specifically for motors, with the associated Kirchhoff-law relations among line current I_L, armature current I_a, and shunt-field current I_f shown on them. It is in this respect that Fig. 15-6 differs from Fig. 15-4. From these relations, together with Eqs. (15-1) or (15-4) and (15-7) to (15-9) and the magnetization curve, we can obtain the principal motor characteristics.

The difference in connections between the shunt and separately excited motor is a matter of detail. The same voltage source is impressed on both field and armature of the shunt motor. Two different sources are used for the separately excited motor, so that the two impressed voltages can be varied independently. This last feature is of importance in obtaining a wide range of speed control.

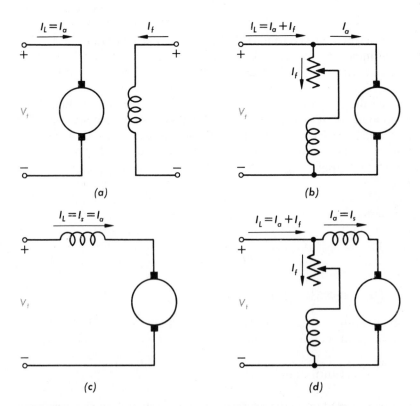

FIGURE 15-6 Types of dc motors, showing relations among currents: (a) separately excited motor; (b) shunt motor; (c) series motor; and (d) compound motor.

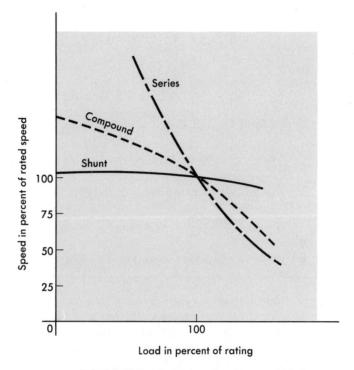

FIGURE 15-7 Direct-current motor speed-load curves.

In a shunt motor, the field current and terminal voltage are directly related by Ohm's law. With constant terminal voltage, the field excitation and hence Φ remain constant unless a rheostat in the motor field is adjusted. The normal characteristic of a shunt motor at constant V_t and R_f shows a slight decrease in speed with added load, as indicated by the solid curve of Fig. 15-7. This decrease results from increased $I_a R_a$ drop, and hence somewhat lower E_a, as load is added.

As will be seen in Sec. 15-5, an outstanding advantage of shunt motors is the relative ease with which speed control can be obtained. Shunt motors are widely used where such control is desired, as in the machine-tool industry.

Example 15-6

A 10-hp 230-V shunt motor takes an armature current of 6.0 A from the 230-V line at no load and runs at 1,200 r/min. The armature resistance is 0.20 Ω.

Determine the speed and electromagnetic torque with 37.0-A armature current and the same flux.

SOLUTION

$$E_a = 230 - (6.0)(0.2) = 229 \text{ V at no load}$$

and

$$E_a = 230 - (37.0)(0.2) = 223 \text{ V under load}$$

Hence from Eq. (15-4), Φ being constant,

$$n = \frac{223}{229} \, 1,200 = 1,165 \text{ r/min}$$

$$P_m = E_a I_a = (223)(37.0) = 8,260 \text{ W}$$

$$T_m = \frac{(60)(8,260)}{2\pi(1,165)} = 67.8 \text{ N-m}$$

$$T_m = (67.8)(0.738) = 50.0 \text{ lb-ft}$$

When a series field is present in a motor, the field MMF is necessarily dependent on the armature current and hence on the motor load. In the series motor, increase in load is accompanied by increases in both the armature current and the flux (provided the iron is not completely saturated). Because flux increases with load, speed must drop in order to maintain the balance between impressed voltage and counter EMF; furthermore, the increase in armature current caused by increased torque is smaller than in the shunt motor because of the increased flux. The series motor is therefore a varying-speed motor with a markedly drooping speed-load characteristic of the type shown dotted in Fig. 15-7. For applications requiring heavy torque overloads, this characteristic is particularly advantageous because the corresponding power overloads are held to more reasonable values by the associated speed drops. Very favorable starting characteristics also result from the increase in flux with increased armature current. Series motors must always be operated under load unless provisions are made in the control circuit to circumvent the destructively high no-load speeds. They are especially suited to traction-type drives, such as cranes and winches.

Example 15-7

When running at its rated load, a 50-hp 550-V 750-r/min series motor takes a current of 74.0 A at 550 V. Its armature resistance is 0.35 Ω, and series-field resistance is 0.15 Ω. When the load torque is double the rated value, the current is 110 A.

Determine the speed and power output at the 100 percent torque overload. For this purpose, consider that doubling the shaft torque doubles the electromagnetic torque.

SOLUTION   At rated load,

$$E_{a1} = 550 - 74(0.35 + 0.15) = 513 \text{ V}$$

At the overload,

$$E_{a2} = 550 - 110(0.35 + 0.15) = 495 \text{ V}$$

From Eq. (15-11), at rated load,

$$T_1 = K_a\Phi_1 I_{a1}$$

and at the overload,

$$T_2 = K_a\Phi_2 I_{a2}$$

Hence

$$\frac{\Phi_2}{\Phi_1} = \frac{T_2}{T_1}\frac{I_{a1}}{I_{a2}} = 2\left(\frac{74}{110}\right) = 1.346$$

From Eq. (15-4), at rated load,

$$E_{a1} = K_a\Phi_1\omega_{m1}$$

and at the overload,

$$E_{a2} = K_a\Phi_2\omega_{m2}$$

Hence

$$\frac{\omega_{m2}}{\omega_{m1}} = \frac{E_{a2}}{E_{a1}}\frac{\Phi_1}{\Phi_2} = \frac{495}{513}\frac{1}{1.346}$$

and

$$n_2 = \frac{495}{513}\frac{1}{1.346}(750) = 537 \text{ r/min}$$

Since the torque is doubled and the speed reduced by $\frac{537}{750}$, the new output is

$$\tfrac{537}{750}(2)(50) = 71.6 \text{ hp}$$

or a 143 percent power load.

In the compound motor, the series field may be connected either cumulatively, so that its MMF adds to that of the shunt field, or differentially, so that it opposes. The differential connection is very rarely used. As shown by the dashed curve in Fig. 15-7, a cumulatively compounded motor will have a speed-load characteristic intermediate between those of a shunt and a series motor, the drop of speed with load depending on the relative number of turns in the shunt and series fields. It does not have the disadvantage of the series motor of very high light-load speed but retains to a considerable degree the advantage of series excitation.

Example 15-8

A compound motor has an armature resistance of 0.10 Ω, series-field resistance 0.10 Ω, shunt-field resistance (including rheostat) 200 Ω, 2,000 shunt-field turns/pole, 10 series-field turns/pole, and the magnetization curve of Fig. 15-2.

Determine the speed and electromagnetic torque when the armature draws 100 A from a 220-V line.

SOLUTION

$$\bullet \quad I_f = \frac{220}{200} = 1.10 \text{ A}$$

Total MMF = (2,000)(1.10) + (10)(100) = 3,200 At/pole.

From Fig. 15-2, this MMF corresponds to an E_a of 252 V at 1,200 r/min. But actually,

$$E_a = 220 - 100(0.10 + 0.10) = 200 \text{ V}$$

so that

$$n = \tfrac{200}{252} 1,200 = 954 \text{ r/min}$$

From Eq. (15-10),

$$T_m = \frac{60}{2\pi} \frac{200}{954} (100) = 201 \text{ N-m}$$

15-5   MOTOR SPEED CONTROL

By combining Eqs. (15-4) and (15-7) and solving for ω_m, we see that the speed in mechanical radians per second is given by

$$\omega_m = \frac{V_t - I_a R_a}{K_a \Phi} \tag{15-13}$$

where V_t is armature terminal voltage, I_a armature current, R_a armature resistance, Φ flux per pole, and K_a a constant fixed by the design of the armature winding. Alternatively, the speed in revolutions per minute is

$$n = \frac{60}{2\pi} \frac{V_t - I_a R_a}{K_a \Phi} \tag{15-14}$$

Inspection of Eq. (15-13) or (15-14) shows that control of speed can be obtained by adjustment of any of the three quantities R_a, Φ, and V_t. The first two methods will be discussed briefly; the third method will be considered more extensively because of its widespread use in control systems.

Armature-resistance control   The speed can be reduced by the insertion of external resistance in series with the armature circuit. It can be used with series, shunt, and compound motors. For the last two types, the series resistor must be connected between the shunt field and the armature, not between the line and the motor. It is the common method of speed control for series motors.

For a fixed value of series armature resistance, the speed will vary widely with load since the speed depends on the voltage drop in this resistance and hence on the armature current demanded by the load. Also, the power loss in the external resistor is large, especially when the speed is greatly reduced. On the other hand, the control equipment is relatively inexpensive. The series-resistance method is often used for short-time or intermittent slowdowns.

Constant armature voltage, controlled field excitation This method is simple and satisfactory for speed control of a shunt or compound motor over a speed range of about 4 or 5 to 1. The field excitation is controlled by variation of the voltage applied to the field circuit or by means of an adjustable resistor in series with the shunt field. The power loss in the resistor is relatively small because the field current is small compared with the armature current. The maximum permissible torque is limited by the permissible armature current and the maximum flux. The latter is limited by magnetic saturation or by heating of the field winding. The maximum speed is determined by commutation and mechanical considerations.

Constant field current, controlled armature voltage This system is the most commonly used when manual or automatic control of speed is required over a wide range and in both directions of rotation. The controlled armature voltage may be obtained from controlled rectifiers receiving input power from an ac source, or it may come from a separately excited dc generator. Because of its special importance, the following discussion is devoted to the separately excited generator-and-motor system, commonly called the *Ward Leonard system*. It is the basic scheme for a wide variety of feedback control systems.

The basic system is shown diagrammatically in Fig. 15-8. The generator is assumed to be driven at constant speed ω_g. Its field is separately excited from a voltage source v_{fg}. The motor is separately excited with constant field current I_{fm}. The combined resistance of the generator and motor armature circuits in series is R_a. Usually, the armature inductances are negligible; they are omitted in Fig. 15-8. The currents i_{fg} and i_a and the internal EMFs e_{eg} and e_{am} are shown with lowercase symbols because all of them may vary with time as adjustments of motor speed are called for.

With its field flux constant, the speed of the motor is directly proportional to the counter EMF e_{am} [see Eq. (15-4)]. The value of e_{am}, in turn, is determined by the generated voltage e_{ag} with the $i_a R_a$ drop deducted from it. The voltage e_{ag} can readily be controlled by adjusting the field voltage v_{fg} or current i_{fg}. Thus, adjustments of the relatively low level of power in the generator field provide excellent control of the output speed of the motor over a

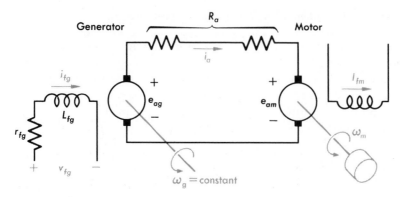

FIGURE 15-8 Schematic diagram for adjustable-armature voltage, or Ward Leonard, control.

wide range. Reversal of polarity of v_{fg} will cause reversal of the motor rotation. Changes in speed will not follow instantaneously upon adjustment in the generator field. Time is required both to change the current i_{fg} in the inductive field winding and to accelerate or decelerate the inertia of the motor armature and driven equipment to the new speed. Often a considerable portion of the design effort is devoted to minimizing these time lags.

Example 15-9

A dc generator and a dc motor have their armatures directly connected to form a Ward Leonard system. The generator is driven at a constant speed, and the motor field is excited by a constant current. The magnetization curve of the generator is assumed to be linear. The motor, rated at 3 hp, has an armature resistance of 0.75 Ω and a torque per unit armature current of 1.25 N·m/A. The generator has an armature resistance of 0.25 Ω, field resistance of 200 Ω, and generated voltage per unit field current of 1,500 V/A.

a A constant voltage of 36.6 V is impressed on the generator field, and the motor load requires a torque of 10.8 N·m. Find the steady operating speed.

b Find the generator-field power required in part **a**. Compare this value of control power with the output of the motor being controlled.

SOLUTION **a** Although we are dealing with steady values throughout this example, we shall retain the lowercase symbols of Fig. 15-8 for ease of reference.

$$i_{fp} = \frac{36.6}{200} = 0.183 \text{ A}$$

$$e_{ag} = 1,500 \times 0.183 = 275 \text{ V}$$

$$i_a = \frac{10.8}{1.25} = 8.63 \text{ A}$$

$$e_{am} = 275 - 8.63(0.25 + 0.75) = 266 \text{ V}$$

Note from Eqs. (15-4) and (15-11) that the same numerical constant K_a enters into the voltage-speed and torque-speed relations when MKS units are used. Since Φ is the same in the two relations, the counter EMF per unit speed of the motor is the same as the torque per unit armature current. It is therefore 1.25 V/mechanical rad/s. Hence the motor speed is

$$\omega_m = \frac{266}{1.25} = 213 \text{ rad/s}$$

or

$$n = \frac{60}{2\pi} \omega_m = \frac{60}{2\pi} 213 = 2,040 \text{ r/min}$$

b The generator-field power is

$$v_{fg}i_{fg} = (36.6)(0.183) = 6.7 \text{ W}$$

The motor power output is

$$T\omega_m = (10.8)(213) = 2,300 \text{ W}$$

The power output being controlled is thus 2,300/6.7, or 343, times the power required to exert control at this speed.

The generator is obviously an important element in Fig. 15-8, for it is there that the power amplification referred to in Example 15-9b takes place. The generator field in this example requires a source with nominal power rating of 5 to 10 W. If this level is too high, then additional power amplification must be inserted between the generator field and the ultimate source of control power. Such additional amplification may be obtained electronically or by means of additional rotating machines.

15-6 TRANSIENT AND DYNAMIC RESPONSES

Many modern applications of dc machines require knowledge of the response of the machine to time-varying control signals rather than simply its performance in the steady state. To introduce the concepts involved, we shall, in this section and the following one, confine our attention to separately excited dc machines. We shall be concerned primarily with two typical problems: (1) the electric transients in generators resulting from changes in excitation, and (2) the dynamics of motors with constant field excitation but time-varying armature voltages. Both problems arise in the control system of Fig. 15-8.

It will be assumed that there is no saturation in the magnetic circuit. The air-gap flux Φ is then directly proportional to the field current i_f, and the magnetization curve is a straight line. Field-circuit inductance L_f is constant. The voltage equation for the field circuit is then

$$e_f = L_f \frac{di_f}{dt} + R_f i_f \tag{15-15}$$

where e_f and R_f are the field voltage and resistance, respectively.

By virtue of Eqs. (15-4) and (15-11), the relations for armature EMF e_a and electromagnetic torque T_m can now be written as

$$e_a = k_f i_f \omega_m \tag{15-16}$$

and

$$T_m = k_f i_f i_a \tag{15-17}$$

i_a being armature current and k_f a constant. Lowercase symbols are used for voltages and currents to indicate that they may all be functions of time. The capital-letter symbol is retained for torque, even though it, too, may be a time function, to avoid confusion with the symbol t for time.

For a generator the armature terminal voltage is

$$v_{ta} = e_a - R_a i_a - L_a \frac{di_a}{dt} \tag{15-18}$$

where R_a and L_a are the armature resistance and inductance. Usually, the armature-inductance term in Eq. (15-18) can be neglected, and sometimes the $R_a i_a$ term can also be omitted. These approximations, of course, lead to greater simplification.

In the particular case of a generator operating at constant speed, Eq. (15-16) can be rewritten as

$$e_a = k_e i_f \tag{15-19}$$

where k_e is a constant equal to the product of k_f and the specified constant value of ω_m.

Example 15-10

A 200-kW 250-V dc generator has the following constants: $R_f = 33.7 \ \Omega$, $R_a = 0.0125 \ \Omega$, $L_f = 25 \ \text{H}$, $L_a = 0.008 \ \text{H}$, and $k_e = 38 \ \text{V/field A}$ at rated speed. The armature circuit is connected to a purely resistive load $R_L = 0.313 \ \Omega$.

The generator is initially unexcited but is rotating at rated speed. A 230-V dc source is suddenly connected to the field terminals. As the terminal voltage builds up and the generator takes on load, its speed does not change appreciably.

Find the terminal voltage as a function of time after connection of the excitation source. Ignore saturation.

SOLUTION   The buildup of field current described by Eq. (15-15) is that of a simple RL-circuit transient, or

$$i_f = \frac{E_f}{R_f}(1 - \epsilon^{-R_f t/L_f})$$

$$= \frac{230}{33.7}(1 - \epsilon^{-33.7t/25}) = 6.83(1 - \epsilon^{-1.35t}) \qquad \text{A}$$

Because e_a and i_f are linearly related, the buildup of armature internal voltage is

$$e_a = (38)(6.83)(1 - \epsilon^{-1.35t}) = 260(1 - \epsilon^{-1.35t}) \qquad \text{V}$$

Conditions in the armature circuit are described by Eq. (15-18) for the machine, and

$$v_{ta} = R_L i_a$$

for the external resistive load. By combining these two relations,

$$\frac{L_a}{R_L}\frac{dv_{ta}}{dt} + \frac{R_L + R_a}{R_L}v_{ta} = e_a$$

Upon substitution of numerical values,

$$0.0255\frac{dv_{ta}}{dt} + 1.04v_{ta} = 260 - 260\epsilon^{-1.35t}$$

This equation is subject to the boundary condition that v_{ta} must be zero at $t = 0^+$ because the armature current cannot change suddenly. Solution by the methods of Chap. 3 yields

$$v_{ta} = 250 - 257\epsilon^{-1.35t} + 7\epsilon^{-40.7t} \qquad \text{V}$$

If armature inductance were ignored, the solution would be

$$v_{ta} = 250 - 250\epsilon^{-1.35t} \qquad \text{V}$$

Comparison of the two solutions will show the influence of armature inductance to be very small.

For a motor, the armature terminal voltage is given by

$$v_{ta} = e_a + R_a i_a + L_a\frac{di_a}{dt} \qquad (15\text{-}20)$$

Again the armature-inductance term can often be neglected. Under dynamic conditions, the electromagnetic torque not only applies to shaft-load requirements T_L but also serves to change the motor speed; that is,

$$T_m = J\frac{d\omega_m}{dt} + T_L \qquad (15\text{-}21)$$

J being the combined moment of inertia of the armature and shaft load. All quantities are measured in MKS units. Rotational losses are either ignored or included in T_L.

In the particular case of a motor operating at constant field current, Eqs. (15-16) and (15-17) can be rewritten as

$$e_a = k\omega_m \tag{15-22}$$

and

$$T_m = ki_a \tag{15-23}$$

where the constant k is the same whether measured in volts per mechanical radian per second or in newton-meters per ampere.

Example 15-11

A small dc motor is directly coupled mechanically to a pure-inertia load (that is, $T_L = 0$), as indicated schematically in Fig. 15-9. It is excited with a constant field current. Both armature inductance and rotational losses are negligible. With the motor at rest, the constant direct voltage V_{ta} is suddenly impressed on its armature terminals.

a Obtain an expression for the armature current as a function of time.

b Obtain an expression for the angular velocity of the shaft as a function of time.

c Consider that the resistance $R_a = 34$ Ω, the moment of inertia $J = 5.85 \times 10^{-5}$ kg·m², and the constant $k = 0.328$ V/rad/s or N·m/A. Determine the time for the motor to come to within approximately 2 percent of its final speed.

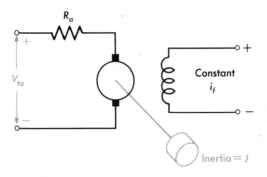

FIGURE 15-9 Direct-current motor with pure-inertia load for Example 15-11.

SOLUTION **a** For the circumstances described, Eqs. (15-20) and (15-21) become

$$V_{ta} = e_a + R_a i_a = k\omega_m + R_a i_a$$

$$T = k i_a = J \frac{d\omega_m}{dt}$$

By combining these equations and eliminating ω_m, we obtain

$$\frac{JR_a}{k^2} \frac{di_a}{dt} + i_a = 0$$

When it is recognized that at $t = 0^+$, $\omega_m = 0$, $e_a = 0$, and $i_a = V_{ta}/R_a$, the solution becomes

$$i_a = \frac{V_{ta}}{R_a} \epsilon^{-(k^2/JR_a)t}$$

b To obtain the angular velocity, the first two equations above are again conbined but with i_a eliminated, yielding

$$\frac{JR_a}{k^2} \frac{d\omega_m}{dt} + \omega_m = \frac{V_{ta}}{k}$$

the solution of which, for zero angular velocity at $t = 0^+$, is

$$\omega_m = \frac{V_{ta}}{k} (1 - \epsilon^{-(k^2/JR_a)t})$$

$$\frac{JR_a}{k^2} = \tau$$

is a time constant in the usual meaning of the term for an electric circuit.

c From this last equation, the time constant is

$$\tau = \frac{(5.85)(10^{-5})(34)}{(0.328)^2} = 0.0185 \text{ s}$$

Now, a simple exponential will decay to 0.0183 or approximately 2 percent of its initial value in a time equal to four time constants. For the motor to come up to speed in this case then requires $(4)(0.0185)$, or 0.074 s.

15-7 TRANSFER FUNCTIONS AND FREQUENCY RESPONSE

The generator in the speed-control system of Fig. 15-8 will be in the steady state only when the desired motor speed and the motor load are constant for appreciable periods of time. The more usual conditions will be that speed changes are desired and that e_f will be changing with time in a more or less random manner which cannot readily be predicted. The voltage e_f may, for example, have a value of 50 V at one instant and -30 V a little later, indicating that reversal of motor torque is desired quickly in order to decelerate the motor. An ideal amplifier would produce at its output a voltage waveform at all times directly proportional to the input waveform. The dc generator is hampered from attaining this ideal by the inductance in its field winding. As has been illustrated in Example 15-10, a time lag is interposed by the field time constant L_f/R_f between a change of e_f and the associated change in i_f and hence in v_{ta}.

A measure of the relative importance of this time lag is furnished by the frequency response of the dc generator. The voltage e_f is considered to vary sinusoidally with time, and the ratio $\mathbf{V}_{ta}/\mathbf{E}_f$ of output to input voltage is evaluated for a range of frequency f or electrical angular velocity ω of the input signal $\mathbf{E}_f$. If this ratio remains substantially constant over a reasonable range of frequencies, the dc generator is judged to be better able to follow rapid variations in the control signal e_f than if the ratio drops off rapidly with frequency. This process is simply application of the frequency-response techniques used for circuits and electronic components.

Boldface symbols for voltages and currents are now used once more to denote that phasor quantities having both phase and magnitude are being dealt with. Consider the generator to have negligible armature inductance and resistance and to be driven at constant speed. Equations (15-15) and (15-19) then become

$$j\omega L_f \mathbf{I}_f + R_f \mathbf{I}_f = \mathbf{E}_f \tag{15-24}$$

and

$$\mathbf{V}_{ta} = k_e \mathbf{I}_f \tag{15-25}$$

By combining Eqs. (15-24) and (15-25), the ratio of output to input voltage is seen to be

$$\mathbf{G}_g = \frac{\mathbf{V}_{ta}}{\mathbf{E}_f} = \frac{k_e/R_f}{1 + j\omega(L_f/R_f)} \tag{15-26}$$

This expression gives the voltage gain as a function of frequency. It is the transfer function relating output to input for the generator as used in Fig. 15-8, and as such, it can be combined with the transfer functions of other cascaded equipment in studies of overall systems performance.

The transfer function, Eq. (15-26), is characterized by two constants: the steady-state dc gain k_e/R_f and the generator field time constant L_f/R_f. Many control amplifiers can be similarly characterized. The steady-state gain is a measure of the output-to-input-voltage ratio under fixed operating conditions, and the time constant is a measure of the time lag in response to varying conditions. The transfer function of such an element containing a single time lag is given by

$$\mathbf{G} = \frac{\text{steady-state gain}}{1 + j\omega \, (\text{time constant})} \tag{15-27}$$

Curves showing the amplitude and phase of the transfer function $\mathbf{G}_g$ as functions of frequency are given in Fig. 15-10. The amplitude drops off and the phase shift increases with frequency. The significance of the response drop is that the dc generator will behave less and less ideally as the frequency of the control-signal voltage variations at the field input increases. It can readily be shown that for $\omega = R_f/L_f$, the amplitude G_g has been reduced to $1/\sqrt{2} = 0.707$ of its value under steady-state dc conditions and the phase shift has become 45°. This value of ω corresponds to the half-power frequency.

This application of frequency-response and transfer-function techniques to a dc generator is a step in bringing dc-machine analysis in line with that for other equipment, such as circuits and electronic amplifiers, to be found in control systems. The use of such techniques, with their implications of ac input and output quantities, for dc machines may appear initially confusing. We know, for instance, that the conventional dc generator in Fig. 15-8 cannot be expected to operate successfully with a 60-Hz voltage impressed on its field. But we also know that the output voltage of the generator will faithfully follow slow variations, including reversal of the field-voltage—of the order of a small fraction of a hertz, say. To present a quantitative measure of the ability of the generator output to follow such variations as their rates of change are increased from very low values is exactly the object of both the frequency-response characteristics and the transfer function. In accordance with the foregoing statements, we expect the output-input amplitude ratio for the generator to be close to its dc value at very low frequencies and to be practically zero at 60 Hz. Numerical computation of frequency-response characteristics such as those in Fig. 15-10 confirms this expectation and enables us to draw an approximate line of frequency demarcation above which the generator is not suitable. For a conventional generator, this line will fall at a low number of cycles per second.

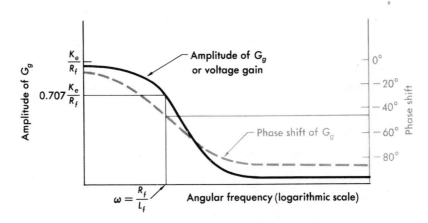

FIGURE 15-10 Frequency-response characteristics of dc generator.

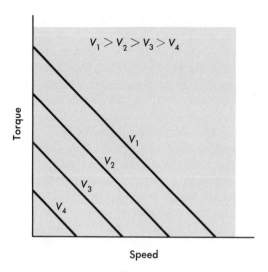

FIGURE 15-11 Typical torque-speed characteristics for dc control motor. The motor is a shunt-wound machine with constant field current; the voltages are armature terminal voltages.

Similar considerations may be applied to dc motors, which are widely used in control systems for positioning and driving shafts. The transfer functions of interest are often those between armature voltage as input and either shaft speed or shaft position as output.

For many applications, particularly those involving large values of power, standard machines are used. In these cases, the control system must be designed to compensate for any shortcomings in the motor characteristics. In other instances, however, the motors used are designed especially to meet the requirements of control systems, chief among which are a high starting torque and low rotor inertia. Such motors are called *control motors*. The principles of operation are identical with those already discussed.

Direct-current control motors are usually shunt motors with the armature and field supplied from separate sources. Control is effected by either armature voltage or field control. For example, the motor in the control system of Fig. 15-8 may be a dc control motor.

The torque-speed characteristics of a typical dc control motor with armature-voltage control are shown in Fig. 15-11. These curves are characterized by high torques in the zero-speed range and a drop-off of torque as the motor speed increases. The drop in the torque serves as a stabilizing feature for the control system. Direct-current control motors with field control usually have their armatures excited from constant-current sources. These motors produce a torque which is independent of speed but directly dependent on the field flux. Care must be exercised in their use to prevent the motor from attaining excessive speeds.

PROBLEMS

15-1 A speed-indicating device consists of a two-pole permanent-magnet dc generator with a dc voltmeter connected to its terminals. The generator has an armature with 1,000 conductors having two parallel paths and a resistance of 1,000 Ω between brushes. The air-gap flux is 36,000 lines/pole. The dc voltmeter, which is connected directly across the brushes, has a resistance of 2,000 Ω.

What speed in revolutions per minute will be indicated by a reading of 5 V on the voltmeter?

15-2 A dc series motor operates at 750 r/min with a line current of 80 A from the 230-V mains. Its armature-circuit resistance is 0.14 Ω, and its field resistance is 0.11 Ω.

Assuming that the flux corresponding to a current of 20 A is 40 percent of that corresponding to a current of 80 A, find the motor speed at a line current of 20 A at 230 V.

15-3 A dc shunt motor is driving a mechanical load which requires a constant torque regardless of speed. The armature current at a speed n_1 is I_{a1}. By ignoring all losses, estimate the armature current as a multiple of I_{a1} when the speed is increased to $2n_1$:
a By changing the armature terminal voltage
b By changing the field flux

15-4 A 25-kW 250-V dc machine has an armature resistance of 0.10 Ω. Points on its magnetization curve at a constant speed of 1,200 r/min are as follows:

Field current, A	0.5	1.0	1.5	2.0	2.5
Induced voltage, V	70	140	195	235	260

Its field is separately excited, and it is driven by a synchronous motor at a constant speed of 1,200 r/min.

Plot a family of curves of armature terminal voltage versus armature current for constant field currents of 2.5, 2.0, 1.5, and 1.0 A.

15-5 The magnetization curve of a dc generator contains the following points, all taken at a speed of 1,000 r/min:

Field current, A	1.50	1.25	1.00	0.50
Induced voltage, V	250	230	200	100

a If the field current is adjusted to 1.25 A, how fast must the generator be driven to generate 250 V at no load?
b What must be the field current to generate 200 V at 800 r/min, no load?
c If the machine is connected as a motor to a 230-V line and the field cur-

rent is adjusted to 1.0 A, how fast will it run at no load? Neglect rotational losses.

15-6a Dc generators whose service requirements demand operation over a wide range of adjustable voltages are usually separately excited rather than self-excited. Why?
b The steady-state current of a shunt generator feeding into a short circuit is frequently lower than the full-load current. Explain briefly.
c How would you convert an overcompounded generator into a flat-compounded generator?
d A flat-compounded self-excited generator is operating at no load when the prime-mover speed is decreased by 10 percent. No other changes are made. Will the change in terminal voltage be equal to, more than, or less than 10 percent?

15-7 A 15-kW 230-V 65-A 1,800-r/min separately excited dc generator supplies a load which is 1,500 ft away. The resistance of the transmission line is 0.075 ohm.

It is desired that the voltage at the load be 230 V at both no load and full load. Tests show that this requires a field current of 1.34 A at no load and 1.48 A at the full load of 65 A. Rather than adjust the field current manually, however, it is proposed to add a series field. The added resistance of the series field is negligible, and the number of turns on the separately excited field is 1,820.

Calculate the necessary number of series-field turns.

15-8 The dc machine of Prob. 15-4 is operated as a separately excited motor.
a For a constant field current of 2.0 A, plot a family of curves of speed in revolutions per minute versus torque in newton-meters for applied armature terminal voltages of 250, 200, 150, and 100 V.
b For a constant applied armature terminal voltage of 200 V, plot a family of curves of speed versus torque for field currents of 2.5, 2.0, 1.5, and 1.0 A.

15-9 A 230-V dc shunt motor has an armature-circuit resistance of 0.1 Ω. This motor operates on the 230-V mains and takes an armature current of 100 A. An external resistance of 1.0 Ω is now inserted in series with the armature, and the electromagnetic torque and field-rheostat setting are unchanged.
a Give the percentage of change in the total current taken by the motor from the mains.
b Give the percentage of change in the speed of the motor, and state whether this will be an increase or a decrease.

15-10 A dc series motor develops its rated power output of 50 hp at a terminal voltage of 240 V and a speed of 1,000 r/min. Under this condition, the armature current is 183 A. The combined armature-circuit and series-field resistance is 0.131 Ω. For this problem, assume that the motor operates with no saturation and that the rotational losses can be neglected.

The motor is used to drive a load having a speed-torque characteristic expressed by

$$T = 11.3\sqrt{n}$$

where T is in newton-meters and n in revolutions per minute.

Find the operating speed of the motor and load when the applied voltage is 240 V.

15-11   A 10-hp 230-V 1,150-r/min shunt motor delivers its rated load at rated speed when taking 38.5 A from a 230-V line. The resistance of the shunt-field circuit is 128 Ω, and the resistance of the armature circuit including brushes is 0.30 Ω.

For some operations, it is desired to have this motor rotate at 450 r/min and deliver 120 percent of full-load electromagnetic torque. The method used to reduce the speed is to introduce series resistance in the armature circuit. With the supply-line potential constant at 230 V, what value of resistance must be inserted in the armature circuit for the desired operation?

15-12a   Two adjustable-speed dc shunt motors have maximum speeds of 1,650 r/min and minimum speeds of 450 r/min. Speed adjustment is obtained by field-rheostat control. Motor A drives a load requiring constant horsepower over the speed range; motor B drives one requiring constant torque. All losses can be neglected.
(1) If the power outputs are equal at 1,650 r/min and the armature currents are each 100 A, what will be the armature currents at 450 r/min?
(2) If the power outputs are equal at 450 r/min and the armature currents are each 100 A, what will be the armature currents at 1,650 r/min?
b   Answer part **a** for speed adjustment by armature-voltage control with conditions otherwise the same.

15-13   A Ward Leonard system of dc-motor speed control is to be used to drive a drum on which cloth is being wound. The motor shaft is coupled directly to the drum. In order to produce a constant winding rate at constant tension on the material being wound, the electromagnetic torque varies from 50 to 100 lb-ft while the speed varies from 1,100 to 550 r/min.

The motor that is driving the generator operates at a constant speed of 1,200 r/min. The generator is a 15-kW 230-V machine with an armature resistance of 0.14 Ω and the magnetization curve of Fig. 15-2. The dc motor is rated 15 hp and 230 V and has an armature resistance of 0.24 Ω and a constant field current of 1.3 A. The magnetization curve of Fig. 15-2 also applies to this motor. When the electromagnetic torque is 50 lb·ft at 1,100 r/min, the motor armature current is 40 A.

Find the range over which the generator shunt-field current must be varied.

15-14   Two identical 5-hp 230-V 17-amp dc shunt machines are to be used as the generator and motor, respectively, in a Ward Leonard system. The generator is driven by a synchronous motor with a speed that is constant at 1,200 r/min. The armature-circuit resistance of each machine is 0.47 Ω (including brushes). Data for the magnetization curve of each machine at 1,200 r/min are as follows:

I_f, A	0.2	0.4	0.6	0.8	1.0	1.2
E_a, V	180	183	230	254	267	276

Compute the maximum and minimum values of generator-field current needed to give the motor a speed range of 300 to 1,500 r/min at full-load armature current (17.0 A) with the motor-field current held constant at 0.50 A.

15-15   A separately excited dc generator has the following constants:
Field-winding resistance = 100 Ω
Field-winding inductance = 50 H
Armature resistance = 0.05 Ω
Armature inductance = 0.5 mH
Generated emf constant = 100 V/field A at 1,200 r/min
The generator is driven at a constant speed of 1,200 r/min. Its field and armature circuits are initially open.
a   At $t = 0$, a constant-voltage source of 250 V is suddenly applied to the terminals of the field winding. Find the equation for the armature terminal voltage as a function of time, and sketch the curve.
b   After steady-state conditions have been established in part **a**, the armature is suddenly connected to a load of resistance 1.20 Ω and inductance 1.5 mH in series. Find the equations for (1) the armature current and (2) the armature terminal voltage as functions of time. Include the effect of the armature inductance and resistance. Sketch the curves.

15-16   A separately excited dc generator has a field-circuit resistance of 100 Ω and an inductance of 50 H. It generates 100 V/field A at 1,200 r/min.
Give the expression for the transfer function relating generated voltage to field voltage when the speed is 1,200 r/min.

15-17   Plot the amplitude of the transfer function for the generator of Prob. 15-16 as a function of frequency in cycles per second.

15-18   A small dc motor operating with constant field flux drives a pure-inertia load. It has an armature resistance of 6 Ω, a speed-voltage constant of 0.9 V/rad/s, and negligible armature inductance. The inertia of the armature plus load is 0.005 kg·m².

Give the expression for the transfer function relating motor angular velocity and armature terminal voltage. Plot the magnitude of this transfer function as a function of frequency.

15-19 Repeat Prob. 15-18 for a significantly larger motor having the following constants:
Motor armature resistance = 0.5 Ω
Motor torque per unit armature current = 2.00 N·m/A
Motor plus load inertia = 10 kg·m²

15-20 Give the expression for the transfer function relating terminal voltage to field voltage for the dc generator of Example 15-10. Only numerical values and the angular frequency ω of the field voltage should appear.

15-21a For the motor of Example 15-11, give a literal expression for the transfer function relating the motor-output angular velocity to the armature input voltage and the angular frequency ω of that voltage.
b Substitute in this expression the numerical values given in Example 15-11c.
c Plot the amplitude of the transfer function of part **b** as a function of frequency.

15-22 The rated line current of a 10-hp 230-V 1,350 r/min shunt motor is 37.5 A. The field current is 0.75 A, and the armature resistance is 0.38 Ω. Determine:
a The full-load efficiency
b The internal torque in newton-meters at rated load
c The internal torque in newton-meters when the line current is 18 A and the field current remains at 0.75 A.
d The speed in part **c**

15-23 A 20-hp 220-V shunt motor has the magnetization curve of Fig. 15-2. There are 2,000 turns/pole on the shunt field, and the armature-circuit resistance is 0.14 Ω.

When the shunt-field rheostat is set for a motor speed of 1,200 r/min at no load, the armature current is 3.5 A. How many series-field turns per pole must be added if the speed is to be 950 r/min for a load requiring an armature current of 72.5 A? Neglect the added resistance of the series turns.

15-24 The dc generator of Prob. 15-4 is operated at a field current of 1.5 A.
a What is the generator-driving speed required for a no-load voltage of 235 V?
b What is the field current required to generate 195 V at a drive speed of 950 r/min.

15-25 The characteristics of a separately excited dc generator are:

Field-winding resistance = 225 Ω
Field-winding inductance = 75 H
Armature resistance = 0.10 Ω
Armature inductance = 1.0 mH
Generator constant = 60 V/field A at 1,200 r/min
The generator is driven at 1,200 r/min and the field is connected to a 120-V source.

a Find the armature terminal voltage at no load.
b Determine the transfer function of the system when the armature is connected to a 2:3-Ω load.
c The field voltage suddenly drops to 110 V. Determine the armature current and terminal voltage as functions of time.

CURRENT MACHINES □ **ALTERNATING-**
ACHINES □ ALTERNATING-**CURRENT**
ALTERNATING-CURRENT MACHINES
ALTERNATING-CURRENT MACHINES
-CURRENT MACHINES □ ALTERNATI
MACHINES □ ALTERNATING-CURRE
ALTERNATING-CURRENT MACHINES
-CURRENT MACHINES □ ALTERNATI
MACHINES □ ALTERNATING-CURRE
ALTERNATING-CURRENT MACHINES □ ALTERNATING CURRENT MA
-CURRENT MACHINES □ ALTERNATING-CURRENT MACHINES □ ALT
MACHINES □ ALTERNATING-CURRENT MACHINES □ ALTERNATING·
ALTERNATING-CURRENT MACHINES □ ALTERNATING CURRENT MA
-CURRENT MACHINES □ ALTERNATING-CURRENT MACHINES □ ALT
MACHINES □ ALTERNATING-CURRENT MACHINES □ ALTERNATING·
ALTERNATING-CURRENT MACHINES □ ALTERNATING CURRENT MA
-CURRENT MACHINES □ ALTERNATING-CURRENT MACHINES □ ALT
MACHINES □ ALTERNATING-CURRENT MACHINES □ ALTERNATING
ALTERNATING-CURRENT MACHINES □ ALTERNATING CURRENT MA
-CURRENT MACHINES □ ALTERNATING-CURRENT MACHINES □ ALT
MACHINES □ ALTERNATING-CURRENT MACHINES □ ALTERNATING·
ALTERNATING-CURRENT MACHINES □ ALTERNATING CURRENT MA
-CURRENT MACHINES □ ALTERNATING-CURRENT MACHINES □ ALT
MACHINES □ ALTERNATING-CURRENT MACHINES □ ALTERNATING·
ALTERNATING-CURRENT MACHINES □ ALTERNATING CURRENT MA
-CURRENT MACHINES □ ALTERNATING-CURRENT MACHINES □ ALT
MACHINES □ ALTERNATING-CURRENT MACHINES □ ALTERNATING
ALTERNATING-CURRENT MACHINES □ ALTERNATING CURRENT MA
-CURRENT MACHINES □ ALTERNATING-CURRENT MACHINES □ ALT
MACHINES □ ALTERNATING-CURRENT MACHINES □ ALTERNATING
ALTERNATING-CURRENT MACHINES □ ALTERNATING CURRENT MA
-CURRENT MACHINES □ ALTERNATING-CURRENT MACHINES □ ALT
MACHINES □ ALTERNATING-CURRENT MACHINES □ ALTERNATING
ALTERNATING-CURRENT MACHINES □ ALTERNATING CURRENT MA
-CURRENT MACHINES □ ALTERNATING-CURRENT MACHINES □ ALT
MACHINES □ ALTERNATING-CURRENT MACHINES □ ALTERNATING
ALTERNATING-CURRENT MACHINES □ ALTERNATING CURRENT MA
-CURRENT MACHINES □ ALTERNATING-CURRENT MACHINES □ ALT
MACHINES □ ALTERNATING-CURRENT MACHINES □ ALTERNATING
ALTERNATING-CURRENT MACHINES □ ALTERNATING CURRENT MA

16

he basis of operation of polyphase synchronous and induction machines will be explored more comprehensively here than was done in the introductory treatment of Chap. 14. Important steady-state and dynamic characteristics will be discussed, but to avoid lengthy analysis, serious quantitative investigations will not be undertaken.

16-1 ROTATING MAGNETIC FIELDS

The ac windings used in polyphase induction and synchronous machines produce magnetic fields of constant amplitude rotating at a uniform speed around the air-gap circumference when the windings carry polyphase currents. This fundamental fact may be demonstrated physically by considering the elementary three-phase two-pole winding of Fig. 16-1. It is the same winding as was used to examine the production of three-phase voltages in Fig. 14-10 and consists of three coils 120° apart in space, each coil forming one phase of a three-phase system. The currents in the three coils as functions of time are shown in Fig. 16-2. To illustrate the field produced by this winding, the cylinder forming the stator is developed in the descriptive-geometry sense in Fig. 16-3; that is, the cylinder is cut and laid out flat, conductor *a* being shown twice in order to identify the ends clearly.

That a uniformly rotating field is produced may be seen by plotting the component and resultant field distributions for several successive instants of time. This is done in Fig. 16-3. It is recommended that Fig. 16-3 be resketched step by step while reading these paragraphs, the figure being clearer when considered piece by piece than when first viewed in finished form. The flux distribution from any one phase is assumed to be sinusoidal, although

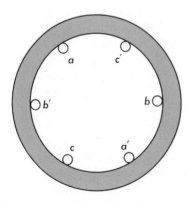

FIGURE 16-1 Simplified three-phase two-pole stator winding.

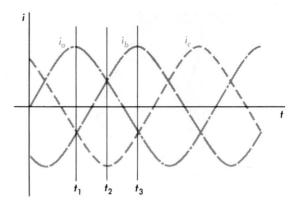

FIGURE 16-2 Instantaneous phase currents in the winding of Fig. 16-1.

such will not be the case if the phase consists of just one concentrated coil, as in Figs. 16-1 and 16-3. Practical windings, however, consist of a group of coils for each phase distributed over the stator surface and so designed that the flux distribution is very nearly sinusoidal.

Consider first the instant t_1 (Fig. 16-2) when the current in phase a is a maximum and the currents in phases b and c are a negative half maximum. The flux distribution produced by phase a alone in Fig. 16-3a then has its time-maximum amplitude. It is centered in space about the axis of coil a and is arbitrarily drawn positive from a to a' to correspond to positive current. The flux distributions produced by phases b and c each, respectively, have half the amplitude of the contribution from phase a, are centered in space about the axis of coil b or c, and are drawn negative from b to b' or c to c' to correspond to negative currents. As shown in Fig. 16-3a, the resultant flux distribution, obtained by adding the individual contributions of the three phases, is a sinusoid centered on phase a and having an amplitude $\frac{3}{2}$ times the maximum contribution of phase a.

For a later instant t_2 (Fig. 16-2), the current in phase c is a negative maximum and those in a and b are positive half maxima. As shown in Fig. 16-3b, the phase-a flux contribution has half its previous amplitude. The phase-b contribution has the same amplitude but the reverse polarity to that at t_1 because its current is reversed. The phase-c contribution has the same polarity as at t_1 but twice the amplitude. The same resultant flux distribution is obtained, but the wave has moved to the right. For a third instant t_3 (Fig. 16-2), similar reasoning leads to the same resultant distribution, but as shown in Fig. 16-3c, it has moved still farther to the right and is centered on phase b. The resultant flux wave corresponds to a magnetic field rotating around the cylindrical stator of Fig. 16-1 at a uniform speed. Results consistent with this conclusion may be obtained by sketching the distribution for any arbitrary instant of time.

One cycle after time t_1 (Fig. 16-2), the resultant field must be back in the

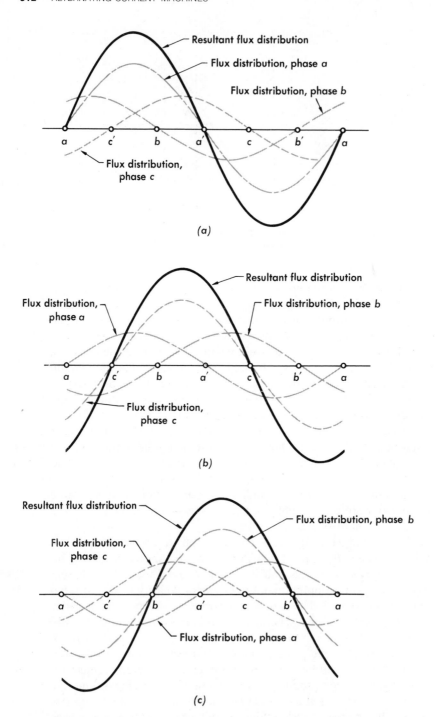

FIGURE 16-3 Component and resultant field distributions caused by the currents of Fig. 16-2 in the winding of Fig. 16-1. (a) For time t_1, Fig. 16-2; (b) for time t_2; (c) for time t_3.

position of Fig. 16-3a. The speed of the field must then be f revolutions per second or 60f revolutions per minute, f being the stator frequency. In many cases, the winding has more than the one group of coils per phase shown in Fig. 16-1; in other words, more than two poles. Figure 14-12 shows a simplified schematic layout of a four-pole winding. Successive phase groups are displaced by 120 *electrical degrees* in this case, rather than 120 *mechanical degrees* as in Fig. 16-1, 2 electrical degrees in Fig. 14-12 equaling 1 mechanical degree. If the winding has p poles, the rotating field travels 2/p revolution per cycle, and p/2 electrical degrees equals 1 mechanical degree. The speed of the stator field, or the *synchronous speed,* is then

$$n_{sf} = \frac{120f}{p} \quad \text{r/min} \tag{16-1}$$

which is the same relation between speed and frequency as was given by Eq. (14-7) for generated ac voltages.

It should be noted that these windings are quite capable of acting as the primary of a three-phase transformer when a second set of windings is introduced and is inductively coupled to the first set. An ac machine with the rotor winding open-circuited is simply an open-circuited transformer with an air gap in the magnetic circuit. In fact, one viewpoint sometimes adopted is to regard an ac machine as a transformer with an additional degree of freedom, that of rotation of one of the windings. As in the transformer, the air-gap flux produced by these stator windings is determined essentially by the magnitude of the impressed voltage. The equivalent statement for a transformer is that the core flux is determined essentially by the impressed primary voltage, a statement which neglects the usually small leakage-impedance drop in the primary. Analogously, the foregoing statement for a machine neglects the corresponding leakage-impedance drop in the stator.

This general type of winding is used in induction motors, synchronous motors, and synchronous generators. The windings are embedded in slots in the stator iron, several slots per pole being devoted to each phase.

16-2 INDUCTION-MOTOR ACTION

A three-phase induction motor consists of a stator winding of the type described in the preceding section and a rotor of one of the two forms shown in Figs. 14-23 and 14-24. Figure 14-23 shows a squirrel-cage rotor with a winding consisting of conducting bars embedded in slots in the rotor iron and short-circuited at each end by conducting end rings. Figure 14-24 shows a wound rotor with a winding similar to and having the same number of poles as the stator winding, the terminals of the winding being connected to the slip rings or collector rings on the left end of the shaft. Carbon brushes bearing on these rings make the rotor terminals available at points external to the motor so that additional resistance can be inserted in the rotor circuit if desired.

To examine the action of an induction motor, recall from Sec. 14-4 that for the production of torque, the rotor and stator magnetic fields must be stationary with respect to each other. Consider a squirrel-cage rotor to be inserted in the rotating magnetic field of the stator. With the rotor stationary, the motor is the equivalent of a short-circuited three-phase transformer. Three-phase voltages of stator frequency are induced in the rotor, and the accompanying currents are determined by voltage magnitude and rotor impedance. Because they are induced by the rotating stator field, these rotor currents inherently produce a rotor field with the same number of poles as the stator and rotating at the same speed with respect to the stationary rotor. Rotor and stator fields are thus stationary with respect to each other in space, and a starting torque is produced. If this torque is sufficient to overcome the opposition to rotation created by the shaft load, the motor will come up to its operating speed. The operating speed can never equal the synchronous speed of the stator field, determined by Eq. (16-1), for the rotor conductors would then be stationary with respect to this field and no voltage would be induced in them.

To determine how rotation is sustained, consider the rotor to be turning at the steady speed n revolutions per minute in the forward direction. The rotor is then traveling at a speed of $(n_{sf} - n)$ revolutions per minute in the backward direction with respect to the rotating field, or the *slip speed* of the rotor is $(n_{sf} - n)$ revolutions per minute. Slip is more usually expressed as a fraction of synchronous speed; that is,

$$s = \frac{n_{sf} - n}{n_{sf}} \tag{16-2}$$

or

$$n = n_{sf}(1 - s) \tag{16-3}$$

This relative motion of flux and rotor conductors causes voltages of frequency equal to sf, called *slip frequency,* to be induced in the rotor. The accompanying rotor currents are determined by the magnitude of the rotational voltage and the rotor impedance at slip frequency. Since the frequency of these currents is now only the fraction s of what it was with the rotor stationary, the rotor field set up by them will travel *with respect to the rotor structure* at only the fraction s of its former speed; in other words, it will travel at $n_{rf} = sn_{sf}$ revolutions per minute in the forward direction with respect to the rotor. But superimposed on this is the mechanical rotation of the rotor [Eq. (16-3)]. The speed of the rotor field in space is the sum of these two terms,

$$sn_{sf} + n_{sf}(1 - s)$$

or n_{sf}. The stator and rotor fields are therefore stationary with respect to each other, torque is produced, and rotation is maintained.

Three of the important practical characteristics of any motor are the starting torque, the maximum possible torque which can be obtained, and the torque-speed curve showing the behavior of the motor under changing load. The squirrel-cage motor is substantially a constant-speed motor having about 3 to 10 percent drop in speed from no load to full load. Because of its ruggedness and simplicity, the squirrel-cage induction motor is very widely used. A wound rotor is generally more expensive and is used only for severe starting duties or when speed control is desired. Speed variation may be obtained by inserting external resistance in the wound rotor. In the normal operating range, this external resistance simply increases the rotor impedance, necessitating a higher slip for a desired rotor field strength and torque.

Example 16-1

A six-pole three-phase wound-rotor induction machine is driven by another machine on its shaft at 1,800 r/min. The rotor is connected to a three-phase 60-Hz system. Describe the magnetic field within the induction machine and the nature of the voltage, if any, produced at the stator terminals.

SOLUTION   The rotor winding will produce a sinusoidal flux distribution in space which, in accordance with Eq. (16-1), will rotate at a speed of 1,200 r/min with respect to the rotor. If the mechanical rotation of the rotor is in the same direction, this field will rotate with a speed of 1,800 + 1,200, or 3,000 r/min with respect to the stator. From Eq. (16-1), the net rotation will induce a sinusoidal voltage of 150 Hz in the stator. If rotation is in the opposite direction, the speed of the field with respect to the stator is 1,800 − 1,200, or 600 r/min, and the stator frequency is 30 Hz.

This example describes the general principle of the *induction frequency converter*.

16-3   INDUCTION-MOTOR PERFORMANCE

One of the important performance characteristics of any motor is the variation of speed as load is added. Conventionally, this characteristic is shown for an induction motor as a plot of torque as a function of slip. Such a torque-slip curve for an induction motor is given by the solid curve of Fig. 16-4; a speed scale is also added on the horizontal axis.

The normal steady operating region for the motor is the right-hand portion of this curve, corresponding to small values of slip. The curve is approximately linear in this region. The starting torque with normal line voltage impressed is the ordinate of the curve at $s = 1$. As the motor comes up to speed, the operating point follows the curve, moving to the right and settling down in the normal region at the torque value required by the load.

An approximate expression for the torque-slip curve at small values of slip can be written by assuming that it is linear over a reasonable range. At the same time it should be recognized that the torque produced by an induction motor also varies as the square of the stator impressed voltage. On this basis, the torque is

$$T = k_T V_t^2 s \tag{16-4}$$

where k_T is a constant for a particular torque-slip curve and V_t is the terminal

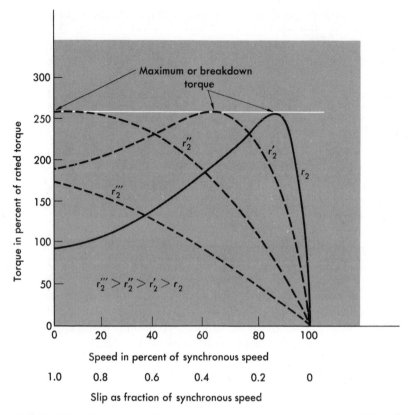

FIGURE 16-4 Typical induction-motor torque-speed curves, showing effect of increasing rotor resistance.

voltage. The expression is useful for approximately evaluating the effects of changing impressed voltage or changing torque requirements on the speed of the motor.

The maximum ordinate of the torque-slip curve (Fig. 16-4) is called the *breakdown torque*. It corresponds to the maximum power which can be transferred across the air gap to the rotor. If the load torque should be increased so that it exceeds the breakdown torque, the motor will stop.

The effects of adding external resistance in the rotor circuit of a wound-rotor motor can be seen from the dotted torque-slip curves of Fig. 16-4. As the rotor-circuit resistance is increased, the maximum torque remains constant. The slip at which it occurs, however, becomes progressively greater. Hence the speed at which any desired value of torque is produced becomes lower as rotor resistance is increased. Variation of starting torque with rotor resistance can also be seen from these curves by noting the zero-speed ordinates. Wound-rotor motors are chosen for applications with severe starting requirements or where control of speed is desired. In fact, the wound-rotor motor offers one of the few possibilities of speed adjustment in a motor supplied with constant-frequency alternating current.

Squirrel-cage motors are classified in accordance with their starting characteristics in the following manner:

Class A: Normal starting torque, normal starting current
Class B: Normal starting torque, low starting current
Class C: High starting torque, low starting current
Class D: High starting torque, high slip

Average torque-slip curves for these four classes of motors are given in Fig. 16-5. These curves show the order of magnitude of starting torque with full voltage impressed. If the motors are started at a reduced voltage, the starting torque is reduced as the square of the voltage in accordance with Eq. (16-4). The curves also show the sacrifices in breakdown torque made to improve starting conditions. Normal starting current (for starting at rated voltage) is 500 to 800 percent of full-load current; low starting current is about three-fourths of this value.

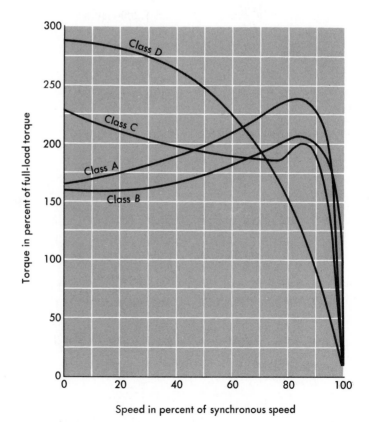

FIGURE 16-5 Typical torque-speed curves for various classes of induction motors.

Example 16-2

A 10-hp 230-V three-phase 60-Hz six-pole squirrel-cage induction motor operates at a full-load slip of 4.0 percent when rated voltage and frequency are impressed.

a What is the full-load speed in revolutions per minute?

b Assume that the torque-slip curve is linear in the normal operating region. Determine the constant k_T in Eq. (16-4) and hence the equation of this part of the torque-slip curve when T is measured in newton-meters and V_t in line-to-line volts.

c According to the expression in part **b**, what is the no-load speed?

d Consider that the load torque is increased to 125 percent of rated while the voltage is reduced to 80 percent of normal. What is the motor speed?

SOLUTION   **a**   From Eq. (16-1), the synchronous speed is

$$n_{sf} = \frac{120f}{p} = \frac{(120)(60)}{6} = 1{,}200 \text{ r/min}$$

From Eq. (16-3), the full-load speed is

$$n = n_{sf}(1 - s) = (1{,}200)(1 - 0.040) = 1{,}152 \text{ r/min}$$

b   The full-load power output is 10 hp or 7,460 W. Hence

$$T = \frac{(60)(7{,}460)}{2\pi(1{,}152)} = 62.0 \text{ N·m}$$

By substitution in Eq. (16-4),

$$62.0 = k_T(230)^2(0.040)$$

or

$$k_T = 0.0293$$

Hence

$$T = 0.0293\,V_t^2 s$$

c   From the last equation, $s = 0$ and n would be 1,200 r/min. Actually, the speed would be slightly lower (perhaps 1,195 r/min) because of friction and windage losses.

d   By substituting in the empirical equation derived in part **b**,

$$(1.25)(62.0) = (0.0293)(0.80 \times 230)^2 s$$

or

$$s = 0.0781$$

and

$$n = (1 - 0.0781)(1{,}200) = 1{,}107 \text{ r/min}$$

16-4 SYNCHRONOUS-GENERATOR PERFORMANCE

Three-phase synchronous machines have ac windings on the stator which are exactly like those described in Sec. 16-1. The field produced by the stator therefore rotates at synchronous speed as determined by Eq. (16-1). Under normal steady-state conditions for either synchronous-generator or synchronous-motor action, no voltage is induced in the rotor by the stator field because, with the rotor winding traveling at the same speed as the stator field, there is no change of rotor flux linkages. Only the impressed direct cur-

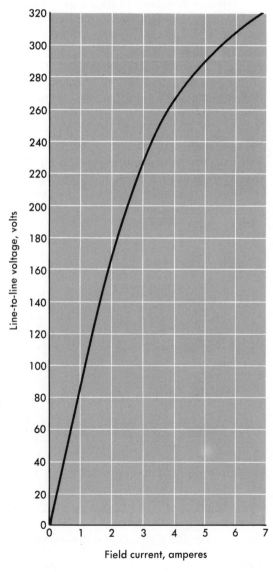

FIGURE 16-6 Open-circuit characteristic of small synchronous machine.

rent is present in the rotor winding, and the ohmic resistance R_f is the only opposition to it.

Except for the commutator action in the dc machine, the broad picture is not unlike that for a separately excited dc generator. As a result, it is not surprising to find that the analysis of performance is also similar. We must remember, however, to use impedance rather than simply resistance in computing the voltage drop in the armature winding.

The armature-generated voltage of a synchronous generator is often called the *excitation voltage*. Like the EMF E_a in a dc machine, it may be related to field current by means of the magnetization curve (Fig. 16-6). The magnetization curve is a plot of the open-circuit terminal voltage of the machine for a series of values of field current, the speed being kept constant at synchronous speed.

The impedance per phase of the armature winding is called the *synchronous impedance* $\mathbf{z}_s$. It is composed of the *effective resistance* r_a of the winding and its reactance x_s, called the *synchronous reactance*. That is,

$$\mathbf{z}_s = r_a + jx_s \quad \Omega \tag{16-5}$$

For large machines, r_a is usually small compared with x_s and can be ignored except when computing losses. Since the iron of the machine is subjected to varying saturation as operating conditions change, x_s is not really constant. For approximate purposes, a value adjusted for a typical degree of saturation can be used.

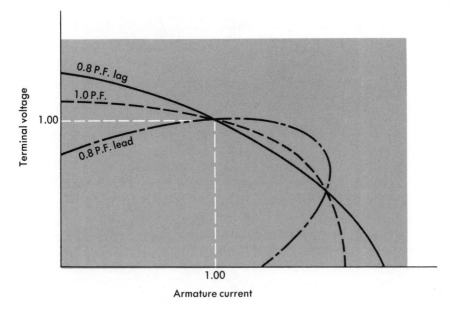

FIGURE 16-7 Terminal voltage as a function of armature current in a synchronous generator.

Synchronous generators are usually rated in terms of the maximum kilovolt-ampere load at a specified voltage, frequency, and power factor (often 80, 85, or 90 percent lagging) which they can carry continuously without overheating. If the field current is held constant while the load varies, the terminal voltage will vary. Characteristic curves of terminal voltage plotted against armature current for three constant power factors are shown in Fig. 16-7. Each curve is drawn for a different value of constant field current. In each case, the field current equals the value required to give rated terminal voltage at rated armature current. Normally, the generator is equipped with a voltage regulator to keep the terminal voltage constant by adjusting the field current as the load changes.

Example 16-3

A 40-kVA 120 (line-to-neutral)/208 (line-to-line)-V 60-Hz three-phase engine-driven generator is to be used as an emergency power source. Its magnetization curve is that of Fig. 16-6, the synchronous reactance is 0.45 Ω per phase, and the armature resistance is negligible.

The terminal voltage is to be held at rated value from no load to 150 percent of load at 0.80 power-factor lagging. Determine the range of field-current adjustment which must be made by a voltage regulator.

SOLUTION   From Fig. 16-6, the no-load field current is 2.6 A. At 150 percent of load, the armature current is

$$I_a = \frac{(40,000)(1.5)}{\sqrt{3}\,(208)} = 166.4 \text{ A}$$

The excitation voltage (per phase for a Y connection) is then

$$\mathbf{E} = \mathbf{V}_t + j\mathbf{I}_a x_s = 120 + j(166.4\underline{/-\cos^{-1} 0.80})(0.45)$$
$$= 164.9 + j59.8 = 175.5\underline{/19.9°}$$

The line-to-line value of excitation voltage is $\sqrt{3}$ (175.5), or 304 V. From Fig. 16-6, the corresponding field current is 5.7 A.

16-5   SYNCHRONOUS MOTORS

The direct current to the rotor of a synchronous machine is usually supplied from a small dc generator, called an *exciter,* which is frequently mounted on the same shaft as the motor and is supplied with mechanical power from the motor. In the synchronous motor pictured in Fig. 14-21, for example, the shaft-driven exciter may be seen to the left of the motor. In order that the rotating fields of stator and rotor will be stationary with respect to each other, the rotor must turn at precisely synchronous speed as determined by Eq. (16-1). Only under these conditions can synchronous-motor action be effective, and a synchronous motor is thus a constant-speed motor regardless of load.

As in the induction motor, the flux per stator pole is essentially determined by the impressed voltage and is therefore approximately constant. The rotor field strength is determined by the rotor or field current, which for normal operation is also kept constant. In terms of the interaction of magnetic fields described in Sec. 14-4, variation in the torque requirements of the load must be taken care of by variation of the power or torque angle δ. At no load, δ is very small, and the torque is just sufficient to overcome rotational losses. When shaft load is added, the rotor drops back in space phase with respect to the rotating stator field just enough so that δ assumes the value required to supply the necessary torque. When δ becomes 90°, the maximum possible torque, called the *pull-out torque,* for a fixed terminal voltage and field current is reached. If this value is exceeded, the motor slows down under the influence of the excessive shaft torque and synchronous-motor action is lost because rotor and stator fields are no longer stationary relative to each other. (Under these conditions, the motor is usually taken off the line by the action of automatic circuit breakers.) This phenomenon is known as *pulling out of step* or *losing synchronism.* Pull-out torque limits the short-time overload that can be placed on the motor.

With the rotor stationary, the rotating stator field is traveling at synchronous speed with respect to the rotor field, and the torque varies sinusoidally with time, reversing during each cycle. Hence a synchronous motor *per se* has no net starting torque. To make the motor self-starting, a squirrel-cage winding, called an *amortisseur* or *damper winding,* is inserted in the rotor pole faces. The rotor then comes up almost to synchronous speed by induction-motor action. If the field winding is energized at this point, the rotor and stator fields are still not quite stationary with respect to each other but move at a slow relative speed equal to the slip speed of the induction-motor action—perhaps at about 5 to 10 r/min, for example. The synchronous-motor torque still varies sinusoidally but only at a very low frequency equal to slip frequency. If the load and inertia are not too great, the positive half cycle of synchronous-motor torque lasts long enough to pull the rotor into synchronism. The maximum torque so obtained is called the *pull-in torque* of the motor; it is the most difficult torque for which to obtain large values in synchronous-motor design.

The fact that damper windings must be inserted to obtain starting torque should not be considered a disadvantage of synchronous motors. At times it may be a decided advantage, for the motor starts on one set of rotor windings and runs on another. Special design of the damper windings for high starting torque therefore need not penalize operating conditions, particularly the efficiency. Amortisseur windings are also effective in damping out hunting (small periodic variations in speed under operating conditions).

The power factor at which a synchronous motor operates can be controlled by varying the field current. This situation is summarized graphically in Fig. 16-8. The solid curves, called *V curves* because of their shape, are

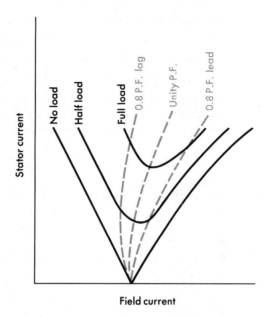

FIGURE 16-8 *V* curves (solid) and compounding curves (dotted) for a synchronous motor.

plots of stator current versus field current for different constant loads. The dotted curves, called *compounding curves,* are loci of constant-power-factor points on the *V* curves. General-purpose synchronous motors are available commercially in two power-factor ratings, 100 and 80 percent (leading). The latter are constructed to withstand the higher field and armature currents that go with leading operation.

Example 16-4

The magnetization curve of a 40-hp 1,200-r/min three-phase 60-Hz 220-V 80 percent power-factor (leading) synchronous motor is given in Fig. 16-6. This motor has an effective stator resistance of 0.05 Ω per phase, a field resistance of 30.0 Ω, a synchronous reactance of 0.45 Ω per phase, and rotational losses of 1,600 W.

For a stator input of 42.0 kVA at 0.80 power-factor leading and rated stator voltage, compute the shaft power output, field current, and efficiency.

SOLUTION   The motor will be treated as though it were Y-connected. The armature current input per phase is

$$I_a = \frac{42,000}{\sqrt{3}\,(220)} = 110 \text{ A}$$

The mechanical-power output is found by deducting the stator copper losses and the rotational losses from the stator power input; that is,

$$\text{Output} = (42{,}000)(0.80) - 3(110)^2(0.05) - 1{,}600$$
$$= 30{,}200 \text{ W} = 40.5 \text{ hp}$$

The excitation voltage will be found as in Example 16-3 except that the sign of the internal-voltage drop will be reversed because we are dealing with a motor (and hence reversed power flow compared with that in a generator).

$$\mathbf{E} = \mathbf{V}_t - \mathbf{I}_a \mathbf{z}_s = \frac{220}{\sqrt{3}} - (110\underline{/\cos^{-1} 0.80})(0.05 + j0.45)$$

The resulting magnitude E is 158 V line to neutral or 274 V line to line. From Fig. 16-6,

$$I_f = 4.3 \text{ A}$$

The total motor input is that to the stator plus that to the field, or

$$(42{,}000)(0.80) + (4.3)^2(30) = 34{,}200 \text{ W}$$

and

$$\text{Efficiency} = \frac{30{,}200}{34{,}200} = 88.3\%$$

16-6 ALTERNATING-CURRENT MOTOR SPEED CONTROL

To change the speed of an induction motor, we can change either the slip or the synchronous speed. To change the speed of a synchronous motor, our only choice is to change the synchronous speed, which means changing either the number of poles or the line frequency. Four methods of speed control are outlined below; the first two depend on change in slip, and the second two on change in synchronous speed.

1 *Line-voltage control.* The torque developed by an induction motor depends on the square of the voltage applied to its stator terminals. This dependence is illustrated by the two torque-speed characteristics in Fig. 16-9. If the load has the torque-speed requirements shown by the dashed line, the speed will be reduced from n_1 to n_2 when the voltage is reduced. This method of speed control is commonly used with small squirrel-cage motors driving fans.

2 *Rotor-resistance control.* The possibility of speed control of a wound-rotor induction motor by changing its rotor-circuit resistance has already been pointed out in Sec. 16-2. The torque-speed characteristics for three different values of rotor resistance are shown in Fig. 16-10. If the load has the torque-speed characteristics shown by the dashed line, the speeds corresponding to each of the values of rotor resistance are n_1, n_2, and n_3.

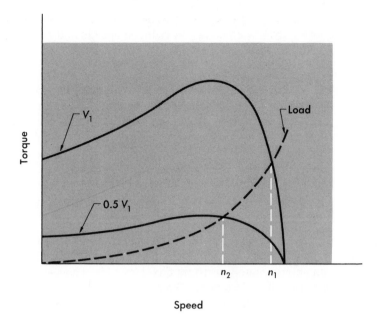

FIGURE 16-9 Induction-motor speed control by means of line voltage.

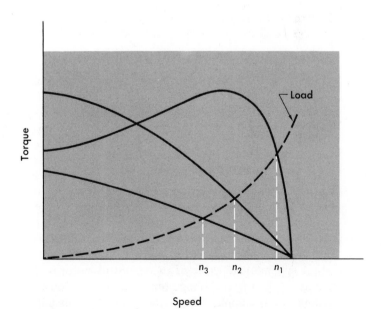

FIGURE 16-10 Induction-motor speed control by means of rotor resistance.

The principal disadvantages of both line-voltage and rotor-resistance control are low efficiency at reduced speeds and poor speed regulation with respect to change in load.

3 *Pole-changing motors.* The stator winding can be designed so that by simple changes in coil connections, the number of poles can be changed in the ratio 2:1. Either of two synchronous speeds can thereby be selected. When this method is applied, it is almost always to a squirrel-cage motor because costly structural complications are involved in changing the number of poles on the rotor of a wound-rotor or synchronous motor. With two independent sets of stator windings, each arranged for pole changing, as many as four synchronous speeds can be obtained in a squirrel-cage motor—for example, 600, 900, 1,200, and 1,800 r/min. The disadvantage, of course, is that more or less discrete speeds are produced rather than a range of speeds.

4 *Line-frequency control.* The synchronous speed of an induction or synchronous motor can be controlled by varying the line frequency. To maintain approximately constant flux density in the motor, the line voltage should be varied directly with the frequency. The motor performance then is similar to that of the dc motor in the Ward Leonard system.

The major problem has been to obtain an adjustable-frequency source. The conventional method has been to use either a synchronous generator or a wound-rotor induction machine as a frequency changer, either of which would require their own adjustable-speed drive. Recent developments in silicon-controlled rectifiers, however, provide much better solutions, as will be indicated in Chap. 17.

16-7 ALTERNATING-CURRENT CONTROL MOTORS

In low-power-level control systems, where the maximum outputs required from the motor range from a fraction of a watt up to a few hundred watts, two-phase induction motors are often used. The two-phase induction motor consists of a stator with two windings displaced 90 electrical degrees from each other in space and a squirrel-cage rotor or the equivalent. The ac voltages applied to the two windings are generally phase-displaced from each other 90° in time. When the voltage magnitudes are equal, the equivalent of balanced two-phase voltages is applied to the stator. The resultant stator flux is then similar to that in a three-phase induction motor. The motor torque-speed curves are also similar to those of a three-phase motor. The two-phase control motor is usually built with a high-resistance rotor to give a high starting torque and a drooping torque-speed characteristic.

A schematic diagram of an ac control motor is shown in Fig. 16-11. The voltage V_1 is a fixed voltage obtained from a constant-voltage source. The voltage V_2 is supplied from the controller, usually from an amplifier at the controller output. The two voltages must be in synchronism; thus they must be derived from the same ultimate ac source. They must also

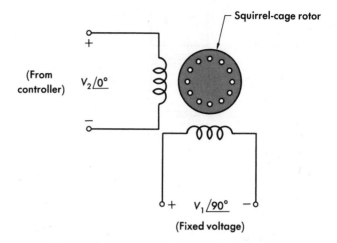

FIGURE 16-11 Schematic diagram of two-phase control motor.

be made to be approximately in time quadrature by introducing a 90° phase shift either in the amplifier or in the source of V_1. If V_2 has a nonzero value leading V_1 by approximately 90°, rotation in one direction is obtained; if V_2 has a nonzero value lagging V_1, rotation in the other direction results. Since the torque is a function of both V_1 and V_2, changing the magnitude of V_2 changes the developed torque of the motor. Torque-speed characteristics of a typical ac control motor are shown in Fig. 16-12. They are similar to those of the dc control motor with armature control. Comments made regarding starting-torque and stability considerations apply equally well to either type.

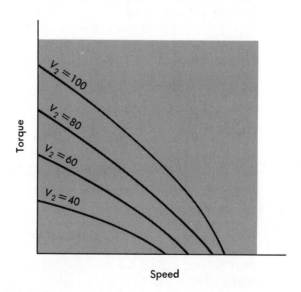

FIGURE 16-12 Torque-speed characteristics of two-phase control motor.

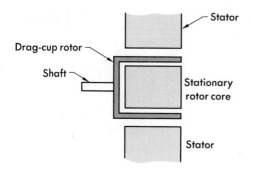

FIGURE 16-13 Cross section of a drag-cup rotor.

The principal disadvantage of the two-phase control motor is the inherent inefficiency of a squirrel-cage induction motor running at a large slip. The ruggedness and simplicity of the cage rotor are great advantages, however, for both economic and technical reasons. There are no brushes riding on sliding contacts. Because the rotor windings do not require insulation, the rotor temperature is limited only by mechanical considerations and indirectly by its effect on the stator-winding temperature. Because there is relatively little inactive material, the inertia of a squirrel-cage rotor can be made less than that of a correspondingly rated dc motor. When the maximum power output is below a few watts, inertia can be minimized by using a thin metallic cup as the rotor. As shown in the simplified sketch of Fig. 16-13, the rotating member is then like a can with one end removed. A stationary iron core, like a plug inside the cup, completes the magnetic circuit. This construction is known as a *drag-cup rotor*.

Example 16-5

For approximate evaluation of the performance of two-phase control motors, two assumptions are sometimes made concerning the torque-speed curves: (1) The stalled torque is assumed to be linearly proportional to the rms control-winding voltage. (2) The torque-speed curves are considered to have a negative slope which is approximately constant and substantially the same for varying values of control-winding voltage. Under these conditions, the torque-speed curves can be represented empirically by the equation

$$T = k_1 V_2 - k_2 \omega_m \tag{16-6}$$

where ω_m is the shaft mechanical velocity, V_2 is the rms control-winding voltage, and k_1 and k_2 are constants.

Consider that such a motor is driving a pure-inertia load. The combined inertia of rotor and load is J kg·m².

a Determine the transfer function relating velocity and control voltage.

b Let the motor be a small 115-V two-pole 60-Hz motor rated at 4 W. With rated voltage impressed, the blocked torque is 0.03 N·m and the torque at half synchronous speed is 0.02 N·m. The rotor moment of inertia is 2.0×10^{-6} kg·m². Find the transfer function for the motor with negligible load inertia and load torque.

SOLUTION   **a**   By equating the motor torque from Eq. (16-6) to the torque required to accelerate the rotor, we have

$$T = k_1 V_2 - k_2 \omega_m = J \frac{d\omega_m}{dt}$$

or

$$J \frac{d\omega_m}{dt} + k_2 \omega_m = k_1 V_2 \tag{16-7}$$

When Eq. (16-7) is rewritten in phasor form,

$$(j\omega) J \Omega_m + k_2 \Omega_m = k_1 \mathbf{V}_2$$

or

$$\mathbf{G}_m = \frac{\Omega_m}{\mathbf{V}_2} = \frac{k_1/k_2}{1 + j\omega(J/k_2)} \tag{16-8}$$

By analogy with Eq. (15-27), the motor time constant is

$$\tau_m = \frac{J}{k_2} \tag{16-9}$$

b   When $T = 0.03$, $V_2 = 115$, and $\omega_m = 0$ are substituted in Eq. (16-6),

$$k_1 = \frac{0.03}{115} = 2.61 \times 10^{-4} \text{ N·m/V}$$

The angular velocity ω_m at half synchronous speed is

$$\frac{1}{2} \frac{4\pi f}{\text{poles}} = \frac{1}{2} \frac{4\pi(60)}{2} = 188.4 \text{ rad/s}$$

When $T = 0.02$, $k_1 V_2 = 0.03$, and $\omega_m = 188.4$ are substituted in Eq. (16-6),

$$k_2 = \frac{0.03 - 0.02}{188.4} = 0.53 \times 10^{-4} \text{ N·m/rad/s}$$

The time constant and transfer function are then, respectively,

$$\tau_m = \frac{2.0 \times 10^{-6}}{0.53 \times 10^{-4}} = 0.038 \text{ s}$$

and

$$\mathbf{G}_m = \frac{4.9}{1 + j\omega(0.038)} \text{ rad/s/V}$$

16-8 FRACTIONAL-HORSEPOWER AC MOTORS

Although three-phase motors are available in ratings down to 1/6 hp (for some types, even lower ratings can be obtained), the majority of ac fractional-horsepower motors operate on single-phase alternating current. One important contributing factor is that only single-phase power is available in a great many cases where small motors must be used. Wide variations in starting- and maximum-torque requirements are encountered, so that many different types of single-phase motors have been developed. Selection of the lowest-priced motor which satisfies the conditions of a particular application is thereby made possible.

Many types of single-phase motors are basically induction motors which differ in their starting methods. Structurally, single-phase induction motors resemble the three-phase squirrel-cage motors considered earlier except that the stator winding is a single-phase winding. They may therefore be represented schematically as in Fig. 16-14, with the understanding that the stator winding is distributed in slots over the stator surface instead of being a concentrated coil. The operation of the motor may be explained in terms of conditions already established for three-phase motors by showing that component rotating fields are produced by the stator winding.

From Fig. 16-14, however, it is evident that the axis of the stator field remains fixed along the coil axis. Also, with alternating current in the coil, the field strength pulsates sinusoidally, the poles alternating in polarity and varying in strength sinusoidally with time. Such a pulsating field may be represented graphically by an arrow of varying length pointing up half the time, down the other half, and having a magnitude and direction determined by the instantaneous magnitude and direction of the coil current, as shown in Fig. 16-15. But it will be seen from Fig. 16-15b and c that such an arrow can be considered as the sum of two equal arrows rotating in opposite directions, each component arrow having a constant length equal to half the maximum

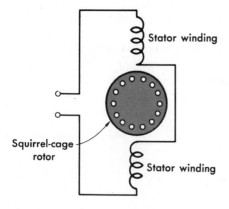

FIGURE 16-14 Schematic diagram of single-phase induction motor.

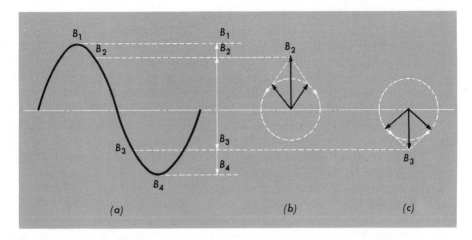

FIGURE 16-15 Sinusoidally pulsating field and its representation by two equal rotating fields revolving in opposite directions.

length of the original arrow. Consequently the pulsating field of the stator can be divided into two rotating fields of equal magnitudes. These component fields rotate in opposite directions at synchronous speed.

Each of these component rotating fields will produce induction-motor action, but the corresponding torques will be in opposite directions. In Fig. 16-16 each of the component curves, shown dotted, is similar to the torque-speed curves of Fig. 16-4 except extended into the other quadrant.

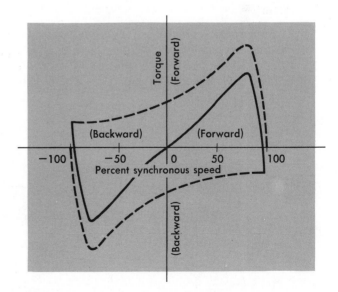

FIGURE 16-16 Resultant torque-speed curve (solid line) and its two components (dotted lines) in a single-phase induction motor.

The resultant curve, which is the sum of the two component curves, shows zero torque at starting but a definite value at any other speed. If, therefore, a single-phase induction motor is started by some other means, it will continue to rotate in whatever direction it was started and will give the same type of performance as a three-phase induction motor. Actually, when the rotor is in motion, the induced rotor currents increase the forward flux wave and decrease the backward flux wave, both relative to their magnitude at standstill. The net torque is thereby increased. This torque is a pulsating torque, a result that is unavoidable in any single-phase motor because of the pulsations in instantaneous power input inherent in a single-phase circuit. The torque referred to on the torque-speed curves for a single-phase motor is the time average of the instantaneous torque.

Single-phase induction motors are classified in accordance with the methods of starting and are usually referred to by names descriptive of those methods. As with integral-horsepower motors, selection of the appropriate type is made from the starting and running characteristics and comparative economics. Starting methods and the resulting characteristics will be considered individually.

1 *Shaded-pole.* As illustrated schematically in Fig. 16-17*a*, the shaded-pole motor has salient poles with one-half of each pole surrounded by a heavy short-circuited winding called a *shading coil.* Induced currents in the shading coil cause the flux in that half of the pole to lag the flux in the other half in building up. The result is a periodic shift in flux from the unshaded to the shaded half of the pole, producing a low starting torque. Typical characteristics are given in Fig. 16-17*b*. This principle is used only in very small motors.

2 *Split-phase.* Split-phase motors have two stator windings, a main winding and an auxiliary winding, with their axes displaced 90 electrical degrees in space. They are connected as shown in Fig. 16-18*a*. The auxiliary winding has a higher resistance-to-reactance ratio than the main winding, so

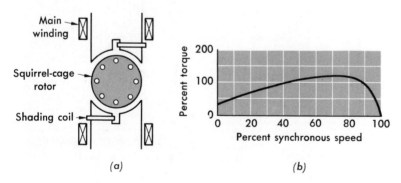

FIGURE 16-17 Shaded-pole motor. (*a*) Schematic representation; (*b*) typical torque-speed curve.

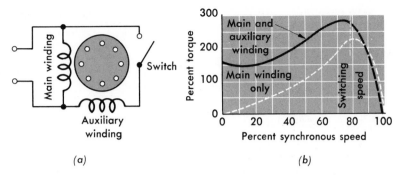

FIGURE 16-18 Split-phase motor. (a) Schematic representation; (b) typical torque-speed curves.

that the two currents are out of phase. The stator field thus first reaches a maximum about the axis of one winding and then somewhat later in time (about 80 to 85 electrical degrees) reaches a maximum about the axis of the winding 90 electrical degrees away in space. The result is a rotating stator field which causes the motor to start. At about 75 percent of synchronous speed, the auxiliary winding is cut out by a centrifugal switch. The general characteristics are shown in Fig. 16-18b.

3 *Capacitor-start induction-run.* This type is also a split-phase motor, but phase displacement between the two currents is obtained by means of a capacitor in series with the auxiliary winding. Again the auxiliary winding is switched out at about 75 percent of synchronous speed. Operating characteristics are given in Fig. 16-19, high starting torque being an outstanding feature.

4 *Single-value-capacitor.* If the capacitor and auxiliary winding of the capacitor-start motor are not cut out after starting, the power factor and run-

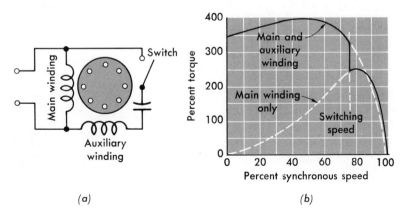

FIGURE 16-19 Capacitor-start induction-run motor. (a) Schematic representation; (b) typical torque-speed curves.

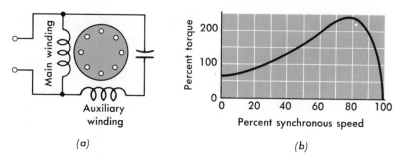

FIGURE 16-20 Single-value-capacitor motor. (a) Schematic representation; (b) typical torque-speed curve.

ning performance may be improved. Such improvement is the general objective of the single-value-capacitor motor. Starting torque must be sacrificed, however, because the capacitance is necessarily a compromise between the best starting and the best running values. The resulting characteristics, together with a schematic diagram, are given in Fig. 16-20.

5 *Capacitor-start capacitor-run.* If two capacitors are used, one for starting and one for running, a compromise need not be made, and optimum starting and running performance can be obtained. As a result, the operating efficiency and power factor are comparatively high. The general features of this motor are given in Fig. 16-21.

An important additional type of small motor is the *series ac motor,* also termed the *universal motor* because it can operate on either alternating or direct current. If alternating current is supplied to a series motor, the stator and rotor field strengths will vary in magnitude sinusoidally with time, but they will vary in exact time phase. Consequently, a torque will be produced, and the performance of the motor will be generally similar to that with direct current. Such will not be true for a shunt motor because the time variations will be almost 90° out of phase, resulting in almost no torque. Commutation

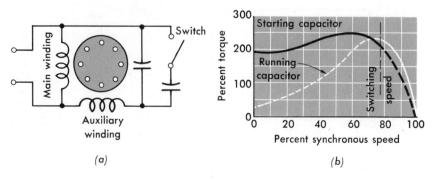

FIGURE 16-21 Capacitor-start capacitor-run motor. (a) Schematic representation; (b) typical torque-speed curves.

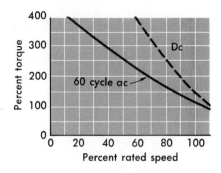

FIGURE 16-22 Typical torque-speed curves of a universal motor.

difficulties will be more severe than with direct current, however, limiting heavy-power usage to low frequencies such as 25 Hz. In the larger sizes, series ac motors are used principally for traction purposes.

In the fractional-horsepower and small integral-horsepower sizes, the commutation difficulties at 60 Hz can be overcome. Typical characteristics of a universal motor are given in Fig. 16-22. They are fundamentally high-speed motors (1,500 to 15,000 r/min). These motors are frequently used with solid-state devices for speed control.

PROBLEMS

16-1 Draw approximately to scale in the manner of Fig. 16-3 the stator flux distributions for a two-pole three-phase induction or synchronous machine at the following instants of time:

a When the current in phase *b* is zero

b An arbitrarily chosen instant not corresponding to zero or maximum current in any phase

Show the contribution from each of the three phases as well as the total stator flux distribution.

16-2 Draw approximately to scale the stator flux distributions corresponding to the three given in Fig. 16-3 but for the reverse phase sequence, i.e., with the applied voltages reaching their time maxima in *cba* order instead of *abc* order.

Basing your answer on comparison of these results with Fig. 16-3, state how you would reverse the rotation of a three-phase synchronous or induction motor.

16-3 Draw approximately to scale the counterpart of Fig. 16-3 for the windings of parts **a** and **b** below.

a A winding consisting of two coils 90° apart in space with impressed voltages which are equal in magnitude but phase-displaced 90° in time. Such a winding would be used in a two-pole motor for three-wire two-phase service.

b A winding consisting of four coils 90° apart in space with impressed voltages which are equal in magnitude but phase-displaced 90° in time. Such a winding would be used in a four-phase motor.

c If the supply frequency in part **b** is 60 Hz, what is the synchronous speed?

d Can the resultant fields in parts **a** and **b** be distinguished from that of a three-phase winding?

e State the effect on the field in part **a** if the two voltages are only about 80° phase-displaced in time. This situation arises during the starting of some single-phase induction motors.

16-4   Sketch curves of speed and rotor frequency as a function of slip for an eight-pole 60-Hz induction motor.

16-5   The twelve-pole three-phase stator of a squirrel-cage induction motor is connected to a source of balanced 60-Hz voltages. For each of three conditions—(1) at the instant of starting, (2) when the rotor mechanical speed is three-fourths of synchronous speed, (3) when the motor is operating at a full-load slip of 4 percent—give the following speeds in revolutions per minute:

a Speed of stator field with respect to stator
b Speed of stator field with respect to rotor
c Speed of rotor field with respect to stator
d Speed of rotor field with respect to rotor
e Speed of rotor field with respect to stator field

16-6   A three-phase induction motor runs at almost 1,200 r/min at no load and at 1,140 r/min at full load when supplied with power from a 60-Hz three-phase line.

a How many poles has the motor?
b What is the percentage of slip at full load?
c What is the corresponding frequency of the rotor voltages?
d What is the corresponding speed of the rotor field with respect to (1) the rotor, (2) the stator, and (3) the stator field?
e What speed would the rotor have at a slip of 10 percent?
f What is the rotor frequency at this speed?
g Repeat parts **c** and **d** for a slip of 10 percent.

16-7   A 50-hp 440-V three-phase 60-Hz eight-pole squirrel-cage induction motor is operated at rated load with rated voltage at rated frequency applied. The slip is 3.0 percent, the efficiency is 91.0 percent, and the line current is 63.0 A.

a Find the speed in revolutions per minute of the rotor field with respect to the rotor.

b Find the shaft torque in pound-feet delivered to the load.

c Find the power factor at the stator terminals.

16-8 A three-phase squirrel-cage induction motor is started by reducing the voltage applied to the motor by a factor of $\frac{1}{2}$. Exciting current can be ignored.

a By what factor is the starting torque reduced?

b By what factor is the current in the motor windings reduced?

16-9 A motor is to be selected for driving a conveyor belt in a plant having a 440-V three-phase 60-Hz supply. The belt is to operate continuously during each working day and will require a normal running torque of 150 lb-ft at a motor speed of approximately 1,750 r/min. Across-the-line starting is to be used, and occasionally the motor must start under a heavy load requiring a torque of 200 lb-ft. It is also essential that the motor be capable of starting with the supply voltage as low as 80 percent of normal.

a What must be the continuous horsepower rating of the motor?

b Will a 440-V Class B induction motor satisfy these conditions? The characteristics of Fig. 16-5 can be considered applicable.

c Will a 440-V Class C induction motor satisfy these conditions? The characteristics of Fig. 16-5 can be considered applicable.

16-10 A 4,160-V 2,500-hp two pole three-phase 60-Hz squirrel-cage induction motor is driving a boiler-feed pump in an electric generating plant. The table below lists points on the motor torque-speed curve at rated voltage, speed being in percentage of synchronous speed and torque in percentage of rated torque. Also listed are the torque requirements of the boiler-feed pump expressed in percentage of motor-rated torque. The drive is started at rated voltage with the discharge valve open but working against a check valve until the pump head equals the system head. The straight-line portion of the pump characteristic between 0 and 10 percent speed represents the torque required to overcome static friction in the bearings. At 92 percent speed, the check valve opens, and there is a discontinuity in the slope of the pump curve. At 98 percent speed, the motor produces its maximum torque.

Percent speed	0	10	30	50	70	90	92	98	99.5
Percent motor torque	75	75	75	75	80	125	155	240	100
Percent pump torque	15	0+	4	12	26	42	44	87	100

a Consider that the motor must occasionally operate with its voltage at 80 percent of its rated value. Determine the pump speed under these conditions.

b To what value can the motor voltage be reduced without making it impossible to reach a speed at which water is delivered to the system?

16-11 A linear motor based on the induction-motor principle consists of a car riding on a track. The track is a developed squirrel-cage winding, and the car, which is 12 ft long, $3\frac{1}{2}$ ft wide, and only $5\frac{1}{2}$ in high, has a developed three-phase eight-pole winding. The center-line distance between adjacent poles $12/8 = 1\frac{1}{2}$ ft. Power at 60 Hz is fed to the car from arms extending through slots to rails below ground level.

a What is the synchronous speed in miles per hour?
b Will the car reach this speed? Explain your answer.
c To what slip frequency does a car speed of 75 mi/h correspond?

(*Note:* Linear induction motors have been proposed for a variety of applications including high-speed ground transportation.)

16-12 A synchronous motor is operating at half load. An increase in its field excitation causes a decrease in armature current. Before the increase, was the motor delivering or absorbing lagging reactive kVA?

16-13 A polyphase synchronous motor is operating at rated frequency and voltage at 0.80 power factor. When the field current is increased, the motor current decreases.

Was the motor operating with a leading or a lagging current? Give reasons for your answer.

16-14 Plot a family of torque-speed curves for the two-phase control motor of Example 16-5**b**. Assume that these characteristics can be represented by Eq. (16-6). Plot curves for 100, 80, 60, 40, and 20 percent of rated control-winding voltage.

16-15 Plot as a function of frequency the amplitude of the transfer function relating velocity and control voltage for the motor of Example 16-5**b**.

16-16 Electric power is supplied to an electrochemical plant at 13.2 kV, three-phase, 60 Hz at an incremental cost of 4 cents per kilowatthour. One of the plant processes requires 4,000 kW of direct current at 600 V. Several methods to provide the dc energy are being considered. One of them is to have a 4,000-kW, 600-V dc generator driven by a 6,000-hp, 2,300-V, three-phase synchronous motor. The power source for the motor is the output of a 13.2- to 2.3-kV, three-phase step-down transformer. The load on the motor generator is assumed constant at 4,000 kW, 24 h/day, 7 days a week. The following data are available for the system components:

Generator losses at full load: 256 kW
Synchronous motor efficiency at full load: 96.2 percent
Transformer efficiency at full load: 99.1 percent

The installed cost of the system, including associated control and protective equipment is $840,000. Maintenance and repairs are estimated to average $4,200 per year. Fixed charges, including the cost of money, are 25 percent per year. The estimated life of the system is 10 years. Using straight-line depreciation, compute the cost per kilowatthour of generating the required dc energy.

16-17 A machine shop uses five boring mills which are driven at constant speed. Two possible arrangements are to be compared on the basis of annual cost. The first is group drive and the second is individual drive for the mills. Group drive requires one 20-hp, three-phase motor and individual drive requires five 5-hp motors.

The following data are available:

	20-HP MOTOR	5-HP MOTOR
Input, full load	16.2 kW	3.65 kW
Input, no load	2.0 kW	0.70 kW
Installed cost	$3,500	$1,600
Life expectancy	10 years	10 years
Yearly maintenance	$ 200	$ 100

Full-load input to the 20-hp motor is estimated for four of the five mills in operation.

Associated equipment, such as line hangers and pulleys, costs $3,000 installed and has expected yearly maintenance costs of $150. Electric power costs 4.5 cents per kilowatthour.

The plant runs 8 h/day, 240 days per year, and each mill operates 80 percent of the time. If individual drives are used, the mills are shut down when they are not in operation. Assume that straight-line depreciation and fixed charges, including short- and long-term interest, are 20 percent.

Compute the average annual cost of operation for each method.

TE POWER PROCESSING □ **CONTROL**
R PROCESSING □ CONTROL **SYSTEMS**
)CESSING □ CONTROL SYSTEMS **AND**
ONTROL SYSTEMS AND **SOLID-STATE**
. SYSTEMS AND SOLID-STATE **POWER**
D SOLID-STATE POWER **PROCESSING**
PROCESSING □ CONTROL SYSTEMS
CONTROL SYSTEMS AND SOLID-STAT
SYSTEMS AND SOLID-STATE POWER I
AND SOLID-STATE POWER PROCESSING □ CONTROL SYSTEMS ANI
SOLID-STATE POWER PROCESSING □ CONTROL SYSTEMS AND SOI
-STATE POWER PROCESSING □ CONTROL SYSTEMS AND SOLID-ST
POWER PROCESSING □ CONTROL SYSTEMS AND SOLID-STATE PO'
PROCESSING □ CONTROL SYSTEMS AND SOLID-STATE POWER PR
CONTROL SYSTEMS AND SOLID-STATE POWER PROCESSING □ COI
SYSTEMS AND SOLID-STATE POWER PROCESSING □ CONTROL SYS
AND SOLID-STATE POWER PROCESSING □ CONTROL SYSTEMS ANI
SOLID-STATE POWER PROCESSING □ CONTROL SYSTEMS AND SOI
-STATE POWER PROCESSING □ CONTROL SYSTEMS AND SOLID-ST/
POWER PROCESSING □ CONTROL SYSTEMS AND SOLID-STATE PO'
PROCESSING □ CONTROL SYSTEMS AND SOLID-STATE POWER PR
CONTROL SYSTEMS AND SOLID-STATE POWER PROCESSING □ COI
SYSTEMS AND SOLID-STATE POWER PROCESSING □ CONTROL SYS
AND SOLID-STATE POWER PROCESSING □ CONTROL SYSTEMS ANI
SOLID-STATE POWER PROCESSING □ CONTROL SYSTEMS AND SOI
-STATE POWER PROCESSING □ CONTROL SYSTEMS AND SOLID-ST/
POWER PROCESSING □ CONTROL SYSTEMS AND SOLID-STATE PO\
PROCESSING □ CONTROL SYSTEMS AND SOLID-STATE POWER PR(
CONTROL SYSTEMS AND SOLID-STATE POWER PROCESSING □ COI
SYSTEMS AND SOLID-STATE POWER PROCESSING □ CONTROL SYS
AND SOLID-STATE POWER PROCESSING □ CONTROL SYSTEMS ANI
SOLID-STATE POWER PROCESSING □ CONTROL SYSTEMS AND SOI
-STATE POWER PROCESSING □ CONTROL SYSTEMS AND SOLID-ST/
POWER PROCESSING □ CONTROL SYSTEMS AND SOLID-STATE PO\
PROCESSING □ CONTROL SYSTEMS AND SOLID-STATE POWER PR(
CONTROL SYSTEMS AND SOLID-STATE POWER PROCESSING □ COI
SYSTEMS AND SOLID-STATE POWER PROCESSING □ CONTROL SYS
AND SOLID-STATE POWER PROCESSING □ CONTROL SYSTEMS ANI
SOLID-STATE POWER PROCESSING □ CONTROL SYSTEMS AND SOI
-STATE POWER PROCESSING □ CONTROL SYSTEMS AND SOLID-ST/
POWER PROCESSING □ CONTROL SYSTEMS AND SOLID-STATE POV
PROCESSING □ CONTROL SYSTEMS AND SOLID-STATE POWER PR(
CONTROL SYSTEMS AND SOLID-STATE POWER PROCESSING □ CON
SYSTEMS AND SOLID-STATE POWER PROCESSING □ CONTROL SYS
AND SOLID-STATE POWER PROCESSING □ CONTROL SYSTEMS ANI
SOLID-STATE POWER PROCESSING □ CONTROL SYSTEMS AND SOL

lectrical energy is widely used because it can be made available in an appropriate form at a desired place, thereby enabling system and device performance to be reliably controlled. In this chapter, two broad aspects associated with system control are described.

One of these is to introduce the different types of control systems and some elementary methods for studying their behavior. Three classes of control systems described are:

1 Motor drive systems, in which the electrical input to a motor is adjusted to control performance.

2 Feedback control systems, in which a measure of the actual performance must be known in order to effect control.

3 Digital control systems, in which a digital processor is an essential element in the system. The output resulting from the data processing forms the basis for system adjustment and control.

Many of the techniques and concepts employed are similar to those developed earlier in the book. Indeed, the discussion of control methods integrates much of the material in circuit theory, electronic devices and circuits, and electromechanical energy-conversion devices. These control techniques are also used in business and social systems and are applied to problems related to inventory control, economic models, health-care delivery systems, and urban planning.

Solid-state power processing circuits are treated to indicate how the appropriate form of energy is provided. In particular, circuits to convert the fixed-frequency alternating current generally available to the direct current or adjustable-frequency alternating current required to operate most electronic circuits and motor drives are described. Also introduced is the silicon-controlled rectifier, one of the principal devices employed to effect this control and conversion.

17-1 RECTIFIERS

Rectification, the conversion of ac to dc, makes use of the nearly unilateral circuit behavior of diodes. As such, the circuit symbol generally used for rectifiers and shown in Fig. 17-1 is the same symbol used for semiconductor

Anode   Cathode

FIGURE 17-1 Rectifier symbol.

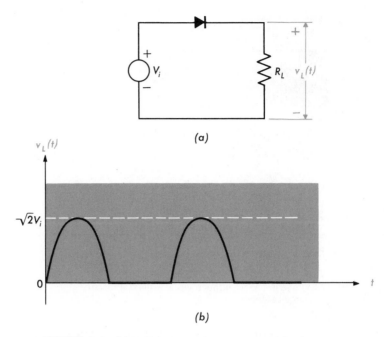

FIGURE 17-2 *(a)* Half-wave rectifier circuit and *(b)* voltage waveform.

diodes (see Sec. 7-5). The arrowhead indicates the forward direction, i.e., the current direction.

The circuit in Fig. 17-2a is a simple, half-wave rectifier which supplies an equivalent load resistance R_L. Power is supplied to the load only during alternate half cycles as illustrated in Fig. 17-2b, in which the diode is assumed ideal. The behavior of the diode in the circuit of Fig. 17-2a is that of a switch which is activated by the voltage across the diode. In many control applications, the use of a controlled switch is required, with control exerted in a circuit external to the power circuit. The principal device which performs this function in solid-state power processing is the silicon-controlled rectifier.

The *silicon-controlled rectifier (SCR)*, or *thyristor*, is a four-layer, three-junction device, and is depicted in Fig. 17-3a. The circuit symbol used for its representation is shown in Fig. 17-3b. The outer *p* and *n* layers act as in the *p-n* junction and are called the *anode* and *cathode*, respectively. The interior *p* and *n* regions serve as the *gate* which acts as the control element. When a small voltage is impressed which makes the anode positive with respect to the cathode, the gate is reverse-biased. Current flow from anode to cathode (the positive direction) is inhibited, the process being called *forward blocking*. The device then appears to have high resistance in both forward and reverse directions. The usual mechanism for starting conduction is to apply a current pulse in the gate-cathode circuit. The pulses neutralize the blocking effect of the gate. The anode current increases markedly, with a

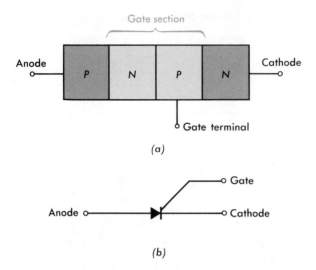

FIGURE 17-3 Silicon-controlled rectifier (SCR) (a) representation and (b) circuit symbol.

simultaneous reduction in the voltage drop across the SCR. The point at which conduction occurs is called *breakover* and is labeled V_{BO} on the volt-ampere characteristics displayed in Fig. 17-4. Once conduction has started, the gate loses all control until the current has been reduced to approximately zero. The SCR, in the conducting region, behaves in a manner similar to a junction diode. As shown in Fig. 17-4, the gate current I_G exerts control beyond that of overcoming forward blocking. The magnitude of I_G also controls the value of the breakover voltage V_{BO}.

The control of V_{BO} is one major reason for the widespread use of SCRs in power-control applications. The type of control exhibited by silicon-controlled rectifiers permits wide adjustment of direct current and voltage magnitudes with a minimum expenditure of energy.

The circuit depicted in Fig. 17-5a is a single-phase SCR half-wave rectifier. When the input voltage exceeds the breakover voltage, the SCR conducts and acts in much the same manner as does the diode in the circuit of Fig. 17-2a. The value of breakover voltage is controlled by the application of a low-power signal to the gate. By means of the gate control pulse, SCR conduction and consequently the direct current in the load are controlled. A typical load current waveform is shown in Fig. 17-5b. The angle α is called the *firing* or *ignition angle,* whose value is determined by the breakover voltage. The sources of the gate control pulses are usually multivibrators and wave-shaping circuits of the type described in Chap. 11.

17-2 AC TO DC CONVERSION

Fixed-amplitude, fixed-frequency alternating current is the primary source of electrical energy. (In the United States, it is 110/220 V rms, 60-Hz sinu-

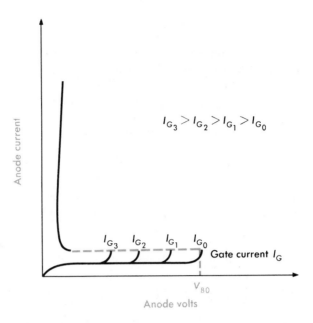

FIGURE 17-4 Volt-ampere characteristic of a silicon-controlled rectifier.

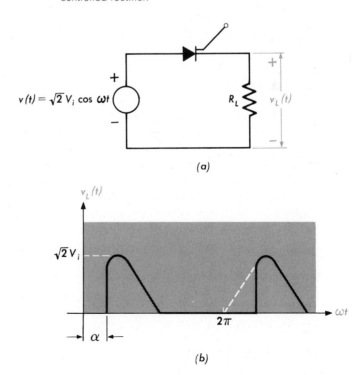

FIGURE 17-5 (a) Half-wave SCR (thyristor) circuit and (b) voltage waveform.

soid; in many parts of Europe 220 V rms, 50-Hz sinusoids are available.) The overwhelming majority of electronic circuits require constant voltages to ensure appropriate operation. For example, many minicomputers require 5-V sources capable of providing a current of 100 A. Other signal processing systems often require 12 V and 15 V supplies in which the current produced varies with load conditions. In addition, many of the motor drives described later in this chapter require dc supplies whose voltage levels can be adjusted to meet desired operating conditions. The rectifiers treated in the preceding section form the basis for the necessary ac-to-dc conversion.

The block diagram for a dc power supply obtained from a primary ac source is depicted in Fig. 17-6. With the exception of the rectifier, whether each of the remaining circuit functions is employed depends on the application. As indicated by the waveforms in Fig. 17-6, the functions of the various circuits are as listed.

1 Transformer: Adjusts the ac level so that the appropriate dc amplitude is achieved.

2 Rectifier: Converts the sinusoidal voltage to a pulsating dc signal.

3 Filter: "Smooths" the waveform by eliminating the ac components from the rectifier output.

4. Regulator: Maintains a constant voltage level independent of load conditions or variations in the amplitude of the ac supply.

The transformer (Sec. 13-10) can be of the step-up or step-down type and its power-handling capacity must be sufficient to supply the load and account for losses in the rectifier, filter, and regulator. Often, the transformer is center-tapped as shown in Fig. 17-7 so that two equal but out of phase voltages are obtained. The turns ratio is determined by the output level required relative to the ac input amplitude.

For improved efficiency of operation, full-wave rectifiers, such as those depicted in Figs. 17-7 and 17-8 are often employed. In the full-wave circuit

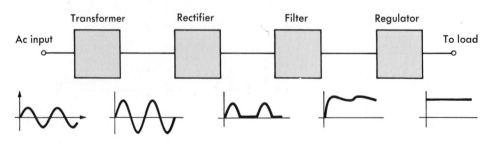

FIGURE 17-6 Block diagram of power supply.

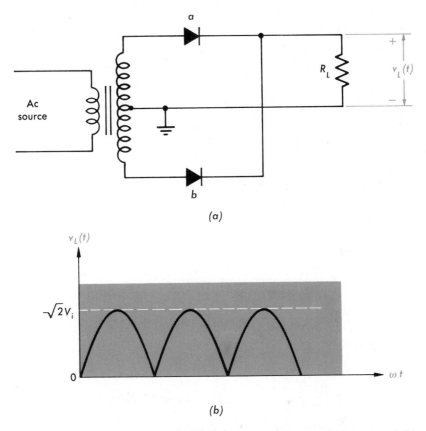

(a)

(b)

FIGURE 17-7 (a) Full-wave rectifier circuit and (b) load-voltage waveform.

(Fig. 17-7a), diode a conducts during alternate half cycles, diode b conducting during the other half cycles. The resulting load voltage, again with the diode considered ideal, is given in Fig. 17-7b, from which the improvement in waveform and power-delivering ability is evident. The bridge circuit of Fig. 17-8 is also a full-wave circuit, having the waveform of Fig. 17-7b. During one half cycle, the current path is through rectifier a, load R_L, and rectifier c; during the other half cycle, it is through rectifier b, load R_L, and rectifier d. Bridge-type rectifiers are frequently used with dc instruments to read alternating voltages.

When rectifiers supply a purely resistive load, the current waveforms are the same as those of Figs. 17-2 and 17-7, with allowance made for the voltage drop across the diode. For the cases where the diode voltage drop can be represented by a constant rectifier resistance R_r, the dc or average value of output current for a half-wave rectifier is

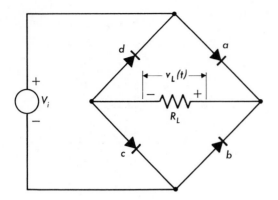

FIGURE 17-8 Full-wave bridge rectifier.

$$I_{dc} = \frac{1}{2\pi} \int_o^\pi \frac{\sqrt{2}\,V_i}{R_r + R_L} \sin \omega t\; d(\omega t) \tag{17-1}$$

$$I_{dc} = \frac{\sqrt{2}}{\pi} \frac{V_i}{R_r + R_L} \qquad \text{(half-wave rectifier)} \tag{17-2}$$

where R_L is the load being supplied, and V_i is the rms value of the impressed voltage. Similarly, for a full-wave circuit

$$I_{dc} = \frac{2\sqrt{2}}{\pi} \frac{V_i}{R_r + R_L} \qquad \text{(full-wave rectifier)} \tag{17-3}$$

When the rectifier interposes a constant voltage drop (as is approximated by the voltage drop in an SCR), conduction does not take place over the entire half cycle, for the anode voltage must become sufficiently positive to overcome the rectifier drop. The same situation holds when the load is a *counter-EMF load,* i.e., one which can be represented (as in Fig. 17-9a) by a constant resistance R_L and a back voltage V_L. Batteries being charged and the armatures of dc motors being supplied from the ac mains through rectifi-

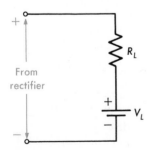

FIGURE 17-9 Representation of counter-EMF load.

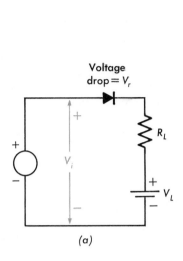

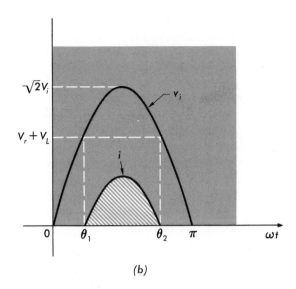

FIGURE 17-10   (a) Half-wave rectifier with counter-EMF load and (b) voltage waveforms.

ers are representative of counter-EMF loads. An example is given in Fig. 17-10a, which shows a half-wave rectifier having a constant-voltage drop V_r supplying such a load.

The resulting voltage and current waveforms are given in Fig. 17-10b. The rectifier will not conduct until the firing angle θ_1 is reached, when the impressed voltage attains the value $V_r + V_L$. Thus,

$$\theta_1 = \sin^{-1} \frac{V_r + V_L}{\sqrt{2} \, V_i} \tag{17-4}$$

Moreover, conduction will cease at the *extinction angle* θ_2, when the impressed voltage drops below $V_r + V_L$. Evidently,

$$\theta_2 = \pi - \theta_1 \tag{17-5}$$

The dc or average value of current is then

$$I_{\text{dc}} = \frac{1}{2\pi} \int_{\theta_1}^{\theta_2} \frac{\sqrt{2} \, V_i \sin \omega t - (V_r + V_L)}{R_L} \, d(\omega t) \tag{17-6}$$

$$I_{\text{dc}} = \frac{\sqrt{2} \, V_i}{\pi \, R_L} \cos \theta_1 - \frac{\pi - 2\theta_1}{2\pi} \frac{V_r + V_L}{R_L} \qquad \text{(half-wave rectifier)} \tag{17-7}$$

Similarly, for a full-wave circuit,

$$I_{\text{dc}} = \frac{2\sqrt{2}}{\pi} \frac{V_i \cos \theta_1}{R_L} - \frac{\pi - 2\theta_1}{2\pi} \frac{V_r + V_L}{R_L} \qquad \text{(full-wave rectifier)} \tag{17-8}$$

For circuits which employ SCRs, with no counter-EMF load, the extinction angle, as shown in Fig. 17-5*b*, is zero, for the SCR cannot be refired until the current returns to zero.

Example 17-1

A storage battery is charged by a full-wave silicon-diode rectifier. The supply voltage V_i is 60 V. The rectifier-voltage drop during conduction is 1 V, and the battery offers a constant counter voltage of 12 V and negligible resistance.

a What resistance must be connected in series with the battery to limit the charging rate to 10 A?

b What power dissipation must the series resistor be able to stand?

c What is the power supplied to the battery?

SOLUTION **a** From Eq. (17-4), the ignition angle is

$$\theta_1 = \sin^{-1} \frac{1 + 12}{\sqrt{2}\,(60)} = 8.81° = 0.154 \text{ rad}$$

From Eq. (17-8),

$$10 = 0.90 \frac{60}{R_L} \cos 8.81° - \frac{\pi - (2)(0.154)}{\pi} \frac{1 + 12}{R_L}$$

or $R_L = 4.16 \ \Omega$

b Power loss in the resistor is determined by the rms current rather than by the average current. From the basic concept (Eq. 3-45),

$$I_{\text{rms}} = \sqrt{\frac{1}{\pi} \int_{8.81°}^{171.2°} \left[\frac{\sqrt{2}(60 \sin \omega t) - (1 + 12)}{4.16} \right]^2 d(\omega t)} = 10.8 \text{ A}$$

Hence the power dissipated in the resistor is

$$(10.8)^2(4.16) = 484 \text{ W}$$

c The battery can utilize only the average or direct current, so that

$$P_b = (10)(12) = 120 \text{ W}$$

In many control applications the load presented to the rectifier is not a pure resistance. The armature circuit of a dc motor contains both resistance and inductance. The performance of rectifiers with inductive loads differs from that of resistively loaded rectifiers because of the phase lag introduced by the inductor. The waveforms in Fig. 17-11*b* are those for one cycle of the impressed voltage v_i and the resistor voltage v_R in the half-wave circuit of

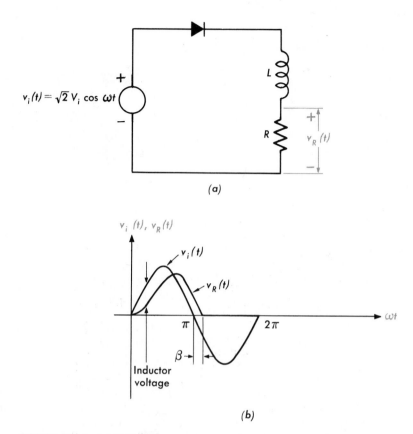

(a)

(b)

FIGURE 17-11   (a) Half-wave rectifier circuit with inductive load and (b) voltage waveform.

Fig. 17-11a. The phase shift and the distortion in the waveform of v_R are attributed to the behavior of the inductor.

The waveforms of all the rectifiers of Figs. 17-2, 17-7, and 17-8 are those of unidirectional voltage but obviously do not represent a constant direct voltage. The ripple or harmonics in the direct voltage are objectionable in

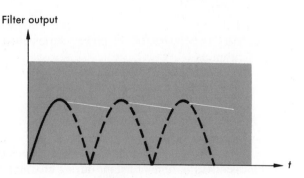

Filter output

FIGURE 17-12   Waveform of filter output.

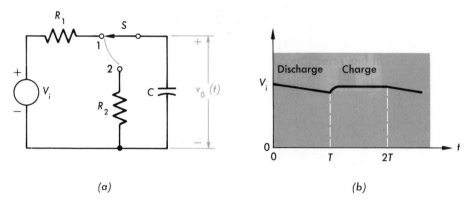

FIGURE 17-13   (a) Circuit representation of filter process and (b) output waveform.

many cases. When rectifiers are used to produce the dc supplies of electronic systems, an unduly large ripple introduces undesirable noise and may even mask the effect of the signal voltage. Filters consisting of shunt capacitors to bypass the ripple and series inductors to offer high series impedance to ripple frequencies are then added to smooth the output-voltage wave to the required degree. The effect of filtering the rectifier output is shown in Fig. 17-12. The dashed curve represents the output voltage of a full-wave rectifier which is the input voltage to the filter; the solid curve is the output of the filter. This output waveform can be considered to result from the alternate charging and discharging of a capacitor, with the discharge process being of considerably longer duration than the period of the full-wave rectifier voltage. To demonstrate the process, the circuit of Fig. 17-13a is useful in representing the behavior of a simple shunt-capacitor filter.

Consider that the switch S has been in position 1 for a long time and beginning at $t = 0$ it is periodically switched from position 1 to position 2 at intervals of T seconds. At $t = 0^-$, the capacitor is charged to the voltage V_i. Placing the switch in position 2 causes the capacitor to discharge through R_2 so that $v_o(t)$ is given as

$$v_o(t) = V_i \epsilon^{-t/R_2C} \tag{17-9}$$

For $R_2C \gg T$, $v_o(t)$ in Eq. (17-9) will be nearly V_i. If ϵ^{-t/R_2C} is expressed by the first two terms of its power series as $1 - t/R_2C$, Eq. (17-9) can be rewritten as

$$v_o(t) = V_i \left(1 - \frac{t}{R_2C} \right) \tag{17-10}$$

The output voltage for the first T seconds is a straight line decreasing from V_i to $V_i(1 - T/R_2C)$.

At $t = T$, the switch is returned to position 1 and the capacitor is recharged. If the time constant during charging (R_1C) is much smaller than T, the capacitor voltage is again V_i at $t = 2T$ seconds (when the switch is re-

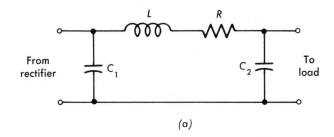

(a)

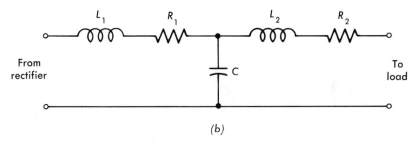

(b)

FIGURE 17-14 (a) Capacitor-input and (b) choke-input filters.

turned to position 2 once again). The resultant output for the initial two intervals is shown in Fig. 17-13b, for which the similarity with the waveform in Fig. 17-12 is evident.

Often, a simple shunt capacitance does not suffice in reducing the ripple when filter sections of the type shown in Fig. 17-14 are used. The circuit shown in Fig. 17-14a is referred to as a *capacitor-input filter,* that in Fig. 17-14b is called a *choke-input filter.* The resistors R, R_1 and R_2 in the filter circuits indicate the coil resistance associated with each inductor.

The circuit depicted in Fig. 17-15 is a simple regulator in which the Zener diode (see Sec. 11-2) is used to maintain a constant voltage across a load resistance R_L when the ac supply is unregulated or if the load is variable. The voltage source V_s and series resistance R_s in Fig. 17-15 represent

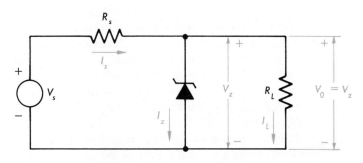

FIGURE 17-15 Representation of Zener-diode regulator.

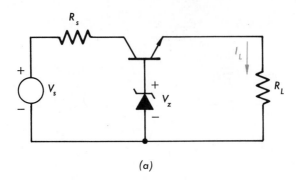

(a)

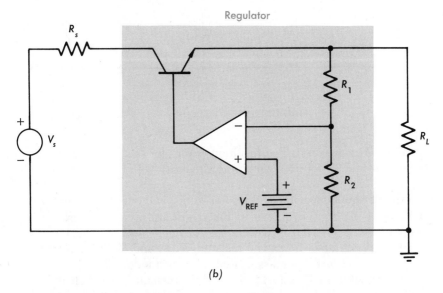

(b)

FIGURE 17-16   Series regulators using (a) a Zener diode and (b) an operational amplifier.

the output of the rectifier and filter. The circuit behaves as a regulator only when V_s is greater than the Zener voltage V_z.

With V_s and V_z specified, the current I_s, equal to $(V_s - V_z)/R_s$, is also fixed. Increasing the value of R_L decreases the load current I_L but increases I_z. To prevent the Zener diode from exceeding its maximum current rating, the circuit of Fig. 17-16a, called a *series regulator,* is employed. Its operation is essentially the same as that of the circuit in Fig. 17-15. The exception is that the diode current is the base current of the transistor and not the emitter current which feeds the load. Further improvement in performance is obtained by use of the circuit of Fig. 17-16b in which the operational amplifier, being part of a feedback system, is more sensitive to changes in load conditions. The enclosed portion of the circuit in Fig. 17-16b is available commercially for a variety of voltage levels as a single monolithic circuit.

17-3 DC TO AC CONVERSION

Inverters are electric circuits used to convert direct voltages into alternating waves. A principal use of inverters is to provide a controlled excitation for adjustable-frequency ac motors. The fundamental principle of operation of an inverter is the periodic interruption of a direct voltage to generate a square wave and is illustrated by the circuit of Fig. 17-17. In most power applications, SCRs perform the function of the periodically operated switch S. The frequency of the resultant square wave is a function only of the switching rate, so that the ac frequency in the load can be varied. If a pure sinusoidal load voltage is desired, the fundamental frequency of the square wave may be obtained by an appropriate filter.

The circuit shown in Fig. 17-18 is called a capacitor-commutated parallel inverter. The two SCRs act as the switching elements, each conducting for one-half of the cycle and blocked for the other half cycle. The function of the capacitor is to keep one SCR in a nonconducting state while the second SCR is conducting. The inductor L is called a choke and serves to isolate the alternating voltages generated by the switching action from the dc supply. Inverters of this type are used as the basic elements in many large and complex inverter systems employed in industry.

17-4 MOTOR DRIVE SYSTEMS

In many industrial applications it is important to adjust the operating characteristics of a motor to particular load conditions. Often, the mechanical load presented to the motor is variable, and to achieve the desired performance, a control system is required. One class of motor controls utilizes the SCR circuits discussed in the previous sections of this chapter. A *motor drive system,* or simply *drive system,* refers to the combination of the motor and its associated control elements and circuits.

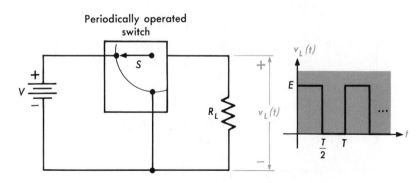

FIGURE 17-17 Schematic representation of an inverter.

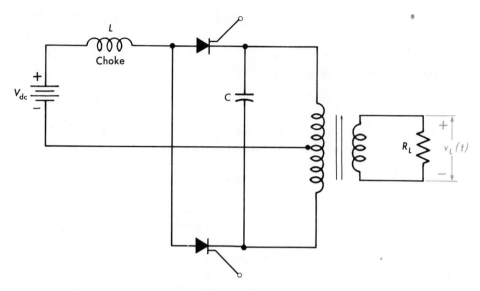

FIGURE 17-18 Capacitor-commutated parallel inverter.

Speed control in dc motors is achieved in primarily one of three ways. The first of these is the control of the dc armature voltage by means of an SCR circuit shown in Fig. 17-19. The motor representation of R_a, L, and the counter-EMF e_a correspond to load conditions discussed in Sec. 17-2. The control signal applied to the gate can be used to adjust the firing angle and thus to control the average voltage applied to the motor.

Control of the current in the field winding is the second method used. Usually, the field acts under constant excitation provided by an SCR circuit. To increase speed, the field current must be decreased. This condition is obtained when the firing angle of the circuit is increased by the control exhibited from the SCR gate circuit.

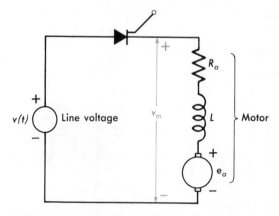

FIGURE 17-19 Half-wave SCR dc-motor drive.

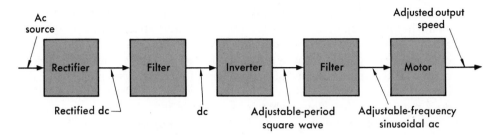

FIGURE 17-20 Block diagram representation of adjustable-frequency ac drive.

The third method of speed control utilizes the SCR as a switch to effectively change the armature resistance. The SCR switch, called a *chopper,* operates in a manner similar to the switching function performed in an inverter. The control of the armature resistance is effected by the chopper circuit by switching the applied voltage between the armature and the switch at a rapid rate. Control of the SCR switch is exerted from the gate circuit so that the average armature voltage is adjusted. The change of armature voltage in this manner has the effect of adding a resistance in series with the armature resistance.

Adjustable-frequency ac drives are used to control the speed of induction and synchronous motors (see Sec. 16-6). As the synchronous speed of these machines is proportional to the applied frequency, control of frequency corresponds to speed control. One method for generating an adjustable-frequency alternating current is by use of the inverters discussed in Sec. 17-3. When the basic source of electrical energy for the motor system is a fixed-frequency alternating current, the system is designed as shown in Fig. 17-20. The rectifier unit converts the ac supply to a dc supply (filtering of the type described in Sec. 17-2 is sometimes used). The inverter circuit reconverts the dc supply to an ac wave, the frequency of the wave being controlled by the pulses applied in the gate circuits of the SCRs used in the inverters.

The motor drives of the type described in this section are widely used in industrial plants, transportation systems, and in large data-processing facilities. They represent, however, only one of a number of different methods by which control of electric power and electrical devices may be effected. The remaining sections of this chapter are used to survey some of the other methods and their underlying concepts.

17-5 INTRODUCTION TO FEEDBACK CONTROL SYSTEMS

Control systems naturally fall into two types, depending on the relationship between the controlled element and the controlling element. An *open-loop* or *open-cycle* control system is one in which the controlling element is unaware of the effect it is producing on the controlled element. A simple example of an open-loop system is an automobile which has no speedometer but which does have a calibrated throttle. The driver can set the throttle for 40 mi/h, for example, and be assured that the actual speed is in that vicinity. Such a control system does not measure, and hence cannot correct for, errors caused by wind velocity, road conditions and slope, motor condition, or any of the many factors affecting operation. The system then has the inherent failing of any system which does not sample the final product: It does not know if it is doing what is expected of it.

Presumably, refinements could be built into the system; a slope-measuring device could correct for road slope, for example. Refinements of this nature, while requiring more complicated control mechanisms, would still not assure the desired speed under all conditions. A closer approach to the ideal can be made by a mechanism which measures the actual speed and makes correction for any deviation of it from the desired value. Perhaps the most obvious means for measuring such a deviation is a speedometer visible to the eye of the driver. By reading the actual speed on the dial and mentally comparing this value with the desired speed, then changing the throttle setting to reduce the error, the driver acts to close the loop between the output speed and input throttle setting. The result is a *closed-loop, closed-cycle,* or *feedback control* system.

In this closed-loop system, the closeness with which the actual output meets the desired condition depends more on the behavior of the feedback element (i.e., the mind of the operator) than on the actual controller (throttle valve, engine, gears, etc.). A feedback control system is in general inherently more accurate than an open-loop system constructed of essentially the same elements. Note particularly that the automobile speed-control system is composed of the same carburetor, engine, transmission, etc., regardless of whether an open- or closed-cycle control system is used. The feedback link which so improves the performance of the system is added more or less as an external appendage. This form of addition is very common, since many systems are designed around existing heavy equipment to which feedback control is added to improve performance.

There are five elements common to all feedback control systems for which Fig. 17-21 is a typical block diagram (see Sec. 6-6). The first of these is the input signal, or reference variable R, which determines the desired value of the output or controlled variable C (the second element). The third element, H, involves measurement of the output and feeding it back to the input, either directly or in a modified but proportional form. Comparison, or summation, is the fourth element; its function is to compare the input signal with the signal fed back from the output. The result of comparison is a difference, or error, signal ε which in turn drives the controller G. The function of the controller, the fifth element, is to produce the output signal. In general, the process represented by the controller contains the system or device whose performance is to be controlled. Note that the block diagram in Fig. 17-21 is identical to the signal-flow graph in Fig. 9-28 for the feedback amplifier if G and H are identified with $a(s)$ and $-f(s)$, respectively.

The major distinction between open-loop and feedback control systems

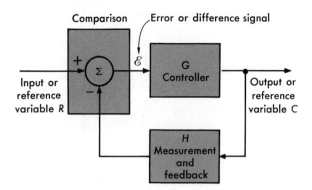

FIGURE 17-21   Block diagram of feedback control system.

is manifested in the comparison of controller elements. In feedback systems the controller processes a signal which contains information regarding the nature of the output, i.e., the difference signal. The controller in an open-loop system responds only to the input signal R.

The input signal to a feedback control system is fundamentally a piece of information stating the performance desired at the output. Thus, to control the speed of a dc motor, the input might indicate that an output speed of 800 r/min is desired. As the speed of the motor is dependent on armature excitation, the input armature voltage is determined from the desired output speed and the motor parameters (see Chap. 15). The input signal establishes the output exactly if all the components in the system are ideal and no external disturbances are present. The necessity for control arises principally from the fact that the motor characteristics do vary. Environmental changes such as temperature variations and component aging, external disturbances such as load variations, and component tolerances in the manufacturing process are among the factors which determine the degree of control required. In addition, feedback techniques are used to minimize inherent non-linearities such as saturation.

17-6 TRANSDUCERS

Position, speed, acceleration, force, pressure, temperature, and volume flow are just some of the measures of the outputs of systems whose performance is electrically controlled. Indeed, if we use a microprocessor to minimize the fuel consumption of an automobile under all driving conditions, a quantity of each type listed must be known. This information is processed and the results used to adjust automobile performance. Evidently, as the microprocessor input and output are electrical signals, it is assumed that the values of the physical quantities which describe the operation can be converted to and from electrical energy.

This assumption also applies to control systems characterized by the block diagram in Fig. 17-21. The measurement and feedback portion of the

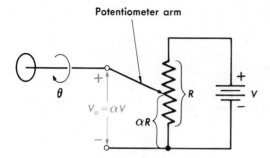

FIGURE 17-22 Position-indicated potentiometer.

system must take the output, whatever its form, and convert it to an electrical signal suitable for comparison. *Transducers* are devices which convert nonelectrical physical quantities to and from electrical energy. There are many types of transducers available for this purpose. The intent of this section is to introduce and briefly describe a few of these and, in so doing, indicate the variety of techniques employed.

A *position-indicating potentiometer,* illustrated in Fig. 17-22, is a device whose output voltage is proportional to the position of the potentiometer arm. The position of the potentiometer arm is, in turn, proportional to the angular displacement θ of the shaft.

The speed of rotation of a shaft is converted into an electrical signal by means of an electric *tachometer.* The shaft rotation is the prime mover for the tachometer which is a dc generator. As described in Sec. 15-1, the generated voltage is directly proportional to the angular speed of the shaft.

Force, torque, and stress can be converted to an electrical signal by means of the *resistance strain-gauge* shown in Fig. 17-23. The circuit arrangement in Fig. 17-23 is that of a Wheatstone bridge (see Example 2-6). In an unstressed state, the bridge is balanced and the output voltage v_o is zero. When subjected to a force, the resistance of the sensing element R_4 changes, causing an imbalance in the bridge circuit. Consequently, an output voltage is produced proportional to the change in resistance, which, in turn, depends on the elongation of the sensing element.

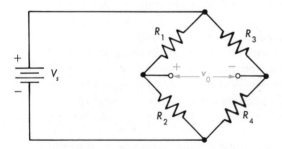

FIGURE 17-23 Resistance strain-gauge.

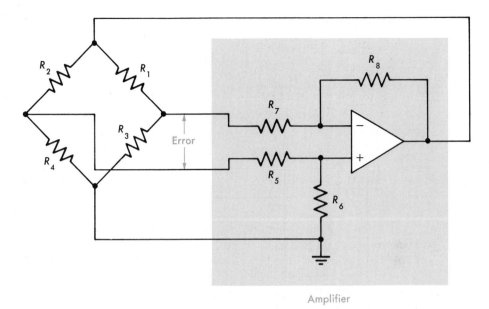

FIGURE 17-24 Hot-wire anemometer circuit.

Hot-wire anemometers used to measure volume and mass flow also employ the Wheatstone-bridge arrangement of Fig. 17-23. The sensing element is a hot wire placed perpendicular to the flow and is kept at constant temperature. To account for heat loss to the surrounding medium, which is proportional to the flow, the current in R_4 must increase. Most commercial anemometers use a feedback circuit as shown in Fig. 17-24 to adjust the current in R_4 and maintain it at constant temperature. The output of the amplifier is calibrated to indicate the flow.

Thermocouples are thermoelectric transducers in which two junctions are formed when two wires of dissimilar metals are connected. When the two junctions are at different temperatures a potential difference exists between the junctions (or a current exists in the wires). Similarly a current in the wires establishes a temperature difference.

The *piezoelectric effect* is a reversible process in which a voltage applied to a crystal produces a mechanical deformation. Piezoelectric transducers are often used to indicate force, acceleration, and pressure. The same effect is used to control oscillator frequencies as described in Sec. 10-4.

Many transducers utilize the change in inductance and capacitance to produce electrical signals just as the strain gauge uses resistance variation. A *capacitor microphone* is an example of a pressure-sensitive electroacoustic transducer.

Another transducer useful in converting acceleration, velocity, and displacement to electrical signals that utilizes changes in inductance is the *linear variable differential transformer* (*LVDT*). A schematic representation of the differential transformer is displayed in Fig. 17-25, in which the two

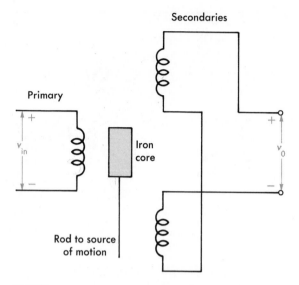

FIGURE 17-25 Circuit representation of a linear variable differential-transformer (LVDT).

output windings are connected series-opposing. When the iron core is in its equilibrium position, the voltage induced in each secondary winding is equal and the output voltage v_o is zero. As the coil is moved, the mutual coupling between the primary and one secondary winding increases and it decreases for the other secondary winding. As a consequence, the voltages induced in each coil differ and produce an output voltage. Construction of LVDT is such that the output amplitude is a linear function of displacement. The relative phase of v_o with respect to the primary voltage indicates whether the core is nearer the upper or lower secondary winding. The utility of differential transformers that are commercially available is enhanced because linear operation is achieved over a wide range of excitation frequencies (usually 25 to 25,000 Hz).

Each of the transducers described to this juncture are analog devices, i.e., the electrical output at each instant of time is proportional to the input quantity at the corresponding instant of time. The use of these transducers in digital control systems requires the conversion of the output, by means of an analog-to-digital converter (see Sec. 12-8). One type of widely used digital transducer is the *optical shaft encoder*. The shaft is filled with a disc which is coded as shown in Fig. 17-26. The darker segments are nonreflective, the lighter segments are reflective. An optical source, usually light-emitting diodes (LEDs), is electrically pulsed at a clock frequency. The reflections from each radial segment are detected by a solar cell and form a pulse train which is binary-coded. Often, each set of four tracks, as shown in Fig. 17-26, is used for a single-decimal digit in a binary-coded decimal output of position; dual channel outputs can also be used to indicate speed. Most commercially available optical shaft encoders are compatible with the TTL and

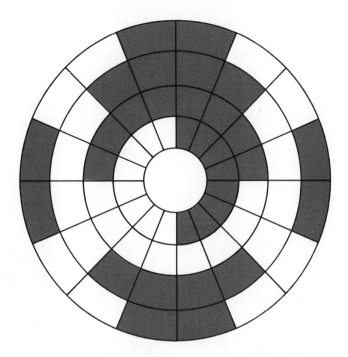

FIGURE 17-26 Optical shaft encoder disc.

CMOS logic gates described in Sec. 11-9. Incremental shaft encoders use a single-track coded disc and are used to measure changes in position or speed. The optical-shaft encoder can also be used to transduce any quantity which is capable of altering the speed or position of a shaft. In addition, shaft encoders often form the input signals for the stepper motors discussed later in this chapter.

17-7 SYSTEM ASPECTS AND CLASSIFICATION

There are two major types of feedback control systems. The first type is called a *regulator,* whose function is to keep the output or controlled variable constant in the face of load variations, parameter changes, etc. Typical applications of regulators include control of the output voltage of a dc generator, control of the rotational speed of a radar antenna used to determine the location of aircraft in the vicinity of an airport, and the control of the temperature at which a chemical reaction occurs. The input reference variable for such a system is a constant, changed only when it is desired to change the output level. In such a system, it is not important that the output exactly equal the input. What is important is that the output stay constant at the desired value. Analysis of such systems necessarily involves evaluation of the changes which the output makes from the desired value when the load or system parameters change. For the purposes of such evaluation, the system input is frequently considered to be the load disturbance. The output resulting from the disturbance can be evaluated, and superposition (Sec. 2-7) used to give the total system response.

Servomechanism, the second type of system, is the name usually given to feedback control systems whose inputs are time-varying. The function of a servomechanism is to provide a one-to-one correspondence between input and output. One of the most commonly used types of servomechanism is a position-control system in which the output position, often the shaft position on a motor, must exactly follow the variations of the input position. Among the more obvious applications are power steering in automobiles, the positioning of airfoils and rudders in aircraft and ships, the directing of radar antennas used for tracking in instrument landing systems, and the positioning of machine tools such as drills, punches, and spot-welders in automated manufacturing facilities. In this type of system, the input shaft position will vary with time in some prescribed manner as in a tracking radar, or will vary randomly as in the case of positioning aircraft control surfaces. Analysis of servomechanisms involves concern with maintaining the one-to-one output-input correspondence. Load disturbances may again play an important role in system behavior but are generally considered separately as a peripheral problem.

The possibility of unwanted oscillations exists in any system in which all or a portion of the output is fed back into the input circuit. Chapter 9 discusses electronic circuits with feedback; certain criteria are established in Sec. 9-9 for ensuring the stability of these circuits. These criteria can also be used to determine the stability of feedback control systems.

Another important consideration in the design of feedback control systems is the system speed of response, or the rate at which a servomechanism will respond to a change in system input or the rate at which a regulator will recover from a load disturbance. Information about response time can be gained from the frequency-response characteristic of the system (see Chaps. 4 to 6). The stability criteria of Sec. 9-9 are also developed on the basis of the frequency-response characteristic, which is then a powerful tool in the evaluation of feedback-control-system performance.

The basic mechanism of a regulator-type feedback system is illustrated by the motor-speed control system shown in Fig. 17-27. The basic function of the system is to maintain a constant angular rotation of the output shaft, which is driving a pure inertia load, by controlling the armature voltage. The five elements of a feedback control system are identified as:

1 The input or reference level is the dc source, V_R.

2 The output is the angular speed of the motor shaft.

3 The tachometer measures the actual output speed and converts this to a proportional voltage.

4 The difference signal ε is obtained from the comparison of V_R and the tachometer output voltage V_T. (By Kirchhoff's voltage law, $\varepsilon = V_R - V_T$.)

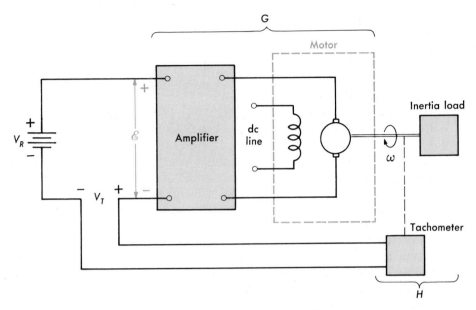

FIGURE 17-27 Speed-control system.

5 The difference signal is amplified and the amplifier output is used to drive the armature of the motor.

The controller G in Fig. 17-21 can be considered as the amplifier-motor combination; the feedback network H is the tachometer.

When the system is operating at the desired speed, the difference signal, when amplified, provides the exact armature voltage necessary to cause the desired angular rotation. In this type of system, as in most regulators, the difference signal must exist; without it there is no signal to drive the motor and produce the output. If, for some reason, the speed increases, the feedback signal V_T also increases. Since the reference level V_R is a constant, the comparison process causes the difference signal ε to decrease, thus tending to decrease the motor speed. Similarly, a decrease in motor speed causes ε to increase, which tends to increase the output speed to the desired value. This principle of self-adjustment is fundamental to all feedback systems, including the feedback amplifiers discussed in Chap. 9.

The position-control system of Fig. 17-28, used to position a pure-inertia load, is a typical servomechanism. The output position is to follow the variations in the input position; by adjusting the armature voltage of the motor, the shaft of the motor rotates to produce the output position. For convenience, the five basic feedback system elements are identified as follows:

1 The input position θ_R is varied by means of the input dial (the steering wheel in a power steering system).

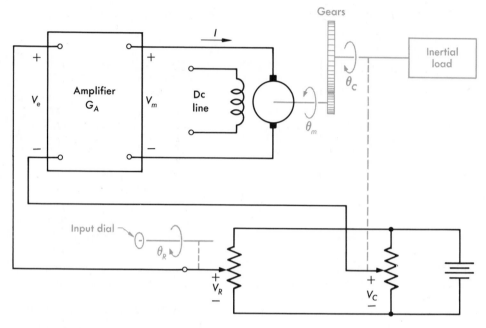

FIGURE 17-28 Position-control servomechanism.

2 The output is the position of the inertial load θ_C (the position of the front tires).

3 Both the output and input positions are converted to electrical signals by means of the position-indicating potentiometers.

4 The generation of the error signal V_e is obtained from the comparison of the output voltages of the two position-indicating potentiometers ($V_e = V_R - V_C$).

5 The amplifier and motor combination make up the controller.

When the output position equals the input position, no error signal is developed. Therefore, the amplifier output is zero, so that no armature voltage is applied to the motor. Consequently, the output shaft does not rotate and maintains its desired position. Many servomechanisms display this characteristic of maintaining a desired output when the difference signal is zero. A change in the input position causes the error signal to be nonzero. The amplified error signal V_m is applied to the armature and causes the motor to rotate, thereby changing the output position. The output shaft continues to rotate until the error voltage again becomes zero, indicating the correspondence between input and output positions.

Comparison of the speed-control (regulator) and position-control (servomechanism) systems leads to a second classification of feedback control systems based on the manner in which the system operates to perform its function rather than on the function which the system performs. The input to a regulator is usually a constant value related in some fashion to the desired output which must remain constant regardless of load fluctuations or variation in the components. Performance criteria for regulators must include (1) the allowable output deviation from the ideal with load changes and in the presence of unwanted disturbance signals, and (2) the speed with which the system can respond to such changes. In the operation of the system of Fig. 17-27, one important fact emerges: As long as the amplifier gain G_A is finite, the difference signal can never be made zero. What this means practically is that some error voltage is required for the system to operate; that is, there must be a controller input to have a controller output. Systems of this type, which require a steady-state error (difference signal) for operation, are called *type 0* systems.

In a servomechanism, the output is to correspond with the time variations of the input. The performance criterion is then the faithfulness with which the system follows the input, and both the steady-state error and speed of response factors must be considered. The position-control system of Fig. 17-28 is one in which a steady-state output can be produced with no difference signal applied to the controller. The condition of zero error in the steady state is typical of many servomechanisms. If a constant error signal existed, the armature voltage would also be a constant and cause the output shaft to rotate at a constant angular velocity. Since the output of the system is the output shaft position, a constant error is seen to produce an output which grows linearly with time. (Angular position is the integral of angular velocity.) Such a system is called a *type 1* system. Note that in this case, when the input signal is removed, the motor will slow down and ultimately stop. However, its position after it has come to rest will differ from the position it had before the error voltage was applied to the controller. The motor shaft output is then a function not only of the error voltage applied to the controller but also of the entire past history of applied errors. The system is capable of remembering and summing up, or integrating, past errors to produce an output even when the instantaneous error is zero.

A classification of feedback control systems based on the nature of the system response can now be developed more formally. The terminology used here is that of a position-control system. Thus the input and output functions are positions; their first time derivatives are velocities; their second time derivatives are accelerations. Position-control-system terminology is particularly useful because well-established names (velocity and acceleration) have been attached to first and second time derivatives of the output quantity as well as to the output itself (position). Table 17-1 shows the type of steady-state error to be expected of each of two types of feedback control systems with each of two specific inputs. The type 0 system, for example, will have a steady-state position error with a constant position input. The type 1 system will have no steady-state error with a constant position input; it will have a position error with a constant velocity input (the two shafts will run at the same velocity, but with an angular displacement between them).

TABLE 17-1 STEADY-STATE ERRORS IN
POSITION-CONTROL SYSTEMS

	SYSTEM	
INPUT	TYPE 0	TYPE 1
Constant position	Position	None
Constant velocity	Velocity	Position

In addition to the types 0 and 1 systems discussed in this section, higher-order systems can be developed. In general, feedback control systems used as regulators are type 0 systems. Servomechanisms are usually type 1 or higher-order systems. In this book, we shall limit our discussion to types 0 and 1 systems.

17-8 ANALYSIS OF FEEDBACK CONTROL SYSTEMS

In the two immediately preceding sections of this chapter the properties of feedback systems and various aspects related to system design and performance are described in qualitative terms. The function of this section is to introduce the quantitative methods used to determine the performance of feedback control systems. What is required is to describe mathematically the effects on the system of load disturbances, steady-state error, amplifier gain, and speed of response. In other words, the analysis is used to quantitatively measure how well the system performs its intended function.

Both regulators and servomechanisms can be represented by the block diagram of Fig. 17-21, which, for convenience, is redrawn in Fig. 17-29. The input signal (reference signal) is indicated by R, the output variable by C. The signal ε, which drives the controller G, is the difference or error signal resulting from the comparison made at the summing mode. The feedback network H provides the measure of the output that is fed back for the purpose of comparison. Mathematically, the block diagram can be used to ana-

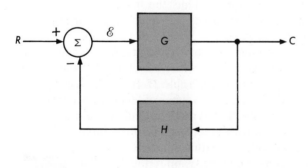

FIGURE 17-29 Basic block diagram of feedback system.

lyze the performance of the system by use of the methods described in Sec. 10-8. For the system of Fig. 17-29, the error signal is

$$\varepsilon = R - HC \tag{17-11}$$

and the output variable, produced by the controller, is

$$C = G\varepsilon \tag{17-12}$$

Combination of Eqs. (17-11) and (17-12) results in

$$\frac{C}{R} = \frac{G}{1 + GH} \tag{17-13}$$

The quantity C/R in Eq. (17-13) is the transfer function of the feedback system. In general, G and H are functions of frequency so that C/R is also a function of frequency. The methods for determination of the system response C caused by an excitation R are those described in Chaps. 4 and 5. For the special case where the excitation is sinusoidal, the techniques developed in Chap. 3 and the Bode diagrams of Secs. 6-2 and 6-3 are particularly useful. All of the pertinent data needed to describe system performance can be evaluated from the system transfer function and the known input variable. The following examples are used to describe, in numerical terms, how the self-adjusting nature of a feedback system corrects for changes in components and load disturbances.

Example 17-2

The block diagram shown in Fig. 17-30*a* is used to represent the speed-control system shown in Fig. 17-27. The input R is the reference dc supply; the output C is the speed of the motor. The controller G_1 is the amplifier-motor combination whose value depends on both the motor constant and gain of the amplifier. The block H represents the tachometer used to measure the output speed and feed the electrical signal back for comparison with the reference. The second input R_D represents an external disturbance which affects the system. The presence of an external disturbance is quite common; consider this system to be the speed control for a radar antenna (the inertia load) used in an automatic landing approach system. Because the radar is used to determine the position, altitude, and speed of aircraft in the vicinity of the airport, it is essential that the antenna rotate at a constant angular speed. The antenna, however, is subject to the effects of the wind. The additional torque produced by the wind tends to slow down or speed up the rotation of the antenna. The action of the wind and its effect on the angular rotation can be considered as the disturbance.

The block diagram in Fig. 17-30*b* is a nonfeedback system which represents the motor driving the antenna directly. The disturbance R_D is also applied to this system.

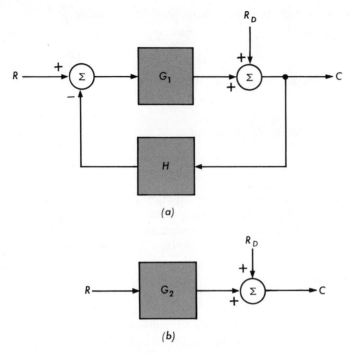

FIGURE 17-30 Systems in presence of disturbance: (a) feedback
system; (b) nonfeedback system.

a For $G_1 = 100$ and $H = 0.49$, determine G_2 so that the transfer function
C/R (in the absence of the disturbing signal) is the same for the systems in
Fig. 17-30a and b.

b For the conditions in a, determine the component of the output caused
by R_D for both systems.

c Determine the value of C/R for each system if the gains (G_1 and G_2) decrease by 10 percent.

SOLUTION **a** For the system of Fig. 17-30a, with $R_D = 0$, the transfer
function is given by Eq. (17-13) as

$$\frac{C}{R} = \frac{100}{1 + (100)(0.49)} = 2.0$$

For $R_D = 0$, the transfer function of the system in Fig. 17-30b is $C/R = G_2$,
so that $G_2 = 2.0$.

b The disturbance component of the output in each case is obtained by superposition. Thus, when $R = 0$, the output is just that produced by R_D. For the system of Fig. 17-30a, this output C_{1D} is

$$C_{1D} = G_1 \varepsilon + R_D$$

where $\varepsilon = -HC_{1D}$

so that $$C_{1D} = \frac{R_D}{1 + G_1 H} = \frac{R_D}{1 + (100)(0.49)} = \frac{R_D}{50}$$

For the system in Fig. 17-30b, when $R = 0$, the output C_{2D} equals R_D.

c A 10 percent decrease in the gain causes both G_1 and G_2 to decrease 10 percent and become 90 and 1.80, respectively. Therefore, for the system of Fig. 17-30a,

$$\frac{C}{R} = \frac{90}{(90)(0.49)} = 1.996$$

and for the system in Fig. 17-30b,

$$\frac{C}{R} = G_2 = 1.80$$

Comparison of the results for each system shows that the effect of the disturbing signal in the feedback system is reduced by a factor of 50, which is the value of $1 + G_1 H$. Similarly, in c, the 10 percent change in the gain produces only a 0.2 percent change in the transfer function of the feedback system. In the nonfeedback system, the value of the transfer function changes by 10 percent. The 0.2 percent change also reflects the reduction by a factor $1 + G_1 H$ of the 10 percent change in G_1. As $G_1 H$ is generally much larger than unity, $1 + G_1 H$ is approximately $G_1 H$. The quantity $G_1 H$ is called the *loop gain* or *return ratio* and is an important measure of the effectiveness of a feedback system. (Further discussion is presented in Sec. 9-7.) The return ratio is also used to determine whether the feedback system is stable. That is, does the system response go to zero for input signals which go to zero with time? Possible instability is often a serious disadvantage of feedback systems. The analysis of system stability is presented in Sec. 9-9 for feedback amplifiers. As feedback control systems are mathematically described in the same manner as are feedback amplifiers, the stability criteria are identical.

Load disturbances in regulators need not be introduced to the system externally. Often, inherent limitations of the devices whose performance is to be controlled can be represented by load disturbances. This feature is illustrated in Example 17-3, whose purpose also outlines the method by which a system can be represented by the block diagram in Fig. 17-29.

Example 17-3

The circuit shown in Fig. 17-31 is a simple voltage regulator for a separately excited dc generator. The voltage V_R is a constant voltage and serves as the reference level. The generator EMF constant K_G is dependent on generator geometry and the speed of the prime mover. The amplifier G_A provides the field voltage V_f required to obtain the output voltage V_o. The purpose of a voltage regulator is to maintain a constant output voltage for all load conditions between no-load and the full-load rating of the generator. Control is achieved by measurement of the output V_o; comparison with the reference signal V_R produces a difference signal V_e. Amplification of V_e by G_A provides the field excitation V_f. A change in V_o causes a change in V_e in the opposite direction; i.e., a negative change in V_o results in a positive change in V_e. However, positive changes in V_e tend to cause positive changes in V_o, thus counteracting the effect of an initial negative change in the output voltage.

a Draw the block diagram of the system.

b For a generator constant $K_G = 1.0$, an amplifier gain $G_A = 50$, a full-load armature drop $I_L R_a$ equal to 10 V, and $V_R = 235$ V, determine the output voltage.

c Repeat **b** for $K_G = 0.9$ with all other parameters having their values as given in part **a**.

d Repeat **b** and **c** for no-load conditions.

SOLUTION **a** Before identifying the various blocks, the basic equations governing each element in the process need be stated. The input to the amplifier is the error signal V_e and is

$$V_e = V_R = V_o$$

The field excitation V_f is

$$V_f = G_A V_e$$

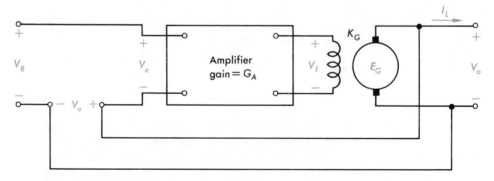

FIGURE 17-31 Voltage regulator for dc generator.

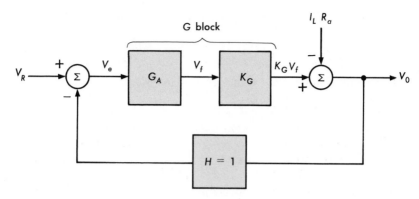

FIGURE 17-32 Block diagram of voltage regulator of Fig. 17-31.

and the output voltage, accounting for the armature drop, is

$$V_o = K_G V_f - I_L R_a$$

To identify the blocks, the equations for the system components are compared with Eqs. (17-11) and (17-12). Comparison of the difference signals results in $H = 1$. This is observed in Fig. 17-31 as the entire output is fed back. Equation (17-12) indicates that the block G processes only the difference signal. The expression for V_o shows that it contains two components, one due to the field excitation V_f, the other due to the armature drop. As $V_f = G_A V_e$, V_o becomes

$$V_o = K_G G_A V_e - I_L R_a$$

The portion of V_o proportional to V_e must represent the block G, so that $G = K_G G_A$. The $I_L R_a$ component of V_o acts as a load disturbance in much the same fashion as the input R_D in Example 17-2. The complete block diagram must therefore contain two excitations, as well as the G and H blocks, as depicted in Fig. 17-32.

b To determine the output voltage, the two components of V_o must be found. The first component is the output produced by V_R, the second that caused by the load disturbance $I_L R_a$. Convenience dictates the use of superposition to determine each component. From Eq. (17-13), the component V_{o1} due to V_R is

$$V_{o1} = \frac{G}{1 + GH} V_R$$

The component V_{o2} due to the armature drop is, from the analysis in Example 17-2,

$$V_{o2} = \frac{-I_L R_a}{1 + GH}$$

The value of V_o is the superposition of V_{o1} and V_{o2} and is

$$V_o = V_{o1} + V_{o2} = \frac{G}{1 + GH} V_R - \frac{I_L R_a}{1 + GH}$$

Substitution of given values results in

$$G = K_G G_A = (1)(50) = 50$$

and

$$V_o = \frac{50}{1 + (50)(1)} 235 - \frac{10}{1 + (50)(1)} = 230.2 \text{ V}$$

c For $K_G = 0.9$, only the value of G changes to $(50)(0.9) = 45$. Substitution of this value of G yields a value of

$$V_o = \frac{45}{1 + (45)(1)} 235 - \frac{10}{1 + (45)(1)} = 229.7 \text{ V}$$

d Under no-load conditions, the value of I_a is zero, so that the armature drop vanishes. The output voltage is then equal to the component produced by V_R. For the conditions of part **b**, the no-load output voltage is

$$V_o = \frac{50}{1 + (50)(1)} 235 = 230.4 \text{ V}$$

and for the conditions of part **c**

$$V_o = \frac{45}{1 + (50)(1)} 235 = 229.9 \text{ V}$$

The results of Example 17-3 reaffirm the fundamental self-regulation property of feedback control systems. For both $K_G = 1$ and $K_G = 0.9$, the difference between the no-load and full-load output voltages is 0.2 V, or less than 0.1 percent. This result indicates that the feedback loop significantly decreases the effect of the load disturbance. The 10-V armature drop affects the output by only 0.2 V, a reduction of 50 times. Significantly, this is the value of GH, the loop gain of the system. Similarly, feedback control reduces the effect of component variations in the system. Comparison of both full-load and no-load voltages for the two values of K_G indicate a difference of 0.5 V, or approximately 0.2 percent. This change is the result of a 10 percent change in K_G, again reflecting a reduction by a factor of 50, the value of loop gain.

In both Examples 17-2 and 17-3 the value of loop gain, and consequently the amount of amplification needed, depend on the allowable deviation that can be tolerated. In the voltage regulator, this is easily seen in that a desired output of 230 V is achieved within only ± 0.5 V, which is generally an allowable error in such systems. Other performance criteria are based on the requirements of system stability and satisfactory transient behavior (e.g., rapid speed of response to varying inputs or load disturbances) of the system. The importance of transient response becomes evident when it is recognized that servomechanism input signals (or regulator load disturbances) may vary in an unpredictable or random fashion. Since it is obviously impossible to solve the transient equations of the system for every possible input or combination of inputs to which it will be subjected, some alternate form of analysis must be used which will enable the transient response to be judged. A ready-made analytical tool is found in the analysis techniques used in electric networks and amplifiers. Both the steady-state frequency response and the response to a step function (a suddenly applied constant excitation) are used to analyze the performance of feedback systems. As exponential excitation functions can be used to represent both sinusoidal and constant excitations, the methods developed in Secs. 3-1 to 3-4 are applicable. These methods are extended and developed in greater depth in the remainder of Chap. 3 and in Chaps. 4, 5 and 6.

Correlation exists between the transient and steady-state frequency response; the analysis of electronic amplifier circuits discussed in Chaps. 8 and 9 serves to illustrate the relationship. One criterion of good amplifier transient performance is a satisfactory frequency-response characteristic. An amplifier which exhibits a constant gain and zero phase shift over the entire frequency spectrum is ideal in that it will faithfully reproduce any given input. Departures from this ideal are necessitated by the characteristics of actual amplifiers. The extent of these departures determines the suitability of the amplifier for a particular application. These remarks apply with equal force to feedback control systems. A system which will respond equally well to inputs of all frequencies can be regarded as ideal. Departures from this ideal must occur, and the overall transient response of the system will depend on the nature of these departures.

Determination of the frequency-response characteristics of physical systems involves the extension of the techniques for electric networks to systems comprising electrical devices, electromechanical transducers, etc. The response is determined from the system transfer function. Since both sinusoidal ($s = j\omega$) and constant excitations ($s = 0$) are useful, the transfer function is developed in terms of the complex-frequency variable s. The methods used are illustrated by the determination of the transfer function of the type 1 position-control servomechanism.

The position-control system of Fig. 17-33 is that shown in Fig. 17-28 with the exception that the load contains both inertia J and friction σ_f. The frictional component accounts for both losses in the motor and mechanical power dissipation in the load. To obtain the transfer function of the system, the characteristics of each component must be determined. One method is to use the block-diagram concept for each element in the system and then combine these elements to obtain a block diagram of the form of Fig. 17-21. The components whose individual block diagrams are to be determined are: the gears which couple the motor to the load, the position-indicating potentiometer, the amplifier, and the motor and load combination.

The position-indicating potentiometer is used to convert the displace-

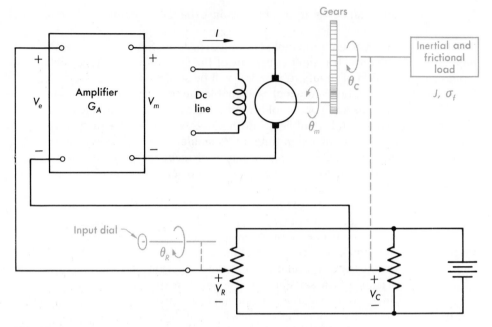

FIGURE 17-33 Position-control system with frictional and inertial load.

ment error $\theta_R - \theta_C$ to the error voltage V_e. As V_r and V_C are proportional to the mechanical angles θ_R and θ_C, respectively,

$$V_e = V_R - V_C = K_p(\theta_R - \theta_C) \tag{17-14}$$

where K_p is the potentiometer constant. The block diagram based on Eq. (17-14) is shown in Fig. 17-34a.

The amplifier can be represented by the block diagram of Fig. 17-34b, in

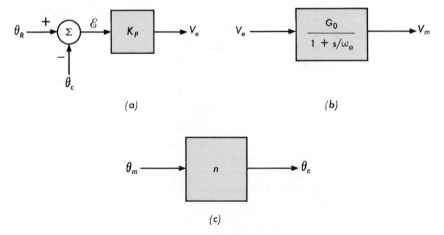

FIGURE 17-34 Component blocks for position-control system: (a) position-indicating potentiometer and error signal; (b) amplifier; (c) motor-to-load gear system.

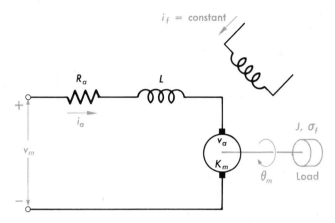

FIGURE 17-35 Representation of a dc motor.

which G_o is the dc value of gain and ω_a is its frequency response (see Sec. 8-1). The amplifier input is the error voltage, and its output V_m is the armature excitation. The gears act so that the motor-shaft position is translated by the gear ratio to the load position as indicated in Fig. 17-34c. The translation process ideally occurs with no power loss so that the gears are a mechanical analog of the transformer.

The dc motor is shown in Fig. 17-35 with the load inertia and frictional components, J and σ_f, reflected to the motor shaft. Reflection of the mechanical load from one side of the gears to the other is accomplished in much the same fashion as reflecting the load from the secondary to the primary in an electric transformer. The motor equations are

$$T_m = K_m i_a \tag{17-15}$$

$$v_a = K_m \omega_m \tag{17-16}$$

and

$$v_m = v_a + R_a i_a + L \frac{di_a}{dt} \tag{17-17}$$

The motor torque T_m is balanced by the inertia and friction torques of the load; the balance equation is

$$T_m = J \frac{d\omega_m}{dt} + \sigma_f \omega_m \tag{17-18}$$

The shaft velocity ω_m is the time rate of change of the angular position, or

$$\omega_m = \frac{d\theta_m}{dt} \tag{17-19}$$

Combining Eq. (17-19) into Eqs. (17-16) and (17-18) gives

$$v_a = K_m \frac{d\theta_m}{dt} \tag{17-20}$$

and

$$T_m = J \frac{d^2\theta_m}{dt^2} + \sigma_f \frac{d\theta_m}{dt} \tag{17-21}$$

Equations (17-15), (17-17), (17-20), and (17-21) can be transformed to the frequency domain by replacing d/dt by s following the method shown in Table 3-1 with exponential excitation implied. The results are

$$T_m = K_m I_a \tag{17-22}$$

$$V_a = s K_m \Theta_m \tag{17-23}$$

$$V_m = V_a + I_a R_a + Ls I_a \tag{17-24}$$

and

$$T_m = Js^2\Theta_m + \sigma_f s\Theta_m \tag{17-25}$$

In these equations, the capital symbols I and Θ denote the amplitude of exponential functions of the form $A\epsilon^{st}$. The block diagram of the motor, based on Eqs. (17-22) to (17-25), is shown in Fig. 17-36a. As is evident in Fig. 17-36a, the counter EMF in the armature circuit indicates an inherent feedback loop in the motor. In addition to motors, many physical devices such as transistors and electronic amplifiers exhibit inherent feedback; this type of feedback is part of the internal mechanism of the device and, in general,

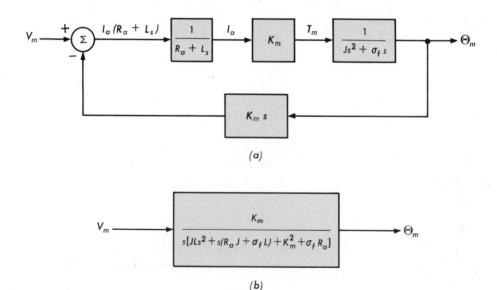

(a)

(b)

FIGURE 17-36   (a) Block diagram representation and (b) equivalent block diagram of dc motor.

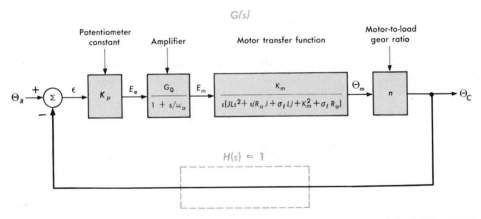

FIGURE 17-37 Standard feedback-system representation of Fig. 17-36.

cannot be used to control system performance. By reduction of the block diagram of Fig. 17-36a, or by the simultaneous solution of Eqs. (17-22) to (17-25), the transfer function relating Θ_m, the shaft position, to V_m, the motor voltage, is determined as

$$\frac{\Theta_m}{V_m} = \frac{K_m}{s[JLs^2 + s(JR_a + \sigma_f L) + K^2 + \sigma_f R_a]} \tag{17-26}$$

The equivalent block diagram of the motor, based on Eq. (17-26), is shown in Fig. 17-36b.

The block-diagram representation of the position-control system is the composite of Figs. 17-34 and 17-36b; it is shown in Fig. 17-37. By combining the cascaded blocks for the amplifier, position-indicating potentiometer, motor, and gears in the block diagram of Fig. 17-37, the resultant structure is that of the standard feedback system. Thus, the components in the position-control system that make up the G and H blocks are readily available.

By use of Eq. (17-13), the transfer function for the complete system is obtained. From this function and by letting $s = j\omega$, the frequency-response characteristic is determined; similarly, the loop gain $\mathbf{GH}(j\omega)$ can also be evaluated. The use of the Bode diagram of Sec. 6-2 is often a convenient method for displaying the frequency response. In particular, the stability analysis in Sec. 9-9, which is applicable to all feedback systems, makes use of the Bode diagram.

The dc steady-state response can also be determined by allowing s to be zero in the transfer function. By use of the methods in Chaps. 4 and 5, the dc steady-state response becomes one component of the step response of the control system. This step response can be used as a measure of the speed of response of the control system. Thus, the transfer function obtained from the block diagram can be used to describe both the steady-state and transient response of a feedback control system.

The motor-drive systems discussed in Sec. 17-4, shown as open-loop control systems, can be converted to feedback control systems. Each of the techniques described in Sec. 17-4 utilizes SCR circuits in which the gate pulses control performance. Measurement of the desired response coupled with appropriate comparison can control the electronic circuits used to generate the gate pulses. Closing the loop by means of the SCR gate circuit results in feedback-type motor-drive systems. Caution should be exercised, however, for the SCR motor drives are nonlinear so that superposition and block-diagram techniques are not directly applicable.

17-9 DIGITAL CONTROL SYSTEMS

Speed, accuracy, and availability of digital computers are the underlying reasons for the development of digital control systems. As presented in Sec. 11-1, a digital system utilizes discrete signals in contrast to the continuous (analog) signals utilized by the systems described earlier in this chapter. Two types of discrete signals are used; these are shown in Fig. 17-38. The sampled-data signal, depicted in Fig. 17-38a, is a sequence of pulses in which the amplitude of each pulse is the same as the amplitude of the analog signal (the dashed curve) at the time of sampling. In Fig. 17-38b the pulse trains in each time interval represent the numeric value, usually expressed in binary terms (see Sec. 11-7), of the amplitudes of the corresponding sampled-data pulse. The binary representation of a number is the basis of digital-computer operation; thus signals of the type in Fig. 17-38b can be fed directly into a computer. This compatibility is the key feature for incorporating electronic digital computers as the controller element in many control systems.

The block diagram in Fig. 17-39 is a functional representation of a type of digital control system. The blocks G and H serve the same function as in

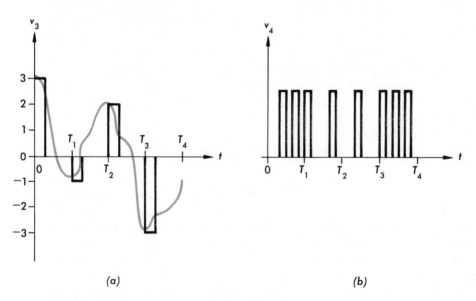

(a) (b)

FIGURE 17-38 Pulse trains: (a) sampled-data signal; (b) digital or numeric signal.

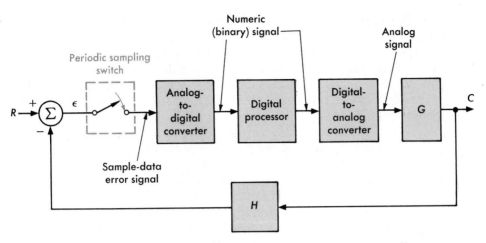

FIGURE 17-39 Representation of a digital control system.

any feedback system; the major differences between the systems in Figs. 17-21 and 17-39 are that the error signal is sampled and the use of a digital processor. The digital processor serves as the controller in this system in that its output, often reconverted to an analog signal, is the excitation for the block G. (Recall that G is the subsystem which provides the output to be controlled.) The digital processor may be either a central computer which controls several functions or it may be a special-purpose computer, such as a microprocessor, designed for the particular control function. The function of the digital processor is twofold. First, it analyzes the difference signal by comparison with the performance of the system stored in the memory unit of the computer. The second function, based on the information obtained from the analysis of the difference signal, is to generate, in numeric form, the excitation signal applied to G required to establish the desired output.

The three characteristics of digital computers used in digital control are the speed and capability of computation, their memories, and the fact they can be programmed. Each of these features, often in combination, is used in particular types of control systems. Large high-speed computers are used for central control in large automated manufacturing facilities. In such systems all three characteristics are used, as one computer controls all of the operations required in the manufacturing process. The computer is programmed to indicate the sequence of operations required and is constantly fed data from various monitoring stations on the progress of the process. Computation is then used to compare actual and desired performance, and the computer generates a new set of instructions to correct for deviations. The new set of instructions is often converted to an analog signal so that the appropriate excitations are applied to the machines which actually do the manufacturing. For example, in a factory whose machines are driven by electric motors, the motor-drive systems described in Sec. 17-4 can be controlled by computer. The basic method is to use the computer to determine the nature of the control pulses applied to the SCRs used in the motor drives.

Computer-control-system operation can be illustrated by considering a machine-tool process used to drill or punch holes in a rectangular piece of material. This operation is common in many manufacturing processes, such as the fabrication of printed circuits, the manufacture of key rings, and in the fabrication of mechanical mountings. An elementary system is depicted in Fig. 17-40; the essential elements are the input, the digital processor, the drill-positioning mechanism, and the sampled-data position-control system.

The input represents the numeric data indicating where the holes are to be drilled. To accomplish this, each point on the board must be given coordinates, (x, y) or (r, θ), and the location of the desired holes must be coded in terms of these coordinates. Usually the input data are entered into the system by means of punched cards or punched tape in much the same fashion as data for a given problem are supplied to a computer.

The drill-positioning mechanism is an electromechanical drive which causes the drill to move to the desired hole location. The sampled-data position-control system can be used to physically position the drill at the desired coordinates. Once the drill is properly positioned, the hole is drilled through the material.

The digital processor has several functions in the system. First, it must convert the input data to electrical signals and store the information. If the input data do not contain the sequence in which the holes are to be drilled, the processor must generate the sequence. This second function may be achieved by an internal program which sorts the data so that the drilling sequence generated requires a minimum displacement between successive

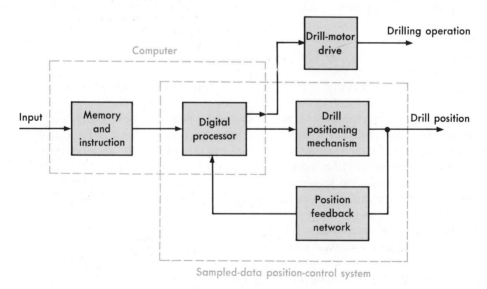

FIGURE 17-40 Functional representation of digitally controlled drill press.

operations. The third function is that the processor is part of the sampled-data system. As such, it acts as a comparative element in positioning the drill and also, after the digital data are converted to an analog signal, controls the drill-positioning drive. After coincidence between actual and desired position is achieved, the fourth function of the processor is that of causing the actual drilling process to begin and end.

The process of control may then be summarized as follows:

1 Upon receiving the input data, the processor determines the sequence of operations and stores both the sequence and hole positions in its memory.

2 The drill is positioned at the first hole location by means of the sampled-data system.

3 When the actual drill position coincides with the desired location stored in the memory, the processor generates a signal which causes the drilling process to begin.

4 At the conclusion of the first drilling operation, the process is repeated for each additional hole.

Additional sophistication can be added to the basic system by requiring other related functions to be controlled. Typical are operations which control the speed and time duration of the drilling process. Speed control adds flexibility to the system by allowing different kinds of materials to be drilled. The applicability of the system to a variety of materials usually requires different bits, each of which has a different optimum cutting speed. Control of the duration of time the bit is cutting effectively controls the depth of the hole. This additional flexibility is manifested in the applicability of the system to materials of differing thicknesses. Both operations can easily be accommodated in the system: speed control can be achieved by a sampled-data version of the speed-control system of Fig. 17-27; timing control, by several of the waveform-generating circuits discussed in Chap. 11. In addition, the SCR circuits used in the motor drives described in Sec. 17-4 can also be adapted for use.

The positioning mechanism for the drill need not be a feedback control system but it can, instead, be derived from an optical shaft encoder (see Sec. 17-6) driving a stepper motor. A *stepper motor* can be thought of as a "digital" motor; it is, in fact, a synchronous motor (see Sec. 16-5) designed to rotate a specific number of degrees for each input pulse applied. Generally, a variable reluctance rotor and multipole, multiphase stator winding are used. The output pulses from the encoder, or other digital device, form the input to a drive circuit whose output delivers the appropriate current to the stator windings. The axis of the air-gap field established by the stator winding field is thus incremented to be in coincidence with the input pulses. The reluctance torque causes the rotor to follow the axis of the air-gap field. The angular change that results is dependent on the number of poles of the motor. Most commercial stepper motors are available to give angular incre-ments between 1° and 30° per input pulse.

17-10 MICROPROCESSOR CONTROL

The microprocessor is rapidly becoming the key component in digital con-trol systems. In essence, a microprocessor is the central processing unit (CPU) of a computing system which is used in conjunction with memory and appropriate input-output (I/O) devices. When the memory and clock genera-tor are integrated on the same chip, the microprocessor becomes a self-contained digital computer (see Sec. 12-7).

In control systems applications, the microprocessor and its associated circuits function as the digital processor shown in Fig. 17-39. Often, it is used as a dedicated computer, i.e., one in which the chip itself is pro-grammed to perform a specific function. Digital watches and automobile fuel injection systems are examples of this type of operation. In other applica-tions where a variety of tasks are to be performed, the microprocessor pro-gram can be altered to fit specific task requirements.

To illustrate the use of the microprocessor as a controller, the tempera-ture control system in Fig. 17-41 is considered. The purpose of the system is to control the temperature of the chemical bath at 90°C ± 2°C. The bath temperature is transduced by the thermocouple and converted to an 8-bit digital signal by the analog-to-digital converter. The timer generates a pulse, whose duration is two clock periods, once each minute. The pulse is used to interrupt the computer by means of the interrupt logic and simultaneously

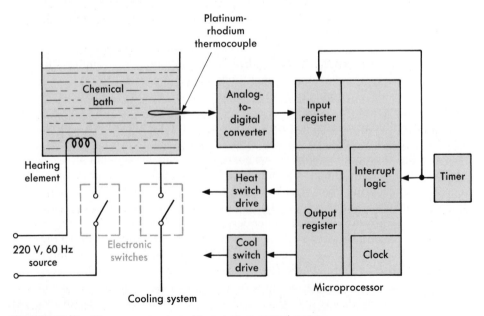

FIGURE 17-41 Microprocessor-based temperature control system.

load the temperature data into the input register. The appearance of logical 1 in bit one of the output register actuates the heat-driver circuit, which causes the heating switch to close. Similarly, logical 1 in bit two of the output register triggers the cooling switch. It is assumed that a 1-min application of heat or cold produces only a small change in the bath temperature. Interrupt logic is used to effect control because the data needed for temperature control is intermittently loaded into the processor. The interrupt logic halts the execution of the program in the processor and performs the subroutine for temperature control. The program in progress is generally controlling some other aspect of system performance. In this case it may be positioning some object that is to be immersed in the bath for a specified time.

The internal operation of the processor is such that the digitized input temperature is compared with the desired value of bath temperature which may be stored in memory. As a result of comparison, if the temperature is greater than 92°C, the program causes a logical 1 to appear at bit two of the output register. In like manner, the program is designed to produce logical 1 in bit one of the output register if the temperature drops below 88°C.

Alternatively, the desired value of bath temperature need not be stored in memory but can be made part of the program. This permits additional system flexibility in that the desired value of bath temperature can be changed. Under these conditions, it is possible to alter the program so that temperature cycling is possible. That is, after the bath temperature has been maintained at 90°C for 30 min, the bath is to be heated to 100°C and allowed to remain at 100°C for 90 min. At this time, cold is to be applied to return the bath temperature to 90° and the cycle is then repeated.

With some hardware additions, a fail-safe protection can also be programmed into the system. Another bit or pair of bits in the output register can be used both to indicate extreme temperatures and to shut down the system.

Two principal advantages of microprocessor control are that the same hardware can be used to perform a variety of tasks and that distributed computation is possible. The first advantage is gained because the microprocessor and its peripheral equipment (sensors, driver circuits, etc.) can be reprogrammed. Therefore, to alter the control function only the program need be changed, obviating the need for new hardware. Distributed computation refers to the interconnection of several microprocessors, each of which performs a given function. This communication between processors, each responsible for a localized control function, can perform many of the functions that are usually performed by a central computer. Often, distributed computation results in less expensive, more reliable operation than centralized control can achieve.

17-11 OTHER TYPES OF DIGITAL CONTROL

A third type of control system which makes use of the computer is called *adaptive control* and is functionally represented in Fig. 17-42. The block *G* is the system or device used to provide the system output for the system input.

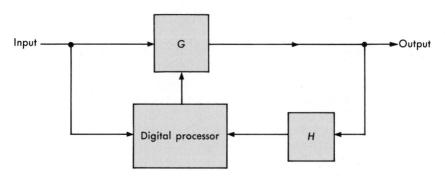

FIGURE 17-42 Functional representation of adaptive control system.

The feedback is provided by H and is converted to a digital signal. This digital signal serves as one input to the digital processor whose function is to effect the necessary control.

The manner by which control is achieved is markedly different in an adaptive system than in the control systems described earlier in this chapter. The adaptive control mechanism is essentially one in which the characteristics of the block G are changed so that for a given input the desired output is obtained. Thus, the parameters of the process or device represented by G are not constant but vary with time. How G changes is determined by the digital processor; the technique used is generally based on the prediction of the performance of the system. The digital processor, often a computer, has other inputs which are stored in its memory unit. These inputs are based on expected values of system components and, from them, the computer calculates the predicted response. The input from the block H is compared with the predicted response. The processor then computes the parameters that G should have for the given value of the actual input signal which is also fed into the computer. The signals representing the parameters of G are then converted, in general, to analog signals whose function is to physically adjust these parameters.

The adaptive control process can be illustrated by a motor-speed-control system based on adjustment of the armature resistance and is shown in Fig. 17-43. The function of the computer is to utilize all of the data inputs and compute the optimum value of R_a so that the motor speed is appropriate to the load. The signal representing the computed value of R_a is converted to one which can be used to position the potentiometer arm of the variable resistance. The actual armature resistance is then controlled by the position of the potentiometer arm, which in turn may be controlled by a sampled-data version of the position-control system of Fig. 17-33.

A major reason for the use of adaptive systems in industrial control is the speed and power of the digital computer. Relative to the time necessary for moving objects, the time required for digital processing is extremely small. Typically, one thinks of the time scale for mechanical motion in seconds or, at best, fractions of a second; electronically the time scale in a com-

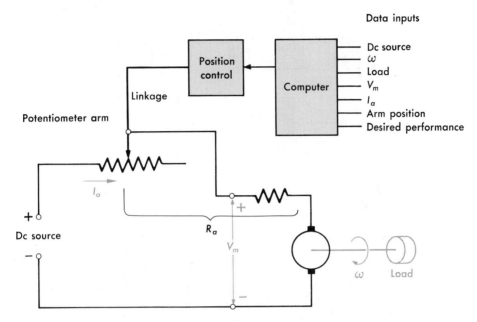

FIGURE 17-43 Adaptive speed-control system.

puter is in the order of nanoseconds (10^{-9} s). In fact, many modern electronic computers are capable of millions of multiplications in one second. Thus, the process of predicting the desired response and comparison with the actual response can be accomplished in a relatively short time compared to the response time of the system to be controlled.

Not all adaptive systems operate digitally; the concept is often used in analog systems (without employing a computer). One example is the *automatic gain control (AGC)* used in commercial AM radios. Because the signals received from different stations vary in strength, it becomes necessary to control the volume so that components such as the speaker are not overdriven. This is desired for both device protection and listener pleasure. Figure 17-44 is a functional representation of such a system. The RF amplifier is used for tuning and to provide some of the amplification required (see Sec. 10-1). The rectifier and filter in the feedback loop provide a dc signal whose amplitude is proportional to the amplitude of the audio output signal. The dc signal is used to control the bias of the RF amplifier. When the audio signal becomes too large, the bias is changed to reduce the gain of the amplifier, thereby tending to reduce the output level. The adaptive nature of the system is manifested in the control of the audio signal at the speaker by changing the gain of the amplifier.

Control-system concepts and techniques have far wider applicability than to the physical systems described in this chapter. This is particularly true for systems which incorporate the digital computer as one element. Inventory control and cash-flow control in business, control of the number

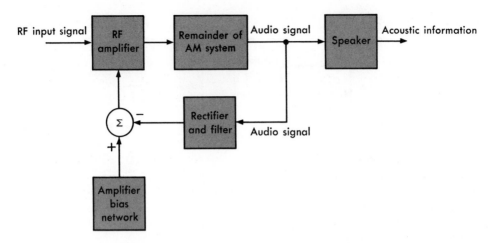

FIGURE 17-44 Representation of automatic-gain-control (AGC) system.

and location of ambulances for health-care emergencies, and behavior modification in the psychology of learning are examples of the use of feedback control. One essential feature in each case is the speed and capacity of the digital computer in predicting the response to the wide range of changes that affect each system. In essence the computer is used to simulate the performance of the system and thus predict system response to each of the excitations. Large-scale systems of this type usually have multiple inputs simultaneously applied; without the benefit of the computer which can predict the responses faster than they actually occur, control cannot be successfully achieved.

For example, in determining the number and storage locations of ambulances in a metropolitan area, several factors come into play. These include population distribution, statistical data on the kind, number, and time of occurrence of emergencies requiring ambulance service, traffic patterns in the area as a function of the time of day, and knowledge of available health-care facilities in the region. In addition, knowledge of possible catastrophes such as airplane disasters, with their potentially large number of emergency patients, will also affect the number and location of the ambulances. The desired response of the system is specified beforehand; i.e., an ambulance must arrive at the scene within 15 min of an emergency. From all of the data, the computer simulates the urban system and determines the total number of ambulances needed and their optimum location. To keep the system current, the data can be continually updated to reflect changing traffic patterns, shifting population, etc. Thus, the initial state of the system becomes the starting point for a continuing adaptive control process.

Because of the computer's ability to store and process a large volume of data and because extensive electronic equipment for communication and control exist, the techniques and concepts of feedback control systems will be even more widely applied in the future.

PROBLEMS

17-1 A storage battery is charged by a rectifier supplied from a 220-V, 50-Hz alternating source. The battery offers a constant counter EMF of 12 V and the combined resistance of the battery and current-limiting resistor is 3Ω. Determine the dc component of the battery-charging current for
a Half-wave rectification
b Full-wave rectification

17-2 A storage battery is charged by a rectifier supplied by a 110-V, 60-Hz alternating source through an $n:1$ transformer. The counter EMF of the battery is 12 V and the combined resistance of the battery and current-limiting resistor is 1.5 Ω. Determine the turns ratio of the transformer required if the battery-charging current is 15 A for
a Half-wave rectification
b Full-wave rectification

17-3 A measure of the effectiveness of a rectifier is provided by the *ripple factor,* defined as the ratio of the RMS value of the ac component of the output voltage V_{ac} to the average value of the output voltage V_{avg}. Determine the ripple factors for the half-wave and full-wave rectified waveforms shown in Figs. 17-2 and 17-7, respectively. *Note:* The rms value of the waveform can be expressed as

$$V_{rms} = \sqrt{V_{ac}^2 + V_{avg}^2}$$

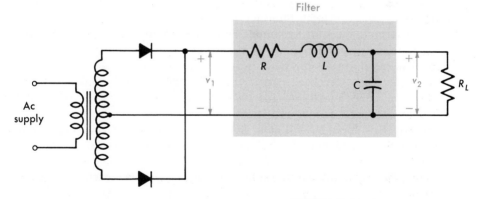

FIGURE 17-45 Circuit for Prob. 17-4.

17-4 A full-wave rectifier with an inductor-input filter is shown in Fig. 17-45. The load is represented by the resistance R_L.
a Show that the filter transfer function for sinusoidal excitation is

$$\frac{\mathbf{V_2}}{\mathbf{V_1}} = \frac{1}{1 + \dfrac{R}{R_L} - \omega^2 LC + j\omega(C + L/R_L)}$$

b One can study the behavior of the filter by considering that its input voltage $v_1(t)$ has the Fourier-series representation

$$v_1(t) = \sqrt{2}\, V_1 \left(\frac{2}{\pi} - \frac{4}{3\pi} \cos 2\omega_o t - \frac{4}{15\pi} \cos 4\omega_o t - \cdots \right)$$

in which V_1 is the rms value of the transformer output voltage and ω_o is its angular frequency. For $V_1 = 20$ V, $\omega_o = 377$ rad/s, $R = 50\ \Omega$, $R_L = 500\ \Omega$, $C = 100\ \mu$F, and $L = 10$ H, determine the dc second- and fourth-harmonic components of the output voltage v_2. It can be assumed that the voltage drop across the rectifier is negligible and that the filter input current never becomes zero.

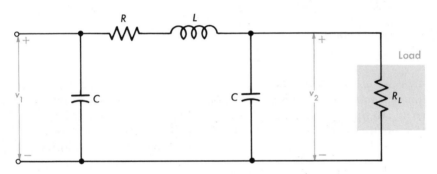

FIGURE 17-46 Filter for Prob. 17-5.

17-5 A capacitor-input filter and load is depicted in Fig. 17-46.
a Determine the transfer function V_2/V_1 assuming a sinusoidal input.
b Find the dc component of load current.
c Assume that $v_1(t)$ can be represented as

$$v_1(t) = \frac{110\sqrt{2}}{\pi} \left(1 - \frac{4}{3} \cos 200\pi t - \cdots \right)$$

and that $R = 25\ \Omega$, $R_L = 100\ \Omega$, $L = 20$ H, and $C = 200\ \mu$F. Determine the ripple factor, as defined in Prob. 17-3, of the output voltage v_2.

17-6 The input voltage to the filter in Fig. 17-46 is represented by

$$v_1(t) = \frac{35\sqrt{2}}{\pi} \left(1 - \frac{4}{3} \cos 754t - \cdots \right)$$

The element values are: $L = 10$ H, $R = 30\ \Omega$, and $R_L = 270\ \Omega$. Determine the value of C for a ripple factor of 0.01.

17-7 In the circuit of Fig. 17-5, the firing angle of the SCR is 60°. Neglecting the voltage drop across the SCR, determine the average current in a 10-Ω load resistance for $v(t) = 440\sqrt{2} \sin 377t$.

17-8 The armature circuit of a dc motor is to be driven by the SCR circuit in Fig. 17-5. The armature resistance is 5 Ω and the counter EMF of the motor is 110 V. The firing angle of the SCR is 90° and the ac input is $\sqrt{2}\ V_1$ cos 377t. Find the value of V_1 required to provide an average voltage of 120 V.

17-9 The SCR circuit in Fig. 17-5 is used to drive the armature of a dc motor at an average voltage of 240 V. The counter EMF is 200 V and the armature resistance is 10 Ω. If $v(t) = 440\sqrt{2} \sin 100\pi t$, find the firing angle of the SCR.

17-10 The circuit in Fig. 17-15 is used as a regulator in a 15-V power supply required to provide 200 mA to a resistive load. The unregulated input voltage V_s varies between 18 and 21 V. Determine the values of
a R_s
b The maximum current in the Zener diode

17-11 The resistor values in the circuit in Fig. 17-15 are $R_s = 10\ \Omega$ and $R_L = 50\ \Omega$. The Zener diode is rated at 6 V and has a maximum allowable current rating of 300 mA. What are the maximum and minimum values of V_s for which regulation occurs?

17-12 A 12-V Zener diode is used in the simple regulator of Fig. 17-15. The current-limiting resistor $R_s = 20\ \Omega$ and V_s varies between 15 and 20 V. Determine the range of values of load resistance R_L and load current I_L for these conditions.

17-13 The transistor in the series regulator of Fig. 17-16a is operated just at the edge of saturation or is saturated. At the edge of saturation the dc characteristics are: $V_{BE} = 0.6$ V, $V_{CE} = 0.3$ V, and $h_{FE} = 25$. The Zener diode is rated at 5.6 V. The load resistance is 5 Ω and the current-limiting resistance is 1.0 Ω.
a What is the minimum value of V_s required for the transistor to operate just at saturation?
b For $V_s = 7.5$ V, determine the current in the Zener diode.

17-14 The system in Fig. 17-47 is to be used as a voltage regulator. The generator is rated at 300 kW at 600 V and its armature and field resistances are 0.05 and 24 Ω, respectively. The generator, driven at constant speed, generates 48 V/field A. The amplifier voltage gain V_f/V_e is 540 and the reference voltage V_R is 12 V. The potentiometer P can be assumed to draw negligible current.
a Determine the amplifier input voltage (error signal) V_e required to provide rated terminal V_t at no load (open-circuit).
b What is the value of H, the potentiometer setting, to achieve the no-load conditions in **a**?

c Using the value of H obtained in **b**, determine the value of V_t at rated armature current.

d Repeat **c** for the case where aging causes the amplifier gain decreases to 480.

e Repeat **c** for the case where the field resistance changes to 28 Ω.

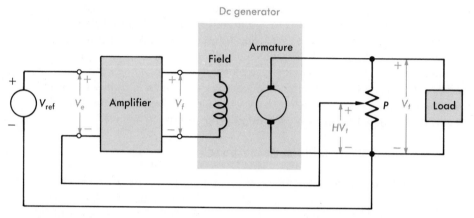

FIGURE 17-47 Voltage regulator for Prob. 17-14.

17-15 The speed-control system depicted in Fig. 17-27 is designed to drive a mechanical load at a constant angular speed of 60 r/min. The motor characteristics are: $K_m = (110/\pi)$ V/rad/s $= (110/\pi)$ N·m/A, $R_a = 2.0$ Ω, and at rated speed the armature current is 3.0 A. The reference voltage V_R is 12 V and the tachometer generates 1.8 V/(rad)(s).

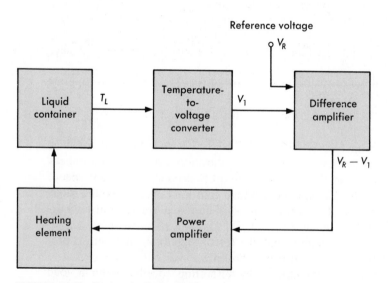

FIGURE 17-48 System for Prob. 17-16.

a Determine the value of the amplifier gain required to maintain rated output speed.
b Using the result in **a**, determine the values of G and H.
c If the amplifier gain decreases by 10 percent, what is the percentage change in output speed?

17-16 The functional representation of a temperature-control system is shown in Fig. 17-48. Such systems are used to control the temperature of the liquid in the container and are used in chemical processes. The temperature-to-voltage converter, usually a thermocouple, acts as the sensing element.
a Explain how this system operates to maintain a constant liquid temperature T_L.
b By what mechanism can the temperature be adjusted to a new value?

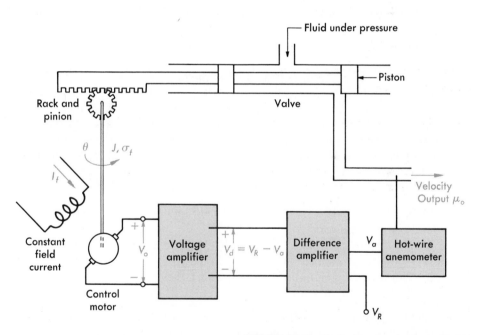

FIGURE 17-49 Control system for Prob. 17-17.

17-17 Control of the flow rate of a fluid is a necessary operation in many industrial processes. In the system shown in Fig. 17-49, the fluid whose flow is to be controlled is supplied to a valve at constant pressure. The output velocity u_o of the fluid is a linear function of the position of the valve piston and, consequently, of the angular shaft position θ of the control motor. That is, $u_o = K_o\theta$. The flow rate is measured by use of a hot-wire anemometer whose output is a voltage $V_a = K_a u_o$. The difference amplifier provides the error signal which, after amplification by A, drives the motor. The motor characteristics are K_m and R_a; J and σ_f are the system moment of inertia and friction referred to the shaft.

a Explain how this system acts to control the rate of fluid flow.

b Draw a block diagram of the system so that the transfer function u_o/V_R can be determined.

c Determine the loop gain GH of the system.

17-18 The numeric values of the system components in the system described in Prob. 17-17 and Fig. 17-49 are:

$$K_o = 50 \text{ m}^3/\text{deg·s} \qquad\qquad R_a = 5.0 \text{ }\Omega$$
$$K_a = 2.7 \text{ }\mu\text{V}/(\text{m}^3)(\text{s}) \qquad J = 2.6 \times 10^{-2} \text{ kg·m}^2$$
$$K_m = 2 \text{ V}/(\text{rad})(\text{s}) = 2 \text{ N·m/s} \qquad \sigma_f = 2.5 \text{ N·m/(rad)(s)}$$

a Evaluate the transfer function u_o/V_R.

b Sketch the frequency response characteristics of GH over the range 1 to 1,000 rad/s.

c Determine the closed-loop response to a unit-step function.

17-19 The temperature of the fluid in the control system in Fig. 17-48 and Prob. 17-16 is to be raised from 20 to 100°C in one hour. The linear heating characteristic of the fluid is such that an average of 1 kW of power for one hour raises the temperature 10°C. The power amplifier transfer ratio is 20 kW/V and the temperature-to-voltage converter generates 50 mW/°C. Determine the value of the reference voltage V_R needed to produce the desired temperature increase.

17-20 In Prob. 17-14, armature and field inductances, L_a and L_f, respectively, were neglected.

a Determine the transfer function V_t/V_R for element values given in Prob. 17-14 and for $L_a = 0.12$ H and $L_f = 24$ H.

b Plot a Bode diagram of the loop gain **GH**.

17-21 A Ward-Leonard control system (see Sec. 15-5) is shown in Fig. 17-50. The following symbols define the system parameters:

R_f, L_f = generator field resistance and inductance, respectively
$\qquad K_g$ = generator constant (induced voltage/field current)
R_g and R_m = armature resistances of motor and generator
$\qquad K_m$ = motor constant (induced voltage/speed or torque/armature current)
$\qquad J$ = moment of inertia of motor and load
$\qquad \sigma_f$ = friction coefficient of motor and load.

Armature inductances have negligible effects.

a Construct a block diagram (or signal-flow graph) in which the angular velocity Ω is the output and V_f, the generator field voltage, is the input.

b Show that the transfer function Ω/V_f is

$$\frac{\Omega}{V_f}(s) = \frac{\dfrac{K_g K_m}{R_f[K_m^2 + \sigma_f(R_g + R_m)]}}{(1 + sL_f/R_f)\left[1 + s\,\dfrac{J(R_g + R_m)}{K_m^2 + \sigma(R_g + R_m)}\right]}$$

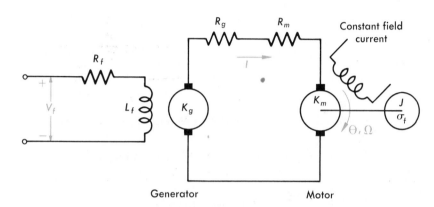

FIGURE 17-50 Ward-Leonard system for Prob. 17-21.

17-22 The machine constants for the generator and motor in the Ward-Leonard system shown in Fig. 17-50 are:

Generator	*Motor*
$R_g = 0.50\ \Omega$	$R_m = 1.5\ \Omega$
$R_f = 400\ \Omega$	$K_m = 1.5625$ N·m/A or V/(rad)(s)
$L_f = 20$ H	$J = 3.0$ kg·m²
$K_g = 1{,}500$ V/field A	$\sigma_f = 0.06$ N·m/(rad)(s)

Armature inductances have negligible effect.

a Draw a block diagram or signal-flow graph for which θ, the angular position of the motor shaft, is taken as the output. The input is the generator field voltage V_f.

b Find the transfer function θ/V_f.

c For $s = 0$, identify the G and H portions of the system.

d Sketch an asymptotic Bode diagram for θ/V_f.

17-23 Feedback systems often employ RC networks to improve the frequency response characteristics over specific bands of frequencies. The network shown in Fig. 17-51a is referred to as a *lead network*.

a Determine the transfer function V_o/V_i and sketch an asymptotic Bode diagram (magnitude and phase).

b Evaluate the dc gain of the network.
c Evaluate the gain as $s \to \infty$.

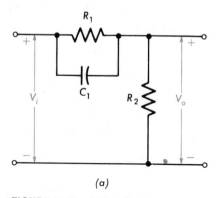 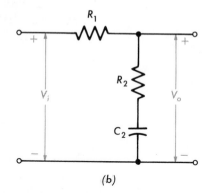

(a)　　　　　　　　　　　　　　　　(b)

FIGURE 17-51　(a) Lead network for Prob. 17-23 and (b) lag network for Prob. 17-24.

17-24　The circuit shown in Fig. 17-51b is referred to as a *lag* network. Repeat Prob. 17-23 for this circuit.

17-25　The element values in Fig. 17-51a are $R_1 = 20$ kΩ, $R_2 = 5$ kΩ, and $C_1 = 0.5$ μF.
a　Sketch the asymptotic Bode diagram for the function V_o/V_i.
b　Why is this circuit called a lead network?
c　Determine the angular frequency at which the phase shift is maximum. *Hint:* It is useful to recall that

$$\tan(\theta_1 \pm \theta_2) = \frac{\tan \theta_1 \pm \tan \theta_2}{1 \mp \tan \theta_1 \tan \theta_2}$$

in the solution.

17-26　The element values in the circuit of Fig. 17-51b are $R_1 = 40$ kΩ, $R_2 = 10$ kΩ, and $C_2 = 0.25$ μF.
a　Sketch the asymptotic Bode diagram for V_o/V_i.
b　Why is this circuit called a lag network?
c　Determine the frequency at which the magnitude of the phase shift is maximum. (See *Hint* in Prob. 17-25.)

17-27　Lead and lag characteristics are often combined in a single network. One such network is depicted in Fig. 17-52 for which the transfer function is expressible as

$$\frac{V_o}{V_i} = \frac{(1 + s/\omega_1)(1 + s/\omega_2)}{(1 + s/\omega_3)(1 + s/\omega_4)}$$

What relations exist between ω_1, ω_2, ω_3, and ω_4 for this transfer function to exhibit lead-lag characteristics?

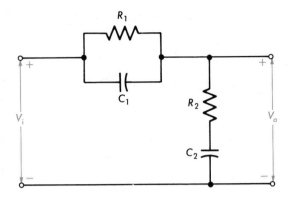

FIGURE 17-52 Network for Prob. 17-27.

17-28 The element values in the circuit of Fig. 17-52 are $R_1 = 100$ kΩ, $R_2 = 25$ kΩ, and $C_1 = C_2 = 1$ μF. Determine the transfer function and sketch the asymptotic Bode diagram.

17-29 The loop-gain GH of an uncompensated feedback system is

$$GH(s) = \frac{10}{s(1 + s/2)}$$

a Sketch the Bode diagram for the system.
b A compensation network is added to the system which makes the transfer function

$$GH(s) = \frac{10(1 + s)}{s(1 + s/2)(1 + s/5)}$$

Sketch the Bode diagram and compare the response with that obtained in **a**.

17-30 The uncompensated and compensated loop-gains GH_1 and GH_2, respectively, of a control system are

$$GH_1(s) = \frac{20}{(1 + s)(1 + s/10)}$$

$$GH_2(s) = \frac{20(1 + 2s)}{(1 + s)(1 + s/10)^2}$$

Sketch the Bode diagrams for GH_1 and GH_2 and compare their responses.

17-31 In many manufacturing processes conveyor belts are used to transport materials between the stations at which the various operations in the process are performed. Assuming that the conveyor belt is driven by a dc motor, describe a computer-based system which can be used to control the speed of the belt. Your discussion should include the hardware needed, such as transducers, amplifiers, A/D converters, etc. Indicate, as well, the data that must be stored in the computer or entered into the computer during the process.

17-32 Describe the operation of a computer control system which can be used in the system of Fig. 17-49. Indicate any additional hardware required and what data is to be stored in the computer.

17-33 **a** Describe the operation of a digital system which can control the position of a shaft. Include both hardware and software requirements of the system.
b Briefly indicate an alternative digital control system which can perform the functions described in **a**.

**Abbreviations
for
Units
and
Technical
Terms**

Term	Abbreviation	Term	Abbreviation
alternating current	ac	henry	H
ampere	A	hertz	Hz
ampere-turn	At	horsepower	hp
amplitude modulation	AM	hour	h
analog-to-digital converter	A/D	inch	in.
		inductance-capacitance	LC
arithmetic logic unit	ALU	input-output	I/O
automatic frequency control	AFC	integrated circuit	IC
		intermediate frequency	IF
automatic gain control	AGC	joule	J
bandwidth	BW	junction field effect transistor	JFET
bel	B		
binary-coded decimal	BCD	kilogram	kg
centimeter	cm	kilohertz	kHz
common-mode rejection ratio	CMRR	kilometer	km
		kilovar	kvar
coulomb	C	kilovolt	kV
cubic centimeter	cm³	kilovoltampere	kVA
cubic meter	m³	kilowatt	kW
decibel	dB	kilowatthour	kWh
degree Celsius	°C	Kirchhoff's current law	KCL
degree Fahrenheit	°F	Kirchhoff's voltage law	KVL
degree Kelvin	°K	large-scale integration	LSI
degree (plane angle)	. . . °	logarithm, base 10	log
digital-to-analog converter	D/A	logarithm, natural	ln
		magnetomotive force	MMF
direct current	dc	megahertz	MHz
electromotive force	EMF	megavolt	MV
farad	F	megawatt	MW
field-effect transistor	FET	metal-oxide semiconductor	MOS
foot	ft		
frequency modulation	FM	meter	m

Term	Abbreviation	Term	Abbreviation
meter-kilogram-second	MKS	pulse-width modulation	PWM
mho	℧	radian	rad
microampere	μA	radio frequency	RF
microfarad	μF	Random-access memory	RAM
microhenry	μH	resistance-capacitance	RC
microsecond	μs	resistance-inductance	RL
mile per hour	mi/h	resistance-inductance-	RLC
mile (statute)	mi	capacitance	
milliampere	mA	revolution per minute	r/min
millihenry	mH	revolution per second	r/s
millimeter	mm	Read-only memory	ROM
millisecond	ms	root-mean-square	rms
millivolt	mV	second (time)	s
milliwatt	mW	silicon-controlled	SCR
minute (time)	min	rectifier	
nanoampere	nA	square foot	ft²
nanosecond	ns	square inch	in²
newton	N	square meter	m²
newton-meter	N-m	var	var
ohm	Ω	volt	V
operational amplifier	OP-AMP	voltage-controlled	VCO
picofarad	pF	oscillator	
pound	lb	voltampere	VA
pulse-amplitude	PAM	watt	W
modulation		watthour	Wh
pulse-code modulation	PCM	weber	Wb

PREFIXES FOR UNITS

Prefix	Multiple	Abbreviation	Prefix	Multiple	Abbreviation
tera	10^{12}	T	centi	10^{-2}	c
giga	10^{9}	G	milli	10^{-3}	m
mega	10^{6}	M	micro	10^{-6}	μ
kilo	10^{3}	k	nano	10^{-9}	n
hecto	10^{2}	h	pico	10^{-12}	p
deka	10	da	femto	10^{-15}	f
deci	10^{-1}	d	atto	10^{-18}	a

The study of electrical engineering requires many mathematical operations for the description and evaluation of circuit or system performance. One of these operations is the manipulation of complex numbers. This appendix presents the necessary background of the algebra of complex numbers commensurate with the requirements of the test material.

B-1 COMPLEX NUMBER REPRESENTATION

Historically, imaginary numbers arose in the solution of certain algebraic equations. For example, in finding the roots of the equation $x^2 + 1 = 0$ it becomes necessary to introduce $x = \pm\sqrt{-1}$ as solutions. Unfortunately the square root of a negative number was called an *imaginary number,* a fact that caused great consternation at the time (and sometimes still does among students of electrical engineering).

> The seemingly preposterous assumption that there was a square root of -1 was justified on pragmatic grounds: it simplified certain calculations and so could be used as long as "real" values were obtained at the end. The parallel with the rules for using negative numbers is striking. If you are trying to determine how many cows there are in a field (that is, if you are working in the domain of positive integers), you may find negative numbers useful in the calculation, but of course the final answer must be in terms of positive numbers because there is no such thing as a negative cow.[1]

We are already familiar with the use of negative values of current, voltage, and force. Yet we are quite aware that these are real, physical quantities. The use of negatives is a convenience in problem solving; that convenience is heightened because we can give physical interpretation to the results. So it is with imaginary numbers when they are used in the description of physical systems. Imaginary numbers represent an efficient, mathematical method for solving problems; a method which readily allows the results obtained to be correlated with the physical world. As with all mathematical

[1] Martin Gardner, "Mathematical Games," *Scientific American,* August 1979, p. 18.

techniques, the rules governing their use must be defined. The approach followed is based on a geometric representation of complex numbers.

B-2 REPRESENTATION OF COMPLEX NUMBERS

The arrow in Fig. B-1 is used to describe a directed line segment in a plane and has two characteristics: it has a length A from the origin; its direction is θ, measured from the horizontal axis in a counterclockwise sense. The segment also can be described as having a horizontal component a of value $A \cos \theta$, and a vertical component b of value $A \sin \theta$. The directed arrow can be described analytically by three equivalent forms, each of which contains information regarding length and direction.

The first is simply to state its length and direction; thus the expression

$$A\underline{/\theta}$$

is a shorthand form of saying the directed line segment has length A and an angle θ. This is called the *polar form* of representation. The second method of presenting the desired information is to give the size of its horizontal and vertical components:

$$a + jb = A \cos \theta + j A \sin \theta$$

called *rectangular form*. Note that the plus sign does not indicate addition in the normal sense but is used to separate the horizontal and vertical components. The significance of the j is to identify that the number b is measured along an axis rotated by 90° counterclockwise from the horizontal. Thus j can be thought of as an operator which rotates a real number by 90° in the positive sense. If the real number b is rotated by 90° and again by 90° for a

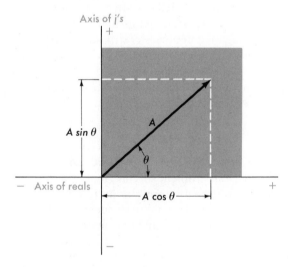

FIGURE B-1 A directed arrow.

total of 180°, the resultant value lies on the negative horizontal axis. There-fore,

$$j \cdot jb = j^2 b = -b$$

implies that

$$j^2 = -1 \qquad \text{or} \qquad j = \sqrt{-1}$$

It is this relation that leads to labeling the vertical axis the *imaginary axis* and the horizontal axis the *real axis*. Similarly, a and b are referred to as the *real* and *imaginary components*, respectively. (Mathematicians use i to denote $\sqrt{-1}$; electrical engineers use j so that confusion with current i is eliminated.)

The *exponential form* is based on Euler's relation which states that

$$\epsilon^{j\theta} = \cos \theta + j \sin \theta$$

so that as an alternative to rectangular form,

$$A \cos \theta + j A \sin \theta = A\epsilon^{j\theta}$$

Since all the expressions represent the same directed arrow, they are equal. The several expressions

$$A\underline{/\theta} = a + jb = A\epsilon^{j\theta} \tag{B-1}$$

all describe a directed segment of length A and angle θ measured counter-clockwise from the positive real axis. By use of the right-triangle relations, these forms are converted from one to another as given in Eqs. (B-2) and (B-3). The conversions described by these relations are internally pro-grammed in many handheld electronic calculators.

$$a = A \cos \theta \qquad b = A \sin \theta \tag{B-2}$$

$$A = \sqrt{a^2 + b^2} \qquad \theta = \tan^{-1} \frac{b}{a} \tag{B-3}$$

Because they contain real and imaginary parts, the expressions of the form of Eq. (B-1) are called *complex numbers*. They are represented by a boldface symbol **A** to differentiate them from the italic A, which is a magnitude.

Directed arrows are known by many names. In the context of this book, directed arrows are called *phasors* when they are used to represent sine and cosine waves. Complex numbers used to change the magnitude and angle of a phasor are called *phasor operators*. Phasor quantities should not be con-fused with vector quantities. Although both types are represented by directed arrows, phasors follow rules analogous to the algebra of real numbers, whereas vectors do not.

B-3 COMPLEX NUMBER ARITHMETIC

Addition and subtraction of complex numbers are most easily accomplished in rectangular form. Let A_1 and A_2 be represented as

$$A_1 = a_1 + jb_1 \quad \text{and} \quad A_2 = a_2 + jb_2$$

Then,

$$A_1 \pm A_2 = (a_1 + jb_1) \pm (a_2 + jb_2) = a_1 \pm a_2 + j(b_1 \pm b_2) \tag{B-4}$$

The result given in Eq. (B-4) states that the real part of the sum of complex numbers is the sum of the real parts of the individual terms. Often, Re **A** is used to identify the real part of **A** and Im **A**, the imaginary component. The previous statement can then be expressed as

$$\text{Re}\{A_1 + A_2\} = \text{Re } A_1 + \text{Re } A_2$$

and

$$\text{Im}\{A_1 + A_2\} = \text{Im } A_1 + \text{Im } A_2 \tag{B-5}$$

Equation (B-5) indicates that the operations of addition (subtraction) and taking the real (imaginary) part of complex numbers can be interchanged.

The process of addition can also be performed graphically in a manner similar to vector addition. The "tip-to-tail" method is illustrated in Fig. B-2 for the complex numbers given in Example B-1.

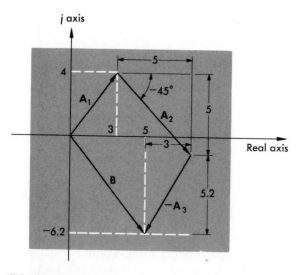

FIGURE B-2 Graphical representation for Example B-1.

Example B-1

For the three complex numbers

$$A_1 = 3 + j4 \qquad A_2 = 7.07\epsilon^{-j\pi/4} \qquad A_3 = 6\underline{/+60°}$$

find

B = **A**₁ + **A**₂ − **A**₃

Express the result in all three forms.

SOLUTION   The numbers **A**₂ and **A**₃·, converted to rectangular form using Eq. (B-2), are

$$\mathbf{A_2} = 7.07 \cos\left(-\frac{\pi}{4}\right) + j\,7.07 \sin\left(-\frac{\pi}{4}\right) = 5 - j5$$

$$\mathbf{A_3} = 6 \cos\,(+60°) + j6 \sin\,(+60°) = 3 + j\,5.20$$

The necessary manipulation to obtain **B** is

$$\mathbf{A_1} = 3 + j4$$
$$\mathbf{A_2} = 5 - j5$$
$$\overline{\mathbf{A_1} + \mathbf{A_2} = 8 - j1}$$
$$-\mathbf{A_3} = -3 - j5.20\,'$$
$$\overline{\mathbf{B} = 5 - j6.20}\qquad \text{in rectangular form}$$

and, from Eq. (B-3),

$$\mathbf{B} = \sqrt{(5)^2 + (6.2)^2}\ \bigg/\ \tan^{-1}\frac{-6.2}{5}$$

$$= 7.96\underline{/-51.1°}\qquad \text{in polar form}$$
$$= 7.96\epsilon^{-j51.1°}\qquad \text{in exponential form}$$

While multiplication can be performed with complex numbers in any form, it is usually more convenient if the numbers are expressed in exponential or polar form. Let **A**₁ and **A**₂ be represented as

$$\mathbf{A_1} = A_1\epsilon^{j\theta_1}\qquad \text{and}\qquad \mathbf{A_2} = A_2\epsilon^{j\theta_2}$$

Then,

$$\mathbf{A_1} \times \mathbf{A_2} = (A_1\epsilon^{j\theta_1})(A_2\epsilon^{j\theta_2}) = A_1 A_2\epsilon^{j(\theta_1+\theta_2)} \qquad\text{(B-6)}$$

following the usual methods for multiplying exponentials. Similarly,

$$(A_1\underline{/\theta_1})(A_2\underline{/\theta_2}) = A_1 A_2\underline{/\theta_1 + \theta_2} \qquad\text{(B-7)}$$

In rectangular form, for **A**₁ = $a_1 + jb_1$ and **A**₂ = $a_2 + jb_2$,

$$\mathbf{A_1 A_2} = (a_1 + jb_1)(a_2 + jb_2) = (a_1 a_2 - b_1 b_2) + j(a_1 b_2 + a_2 b_1) \qquad\text{(B-8)}$$

Note that in obtaining Eq. (B-8), the identity $j^2 = -1$ is used.

Division is the inverse of multiplication so that in exponential or polar form

$$\frac{\mathbf{A}_1}{\mathbf{A}_2} = \frac{A_1 \epsilon^{j\theta_1}}{A_2 \epsilon^{j\theta_2}} = \frac{A_1 \underline{/\theta_1}}{A_2 \underline{/\theta_2}} = \frac{A_1}{A_2} \epsilon^{j(\theta_1 - \theta_2)} = \frac{A_1}{A_2} \underline{/\theta_1 - \theta_2} \tag{B-9}$$

Division in rectangular form is performed by multiplying numerator and denominator by the conjugate of the denominator. The *conjugate* of a complex number is formed by reversing the sign of the j-term. Therefore, for a complex number $\mathbf{A}$ and its conjugate, denoted by $\mathbf{A}^*$,

$$\mathbf{A} = a + jb = A\underline{/\theta} \qquad \text{and} \qquad \mathbf{A}^* = a - jb = A\underline{/-\theta} \tag{B-10}$$

Then,

$$\frac{\mathbf{A}_1}{\mathbf{A}_2} = \frac{a_1 + jb_1}{a_2 + jb_2} \times \frac{a_2 - jb_2}{a_2 - jb_2} = \frac{(a_1a_2 + b_1b_2) + j(a_2b_1 - a_1b_2)}{a_2^2 + b_2^2} \tag{B-11}$$

The process of eliminating the j-term in the denominator of the fraction $\mathbf{A}_1/\mathbf{A}_2$ is called *rationalization*.

Example B-2

For the three complex numbers

$$\mathbf{A}_1 = 10\underline{/26.6°} \qquad \mathbf{A}_2 = 5\underline{/71.6°} \qquad \mathbf{A}_3 = -2 - j3$$

find $\mathbf{A}_1\mathbf{A}_2/\mathbf{A}_3$.

SOLUTION   The result will first be obtained using polar form. $\mathbf{A}_3$ must be converted and from Eq. (B-3)

$$\mathbf{A}_3 = \sqrt{(-2)^2 + (-3)^2} \underline{/\tan^{-1} \frac{-2}{-3}} = 3.61\underline{/236.3°}$$

$$\frac{\mathbf{A}_1\mathbf{A}_2}{\mathbf{A}_3} = \frac{(10\underline{/26.6°})(5\underline{/71.6°})}{3.61\underline{/236.3°}} = \frac{10 \times 5}{3.61} \underline{/26.6° + 71.6° - 236.3°}$$

$$= 13.9\underline{/-138.1°} = 13.9\underline{/221.9°}$$

To perform the division in rectangular form, $\mathbf{A}_1 \times \mathbf{A}_2$ is obtained first and converted to rectangular form.

$$\mathbf{A}_1\mathbf{A}_2 = (10\underline{/26.6°})(5\underline{/71.6°}) = 50\underline{/98.2°} = -7.13 + j49.5$$

Then, forming the quotient and rationalizing gives

$$\frac{\mathbf{A}_1\mathbf{A}_2}{\mathbf{A}_3} = \frac{-7.13 + j49.5}{-2 - j3} \times \frac{-2 + j3}{-2 + j3} = \frac{-134.3 - j120.4}{4 + 9} = -10.33 - j9.26$$

Converting the result to polar form gives $13.9\underline{/221.9°}$

The algebra of complex numbers satisfy the associative, commutative and distributive laws. These are summarized, respectively, in Eq. (B-12).

$$\mathbf{A} + (\mathbf{B} + \mathbf{C}) = (\mathbf{A} + \mathbf{B}) + \mathbf{C} \quad \text{and} \quad (\mathbf{AB})\mathbf{C} = \mathbf{A}(\mathbf{BC})$$

$$\mathbf{A} + \mathbf{B} = \mathbf{B} + \mathbf{A} \quad \text{and} \quad \mathbf{AB} = \mathbf{BA} \tag{B-12}$$

$$\mathbf{A}(\mathbf{B} + \mathbf{C}) = \mathbf{AB} + \mathbf{AC}$$

B-4 SOME USEFUL PROPERTIES

Complex numbers, when they arise in dealing with physical systems, appear in conjunction with their conjugate values. Three properties of complex numbers and their conjugates are of particular interest. The first two, stated in Eqs. (B-13) and (B-14), are direct consequences of the definition in Eq. (B-10).

$$\mathbf{A} + \mathbf{A}^* = a + jb + a - jb = 2a = 2Re \ \mathbf{A} \tag{B-13}$$

$$\mathbf{A} \times \mathbf{A}^* = A\epsilon^{j\theta} \times A\epsilon^{-j\theta} = A^2 \tag{B-14}$$

The results in Eqs. (B-13) and (B-14) indicate that both the sum and product of a number and its conjugate are real.

A third property involves complex numbers in functional relationships. Let $\mathbf{z}$ be a variable which takes on complex values and $f(\mathbf{z})$ be a function which can be expressed as a polynomial in $\mathbf{z}$ with real coefficients. Then, for

$$f(\mathbf{z}) = A + jB$$

if $\mathbf{z}$ is replaced by its conjugate $\mathbf{z}^*$, it is true that

$$f(\mathbf{z}^*) = A - jB = \overline{f(\mathbf{z})^*} \tag{B-15}$$

Thus, as expressed in Eq. (B-15), the function of the conjugate of the variable is the conjugate of the given function.

These properties are useful in determining the roots of algebraic equations. Consider the equation $x^2 + bx + c = 0$ whose roots are z_1 and z_2. (These can easily be determined by the quadratic formula.) Therefore,

$$x^2 + bx + c = (x - z_1)(x - z_2) = x^2 + x(-z_1 - z_2) + z_1 z_2$$

Coefficient comparison gives

$$b = -(z_1 + z_2) \quad \text{and} \quad c = z_1 z_2$$

As b and c are real, the solutions require that both the sum of the roots and the product of the roots be real. From Eqs. (B-13) and (B-14), it is apparent that z_1 and z_2 are conjugates. In general, any complex roots of a polynomial with real coefficients must appear in conjugate pairs. Keep in mind that the equations whose roots we seek describe physical systems whose elements $(L, R, C, \text{etc.})$ are real. We can now appreciate, with the aid of the preceding discussion, that the use of complex numbers does give rise to physically meaningful solutions.

PROBLEMS

B-1   **a**   Express each of the following in rectangular form: $2.24\underline{/26.4}$, $10\epsilon^{j30°}$, $15\underline{/-72°}$, $20\epsilon^{-j126.9°}$, and $0.3\underline{/140°}$

b   Express each of the following in polar and exponential form: $3 - j4$, $-15 + j5$, $10 + j20$, $-10 - j24$, and $0.25 + j0.433$.

B-2   For the three complex numbers $\mathbf{A} = 8 - j6$, $\mathbf{B} = 20 + j10$, and $\mathbf{C} = 0.866 - j0.5$, perform the following operations:

a $\mathbf{A} - \mathbf{B} + \mathbf{C}$
b $\mathbf{BC} - \mathbf{A}$
c $(\mathbf{ABC})^2$
d $\mathbf{AC}/\mathbf{B} - \mathbf{BC}/\mathbf{A}$

B-3   For the three complex numbers $\mathbf{A} = 10\epsilon^{-j45°}$, $\mathbf{B} = -4 + j2$, and $\mathbf{C} = 12\underline{/30°}$, perform the following operations:

a $\mathbf{AB} - \mathbf{C}/\mathbf{B}$
b $\mathbf{C} + \mathbf{A}/(\mathbf{B} - \mathbf{C})$
c $(\mathbf{A} + \mathbf{BC})/(\mathbf{C} - \mathbf{AB})$
d $\sqrt{\mathbf{C}(\mathbf{A} - \mathbf{B})}$

ANSWERS TO SELECTED PROBLEMS

ANSWERS TO SELECTED PROBLEMS

Chapter 1

1-1a 0.833 A   **b** 3,000 C   **c** $21.90

1-3b 7.5 W

1-5b 50 W

1-7a 5 W   **b** 0

1-9a $W(\frac{1}{3}) = 0.516$ J, $W(1) = 2.01$ J   **b** $W(\frac{1}{3}) = 0.892$ J, $W(1) = 5.28$ J

1-11a 7.07 mA   **b** $p(t) = 50 + 40\sqrt{2} \sin 10^6 t + 10 \sin 2 \times 10^6 t$ mW   **c** 50 mW

1-13a $6.40 \cos 377t$ A   **c** $543 \sin 754t$ W

1-15 $i_C(t) = 5$ mA $= -2.5$ mA $= 0$; $0 \le t \le 50$ ns, 50 ns $\le t \le 150$ ns, $t \ge 150$ ns; $p(t) = 5 \times 10^5 t$ W $= -18.75 \times 10^{-3} + 12.5 \times 10^4 t$ W $= 0$

1-17a $10\epsilon^{-10^6 t}$ V   **b** $10^{-2}\epsilon^{-10^6 t}(1 - \epsilon^{-10^6 t})$ W

1-19 $v_1(t) = -47.1 \times 10^3 \sin 377$ t V, $v_2(t) = -28.3 \times 10^3 \sin 377t$ V

1-21a 2 A   **b** 31 V   **c** -90 W   **d** 50 W

1-23a 9.14 V   **b** $I_1 = 4.57$ A, $I_2 = 2.29$ A, $I_3 = 1.14$ A   **c** 45.7 W

1-25 $4.5\epsilon^{-2t}$ A

1-27   12 W (dissipated)

1-29 $-5\epsilon^{-t}$ A

1-31 $20\epsilon^{-2t}$ V

1-33   0.5 A

Chapter 2

2–2a 300 kΩ   b 0.0025 Ω

2–4a For Fig. 2–40a: V_T = 10.0 V, I_A = 0.909 A; for Fig. 2–40b: V_T = 9.09 V, I_A = 0.910 A
   b For Fig. 2–40a: the measured resistance = 11.0 Ω; for Fig. 2–40b: the measured resistance
     = 9.99 Ω

2–6a For Fig. 2–40a: V_T = 10.0 V, I_A = 99 mA; for Fig. 2–40b: V_T = 9.90 V, I_A = 100 mA
   b In both circuits, the measured resistance is 99 Ω

2–8a 14.95 kΩ   c The error in any reading is ±3μA. This error in the meter reading produces an
     error in determining R_x that is greater than 10 percent

2–10a 2,818 Ω   b 94.3 Ω

2–12   5.0 V

2–14a I_1 = 0.2 mA, I_2 = −37.5 mA   b 60 μW   c 5.63 W

2–16   1.90 V

2–18   2 A

2–20a 11.0 V   b 0.5 A

2–22 V_{cd} = 0

2–24   8/3 A

2–28   −4 A

2–30   20V

2–32   0.749 A

2–34a 484 W   b 155 W

2–36   −0.257 V

2–38   0.5 A

2–42a V_O = 9.18 V, R_O = 1.76 Ω   b 8 V

2–44   −1.37 mV

2–46a V_O − 0.257 V, R_O = 8.15 Ω   b 2.03 mW

2–48   −0.231 A

2–50   6.29 kΩ

Chapter 3

3–1a −2 Ω   b −0.5 Ω   c −10ε^{-t} V

3–3a 2ε^{-2t} A   b −2ε^{-2t} A   c −0.2 Ω

3–5b $v(t) = I(R + Ls + 1/Cs)\epsilon^{st}$ V

3–7 $-15\epsilon^{-t}$ A

3–9 $-4\epsilon^{-3t}$ V

3–11b $-13.3^{-t/4}$ m/s

3–13a 50 kHz   **b** 1.67 μs   **c** 11.7 μs   **d** 20 μs

3–15b 1.94 cos $(t/3 + 31°)$ A

3–17   1.97 cos $(3t + 144.5°)$ A

3–19a $50 + j50$ Ω **b** $0.707\underline{/45°}$ V

3–21 $V_{avg} = V_m/4$ V, $V_{rms} = V_m/2$ V

3–23 $I_{avg} = 0$, $I_{rms} = I_m/\sqrt{3}$ A

3–25a 500 W, -866 vars, 1 kVA   **b** $10 \underline{/-60°}$ Ω **c** 115 μF

3–27a 1.10 kW, 1.10 kvars, 1.56 kVA   **b** $R = 22$ Ω, $X = 22$ Ω **c** 58.4 mH

3–29 $8.83\sqrt{2}$ cos $(t - 6.31°)$ V

3–31 $0.969 \underline{/78.4°}$ A

3–33 $4.78\sqrt{2}$ cos $(2t - 66.0°)$ V

3–35 $\mathbf{Z_1 Z_4 = Z_2 Z_3}$

3–37a $R_p = 33.3$ kΩ, $C_p = 0.02$ μF   **b** 3.25 V   **d** $k = 28.5 \times 10^{-10}$ H·m^2/N

3–39 $V_2 = 13.7 \underline{/-249°}$ V

3–41a $v_0(t) = -0.354\sqrt{2}$ cos $(4 \times 10^7 t - 22.3°)$ **b** $\mathbf{Z_{in}} = 146 \underline{/-105.5°}$ Ω

3–43a 1,629 W   **b** $P_{40} = 1,313$ W, $P_{60} = 315$ W

3–45a 2,485 W   **b** 833 W   **c** The $100 \underline{/0°} - $ V generator supplies 3,167 W; the $100 \underline{/-30°} - $ V generator absorbs 573 W

3–47a $\mathbf{V_o} = 8.83\underline{/-6.31°}$ V, $\mathbf{Z_o} = 1.09\underline{/-51.5°}$ Ω **b** $\mathbf{Z_L} = 1.09\underline{/51.5°}$ Ω **c** 28.6 W

3–49a $\mathbf{V_o} = 9.70 \underline{/71.9°}$ V, $\mathbf{Z_o} = 7.97 - j1.12$ Ω **b** $\mathbf{I_2} = 0.967 \underline{/78.3°}$ A

3–51a $\mathbf{V_o} = 89.4 \underline{/-38.0°}$ V, $\mathbf{Z_o} = 0.894 \underline{/28.8°}$ Ω **b** 3,168 W

3–53 $v_0(t) = -0.355\sqrt{2}$ cos $(4 \times 10^7 t - 22.3°)$

3–55a 14.14 A   **b** $200.2 \underline{/2.86°}$ V   **c** 2,831 VA, 2,099 W, $-1,899$ vars

3–57a $P_1 = 28$ kW, $Q_L = 20$ kvars, $VA = 34.4$ kVA, $I_L = 78.2$ A   **b** 329 μF   **c** 552 μF

3–59a $\mathbf{V_1} = 477 \underline{/0.9°}$ V, $\mathbf{V_2} = 466 \underline{/0.7°}$ V   **b** 5.01 kVA, 3.90 kW, 3.15 kvars

3–61a $v(t) = 107.6 + 215.2 \sum\limits_{1}^{\infty} \dfrac{(-1)^k}{4k^2 - 1}$ cos $(2\pi k \times 120t)$ **b** the third harmonic

3-63 $i(t) = \dfrac{2I_m}{\pi} \displaystyle\sum_{1}^{\infty} \dfrac{(-1)^{n+1}}{n} \sin \dfrac{2\pi n t}{T}$ A

3-65 $v_O(t) = 86.1 + 1.03 \cos (2\pi \times 120t + 26.2°) + 0.271 \cos (2\pi \times 240t + 13.5°)$ V

3-67 $v_O(t) = 8.95 \sin (2{,}000\pi t - 62.1°) - 2.45 \sin (4{,}000\pi t - 75.1°) + 1.11 \sin (6{,}000\pi t$ $- 80.0°) + 0.628 \sin (8{,}000\pi t - 82.4°)$

Chapter 4

4-2a $+5\epsilon^{-2t}$ A   **b** 0   **c** $+5\epsilon^{-2t}$ A

4-4a -0.9 V   **b** $-1.8\epsilon^{-10^5 t} \sin 10^5 t$ V

4-6a $-3 \cos t$ V   **b** $1.58 \cos (t - 161.6°)$ V   **c** $1.58 \cos (t - 161.6°)$ V

4-8a $k\epsilon^{-2t/3}$ **b** -2 A

4-10a $k\epsilon^{49 \times 10^6 t}$ **b** 98 V

4-12 $v(0^+) = 14$ V, $i_L(0^+) = 5$ A, $(dv/dt)(0^+) = -80$V/s, $(di_L/dt)(0^+) = 28$ A/s

4-14 $v_O(0^+) = 0$, $(dv_O/dt)(0^+) = 0$

4-16 $-8\epsilon^{-t} + 22\epsilon^{-4t}$ V

4-18 $0.1\epsilon^{49 \times 10^6 t}$, $-2.04 \times 10^{-8}(1 - \epsilon^{49 \times 10^6 t})$, $2.04 \times 10^{-8} + 0.1\epsilon^{49 \times 10^6 t}$

4-20a For $0 \le t \le 0.5$ s: case 1: $1.414[\epsilon^{-0.172t} - \epsilon^{-5.83t}] = 1.414f(t)$; case 2: $0.927f(t)$; case 3: $1.857f(t)$
       For $0.5 \text{ s} \le t$: case 1: $-0.322\epsilon^{-(t-0.5)/3} + 1.542\epsilon^{-3(t-0.5)}$; case 2: $-0.288\epsilon^{-(t-0.5)/2}$ $+ 1.088\epsilon^{-4(t-0.5)}$; case 3: $-0.397\epsilon^{-(t-0.5)/4} + 2.00\epsilon^{-2(t-0.5)}$
   b Case 1: 1.29 W; case 2: 1.30 W; case 3: 1.40 W

4-22 $6.71\epsilon^{-4t} \cos (2t + 63.4°)$ A

4-24a $10\epsilon^{-10t}(1 + 5t)$ V   **b** 0.6 s

4-26a $-9(1 - \epsilon^{-10^5 t})u(t)$ V   **b** $-9[1 - \epsilon^{-10^5(t - 10^{-3})}]u(t - 10^{-3})$ V

4-28 $8(\epsilon^{-2t} - \epsilon^{-t/4})u(t) - 8[\epsilon^{-2(t-1)} - \epsilon^{-(t-1)/4}]u(t - 1)$ V

4-30 $2 \displaystyle\sum_{0}^{4} u(t - n) - 2 \sum_{5}^{9} u(t - n)$

4-32 $(20\epsilon^{-t/2} \sin t/2)u(t) - [40\epsilon^{-(t-1)/2} \sin (t-1)/2]u(t - 1) + [20\epsilon^{-(t-2)/2} \sin (t-2)/2]$ $u(t - 2)$

4-34a $v_C(t) = 2\epsilon^{-2t/3}$ **b** $2\epsilon^{-2(t-3)/3}u(t - 3)$

4-36a $i(t) = 0.25(1 - \epsilon^{-2{,}000t})$ $0 \le t \le T_1$
       $= i(T_1)\epsilon^{-2{,}000(t - T_1)}$ $t \le T_1$
   b $T_1 < 50 \ \mu$s, $T_2 \ge 2.5$ ms

4-38 $(s + 4)/2s$

4-40a $8 \times 10^4/(0.9s^2 + 570s + 10^5)$ **b** $(9s^2 + 5,700s + 10^6)/(0.9s^2 + 570s + 10^4)$
  c $ke^{-317t} \cos(106t + \theta)$ **d** 9.61 V

4-42 $12(1 - \epsilon^{-10^5 t})u(t)$

4-44b $10^6(-0.2 + 2 \times 10^{-12}s)/(5 + 5.34 \times 10^{-7}s + 10^{-16}s^2)$ **c** $[-40 + 40.0743\epsilon^{-9.38 \times 10^6 t}$
  $- 0.0743\epsilon^{-53.3 \times 10^8 t}]u(t)$

4-46a $(100s + 10)/(250s^2 + 269s + 260)$ **b** $-(100s^2 + 70s + 100)/(250s^2 + 269s + 260)$
  c $ke^{-0.538t} \cos(0.866t + \theta)$ V   **d** $-5.39\epsilon^{-t}$ V

4-48 $\dfrac{[25s(s+2)]\,V + 200I(s+5)}{27s^2 + 114s + 270}$

4-50 $V_0 = \dfrac{50sV}{27s^2 + 114s + 270}$, $Z_0 = \dfrac{8s^3 + 140s^2 + 200s + 1,000}{27s^2 + 114s + 270}$

4-52a $2\,\Omega$ in parallel with $1/3$ H   **b** $0.5\epsilon^{-3t/2}$

4-54b $2 - (6 + \sqrt{2})/4\epsilon^{-(2-\sqrt{2})t} + (\sqrt{2} - 2)/4\epsilon^{-(2+\sqrt{2})t}$

4-56 $3(s + 1)(s + 2)/4(3s^2 + 6s + 1)$

Chapter 5

5-5 $1/(s^2 - 1)$

5-7a $-4/(s+1) + 8/(s+2)$ **b** $24/s - 60/(s+1) + 48/(s+2)$ **c** $1 + 2/(s+3) - 6/s+4)$

5-9a $5(1 - \epsilon^{-2t})u(t)$ **b** $5(1 + \epsilon^{-2t})u(t)$ **c** $5/2[1 - (2t-1)\epsilon^{-2t}]u(t)$ **d** $10\delta(t) + 5(1 - \epsilon^{-2t})u(t)$

5-11a $12\sqrt{2}\epsilon^{-2t} \sin(2t - 3\pi/4)$ **b** $(3\epsilon^{-2t})/4[4t \cos 2t - 2(1 - 2t) \sin 2t]$ **c** $3[1 - \sqrt{2}\epsilon^{-2t} \sin(2t + 3\pi/4)]$ **d** $1 - 3\epsilon^{-3t} \sin 4t$

5-13 $2 - 3\epsilon^{-t} + \epsilon^{-3t}$

5-15 $-3\epsilon^{-2t} - 5.37 \cos(t + 26.6°) + 9.49\epsilon^{-t} \cos(t - 34.7°)$

5-17a $-\dfrac{1}{2}\ln[s^2/(1 + s^2)]$ **b** $2/(s + 1)^3$

5-19 $-\{s/[s^2 + (5.01 \times 10^6)^2] + s/[s^2 + (4.99 \times 10^6)^2]\}$

5-21 For single pulse: $(2 \times 10^6)/s^2[1 - \epsilon^{-10^{-8}s} - \epsilon^{-5 \times 10^{-8}s} + \epsilon^{-6 \times 10^{-8}s}]$
  For periodic pulse train: $(2 \times 10^6)/s^2[(1 - \epsilon^{-10^{-8}s})(1 - \epsilon^{-5 \times 10^{-8}s})/(1 - \epsilon^{-10^{-7}s})$

5-23a $\epsilon^{-(t-1)}u(t-1)$ **b** $(t-2)\epsilon^{-(t-2)}u(t-2)$

5-25a $f_1(0^+) = 4, f_1(\infty) = 0; f_2(0^+) = 12, f_2(\infty) = 24; f_3(0^+)$ undefined, $f_3(\infty) = 0$;
  $f_4(0^+) = 0, f_4(\infty) = 0$

5-27a $f_1(0^+) = 12, f_1(\infty) = 0; f_2(0^+) = 0, f_2(\infty) = 0; f_3(0^+) = 0, f_3(\infty) = 3$;
  $f_4(0^+) = 1, f_4(\infty) = 1$

5-29 $40/\sqrt{3}\epsilon^{-t/2} \sin(\sqrt{3}t/2)$ V

5-31 $6\sqrt{2}\epsilon^{-t} \cos(t - \pi/4)$ V

5-33 $\frac{50}{3}\epsilon^{-6} - \frac{20}{3}t\epsilon^{-t} - \frac{50}{3}\epsilon^{-2t}$ V

5-35 $2.04 \times 10^{-5}[0.1t\epsilon^{-0.1t} - \epsilon^{-0.1t} + \epsilon^{49 \times 10^{5}t}]$ V

5-37 $[-2 + 2\epsilon^{-9.31 \times 10^{6}t} + 0.004\epsilon^{-53.7 \times 10^{8}t}]u(t) - [-2 + 2\epsilon^{-9.31 \times 10^{6}(t-10^{-5})}$
       $+ 0.004\epsilon^{-53.7 \times 10^{8}(t-10^{-5})}]u(t-10^{-5})$

5-39 $1 - 2\epsilon^{-t/RC}$ V

5-41a $V_O/V_S(s) = 2s^2/(s^2 + 3.5s + 1)$ b $2.219\epsilon^{-3.186t} - 0.219\epsilon^{-0.3138t}$ V
       c $-0.0196\epsilon^{-0.3138t} + 2.02\epsilon^{-3.186t} + 0.517 \cos(t - \pi/2)$ V

5-43 $\sqrt{2}s^2/(s^2 + \sqrt{2}s + 1)$ Ω

5-45a $8(1 - 3\epsilon^{-2t} + 1\ 2\epsilon^{-3t})$ V   b $2.85s$

5-49 $r_\pi/[1 + sr_\pi(C_\pi + C_u)]$ Ω

5-51a $0.125s$ (circuit time)   b $v_0(0.5) = 2.79$ mV, $v_0(0.05) = 133$ mV

5-53a $H(s) = (s^2 + 4)/[(s+1)(s+2)]$ b 2 rad/s

5-55a   Poles at $s = -1, -1$; zeros at $s = -2, -2$ b $K_0\delta(t) + K_1\epsilon^{-t} + K_2t\epsilon^{-t}$
       c $[(t-2)\epsilon^{-(t-2)} + \epsilon^{-(t-2)}]u(t-2)$

5-57a $1 - \epsilon^{-t} - t\epsilon^{-0.5t}$ b $1 - \epsilon^{-t}$

Chapter 6

6-2a $(1 + j\omega/0.5)/(1 + j\omega/0.25)$

6-4a $H(j\omega) = (4.444 \times 10^4)/(4.938 \times 10^4 - \omega^2 + j\omega \times 140.47)$ b $44.6 + 16.2 \cos(377t - 150.2°)$ V
       d The response at higher frequencies is attenuated.

6-6a $H(j\omega) = -2\omega^2/(1 - \omega^2 + j3.5\omega)$ c High-pass

6-8a $F_1(j2) = 32$ dB, $\angle F_1(j2) = -90°$ b $F_2(j2) = 33$ dB, $\angle F_2(j2) = -121.5°$ c $F_3(j2) = 33$ dB,
       $\angle F_3(j2) = -121.5°$ These values are obtained from the "smoothed" asymptotic Bode
       diagrams.

6-10b   By use of the asymptotic Bode diagram,
       For H_1: $v_2(t) = 7.94 \cos(4t - 117°)$ mV
       For H_2: $v_2(t) = 1.58 \cos(4t - 207°)$ mV
       For H_3: $v_2(t) = 6.30 \cos(4t - 139.5°)$ mV
       c For H_1, $\omega = 6.31$ rad/s; for H_2, $\omega = 2.24$ rad/s; for H_3, $\omega = 5.00$ rad/s

6-12b   For T_1, $\omega = 12.6$ rad/s; for T_2, $\omega = 15.8$ rad/s   c For T_1, $\omega = 10$ rad/s; for T_2, $\omega = 24.6$ rad/s

6-14b   No value exists if the asymptotic Bode diagram is used. Use of the actual characteristic
       gives $\omega = 0.162$ rad/s and $\omega = 0.771$ rad/s.

6-16 $\omega = 1.30 \times 10^9$ rad/s

6-18b   100 kHz   c $2\pi \times 10^6$ rad/s

6-20a $R = 10\ \Omega$, $L = 20$ mH, $C = 0.5\ \mu$F **b** 10 V

6-22a $G = 2.5 \times 10^{-6}\ \Omega$, $C = 1,990$ pF, $L = 12.7$ H **b** $f = 3.967$ kHz and 4.033 kHz

6-24a $R = 40\ \Omega$, $L = 20$ mH, $C = 5,000$ pF **b** $\omega = 1.0075 \times 10^5$ rad/s and 0.9925×10^5 rad/s

6-26a $Z_{ab} = (LC_s s^2 + RC_s s + 1)/\{s(C_s \pm C_p)[1 + (RC_p C_s/(C_p + C_s)s + (LC_p C_s)/(C_p + C_s)s^2]\}$

 b The two resonant frequencies are, approximately, $\omega_1 = \omega_s \sqrt{1 - 1/Q_s^2}$ and
 $\omega_2 = \omega_p \sqrt{1 + C_p/C_s(1 + 2/Q_s^2)}$
 where $\omega_s = 1/\sqrt{LC_s}$, $\omega_p = 1/\sqrt{LC_p}$, $Q_s = \omega_s L/R$

6-28 $R = 2.75\ \Omega$, $L = 45\ \mu$H, $c = 563$ pF

6-30a 2.5 mV **b** 64×10^3 rad/s

6-32a $\dfrac{V_2}{V_1}(s) = (-A_s/R_2 C_2)/[s^2 + s(\dfrac{1}{R_1 C_1} + \dfrac{1}{R_2 C_2} + \dfrac{1}{R_2 C_1}) + (1 + A)/R_1 R_2 C_1 C_2]$
 b $\omega_0 = 10^4$ rad/s, $BW = 10^3$ rad/s

6-34 $z_{11} = (100 + 0.318s)/(1.01 \times 10^{-8}s + 3.18 \times 10^{-2}s + 1) = N(s)/D(s)\ \Omega$
 $z_{12} = z_{21} = 90/D(s)\ \Omega$; $z_{22} = 90(1.01 \times 10^{-3}s^2 + 3.18 \times 10^{-3}s + 1)/D(s)\ \Omega$

6-36a $z_{11} = 6 \times 10^3 + [2 \times 10^3(1 + 1.8 \times 10^{-4}s)]/(1.2 \times 10^{-8}s^2 + 1.41 \times 10^{-4}s + 1)\ \Omega$
 $z_{22} = 0.612s/(1.2 \times 10^{-8}s^2 + 1.41 \times 10^{-4}s + 1)\ \Omega$, $z_{12} = z_{22} = 0$
 b $(V_2/V_1)(s) = 0.612s/(7.2 \times 10^{-5}s^2 + 1.20s + 8 \times 10^3)$

6-38a $y_{11} = 1/h_{11}$; $y_{12} = -h_{12}/h_{11}$; $y_{21} = h_{21}/h_{11}$; $y_{22} = (h_{11}h_{22} - h_{12}h_{21})/h_{11}$
 b $V_2/V_1 = h_{21}/h_{11}h_{22} - h_{12}h_{21})$

6-40a $h_{11} \cdot 2/(s + 1)$; $h_{21} = -1$, $h_{12} = 1$, $h_{22} = 0.5\ \Omega$ $\dfrac{V_2}{V_1} = (s + 2)/(s + 1)$

6-42 $h_{11} = (10 + 0.318s)/(1.01 \times 10^{-4}s^2 + 3.18 \times 10^{-3}s + 1) = [N(s)/D(s)]h_{21} = -1/D(s)$;
 $h_{12} = 1/D(s)$; $h_{22} = 3.18 \times 10^{-4}s/D(s)$

6-44 $z_{11} - z_{12} \rightarrow R = 2$; $z_{12} \rightarrow C = 1$ F; $z_{22} - z_{12} \rightarrow 1\ \Omega$ in series with 1 F

6-46a $X_O/X_s = G_1 G_2 G_3/(1_2 G_3 H_2 + G_1 G_2 G_3 H_1)$ **b** $1/H_1$

6-48b $z_{21} Z_L/[z_{11}(z_{22} + Z_L) - z_{12} z_{21}]$

6-50a $G = a_1 a_2/(1 - a_1 f_1 - a_1 a_2 f_2)$ **b** $a_1 f_1 = 1$

6-52 $[a_0(a_1 a_2 + a_3)(1 - f_2)]/[1 - f_2 - f_1 f_3(a_1 a_2 + a_3)]$

6-54b $-5/4$ A/V

Chapter 7

7-3a $\dfrac{V_2}{I_s} = -g\ \dfrac{R_s R_i}{R_s + R_i}\ \dfrac{R_L R_o}{R_L + R_o}$ **b** $G_p = g^2 \dfrac{R_L R_s R_i R_o^2}{(R_s + R_i)(R_L + R_o)^2}$

 d and **e** Either $\left|\dfrac{I_2}{I_s}\right|$ or $\left|\dfrac{V_2}{V_1}\right|$ must be greater than one.

7-5a $212\ \Omega$ **b** 4,850 **c** 485

7-7   Switch closed for $1 \le t \le 2$ μs; $v_o = 1$ V, $i_L = 20$ mA
       Switch open for $0 \le t \le 1$ μs; $v_o = 5$ V, $i_L = 0$

7-9   Switch open for $0 \le t \le 1.12$ μs; $v_o = 5$ V, $i_L = 0$
       Switch closed for $1.12 \le t \le 2$ μs; $v_o = 1$ V, $i_L = 20$ mA

7-11a Switch open for $0 \le t \le 1.12$ μs; $v_o = 14.9$ V, $i_L = 0.149$ mA
       Switch closed for $1.12 \le t \le 2$ μs; $v_o = 0.294$ V, $i_L = 14.7$ mA
    b 3.145 mW

7-13b $V_D = 0.65$ V, $I_D = 95$ mA

7-15a $I_{DQ} = 4.5$ mA, $V_{DQ} = 0.6$ V   b 2.70 mW   c 2.10 mA, 0.90 mA, 8.5 mA, 22 mA

7-17   34.7 mV, -14.9 V

7-19a 10 mA   b 7.5 mA

7-21 $v_o = 6$ V, $0 \le t \le 6$ ms; $v_o = 8$ V at $t = 10$ ms

7-23 $v_o = 5$ V for all $t < 5$ ms

7-25   One diode conducts for $36.9° \le \omega t \le 143.1°$;
       the second diode conducts when $210° \le \omega t \le 330°$.

7-27a $v_o = 2v_i/3$ for $v_i > 0$; $v_o = 0$ for $v_i < 0$
    b $v_o = \frac{2}{3}v_i - 0.4$ for $v_i \ge 0.6$ V, $v_o = 0$ for $v_i < 0.6$ V

7-29a 0.98   b 2.61×10^{-16} A for $\eta = 1$, 0.115 nA for $\eta = 2$

7-31a $I_{CQ} = 1.9$ mA, $V_{CQ} = 6.2$ V   b 11.8 mW   c 87 kΩ

7-33a $I_{CQ} = -1.15$A, $V_{CQ} = -25$ V   b 69.6 W

7-35a $R_C = 700$ Ω, $R_B = 230$ kΩ

7-37a $I_{CQ} = 1.98$ mA, $V_{CEQ} = 4.60$ V   b $I_{CQ} = 1.986$ mA, $V_{CEQ} = 4.61$ V

7-39a 4.1 mA, 3.8 V   c 42.5 μA

7-41a $R_C = 250$ Ω, $R_E = 250$ Ω, $R_B = 48$ kΩ b 5.18 mA, 2.41 V

7-43a 2 mA, 5 V   b 40 μA, -12.4 V

7-45a 10 mA, 15 V   b 11.8 mA, 12.3 V   c 7.66 mA, 18.5 V

7-47a 0.135 mA, 8.60 V   b 0.127 mA, 8.76 V   c 0.138 mA, 8.54 V

7-49a 3.6 mA, 11 V

7-51a 161 Ω, 17.4 V   b 161 Ω, 36 V

7-53a 9 mA, 3V   b $R_S = 250$ Ω, $R_D = 2.5$ kΩ

7-55   3 mA, 0 V

7-57a 5,000 $\Omega/\square$ b 50 kΩ

Chapter 8

8-2 $G_o = 3$, $F_H = 600$ kHz

8-4a 1 MHz   **b** 18 kΩ

8-6a 1 MHz   **b** 19.9 cos($5 \times 10^5 t - 4.6°$)

8-8 $R_2 = 20$ kΩ, $R_4 \to \infty$

8-10a 0.823 mA   **b** −2.07 percent   **c** 1.10 percent

8-12a 18.5 kΩ   **b** 42.7°C

8-14a 0.466 mA   **b** min $I_C = 0.459$ mA, max $I_C = 0.468$ mA

8-16 $I_C = \dfrac{h_{FE}(h_{FE}+1)}{h^2_{FE} + 2h_{FE} + 2}\ \dfrac{V_{CC} - 2V_{BE}}{R}$

8-18   min $I_C = 2.1193$ mA, max $I_C = 2.1199$ mA

8-20 $R = 19.7$ kΩ, $R_E = 762$ Ω

8-22a 2.03 mA, 6.92 V   **b** 2.29 mA, 6.42 V

8-24a 1.10 mA, 6.20 V for $\Delta T = +50°C$; 0.999 mA, 7.01 V for $\Delta T = -50°C$
      b 1.17 mA, 5.63 V for $\Delta T = +50°C$; 1.06 mA, 6.49 V for $\Delta T = -50°C$

8-26 $R_B = 10$ kΩ, $R = 7.2$ kΩ

8-28a $I_{DQ} = 10$ mA, $V_{DSQ} = 10$ V   **b** 8 V

8-30a $R_K = 2.5$ kΩ   **b** −10 V, −5 V

8-32a 3 kΩ   **b** $(V_{DD} - 9.4)/9.4$

8-36a $v_o(t) = 0.0651$ cos ($10^6 t - 78.7°$) V   **b** $v_o(t) = 14.6$ cos($10^6 t - 20.1°$) μV

8-42a $r_\pi = 1$ kΩ, $g_m = 0.1$ ℧, $C_\pi = 9$ pF   **b** C_x 110 pF

8-44a $g_m 52$ m℧, $g_{ds} = 30$ μ℧, $C_{gd} = 6$ pF, $C_{gs} = 6$ pF   **b** $C_x = 104$ pF

8-46a −182   **b** 12.7 Mrad/s   **c** 610 rad/s

8-48   −462, 4.13 Mrad/s

8-50a $t_2 = 2.303/\omega_H$, $t_1 = 0.105/\omega_H$ **b** $2.2/\omega_H$

8-52a 1,420 Ω   **b** −283   **c** 1.67 μF

8-54a −1.49, 1.03×10^9 rad/s, 15.7 rad/s

8-56a $R_E = 21.7$ Ω, $R_L = 246$ Ω   **b** for $h_{fe} = 240$, $\omega_H = 1.88 \times 10^9$ rad/s;
      for $h_{fe} = 60$, $\omega_H = 1.93 \times 10^9$ rad/s

8-58a −16.3; 5,525 Ω   *b* 5.52 percent   **c** −9.20 percent   **d** 10 MHz   **e** 27.8 μF   **f** 0.398 μF

8-60a 20 kΩ **b** 0.0246 μF **c** 5.77 MHz

8-62 $G_o = \dfrac{(\mu + 1)R_L}{R_k(1 + \mu) + r_d + R_L}$, $Z_{\text{in}} = R_k + \dfrac{R_L + r_d}{1 + \mu}$

8-64a $\omega_H \approx (R_L + r_d)/R_G r_d C_{gd}(1 + g_m R_L)$

8-66a 0.891, 39.3 Ω **b** 42.9 kΩ, 0.0332 μF

8-68 0.991, $R_{\text{in}} = 23$ MΩ, $R_o = 10.3$ Ω

8-70 $\omega_H = 3.67 \times 10^7$ rad/s

8-72a $A_{DM} = -1{,}480$, $A_{CM} = -0.222$ **b** 76.5 dB

8-74a CMRR = 66 dB **b** 4.0 V

8-76a 72.3 dB **b** 61.5 kΩ **c** 171 kΩ

8-78 $\omega_H|_{DM} = 3.20 \times 10^4$ rad/s, $\omega_H|_{CM} = 2.02 \times 10^8$ rad/s

Chapter 9

9-1 $G_{vo} = -40$, $f_{HO} = 190$ kHz (Bode diagram), $f_{HO} = 143$ kHz (dominant-pole)

9-3 $G_{vo} = 160$, $f_{HO} = 98$ kHz (Bode diagram)

9-5 $G_{vo} = -750$, $f_{HO} = 40$ kHz

9-7 $G_{vo} = -750$, $f_{HO} = 50$ kHz

9-9 An infinite number of solutions exist.

9-11 An infinite number of solutions exist.

9-13a 1,210 **b** 47.1 kHz

9-15a 1,000 **b** 50 kHz

9-17a 125 rad/s **b** 2,440 rad/s

9-19a 894 rad/s **b** 26.2 μF

9-21a $G_{vo} = 137$, $\omega_{HO} = 2.77 \times 10^7$ rad/s **b** $G_{vo} = 148$, $\omega_{HO} = 2.72 \times 10^7$ rad/s

9-23a 13.0 Ω **b** 5.62×10^6 rad/s

9-25a 2.44×10^7 rad/s **b** 2.93×10^7 rad/s **c** 1.89×10^7 rad/s

9-27a 2.84×10^7 rad/s **b** 2.82×10^7 rad/s **c** 2.87×10^7 rad/s

9-29 39.7 rad/s

9-31a 4.6 μF **b** 0.0218 μF **c** 0.22 μF **d** 46 μF

9-33a 5.45 kΩ **b** 1.65×10^6 rad/s

9-35   For m = 0.25: $v_o(t) = 1 - t\epsilon^{-2t} - \epsilon^{-2t}$;
       For m = 25/36: $v_o(t) = 1 + 1.042\epsilon^{-0.72t} \cos(0.96t - 163.8°)$

9-37 $G_{vo} = 8 \times 10^4$, $\omega_{HO} = 4.76 \times 10^7$ rad/s

9-39a 950 Ω   b 47.5 Ω

9-43a $|a_o| = 1,275$, $|f_o| = 0.0192$ b 4,080 rad/s

9-47a $G_{DT} = 0$, $T = h_{fe}R_C/(R_s + r_\pi + R_E)$, $K = -R_C/R_E$ b The results are identical

9-49a $K = 6$, $T = 86$ dB   b 4×10^5 rad/s

9-51 $G_{DT} = 6.56 \times 10^{-6}$ Ω, $T = 6,570$, $K = -10^{-3}$ ℧

9-53 $T = 0.0966$, $K = -10^4$ Ω, $G_{DT} = 48.3$ Ω

9-55   13.7 MΩ

9-57 $Z_{in} = 1.34$ kΩ, $Z_{out} = 2.39$ kΩ

9-59   476 MΩ

9-61 $T(s) = (8 \times 10^4)/(1 + s/20\pi)$, $f_{GC} = 800$ kHz

9-63   Barely stable

9-65a 66.9   b 10.3 (from asymptotic plot)

9-67a 100 dB   b 44 dB   c 90°

Chapter 10

10-2a 100 kHz   b $R = 10$ kΩ, $c = 276$ pF

10-4a $\omega = 1/\sqrt{6}RC$ rad/s   b $K = -29$ c $f_{MAX} = 0.1\, GBW/29$

10-6 $R = 6.56$ kΩ, $C = 65$ pF

10-8a $15.9 \le f \le 1,590$ Hz   b 5 kHz $\le BW_{min} \le 32$ kHz

10-10a 13.9 pF for $f_o = 950$ kHz, 3.0 pF for $f_o = 2,050$ kHz   b $h_{fe} = 100$

10-12 $\omega = 1/\sqrt{C(L_1 + L_2)}$, $A_v = -L_2/L_1$

10-14 $\omega = 1/\sqrt{LC_2} \times \sqrt{1 + (C_2/C_1)(1 + R_L/R_{in})}$

$$A_i = C_2/C_1(1 + R_L/R_{in}) + R_L C_2/L \,[R_i(1 + C_2/C_1) + R_L]$$

10-16a $\omega_o = 4 \times 10^5$ rad/s, $Q = 12.5$ b $v_o = 3.0$ V

10-18 $f_o = 159$ kHz, $BW = 100$ kHz

10-20a $s = \pm j(1/\sqrt{R_1 C_1})\sqrt{(8R_2/R_1)}$ b $\omega_o = $ zero of the twin-tee

10-22a $A = 1.586$, $R = 15.9$ kΩ   b $R_3 = 12.3$ kΩ, $R_4 = 15.9$ kΩ   d 1.12 V

10-24a $A_v = 7$, $R = 37.5$ kΩ **b** $|V_o(j\omega_o)| = 0.778$ V, $V_o(0) = 0.778$; $|V_o(j2\omega_o)| = 0.216$ V

10-26 $K = 0.414$; an infinite set of values for R_1, R_2, C_1, and C_2 exist

10-28 $A_v = 3.33$, $Q = \sqrt{3}/3$; element values are arbitrary

10-30 $R_1 = R_2 = 6.21$ kΩ, $R_3 = 2.21$ kΩ

10-32a 100 rad/s   **b** The bandwidth can vary from 1 rad/s to an all-pass sytem

10-34   Bandpass filter

10-36a $\dfrac{V_4}{V_1}(s) = \dfrac{R_6/R_1R_2R_4C_1C_3}{s^2 + s(R_6/R_3R_5C) + 1/R_2R_4C_1C_3}$

b $\dfrac{R_5s/R_1R_4C_3}{s^2 + s/R_3C_3 + R_5/R_2R_4R_6C_1C_3}$

10-38   Stagger-tuning gives better performance.

10-40   For $f + f_{IF}$, the oscillator must be varied from 1 to 2.06 MHz; for $f - f_{IF}$, the oscillator range is 80 kHz to 1.14 MHz; this range is impractical as oscillation frequencies are difficult to achieve over more than two orders of magnitude

10-42   0.8 pF

10-48a $R_2/R_1 = 5$, $R_3 = R_4$ **b** with $v_2 = 0$, $v_1 = 3$ V; with $v_1 = 0$, $v_2 = 5$ V   **c** $V_{max} = 1.875$ V

Chapter 11

11-5b 3.33 mA, 1 mA

11-9   The amplitude and duration of the output pulse decrease; the output begins to more closely approximate a triangle

11-11a 23.5 kΩ **b** $I_C = 20$ mA, $I_B = 0.4$ mA

11-19   644 V

11-21a 1.14 mA   **b** 0.14 mA   **c** 1.2 V

11-23 $R_L = 470$ Ω, $R_1 = 5.2$ kΩ, $R_2 = 15.9$ kΩ

11-25 $R_L = 1$ k$\Omega\times$, $R = 20$ kΩ, $R_1 = 13.6$ kΩ, $R_2 = 54.5$ kΩ, $C = 72.1$ pF

11-29 $38_{10} = 100110$; $82_{10} = 1010010$; $127_{10} = 1111111$; $270_{10} = 100001110$; $360_{10} = 101101000$; $1066_{10} = 10000101010$

11-31   13, 22, 53, 44, 82, 213

11-33a $\overline{C}(\overline{A} + B)$ **b** $AC(B + D) + BD(A + C)$ **c** $B(A + C) + ABC$

11-43 $Y = A \oplus B$

11-45 $Y = \overline{(AB + CD)}$

11-47 $Y_2 = A + B + \overline{C} + D$

11-49 $Y_4 = \overline{D}(\overline{A}\overline{B} + \overline{A}C + \overline{A}\overline{B}\overline{C})$

Chapter 12

12-2a 137; 1492; 1935   **b** no error; three most significant bits

12-8 $D_N = (A_N \oplus B_N) \oplus C_N$; $C_N = A_N B_N + C_{N-1}(A_N \oplus B_N)$

12-10 $ABC + A\overline{B}\overline{C} + A\overline{B}\overline{C} + \overline{A}\overline{B}\overline{C}$

12-12 $13_{10} = 01101$, $-13_{10} = 10011$, $162_{10} = 010100010$, $-162_{10} = 101011110$

12-14   0000011, 1111100, 1000111, 0111001

12-18a   10110110, 10010101101, 101001011100   **b** no error; lose the three most significant bits; lose the four most significant bits

12-20 142_8, 1375_8, 10341_8

12-22 62_{10}, 355_{10}, 1150_{10}

12-24 $10C_{16}$, 27_{16}, 1663_{16}

12-26 47_{10}, 2211_{10}, 3440_{10}

12-28 153_8, 764_8, 5131_8

12-30   0010   1001, 0011   0000   0110, 0101   0111   0100   0001

12-32   1.11011001, 100101.10011, 0.000101110001

12-34a   5 NAND-gates and 5 INVERTERS   **b** 5 NOR-gates and 2 INVERTERS

12-38 $Y_0 = 1$ s, $Y_1 = 10$ s, $Y_2 = 1$ min, $Y_3 = 10$ min, $Y_4 = 1$ h, $Y_5 = 10$ h

12-40   Outputs form an 8:1 divider

12-52   625 mV increment

Chapter 13

13-3a 104   **b** 135   **c** 151   **d** 160

13-5   0.28 A

13-7   1.81 A

13-9a 1.35 Wb/m²   **b** 0.71 Wb/m²   **c** 0.29 Wb/m²

13-13
Ratio	0.1	1.0	4.0	$\sqrt{40}$	8	10
Milliwatts	0.003	0.296	2.56	3.12	2.97	2.56

13–15a Ratio = 6, I_1 = 3.62 A, I_2 = 21.7 A   **b** (1) 3,830 Ω   (2) 106 Ω

13–17a (1) $13.8\underline{/52.6°}$ Ω, (2) $0.138\underline{/52.6°}$ Ω   **b** 2,360 V
  c 2,310 V   **d** 97.3 percent   **e** Same   **f** 171 A

13–19a $15\underline{/-30°}$, $15\underline{/210°}$, $15\underline{/90°}$ **b** $416\underline{/30°}$, $416\underline{/270°}$, $416\underline{/150°}$,   **c** 9,350 W, 5,400 var

13–21a 224 A   **b** 87.9 kW   **c** 0.984 leading

13–23a 12.0 kV   **b** 3,120 kW   **c** 0.728 lagging

Chapter 14

14–1a 1,470 N   **b** 405 N   **c** 68.5 N

14–3   21 lb

14–5a 9.9 V   **b** 9.9 V

14–7   0.785 N•m

14–9a 266 V   **b** 460 V

14–11a 24 poles   **b** 10 poles   **c** 300 r/min

14–13   1,200 Hz

14–15a 140.6 A   **b** 3.6 A   **c** 71.7 A   **d** 5.3 A   **e** 183 lb•ft

14–17   Rms hp = 853; 900-hp motor

Chapter 15

15–1   1,250 r/min

15–5a 1,085 r/min   **b** 1.50 A   **c** 1,150 r/min

15–7   4 turns

15–9a No change   **b** 45.5 percent decrease

15–11   3 Ω

15–13   0.69 A to 1.27 A

15–15a $v_t = 250(1 - \epsilon^{-2t})$ **b** (1) $i_a = 200(1 - \epsilon^{-625t})$, (2) $v_t = 240 - 52\epsilon^{-625t}$

15–19 $\dfrac{0.5}{1 + j1.25\omega}$

15–21b $\dfrac{3.05}{1 + j0.0185\omega}$

Chapter 16

16–3c 3,600 r/min

16–7a 27 r/min   **b** 301 lb•ft   **c** 0.854 lagging

16-9a 50 hp   **b** No   **c** Yes

16-11a 123 mi/h   **c** 23.4 Hz

16-15   Transfer function is $\dfrac{4.9}{1 + j0.038\omega}$

Chapter 17

17-1a 31.0 A   **b** 64.0 A

17-3   Half-wave rectifier: 1.21; full-wave rectifier: 0.484

17-5a $V_2/V_1(j\omega) = (R_L)/(R + R_L - \omega^2 R_L LC + j\omega RR_L C)$ **b** 0.396 A   **c** 0.00996

17-7   14.9 A

17-9   97.7°

17-11 $7.2 \le V_s \le 10.2$ V

17-13a 6.26 V   **b** 38.5 mA

17-15a 327   **b** $H = 1.8$ V/(rad)(s), $G = 9.35$ rad/(s)(V)   **c** -0.0389 rad/s

17-17c $GH = K_o K_a K_m A/[R_a(J_s^2 + \sigma s) + K_m^2 s]$

17-19   3.4 V

17-23a $(V_o/V_i)(s) = [R_2/(R_1 + R_2)]$
$\quad \cdot [(1 + sR_1C_1)]/[(1 + sR_1R_2C_1)/(R_1 + R_2)]$ **b** $R_1/(R_1 + R_2)$ **c** unity

17-25c 223.6 rad/s

17-27   A necessary condition is that $\omega_1 < \omega_2 \le \omega_3 < \omega_4$

Appendix B

B2a $-11.13 - j16.50$ **b** $14.32 + j4.66$ **c** $5 \times 10^4 \underline{/-80.6°}$ **d** $-1.839 - j1.680$